T0189541

The Finite Element Method and Applications in Engineering Using ANSYS®

Erdogan Madenci • Ibrahim Guven

The Finite Element Method and Applications in Engineering Using ANSYS®

Second Edition

Springer is a brand of Springer US

Erdogan Madenci
Department of Aerospace and Mechanical Engineering
The University of Arizona
Tucson, Arizona
USA

Ibrahim Guven
Department of Mechanical and Nuclear Engineering
Virginia Commonwealth University
Richmond, Virginia
USA

Electronic supplementary material can be found at http://link.springer.com/book/10.1007/978-1-4899-7550-8.

ISBN 978-1-4899-7735-9 ISBN 978-1-4899-7550-8 (eBook)
DOI 10.1007/978-1-4899-7550-8

Springer New York Heidelberg Dordrecht London

Softcover reprint of the hardcover 2nd edition 2015

Printed on acid-free paper

Springer is part of Springer Science+Business Media (www.springer.com)

Preface

The finite element method (FEM) has become a staple for predicting and simulating the physical behavior of complex engineering systems. The commercial finite element analysis (FEA) programs have gained common acceptance among engineers in industry and researchers at universities and government laboratories. Therefore, academic engineering departments include graduate or undergraduate senior-level courses that cover not only the theory of FEM but also its applications using the commercially available FEA programs.

The goal of this book is to provide students with a theoretical and practical knowledge of the finite element method and the skills required to analyze engineering problems with ANSYS®, a commercially available FEA program. This book, designed for seniors and first-year graduate students, as well as practicing engineers, is introductory and self-contained in order to minimize the need for additional reference material.

In addition to the fundamental topics in finite element methods, it presents advanced topics concerning modeling and analysis with ANSYS®. These topics are introduced through extensive examples in a step-by-step fashion from various engineering disciplines. The book focuses on the use of ANSYS® through both the Graphics User Interface (GUI) and the ANSYS® Parametric Design Language (APDL). Furthermore, it includes a CD-ROM with the "*input*" files for the example problems so that the students can regenerate them on their own computers. Because of printing costs, the printed figures and screen shots are all in gray scale. However, color versions are provided on the accompanying CD-ROM.

Chapter 1 provides an introduction to the concept of FEM. In Chap. 2, the analysis capabilities and fundamentals of ANSYS®, as well as practical modeling considerations, are presented. The fundamentals of discretization and approximation functions are presented in Chap. 3. The modeling techniques and details of mesh generation in ANSYS® are presented in Chap. 4. Steps for obtaining solutions and reviews of results are presented in Chap. 5. In Chap. 6, the derivation of finite element equations based on the method of weighted residuals and principle of minimum potential energy is explained and demonstrated through example problems. The use of commands and APDL and the development of macro files are presented in Chap. 7. In Chap. 8, example problems on linear structural analysis are worked out in detail in a step-by-step fashion. The example problems related to heat transfer and moisture diffusion are demonstrated in Chap. 9. Nonlinear structural problems are presented in Chap. 10. Advanced topics concerning submodeling, substructuring, interaction with external files, and modification of ANSYS®-GUI are presented in Chap. 11.

There are more than 40 example problems considered in this book; solutions to most of these problems using ANSYS® are demonstrated using GUI in a step-by-step fashion. The remaining problems are demonstrated using the APDL. However, the steps taken in either GUI- or APDL-based solutions may not be the optimum/shortest possible way. Considering the steps involved in obtaining solutions to engineering problems (e.g., model generation, meshing, solution options, etc.), there exist many different routes to achieve the same solution. Therefore, the authors strongly encourage the students/engineers to experiment with modifications to the analysis steps presented in this book.

We are greatly indebted to Connie Spencer for her invaluable efforts in typing, editing, and assisting with each detail associated with the completion of this book. Also, we appreciate the contributions made by Dr. Atila Barut, Dr. Erkan Oterkus, Dr. Abigail Agwai, Dr. Manabendra Das, and Dr. Bahattin Kilic in the solution of the example problems. Last, but not least, we thank Mr. Mehmet Dorduncu for his careful review of the modeling steps and example problems, and for capturing the ANSYS screen shots in this version of the book. The permission provided by ANSYS, Inc. to print the screen shots is also appreciated.

Contents

List of Problems Solved

Chapter 1
Introduction

1.1 Concept

The Finite Element Analysis (FEA) method, originally introduced by Turner et al. (1956), is a powerful computational technique for approximate solutions to a variety of "real-world" engineering problems having complex domains subjected to general boundary conditions. FEA has become an essential step in the design or modeling of a physical phenomenon in various engineering disciplines. A physical phenomenon usually occurs in a continuum of matter (solid, liquid, or gas) involving several field variables. The field variables vary from point to point, thus possessing an infinite number of solutions in the domain. Within the scope of this book, a continuum with a known boundary is called a domain.

The basis of FEA relies on the decomposition of the domain into a finite number of subdomains (elements) for which the systematic approximate solution is constructed by applying the variational or weighted residual methods. In effect, FEA reduces the problem to that of a finite number of unknowns by dividing the domain into elements and by expressing the unknown field variable in terms of the assumed approximating functions within each element. These functions (also called interpolation functions) are defined in terms of the values of the field variables at specific points, referred to as nodes. Nodes are usually located along the element boundaries, and they connect adjacent elements.

The ability to discretize the irregular domains with finite elements makes the method a valuable and practical analysis tool for the solution of boundary, initial, and eigenvalue problems arising in various engineering disciplines. Since its inception, many technical papers and books have appeared on the development and application of FEA. The books by Desai and Abel (1971), Oden (1972), Gallagher (1975), Huebner (1975), Bathe and Wilson (1976), Ziekiewicz (1977), Cook (1981), and Bathe (1996) have influenced the current state of FEA. Representative common engineering problems and their corresponding FEA discretizations are illustrated in Fig. 1.1.
The finite element analysis method requires the following major steps:

- Discretization of the domain into a finite number of subdomains (elements).
- Selection of interpolation functions.

The online version of this book (doi: 10.1007/978-1-4939-1007-6_1) contains supplementary material, which is available to authorized users

E. Madenci, I. Guven, *The Finite Element Method and Applications in Engineering Using ANSYS®*, DOI 10.1007/978-1-4899-7550-8_1

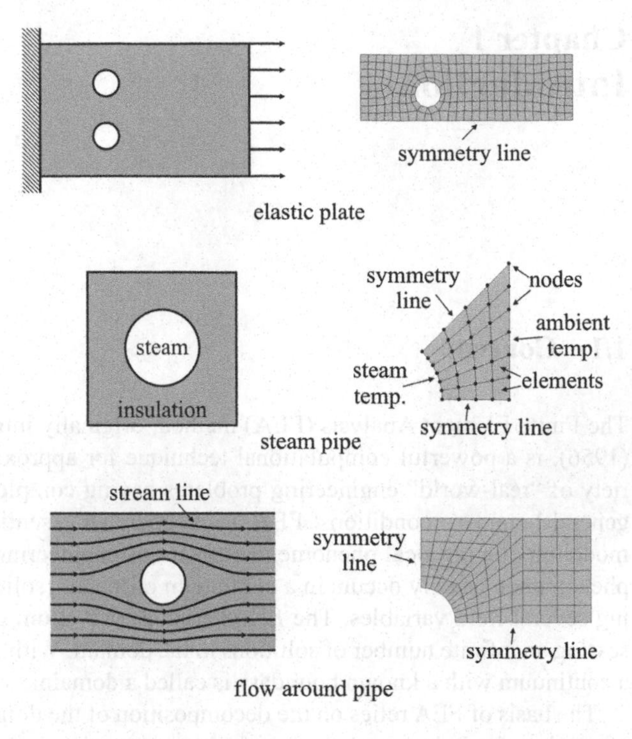

Fig. 1.1 FEA representation of practical engineering problems

- Development of the element matrix for the subdomain (element).
- Assembly of the element matrices for each subdomain to obtain the global matrix for the entire domain.
- Imposition of the boundary conditions.
- Solution of equations.
- Additional computations (if desired).

There are three main approaches to constructing an approximate solution based on the concept of FEA:

Direct Approach This approach is used for relatively simple problems, and it usually serves as a means to explain the concept of FEA and its important steps (discussed in Sect. 1.4).

Weighted Residuals This is a versatile method, allowing the application of FEA to problems whose functionals cannot be constructed. This approach directly utilizes the governing differential equations, such as those of heat transfer and fluid mechanics (discussed in Sect. 6.1).

Variational Approach This approach relies on the calculus of variations, which involves extremizing a functional. This functional corresponds to the potential energy in structural mechanics (discussed in Sect. 6.2).

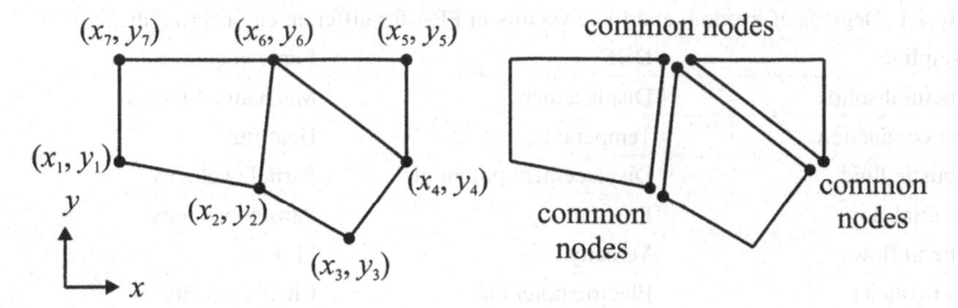

Fig. 1.2 Division of a domain into subdomains (elements)

In matrix notation, the global system of equations can be cast into

$$\mathbf{K}\mathbf{u} = \mathbf{F} \tag{1.1}$$

where $\mathbf{K}$ is the system stiffness matrix, $\mathbf{u}$ is the vector of unknowns, and $\mathbf{F}$ is the force vector. Depending on the nature of the problem, $\mathbf{K}$ may be dependent on $\mathbf{u}$, i.e., $\mathbf{K} = \mathbf{K}(\mathbf{u})$ and $\mathbf{F}$ may be time dependent, i.e., $\mathbf{F} = \mathbf{F}(t)$.

1.2 Nodes

As shown in Fig. 1.2, the transformation of the practical engineering problem to a mathematical representation is achieved by discretizing the domain of interest into elements (subdomains). These elements are connected to each other by their "common" nodes. A node specifies the coordinate location in space where degrees of freedom and actions of the physical problem exist. The nodal unknown(s) in the matrix system of equations represents one (or more) of the primary field variables. Nodal variables assigned to an element are called the degrees of freedom of the element.

The common nodes shown in Fig. 1.2 provide continuity for the nodal variables (degrees of freedom). Degrees of freedom (DOF) of a node are dictated by the physical nature of the problem and the element type. Table 1.1 presents the DOF and corresponding "forces" used in FEA for different physical problems.

1.3 Elements

Depending on the geometry and the physical nature of the problem, the domain of interest can be discretized by employing line, area, or volume elements. Some of the common elements in FEA are shown in Fig. 1.3. Each element, identified by an element number, is defined by a specific sequence of global node numbers. The

Table 1.1 Degrees of freedom and force vectors in FEA for different engineering disciplines

Discipline	DOF	Force vector
Structural/solids	Displacement	Mechanical forces
Heat conduction	Temperature	Heat flux
Acoustic fluid	Displacement potential	Particle velocity
Potential flow	Pressure	Particle velocity
General flows	Velocity	Fluxes
Electrostatics	Electric potential	Charge density
Magnetostatics	Magnetic potential	Magnetic intensity

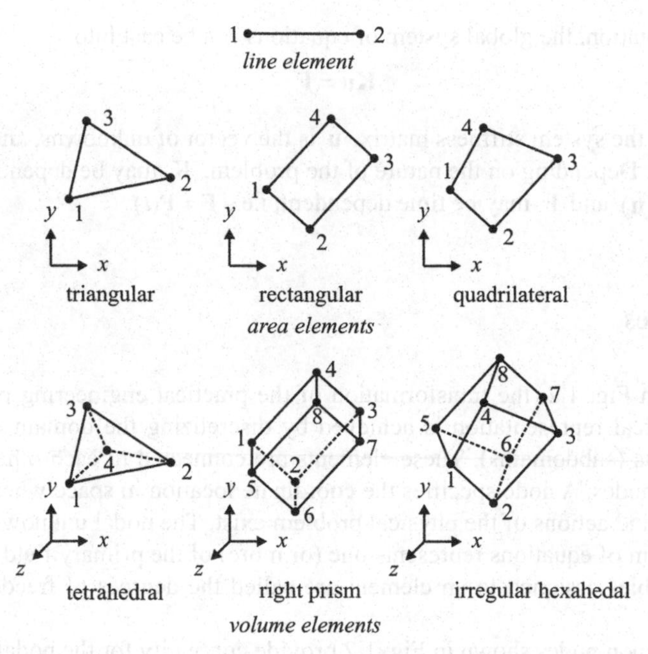

Fig. 1.3 Description of line, area, and volume elements with node numbers at the element level

specific sequence (usually counterclockwise) is based on the node numbering at the element level. The node numbering sequence for the elements shown in Fig. 1.4 are presented in Table 1.2.

1.4 Direct Approach

Although the direct approach is suitable for simple problems, it involves each fundamental step of a typical finite element analysis. Therefore, this approach is demonstrated by considering a linear spring system and heat flow in a one-dimensional (1-D) domain.

Table 1.2 Description of numbering at the element level

Element Number	Node 1	Node 2	Node 3	Node 4
1	1	2	6	7
2	3	4	6	2
3	4	5	6	

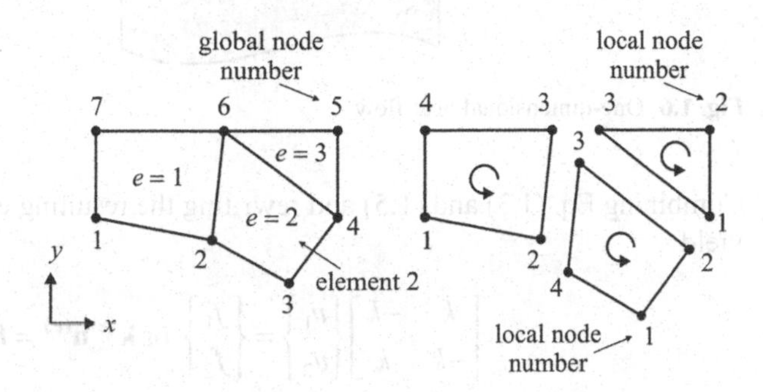

Fig. 1.4 Discretization of a domain: element and node numbering

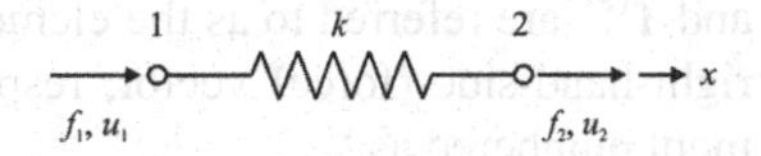

Fig. 1.5 Free-body diagram of a linear spring element

1.4.1 Linear Spring

As shown in Fig. 1.5, a linear spring with stiffness k has two nodes. Each node is subjected to axial loads of f_1 and f_2, resulting in displacements of u_1 and u_2 in their defined positive directions.

Subjected to these nodal forces, the resulting deformation of the spring becomes

$$u = u_1 - u_2 \tag{1.2}$$

which is related to the force acting on the spring by

$$f_1 = ku = k(u_1 - u_2) \tag{1.3}$$

The equilibrium of forces requires that

$$f_2 = -f_1 \tag{1.4}$$

which yields

$$f_2 = k(u_2 - u_1) \tag{1.5}$$

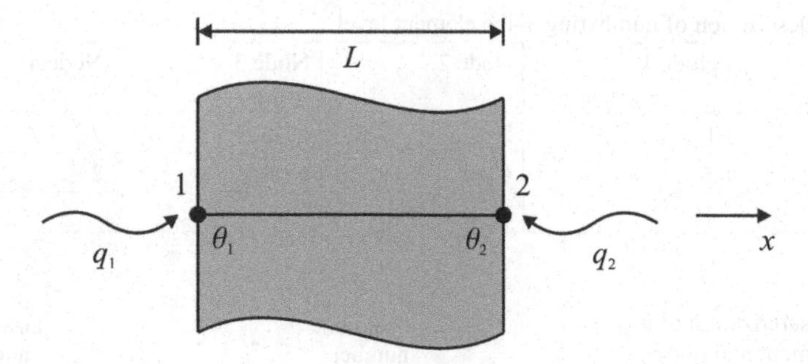

Fig. 1.6 One-dimensional heat flow

Combining Eq. (1.3) and (1.5) and rewriting the resulting equations in matrix form yield

$$\begin{bmatrix} k & -k \\ -k & k \end{bmatrix} \begin{Bmatrix} u_1 \\ u_2 \end{Bmatrix} = \begin{Bmatrix} f_1 \\ f_2 \end{Bmatrix} \text{ or } \mathbf{k}^{(e)}\mathbf{u}^{(e)} = \mathbf{f}^{(e)} \tag{1.6}$$

in which $\mathbf{u}^{(e)}$ is the vector of nodal unknowns representing displacement and $\mathbf{k}^{(e)}$ and $\mathbf{f}^{(e)}$ are referred to as the element characteristic (stiffness) matrix and element right-hand-side (force) vector, respectively. The superscript (e) denotes the element numbered as 'e'.

The stiffness matrix can be expressed in indicial form as $k_{ij}^{(e)}$

$$\mathbf{k}^{(e)} \sim k_{ij}^{(e)} \tag{1.7}$$

where the subscripts i and j $(i, j = 1, 2)$ are the row and the column numbers. The coefficients, $k_{ij}^{(e)}$, may be interpreted as the force required at node i to produce a unit displacement at node j while all the other nodes are fixed.

1.4.2 Heat Flow

Uniform heat flow through the thickness of a domain whose in-plane dimensions are long in comparison to its thickness can be considered as a one-dimensional analysis. The cross section of such a domain is shown in Fig. 1.6. In accordance with Fourier's Law, the rate of heat flow per unit area in the x-direction can be written as

$$q = -kA\frac{d\theta}{dx} \tag{1.8}$$

where A is the area normal to the heat flow, θ is the temperature, and k is the coefficient of thermal conductivity. For constant k, Eq. (1.8) can be rewritten as

$$q = -kA\frac{\Delta\theta}{L} \tag{1.9}$$

in which $\Delta\theta = \theta_2 - \theta_1$ denotes the temperature drop across the thickness denoted by L of the domain.

As illustrated in Fig. 1.6, the nodal flux (heat flow entering a node) at Node 1 becomes

$$q_1 = \frac{kA}{L}(\theta_1 - \theta_2) \tag{1.10}$$

The balance of the heat flux requires that

$$q_2 = -q_1 \tag{1.11}$$

which yields

$$q_2 = -\frac{kA}{L}(\theta_1 - \theta_2) \tag{1.12}$$

Combining Eq. (1.10) and (1.12) and rewriting the resulting equations in matrix form yield

$$\frac{kA}{L}\begin{bmatrix} 1 & -1 \\ -1 & 1 \end{bmatrix}\begin{Bmatrix} \theta_1 \\ \theta_2 \end{Bmatrix} = \begin{Bmatrix} q_1 \\ q_2 \end{Bmatrix} \text{ or } \mathbf{k}^{(e)}\boldsymbol{\theta}^{(e)} = \mathbf{q}^{(e)} \tag{1.13}$$

in which $\boldsymbol{\theta}^{(e)}$ is the vector of nodal unknowns representing temperature and $\mathbf{k}^{(e)}$ and $\mathbf{q}^{(e)}$ are referred to as the element characteristic matrix and element right-hand-side vector, respectively.

1.4.3 Assembly of the Global System of Equations

Modeling an engineering problem with finite elements requires the assembly of element characteristic (stiffness) matrices and element right-hand-side (force) vectors, leading to the global system of equations

$$\mathbf{K}\mathbf{u} = \mathbf{F} \tag{1.14}$$

in which $\mathbf{K}$ is the assembly of element characteristic matrices, referred to as the global system matrix and $\mathbf{F}$ is the assembly of element right-hand-side vectors, referred to as the global right-hand-side (force) vector. The vector of nodal unknowns is represented by $\mathbf{u}$.

The global system matrix, $\mathbf{K}$, can be obtained from the "expanded" element coefficient matrices, $\mathbf{k}^{(e)}$, by summation in the form

$$\mathbf{K} = \sum_{e=1}^{E} \mathbf{k}^{(e)} \tag{1.15}$$

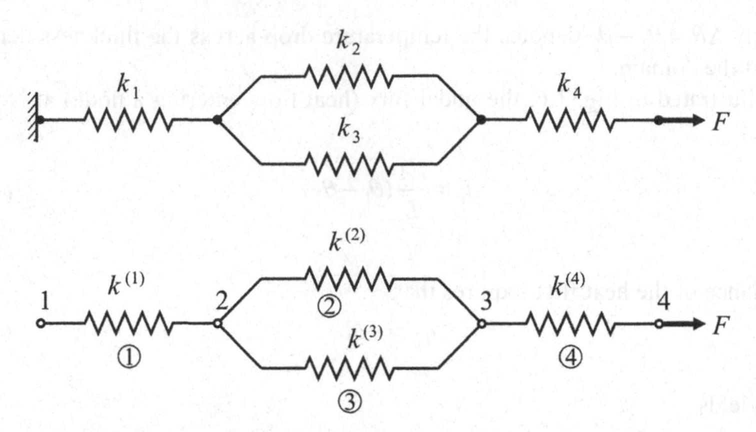

Fig. 1.7 System of linear springs (*top*) and corresponding FEA model (*bottom*)

in which the parameter E denotes the total number of elements. The "expanded" element characteristic matrices are the same size as the global system matrix but have rows and columns of zeros corresponding to the nodes not associated with element (e). The size of the global system matrix is dictated by the highest number among the global node numbers.

Similarly, the global right-hand-side vector, $\mathbf{F}$, can be obtained from the "expanded" element coefficient vectors, $\mathbf{f}^{(e)}$, by summation in the form

$$\mathbf{F} = \sum_{e=1}^{E} \mathbf{f}^{(e)} \tag{1.16}$$

The "expanded" element right-hand-side vectors are the same size as the global right-hand-side vector but have rows of zeros corresponding to the nodes not associated with element (e). The size of the global right-hand-side vector is also dictated by the highest number among the global node numbers.

The explicit steps in the construction of the global system matrix and the global right-hand-side-vector are explained by considering the system of linear springs shown in Fig. 1.7. Associated with element (e), the element equations for a spring given by Eq. (1.6) are rewritten as

$$\begin{bmatrix} k_{11}^{(e)} & k_{12}^{(e)} \\ k_{21}^{(e)} & k_{22}^{(e)} \end{bmatrix} \begin{Bmatrix} u_1^{(e)} \\ u_2^{(e)} \end{Bmatrix} = \begin{Bmatrix} f_1^{(e)} \\ f_2^{(e)} \end{Bmatrix} \tag{1.17}$$

in which $k_{11}^{(e)} = k_{22}^{(e)} = k^{(e)}$ and $k_{12}^{(e)} = k_{21}^{(e)} = -k^{(e)}$. The subscripts used in Eq. (1.17) correspond to Node 1 and Node 2, the local node numbers of element (e). The global node numbers specifying the connectivity among the elements for this system of springs is shown in Fig. 1.7, and the connectivity information is tabulated in Table 1.3.

Table 1.3 Table of connectivity

Element number	Local node numbering	Global node numbering
1	1	1
	2	2
2	1	2
	2	3
3	1	2
	2	3
4	1	3
	2	4

In accordance with Eq. (1.15), the size of the global system matrix is (4×4) and the specific contribution from each element is captured as

$$\text{Element 1:}\quad \overset{\begin{matrix}\boxed{1} & \boxed{2}\end{matrix}}{\begin{bmatrix} k_{11}^{(1)} & k_{12}^{(1)} \\ k_{21}^{(1)} & k_{22}^{(1)} \end{bmatrix}} \begin{matrix}\boxed{1}\\ \boxed{2}\end{matrix} \Rightarrow \overset{\begin{matrix}\boxed{1} & \boxed{2} & \boxed{3} & \boxed{4}\end{matrix}}{\begin{bmatrix} k_{11}^{(1)} & k_{12}^{(1)} & 0 & 0 \\ k_{21}^{(1)} & k_{22}^{(1)} & 0 & 0 \\ 0 & 0 & 0 & 0 \\ 0 & 0 & 0 & 0 \end{bmatrix}} \begin{matrix}\boxed{1}\\ \boxed{2}\\ \boxed{3}\\ \boxed{4}\end{matrix} \equiv \mathbf{k}^{(1)} \tag{1.18}$$

$$\text{Element 2:}\quad \overset{\begin{matrix}\boxed{2} & \boxed{3}\end{matrix}}{\begin{bmatrix} k_{11}^{(2)} & k_{12}^{(2)} \\ k_{21}^{(2)} & k_{22}^{(2)} \end{bmatrix}} \begin{matrix}\boxed{2}\\ \boxed{3}\end{matrix} \Rightarrow \overset{\begin{matrix}\boxed{1} & \boxed{2} & \boxed{3} & \boxed{4}\end{matrix}}{\begin{bmatrix} 0 & 0 & 0 & 0 \\ 0 & k_{11}^{(2)} & k_{12}^{(2)} & 0 \\ 0 & k_{21}^{(2)} & k_{22}^{(2)} & 0 \\ 0 & 0 & 0 & 0 \end{bmatrix}} \begin{matrix}\boxed{1}\\ \boxed{2}\\ \boxed{3}\\ \boxed{4}\end{matrix} \equiv \mathbf{k}^{(2)} \tag{1.19}$$

$$\text{Element 3:}\quad \overset{\begin{matrix}\boxed{2} & \boxed{3}\end{matrix}}{\begin{bmatrix} k_{11}^{(3)} & k_{12}^{(3)} \\ k_{21}^{(3)} & k_{22}^{(3)} \end{bmatrix}} \begin{matrix}\boxed{2}\\ \boxed{3}\end{matrix} \Rightarrow \overset{\begin{matrix}\boxed{1} & \boxed{2} & \boxed{3} & \boxed{4}\end{matrix}}{\begin{bmatrix} 0 & 0 & 0 & 0 \\ 0 & k_{11}^{(3)} & k_{12}^{(3)} & 0 \\ 0 & k_{21}^{(3)} & k_{22}^{(3)} & 0 \\ 0 & 0 & 0 & 0 \end{bmatrix}} \begin{matrix}\boxed{1}\\ \boxed{2}\\ \boxed{3}\\ \boxed{4}\end{matrix} \equiv \mathbf{k}^{(3)} \tag{1.20}$$

$$\text{Element 4:}\quad \begin{matrix} \boxed{3} & \boxed{4} \end{matrix} \begin{bmatrix} k_{11}^{(4)} & k_{12}^{(4)} \\ k_{21}^{(4)} & k_{22}^{(4)} \end{bmatrix} \begin{matrix} \boxed{3} \\ \boxed{4} \end{matrix} \Rightarrow \begin{matrix} \boxed{1} & \boxed{2} & \boxed{3} & \boxed{4} \end{matrix} \begin{bmatrix} 0 & 0 & 0 & 0 \\ 0 & 0 & 0 & 0 \\ 0 & 0 & k_{11}^{(4)} & k_{12}^{(4)} \\ 0 & 0 & k_{21}^{(4)} & k_{22}^{(4)} \end{bmatrix} \begin{matrix} \boxed{1} \\ \boxed{2} \\ \boxed{3} \\ \boxed{4} \end{matrix} \equiv \mathbf{k}^{(4)} \tag{1.21}$$

Performing their assembly leads to

$$\mathbf{K} = \sum_{e=1}^{4} \mathbf{k}^{(e)} = \mathbf{k}^{(1)} + \mathbf{k}^{(2)} + \mathbf{k}^{(3)} + \mathbf{k}^{(4)} \tag{1.22}$$

or

$$\mathbf{K} = \begin{bmatrix} k_{11}^{(1)} & k_{12}^{(1)} & 0 & 0 \\ k_{21}^{(1)} & \left(k_{22}^{(1)} + k_{11}^{(2)} + k_{11}^{(3)}\right) & \left(k_{12}^{(2)} + k_{12}^{(3)}\right) & 0 \\ 0 & \left(k_{21}^{(2)} + k_{21}^{(3)}\right) & \left(k_{22}^{(2)} + k_{22}^{(3)} + k_{11}^{(4)}\right) & k_{12}^{(4)} \\ 0 & 0 & k_{21}^{(4)} & k_{22}^{(4)} \end{bmatrix} \tag{1.23}$$

In accordance with Eq. (1.16), the size of the global right-hand-side vector is (4×1) and the specific contribution from each element is captured as

$$\text{Element 1:}\quad \begin{Bmatrix} f_1^{(1)} \\ f_2^{(1)} \end{Bmatrix} \begin{matrix} \boxed{1} \\ \boxed{2} \end{matrix} \Rightarrow \begin{Bmatrix} f_1^{(1)} \\ f_2^{(1)} \\ 0 \\ 0 \end{Bmatrix} \begin{matrix} \boxed{1} \\ \boxed{2} \\ \boxed{3} \\ \boxed{4} \end{matrix} \equiv \mathbf{f}^{(1)} \tag{1.24}$$

$$\text{Element 2:}\quad \begin{Bmatrix} f_1^{(2)} \\ f_2^{(2)} \end{Bmatrix} \begin{matrix} \boxed{2} \\ \boxed{3} \end{matrix} \Rightarrow \begin{Bmatrix} 0 \\ f_1^{(2)} \\ f_2^{(2)} \\ 0 \end{Bmatrix} \begin{matrix} \boxed{1} \\ \boxed{2} \\ \boxed{3} \\ \boxed{4} \end{matrix} \equiv \mathbf{f}^{(2)} \tag{1.25}$$

$$\text{Element 3:}\quad \begin{Bmatrix} f_1^{(3)} \\ f_2^{(3)} \end{Bmatrix} \begin{matrix} \boxed{2} \\ \boxed{3} \end{matrix} \Rightarrow \begin{Bmatrix} 0 \\ f_1^{(3)} \\ f_2^{(3)} \\ 0 \end{Bmatrix} \begin{matrix} \boxed{1} \\ \boxed{2} \\ \boxed{3} \\ \boxed{4} \end{matrix} \equiv \mathbf{f}^{(3)} \tag{1.26}$$

$$\text{Element 4:} \qquad \begin{Bmatrix} f_1^{(4)} \\ f_2^{(4)} \end{Bmatrix} \begin{matrix} \boxed{3} \\ \boxed{4} \end{matrix} \Rightarrow \begin{Bmatrix} 0 \\ 0 \\ f_1^{(4)} \\ f_2^{(4)} \end{Bmatrix} \begin{matrix} \boxed{1} \\ \boxed{2} \\ \boxed{3} \\ \boxed{4} \end{matrix} \equiv \mathbf{f}^{(4)} \tag{1.27}$$

Similarly, performing their assembly leads to

$$\mathbf{F} = \sum_{e=1}^{4} \mathbf{f}^{(e)} = \mathbf{f}^{(1)} + \mathbf{f}^{(2)} + \mathbf{f}^{(3)} + \mathbf{f}^{(4)} \tag{1.28}$$

or

$$\mathbf{F} = \begin{Bmatrix} f_1 \\ f_2 \\ f_3 \\ f_4 \end{Bmatrix} = \begin{Bmatrix} f_1^{(1)} \\ f_2^{(1)} + f_1^{(2)} + f_1^{(3)} \\ f_2^{(2)} + f_2^{(3)} + f_1^{(4)} \\ f_2^{(4)} \end{Bmatrix} \tag{1.29}$$

Consistent with the assembly of the global system matrix and the global right-hand-side vector, the vector of unknowns, $\mathbf{u}$, becomes

$$\mathbf{u} = \begin{Bmatrix} u_1 \\ u_2 \\ u_3 \\ u_4 \end{Bmatrix} = \begin{Bmatrix} u_1^{(1)} \\ u_2^{(1)} = u_1^{(2)} = u_1^{(3)} \\ u_2^{(2)} = u_2^{(3)} = u_1^{(4)} \\ u_2^{(4)} \end{Bmatrix} \tag{1.30}$$

1.4.4 Solution of the Global System of Equations

In order for the global system of equations to have a unique solution, the determinant of the global system matrix must be nonzero. However, an examination of the global system matrix reveals that one of its eigenvalues is zero, thus resulting in a zero determinant or singular matrix. Therefore, the solution is not unique. The eigenvector corresponding to the zero eigenvalue represents the translational mode, and the remaining nonzero eigenvalues represent all of the deformation modes.

For the specific values of $k_{11}^{(e)} = k_{22}^{(e)} = k^{(e)}$ and $k_{12}^{(e)} = k_{21}^{(e)} = -k^{(e)}$, the global system matrix becomes

$$\mathbf{K} = k^{(e)} \begin{bmatrix} 1 & -1 & 0 & 0 \\ -1 & 3 & -2 & 0 \\ 0 & -2 & 3 & -1 \\ 0 & 0 & -1 & 1 \end{bmatrix} \tag{1.31}$$

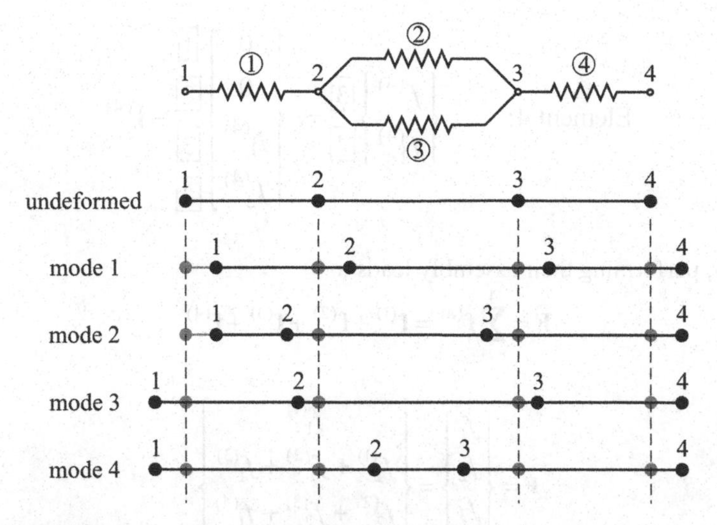

Fig. 1.8 Possible solution modes for the system of linear springs

with its eigenvalues $\lambda_1 = 0$, $\lambda_2 = 2$, $\lambda_3 = 3-\sqrt{5}$, and $\lambda_4 = 3+\sqrt{5}$. The corresponding eigenvectors are

$$\mathbf{u}^{(1)} = \begin{Bmatrix} 1 \\ 1 \\ 1 \\ 1 \end{Bmatrix}, \mathbf{u}^{(2)} = \begin{Bmatrix} 1 \\ -1 \\ -1 \\ 1 \end{Bmatrix}, \mathbf{u}^{(3)} = \begin{Bmatrix} -1 \\ 2-\sqrt{5} \\ -2+\sqrt{5} \\ 1 \end{Bmatrix}, \mathbf{u}^{(4)} = \begin{Bmatrix} -1 \\ 2+\sqrt{5} \\ -2-\sqrt{5} \\ 1 \end{Bmatrix} \tag{1.32}$$

Each of these eigenvectors represents a possible solution mode. The contribution of each solution mode is illustrated in Fig. 1.8.

In order for the global system of equations to have a unique solution, the global system matrix is rendered nonsingular by eliminating the zero eigenvalue. This is achieved by introducing a boundary condition so as to suppress the translational mode of the solution corresponding to the zero eigenvalue.

1.4.5 Boundary Conditions

As shown in Fig. 1.7, Node 1 is restrained from displacement. This constraint is satisfied by imposing the boundary condition of $u_1 = 0$. Either the nodal displacements, u_i, or the nodal forces, f_i, can be specified at a given node. It is physically impossible to specify both of them as known or as unknown. Therefore, the nodal force f_1 remains as one of the unknowns. The nodal displacements, u_2, u_3, and u_4 are treated as unknowns, and the corresponding nodal forces have values of $f_2 = 0$, $f_3 = 0$, and $f_4 = F$.

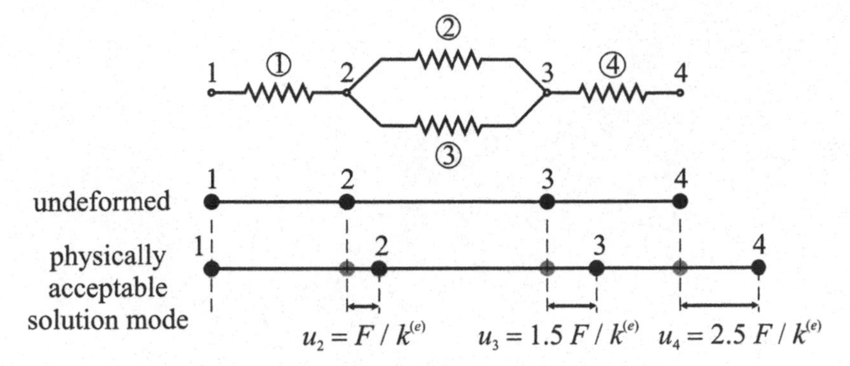

Fig. 1.9 Physically acceptable solution mode for the system of linear springs

These specified values are invoked into the global system of equations as

$$k^{(e)}\begin{bmatrix} 1 & -1 & 0 & 0 \\ -1 & 3 & -2 & 0 \\ 0 & -2 & 3 & -1 \\ 0 & 0 & -1 & 1 \end{bmatrix}\begin{Bmatrix} u_1 = 0 \\ u_2 \\ u_3 \\ u_4 \end{Bmatrix} = \begin{Bmatrix} f_1 \\ f_2 = 0 \\ f_3 = 0 \\ f_4 = F \end{Bmatrix} \tag{1.33}$$

leading to the following equations:

$$k^{(e)}\begin{bmatrix} 3 & -2 & 0 \\ -2 & 3 & -1 \\ 0 & -1 & 1 \end{bmatrix}\begin{Bmatrix} u_2 \\ u_3 \\ u_4 \end{Bmatrix} = \begin{Bmatrix} 0 \\ 0 \\ F \end{Bmatrix} \tag{1.34}$$

and

$$-k^{(e)}u_2 = f_1 \tag{1.35}$$

The coefficient matrix in Eq. (1.34) is no longer singular, and the solutions to these equations are obtained as

$$u_2 = \frac{F}{k^{(e)}},\ u_3 = \frac{3}{2}\frac{F}{k^{(e)}},\ u_4 = \frac{5}{2}\frac{F}{k^{(e)}} \tag{1.36}$$

and the unknown nodal force f_1 is determined as $f_1 = -F$. The final physically acceptable solution mode is shown in Fig. 1.9.

There exist systematic approaches to assemble the global coefficient matrix while invoking the specified nodal values (Bathe and Wilson 1976; Bathe 1996). The specified nodal variables are eliminated in advance from the global system of equations prior to the solution.

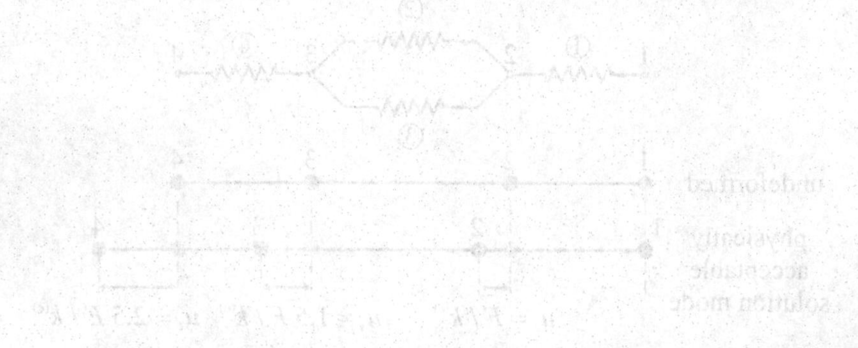

Chapter 2
Fundamentals of ANSYS

2.1 Useful Definitions

Before delving into the details of the procedures related to the ANSYS program, we define the following terms:

Jobname A specific name to be used for the files created during an ANSYS session. This name can be assigned either before or after starting the ANSYS program.

Working Directory A specific folder (directory) for ANSYS to store all of the files created during a session. It is possible to specify the *Working Directory* before or after starting ANSYS.

Interactive Mode This is the most common mode of interaction between the user and the ANSYS program. It involves activation of a platform called *Graphical User Interface* (*GUI*), which is composed of menus, dialog boxes, push-buttons, and different windows. *Interactive Mode* is the recommended mode for beginner ANSYS users as it provides an excellent platform for learning. It is also highly effective for postprocessing.

Batch Mode This is a method to use the ANSYS program without activating the *GUI*. It involves an *Input File* written in *ANSYS Parametric Design Language* (*APDL*), which allows the use of parameters and common programming features such as *DO* loops and *IF* statements. These capabilities make the *Batch Mode* a very powerful analysis tool. Another distinct advantage of the *Batch Mode* is realized when there is an error/mistake in the model generation. This type of problem can be fixed by modifying a small portion of the *Input File* and reading it again, saving the user a great deal of time.

Combined Mode This is a combination of the *Interactive* and *Batch Modes* in which the user activates the *GUI* and reads the *Input File*. Typically, this method allows the user to generate the model and obtain the solution using the *Input File* while reviewing the results using the *Postprocessor* within the *GUI*. This method combines the salient advantages of the *Interactive* and *Batch Modes*.

The online version of this book (doi: 10.1007/978-1-4939-1007-6_2) contains supplementary material, which is available to authorized users

E. Madenci, I. Guven, *The Finite Element Method and Applications in Engineering Using ANSYS®*, DOI 10.1007/978-1-4899-7550-8_2

2.2 Before an ANSYS Session

The construction of solutions to engineering problems using FEA requires either the development of a computer program based on the FEA formulation or the use of a commercially available general-purpose FEA program such as ANSYS. The ANSYS program is a powerful, multi-purpose analysis tool that can be used in a wide variety of engineering disciplines. Before using ANSYS to generate an FEA model of a physical system, the following questions should be answered based on engineering judgment and observations:

- What are the objectives of this analysis?
- Should the entire physical system be modeled, or just a portion?
- How much detail should be included in the model?
- How refined should the finite element mesh be?

In answering such questions, the computational expense should be balanced against the accuracy of the results. Therefore, the ANSYS finite element program can be employed in a correct and efficient way after considering the following:

- Type of problem.
- Time dependence.
- Nonlinearity.
- Modeling idealizations/simplifications.

Each of these topics is discussed in this section.

2.2.1 Analysis Discipline

The ANSYS program is capable of simulating problems in a wide range of engineering disciplines. However, this book focuses on the following disciplines:

Structural Analysis Deformation, stress, and strain fields, as well as reaction forces in a solid body.

Thermal Analysis Steady-state or time-dependent temperature field and heat flux in a solid body.

2.2.1.1 Structural Analysis

This analysis type addresses several different structural problems, for example:

Static Analysis The applied loads and support conditions of the solid body do not change with time. Nonlinear material and geometrical properties such as plasticity, contact, creep, etc., are available.

Modal Analysis This option concerns natural frequencies and modal shapes of a structure.

Table 2.1 Degrees of freedom for structural and thermal analysis disciplines

Discipline	Quantity	DOF
Structural	Displacement, stress, strain, reaction forces	Displacement
Thermal	Temperature, flux	Temperature

Harmonic Analysis The response of a structure subjected to loads only exhibiting sinusoidal behavior in time.

Transient Dynamic The response of a structure subjected to loads with arbitrary behavior in time.

Eigenvalue Buckling This option concerns the buckling loads and buckling modes of a structure.

2.2.1.2 Thermal Analysis

This analysis type addresses several different thermal problems, for example:

Primary Heat Transfer Steady-state or transient conduction, convection and radiation.

Phase Change Melting or freezing.

Thermomechanical Analysis Thermal analysis results are employed to compute displacement, stress, and strain fields due to differential thermal expansion.

2.2.1.3 Degrees of Freedom

The ANSYS solution for each of these analysis disciplines provides nodal values of the field variable. This primary unknown is called a degree of freedom (DOF). The degrees of freedom for these disciplines are presented in Table 2.1. The analysis discipline should be chosen based on the quantities of interest.

2.2.2 Time Dependence

The analysis with ANSYS should be time-dependent if:

- The solid body is subjected to time varying loads.
- The solid body has an initially specified temperature distribution.
- The body changes phase.

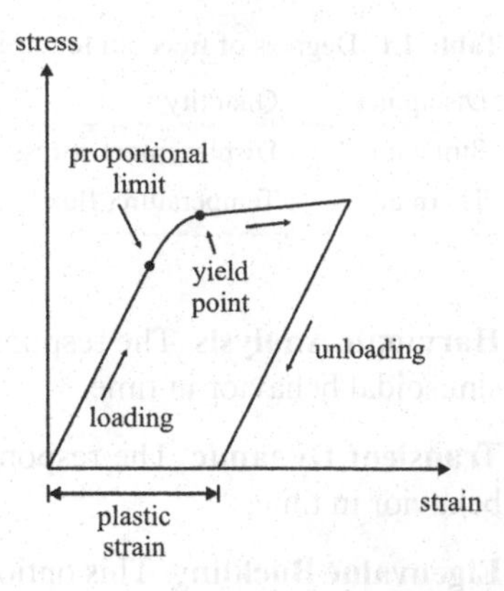

Fig. 2.1 Non-linear material response

2.2.3 *Nonlinearity*

Most real-world physical phenomena exhibit nonlinear behavior. There are many situations in which assuming a linear behavior for the physical system might provide satisfactory results. On the other hand, there are circumstances or phenomena that might require a nonlinear solution. A nonlinear structural behavior may arise because of geometric and material nonlinearities, as well as a change in the boundary conditions and structural integrity. These nonlinearities are discussed briefly in the following subsections.

2.2.3.1 Geometric Nonlinearity

There are two main types of geometric nonlinearity:

Large Deflection and Rotation If the structure undergoes large displacements compared to its smallest dimension and rotations to such an extent that its original dimensions and position, as well as the loading direction, change significantly, the large deflection and rotation analysis becomes necessary. For example, a fishing rod with a low lateral stiffness under a lateral load experiences large deflections and rotations.

Stress Stiffening When the stress in one direction affects the stiffness in another direction, stress stiffening occurs. Typically, a structure that has little or no stiffness in compression while having considerable stiffness in tension exhibits this behavior. Cables, membranes, or spinning structures exhibit stress stiffening.

2.2.3.2 Material Nonlinearity

A typical nonlinear stress-strain curve is given in Fig. 2.1. A linear material response is a good approximation if the material exhibits a nearly linear stress-strain

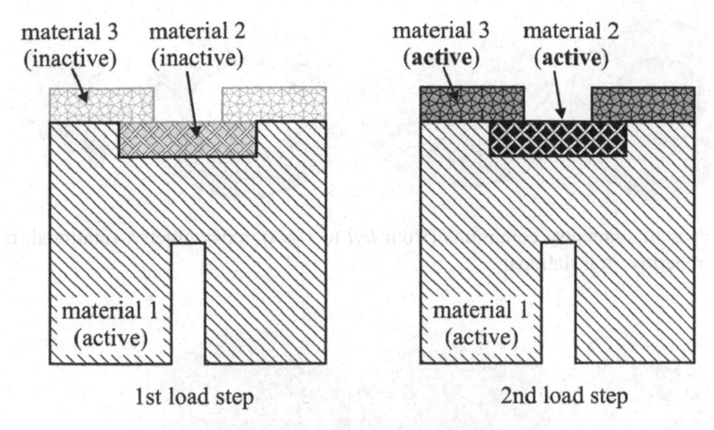

Fig. 2.2 Element birth and death used in a manufacturing problem

curve up to a proportional limit and the loading is in a manner that does not create stresses higher than the yield stress anywhere in the body.

Nonlinear material behavior in ANSYS is characterized as:

Plasticity Permanent, time-independent deformation.

Creep Permanent, time-dependent deformation.

Nonlinear Elastic Nonlinear stress-strain curve; upon unloading, the structure returns back to its original state—no permanent deformations.

Viscoelasticity Time-dependent deformation under constant load. Full recovery upon unloading.

Hyperelasticity Rubber-like materials.

2.2.3.3 Changing-status Nonlinearity

Many common structural features exhibit nonlinear behavior that is status dependent. When the status of the physical system changes, its stiffness shifts abruptly. The ANSYS program offers solutions to such phenomena through the use of nonlinear contact elements and birth and death options. This type of behavior is common in modeling manufacturing processes such as that of a shrink-fit (Fig. 2.2).

2.2.4 Practical Modeling Considerations

In order to reduce computational time, minor details that do not influence the results should not be included in the FE model. Minor details can also be ignored in order to render the geometry symmetric, which leads to a reduced FE model. However, in certain structures, "small" details such as fillets or holes may be the areas of

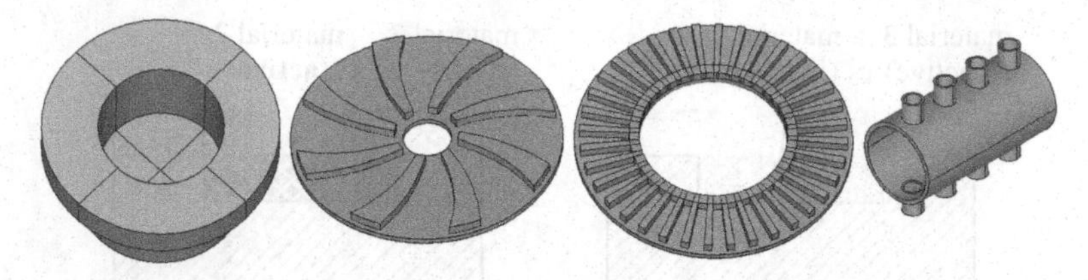

Fig. 2.3 Types of symmetry conditions (from *left* to *right*): axisymmetry, rotational, reflective/planar, and repetitive/translational

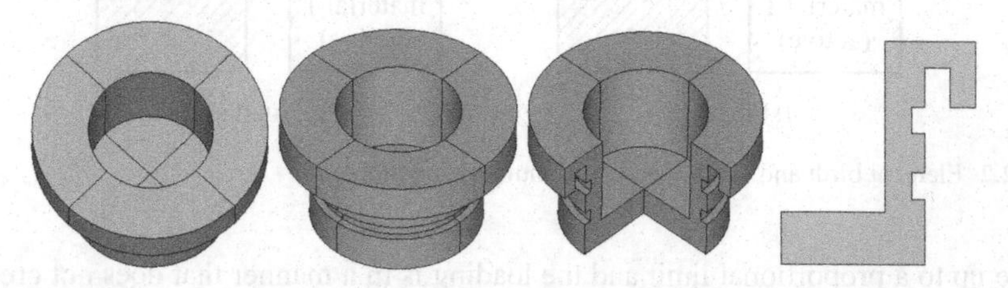

Fig. 2.4 Different views of a 3-D body with axisymmetry and its cross section (far *right*)

maximum stress, which might prove to be extremely important in the analysis and design. Engineering judgment is essential to balance the possible gain in computational cost against the loss of accuracy.

2.2.4.1 Symmetry Conditions

If the physical system under consideration exhibits symmetry in geometry, material properties, and loading, then it is computationally advantageous to model only a representative portion. If the symmetry observations are to be included in the model generation, the physical system must exhibit symmetry in all of the following:

- Geometry.
- Material properties.
- Loading.
- Degree of freedom constraints.

Different types of symmetry are:

- Axisymmetry.
- Rotational symmetry.
- Planar or reflective symmetry.
- Repetitive or translational symmetry.

Examples for each of the symmetry types are shown in Fig. 2.3. Each of these symmetry types is discussed below.

Axisymmetry As illustrated in Fig. 2.4, axisymmetry is the symmetry about a central axis, as exhibited by structures such as light bulbs, straight pipes, cones, circular plates, and domes.

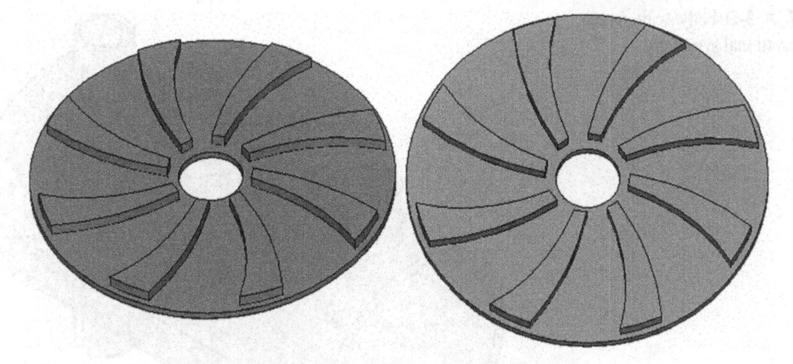

Fig. 2.5 Different views of a 3-D body with rotational symmetry

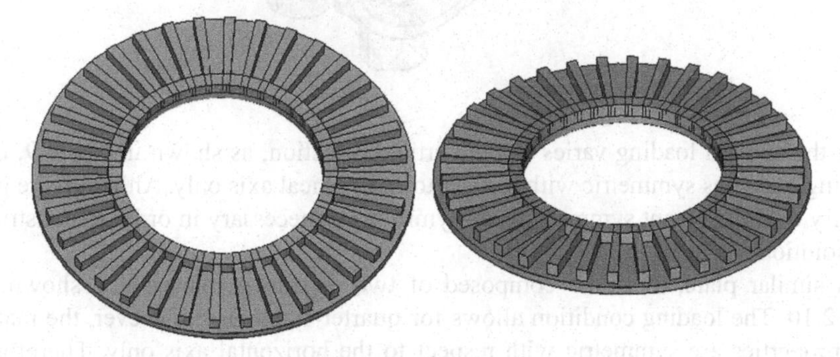

Fig. 2.6 Different views of a 3-D body with reflective/planar symmetry

Rotational Symmetry A structure possesses rotational symmetry when it is made up of repeated segments arranged about a central axis. An example is a turbine rotor (see Fig. 2.5).

Planar or Reflective Symmetry When one-half of a structure is a mirror image of the other half, planar or reflective symmetry exists, as shown in Fig. 2.6. In this case, the plane of symmetry is located on the surface of the mirror.

Repetitive or Translational Symmetry Repetitive or translational symmetry exists when a structure is made up of repeated segments lined up in a row, such as a long pipe with evenly spaced cooling fins, as shown in Fig. 2.7.

Symmetry in Material Properties, Loading, Displacements Once symmetry in geometry is observed, the same symmetry plane or axis should also be valid for the material properties, loading (forces, pressure, etc.), and constraints. For example, a homogeneous and isotropic square plate with a hole at the center under horizontal tensile loading (Fig. 2.8) has octant (1/8th) symmetry in both geometry and material with respect to horizontal, vertical, and both diagonal axes. However, the loading is symmetric with respect to horizontal and vertical axes only. Therefore, a quarter of the structure is required in the construction of the solution.

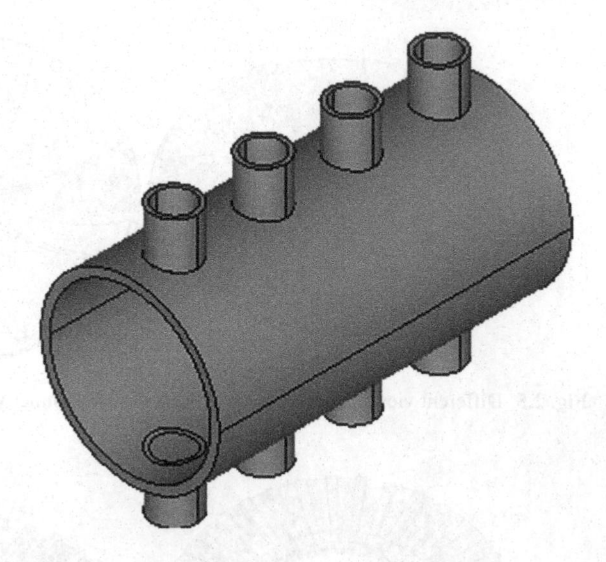

Fig. 2.7 A 3-D body with repetitive/translational symmetry

If the applied loading varies in the vertical direction, as shown in Fig. 2.9, the loading becomes symmetric with respect to the vertical axis only. Although the geometry exhibits octant symmetry, half-symmetry is necessary in order to construct the solution.

A similar plate, this time composed of two dissimilar materials is shown in Fig. 2.10. The loading condition allows for quarter-symmetry; however, the material properties are symmetric with respect to the horizontal axis only. Therefore, it is limited to half-symmetry. If this plate is subjected to a horizontal tensile load varying in the vertical direction, as shown in Fig. 2.11, no symmetry condition is present.

Since a structure may exhibit symmetry in one or more of the aforementioned categories, one should try to find the smallest possible segment of the structure that

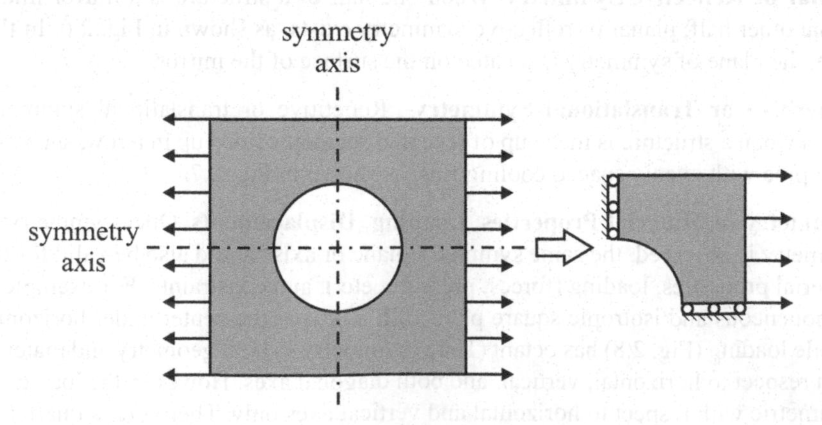

Fig. 2.8 Example of quarter-symmetry

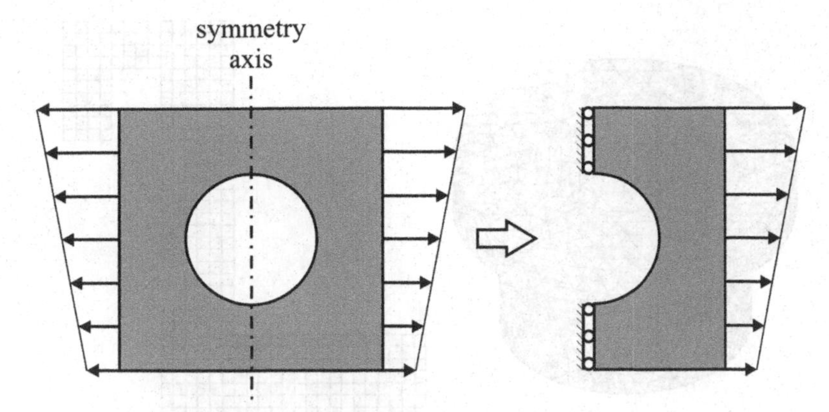

Fig. 2.9 Example of half-symmetry with respect to vertical axis

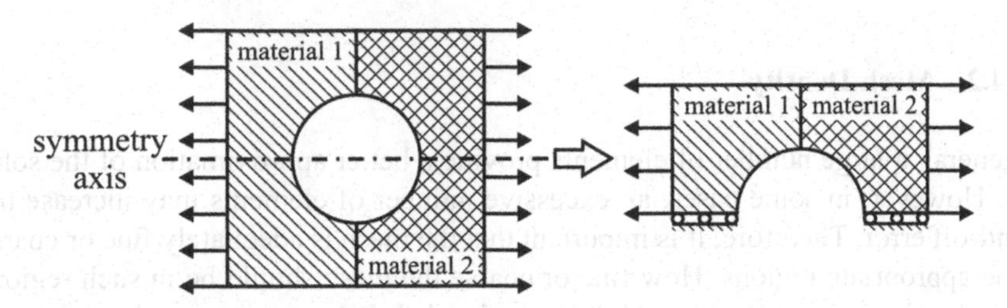

Fig. 2.10 Example of half-symmetry with respect to horizontal axis

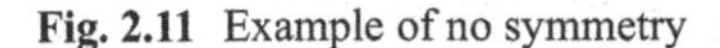
Fig. 2.11 Example of no symmetry

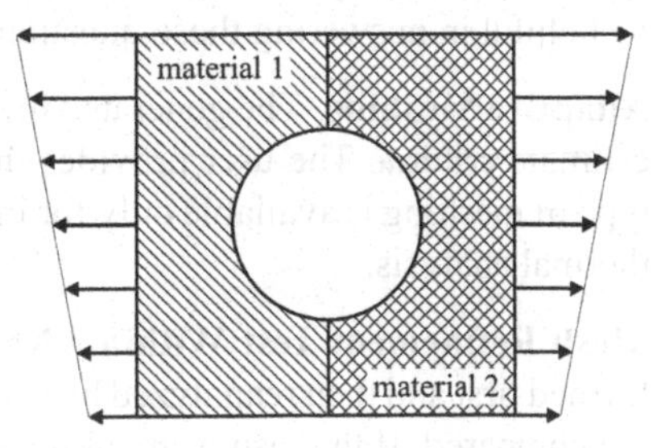

would represent the entire structure. If the physical system exhibits symmetry in geometry, material properties, loading, and displacement constraints, it is computationally advantageous to use symmetry in the analysis. Typically, the use of symmetry produces better results as it leads to a finer, more detailed model than would otherwise be possible.

A three-dimensional finite element mesh of the structure shown in Fig. 2.12 contains 18,739 tetrahedral elements with 5014 nodes. However, the two-dimensional mesh of the cross section necessary for the axisymmetric analysis has 372 quadrilateral elements and 447 nodes. The use of symmetry in this case reduces the CPU time required for the solution while delivering the same level of accuracy in the results.

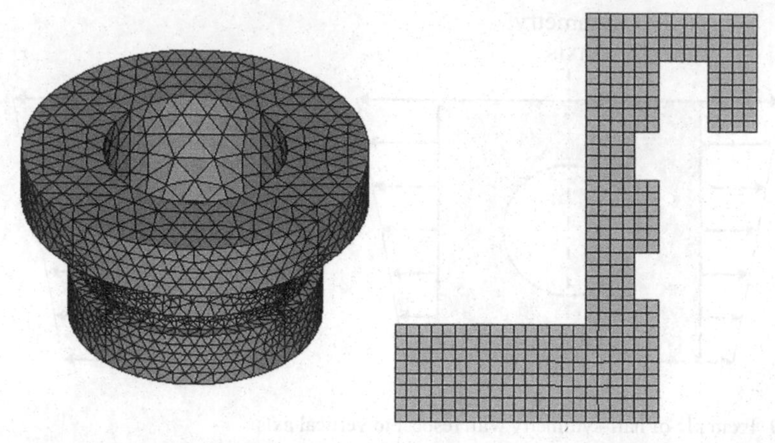

Fig. 2.12 Three-dimensional mesh of a structure (*left*) and 2-D mesh of the same structure (*right*) using axisymmetry

2.2.4.2 Mesh Density

In general, a large number of elements provide a better approximation of the solution. However, in some cases, an excessive number of elements may increase the round-off error. Therefore, it is important that the mesh is adequately fine or coarse in the appropriate regions. How fine or coarse the mesh should be in such regions is another important question. Unfortunately, definitive answers to the questions about mesh refinement are not available since it is completely dependent on the specific physical system considered. However, there are some techniques that might be helpful in answering these questions:

Adaptive Meshing The generated mesh is required to meet acceptable energy error estimate criteria. The user provides the "acceptable" error level information. This type of meshing is available only for linear static structural analysis and steady-state thermal analysis.

Mesh Refinement Test Within ANSYS An analysis with an initial mesh is performed first and then reanalyzed by using twice as many elements. The two solutions are compared. If the results are close to each other, the initial mesh configuration is considered to be adequate. If there are substantial differences between the two, the analysis should continue with a more-refined mesh and a subsequent comparison until convergence is established.

Submodeling If the mesh refinement test yields nearly identical results for most regions and substantial differences in only a portion of the model, the built-in "submodeling" feature of ANSYS should be employed for localized mesh refinement. This feature is described in Chap. 11.

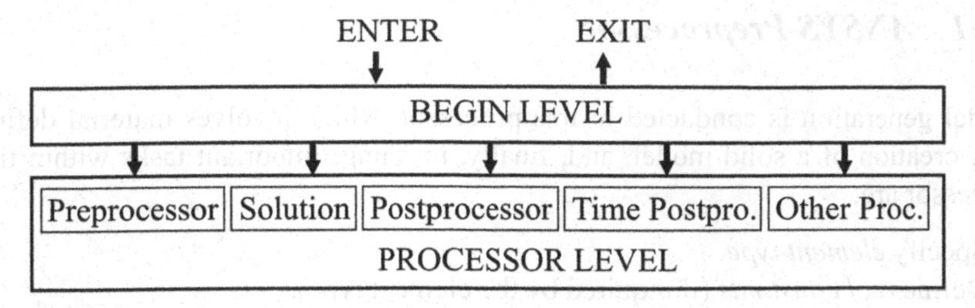

Fig. 2.13 Schematic of ANSYS levels

2.3 Organization of ANSYS Software

There are two primary levels in the ANSYS program, as shown in Fig. 2.13:

Begin Level Gateway into and out of ANSYS and platform to utilize some global controls such as changing the *jobname*, etc.

Processor Level This level contains the processors (preprocessor, solution, postprocessor, etc.) that are used to conduct finite element analyses.

The user is in the *Begin Level* upon entering the ANSYS program. One can proceed to the *Processor Level* by clicking the mouse on one of the processor selections in the ANSYS *Main Menu*.

2.4 ANSYS Analysis Approach

There are three main steps in a typical ANSYS analysis:

- Model generation:
  - Simplifications, idealizations.
  - Define materials/material properties.
  - Generate finite element model (mesh).
- Solution:
  - Specify boundary conditions.
  - Obtain the solution.
- Review results:
  - Plot/list results.
  - Check for validity.

Each of these steps corresponds to a specific processor or processors within the *Processor Level* in ANSYS. In particular, model generation is done in the *Preprocessor* and application of loads and the solution is performed in the *Solution Processor*. Finally, the results are viewed in the *General Postprocessor* and *Time History Postprocessor* for steady-state (static) and transient (time-dependent) problems, respectively. There are several other processors within the ANSYS program. These mostly concern optimization and probabilistic-type problems. The most commonly used processors are described in the following subsections.

2.4.1 ANSYS Preprocessor

Model generation is conducted in this processor, which involves material definition, creation of a solid model, and, finally, meshing. Important tasks within this processor are:

- Specify *element type*.
- Define *real constants* (if required by the element type).
- Define *material properties*.
- Create the model geometry.
- Generate the mesh.

Although the boundary conditions can also be specified in this processor, it is usually done in the *Solution Processor*.

2.4.2 ANSYS Solution Processor

This processor is used for obtaining the solution for the finite element model that is generated within the *Preprocessor*. Important tasks within this processor are:

- Define *analysis type* and *analysis options*.
- Specify boundary conditions.
- Obtain solution.

2.4.3 ANSYS General Postprocessor

In this processor, the results at a specific time (if the *analysis type* is transient) over the entire or a portion of the model are reviewed. This includes the plotting of contours, vector displays, deformed shapes, and listings of the results in tabular format.

2.4.4 ANSYS Time History Postprocessor

This processor is used to review results at specific points in time (if the *analysis type* is transient). Similar to the General Postprocessor, it provides graphical variations and tabular listings of results data as functions of time.

2.5 ANSYS File Structure

Several files are created during a typical ANSYS analysis. Some of these files are in ASCII format while the others are binary. Brief descriptions of common file types are given below.

2.5.1 Database File

During a typical ANSYS analysis, input and output data reside in memory until they are saved in a *Database File,* which is saved in the *Working Directory*. The syntax for the name of the *Database File* is *jobname.db*. This binary file includes the element type, material properties, geometry (solid model), mesh (nodal coordinates and element connectivity), and the results if a solution is obtained. Once the *Database File* is saved, the user can resume from this file at any time. There are three distinct ways to save and resume the *Database File*:

- Use the *Utility Menu.*
- Click on ***SAVE_DB*** or ***RESUM_DB*** button on the *ANSYS Toolbar*.
- Issue the command **SAVE** or **RESUME** in the *Input Field.*

2.5.2 Log File

The *Log File* is an ASCII file, which is created (or resumed) immediately upon entering ANSYS. Every action taken by the user is stored sequentially in this file in command format (ANSYS Parametric Design Language (APDL)). The syntax for the name of the *Log File*, which is also saved in the *Working Directory*, is *jobname.log*. If *jobname.log* already exists in the *Working Directory*, ANSYS appends the newly executed actions instead of overwriting the file. The *Log File* can be utilized to:

- Understand how an analysis was performed by another user.
- Learn the command equivalents of the actions taken within ANSYS.

2.6 Error File

Similar to the *Log File*, the *Error File* is an ASCII file, which is created (or resumed) immediately upon entering ANSYS. This file captures all warning and error messages issued by ANSYS during a session. It is saved in the *Working Directory* with the following syntax for the name: *jobname.err*. If *jobname.err* already exists in the *Working Directory*, ANSYS appends the newly issued warning and error messages instead of overwriting the file. This file is particularly important when ANSYS issues several warning and error messages too quickly during an interactive session. The user can then consult the *Error File* to discover the exact cause(s) of each of the warnings or errors.

2.6.1 Results Files

The results of an ANSYS analysis are stored in a separate *Results File*. This file is a binary file and, depending upon the *Analysis Type*, the file's extension takes a different form. The following syntax applies to the *Results File* name for the selected *Analysis Type*:

Structural analysis	*jobname.rst*
Thermal analysis	*jobname.rth*
Fluids analysis	*jobname.rfl*

2.7 Description of ANSYS Menus and Windows

When using the ANSYS program in *Interactive Mode*, the *Graphical User Interface* (*GUI*) is activated. The *GUI* has six distinct components:

Utility Menu Contains functions that are available throughout the ANSYS session, such as file controls, selecting, graphic controls, and parameters. The ANSYS *Help System* is also accessible through this menu.

Main Menu Contains the primary ANSYS functions organized by processors (*Preprocessor*, *Solution*, *General Postprocessor*, etc.).

Toolbar Contains push-buttons for executing commonly used ANSYS commands and functions. Customized buttons can be created.

Input Field Displays a text field for typing commands. All previously typed commands are stored in a pull-down menu for easy reference and access.

Graphics Window Displays the graphical representation of the models/meshes created within ANSYS. Also, the related results are reviewed in this window.

Output Window Receives text output from the program. This window is usually positioned behind other windows and can be raised to the front when necessary.

Figure 2.14 shows a typical ANSYS *GUI* with each of the preceding components identified.

2.7.1 Utility Menu

The *Utility Menu* contains utility functions that are independent of *ANSYS Levels* (i.e., begin and processor levels), with some exceptions. The *Utility Menu* contains ten items, each of which brings up a pull-down menu of subitems. Clicking the left mouse button on these subitems will result in one of the following:

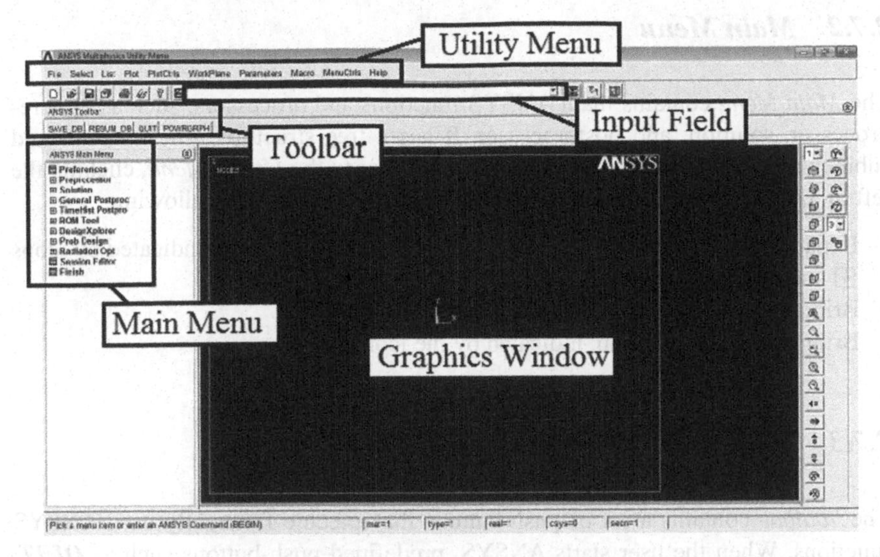

Fig. 2.14 Typical ANSYS GUI with separate components identified

- Bring up a submenu, indicated by the icon ▶ .
- Immediately execute a function.
- Bring up a dialog box, indicated by the icon ... .
- Bring up a picking menu, indicated by the icon + .

Brief descriptions of each of the menu items under the *Utility Menu* are given below.

File item under *Utility Menu*: Contains file- and database-related functions, such as clearing the database, reading an input file, saving the database to a file, or resuming a database from a file. This menu item can be used to exit the program.
Select item under *Utility Menu*: Includes functions that allow the user to select a subset of data and to create *Components*.
List item under *Utility Menu*: This menu item allows the user to list any data stored in the ANSYS database. Also, status information about different areas of the program and contents of files in the system are available.
Plot item under *Utility Menu*: This menu item allows the user to plot ANSYS entities such as keypoints, lines, areas, volumes, nodes, and elements. If a solution is obtained, results can also be plotted through this menu item.
PlotCtrls item under *Utility Menu*: Contains functions that control the view, style, and other characteristics of graphic displays.
WorkPlane item under *Utility Menu*: Use of *WorkPlane* offers great convenience for *Solid Model* generation. This menu item enables the user to toggle the *Working Plane* on and off, and to move, rotate, and maneuver it. Coordinate system operations are also performed under this menu item.
Parameters item under *Utility Menu*: Contains functions to define, edit, and delete scalar and array parameters.
Macro item under *Utility Menu*: This menu item allows the user to execute *Macros* and data blocks. Under this menu item, the user can also manipulate the push-buttons on the *Toolbar*.
MenuCtrls item under *Utility Menu*: Allows the user to format the menus, as well as manipulate the *Toolbar*.
Help item under *Utility Menu*: Brings up the ANSYS *Help System*.

2.7.2 Main Menu

The *Main Menu* contains main ANSYS functions and processors, such as the preprocessor, solution, and postprocessor. It has a tree structure, where menus and submenus can be expanded and collapsed. Similar to the *Utility Menu*, clicking the left mouse button on the Main Menu items results in one of the following:

- Expand or collapse the submenus attached to the menu item, indicated by icons ⊞ and ⊟, respectively.
- Bring up a dialog box, indicated by the icon ▦.
- Bring up a picking menu, indicated by the icon ↗.

2.7.3 Toolbar

The *Toolbar* contains a set of push-buttons that execute frequently used ANSYS functions. When the user starts ANSYS, predefined push-buttons such as ***QUIT***, ***SAVE_DB***, and ***RESUM_DB*** appear in the toolbar. The user can create customized push-buttons and delete or edit the existing ones.

2.7.4 Input Field

This field allows the user to type in commands directly as opposed to the use of menu items. The *Input Field* consists of two main regions:

- Command entry box.
- History buffer.

2.7.5 Graphics Window

All ANSYS graphics are displayed in the *Graphics Window*. Also, the user performs all of the graphical "picking" in this window.

2.7.6 Output Window

All of the text output generated as a result of command responses, warnings, and errors appear in the *Output Window*. It is positioned behind the main ANSYS window, but can be raised to the front when necessary.

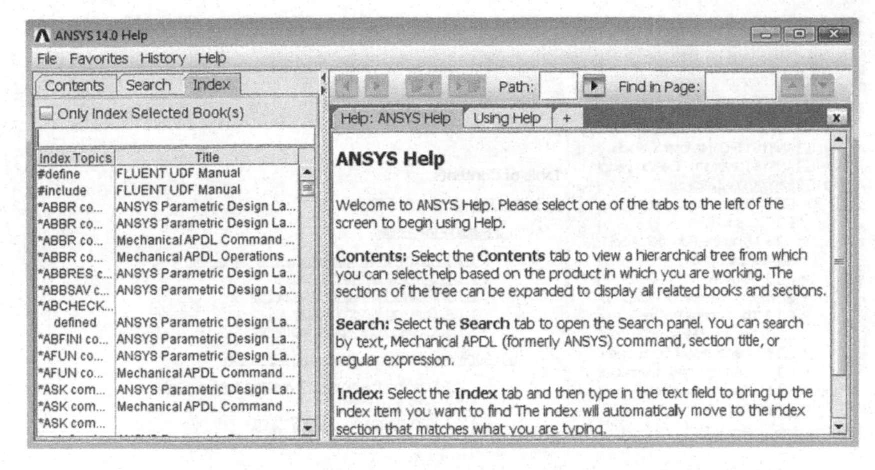

Fig. 2.15 ANSYS *Help System*

2.8 Using the ANSYS Help System

Information on ANSYS procedures, commands, and concepts can be found in the ANSYS *Help System*. The importance of knowing how to use the *Help System* cannot be overemphasized. It can be accessed within the *Graphical User Interface* (*GUI*) in three ways:

- By choosing the *Help* menu item under *Utility Menu*.
- By pressing the ***Help*** button within dialog boxes.
- By entering the **HELP** command directly in the *Input Field*.

The *Help System* is also available as a stand-alone program outside of ANSYS. The user can bring up the desired help topic by choosing it from the system's table of contents or index, through a word search, or by choosing a hypertext link. The *Help System* is built on the HTML platform in the form of web pages. As indicated in Fig. 2.15, there are three tabs on the left of the *Help Window*: *Contents*, *Index*, and *Search*. The help pages are displayed on the right side of the *Help Window*. Selected topics regarding the *Help System* are discussed in the following subsections.

2.8.1 Help Contents

The first tab on the left side of the *Help Window* is the *Contents Tab*, as shown in Fig. 2.16. It is a collection of several different ANSYS *Manuals* containing thousands of pages. The *Contents Tab* is organized in a tree structure for easy navigation. It is recommended that beginner ANSYS users take the time to read the relevant chapters in each *Manual*. Throughout this book, the reader is referred to several specific chapters in these *Manuals* for a thorough understanding of the topics being discussed.

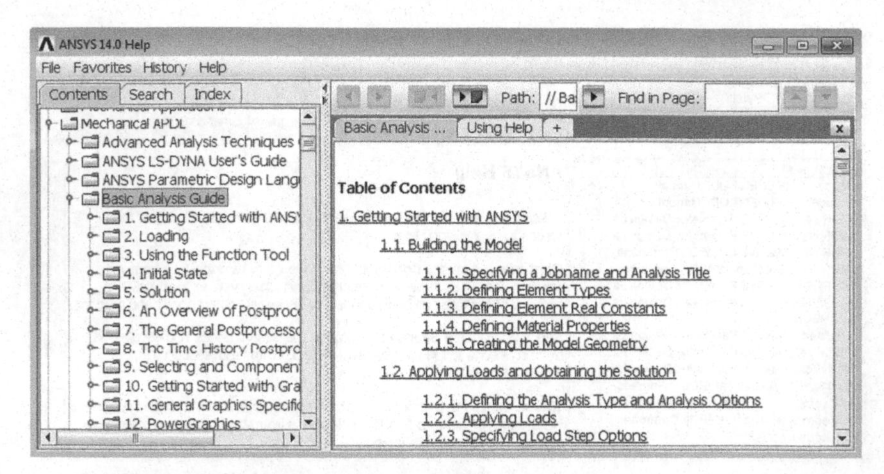

Fig. 2.16 ANSYS *Help System* with *Contents* tab activated

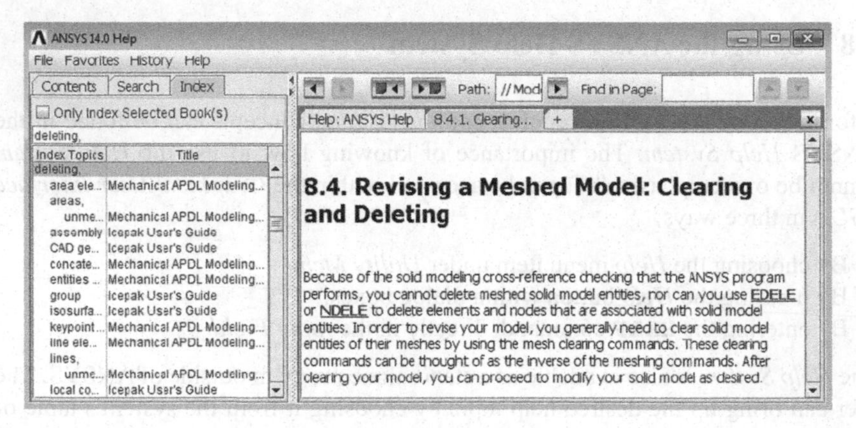

Fig. 2.17 ANSYS *Help System* with *Index* tab activated

2.8.2 Help Index

The *Index Tab.* (Fig. 2.17) is the second tab on the left side of the *Help Window*. Every single help page contained in the ANSYS *Help System* is exhaustively listed under this tab. It is useful for finding which help pages are available for a given topic. Upon typing the topic of interest, a list of help pages appears, giving the user a chance to browse for the most-relevant help page.

2.8.3 Search in Help

The user can perform a word search of the ANSYS *Manuals* through the *Search Tab.* (Fig. 2.18), which is the third tab on the left side of the *Help Window*. As a

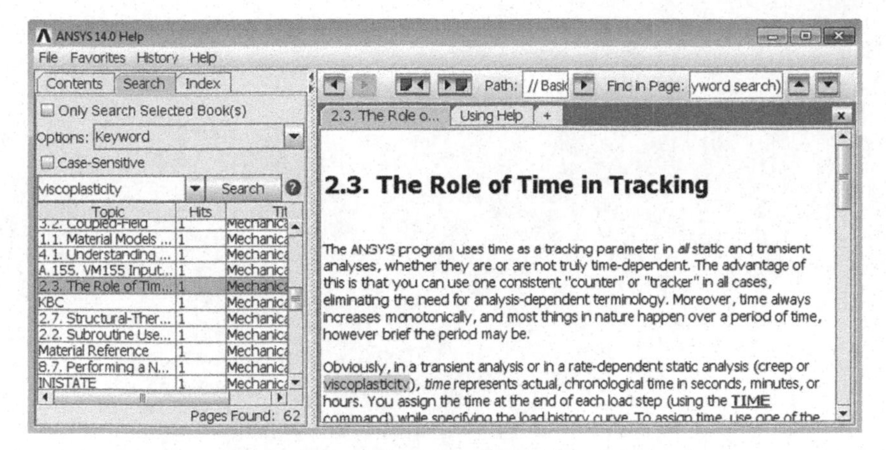

Fig. 2.18 ANSYS *Help System* with *Search* tab activated

result of an inquiry, a list of help pages containing the search word appears and the user can select which pages to display.

2.8.4 Verification Manual

Although all of the ANSYS *Manuals* included under the *Contents Tab* are important sources of information, one particular *Manual* deserves special emphasis, the *Verification Manual*. The purpose of this *Manual* is to demonstrate the capabilities of ANSYS in solving fundamental engineering problems with analytical solutions. Another important feature of the *Verification Manual* is its suitability as an effective learning tool. There is a corresponding *Input File* for each of the verification problems included in this manual (in excess of 200). As mentioned earlier, the input files contain ANSYS commands to be executed sequentially when read from within ANSYS. Each of these commands corresponds to a specific action in the *Interactive Mode*. Once the verification problem that is the closest to the problem at hand is identified, the user can then study the corresponding *Input File* and learn the essential steps in solving the problem using ANSYS. The *Verification Manual* also serves as an excellent tool for learning to use ANSYS in *Batch Mode*.

Chapter 3
Fundamentals of Discretization

3.1 Local and Global Numbering

In solving an engineering problem with the finite element method (FEM), the domain is discretized by employing elements. The characteristics of the problem dictate the dimensionality of the problem, i.e., one, two, or three dimensional. A brief summary of the common element types utilized in a finite element analysis (FEA) is presented in Fig. 3.1. Once the domain of the problem is discretized by elements, a unique element number identifies each element and a unique node number identifies each node in the domain. As illustrated in Fig. 3.2, nodes are also numbered within each element, and are called local node numbers. The unique node numbering within the entire domain is called global node numbering. This is part of the computational procedure in FEA.

3.2 Approximation Functions

The variation of the field variable, $\phi^{(e)}$, over an element is approximated by an appropriate choice of functions, as illustrated in Fig. 3.3. The selection of these functions is the core of the finite element method. The approximation functions should be reliable in the sense that as the mesh becomes more refined, the approximate solution should converge to the exact solution monotonically. Oscillatory convergence is unreliable because it is possible to observe an increase in error with the refined mesh. Oscillatory and monotonic convergences are demonstrated in Fig. 3.4. Common approximation functions are usually polynomials since their differentiation and integration are rather straightforward compared to other functions.

In order to achieve a monotonically convergent solution, the polynomials chosen as approximation functions must satisfy four requirements:

Requirement 1 *Continuous behavior of the approximation function within the element—no kinks or jumps.*

The online version of this book (doi: 10.1007/978-1-4939-1007-6_3) contains supplementary material, which is available to authorized users

E. Madenci, I. Guven, *The Finite Element Method and Applications in Engineering Using ANSYS®*, DOI 10.1007/978-1-4899-7550-8_3

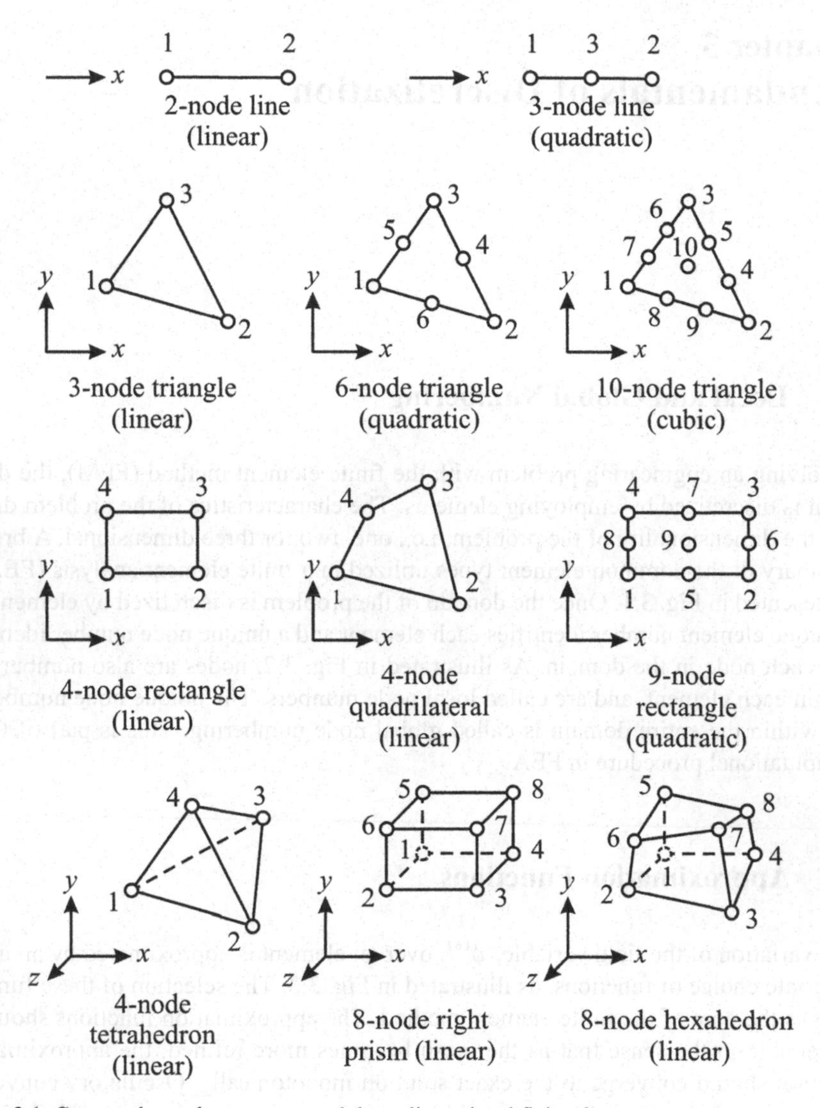

Fig. 3.1 Commonly used one-, two-, and three-dimensional finite elements

Requirement 2 *Compatibility along the common nodes, boundaries or surfaces between adjacent elements—no gaps between elements*

The elements satisfying the continuity and compatibility requirements are called conformal elements (Fig. 3.5).

Requirement 3 *Completeness, permitting rigid body motion of the element and ensuring (constant) variation of ϕ and its derivatives within the element.*

The reason for this requirement is best illustrated by considering a cantilever beam under a concentrated load in the middle (Fig. 3.6). As a result of this loading,

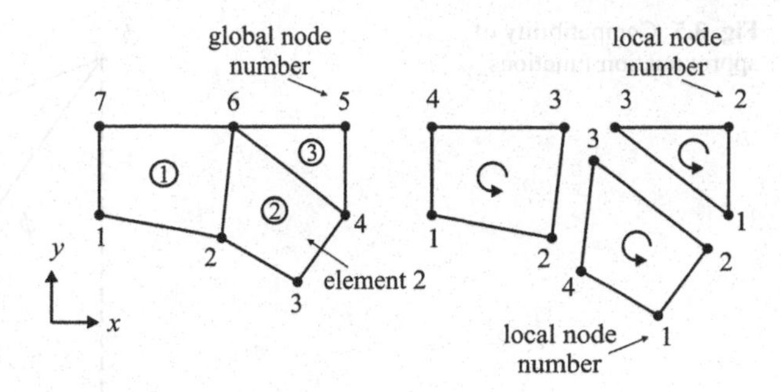

Fig. 3.2 Element numbers, global node numbers, and local node numbers

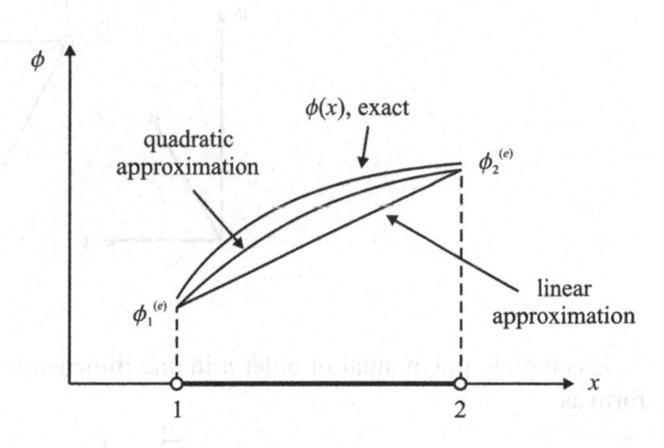

Fig. 3.3 Element approximation functions

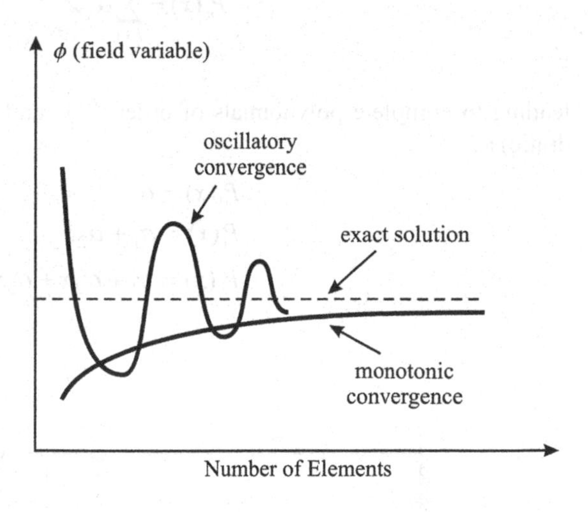

Fig. 3.4 Oscillatory and monotonic convergence of approximate solution

deformation occurs only to the left of the load. The section of the beam to the right of the load experiences only rigid-body translations and rotations (constant displacements and zero strain), i.e., no stresses and strains occur. Therefore, the element approximation functions must permit such behavior. Complete polynomials satisfy these requirements.

Fig. 3.5 Compatibility of approximation functions

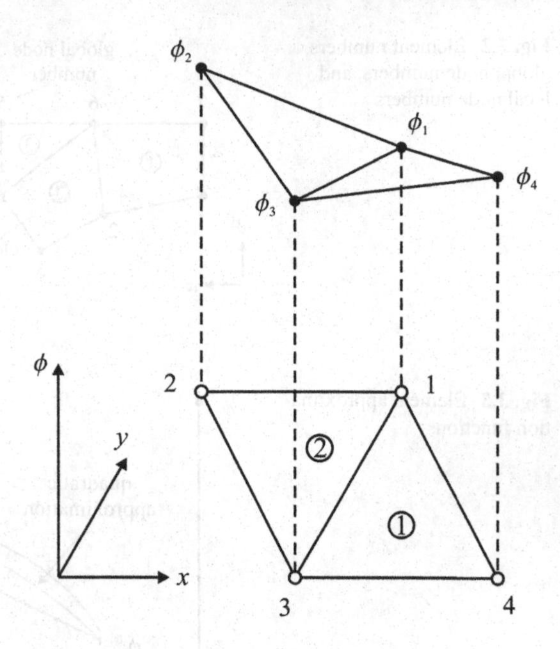

A complete polynomial of order n in one dimension can be written in compact form as

$$P_n(x) = \sum_{k=1}^{n+1} \alpha_k x^{k-1} \tag{3.1}$$

leading to complete polynomials of order 0, 1, and 2 (constant, linear, and quadratic) as

$$\begin{aligned} P_0(x) &= \alpha_1 \\ P_1(x) &= \alpha_1 + \alpha_2 x \\ P_2(x) &= \alpha_1 + \alpha_2 x + \alpha_3 x^2 \end{aligned} \tag{3.2}$$

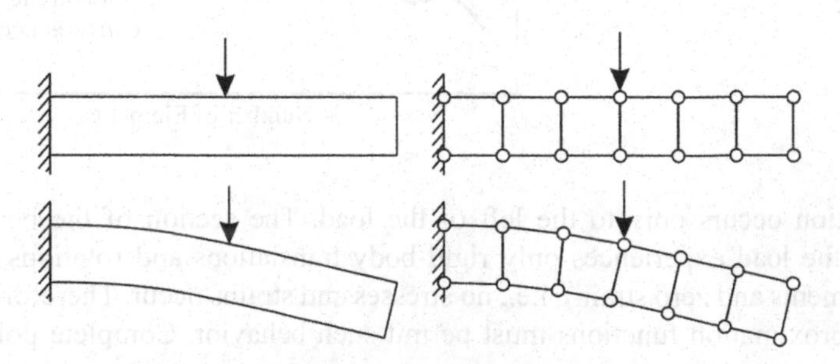

Fig. 3.6 A cantilever beam loaded at the middle and its FEA model

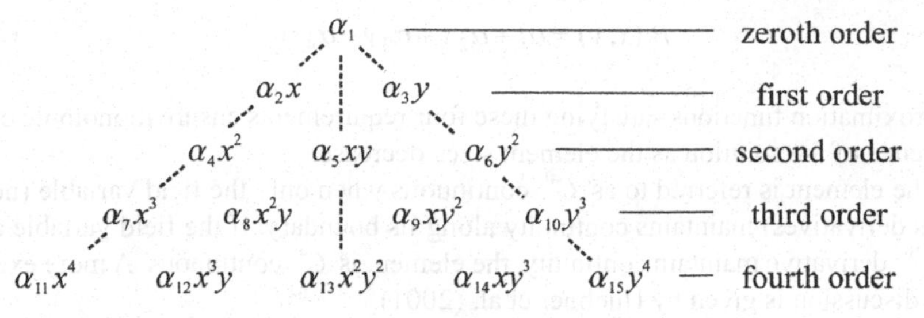

Fig. 3.7 Pascal's triangle for complete polynomials

In two dimensions, the compact form for a complete polynomial of order n can be written as

$$P_n(x) = \sum_{k=1}^{\frac{(n+1)(n+2)}{2}} \alpha_k x^i y^j \quad i + j \leq n \tag{3.3}$$

Constant, linear, and quadratic complete polynomials in two dimensions can be written as

$$\begin{aligned} P_0(x,y) &= \alpha_1 \\ P_1(x,y) &= \alpha_1 + \alpha_2 x + \alpha_3 y \\ P_2(x,y) &= \alpha_1 + \alpha_2 x + \alpha_3 y + \alpha_4 x^2 + \alpha_5 xy + \alpha_6 y^2 \end{aligned} \tag{3.4}$$

The Pascal triangle shown in Fig. 3.7 is useful for including the appropriate terms to obtain complete approximating functions in any order.

The order of the polynomial as an approximation function is dictated by the total number of nodes in an element, i.e., the number of coefficients, α_i, in the approximation function must be the same as the number of nodes in the element.

Requirement 4 *Geometric isotropy for the same behavior in each direction.*

Using complete polynomials satisfies this requirement of translation and rotation of the coordinate system. If the required degree of completeness does not provide a number of terms equal to the number of nodes, then this requirement can be satisfied by disregarding the non-symmetrical terms. In the case of a 4-noded rectangular element, the first-order complete polynomial has three coefficients, one less than the number of nodes. In order to circumvent this deficiency, the order of the polynomial can be increased to "complete" in the second degree, having six coefficients, two more than the number of nodes. As a result, two of the additional higher-order terms, which are $\alpha_4 x^2$, $\alpha_5 xy$, and $\alpha_6 y^2$, must be removed from the approximation function.

In order to satisfy the condition of geometric isotropy, only the term $\alpha_5 xy$ is retained in the approximation function, leading to

$$P_2(x,y) = \alpha_1 + \alpha_2 x + \alpha_3 y + \alpha_4 xy \tag{3.5}$$

Approximation functions satisfying these four requirements ensure monotonic convergence of the solution as the element sizes decrease.

The element is referred to as C^0 continuous when only the field variable (none of its derivatives) maintains continuity along its boundary. If the field variable and its rth derivative maintain continuity, the element is C^r continuous. A more extensive discussion is given by Huebner et al. (2001).

3.3 Coordinate Systems

3.3.1 Generalized Coordinates

The coefficients of the approximation functions, α_i, are referred to as the generalized coordinates. They are not identified with particular nodes. The generalized coordinates are independent parameters that specify the magnitude of the prescribed distribution of the field variable. They have no direct physical interpretation, but rather are linear combinations of the physical nodal degrees of freedom.

3.3.2 Global Coordinates

Global coordinates are convenient for specifying the location of each node, the orientation of each element, and the boundary conditions and loads for the entire domain. Also, the solution to the field variable is generally represented with respect to the global coordinates. However, approximation functions described in terms of the global coordinates are not convenient to use in the evaluation of integrals necessary for the construction of the element matrix.

3.3.3 Local Coordinates

A local coordinate system whose origin is located within the element is introduced in order to simplify the algebraic manipulations in the derivation of the element matrix. The use of natural coordinates in expressing the approximation functions is particularly advantageous because special integration formulas can often be employed to evaluate the integrals in the element matrix. Natural coordinates also play a crucial role in the development of elements with curved boundaries (discussed under isoparametric elements, Sect. 6.2.2.5).

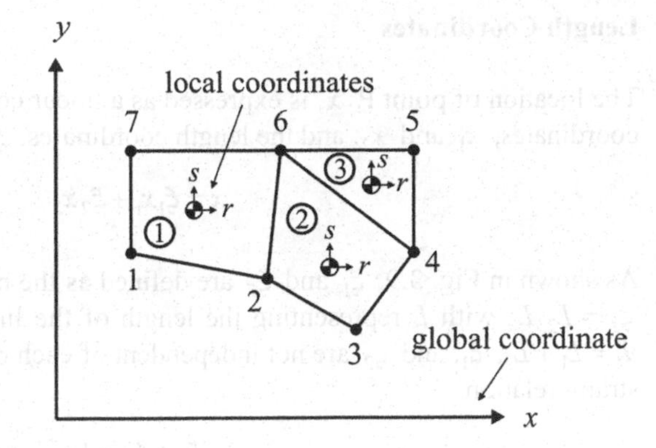

Fig. 3.8 Local and global coordinates in two dimensions

3.3.4 Natural Coordinates

A local coordinate system that permits the specification of a point within the element by a dimensionless parameter whose absolute magnitude never exceeds unity is referred to as a natural coordinate system. Natural coordinates are dimensionless. They are defined with respect to the element rather than with reference to the global coordinates. Also, the natural coordinates are functions of the global coordinates in which the element is defined. As illustrated in Fig. 3.8, the basic purpose of the natural coordinate system is to describe the location of a point inside an element in terms of coordinates associated with the nodes of the element.

3.3.4.1 Natural Coordinates in One Dimension

As shown in Fig. 3.9, within a one-dimensional element (line segment), defined by two nodes (one at each end), the location of a point P denoted by x (global coordinate) on the element can be expressed in terms of length or centroidal coordinates.

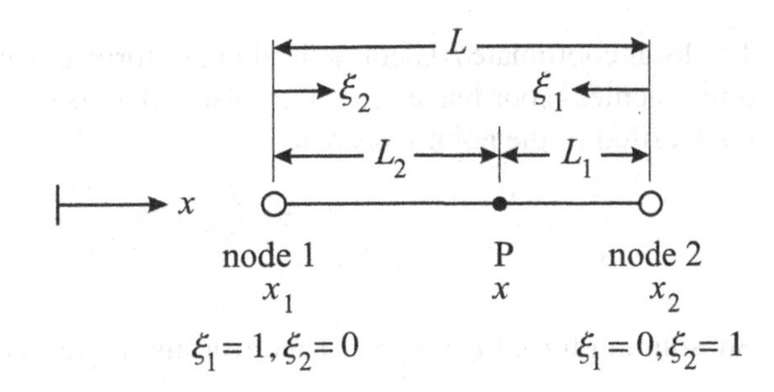

Fig. 3.9 Length coordinates in one dimension

Length Coordinates

The location of point P, x, is expressed as a linear combination of the global nodal coordinates, x_1 and x_2, and the length coordinates, ξ_1 and ξ_2, as

$$x = \xi_1 x_1 + \xi_2 x_2 \tag{3.6}$$

As shown in Fig. 3.9, ξ_1 and ξ_2 are defined as the ratios of lengths $\xi_1 = L_1/L$ and $\xi_2 = L_2/L$, with L representing the length of the line segment, $L = x_2 - x_1$. Since $L = L_1 + L_2$, ξ_1, and ξ_2 are not independent of each other and must satisfy the constraint relation

$$\xi_1 + \xi_2 = 1 \tag{3.7}$$

Solving for ξ_1 and ξ_2 via these equations written in matrix form as

$$\begin{Bmatrix} 1 \\ x \end{Bmatrix} = \begin{bmatrix} 1 & 1 \\ x_1 & x_2 \end{bmatrix} \begin{Bmatrix} \xi_1 \\ \xi_2 \end{Bmatrix} \tag{3.8}$$

results in

$$\xi_1 = \frac{x_2 - x}{x_2 - x_1} \text{ and } \xi_2 = \frac{x - x_1}{x_2 - x_1} \tag{3.9}$$

Such coordinates, whose behavior is shown in Fig. 3.10, have the property that one particular coordinate has a unit value at one node of the element and a zero value at the other node(s), i.e., $\xi_1(x_1) = 1$ and $\xi_1(x_2) = 0$, and $\xi_2(x_1) = 0$ and $\xi_2(x_2) = 1$.

Centroidal Coordinates

As shown in Fig. 3.11, x (the location of point P) with respect to a local coordinate system, r, located at the centroid of the line element becomes

$$x = r + x_1 + \frac{L}{2} \tag{3.10}$$

The local coordinate r is normalized in the form $\xi = r/(L/2)$ in order to achieve a dimensionless coordinate, ξ, and to ensure that its range never exceeds unity. Thus, the location of the point P becomes

$$x = \frac{L}{2}\xi + x_1 + \frac{L}{2} \tag{3.11}$$

Substituting for L ($L = x_2 - x_1$) and rearranging terms leads to

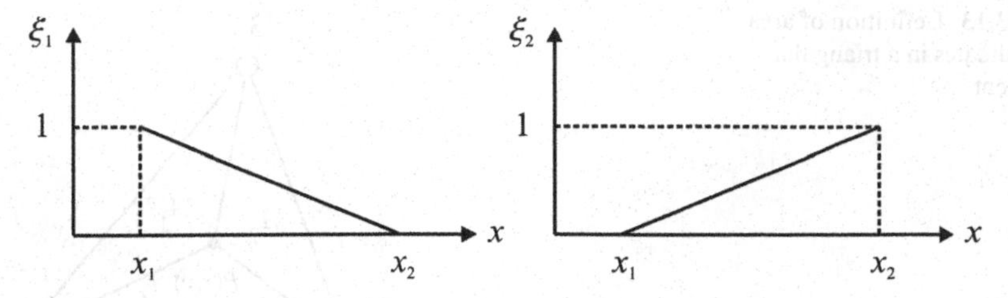

Fig. 3.10 Variation of length coordinates within the element

Fig. 3.11 Centroidal coordinates in one dimension

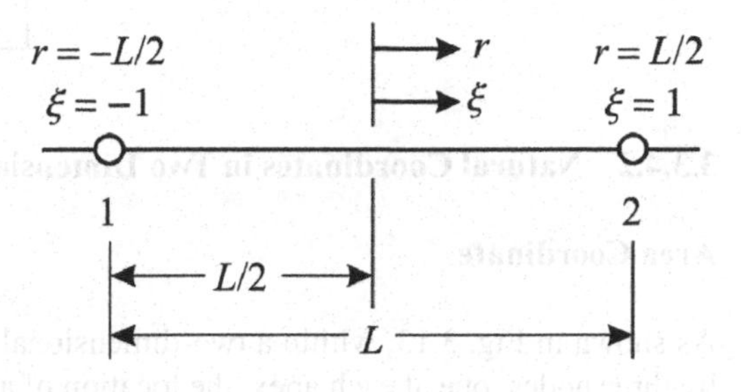

$$x = \frac{1}{2}(1-\xi)x_1 + \frac{1}{2}(1+\xi)x_2 \tag{3.12}$$

or

$$x = \sum_{i=1}^{2} N_i x_i \tag{3.13}$$

with $N_1 = (1-\xi)/2$ and $N_2 = (1+\xi)/2$. As shown in Fig. 3.12, $N_1(-1) = 1$ and $N_1(1) = 0$, and $N_2(-1) = 0$ and $N_2(1) = 1$.

Fig. 3.12 Variation of centroidal coordinates within the element

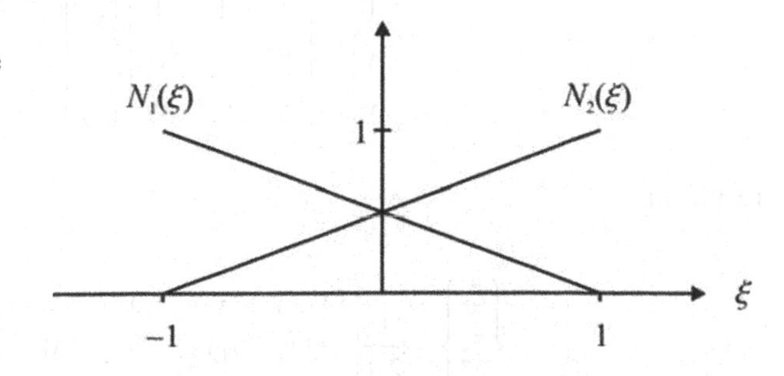

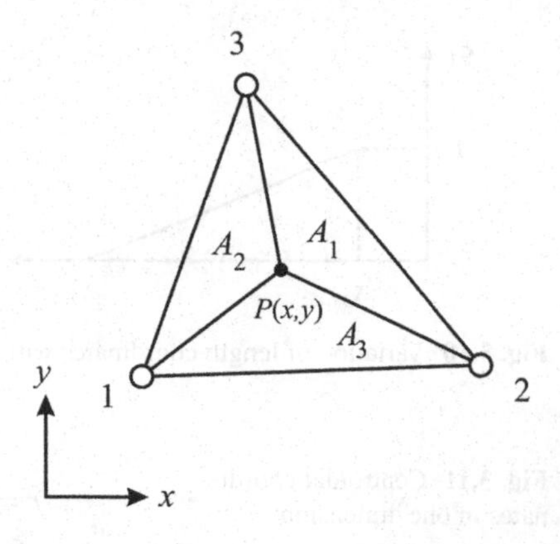

Fig. 3.13 Definition of area coordinates in a triangular element

3.3.4.2 Natural Coordinates in Two Dimensions

Area Coordinates

As shown in Fig. 3.13, within a two-dimensional element (triangular area) defined by three nodes, one at each apex, the location of a point P, denoted by (x, y) (global coordinates), on the element can be expressed as linear combinations of the global nodal coordinates, (x_1, y_1), (x_2, y_2), and (x_3, y_3), and the area coordinates, ξ_1, ξ_2, and x, as

$$\begin{aligned} x &= \xi_1 x_1 + \xi_2 x_2 + \xi_3 x_3 \\ y &= \xi_1 y_1 + \xi_2 y_2 + \xi_3 y_3 \end{aligned} \tag{3.14}$$

As illustrated in Fig. 3.13, ξ_1, ξ_2, and ξ_3 are defined as the ratios of areas $\xi_1 = A_1/A$, $\xi_2 = A_2/A$, and $\xi_3 = A_3/A$, with A representing the area of the triangle. Since $A_1 + A_2 + A_3 = 1$, ξ_1, ξ_2, and ξ_3 are not independent of each other and must satisfy the constraint relation

$$\xi_1 + \xi_2 + \xi_3 = 1 \tag{3.15}$$

Solving for ξ_1, ξ_2, and ξ_3 via Eq. (3.14) and (3.15) written in matrix form as

$$\begin{Bmatrix} 1 \\ x \\ y \end{Bmatrix} = \begin{bmatrix} 1 & 1 & 1 \\ x_1 & x_2 & x_3 \\ y_1 & y_2 & y_3 \end{bmatrix} \begin{Bmatrix} \xi_1 \\ \xi_2 \\ \xi_3 \end{Bmatrix} \tag{3.16}$$

results in

$$\begin{Bmatrix} \xi_1 \\ \xi_2 \\ \xi_3 \end{Bmatrix} = \frac{1}{2A} \begin{bmatrix} (x_2 y_3 - x_3 y_2) & y_{23} & x_{32} \\ (x_3 y_1 - x_1 y_3) & y_{31} & x_{13} \\ (x_1 y_2 - x_2 y_1) & y_{12} & x_{21} \end{bmatrix} \begin{Bmatrix} 1 \\ x \\ y \end{Bmatrix} \tag{3.17}$$

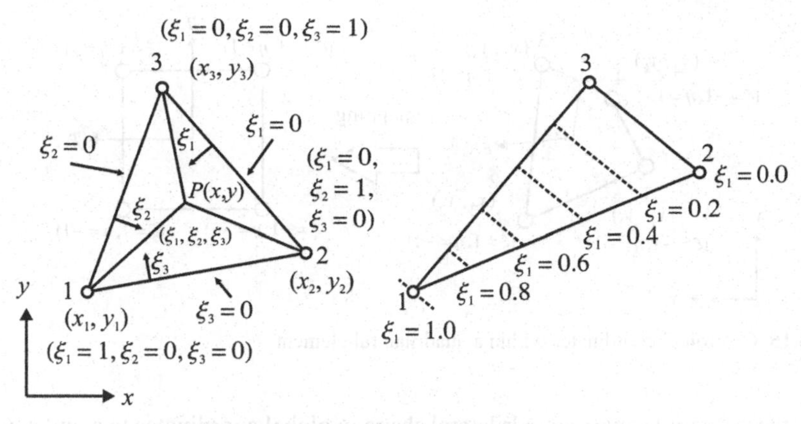

Fig. 3.14 Area coordinates within a triangular element

where $x_{mn} = x_m - x_n$, $y_{mn} = y_m - y_n$, and

$$2A = \begin{vmatrix} 1 & 1 & 1 \\ x_1 & x_2 & x_3 \\ y_1 & y_2 & y_3 \end{vmatrix} \tag{3.18}$$

As shown in Fig. 3.14, one particular area coordinate has a unit value at one node of the element and a zero value at the other node(s); $\xi_i(x_j) = \delta_{ij}$, where $\delta_{ij} = 1$ for $i = j$ and $\delta_{ij} = 0$ for $i \neq j$.

The exact evaluation of the area integrals over a triangle can be obtained by employing the expression

$$I = \int_A \xi_1^m \xi_2^n \xi_3^\ell \, dxdy = \frac{m!n!\ell!}{(m+n+\ell+2)!} 2A \tag{3.19}$$

Centroidal Coordinates

In the case of a two-dimensional element with a quadrilateral shape defined by four nodes, one at each corner, the location of a point P, denoted by (x, y), on the element can be expressed with respect to the centroidal coordinate system (ξ, η) whose origin coincides with the centroid of the quadrilateral area, as shown in Fig. 3.15. The relationship between (x, y) and (ξ, η) can be expressed as

$$\begin{aligned} x &= a_x + b_x\xi + c_x\eta + d_x\xi\eta \\ y &= a_y + b_y\xi + c_y\eta + d_y\xi\eta \end{aligned} \tag{3.20}$$

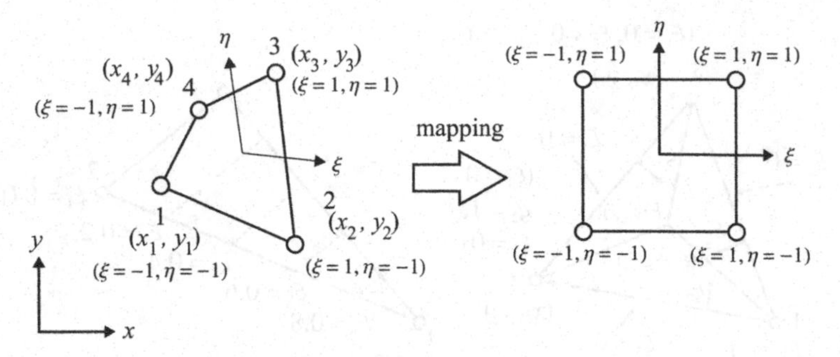

Fig. 3.15 Centroidal coordinates within a quadrilateral element

Also, these relations map a quadrilateral shape in global coordinates to a unit square in natural (centroidal) coordinates. Evaluation of these equations along $\eta = -1$ leads to

$$\begin{aligned} x &= a_x + b_x\xi - c_x - d_x\xi \\ y &= a_y + b_y\xi - c_y - d_y\xi \end{aligned} \tag{3.21}$$

Eliminating the coordinate ξ from the resulting equations yields the linear relationship between the global coordinates

$$y = A + Bx \tag{3.22}$$

in which A and B are known explicitly. Considering the remaining sides of the square in the centroidal coordinates defined by the lines $\eta = 1$, $\xi = 1$, and $\xi = -1$ results in a straight-sided quadrilateral.

Evaluation of x at $\xi = \pm 1$ and $\eta = \pm 1$ (four corners) leads to

$$\begin{aligned} x_1 &= a_x - b_x - c_x + d_x \\ x_2 &= a_x + b_x - c_x - d_x \\ x_3 &= a_x + b_x + c_x + d_x \\ x_4 &= a_x - b_x + c_x - d_x \end{aligned} \tag{3.23}$$

Solving for the coefficients a_x, b_x, c_x, and d_x, substituting back into Eq. (3.20), and collecting the terms multiplying x_i gives

$$\begin{aligned} x = &\frac{1}{4}(1-\xi)(1-\eta)x_1 + \frac{1}{4}(1+\xi)(1-\eta)x_2 \\ &+ \frac{1}{4}(1+\xi)(1+\eta)x_3 + \frac{1}{4}(1-\xi)(1+\eta)x_4 \end{aligned} \tag{3.24}$$

A similar operation performed on y in Eq. (3.20) yields

$$\begin{aligned} y = &\frac{1}{4}(1-\xi)(1-\eta)y_1 + \frac{1}{4}(1+\xi)(1-\eta)y_2 \\ &+\frac{1}{4}(1+\xi)(1+\eta)y_3 + \frac{1}{4}(1-\xi)(1+\eta)y_4 \end{aligned} \tag{3.25}$$

Defining

$$\begin{aligned} N_1 &= \frac{1}{4}(1-\xi)(1-\eta) \qquad & N_2 &= \frac{1}{4}(1+\xi)(1-\eta) \\ N_3 &= \frac{1}{4}(1+\xi)(1+\eta) \qquad & N_4 &= \frac{1}{4}(1-\xi)(1+\eta) \end{aligned} \tag{3.26}$$

allows Eq. (3.24) and (3.25) to be rewritten as

$$x = \sum_{i=1}^{4} N_i(\xi,\eta)x_i \quad \text{and} \quad y = \sum_{i=1}^{4} N_i(\xi,\eta)y_i \tag{3.27}$$

Note that N_i can be written in compact form as

$$N_i = \frac{1}{4}(1+\xi\xi_i)(1+\eta\eta_i) \tag{3.28}$$

with ξ_i and η_i representing the coordinates of the corner nodes in the natural coordinate system. It is worth noting that $N_i(\xi_j,\eta_j)=\delta_{ij}$, where $\delta_{ij}=1$ for $i=j$ and $\delta_{ij}=0$ for $i \neq j$. The variations of N_i within a quadrilateral element are given schematically in Fig. 3.16.

3.4 Shape Functions

Shape functions constitute the subset of element approximation functions. They cannot be chosen arbitrarily. As discussed in the previous section, the element approximation functions are chosen to be *complete polynomials* with unknown generalized coordinates. For a one-dimensional element with m nodes as shown in Fig. 3.17, the element approximation function for the field variable, $\phi(x)$, is assumed as a polynomial of order $(m-1)$

$$\phi^{(e)}(x) = \alpha_1 + \alpha_2 x + \alpha_3 x^2 + \alpha_4 x^3 + \cdots + \alpha_{m-1}x^{m-2} + \alpha_m x^{m-1} \tag{3.29}$$

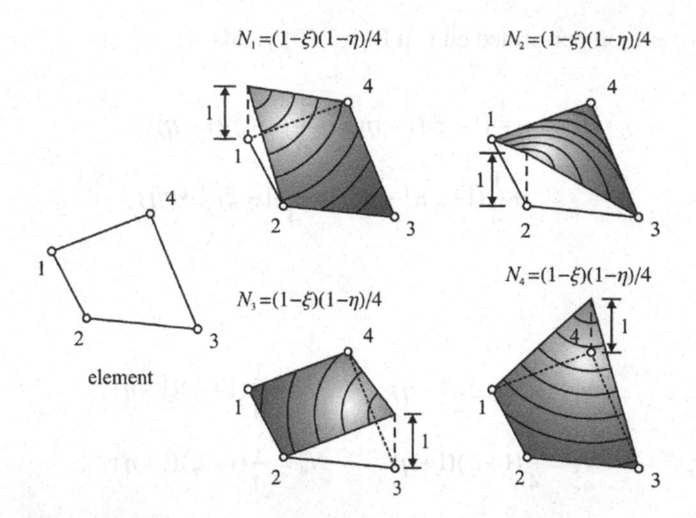

Fig. 3.16 Variation of N_i within a quadrilateral

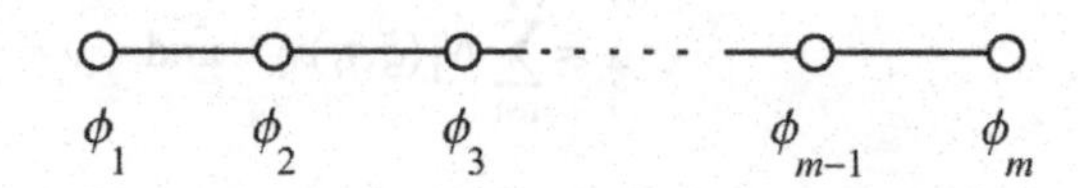

Fig. 3.17 A one-dimensional element with m nodes

or

$$\phi^{(e)}(x) = \mathbf{g}^T \boldsymbol{\alpha} \tag{3.30}$$

where

$$\mathbf{g}^T = \{1 \quad x \quad x^2 \quad \cdots \quad x^{m-1}\} \tag{3.31}$$

and

$$\boldsymbol{\alpha}^T = \{\alpha_1 \quad \alpha_2 \quad \alpha_3 \quad \cdots \quad \alpha_m\} \tag{3.32}$$

Note that the number of generalized coordinates ($\alpha_i, i = 1,2,\ldots,m$) is equal to the number of nodes within the element.

The field variable, $\phi^{(e)}(x)$, can also be expressed within the element through the use of its nodal values, ϕ_i $(i = 1, m)$, in the form

$$\begin{aligned} \phi^{(e)}(x) = {} & N_1\phi_1 + N_2\phi_2 + N_3\phi_3 + N_4\phi_4 + \cdots \\ & + N_{m-1}\phi_{m-1} + N_m\phi_m \end{aligned} \tag{3.33}$$

or

$$\phi^{(e)}(x) = \mathbf{N}^T \boldsymbol{\varphi} \tag{3.34}$$

where

$$\mathbf{N}^T = \{N_1 \quad N_2 \quad N_3 \quad \cdots \quad N_m\} \tag{3.35}$$

and

$$\boldsymbol{\varphi}^T = \{\phi_1 \quad \phi_2 \quad \phi_3 \quad \cdots \quad \phi_m\} \tag{3.36}$$

in which $N_i \ (i = 1, m)$ are referred to as shape functions. These functions are associated with node i and must have a unit value at node i and a zero value at all other nodes. Furthermore, they must have the same degree of polynomial variation as in the element approximation function.

The explicit form of the shape functions can be determined by solving for the generalized coordinates, α_i, in terms of the nodal coordinates, x_i, and nodal values, $\phi_i \ (i = 1, 2, \ldots, m)$, through Eq. (3.29), and rearranging the resulting expressions in the form of Eq. (3.34). At each node, the field variable $\phi^{(e)}(x)$ is evaluated as

$$\begin{aligned}
\phi_1 &= \alpha_1 + \alpha_2 x_1 + \alpha_3 x_1^2 + \alpha_4 x_1^3 + \cdots + \alpha_{m-1} x_1^{m-2} + \alpha_m x_1^{m-1} \\
\phi_2 &= \alpha_1 + \alpha_2 x_2 + \alpha_3 x_2^2 + \alpha_4 x_2^3 + \cdots + \alpha_{m-1} x_2^{m-2} + \alpha_m x_2^{m-1} \\
&\ \ \vdots \\
\phi_m &= \alpha_1 + \alpha_2 x_m + \alpha_3 x_m^2 + \alpha_4 x_m^3 + \cdots + \alpha_{m-1} x_m^{m-2} + \alpha_m x_m^{m-1}
\end{aligned} \tag{3.37}$$

or in matrix form

$$\begin{Bmatrix} \phi_1 \\ \phi_2 \\ \phi_3 \\ \vdots \\ \phi_m \end{Bmatrix} = \begin{bmatrix} 1 & x_1 & x_1^2 & \cdots & x_1^{m-1} \\ 1 & x_2 & x_2^2 & \cdots & x_2^{m-1} \\ 1 & x_3 & x_3^2 & \cdots & x_3^{m-1} \\ \vdots & \vdots & \vdots & \ddots & \vdots \\ 1 & x_m & x_m^2 & \cdots & x_m^{m-1} \end{bmatrix} \begin{Bmatrix} \alpha_1 \\ \alpha_2 \\ \alpha_3 \\ \vdots \\ \alpha_m \end{Bmatrix} \quad or \quad \boldsymbol{\varphi} = \mathbf{A}\boldsymbol{\alpha} \tag{3.38}$$

Solving for the generalized coordinates in terms of nodal coordinates and nodal values of the field variable yields

$$\boldsymbol{\alpha} = \mathbf{A}^{-1} \boldsymbol{\varphi} \tag{3.39}$$

Substituting for the generalized coordinates in Eq. (3.30) results in

$$\phi^{(e)}(x) = \mathbf{g}^T \mathbf{A}^{-1} \boldsymbol{\varphi} \tag{3.40}$$

Comparison of Eq. (3.40) and (3.34) leads to the explicit form of the shape functions N_i as

$$\mathbf{N}^T = \mathbf{g}^T \mathbf{A}^{-1} \tag{3.41}$$

This formulation illustrates the determination of the shape functions for a one-dimensional element; its extension to two dimensions is straightforward. The properties of shape functions are:

1. $N_i = 1$ at node i and $N_i = 0$ at all other nodes.
2. $\sum_{i=1}^{m} N_i = 1$.

3.4.1 Linear Line Element with Two Nodes

3.4.1.1 Global Coordinate

For a line element with two nodes, the field variable, $\phi^{(e)}$, is approximated by a linear function (refer to Fig. 3.18) in terms of the global coordinate, x, as

$$\phi^{(e)}(x) = \alpha_1 + \alpha_2 x \tag{3.42}$$

This element approximation function ensures the inter-element continuity of only the field variable. The nodal values of the function are identified by ϕ_1 and ϕ_2.

Evaluation of the function at each node with coordinates x_1 and x_2 leads to

$$\phi_1 = \alpha_1 + \alpha_2 x_1 \quad \text{and} \quad \phi_2 = \alpha_1 + \alpha_2 x_2 \tag{3.43}$$

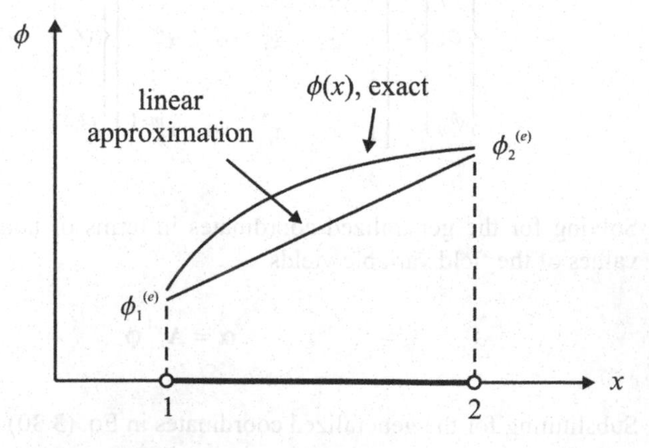

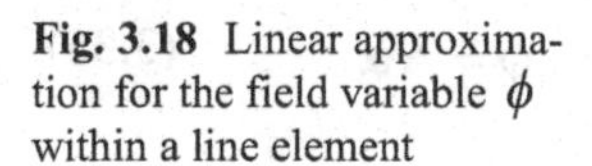
Fig. 3.18 Linear approximation for the field variable ϕ within a line element

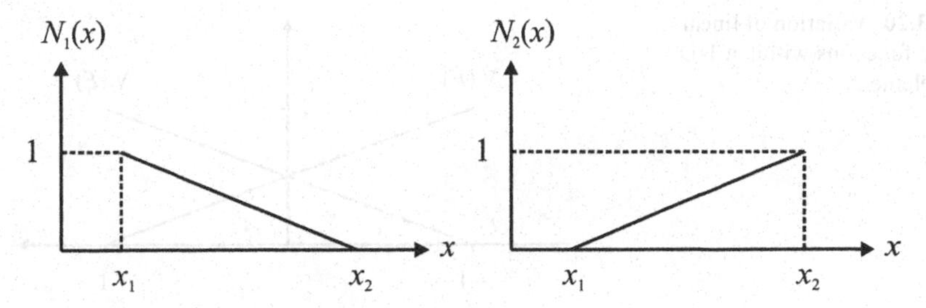

Fig. 3.19 Variation of linear shape functions within a 1-D line element

Solving for α_1 and α_2 and substituting for them in the element approximation function results in

$$\phi^{(e)}(x) = N_1(x)\phi_1 + N_2(x)\phi_2 \tag{3.44}$$

where $N_1 = (x_2 - x)/(x_2 - x_1)$ and $N_2 = (x - x_1)/(x_2 - x_1)$. These functions, referred to as interpolation or shape functions, are the same as the length coordinates, ξ_1 and ξ_2, and they also vary linearly with x (Fig. 3.19), as does the element approximation function. Because $N_i(x_j) = \delta_{ij}$, where $\delta_{ij} = 1$ for $i = j$ and $\delta_{ij} = 0$ for $i \neq j$,

$$1 = \sum_{i=1}^{2} N_i \tag{3.45}$$

3.4.1.2 Centroidal Coordinate

For a line element with two nodes, the field variable, $\phi^{(e)}$, is approximated by a linear function in terms of the natural (centroidal) coordinate, ξ, as

$$\phi^{(e)}(\xi) = \alpha_1 + \alpha_2 \xi \tag{3.46}$$

This element approximation function ensures the inter-element continuity of the field variable. The nodal values of the function are identified by ϕ_1 and ϕ_2. Evaluation of the function at each node with coordinates $\xi = -1$ and $\xi = 1$ leads to

$$\phi_1 = \alpha_1 - \alpha_2 \quad \text{and} \quad \phi_2 = \alpha_1 + \alpha_2 \tag{3.47}$$

Solving for α_1 and α_2 and substituting for them in the element approximation function results in

$$\phi^{(e)}(\xi) = N_1(\xi)\phi_1 + N_2(\xi)\phi_2 \tag{3.48}$$

where $N_1(\xi) = (1 - \xi)/2$ and $N_2(\xi) = (1 + \xi)/2$. These functions, referred to as interpolation or shape functions, vary linearly with ξ (Fig. 3.20), as in the case of the element approximation function.

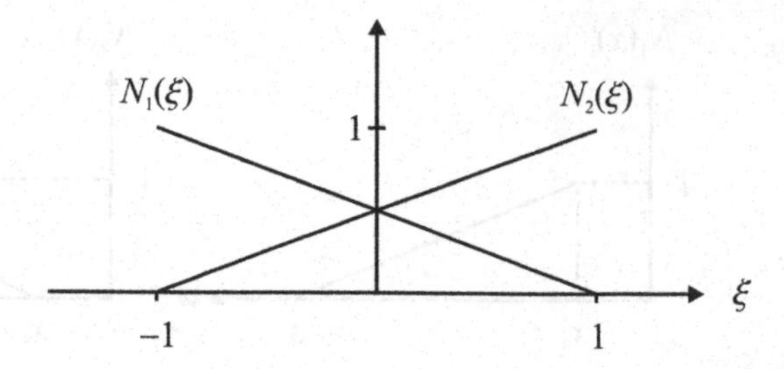

Fig. 3.20 Variation of linear shape functions within a 1-D line element

Also, they have the property

$$1 = \sum_{i=1}^{2} N_i \tag{3.49}$$

because $N_i(\xi_j) = \delta_{ij}$, where $\delta_{ij} = 1$ for $i = j$ and $\delta_{ij} = 0$ for $i \neq j$.

3.4.2 Quadratic Line Element with Three Nodes: Centroidal Coordinate

For a line element with three nodes, the field variable, $\phi^{(e)}$, is approximated by a quadratic function (schematic given in Fig. 3.21) in terms of the natural (centroidal) coordinate, ξ, as

$$\phi^{(e)}(\xi) = \alpha_1 + \alpha_2 \xi + \alpha_3 \xi^2 \tag{3.50}$$

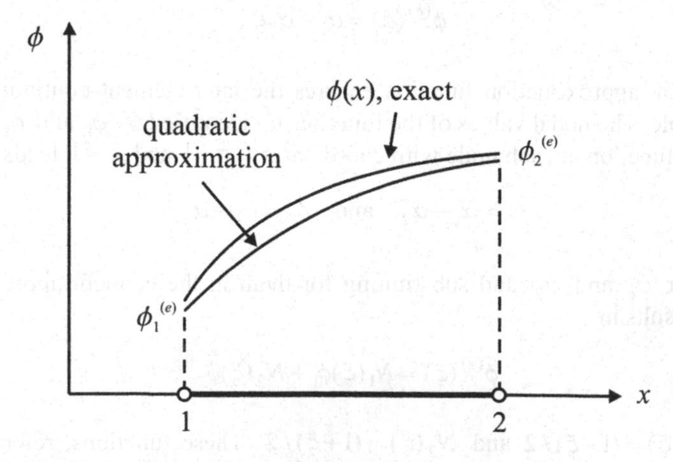

Fig. 3.21 Quadratic approximation for the field variable ϕ within a line element

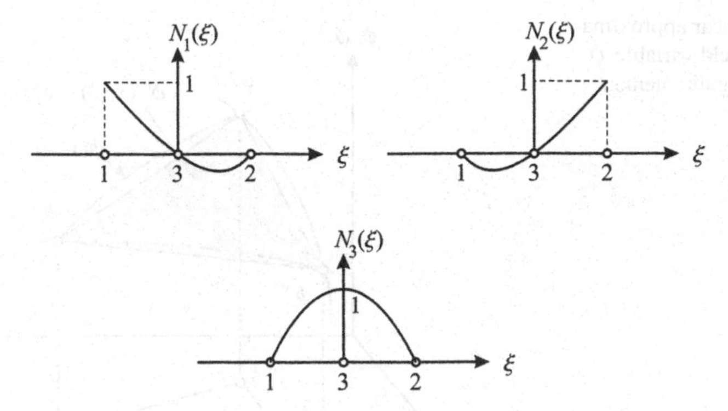

Fig. 3.22 Variation of quadratic shape functions within a 1-D line element

in order to ensure the inter-element continuity of the field variable. The element nodes are identified as 1, 2, and 3, with their nodal values as ϕ_1, ϕ_2, and ϕ_3. The middle node is located at the center of the line element. Evaluation of the function at each node with coordinates $\xi = -1$, $\xi = 0$, and $\xi = 1$ leads to

$$\phi_1 = \alpha_1 - \alpha_2 + \alpha_3 \tag{3.51}$$

Solving for α_1, α_2, and α_3 and substituting for them in the element approximation function results in

$$\phi^{(e)}(\xi) = N_1(\xi)\phi_1 + N_2(\xi)\phi_2 + N_3(\xi)\phi_3 \tag{3.52}$$

where $N_1(\xi) = \xi/[2(\xi - 1)]$, $N_2(\xi) = \xi/[2(\xi + 1)]$, and $N_3(\xi) = -(\xi + 1)(\xi - 1)$. These functions, referred to as interpolation or shape functions, vary quadratically with ξ (Fig. 3.22), as in the case of element approximation function.

Also, they have the property

$$1 = \sum_{i=1}^{3} N_i \tag{3.53}$$

because $N_i(\xi_j) = \delta_{ij}$, where $\delta_{ij} = 1$ for $i = j$ and $\delta_{ij} = 0$ for $i \neq j$.

3.4.3 *Linear Triangular Element with Three Nodes: Global Coordinate*

Within a two-dimensional element (triangular area) defined by three nodes, one at each apex, the variation of the field variable, $\phi^{(e)}(x, y)$, can be approximated by a linear function (as illustrated in Fig. 3.23) of the form

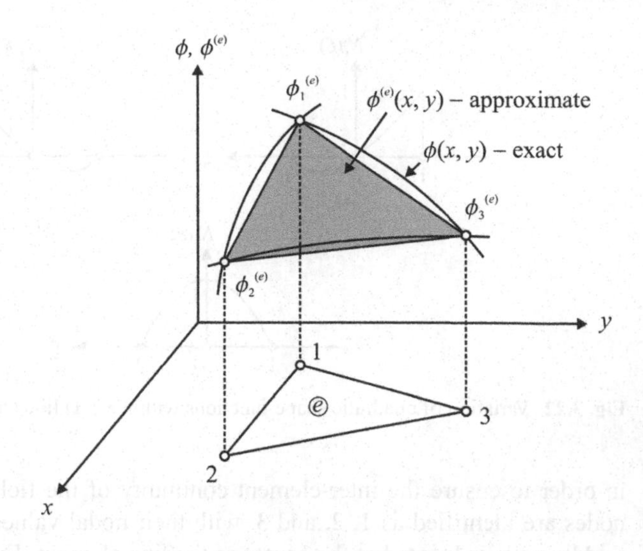

Fig. 3.23 Linear approximation for the field variable ϕ within a triangular element

$$\phi^{(e)}(x, y) = \alpha_1 + \alpha_2 x + \alpha_3 y \tag{3.54}$$

This function ensures the inter-element continuity of the field variable $\phi^{(e)}(x, y)$.

The element nodes are identified as 1, 2, and 3 in a counterclockwise orientation, with their nodal values as ϕ_1, ϕ_2, and ϕ_3. The nodal coordinates are specified by (x_1, y_1), (x_2, y_2), and (x_3, y_3).

The nodal values of the field variable must be satisfied as

$$\begin{aligned} \phi_1 &= \alpha_1 + \alpha_2 x_1 + \alpha_3 y_1 \\ \phi_2 &= \alpha_1 + \alpha_2 x_2 + \alpha_3 y_2 \\ \phi_3 &= \alpha_1 + \alpha_2 x_3 + \alpha_3 y_3 \end{aligned} \tag{3.55}$$

leading to the determination of the generalized coefficients in the form

$$\begin{Bmatrix} \alpha_1 \\ \alpha_2 \\ \alpha_3 \end{Bmatrix} = \frac{1}{2A} \begin{bmatrix} (x_2 y_3 - x_3 y_2) & (x_3 y_1 - x_1 y_3) & (x_1 y_2 - x_2 y_1) \\ (y_2 - y_3) & (y_3 - y_1) & (y_1 - y_2) \\ (x_3 - x_2) & (x_1 - x_3) & (x_2 - x_1) \end{bmatrix} \begin{Bmatrix} \phi_1 \\ \phi_2 \\ \phi_3 \end{Bmatrix} \tag{3.56}$$

where

$$2A = \begin{vmatrix} 1 & x_1 & y_1 \\ 1 & x_2 & y_2 \\ 1 & x_3 & y_3 \end{vmatrix} \tag{3.57}$$

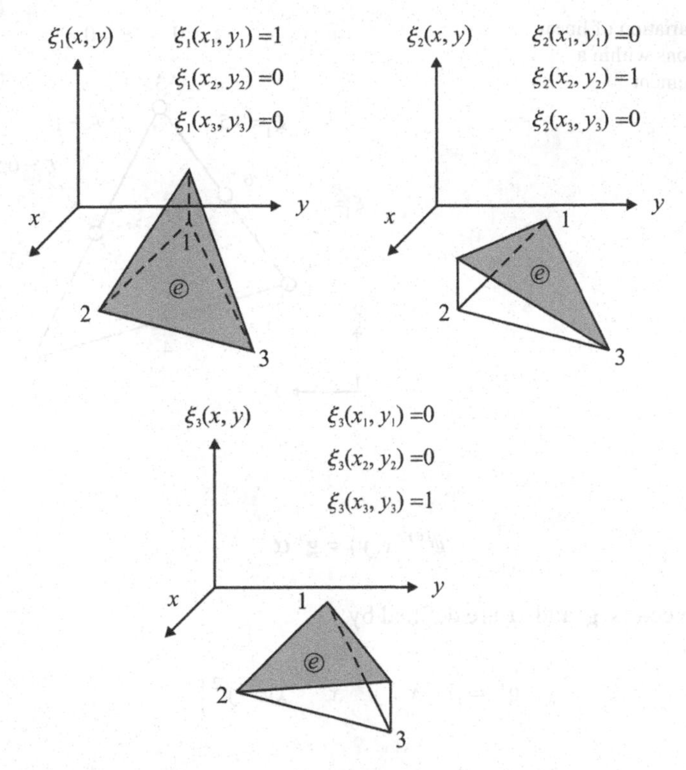

Fig. 3.24 Variation of linear shape functions within a triangular element

Substitution of α_1, α_2, and α_3 into the expression for the element approximation function results in

$$\phi^{(e)}(x,y) = N_1(x,y)\phi_1 + N_2(x,y)\phi_2 + N_3(x,y)\phi_3 \tag{3.58}$$

where the shape functions $N_1 = \xi_1$, $N_2 = \xi_2$, and $N_3 = \xi_3$ are the same as the area coordinates with properties $\xi_i(x_j, y_j) = \delta_{ij}$ and $\sum_{i=1}^{3} \xi_i = 1$. Their variation within the element is given in Fig. 3.24.

3.4.4 Quadratic Triangular Element with Six Nodes

The field variable can be approximated by a complete quadratic function within a triangular element in the form

$$\phi^{(e)}(x,y) = \alpha_1 + \alpha_2 x + \alpha_3 y + \alpha_4 x^2 + \alpha_5 xy + \alpha_6 y^2 \tag{3.59a}$$

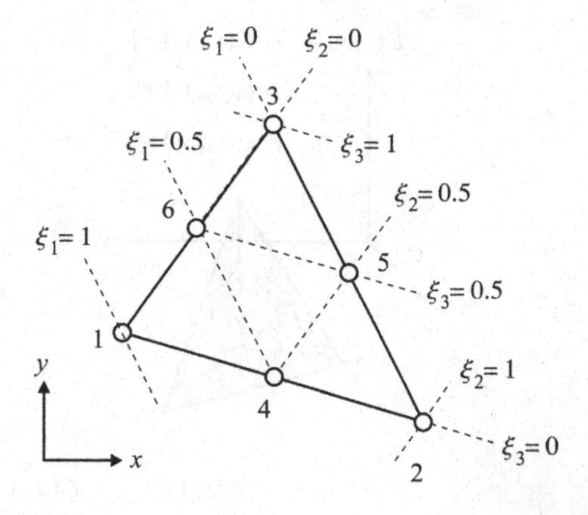

Fig. 3.25 Variation of linear shape functions within a triangular element

or

$$\phi^{(e)}(x,y) = \mathbf{g}^T \boldsymbol{\alpha} \tag{3.59b}$$

where the vectors $\mathbf{g}$ and $\boldsymbol{\alpha}$ are defined by

$$\mathbf{g}^T = \left\{1 \quad x \quad y \quad x^2 \quad xy \quad y^2\right\} \tag{3.60}$$

and

$$\boldsymbol{\alpha}^T = \left\{\alpha_1 \quad \alpha_2 \quad \alpha_3 \quad \alpha_4 \quad \alpha_5 \quad \alpha_6\right\} \tag{3.61}$$

However, this representation requires a triangular element with six nodes, as shown in Fig. 3.25, in order to determine its six unknown coefficients, α_i.

At each node, the field variable, $\phi^{(e)}(x_i, y_i)$, is evaluated as

$$\begin{Bmatrix} \phi_1 \\ \phi_2 \\ \phi_3 \\ \phi_4 \\ \phi_5 \\ \phi_6 \end{Bmatrix} = \begin{bmatrix} 1 & x_1 & y_1 & x_1^2 & x_1 y_1 & y_1^2 \\ 1 & x_2 & y_2 & x_2^2 & x_2 y_2 & y_2^2 \\ 1 & x_3 & y_3 & x_3^2 & x_3 y_3 & y_3^2 \\ 1 & x_4 & y_4 & x_4^2 & x_4 y_4 & y_4^2 \\ 1 & x_5 & y_5 & x_5^2 & x_5 y_5 & y_5^2 \\ 1 & x_6 & y_6 & x_6^2 & x_6 y_6 & y_6^2 \end{bmatrix} \begin{Bmatrix} \alpha_1 \\ \alpha_2 \\ \alpha_3 \\ \alpha_4 \\ \alpha_5 \\ \alpha_6 \end{Bmatrix} \quad \text{or } \boldsymbol{\varphi} = \mathbf{A}\boldsymbol{\alpha} \tag{3.62}$$

Solving for the generalized coordinates in terms of nodal coordinates and nodal values of the field variable yields

$$\boldsymbol{\alpha} = \mathbf{A}^{-1}\boldsymbol{\varphi} \tag{3.63}$$

Substituting for the generalized coordinates in Eq. (3.59) results in

$$\phi^{(e)}(x, y) = \mathbf{g}^T \mathbf{A}^{-1}\boldsymbol{\varphi} \tag{3.64}$$

However, $\phi^{(e)}(x, y)$ can also be expressed within the element through the use of its nodal values ϕ_i as

$$\phi^{(e)}(x, y) = \sum_{i=1}^{6} N_i(x, y)\phi_i \quad \text{or} \quad \phi^{(e)}(x, y) = \mathbf{N}^T \boldsymbol{\varphi} \tag{3.65}$$

where $\mathbf{N}$ is the vector of shape functions, N_i $(i = 1, 6)$. Comparison of the last two equations results in the explicit form of the shape functions N_i as

$$\mathbf{N}^T = \mathbf{g}^T \mathbf{A}^{-1} \tag{3.66}$$

In providing the explicit forms of the shape functions, lengthy expressions are avoided by utilizing the expressions for the area coordinates of ξ_1, ξ_2, and ξ_3, as derived in Eq. (3.17), thus leading to

$$\mathbf{N}^T = \{(2\xi_1 - 1)\xi_1 \quad (2\xi_2 - 1)\xi_2 \quad (2\xi_3 - 1)\xi_3 \quad 4\xi_1\xi_2 \quad 4\xi_2\xi_3 \quad 4\xi_3\xi_1\} \tag{3.67}$$

or

$$\begin{aligned} &N_1 = (2\xi_1 - 1)\xi_1, \quad N_2 = (2\xi_2 - 1)\xi_2, \quad N_3 = (2\xi_3 - 1)\xi_3 \\ &N_4 = 4\xi_1\xi_2, \qquad N_5 = 4\xi_2\xi_3, \qquad N_6 = 4\xi_3\xi_1 \end{aligned} \tag{3.68}$$

Variation of these shape functions within the element is shown in Fig. 3.26.

3.4.5 *Linear Quadrilateral Element with Four Nodes: Centroidal Coordinate*

For a quadrilateral element with four nodes, the field variable, $\phi^{(e)}(x, y)$, is approximated by a linear function (refer to Fig. 3.27) in terms of the natural (centroidal) coordinates, $-1 \le \xi \le 1$ and $-1 \le \eta \le 1$, as

$$\phi^{(e)}(\xi, \eta) = \alpha_1 + \alpha_2\xi + \alpha_3\eta + \alpha_4\xi\eta \tag{3.69}$$

This element approximation function ensures the inter-element continuity of only the field variable. The nodal values of the function are identified by ϕ_1, ϕ_2, ϕ_3, and ϕ_4. Evaluation of the function at each node with coordinates $(\xi_1 = -1, \eta_1 = -1)$, $(\xi_2 = 1, \eta_2 = -1)$, $(\xi_3 = 1, \eta_3 = 1)$, and $(\xi_4 = -1, \eta_4 = 1)$ leads to

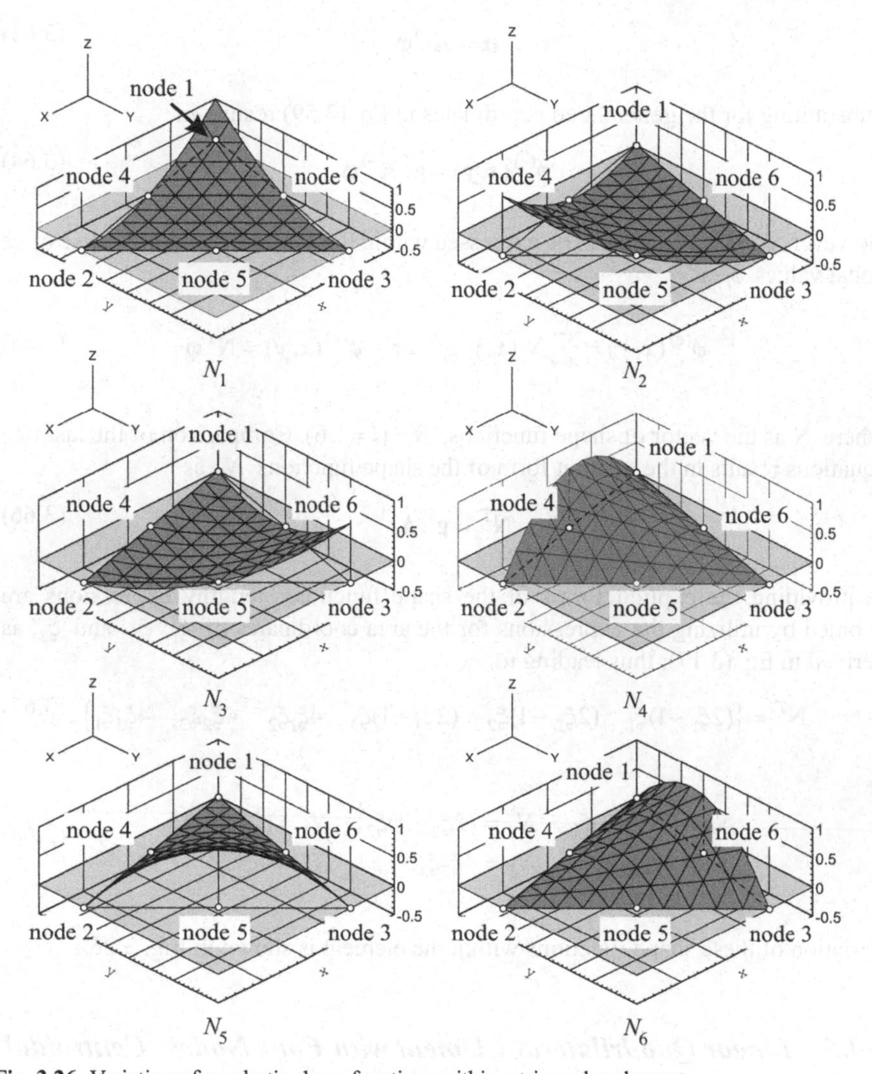

Fig. 3.26 Variation of quadratic shape functions within a triangular element

$$\begin{Bmatrix} \phi_1 \\ \phi_2 \\ \phi_3 \\ \phi_4 \end{Bmatrix} = \begin{bmatrix} 1 & -1 & -1 & 1 \\ 1 & 1 & -1 & -1 \\ 1 & 1 & 1 & 1 \\ 1 & -1 & 1 & -1 \end{bmatrix} \begin{Bmatrix} \alpha_1 \\ \alpha_2 \\ \alpha_3 \\ \alpha_4 \end{Bmatrix} \tag{3.70}$$

Solving for α_1, α_2, α_3, and α_4 results in

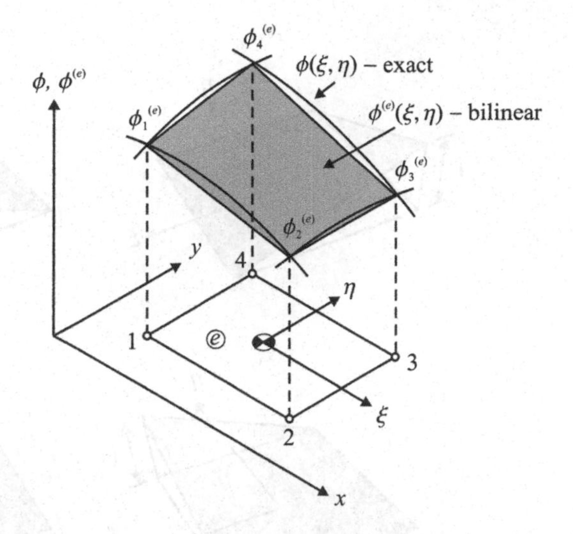

Fig. 3.27 Bi-linear approximation for the field variable ϕ within a quadrilateral element

$$\begin{Bmatrix} \alpha_1 \\ \alpha_2 \\ \alpha_3 \\ \alpha_4 \end{Bmatrix} = \frac{1}{4} \begin{bmatrix} 1 & 1 & 1 & 1 \\ -1 & 1 & 1 & -1 \\ -1 & -1 & 1 & 1 \\ 1 & -1 & 1 & -1 \end{bmatrix} \begin{Bmatrix} \phi_1 \\ \phi_2 \\ \phi_3 \\ \phi_4 \end{Bmatrix} \tag{3.71}$$

and their substitution in the element approximation function yields

$$\phi^{(e)}(\xi,\eta) = \sum_{i=1}^{4} N_i(\xi,\eta)\phi_i \tag{3.72}$$

in which

$$N_i = \frac{1}{4}(1+\xi\xi_i)(1+\eta\eta_i) \tag{3.73}$$

with ξ_i and η_i representing the coordinates of the corner nodes in the natural coordinate system. The shape functions have the property $N_i(\xi_j,\eta_j) = \delta_{ij}$, where $\delta_{ij} = 1$ for $i = j$ and $\delta_{ij} = 0$ for $i \neq j$. They are graphically illustrated in Fig. 3.28.

3.5 Isoparametric Elements: Curved Boundaries

The modeling of domains involving curved boundaries by using straight-sided elements may not provide satisfactory results. However, the family of elements known as "isoparametric elements" is suitable for such boundaries. The shape (or geom-

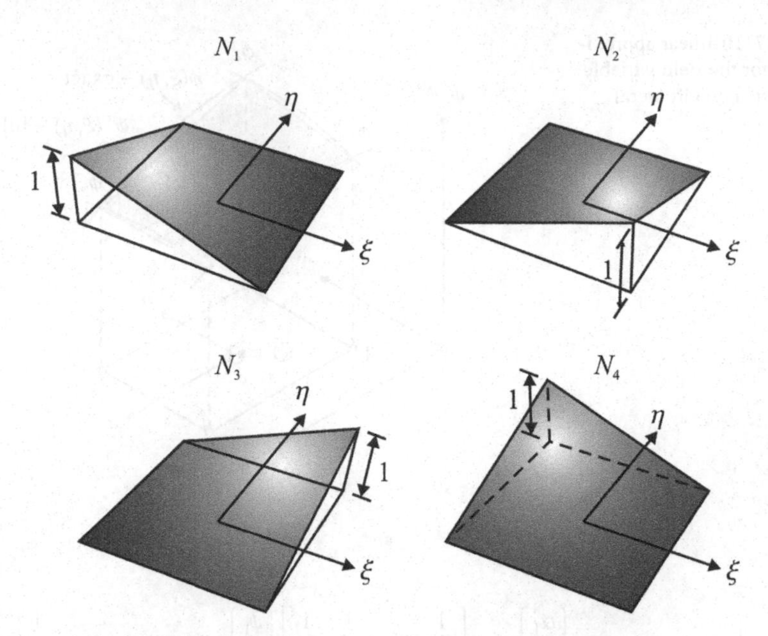

Fig. 3.28 Variation of bi-linear shape functions within a quadrilateral element

etry) and the field variable of these elements are described by the same interpolation functions of the same order. The representation of geometry (element shape) in terms of linear (or nonlinear) shape functions can be considered as a mapping procedure that transforms a square in local coordinates to a regular quadrilateral (or distorted shape) in global coordinates (Fig. 3.29) (Ergatoudis et al. 1968).

The most widely used elements are triangular or quadrilateral because of their ability to approximate complex geometries. An arbitrary straight-sided quadrilateral in global coordinates, (x, y), can be obtained by a point mapping from the "standard square" defined in natural coordinates, (ξ,η). The mapping shown in Fig. 3.29 can be achieved by

$$\begin{aligned} x &= \frac{1}{4}(1-\xi)(1-\eta)x_1 + \frac{1}{4}(1+\xi)(1-\eta)x_2 \\ &\quad + \frac{1}{4}(1+\xi)(1+\eta)x_3 + \frac{1}{4}(1-\xi)(1+\eta)x_4 \\ y &= \frac{1}{4}(1-\xi)(1-\eta)y_1 + \frac{1}{4}(1+\xi)(1-\eta)y_2 \\ &\quad + \frac{1}{4}(1+\xi)(1+\eta)y_3 + \frac{1}{4}(1-\xi)(1+\eta)y_4 \end{aligned} \tag{3.74}$$

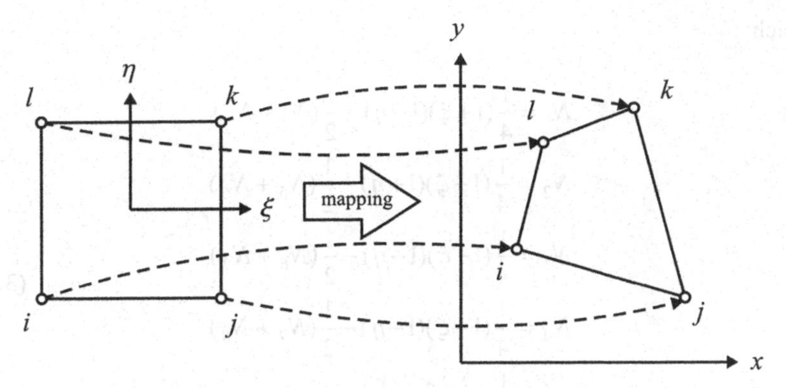

Fig. 3.29 Mapping from a unit square to an arbitrary straight-sided quadrilateral

or

$$x = \sum_{i=1}^{4} N_i(\xi,\eta)x_i \quad \text{and} \quad y = \sum_{i=1}^{4} N_i(\xi,\eta)y_i \tag{3.75}$$

in which

$$N_i = \frac{1}{4}(1+\xi\xi_i)(1+\eta\eta_i) \tag{3.76}$$

with $(\xi_1 = -1,\ \eta_1 = -1)$ $(\xi_2 = 1,\ \eta_2 = -1)$ $(\xi_3 = 1,\ \eta_3 = 1)$, and $(\xi_4 = -1,\ \eta_4 = 1)$.

In the case of an element with curved boundaries in global coordinates, quadratic shape functions can be used to map it on to a unit square in local coordinates, as shown in Fig. 3.30. The mapping can be achieved by

$$x = \sum_{i=1}^{8} N_i(\xi,\eta)x_i \quad \text{and} \quad y = \sum_{i=1}^{8} N_i(\xi,\eta)y_i \tag{3.77}$$

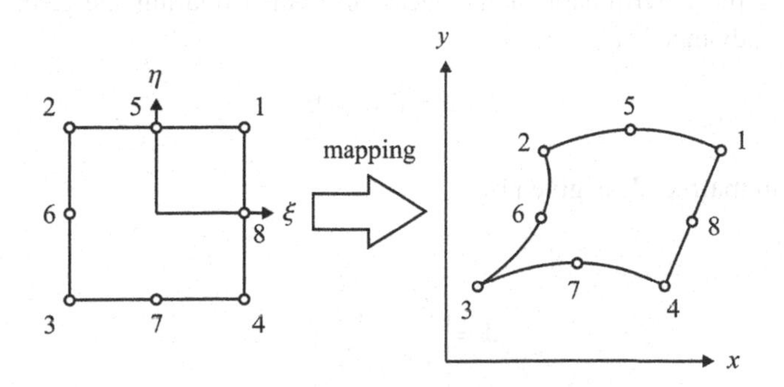

Fig. 3.30 Mapping from a unit square to a quadrilateral with curved sides

in which

$$
\begin{aligned}
N_1 &= \frac{1}{4}(1+\xi)(1+\eta) - \frac{1}{2}(N_5 + N_8) \\
N_2 &= \frac{1}{4}(1-\xi)(1+\eta) - \frac{1}{2}(N_5 + N_6) \\
N_3 &= \frac{1}{4}(1-\xi)(1-\eta) - \frac{1}{2}(N_6 + N_7) \\
N_4 &= \frac{1}{4}(1+\xi)(1-\eta) - \frac{1}{2}(N_7 + N_8) \\
N_5 &= \frac{1}{2}(1-\xi^2)(1+\eta) \\
N_6 &= \frac{1}{2}(1-\xi)(1-\eta^2) \\
N_7 &= \frac{1}{2}(1-\xi^2)(1-\eta) \\
N_8 &= \frac{1}{2}(1+\xi)(1-\eta^2)
\end{aligned}
\tag{3.78}
$$

When the elements have curved boundaries, or arbitrary nodal locations (such as the quadrilaterals), the integrals appearing in the expression for the element matrix are most easily evaluated by using a natural coordinate system. Since it is more advantageous to use natural coordinates, the variables of integration are changed so that the integrals can be evaluated using natural coordinates. In two dimensions, the integral over an arbitrary quadrilateral region of $dxdy$ becomes an integral over a square area of $d\xi d\eta$ in a natural coordinate system in the form

$$\int_A f(x,y)\,dxdy = \int_{-1}^{1}\int_{-1}^{1} g(\xi,\eta)|\mathbf{J}|\,d\xi d\eta \tag{3.79}$$

where $|\mathbf{J}|$ is the determinant of the Jacobian matrix relating the term $dxdy$ to $d\xi d\eta$ from advanced calculus as

$$dx\,dy = |\mathbf{J}|\,d\xi d\eta \tag{3.80}$$

The Jacobian matrix, $\mathbf{J}$, is given by

$$\mathbf{J} = \begin{bmatrix} \dfrac{\partial x}{\partial \xi} & \dfrac{\partial y}{\partial \xi} \\ \dfrac{\partial x}{\partial \eta} & \dfrac{\partial y}{\partial \eta} \end{bmatrix} \tag{3.81}$$

whose determinant is always positive, $|\mathbf{J}| > 0$, for a one-to-one mapping.

It is not necessary to use interpolation or shape functions of the same order for describing both the geometry and field variable of an element. If the geometry is described by a lower-order model (in comparison to that for the field variable), the element is called a "subparametric element." On the other hand, if the geometry is described by a higher-order interpolation function, then the element is termed a "superparametric" element.

3.6 Numerical Evaluation of Integrals

The evaluation of line or area integrals appearing in the finite element equations can be performed numerically by employing the Gaussian integration method (Stroud and Secrest 1966). This method locates sampling points (also called Gaussian points) to achieve the greatest accuracy.

3.6.1 Line Integrals

The line integrals encountered commonly are of the form

$$I = \int_a^b f(x)\,dx \tag{3.82}$$

The limits of this integral can be changed by introducing a new variable as

$$x = \frac{1}{2}[(b-a)\xi + (b+a)] \tag{3.83}$$

Thus, the integral given by Eq. (3.82) can be rewritten as

$$I = \int_{-1}^{1} f(\xi) J d\xi \tag{3.84}$$

in which the variables ξ and J are given by

$$\xi = \frac{2}{b-a}\left[x - \frac{(b+a)}{2}\right] \tag{3.85}$$

and

$$J = \frac{dx}{d\xi} = \frac{b-a}{2} \tag{3.86}$$

Integrals expressed in the form of Eq. (3.84) are almost always evaluated numerically. The most commonly used Gaussian integration technique approximates the integral in the form

$$I = \int_{-1}^{1} f(\xi)\, d\xi \approx \sum_{i=1}^{n} w_i f(\xi_i) \tag{3.87}$$

The weights of the numerical integration are denoted by w_i, and the number of evaluation points, ξ_i (referred to as the Gaussian points), depends on the order of the polynomial approximation of the integrand.

In general, the integrand $f(\xi)$ in Eq. (3.87) can be approximated as

$$f(\xi) = \alpha_1 + \alpha_2 \xi + \alpha_3 \xi^2 + \alpha_4 \xi^3 + \ldots + \alpha_{2n} \xi^{2n-1} \tag{3.88}$$

resulting in

$$I = \int_{-1}^{1} f(\xi)\, d\xi = 2\alpha_1 + \frac{2}{3}\alpha_3 + \ldots + \frac{2}{2n-1}\alpha_{2n-1} \tag{3.89}$$

and

$$\begin{aligned} I = \sum_{i=1}^{n} w_i f(\xi_i) = \alpha_1 \sum_{i=1}^{n} w_i + \alpha_2 \sum_{i=1}^{n} w_i \xi_i + \alpha_3 \sum_{i=1}^{n} w_i \xi_i^2 + \ldots \\ + \alpha_{2n} \sum_{i=1}^{n} w_i \xi_i^{2n-1} \end{aligned} \tag{3.90}$$

Equating the coefficients of the $\alpha_i^{'}$s in Eq. (3.89) and (3.90) leads to

$$\begin{aligned} &\sum_{i=1}^{n} w_i = 2, \ \sum_{i=1}^{n} w_i \xi_i = 0 \\ &\sum_{i=1}^{n} w_i \xi_i^2 = \frac{2}{3}, \ \sum_{i=1}^{n} w_i \xi_i^{2n-2} = \frac{2}{2n-1} \\ &\sum_{i=1}^{n} w_i \xi_i^{2n-1} = 0 \end{aligned} \tag{3.91}$$

providing $2n$ equations in n unknowns for positions ξ_i and n unknowns for weights w_i. Hence, for a polynomial of degree $p = 2n-1$, it is sufficient to use n sampling points for exact integration, i.e., the exact integration is obtained if $n \geq (p+1)/2$. This means that for "n" sampling points, a polynomial of degree $(2n-1)$ can be integrated exactly.

Rewriting Eq. (3.84) in its final form as

$$I = \int_a^b f(x)\,dx = \frac{b-a}{2}\int_{-1}^{1} f\left[\frac{b-a}{2}\xi + \frac{b+a}{2}\right]d\xi \tag{3.92}$$

and assuming a third-order polynomial ($p=3$) approximation for $f(\xi)$ in Eq. (3.92), this integral is approximated with two sampling points ($n=2$) as

$$I \approx w_1 f(\xi_1) + w_2 f(\xi_2) \tag{3.93}$$

where $-1 \leq \xi_1 \; \xi_2 \leq 1$, and $w_1 \;\; w_2$, (Gaussian weights), ξ_1, and ξ_2 are to be determined. For each coefficient of the cubic representation of $f(\xi)$, Eq. (3.91) yields

$$\int_{-1}^{1} \xi^3 d\xi = 0 = w_1\xi_1^3 + w_2\xi_2^3 \tag{3.94a}$$

$$\int_{-1}^{1} \xi^2 d\xi = \frac{2}{3} = w_1\xi_1^2 + w_2\xi_2^2 \tag{3.94b}$$

$$\int_{-1}^{1} \xi d\xi = 0 = w_1\xi_1 + w_2\xi_2 \tag{3.94c}$$

$$\int_{-1}^{1} d\xi = 2 = w_1 + w_2 \tag{3.94d}$$

Multiplying Eq. (3.94c) by ξ_1^2 and subtracting it from Eq. (3.94a) gives

$$w_2\xi_2(\xi_2^2 - \xi_1^2) = w_2\xi_2(\xi_2 - \xi_1)(\xi_2 + \xi_1) = 0 \tag{3.95}$$

For this equality to be valid, the possibilities are:

1. $w_2 = 0 \rightarrow$ one-term formula—reject.
2. $\xi_2 = 0 \rightarrow w_1 = 0$ one-term formula—reject.
3. $\xi_1 = \xi_2 \rightarrow w_1 = 0$ one-term formula—reject.
4. $\xi_2 = -\xi_1 \rightarrow$ ACCEPTED.

Thus, substituting for $\xi_2 = -\xi_1$ in Eq. (3.94) leads to

$$w_1 = w_2 \tag{3.96a}$$

Table 3.1 Positions and weights for Gauss integration

Gauss points	w_i	w_i
$n=1$	0.00	2.00
$n=2$	$\pm\sqrt{1/3}$	1.00
$n=3$	0.00 x	x x
$n=4$	±0.339981 ±0.861136	0.652145 0.347854
$n=5$	0.00 x x	0.568888 0.478628 0.236926

$$2\xi_1^2 = \frac{2}{\sqrt{3}} \rightarrow \xi_1 = \frac{1}{\sqrt{3}}, \xi_2 = -\frac{1}{\sqrt{3}} \tag{3.96b}$$

$$w_1 = w_2 = 1 \tag{3.96c}$$

The numerical integration, Eq. (3.93) becomes

$$I \approx f\left(-\frac{1}{\sqrt{3}}\right) + f\left(\frac{1}{\sqrt{3}}\right) \tag{3.97}$$

The Gaussian points and weights for polynomials of order up to 5 are summarized in Table 3.1. The Gaussian points for higher order polynomial approximation are given by Abramowitz and Stegun (1972).

An example is considered that evaluates the line integral given by

$$I = \int_{-0.25}^{0.25} e^x dx \tag{3.98}$$

This integral can be rewritten as

$$I = \frac{1}{4}\int_{-1}^{1} e^{\xi/4} d\xi \tag{3.99}$$

Applying Gauss's formula with $n=2$ integration points, this integral is approximated as

$$I \approx \frac{1}{4}[e^{-1/4\sqrt{3}} + e^{1/4\sqrt{3}}] = 0.505217 \tag{3.100}$$

The exact solution is $I = 2 \times \sinh(0.25) = 0.505224$.

3.6.2 *Triangular Area Integrals*

The area integrals over a triangular region given in the form

$$I = \int_A f(x,y)dA \tag{3.101}$$

can be rewritten as

$$I = \int_0^1 \int_0^{1-\xi_2} f(\xi_1 \xi_2)|\mathbf{J}|d\xi_1 d\xi_2 \tag{3.102}$$

in which $|\mathbf{J}|$ is the determinant of the *Jacobian* matrix expressed as

$$\mathbf{J} = \begin{bmatrix} \dfrac{\partial x}{\partial \xi_1} & \dfrac{\partial y}{\partial \xi_1} \\ \dfrac{\partial x}{\partial \xi_2} & \dfrac{\partial y}{\partial \xi_2} \end{bmatrix} = \begin{bmatrix} (x_1 - x_3) & (y_1 - y_3) \\ (x_2 - x_3) & (y_2 - y_3) \end{bmatrix} = 2A \tag{3.103}$$

relating the area coordinates (discussed in Sect. 3.3.4.2.1) to Cartesian coordinates

$$\begin{Bmatrix} \dfrac{\partial}{\partial \xi_1} \\ \dfrac{\partial}{\partial \xi_2} \end{Bmatrix} = [\mathbf{J}] \begin{Bmatrix} \dfrac{\partial}{\partial x} \\ \dfrac{\partial}{\partial y} \end{Bmatrix} \tag{3.104}$$

The extent of the triangular area of integration is defined by the coordinates (x_i, y_i) (with $i = 1,2,3$) of the vertices. The Gaussian approximation to the integration is expressed as

$$I = \int_0^1 \int_0^{1-\xi_2} f(\xi_1, \xi_2)|\mathbf{J}|d\xi_1 d\xi_2 \approx 2A \sum_{i=1}^{n} w_i f(\xi_{1i}, \xi_{2i}) \tag{3.105}$$

in which the weights of the numerical integration are denoted by w_i. The number of evaluation points, ξ_{1i} and ξ_{2i}, are referred to as the Gaussian integration points and they depend on the order of the polynomial approximation of the integrand. Depending on the degree of approximation, the weights and the evaluation points are given by Huebner et al. (2001).

An example is considered that evaluates the area integral given by

$$I = \int_A xy dA \tag{3.106}$$

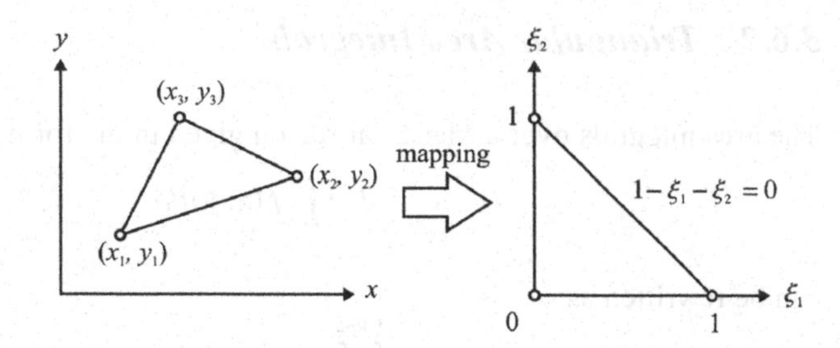

Fig. 3.31 A triangular element and its mapping

in which the area A is defined by a triangle whose vertices are (1,1), (3,2), and (2,3), as shown in Fig. 3.31. This integral can also be evaluated exactly by using Eq. (3.19).

The coordinates (x, y) of a point within a triangular area can be expressed as linear combinations of the nodal coordinates (x_1, y_1) (x_2, y_2), and (x_3, y_3) and the area coordinates ξ_1, ξ_2, and ξ_3 as

$$x = \xi_1 x_1 + \xi_2 x_2 + \xi_3 x_3 = \xi_1 + 3\xi_2 + 2\xi_3 = -\xi_1 + \xi_2 + 2 \tag{3.107a}$$

$$y = \xi_1 y_1 + \xi_2 y_2 + \xi_3 y_3 = \xi_1 + 2\xi_2 + 3\xi_3 = -2\xi_1 - \xi_2 + 3 \tag{3.107b}$$

with

$$\xi_1 + \xi_2 + \xi_3 = 1 \tag{3.107c}$$

Substituting for x and y in the integrand of Eq. (3.106) results in

$$I = 2A \int_0^1 \int_0^{1-\xi_2} (2\xi_1^2 - \xi_2^2 - \xi_1\xi_2 - 7\xi_1 + \xi_2 + 6) d\xi_1 d\xi_2 \tag{3.108}$$

Utilizing $n = 3$ Gaussian points as shown in Fig. 3.32, approximation to the integration by Eq. (3.105) becomes

$$I \approx 2A[w_1 f(\xi_{11}, \xi_{21}) + w_2 f(\xi_{12}, \xi_{22}) + w_3 f(\xi_{13}, \xi_{23})] \tag{3.109}$$

in which $w_1 = w_2 = w_3 = 1/6$, $\xi_{11} = 1/2$, $\xi_{21} = 0$, $\xi_{12} = 1/2$, $\xi_{22} = 1/2$, $\xi_{13} = 0$, and $\xi_{23} = 1/2$. The area of the triangle is obtained from Eq. (3.18) as $2A = 3$. Thus, the Gaussian approximation leads to

$$\begin{aligned} I \approx 2A[&w_1(2\xi_{11}^2 - \xi_{21}^2 - \xi_{11}\xi_{21} - 7\xi_{11} + \xi_{21} + 6) \\ &+ w_2(2\xi_{12}^2 - \xi_{22}^2 - \xi_{12}\xi_{22} - 7\xi_{12} + \xi_{22} + 6) \\ &+ w_3(2\xi_{13}^2 - \xi_{23}^2 - \xi_{13}\xi_{23} - 7\xi_{13} + \xi_{23} + 6)] \end{aligned}$$

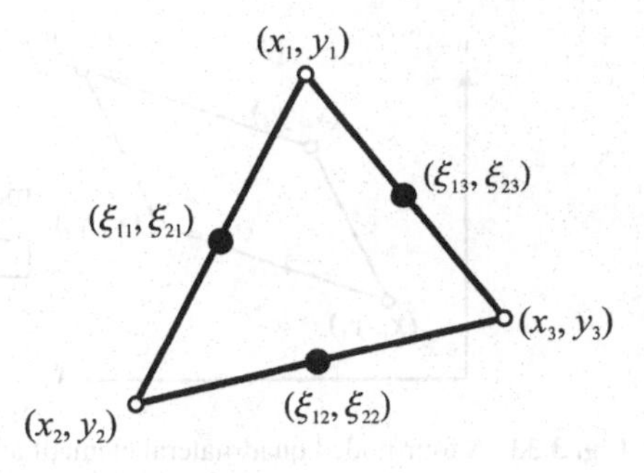

Fig. 3.32 Three Gaussian points, located at mid-sides, for approximate integration

and

$$I \approx 3\frac{1}{6}\left[\left(2\frac{1}{4}-7\frac{1}{2}+6\right)+\left(2\frac{1}{4}-\frac{1}{4}-\frac{1}{4}-7\frac{1}{2}+\frac{1}{2}+6\right)+\left(-\frac{1}{4}+\frac{1}{2}+6\right)\right]$$

and

$$I \approx \frac{1}{2}\left[6+\frac{25}{4}\right] = 6.125 \tag{3.110}$$

For the exact evaluation, substituting for x and y in the integrand results in

$$I = \int_A (\xi_1^2 + 6\xi_2^2 + 6\xi_3^2 + 5\xi_1\xi_2 + 13\xi_2\xi_3 + 5\xi_1\xi_3)dA \tag{3.111}$$

Utilizing the formula of Eq. (3.19) for exact integration results in

$$\begin{aligned} I &= \frac{2A}{4!}(2! + 6\times 2! + 6\times 2! + 5\times 1!\times 1! + 13\times 1!\times 1! + 5\times 1!\times 1!) \\ &= \frac{3}{24}[13\times 2 + 23] = \frac{49}{8} = 6.125 \end{aligned} \tag{3.112}$$

3.6.3 Quadrilateral Area Integrals

The quadrilateral area integrals appearing in the form

$$I = \int_a^b\int_c^d f(x,y)dx\,dy \tag{3.113}$$

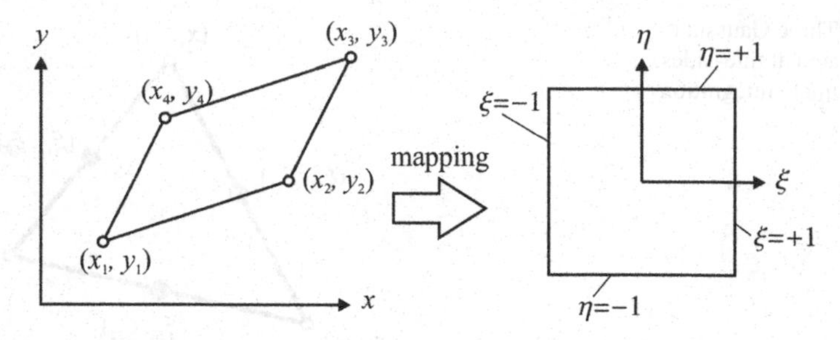

Fig. 3.33 A four-noded quadrilateral element and its mapping

can be rewritten as

$$I = \int_{-1}^{1}\int_{-1}^{1} f(\xi,\eta)\left|\mathbf{J}\right| d\xi d\eta \tag{3.114}$$

in which $|\mathbf{J}|$ is the determinant of the Jacobian matrix expressed as

$$\mathbf{J} = \begin{bmatrix} \dfrac{\partial x}{\partial \xi} & \dfrac{\partial y}{\partial \xi} \\ \dfrac{\partial x}{\partial \eta} & \dfrac{\partial y}{\partial \eta} \end{bmatrix} \text{ relating } \begin{Bmatrix} \dfrac{\partial}{\partial \xi} \\ \dfrac{\partial}{\partial \eta} \end{Bmatrix} = [\mathbf{J}] \begin{Bmatrix} \dfrac{\partial}{\partial x} \\ \dfrac{\partial}{\partial y} \end{Bmatrix} \tag{3.115}$$

These integrals can be evaluated first with respect to one variable and then with respect to the other leading to

$$I = \int_{-1}^{1}\int_{-1}^{1} f(\xi,\eta)\left|\mathbf{J}\right| d\xi d\eta \approx \sum_{i=1}^{n}\sum_{j=1}^{n} w_i w_j f(\xi_i,\eta_j)\left|\mathbf{J}(\xi_i,\eta_j)\right| \tag{3.116}$$

in which w_i represent the weights of the numerical integration, and ξ_i and η_i are the Gaussian integration points. They are given by Abramowitz and Stegun (1972) and depend on the order of the polynomial approximation of the integrand.

An example is considered that evaluates the area integral given by

$$I = \int_A xy\, dA \tag{3.117}$$

in which the area A is defined by a quadrilateral whose vertices are (1,1), (3,2), (4,4), and (2,3) as shown in Fig. 3.33.

The coordinates (x, y) of a point within a quadrilateral area can be expressed as linear combinations of the nodal coordinates (x_1, y_1) (x_2, y_2) (x_3, y_3), and (x_4, y_4) and the natural coordinates ξ and η as

Fig. 3.34 Two Gaussian points, in each direction, for approximate integration

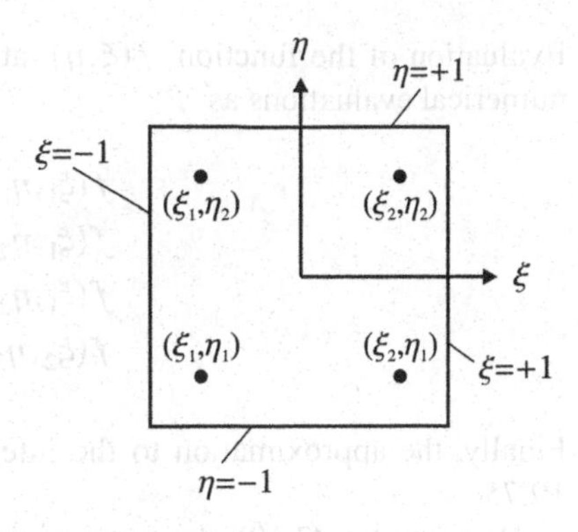

$$x = \frac{1}{4}[(1-\xi)(1-\eta) + 3(1+\xi)(1-\eta) \\ +4(1+\xi)(1+\eta) + 2(1-\xi)(1+\eta)] \tag{3.118}$$

$$y = \frac{1}{4}[(1-\xi)(1-\eta) + 2(1+\xi)(1-\eta) \\ +4(1+\xi)(1+\eta) + 3(1-\xi)(1+\eta)] \tag{3.119}$$

The Jacobian matrix is obtained as

$$\mathbf{J} = \begin{bmatrix} 1 & 1/2 \\ 1/2 & 1 \end{bmatrix}$$

with its determinant $|\mathbf{J}| = 3/4$.

Utilizing two Gaussian points as shown in Fig. 3.34, the approximation to the integration becomes

$$I \approx \frac{3}{4}[w_1 w_1 f(\xi_1, \eta_1) + w_1 w_2 f(\xi_1, \eta_2) \\ +w_2 w_1 f(\xi_2, \eta_1) + w_2 w_2 f(\xi_2, \eta_2)] \tag{3.120}$$

in which $w_1 = w_2 = 1$ $\xi_1 = -1/\sqrt{3}$ $\xi_2 = 1/\sqrt{3}$ $\eta_1 = -1/\sqrt{3}$ and $\eta_2 = 1/\sqrt{3}$

The function $f(\xi, \eta)$ is expressed as

$$f(\xi,\eta) = \frac{1}{16}[(1-\xi)(1-\eta) + 3(1+\xi)(1-\eta) \\ +4(1+\xi)(1+\eta) + 2(1-\xi)(1+\eta)] \\ \times[(1-\xi)(1-\eta) + 2(1+\xi)(1-\eta) \\ +4(1+\xi)(1+\eta) + 3(1-\xi)(1+\eta)] \tag{3.121}$$

Evaluation of the function $f(\xi,\eta)$ at Gaussian integration points results in their numerical evaluations as

$$f(\xi_1,\eta_1)=11.33012702$$
$$f(\xi_1,\eta_2)=6.16666667$$
$$f(\xi_1,\eta_2)=6.16666667$$
$$f(\xi_2,\eta_2)=2.66987298$$

Finally, the approximation to the integral from Eq. (3. 120) is determined to be 19.75.

By using Eq. (3.19), the exact evaluation of this integral can be obtained by integration over two triangular regions defined by the vertices (1,1), (3,2), and (2,3) and (3,2), (4,4), and (2,3). The exact integration over these two regions are obtained as 6.125 and 12.125. Their summation provides the exact integration over a quadrilateral defined by vertices (1,1), (3,2), (4,4), and (2,3). Thus, the exact integration becomes 19.75.

3.7 Problems

3.1. The *completeness criterion* for convergence of finite element solutions requires that the interpolating function must be able to reproduce exactly (that is, interpolate to the exact value at every point in the element). In particular, the approximation function $\phi(x,y)$ is specified as

$$\phi(x,y)=a+bx+cy=\sum N_i\phi_i$$

where a, b, and c are arbitrary constants, ϕ_i are the nodal values, and $N_i(x,y)$ are the interpolating functions.

(a) Derive a set of three equations that the interpolating functions $N_i(x,y)$ must satisfy for completeness.

(b) Show that the standard and quadratic linear interpolation functions for a triangular domain satisfy these requirements.

3.2. Using the coordinate transformation equations given in Sect. 3.5 for an 8-noded quadrilateral element, determine the isoparametric element shape whose nodal locations are

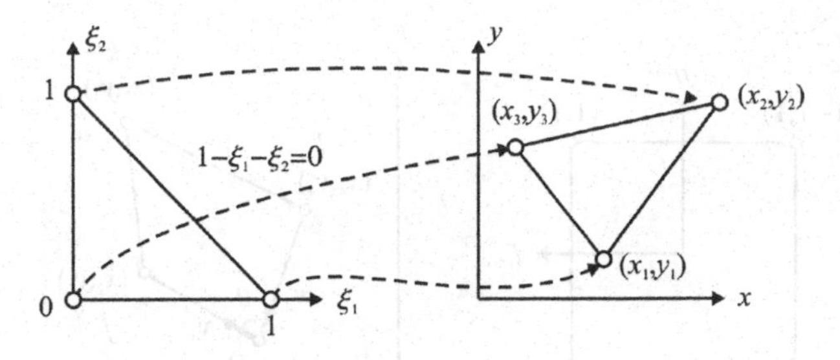

Fig. 3.35 A triangular element and its mapping

Node no.	x	y
1	6.0	3.0
2	−4.0	3.0
3	− 5.0	−3.0
4	4.0	− 3.0
5	1.0	4.0
6	− 3.0	0.5
7	0.0	− 2.0
8	5.0	0.0

3.3. The isoparametric formulation is useful for triangular, as well as for quadrilateral, elements. Also, the area coordinates (ξ_1, ξ_2, ξ_3) are commonly employed for triangular elements instead of using the local coordinates (r, s). However, because only two of these are independent coordinates, one of them, say ξ_3, can be eliminated in favor of ξ_1 and ξ_2. Thus, for a 3-noded triangle, the interpolation functions are $N_i = \xi_i$ $(i = 1, 2, 3)$ and the coordinate transformations, using $\xi_3 = 1 - \xi_1 - \xi_2$, are

$$x = \xi_1 x_1 + \xi_2 x_2 + (1 - \xi_1 - \xi_2) x_3$$
$$y = \xi_1 y_1 + \xi_2 y_2 + (1 - \xi_1 - \xi_2) y_3$$

As illustrated in Fig. 3.35, this clearly maps a triangle with vertices (1,0), (0,1), and (0,0) in the ξ_1 - ξ_2 plane into a triangle with vertices (x_1, y_1), (x_2, y_2), and (x_3, y_3) in the *x-y* plane. Also, the integrals in the *x-y* plane may be related to integrals in the ξ_1 - ξ_2 plane by

$$|\boldsymbol{J}| d\xi_2 d\xi_1$$

Explicitly determine the coordinate transformations and the Jacobian matrix for the 6-noded triangle having the side nodes located at the midpoint of each side. Explain how it is possible to obtain a triangular element in the *x-y* plane with one or more curved sides. What is the form of the curve?

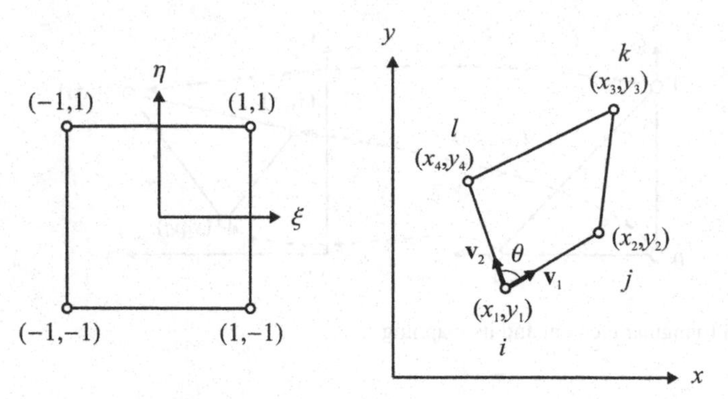

Fig. 3.36 A four-noded quadrilateral element and its mapping

3.4. For a 4-noded element shown in Fig. 3.36, the mapping is achieved by

$$x = \sum_{i=1}^{4} N_i(\xi,\eta)x_i \quad \text{and} \quad y = \sum_{i=1}^{4} N_i(\xi,\eta)y_i$$

where

$$N_1 = \frac{1}{4}(1-\xi)(1-\eta),\; N_2 = \frac{1}{4}(1+\xi)(1-\eta)$$

$$N_3 = \frac{1}{4}(1+\xi)(1+\eta) \quad N_4 = \frac{1}{4}(1-\xi)(1+\eta)$$

(a) For this element explicitly determine the Jacobian determinant, and show that it is strictly linear in the local coordinates ξ and η and that the term proportional to the product $\xi\eta$ vanishes.

(b) Show that the Jacobian determinant becomes

$$|\mathbf{J}| = \frac{1}{4}[(x_4 - x_3)(y_2 - y_3) - (x_2 - x_3)(y_4 - y_3)]$$

for $\xi = \eta = 1$.

(c) Using the definition of the cross-product of the vectors v_1 and v_2 shown in Fig. 3.36, show that at $\xi = \eta = 1$

$$|\mathbf{J}| > 0 \text{ if } 0 < \theta < \pi$$

(d) Based on the results of parts (a) and (c), provide a short argument to show that $|\mathbf{J}| > 0$ throughout the element and, hence, the coordinate transformation $(\xi,\eta) \rightarrow (x,y)$ is unique and invertible if the interior angles at all nodes are less than 180°.

Chapter 4
ANSYS Preprocessor

4.1 Fundamentals of Modeling

The fundamental concepts and the *Begin* and *Processor Levels* of the ANSYS finite element program are described in Chap. 2. Specifications of all the geometric and material properties, as well as the generation of solid and finite element models, are conducted at the preprocessor level.

There are two approaches for creating a finite element model: solid modeling and direct generation. The solid modeling approach utilizes *Primitives* (pre-defined geometric shapes) and operations similar to those of computer-aided design (CAD) tools, and internally generates the nodes and the elements based on user specifications. Solid modeling is the most commonly used approach because it is much more versatile and powerful. However, the user must have a strong understanding of the concept of meshing in order to utilize the solid modeling approach successfully and efficiently.

Direct generation is entirely dependent on user input for the size, shape, and connectivity of each element and coordinates of each node before it creates the nodes and elements one at a time. It requires the user to keep track of the node and element numbering, which may become tedious—sometimes practically impossible—for complex problems requiring thousands of nodes. It is, however, extremely useful for simple problems as one has full control over the model.

A combination of the two approaches is not only possible, but also advantageous in many cases. A comprehensive list of some important advantages and disadvantages is given in Table 4.1.

4.2 Modeling Operations

Within the *ANSYS Preprocessor*, a finite element model is generated by utilizing various operations, which are explained in this section.

The online version of this book (doi: 10.1007/978-1-4939-1007-6_4) contains supplementary material, which is available to authorized users

E. Madenci, I. Guven, *The Finite Element Method and Applications in Engineering Using ANSYS®*, DOI 10.1007/978-1-4899-7550-8_4

Table 4.1 Advantages and disadvantages of solid modeling and direct generation

Advantages	Disadvantages
Solid modeling	
Powerful (sometimes the only feasible way) in modeling three-dimensional solid volumes with complex geometry	If the user does not have a good understanding of meshing, ANSYS may not be able to generate the finite element mesh
User data input is rather low	For simple problems, using solid modeling may be ponderous
Common computer-aided design (CAD)-type operations such as extrusions, dragging, and rotations are utilized which are not possible when working directly with the nodes and elements	
With the basic (primitive) areas and volumes (rectangular, circular etc. areas; cubic, cylindrical, spherical etc. volumes), the Boolean operations (add, subtract, overlap etc.) can be used easily to modify (or tailor) these basic areas or volumes to obtain the desired shape	
Direct generation	
Provides the user with complete control of placement and numbering of nodes and elements	Use of direct generation is extremely tedious for solving real engineering design applications, especially when the problem can not be simplified to a two-dimensional idealization
For simple problems, the direct generation is the shortest way to generate a finite element mesh	

4.2.1 Title

This operation defines the title for the ANSYS analysis. This is an optional but recommended step in a typical ANSYS session. It helps the user to keep track of the problems by appearing in the graphics display and output. It becomes extremely useful when the user conducts a case study that involves the same model with different boundary conditions, different material properties, etc. The following menu path is used to change (or specify) the title:

Utility Menu > File > Change Title

which brings up the dialog box shown in Fig. 4.1. After entering the desired title in the *text box*, clicking on ***OK*** completes the specification of the title.

4.2.2 Elements

The nodes and elements are the essential parts of a finite element model. Before starting meshing, the *element type*(*s*) to be used must be defined (otherwise ANSYS refuses to create the mesh). The ANSYS software contains more than 100 different

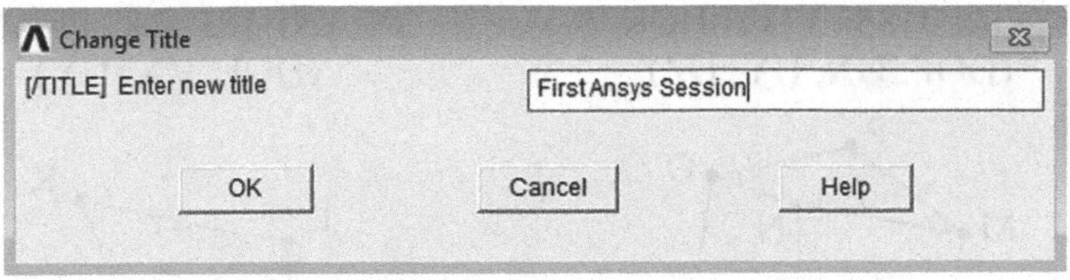

Fig. 4.1 Dialog box for specifying the title

element types in its element library. Each element type has a unique number and a prefix that identifies the element category, such as **BEAM188**, **PLANE182**, **SOLID185**, etc. The elements that are available in ANSYS can be classified according to many different criteria, such as dimensionality, analysis discipline, and material behavior. ANSYS classifies the elements in 23 different groups. In this section, the elements from four of these groups—specifically, structural, thermal, fluid, and FLOTRAN CFD—are considered for different analysis objectives.

1. ***Structural***: For this group of elements, the degrees of freedom at the nodes are displacements. As shown in Fig. 4.2, the structural analysis employs plane, link, beam, pipe, solid, and shell elements. All of the above "subgroups" of elements include several *element types* with different degree-of-freedom (DOF) sets.Consider the entries **Quad 4node 182**, **Quad 8node 183**, and **Brick 8node 185** from the ***Structural Solid*** subgroup. The first two elements types, **Quad182** and **Quad183**, are used for two-dimensional structural problems (plane stress, plane strain, or axisymmetric) whereas the third one is used for three-dimensional structural problems. The difference between **Quad182** and **Quad183** elements is that they have a different number of nodes per element, which implies that they are employing different interpolation functions for the variation of the degrees of freedom along the edges of the element. In this particular case, the variation of displacements along the element edges is assumed to be linear for **Quad 4node 182** and quadratic for **Quad 4node 183**, as shown in Fig. 4.3. The interpolation functions for the **Brick 8node 185** element are linear.
2. ***Thermal***: For this group of elements, the degrees of freedom at the nodes are temperatures. The thermal analysis employs mass, link, solid and shell subgroups. The element types in this group differ from each other with similar considerations as explained for structural discipline. Two commonly used thermal elements are shown in Fig. 4.4.
3. ***Fluid***: For this group of elements, depending on the type, the degrees of freedom appear as a pair, velocity-pressure or pressure-temperature, at the nodes. Included in this group are two- and three- dimensional acoustic, thermal-fluid coupled pipe, and contained-fluid types of elements.
4. ***FLOTRAN CFD***: This group of elements is similar to the previous one, except it is based on the method of computational finite difference.

SOLID185- 3-D BRICK
(DOF: UX, UY, UZ)

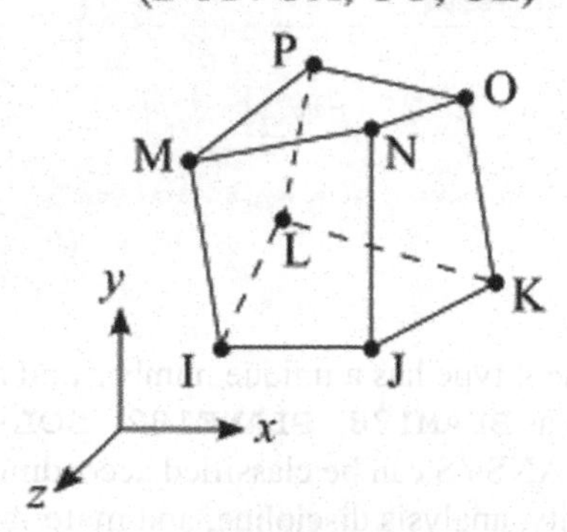

PLANE182- 2-D PLANE
(DOF: UX, UY)

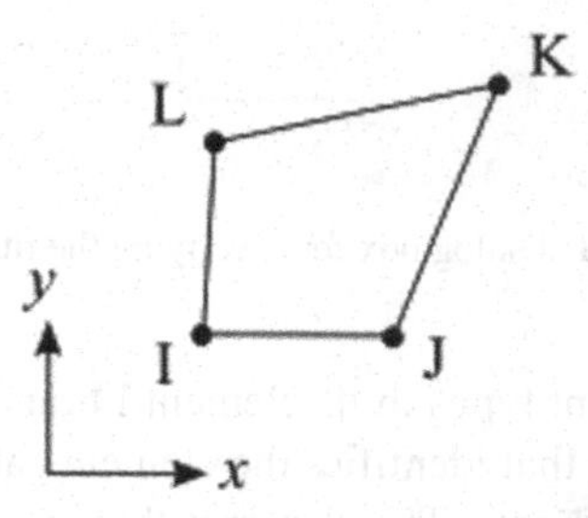

LINK180- 3-D SPAR
(DOF: UX, UY, UZ)

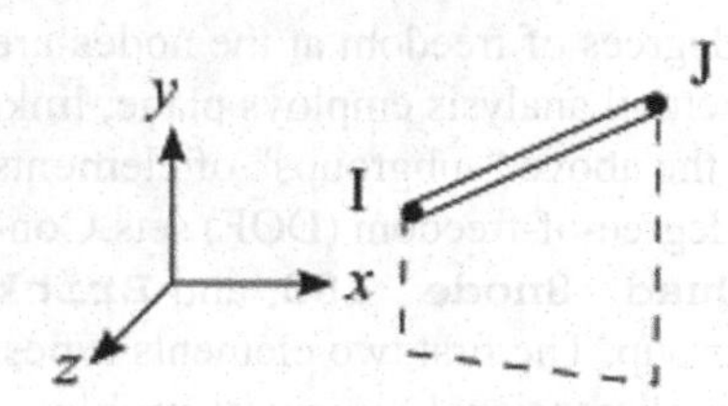

BEAM188- 2-D BEAM
(DOF: UX, UY, ROTZ)

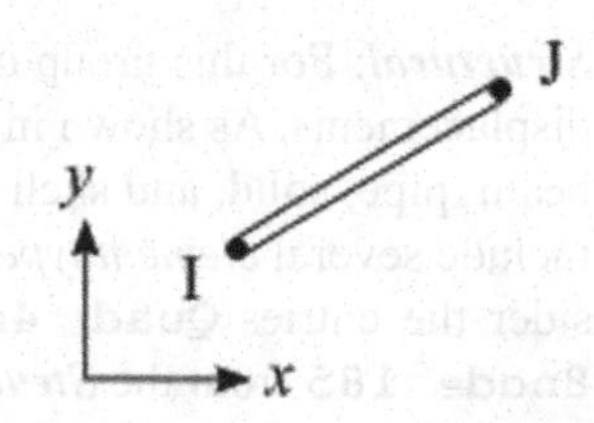

PIPE288- 3-D STRAIGHT PIPE
(DOF: UX, UY, UZ,
ROTX,ROTY, ROTZ)

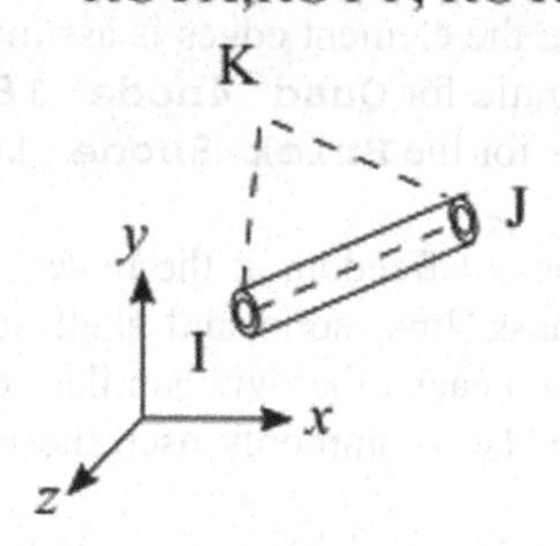

SHELL181- 3-D SHELL
(DOF: UX, UY, UZ,
ROTX,ROTY, ROTZ)

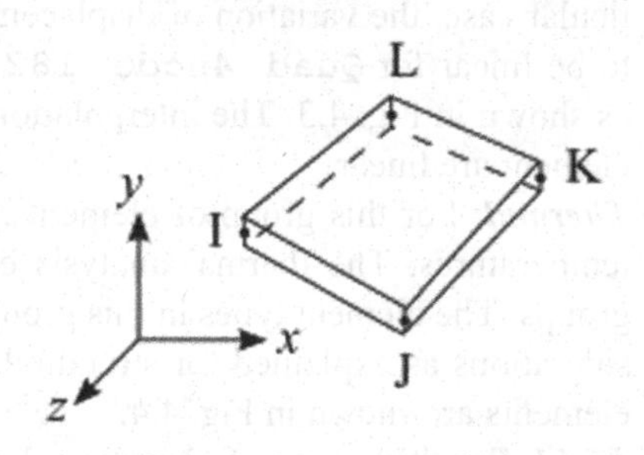

Fig. 4.2 Examples of structural elements in ANSYS

Each discipline requires the use of its own element types because the element type determines the degree-of-freedom set (displacements, temperatures, pressures, etc.) and the dimensionality of the problem (2-D or 3-D).

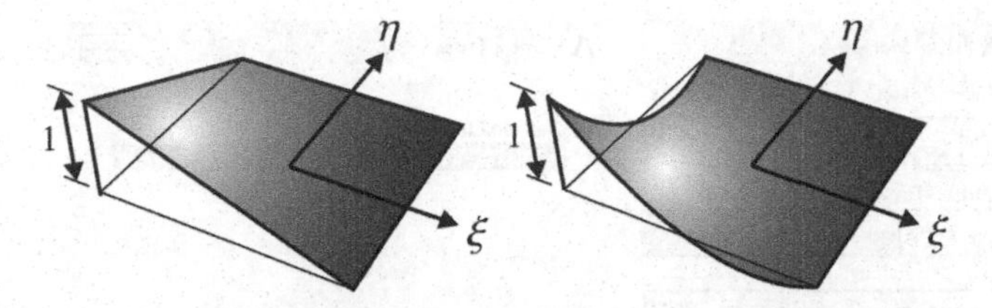

Fig. 4.3 Linear and quadratic variations of displacements within a 2-D element

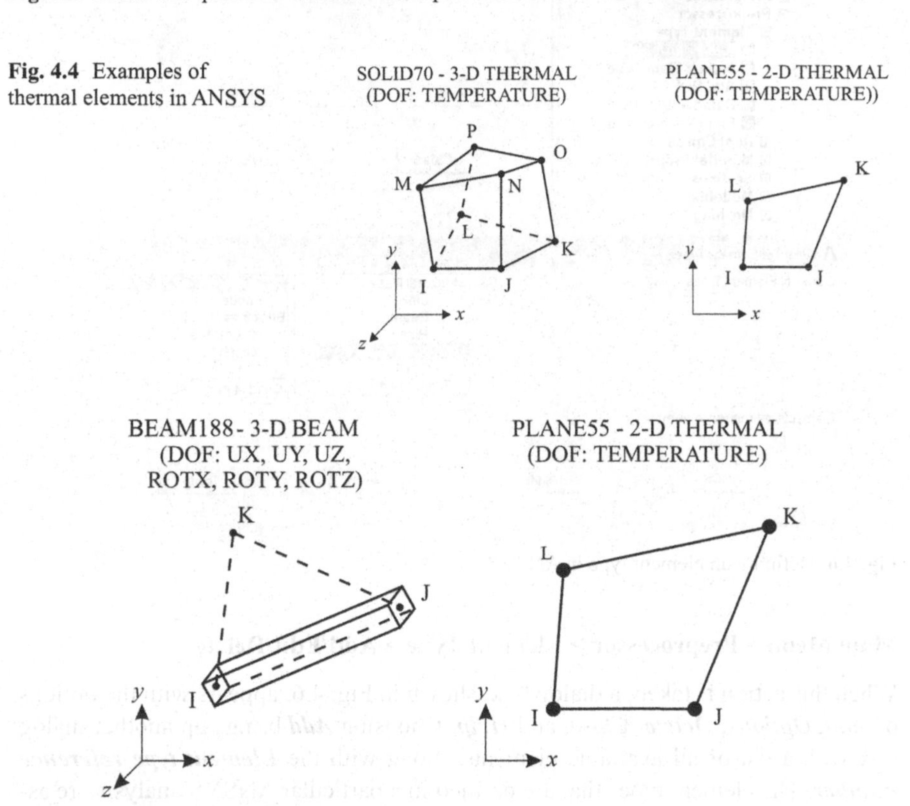

Fig. 4.4 Examples of thermal elements in ANSYS

Fig. 4.5 **BEAM188** element for 3-D problems and **PLANE55** element for 2-D problems

For example, the **`BEAM188`** element, shown in Fig. 4.5, has six structural degrees of freedom (displacements and rotations in and about the *x*-, *y*-, and *z*-directions) at each of the two nodes, is a line element, and can be modeled in 3-D space. The **`PLANE55`** element, also shown in Fig. 4.5, which has a total of four thermal degrees of freedom (temperature at each node), is a 4-noded quadrilateral element, and can be used only for two-dimensional problems.

In order to specify an element type, the user must be in the *Preprocessor*. The menu path for element specification is

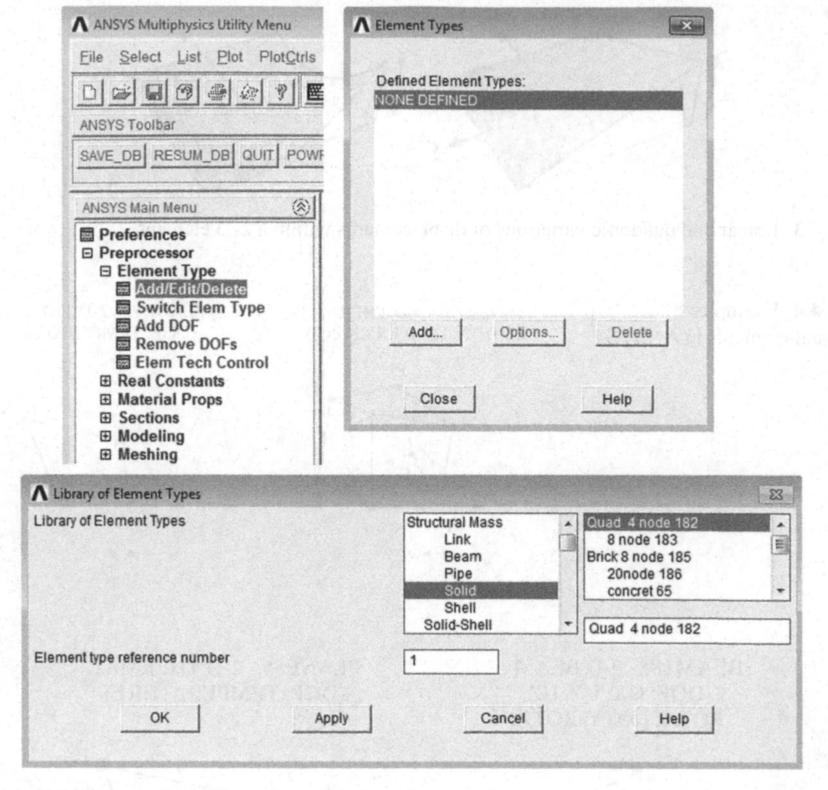

Fig. 4.6 Defining an element type in ANSYS

Main Menu > Preprocessor > Element Type > Add/Edit/Delete

When this action is taken, a dialog box, shown in Fig. 4.6, appears with the options of ***Add***, ***Options***, ***Delete***, ***Close***, and ***Help***. Choosing ***Add*** brings up another dialog box with a list of all available elements, along with the ***Element type reference number***. The element types that are defined in a particular ANSYS analysis are assigned reference numbers. This reference number is used when creating the mesh. If the analysis requires the use of more than one element type, switching from one type to another one is achieved by referring to this number (this point is further explained when discussing *Element Attributes*).

If the user wants to delete an existing element type, it is achieved by using the same *GUI* path and choosing ***Delete***, as shown in Fig. 4.7:

Main Menu > Preprocessor > Element Type > Add/Edit/Delete

Many element types have additional options, known as *keyoptions* (**KEYOPT**), and are referred to as **KEYOPT(1)**, **KEYOPT(2)**, etc. For example, as shown in Fig. 4.8, **KEYOPT(3)** for **SOLID182** (4-noded quadrilateral 2-D structural element) allows the user to specify the type of two-dimensional idealization, i.e., plane stress, plane strain, axisymmetric, or plane stress with thickness.

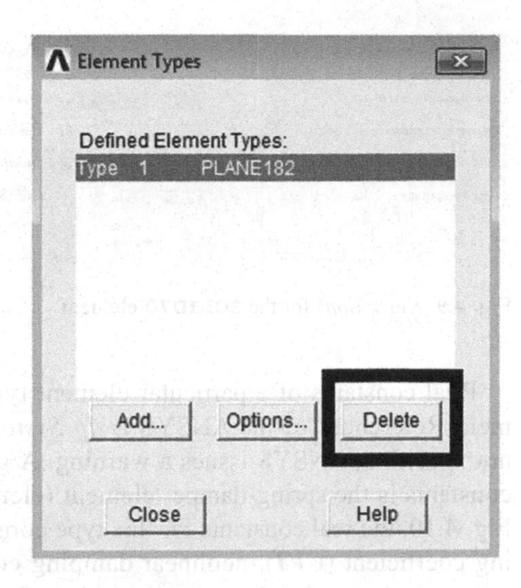

Fig. 4.7 Deleting an element type

Another example is shown in Fig. 4.9, in which **`KEYOPT(7)`** for **`SOLID70`** (8-noded thermal solid element for 3-D problems) permits the specification of a standard heat transfer or a nonlinear steady-state fluid flow through a porous medium.

Keyoptions are specified using the same *GUI* path and choosing ***Options*** from the *Element Types* dialog box.

4.2.3 Real Constants

As described in Chap. 1, the calculation of the element matrices requires material properties, nodal coordinates and geometrical parameters. Any data required for the calculation of the element matrix that cannot be determined from the nodal coordinates or material properties are called "real constants" in ANSYS. Typically, *real constants* are area, thickness, inner diameter, outer diameter, spring constant, damping coefficient etc. Not all element types require real constants.

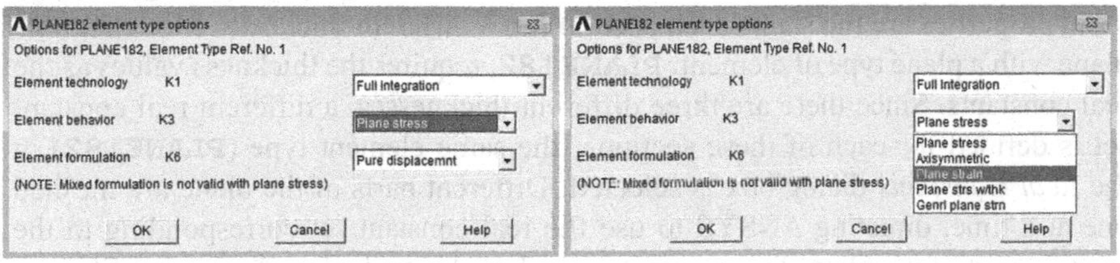

Fig. 4.8 *Keyoptions* for the **`PLANE182`** element

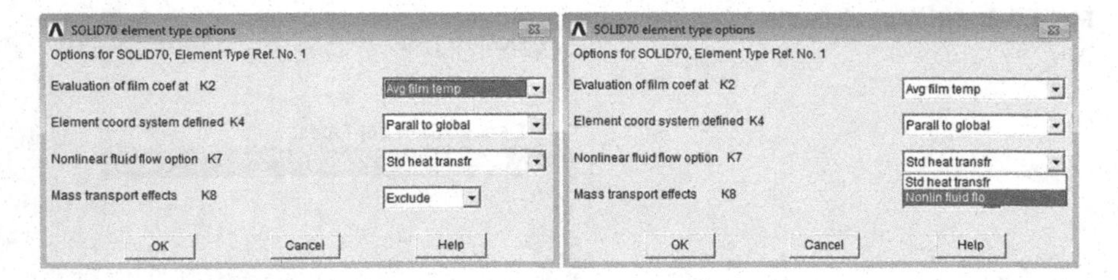

Fig. 4.9 *Keyoptions* for the **SOLID70** element

Real constants of a particular element type are briefly explained in the "Element Reference" of the ANSYS *Help System*. If the required real constants are not specified, ANSYS issues a warning. A good example for describing the real constants is the spring-damper element (element type **COMBIN14**). As shown in Fig. 4.10, the real constants for this type consist of the spring constant (***K***), damping coefficient (***CV1***), nonlinear damping coefficient (***CV2***), etc. In some cases, a complete set of real constants may not be required; in other cases, if the real constants are not specified, ANSYS may use a default value for that particular parameter. It is recommended that the "Element Reference" be consulted for the particular element type.

For each real constant set, ANSYS requires a reference number. If it is not assigned by the user, ANSYS automatically assigns a number, as shown in Fig. 4.10.

Real constants are specified using the following *GUI* path:

Main Menu > Preprocessor > Real Constants > Add/Edit/Delete

This brings up the *Real Constants* dialog box, where clicking on ***Add*** leads to another dialog box having a list of currently defined element types. Choosing the element type for which the real constants are specified (if there are no required real constants for the selected element type, a warning window pops up) and hitting ***OK*** brings up a new dialog box. The real constants for that specific element type appear; after filling in the boxes, hitting ***OK*** completes this operation.

For models having multiple element types, a distinct real constant set (that is, a different reference number) is assigned for each element type. ANSYS issues a warning message if multiple element types are referenced to the same real constant set. However, there are cases where it is necessary to specify several real constant sets for the same element type. This feature is explained further by considering a plate composed of three different sections, as shown in Fig. 4.11. Although the material properties are the same, each section has a different thickness. Modeling this plane with a plane type of element, **PLANE182**, requires the thickness values as the real constants. Since there are three different thicknesses, a different real constant set is defined for each of these sections; the same element type (**PLANE182**) in the *Real Constants* dialog box is selected. Different parts of the plane are meshed one at a time, directing ANSYS to use the real constant set corresponding to the specific part of the plate. This concept is further clarified when discussing *Element Attributes*.

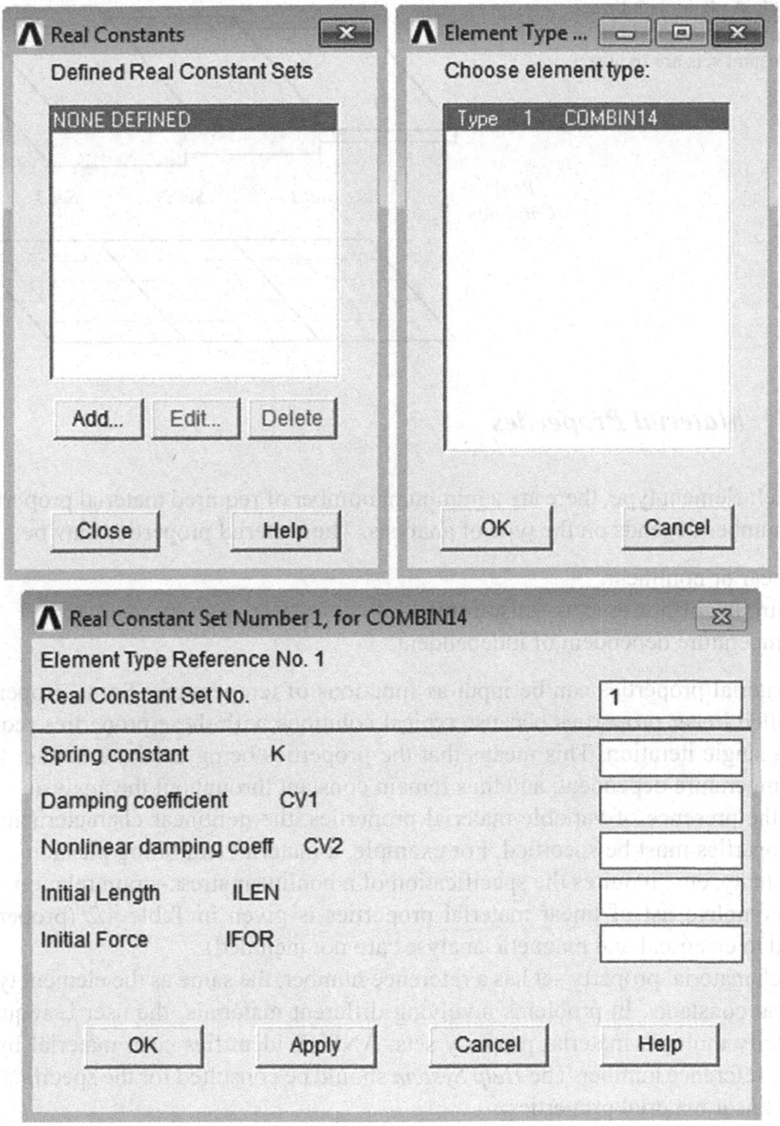

Real Constants
Defined Real Constant Sets
NONE DEFINED
Add... Edit... Delete
Close Help

Element Type ...
Choose element type:
Type 1 COMBIN14
OK Cancel

Real Constant Set Number 1, for COMBIN14
Element Type Reference No. 1
Real Constant Set No. 1
Spring constant K
Damping coefficient CV1
Nonlinear damping coeff CV2
Initial Length ILEN
Initial Force IFOR
OK Apply Cancel Help

Fig. 4.10 Real constants for the **COMBIN14** element

CAUTION It is the user's responsibility to keep track of units that are used in the analysis. The user does not need to give ANSYS the system of units being used. The user should decide which system of units to use and be consistent throughout the analysis (i.e., dimensions of the input, real constants, material properties and loads). ANSYS WILL NOT CONVERT UNITS. Also, the solution quantities are given in terms of the units of the input.

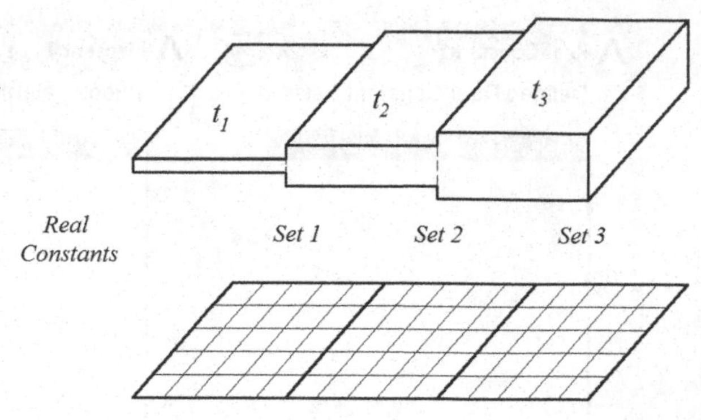

Fig. 4.11 A plane with three different thicknesses; three real constant sets are required

4.2.4 Material Properties

For each element type, there are a minimum number of required material properties. This number depends on the type of analysis. The material properties may be:

- Linear or nonlinear.
- Isotropic, orthotropic, or anisotropic.
- Temperature dependent or independent.

All material properties can be input as functions of temperature. Some properties are called linear properties because typical solutions with these properties require only a single iteration. This means that the properties being used are neither time nor temperature dependent, and thus remain constant throughout the analysis.

In the presence of variable material properties, the nonlinear characteristics of the properties must be specified. For example, a material exhibiting plasticity, viscoplasticity, etc., requires the specification of a nonlinear stress-strain relation.

A complete list of linear material properties is given in Table 4.2 (properties related to electrical and magnetic analyses are not included).

Each material property set has a reference number, the same as the element types and real constants. In problems involving different materials, the user is required to specify multiple material property sets. ANSYS identifies each material by its unique reference number. The *Help System* should be consulted for the specification of nonlinear material properties.

The following menu path is used to specify constant isotropic or orthotropic material properties:

Main Menu > Preprocessor > Material Props > Material Models

This brings up the *Define Material Model Behavior* dialog box, as shown in Fig. 4.12. On the left side of this window, material models are listed based on their material reference numbers. On the right side, available material models are organized based on the analysis type (e.g., structural, thermal, etc.). Figure 4.13 shows an expanded view of the material models available under *Structural* analysis. As observed in the figure, if a linear material response is to be used, then the user double-clicks

Table 4.2 List of material properties for structural, thermal, and fluids disciplines

Label	Units	Description
EX	Force/Area	Elastic modulus, element *x*-direction
EY		Elastic modulus, element *y*-direction
EZ		Elastic modulus, element *z*-direction
ALPX	Strain/Temp	Coefficient of thermal expansion, element *x*-direction
ALPY		Coefficient of thermal expansion, element *y*-direction
ALPZ		Coefficient of thermal expansion, element *z*-direction
REFT	Temp	Reference temperature (as a property)
PRXY	None	Major Poisson's ratio, *x-y* plane
PRYZ		Major Poisson's ratio, *y-z* plane
PRXZ		Major Poisson's ratio, *x-z* plane
NUXY		Minor Poisson's ratio, *x-y* plane
NUYZ		Minor Poisson's ratio, *y-z* plane
NUXZ		Minor Poisson's ratio, *x-z* plane
GXY	Force/Area	Shear modulus, *x-y* plane
GYZ		Shear modulus, *y-z* plane
GXZ		Shear modulus, *x-z* plane
DAMP	Time	K matrix multiplier for damping
MU	None	Coefficient of friction (or, for **FLUID29** element, boundary admittance)
DENS	Mass/Vol	Mass density
C	Heat/Mass × Temp	Specific heat
ENTH	Heat/Vol	Enthalpy
KXX	Heat × Length/ (Time × Area × Temp)	Thermal conductivity, element *x*-direction
KYY		Thermal conductivity, element *y*-direction
KZZ		Thermal conductivity, element *z*-direction
HF	Heat/ (Time × Area × Temp)	Convection (or film) coefficient
EMIS	None	Emissivity
QRATE	Heat/Time	Heat generation rate (**MASS71** element only)
VISC	Force × Time/Length^2	Viscosity
SONC	Length/Time	Sonic velocity (**FLUID29** and **FLUID30**)

on the ***Linear*** option to expand. After double-clicking on the ***Elastic*** option under ***Linear***, three options are available for the user: ***isotropic***, ***orthotropic***, and ***anisotropic***. Upon double-clicking on any of these options, a new dialog box appears. Figure 4.14 (left) shows the dialog box corresponding to the ***isotropic*** option. If the material properties are temperature dependent, the ***Add Temperature*** button is used for adding columns for different temperatures, as shown in Fig. 4.14 (right).

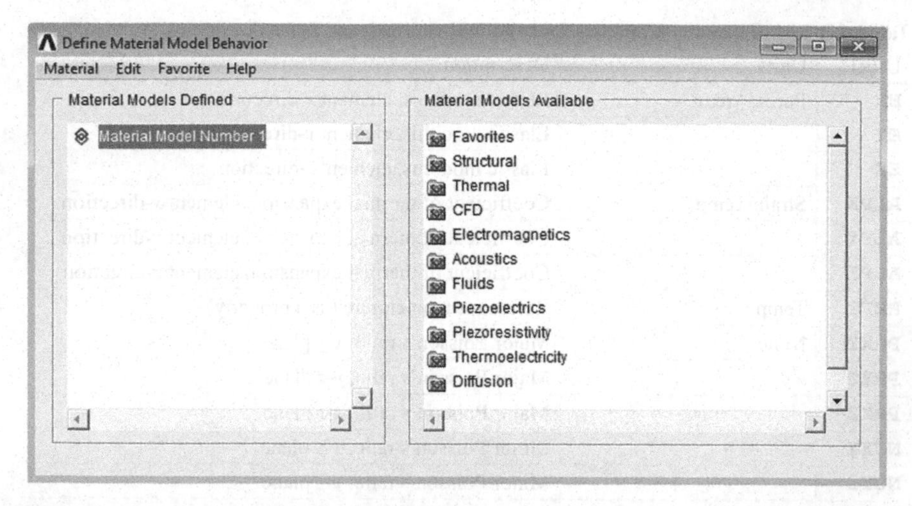

Fig. 4.12 Dialog box for defining material models

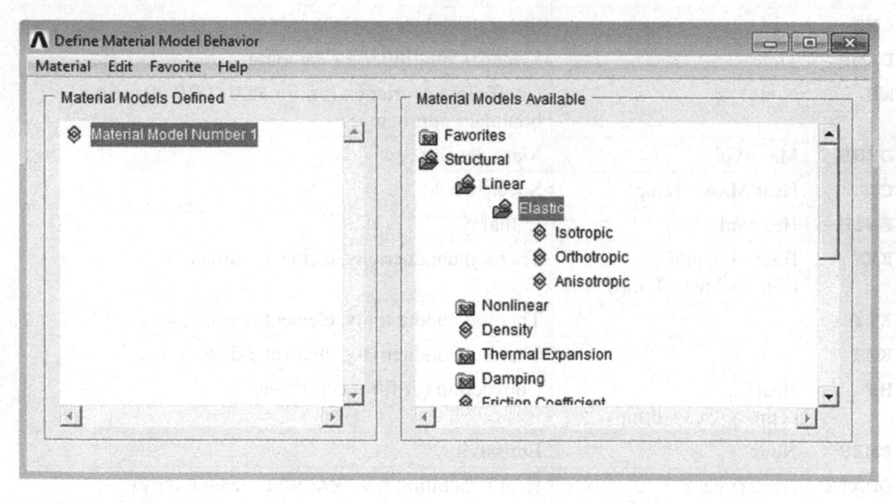

Fig. 4.13 Expanded view of the material models under the *Structural* discipline

4.2.5 Element Attributes

Every element in ANSYS is identified by the element type, real constant set, material property set, and element coordinate system. These are called *element attributes*. In order to create a mesh, the element type(s) must be specified a priori and the material properties (and real constants, depending on the element type) must be specified in order to obtain a solution. The element coordinate system is defined internally.

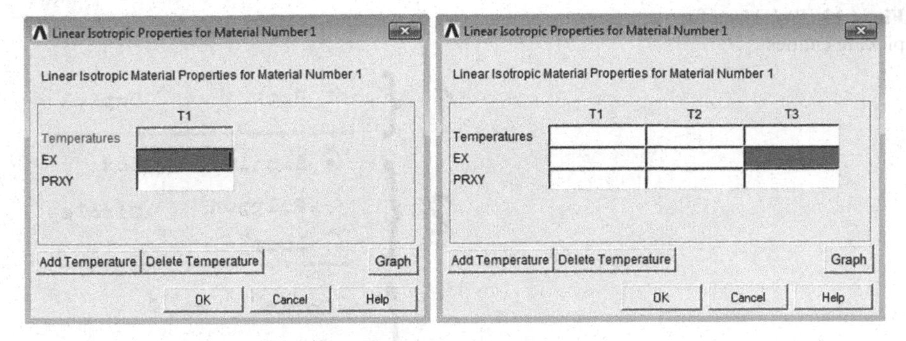

Fig. 4.14 Dialog box for isotropic properties: not temperature dependent (*left*) and temperature dependent (*right*)

4.2.6 Interaction with the Graphics Window: Picking Entities

When using ANSYS through the *GUI*, part of the interaction between the user and the software involves picking entities or locations in the *Graphics Window*. These interactions are performed using the *Pick Menus*. Figures 4.15 and 4.16 show two examples of such menus. Picking operations are performed using the left mouse button.

When picking entities through the *Pick Menu*, there are five distinct fields, as shown in Fig. 4.15:

1. *Pick/Unpick Field*: Using the radio-buttons, the user selects whether the entities are to be picked or unpicked. This feature is useful when the user picks entities other than the intended ones. Instead of using the radio-buttons, the user may use the right mouse button to toggle between the ***Pick*** and ***Unpick*** modes.
2. *Picking Style Field*: By default, the user picks entities one at a time (i.e., radio-button ***Single*** in the *Pick Menu*). However, if the number of entities to be picked is a large number, the ***Single*** picking mode may become tedious, and one of the other modes may be preferable in such situations. Available options include:

   Box: The user draws a rectangle in the *Graphics Window* by holding down the left mouse button; entities located inside the rectangular box are picked.

   Polygon: The user draws a polygon in the *Graphics Window*. Vertices of the polygon are created by single clicks on the left mouse button. The polygon is finalized when the user clicks on the first vertex created. The entities located inside the polygon are picked.

   Circle: When the entities follow a radial pattern, it may be more convenient to pick them through a circular region. This option permits the user to draw a circle in the *Graphics Window* by holding down the left mouse button.

3. *Information Field*: This field provides the user with useful information such as the number of currently picked entities, maximum number of entities that can be picked, and the last entity number picked.

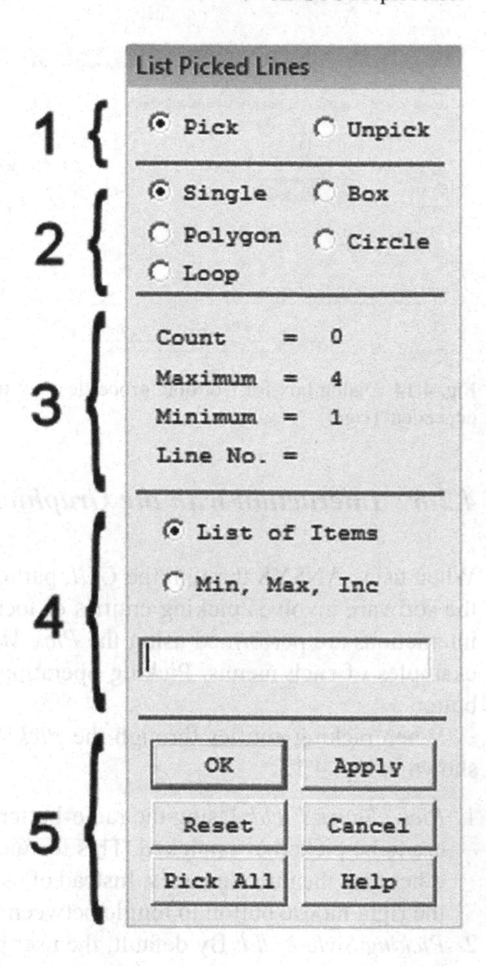

Fig. 4.15 *Pick Menu* for picking entities

4. *Text Field*: Using this option, the user may provide text input for the entities to be picked instead of picking them in the *Graphics Window*. This can be done in two different formats:

   List of Items: When the radio-button next to ***List of Items*** is selected (default), the user may enter a list of the entity numbers to be picked, separated by commas, in the text field.

   Min, Max, Inc: When the radio-button next to ***Min, Max, Inc*** is selected, the user may enter the entity numbers to be picked in the text field in the format Minimum, Maximum, Increment. For example, if the user enters 1, 5, 2, then ANSYS picks entities 1, 3 and 5.

5. *Action Field*:This field involves familiar actions, such as:

   OK: Finishes the picking operation and closes the *Pick Menu*.

   Apply: Applies the picking performed so far while keeping the *Pick Menu* active.

   Reset: The picking operations performed so far are ignored and the configuration is set to the one that existed when the *Pick Menu* appeared.

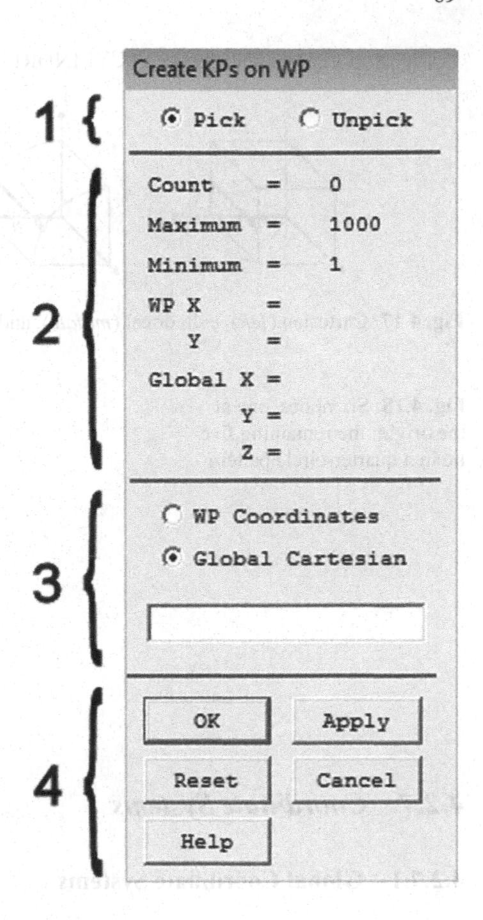

Fig. 4.16 *Pick Menu* for picking locations

Cancel: Closes the *Pick Menu* without performing picking.
Pick All: All of the items under consideration are picked and the *Pick Menu* is closed.
Help: Displays the Help Page related to the current operation.

Picking locations is similar to picking entities, except for slight differences in the *Pick Menu*. This time the menu has four fields, as shown in Fig. 4.16:

1. *Pick/Unpick Field*: This field is the same as explained above.
2. *Information Field*: Similar to the previous case, this field provides the user with useful information such as the number of currently picked locations, maximum and minimum possible picking operations, and the *Working Plane* and *Global Cartesian coordinates* of the last location picked.
3. *Text Field*: Using this option, the user can provide the coordinates of the location to be picked instead of picking them in the *Graphics Window*. This can be done in two different formats: *Working Plane* or *Global Cartesian Coordinates*. In either case, the coordinates are separated by commas.
4. *Action Field*: This field is the same as explained above, with exception of the absence of the ***Pick All*** button.

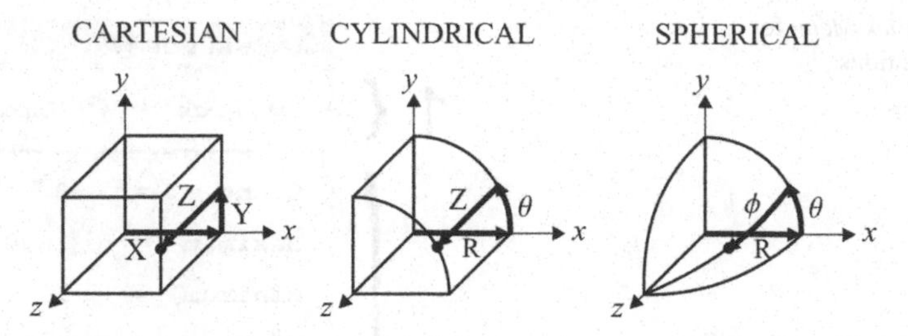

Fig. 4.17 Cartesian (*left*), cylindrical (*middle*), and spherical (*right*) coordinate systems

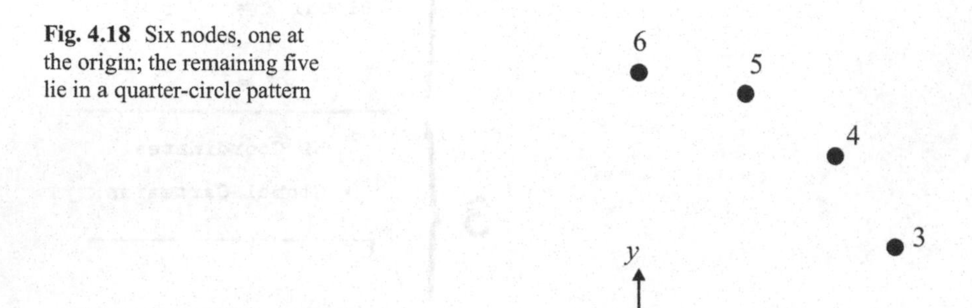

Fig. 4.18 Six nodes, one at the origin; the remaining five lie in a quarter-circle pattern

4.2.7 Coordinate Systems

4.2.7.1 Global Coordinate Systems

When the user starts an ANSYS session, the coordinate system (CS) is Cartesian by default. However, there are many situations where using other coordinate systems (cylindrical or spherical) is more convenient. There are four predefined coordinate systems in ANSYS: Cartesian, cylindrical, spherical, and toroidal; the first three of them are shown in Fig. 4.17.

All of these coordinate systems have the same origin (global origin) and are called global coordinate systems. Although the session starts with the Cartesian CS, the user can switch to one of the other three coordinate systems at any time. The CS currently used is referred to as the *active coordinate system* (active CS); any action referring to the coordinates is performed in the active CS. For example, either a Cartesian or cylindrical CS can be used to create the nodes at the locations shown in Fig. 4.18. The nodes around the unit circle are equally spaced.

In reference to a Cartesian CS, Nodes 1, 2, and 6 can easily be created because the coordinates are explicitly given as **(0, 0, 0)**, **(1, 0, 0)** and **(0, 1, 0)**, respectively. For nodes 3, 4, and 5, trigonometric relations can be used to calculate the x-, y-, and z-coordinates with a desired precision or round-off.

An alternative to the calculation of these coordinates is to change the active CS from Cartesian to cylindrical. In the cylindrical coordinate system, any reference to

Table 4.3 Nodal coordinates in Cartesian and cylindrical coordinate systems

Node	Cartesian			Cylindrical		
	x	y	z	r	θ	z
1	0	0	0	0	0	0
2	1	0	0	1	0	0
3	0.924	0.383	0	1	22.5	0
	0.9239	0.3827				
	0.92388	0.38268				
4	0.707	0.707	0	1	45	0
	0.7071	0.7071				
	0.707106	0.707106				
5	0.383	0.924	0	1	67.5	0
	0.3827	0.9239				
	0.38268	0.92388				
6	0	1	0	1	90	0

x-, y-, and z-coordinates are treated as r, θ, and z. The coordinates of the nodes 3, 4, and 5 in the cylindrical CS are specified as **(1, 22.5, 0)**, **(1, 45, 0)** and **(1, 67.5, 0)**, respectively.

By changing the active CS, unnecessary algebraic calculations and the potential loss of accuracy are avoided.

The coordinates of the nodes in the Cartesian and cylindrical coordinate systems are given in Table 4.3. The "loss of accuracy" can be observed by examining the possible x- and y-coordinates of nodes 3, 4, and 5.

The menu path to change the active CS is given as

Utility Menu > WorkPlane > Change Active CS to

Selection of one of the top three choices in the dialog box,, i.e., global Cartesian, global cylindrical, or global spherical, completes this operation. The CS that is chosen remains active and, in turn, all the coordinates are referenced to that CS, until the user changes it.

4.2.7.2 Local Coordinate Systems

The global coordinate systems all share the same origin (global origin) with a predefined orientation. There are situations where changing one type of global coordinate system to a different global coordinate system does not provide enough convenience or sometimes makes it even more complicated.

It may turn out that what the user really needs is to change the orientation of the CS and/or location of the origin.

In such cases, the user needs to define a CS by offsetting the origin or changing the orientation, or both. Such a coordinate system is called a *local coordinate system* (local CS). A *local CS* can be created by specifying either a location for the origin or

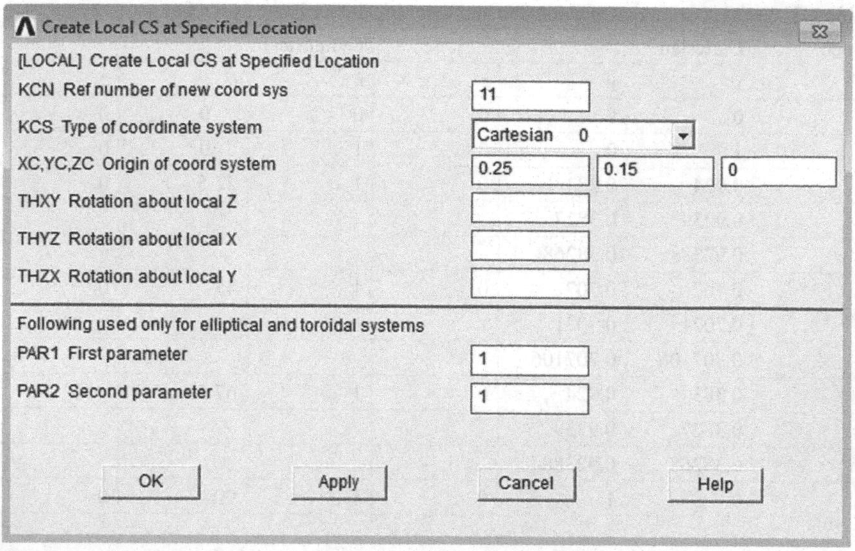

Create Local CS at Specified Location
[LOCAL] Create Local CS at Specified Location
KCN Ref number of new coord sys 11
KCS Type of coordinate system Cartesian 0
XC,YC,ZC Origin of coord system 0.25 0.15 0
THXY Rotation about local Z
THYZ Rotation about local X
THZX Rotation about local Y
Following used only for elliptical and toroidal systems
PAR1 First parameter 1
PAR2 Second parameter 1
OK Apply Cancel Help

Fig. 4.19 Dialog box for creating a *Local CS* at a specified location

three keypoints or nodes. Only one CS can be active at a given time. ANSYS requires that *local coordinate systems* have reference numbers that are *greater than or equal to* 11. The menu path for creating a *local CS* at specified location is given as

Utility Menu > WorkPlane > Local Coordinate Systems > Create Local CS > At Specified Loc +

This brings up a *Pick Menu*, requesting the user to enter the coordinates of the points in the *text field* inside the *Pick Menu*, or to pick the points by clicking the mouse pointer on the *Graphics Window*.

After picking the origin, clicking on ***OK*** brings up the dialog box shown in Fig. 4.19. There are several text boxes to fill out in this dialog box. First is the reference number (by default, it is 11). If a *local CS* was defined previously, as a default, ANSYS assigns the smallest available reference number that is greater than or equal to 11. If this reference number is not desired, the user enters the new reference number for this CS.Below the reference number box, there is a pull-down menu for the CS type: Cartesian, cylindrical, or spherical (toroidal is not discussed herein). The coordinates of the origin of the *local CS* with respect to the global origin should already appear in the CS type menu. Finally, rotation angles with respect to the active CS (not necessarily global Cartesian) are entered.

4.2.8 Working Plane

Within the ANSYS environment, regardless of the dimensionality of the problem (2-D or 3-D), calculations are performed in a 3-D space. If the problem is 2-D, then ANSYS uses the x-y plane, which is the $z=0$ plane.

The *Working Plane* (WP) is a 2-D plane with the origin of a 2-D coordinate system (Cartesian or polar) and a display grid. It is designed to facilitate solid model generation, where many solid model entities are created by referring to the origin of the WP.

In order to view the WP, the menu path is given as

Utility Menu > WorkPlane > Display Working Plane

A checkmark appears on the left of this menu item. Similarly, one can turn the display WP off by using the exact same menu path, resulting in the disappearance of the checkmark. By default, only the triad that is attached to the WP is shown in the *Graphics Window*. Viewing the grid is achieved by the menu path:

Utility Menu > WorkPlane > WP Settings

This brings up the *WP Settings Window*. Clicking on the ***Grid and Triad*** radio-button turns on both the grid and the triad; clicking on the ***Grid Only*** radio-button turns on only the grid. Hitting the ***Apply*** or ***OK*** button activates the new setting.

Using the two radio-buttons at the top permits a switch between the Cartesian and cylindrical (polar) CS. The WP can be placed at any point in the 3-D space with an arbitrary orientation. There can only be one working plane at a time. By default, the WP is the *x-y* plane of the global CS.A working plane can be defined by specifying either three points or nodes or keypoints.

At this point, defining a WP by three points is explained. The menu path is given as

Utility Menu > WorkPlane > Align WP with > XYZ Locations

A *Pick Menu* appears, prompting the user to enter the coordinates of the points in the *text field* or pick the points by clicking the mouse pointer on the *Graphics Window*.

The user needs three noncolinear points to define a plane. The first point is the origin of the WP. The second point defines the WP *x*-axis along the line defined between the first and itself. The third point defines the direction of the positive WP y-axis. Two examples of these operations are illustrated in Fig. 4.20. When all three points are entered, clicking on ***OK*** in the *pick menu* completes the definition of the WP by three points.

As shown in Fig. 4.21, an existing WP can be moved to a new location by providing offset distances in the *x*-, *y*-, and *z*-directions, which yields a WP parallel to its previous orientation. Also, an existing WP can be rotated in all three directions, as shown in Fig. 4.22. If the user rotates the WP about the *z*-axis (which is the direction normal to the WP—not to the global CS), then the WP remains in the same plane but the WP *x*- and *y*-axes rotate within the plane. These movements can be made by the following menu path:

Utility Menu > WorkPlane > Offset WP by Increments

This brings up the *Offset* window. This window requires the offset values in **X, Y,** and **Z.** In the *Offset WP* window, which is used for both translation and rotation, there are six push-buttons for translation and six push-buttons for rotation. These are used for incremental translation and rotation in and around, respectively,

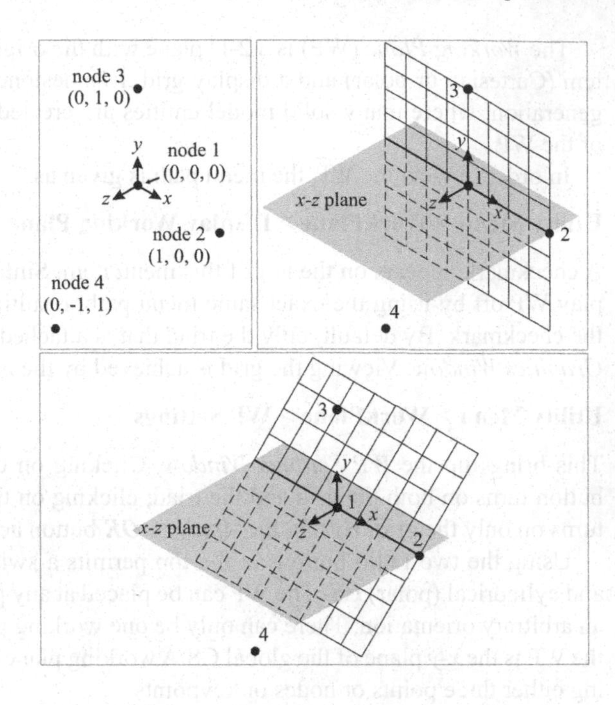

Fig. 4.20 Four nodes in 3-D space (*top left*); WP defined on the plane defined by nodes 1, 2, and 3 (*top right*) and by nodes 1, 2, and 4 (*bottom*)

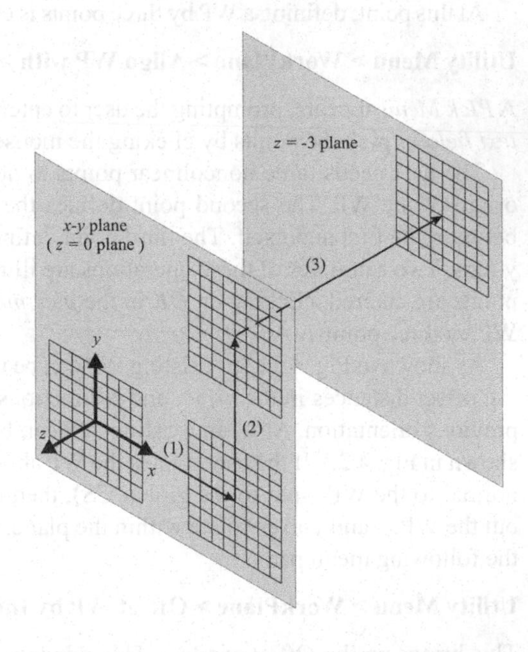

Fig. 4.21 WP first moved 3 units in the x-direction, then 3 units in the y-direction, and, finally, − 3 units in the z-direction

positive and negative x-, y-, and z-axes. The increment is given by a sliding button right below the buttons (one for translation and one for rotation). If the display WP is turned on, the resulting incremental translation or rotation can be observed immediately (without having to hit ***Apply***).

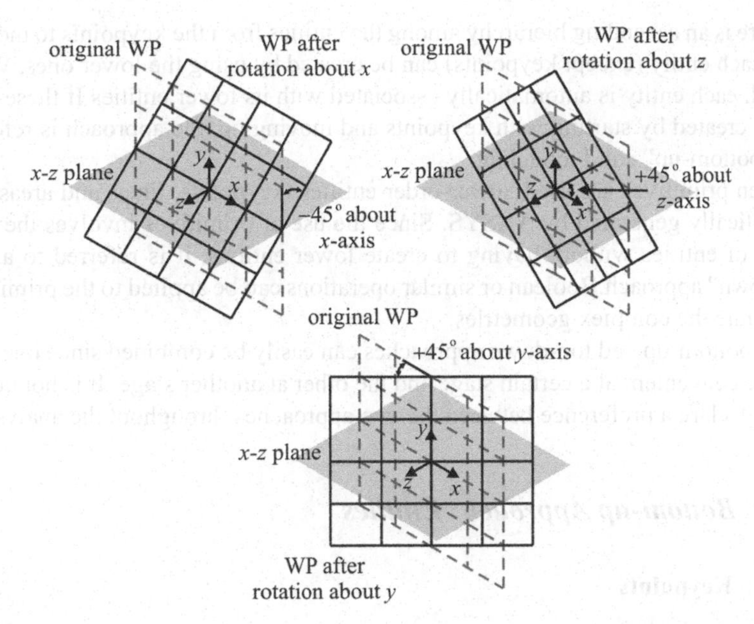

Fig. 4.22 WP rotated −45° about the *x*-axis (*top left*), +45° about the *z*-axis (*top right*), and +45° about *y*-axis (*bottom*)

CAUTION The **X** and **Y** refer to the WP's *x*- and *y*-axes (not global axes) and *z* is the direction normal to the WP (not global CS); the positive direction is established by the right-hand rule.

4.3 Solid Modeling

The geometrical representation of the physical system is referred to as the solid model. In model generation with ANSYS, the ultimate goal is to create a finite element mesh of the physical system. There are two main paths in ANSYS to generate the nodes and elements of the mesh: (1) direct generation and (2) solid modeling and meshing.

In direct generation, every single node is generated by entering their coordinates followed by generation of the elements through the connectivity information. Since most real engineering problems require a high number of nodes and elements (i.e., hundreds or thousands), direct generation is not feasible. Solid modeling is a very powerful alternative to direct generation.

Solid modeling involves the creation of geometrical entities, such as lines, areas, or volumes, that represent the actual geometry of the problem.Once completed, they can be meshed by ANSYS automatically (user still has control over the meshing through user-specified preferences for mesh density, etc.). A solid model can be created by using either *entities* or *primitives*. The *entities* refer to the *keypoints*, *lines*, *areas*, and *volumes*. The *primitives* are predefined geometrical shapes.

There is an ascending hierarchy among the entities from the keypoints to the volumes.Each entity (except keypoints) can be created by using the lower ones. When defined, each entity is automatically associated with its lower entities.If these entities are created by starting with keypoints and moving up, the approach is referred to as "bottom-up" solid modeling.

When primitives are used, lower-order entities (keypoints, lines, and areas) are automatically generated by ANSYS. Since the use of primitives involves the generation of entities without having to create lower entities, it is referred to as the "top-down" approach.Boolean or similar operations can be applied to the primitives to generate the complex geometries.

The bottom-up and top-down approaches can easily be combined since one may be more convenient at a certain stage and the other at another stage. It is not necessary to declare a preference between the two approaches throughout the analysis.

4.3.1 Bottom-up Approach: Entities

4.3.1.1 Keypoints

When the bottom-up solid model generation approach is used, the user starts by generating the keypoints. The higher entities (lines, areas, and volumes) can then be defined by using the keypoints. The keypoints necessary to create a higher-order entity for modeling different parts of the geometry should be generated a priori. When areas or volumes are generated using keypoints, the intermediate entities are generated automatically by ANSYS. The creation of keypoints on the WP and in the *active CS* is explained herein.

The following menu path is suggested to create a keypoint on WP:

Main Menu > Preprocessor > Modeling > Create > Keypoints > On Working Plane

This brings up a *Pick Menu*, where ANSYS expects the user to pick points on the WP.Once the points are picked by clicking on the left mouse key, hitting on the ***Apply*** or ***OK*** button completes this task (***OK*** closes the *Pick Menu*).When using this option, turning on the display WP with the grid visible is highly recommended.

IMPORTANT HINT When using the *Pick Menu*, picking the exact location might become a real challenge and, in turn, result in the generation of unnecessary entities. These "extra" entities might cause confusion and possible errors in the course of the solid model generation. Whenever there is a *Pick Menu*, the user can hold down (no release) the left mouse button and move the pointer on the *Graphics Window*. This action shows the mouse pointer coordinates on the *Pick Menu*. When the target coordinates are found, the user can release the button to finish the picking.

The following menu path is suggested to create keypoint(s) (KP) in the active CS:

Fig. 4.23 A straight line

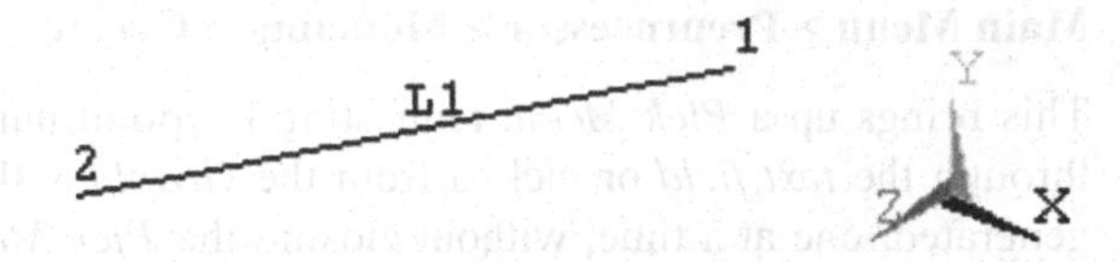

Main Menu > Preprocessor > Modeling > Create > Keypoints > In Active CS

This brings up a dialog box with four input fields for the KP number and the *x*-, *y*-, and *z*-coordinates. Once this information is supplied, hitting ***OK*** creates the KP and exits from this dialog box.Alternatively, the ***Apply*** button can be clicked on and more keypoints can be created.

The coordinates defining a KP can be modified as follows:

Main Menu > Preprocessor > Modeling > Move/Modify > Keypoints > Single KP

This brings up a *Pick Menu*. First, KP is picked from the *Graphics Window*, or its number is typed in the *text field*. Then, the new location is picked or the new coordinates are typed. If a KP is modified, any mesh that is attached to that KP is automatically cleared, and any higher-order entities that are associated with that KP also are modified accordingly.

4.3.1.2 Lines

Lines are used for either creating a mesh with line elements or creating areas and volumes. A straight line, an arc, and a cubic spline can be created, as shown in Fig. 4.23 and 4.24.

Creating a True Straight Line

By using the following menu path, a straight line can be created regardless of the active CS. The only input needed is two keypoints. The menu path is given as

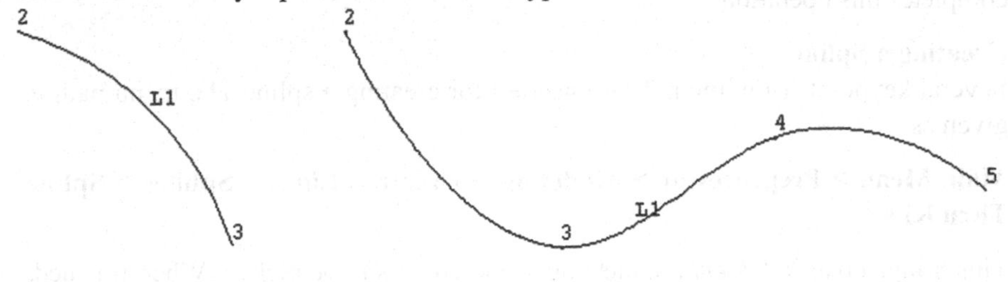

Fig. 4.24 An arc (left) and a cubic spline (right)

Main Menu > Preprocessor > Modeling > Create > Lines > Straight Line

This brings up a *Pick Menu*, requesting keypoint numbers, which can be entered through the *text field* or picked from the *Graphics Window*.Multiple lines can be generated, one at a time, without closing the *Pick Menu* (by using ***Apply*** button). The straight line (L1) is generated by keypoints (KP1) and (KP2).

Creating a Straight Line in the Active CS
This method creates a straight line in the active CS. If the active CS is a Cartesian CS, then the line is a true straight line. If the active CS is a cylindrical CS, then the line is a helical spiral. The menu path is given as

Main Menu > Preprocessor > Modeling > Create > Lines > In Active Coord

This brings up a *Pick Menu,* requesting the keypoints. The keypoint numbers are supplied through either the *text field* or by picking them using the *Graphics Window*. Multiple lines can be generated, one at a time, without closing the *Pick Menu*. It works the same way as creating a true straight line.

Creating an Arc
Creating an arc requires three keypoints. The arc is circular, regardless of the active CS. It is generated between the first and the second keypoints. The third KP defines the plane of the arc, as well as the positive curvature side. It does not have to be at the center of the curvature. The menu path is given as

Main Menu > Preprocessor > Modeling > Create > Lines > Arcs > By End KPs & Rad

This brings up a *Pick Menu*, requesting the two end keypoints. These keypoint numbers are supplied through either the *text field* or by picking them using the *Graphics Window*. Upon hitting ***OK*** in the *Pick Menu*, ANSYS requests the third KP, which defines the positive curvature side. After entering it the same way and hitting ***OK*** in the *Pick Menu*, a dialog box appears. The first field is the radius and the remaining 3 are the keypoints that have already been input. Entering the radius and hitting ***OK*** completes this operation.

Creating a Spline
Several keypoints (minimum 2) are needed for creating a spline.The menu path is given as

Main Menu > Preprocessor > Modeling > Create > Lines > Splines > Spline Thru KPs

This brings up a *Pick Menu*, requesting the keypoints to be picked. When finished, hitting ***OK*** finishes the spline creation. Multiple splines can be generated, one at a time, without closing the *Pick Menu* by hitting the ***Apply*** button instead of ***OK***.

Once the lines are defined, areas can be created by using them.

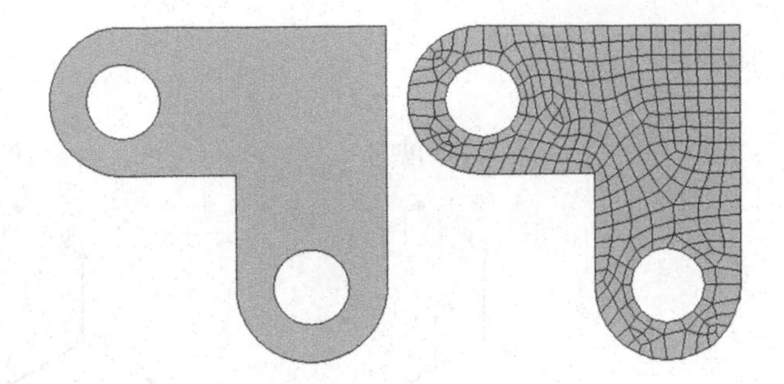

Fig. 4.25 An area in the *x-y* plane (*left*); meshed (*right*)

4.3.1.3 Areas

Areas are used to create a mesh with area elements and to create volumes.If the geometry involves a 2-D domain, the area(s) is (are) required to be flat, lying on the *x-y* plane. If the geometry involves 3-D bodies, then the areas that define the faces of the volume(s) can be flat or curved. A mesh created from a flat area and volumes created from flat and curved areas are shown in Figs. 4.25 and 4.26.

In bottom-up approach, areas can be created by using either keypoints or lines.

Creating an Area Using Keypoints
A minimum of 3 keypoints is required, and the maximum number allowed is 18.If more than 3 keypoints are used, they must lie in the same plane (co-planar), as shown in Fig. 4.27. The menu path is given as

Main Menu > Preprocessor > Modeling > Create > -Areas- Arbitrary > Through KPs

which brings up a *Pick Menu*, requesting the keypoints to be picked. When finished, clicking on ***OK*** creates the area.

CAUTION In the PC version, it is recommended that the input window be used.

Creating an Area Using Lines
In creating an area by lines, a minimum of 3 previously defined lines are required, and the maximum number of lines allowed is 10. If more than 3 lines are used, they

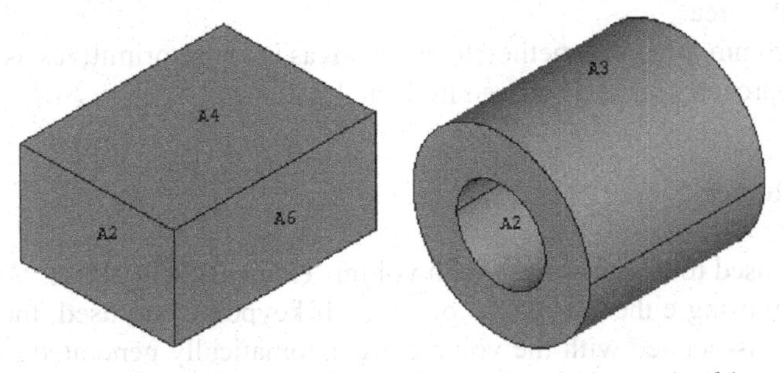

Fig. 4.26 Volume composed of flat areas (*left*) and flat and curved areas (*right*)

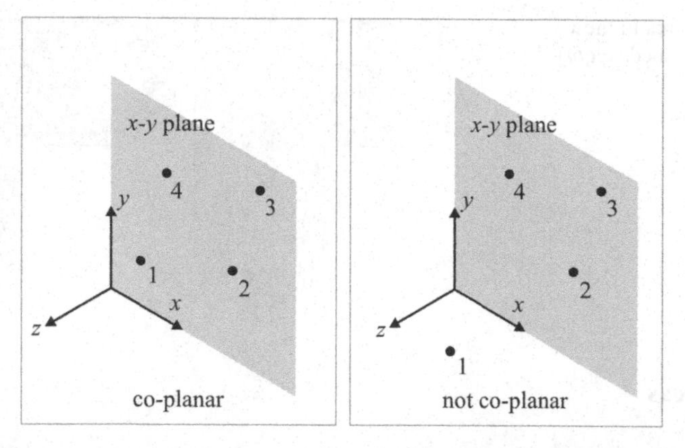

Fig. 4.27 Coplanarity of keypoints: 4 coplanar keypoints (*left*) and 4 noncoplanar keypoints (*right*)

Fig. 4.28 A meshed volume

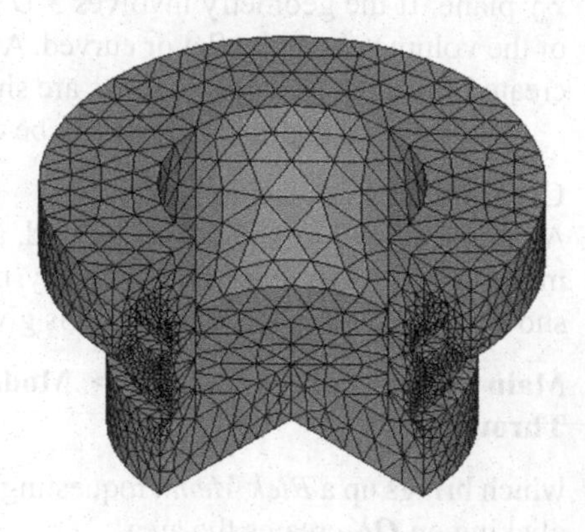

must be co-planar.Lines must be given in a clockwise or counterclockwise order, and they must form a simply connected closed curve. The menu path is given as

Main Menu > Preprocessor > Modeling > Create > -Areas- Arbitrary > By Lines

This brings up a *Pick Menu*, requesting the lines to be picked. When finished, hitting ***OK*** creates the area.

Another commonly used method to create areas is to use primitives as part of the top-down approach; this is discussed in Sect. 4.3.2.

4.3.1.4 Volumes

Volumes are used to create a mesh with volume elements (Fig. 4.28). Volumes can be created by using either keypoints or areas. If keypoints are used, the areas and lines that are associated with the volume are automatically generated by ANSYS. Two basic methods are presented below.

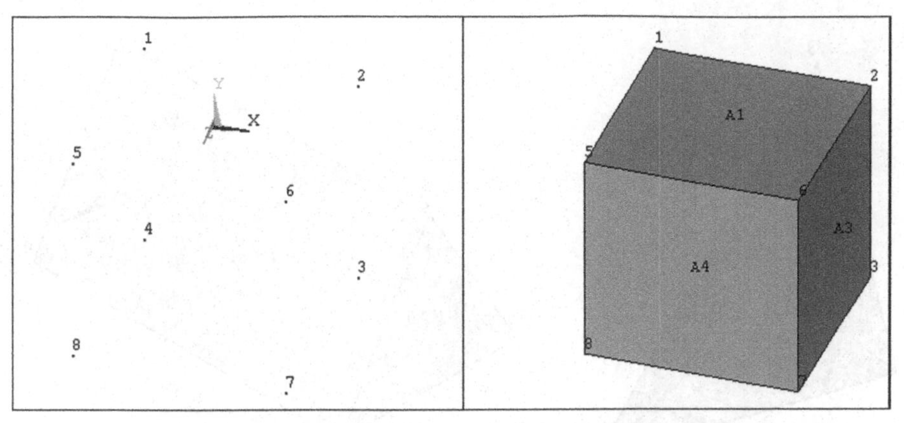

Fig. 4.29 Eight keypoints (*left*); volume created by picking keypoints in 1-2-6-5-4-3-7-8 order (*right*)

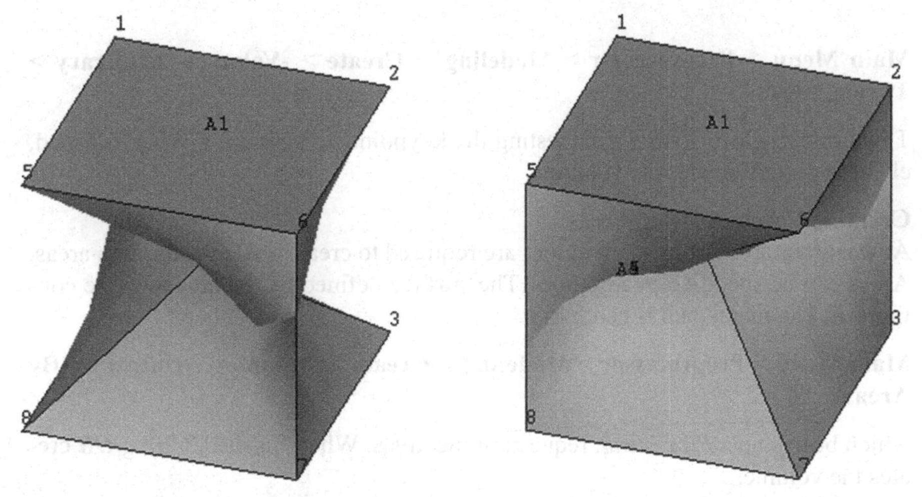

Fig. 4.30 Volumes created by picking keypoints in 1-2-6-5-4-8-7-3 order (*left*) and 1-2-6-5-7-3-4-8 order (*right*)

Creating Volumes Using Keypoints

A maximum of 8 and a minimum of 4 keypoints are required to create a volume using keypoints.Keypoints must be specified in a continuous order.If the volume has 6 faces, two of the opposite faces are required to be specified by the user, and keypoints defining both of these faces should be given in either a clockwise or counterclockwise direction.

For example, a 6-faced volume, shown in Fig. 4.29, requires 8 keypoints.The correct counterclockwise sequence of keypoints is **1-2-6-5-4-3-7-8**. Incorrect sequences, such as **1-2-6-5-4-8-7-3** or **1-2-6-5-7-3-4-8** (Fig. 4.30), have neither a clockwise nor counterclockwise sense and fail to produce the 6-faced volume. Figure 4.31 illustrates volumes with 4 and 5 faces (tetrahedron and triangular prism). The menu path is given as

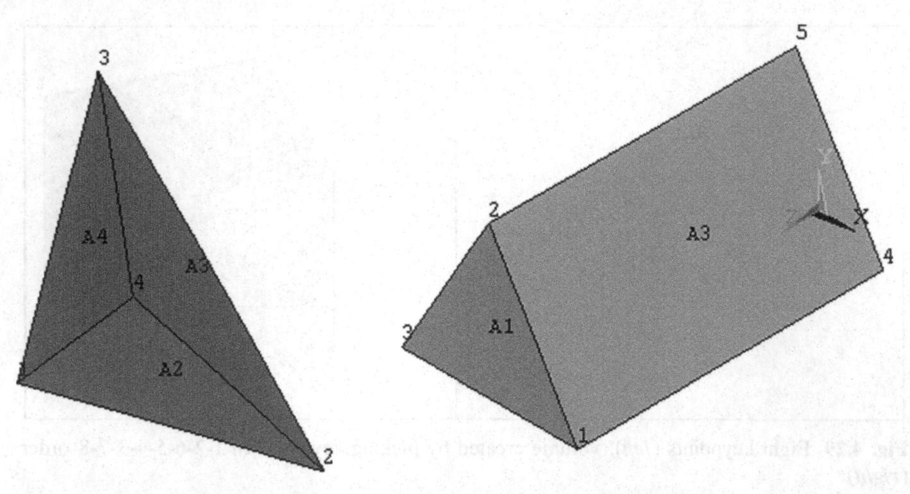

Fig. 4.31 A tetrahedron, 4 faces (*left*), and a triangular prism, 5 faces (*right*)

Main Menu > Preprocessor > Modeling > Create > -Volumes- Arbitrary > Through KPs

This brings up a *Pick Menu*, requesting the keypoints to be picked. When finished, clicking on ***OK*** creates the volume.

Creating Volumes Using Areas
At least four areas (maximum of ten) are required to create a volume through areas. Areas can be specified in any order. The surface defined by the area must be continuous. The menu path is given as

Main Menu > Preprocessor > Modeling > Create > -Volumes- Arbitrary > By Areas

which brings up a *Pick Menu*, requesting the areas. When finished, hitting ***OK*** creates the volume.

4.3.2 Top-Down Approach: Primitives

The primitives are predefined geometrical shapes that enable the user to create a solid model entity (area or volume) with the execution of a single menu item. The user is not required to create keypoints and lines prior to using primitives.

4.3.2.1 Area Primitives

Area primitives are available for the generation of rectangles, circles, and polygons. There are different ways to create each of these primitives. The basic methods are presented here.

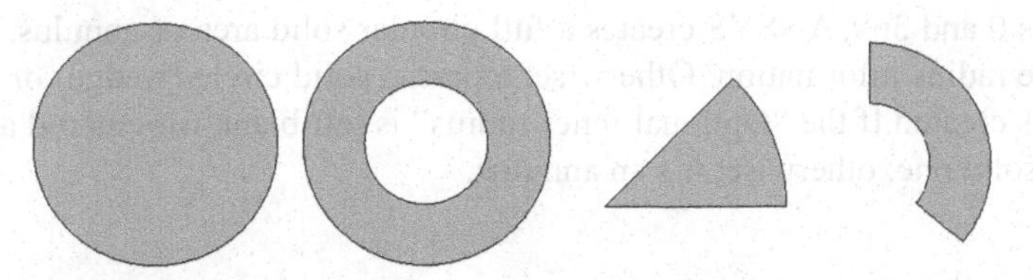

Fig. 4.32 Circular area primitives

Rectangle by Dimension
The menu path is given as

Main Menu > Preprocessor > Modeling > Create > Rectangle > By Dimensions

This brings up a dialog box asking for *Working Plane* ***X*** and ***Y*** coordinates of the two corners of the rectangle.After filling out the four fields in this box, clicking on ***OK*** creates the rectangle in the *Graphics Window*.

Rectangle by 2 Corners
The menu path is given as

Main Menu > Preprocessor > Modeling > Create > Rectangle > By 2 Corners

This brings up a *Pick Menu*. There are two ways to finish this action. One way is to use the four fields in the pick menu to input WP coordinates of one corner and the dimensions of the rectangle. The other method is to use the left mouse button to click on the *Graphics Window* to define one corner. After this, as the mouse pointer is moved, ANSYS displays possible rectangles as outlines, with the dimensions quantitatively indicated. When the user finds the right dimensions, a left-click creates the rectangular area.

Solid Circular Area
The menu path is given as

Main Menu > Preprocessor > Modeling > Create > Circle > Solid Circle

This brings up a *Pick Menu*, requesting *Working Plane* ***X*** and ***Y*** of the center of the circle and its ***radius***. They can be supplied either by filling out the fields in the *Pick Menu* or using the mouse pointer. Picking the center of the circle, moving the pointer to find the desired radius (as the mouse pointer is moved, similar to creating rectangles by dimensions, ANSYS plots the circle's outline with the radius identified), and clicking again finalizes the circle generation.

Circular Area by Dimensions
With this option, a solid circle, annulus, circular segment (wedge) or partial annulus can be generated, as shown in Fig. 4.32. The menu path is given as

Main Menu > Preprocessor > Modeling > Create > Circle > By Dimensions

This brings up a dialog box requesting the outer and inner (optional) radii and starting and ending angles of the circular sector. All four of the geometrical parameters are defined with respect to *Working Plane*.If the starting and ending angles are entered as 0 and 360, ANSYS creates a full circular solid area or annulus, depending on the radius information. Otherwise, a partial solid circle (wedge) or a partial annulus is created.If the "Optional inner radius" is left blank (or entered as 0), the area is a solid one; otherwise, it's an annulus.

Polygon
The menu path is given as

Main Menu > Preprocessor > Modeling > Create > -Areas- Polygon > By Vertices

This brings up a *Pick Menu*, requesting the vertices. All the vertices are on the *Working Plane*. Naturally, the polygon must be closed; this is achieved by picking the first point one more time after picking the last point. If the user does not pick the first point to close the polygon and hits ***OK*** in the *Pick Menu*, ANSYS automatically closes the polygon by defining a line between the last and the first point.

4.3.2.2 Volume Primitives

Volume primitives are available for generation of blocks, cylinders, prisms, spheres, or cones. There are different ways to create each of these primitives. The basic methods are presented here.

Block
A block is a rectangular prism. It is created using the following menu path:

Main Menu > Preprocessor > Modeling > Create > Volumes > Block > By Dimensions

This brings up a dialog box requesting six coordinates, the starting and ending *x*-, *y*-, and *z*-coordinates in the active coordinate system. Figure 4.33 shows the isometric view of a block created using $x_1=y_1=z_1=0$, $x_2=1$, $y_2=2$, and $z_2=3$.

Cylinder The user can create solid or hollow cylinders that encompass either the entire angular range or a part thereof. Cylinders are created using the following menu path:

Main Menu > Preprocessor > Modeling > Create > Volumes > Cylinder > By Dimensions

Six parameters are requested in the dialog box:

RAD1 and ***RAD2***: Outer and inner radii of the cylinder. If ***RAD2*** is not specified (or specified as zero), then the cylinder is solid; otherwise, it's hollow.
Z1 and ***Z2***: Starting and ending z-coordinates.

THETA1 and ***THETA2***:Starting and ending angles, measured in degrees, with the active coordinate system *z*-axis defining the axis of rotation.

Figure 4.34 (left) shows an isometric view of a hollow cylinder created using ***RAD1***=1, ***RAD2***=0.5, ***Z1***=0, ***Z2***=2, ***THETA1***=0 and ***THETA2***=360.When the

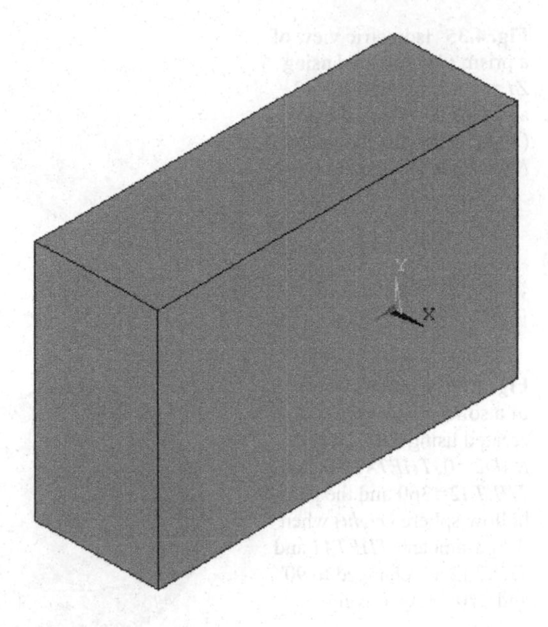

Fig. 4.33 Isometric view of a block created using $x_1=y_1=z_1=0$, $x_2=1$, $y_2=2$, and $z_2=3$

parameter ***THETA1*** is changed to 135° and the other parameters are kept the same, the partial hollow cylinder shown in Fig. 4.34 (right) is created.

Prism

Regular prisms are created using this option. A regular prism is a volume with a constant polygonal cross section in the *Working Plane* *z*-direction. The menu path for creating prisms is given as

Main Menu > Preprocessor > Modeling > Create > Volumes > Prism > By Side Length

This brings up a dialog box requesting the starting and ending *z*-coordinates ***Z1*** and ***Z2***, respectively; the number of sides (***NSIDES***); and the length of each side

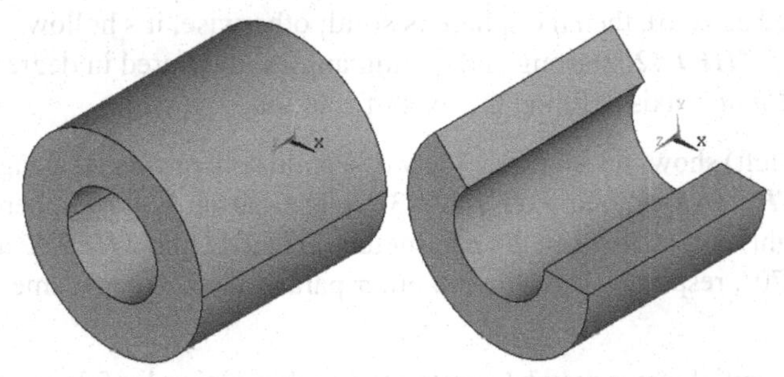

Fig. 4.34 Isometric view of a hollow cylinder (*left*) created using *RAD1*=1, *RAD2*=0.5, *Z1*=0, *Z2*=2, *THETA1*=0, and *THETA2*=360 and a partial hollow cylinder (*right*) when the parameter *THETA1* is changed to 135

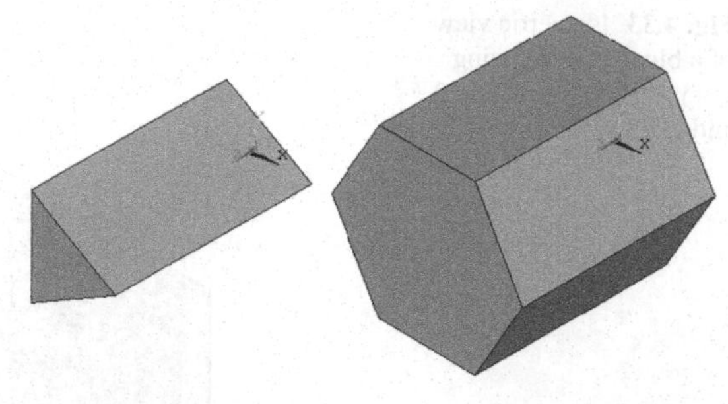

Fig. 4.35 Isometric view of a prism (*left*) created using *Z1*=0, *Z2*=2, *NSIDES*=3, and *LSIDE*=1 and the prism (*right*) when the parameter *NSIDES* is changed to 6

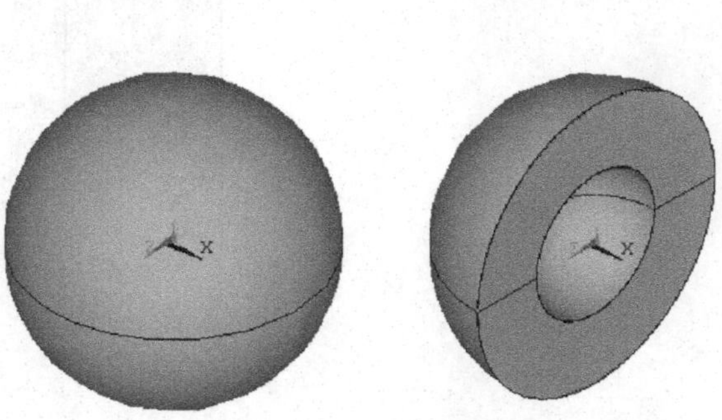

Fig. 4.36 Isometric view of a solid sphere (*left*) created using *RAD1*=1, *RAD2*=0, *THETA1*=0, and *THETA2*=360 and the partial hollow sphere (*right*) when the parameters *THETA1* and *THETA2* are changed to 90° and 270°, respectively

(***LSIDE***).The center of the polygonal area coincides with the *Working Plane* origin. Figure 4.35 (left) shows the isometric view of a prism created using ***Z1***=0, ***Z2***=2, ***NSIDES***=3, and ***LSIDE***=1. Figure 4.35 (right) shows the prism when the parameter ***NSIDES*** is changed to 6.

Sphere
The user can create solid or hollow spheres by using the following menu path:

Main Menu > Preprocessor > Modeling > Create > Volumes > Sphere > By Dimensions

Four parameters are requested in the dialog box:

RAD1 and ***RAD2***: Outer and inner radii of the cylinder. If ***RAD2*** is not specified (or is specified as zero), then the sphere is solid; otherwise, it's hollow.
THETA1 and ***THETA2***: Starting and ending angles, measured in degrees, with the *Working Plane* z-axis defining the axis of rotation.

Figure 4.36 (left) shows an isometric view of a solid sphere created using ***RAD1***=1, ***RAD2***=0, ***THETA1***=0, and ***THETA2***=360. The partial hollow sphere shown in Fig. 4.36 (right) is created when the parameters ***THETA1*** and ***THETA2*** are changed to 90° and 270°, respectively, while the other parameters are kept same.

Cone
Complete or partial cones may be created using this option by following the menu path:

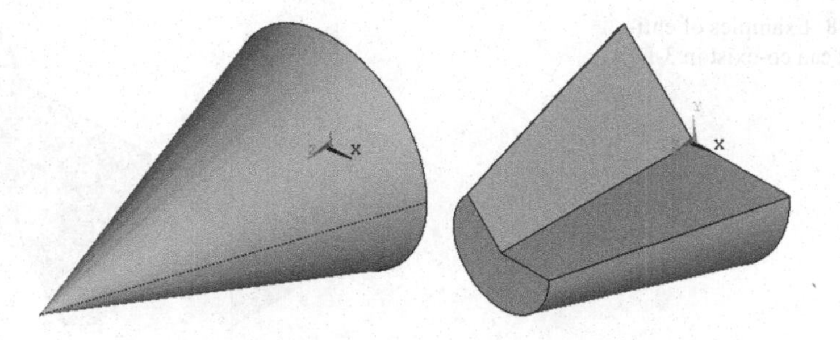

Fig. 4.37 Isometric view of a cone (*left*) created using *RBOT*=1, *RTOP*=0, *Z1*=0, *Z2*=3, *THETA1*=0, and *THETA2*=360 and the partial conical section (*right*) when parameters *RTOP*, *Z2*, and *THETA1* are changed to 0.5, 2, and 135, respectively

Main Menu > Preprocessor > Modeling > Create > Volumes > Cone > By Dimensions

In the dialog box, six parameters are requested:

RBOT and ***RTOP***: Bottom and top radii of the cone.If ***RTOP*** is not specified (or is specified as zero), then a complete cone is generated. If a nonzero ***RTOP*** is specified, then the volume generated is a conical section with parallel top and bottom sides.

Z1 and ***Z2***: Starting and ending *z*-coordinates.

THETA1 and ***THETA2***: Starting and ending angles, measured in degrees, with the *Working Plane* *z*-axis defining the axis of rotation.It is used for creating conical sections.

Figure 4.37 (left) shows an isometric view of a cone created using ***RBOT***=1, ***RTOP***=0, ***Z1***=0, ***Z2***=3, ***THETA1***=0, and ***THETA2***=360.The partial conical section shown in Fig. 4.37 (right) is created when the parameters ***RTOP***, ***Z2***, and ***THETA1*** are changed to 0.5, 2, and 135, respectively, while the other parameters are kept same.

4.4 Boolean Operators

Many engineering problems possess a complex geometry, making model generation a real challenge. However, the solid model entities can be subjected to certain operations that make model generation much easier. These operations, referred to as Boolean operations, utilize logical operators such as add, subtract, divide, etc. The Boolean operators are applied to generate more complex entities using simple entities (see Fig. 4.38).

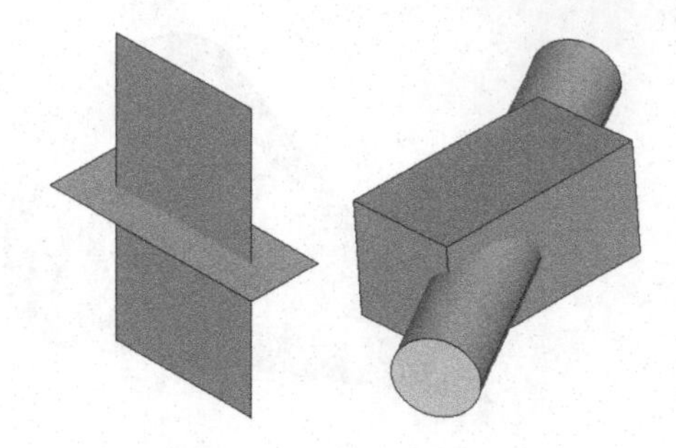

Fig. 4.38 Examples of entities that can co-exist in 3-D space

4.4.1 Adding

The areas to be added must be co-planar (lie in the same plane). As shown in Fig. 4.39, the areas (or volumes) must have either a common boundary or an overlapping region. The original areas or volumes that are added will be deleted unless otherwise enforced by the user. The addition of areas or volumes results in a single (possibly complex geometry) entity, as shown in Fig. 4.40.

Adding entities can be performed by the following menu paths:

Main Menu > Preprocessor > Modeling > Operate > Booleans > Add > Lines

Main Menu > Preprocessor > Modeling > Operate > Booleans > Add > Areas

Main Menu > Preprocessor > Modeling > Operate > Booleans > Add > Volumes

4.4.2 Subtracting

Entities can be subtracted from each other to obtain new entities.Subtracting entities can be executed through the menu paths given below:

Main Menu > Preprocessor > Modeling > Operate > Booleans > Subtract > Lines

Main Menu > Preprocessor > Modeling > Operate > Booleans > Subtract > Areas

Main Menu > Preprocessor > Modeling > Operate > Booleans > Subtract > Volumes

This brings up a *Pick Menu*, requesting the user to pick or enter the base entity from which to subtract. The user picks the entities to be subtracted and clicks on the ***OK*** button to complete the operation.

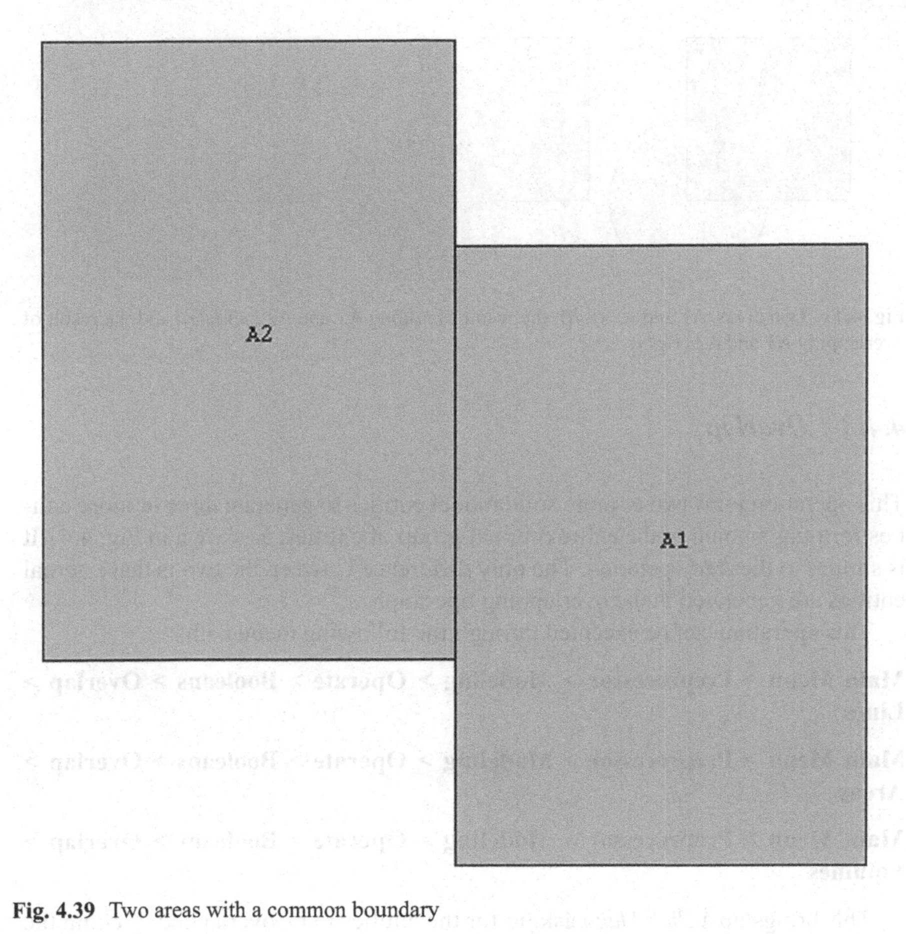

Fig. 4.39 Two areas with a common boundary

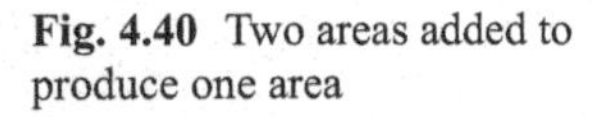

Fig. 4.40 Two areas added to produce one area

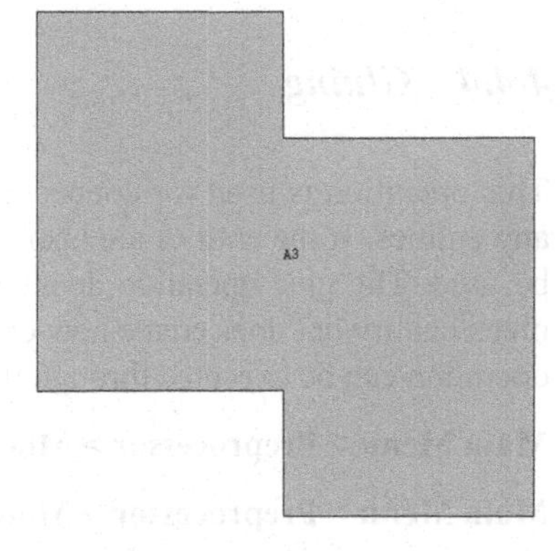

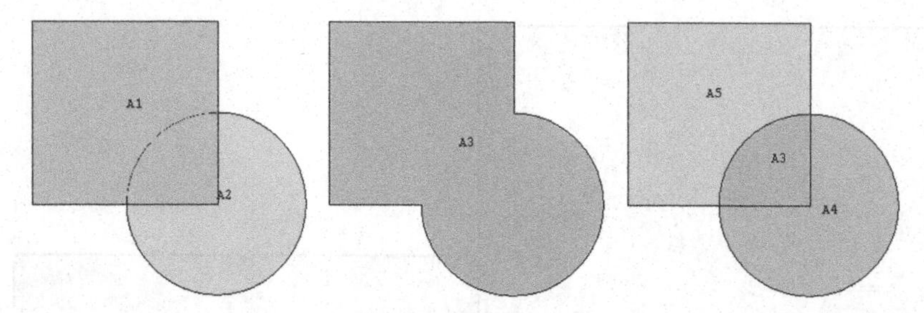

Fig. 4.41 Two areas, A1 and A2 (*left*); the result of adding A1 and A2 (*middle*); and the result of overlapping A1 and A2 (*right*)

4.4.3 Overlap

This operation joins two or more solid model entities to generate three or more entities forming a union of the entire original group of entities, as shown in Fig. 4.41.It is similar to the ***Add*** operation. The only difference between the two is that internal entities are generated in the overlapping operation.

This operation can be executed through the following menu paths:

Main Menu > Preprocessor > Modeling > Operate > Booleans > Overlap > Lines

Main Menu > Preprocessor > Modeling > Operate > Booleans > Overlap > Areas

Main Menu > Preprocessor > Modeling > Operate > Booleans > Overlap > Volumes

This brings up a *Pick Menu* asking for the entities to be overlapped. Picking the entities followed by hitting ***OK*** completes the operation.

4.4.4 Gluing

This operation is used for connecting entities that are "touching" but not sharing any entities. If the entities are apart from or overlapping each other, gluing cannot be used. The glue operation does not produce additional entities of the same dimensionality but does create new entities that have one lower dimensionality. This operation can be executed through the following menu paths:

Main Menu > Preprocessor > Modeling > Operate > Booleans > Glue > Lines

Main Menu > Preprocessor > Modeling > Operate > Booleans > Glue > Areas

Main Menu > Preprocessor > Modeling > Operate > Booleans > Glue > Volumes

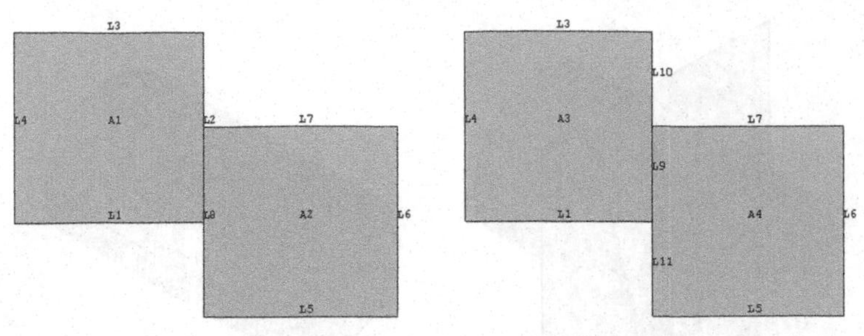

Fig. 4.42 Two areas with a common boundary plotted with line numbers (*left*); they do not share any lines as area 1 (A1) is defined by lines 1 through 4 and area 2 (A2) is defined by lines 5 through 8. After gluing (*right*), the areas share line 9

Before gluing the two areas shown in Fig. 4.42, there are two lines at the interface between Area 1 (A1) and Area 2 (A2).One of these lines is attached to A1, defined by keypoints **2** and **3**, and the other one is attached to A2, defined by keypoints **5** and **8**. Before gluing, these two areas do not "know" of each other's existence because they do not "truly" share any entities. Gluing makes sure that they share entities.After gluing, there are two lines along the right vertical side of A1, and two lines along the left side of A2. The lines along the right vertical side of A1 are defined by keypoints **3** and **8** and keypoints **8** and **2** whereas the lines along the left vertical side of A2 are defined by keypoints **2** and **8** and keypoints **5** and **2**. After gluing, the two areas share one line and two keypoints.

4.4.5 Dividing

A solid model entity can be divided into smaller parts by using other solid model entities. By default, a divided solid model entity is deleted after the operation. There is a wide range of choices for this operation. Some of the available options are presented in Fig. 4.43–4.46.

The menu paths for these operations are given as

Volume by Area

Main Menu > Preprocessor > Modeling > Operate > Booleans > Divide > Volume by Area

Area by Volume

Main Menu > Preprocessor > Modeling > Operate > Booleans > Divide > Area by Volume

Area by Area

Main Menu > Preprocessor > Modeling > Operate > Booleans > Divide > Area by Area

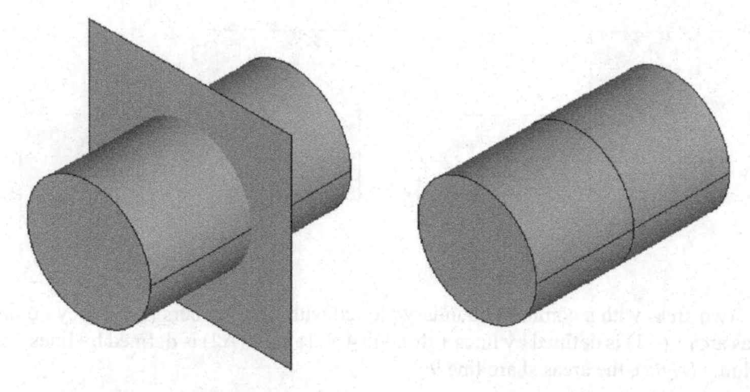

Fig. 4.43 A cylindrical volume is divided into two smaller cylindrical volumes by an area

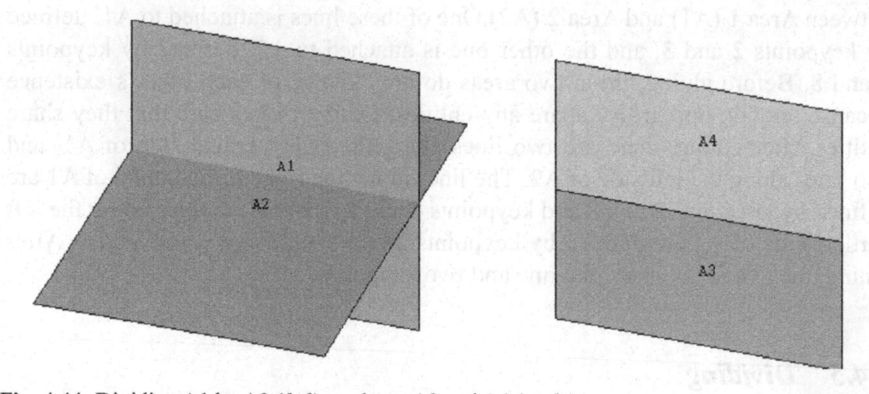

Fig. 4.44 Dividing A1 by A2 (*left*) produces A3 and A4 (*right*)

Area by Line

Main Menu > Preprocessor > Modeling > Operate > Booleans > Divide > Area by Line

Line by Volume

Main Menu > Preprocessor > Modeling > Operate > Booleans > Divide > Line by Volume

Line by Area

Main Menu > Preprocessor > Modeling > Operate > Booleans > Divide > Line by Area

Line by Line

Main Menu > Preprocessor > Modeling > Operate > Booleans > Divide > Line by Line

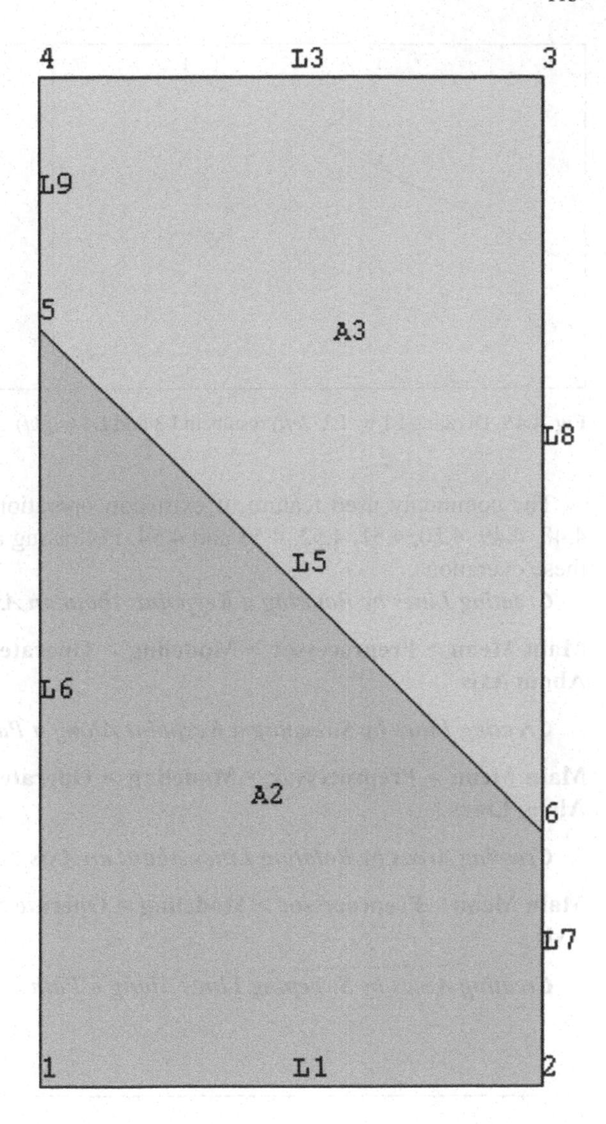

Fig. 4.45 Dividing an area by a line (requires the dividing line to be in the same plane as the area)

4.5 Additional Operations

4.5.1 Extrusion

In addition to Boolean operators, extrusion of the existing entities can be used to generate higher entities. By extruding (dragging) an entity about an axis, one can create a new solid model entity, which is one order higher than the original one (e.g., lines from keypoints, volumes from areas).

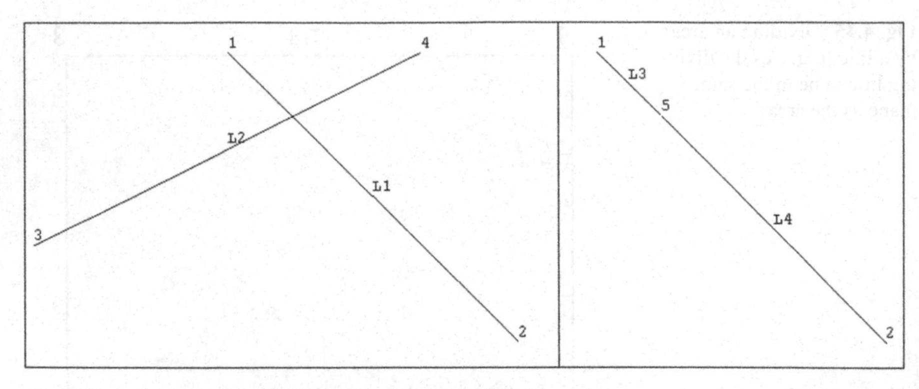

Fig. 4.46 Dividing L1 by L2 (*left*) results in L3 and L4 (*right*)

The commonly used feature of extrusion operation is described in Fig. 4.47, 4.48, 4.49, 4.50, 4.51, 4.52, 4.53 and 4.54. Following are the menu paths used for these operations:

Creating Lines by Rotating a Keypoint About an Axis

Main Menu > Preprocessor > Modeling > Operate > Extrude > -Keypoints- About Axis

Creating Lines by Sweeping a Keypoint Along a Path

Main Menu > Preprocessor > Modeling > Operate > Extrude > -Keypoints- Along Lines

Creating Areas by Rotating Lines About an Axis

Main Menu > Preprocessor > Modeling > Operate > Extrude > -Lines- About Axis

Creating Areas by Sweeping Lines Along a Path

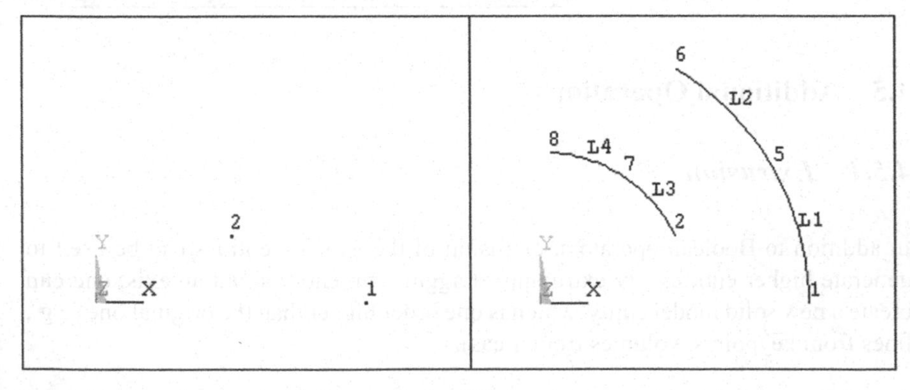

Fig. 4.47 Keypoints 1 and 2 (*left*) are rotated +60° around the *z*-axis in two increments of 30° to create the lines (*right*)

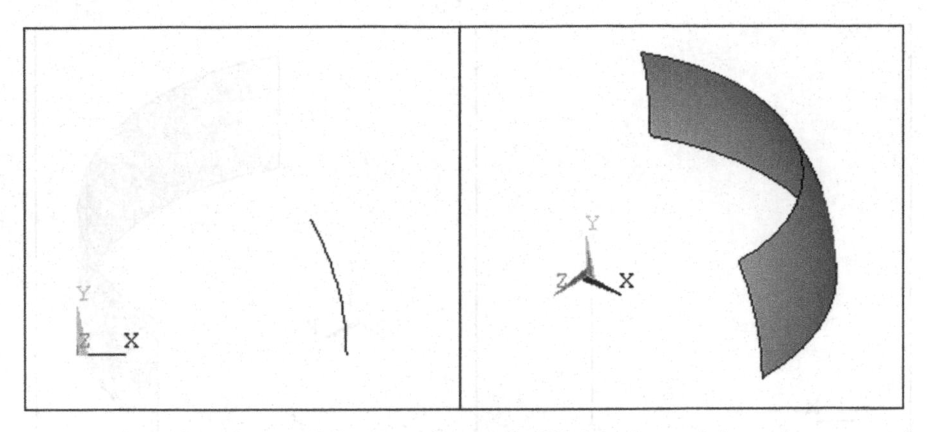

Fig. 4.48 Front view of an arc (*left*); the arc is rotated +60° about the *y*-axis in two increments of 30° to create the curved areas (shown in oblique view on the *right*)

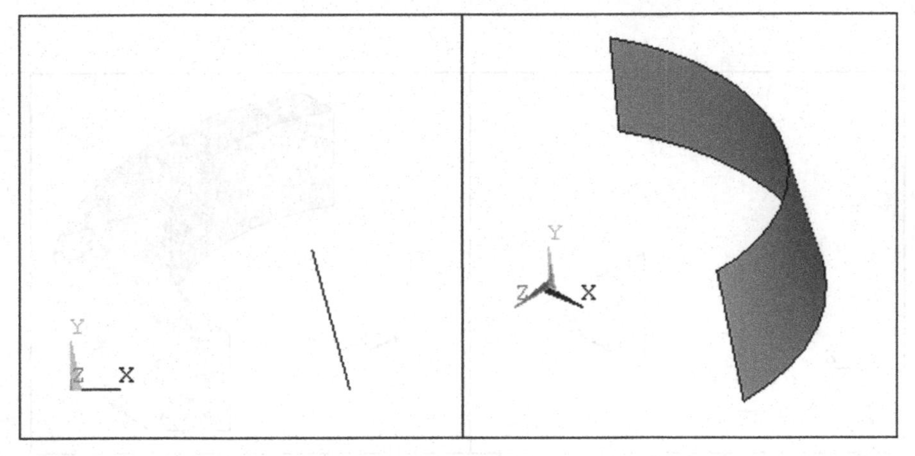

Fig. 4.49 Front view of a straight line (*left*); the line is rotated +60° about the *y*-axis in two increments of 30° to create the curved areas (shown in oblique view on the *right*)

Main Menu > Preprocessor > Modeling > Operate > Extrude > -Lines- Along Lines

Creating Volumes by Rotating Areas About an Axis

Main Menu > Preprocessor > Modeling > Operate > Extrude > -Areas- About Axis

Creating Volumes by Sweeping Areas Along a Path

Main Menu > Preprocessor > Modeling > Operate > Extrude > -Areas- Along Lines

Creating Volumes by Extruding Areas

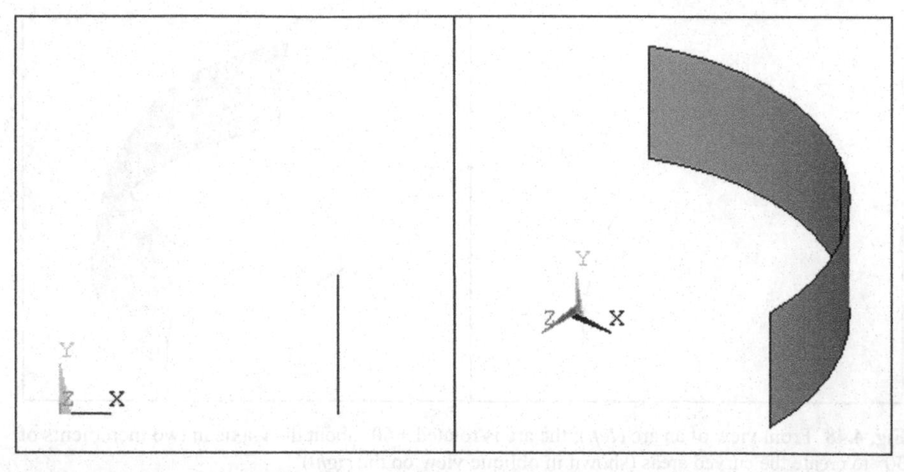

Fig. 4.50 Front view of a straight line (*left*); the line is rotated + 120° about the *y*-axis in two increments of 60° to create the curved areas (shown in oblique view on the *right*)

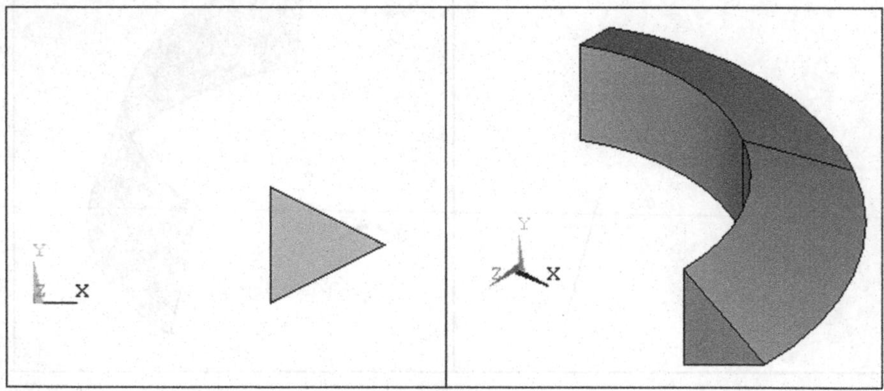

Fig. 4.51 Front view of an area (*left*); the area is rotated + 120° about the *y*-axis in two increments of 60° to create the volumes (shown in oblique view on the *right*)

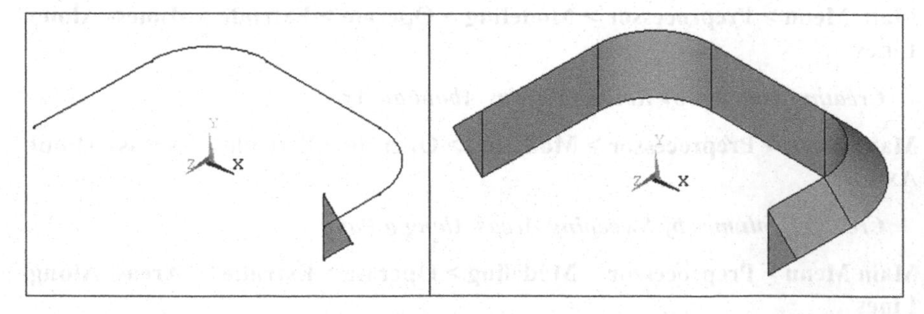

Fig. 4.52 Oblique view of a path defined by lines and an area to be swept along the path (*left*) and the volume created by sweeping the area along the path (*right*)

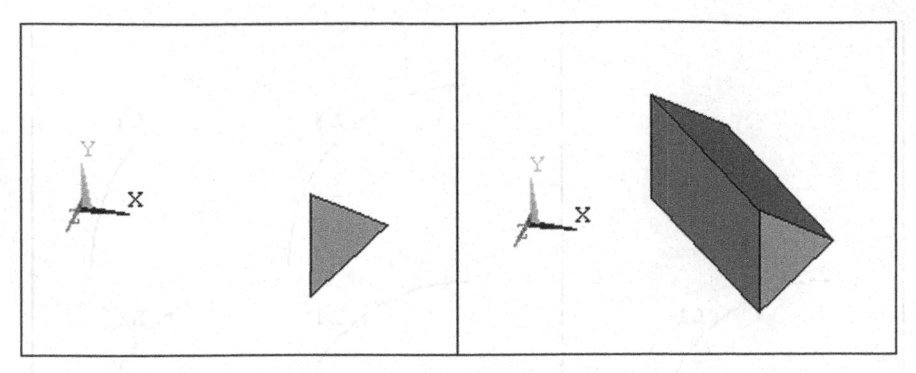

Fig. 4.53 Oblique view of an area to be extruded along its normal (*left*) and oblique view of the volume created by extrusion of the area along its normal (*right*)

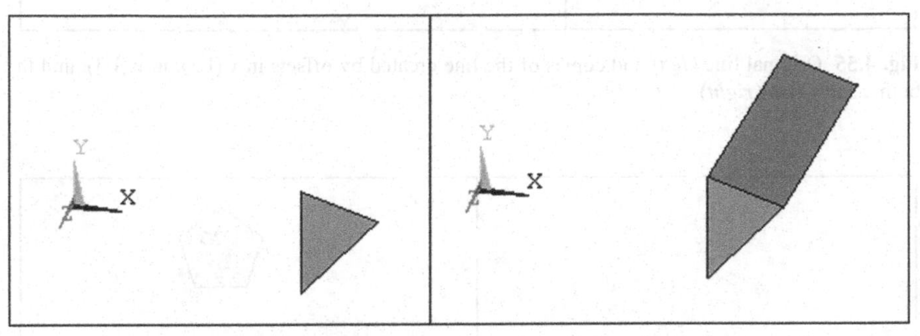

Fig. 4.54 Oblique view of an area to be offset in *x*, *y*, or *z* (*left*) and oblique view of the volume created by offsetting the area in the *z*-direction (*right*)

Main Menu > Preprocessor > Modeling > Operate > Extrude > -Areas- Along Normal

Creating Volumes by Offsetting Areas

Main Menu > Preprocessor > Modeling > Operate > Extrude > -Areas- By XYZ Offset

4.5.2 Moving and Copying

Previously created entities can be moved or copied.Also, if a repeated symmetry or skew-symmetry exists in the geometry, the user can create a representative entity to create the geometry by copying it to a new location.Representativc applications are described in Fig. 4.55, 4.56, and 4.57.

The common menu paths for moving entities are specified as

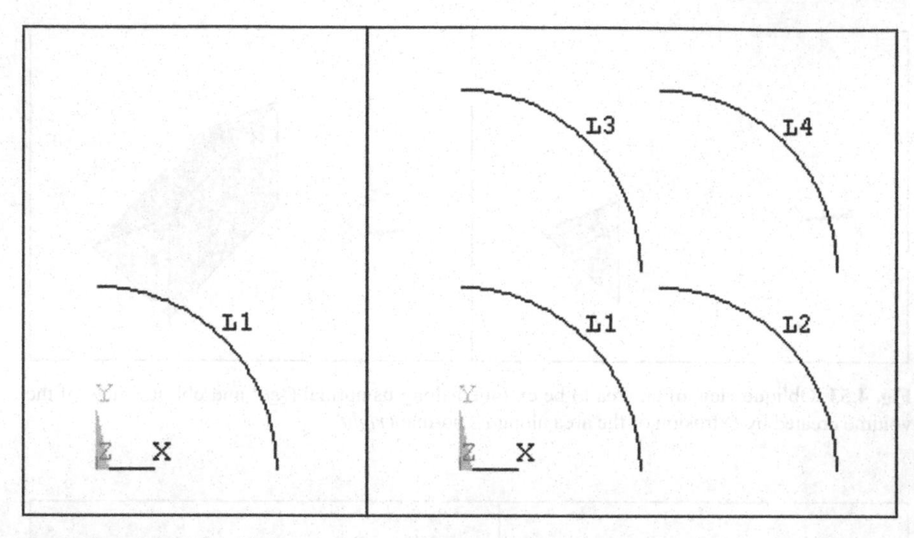

Fig. 4.55 Original line (*left*) and copies of the line created by offsets in x (L2), in y (L3), and in both x and y (L4)(*right*)

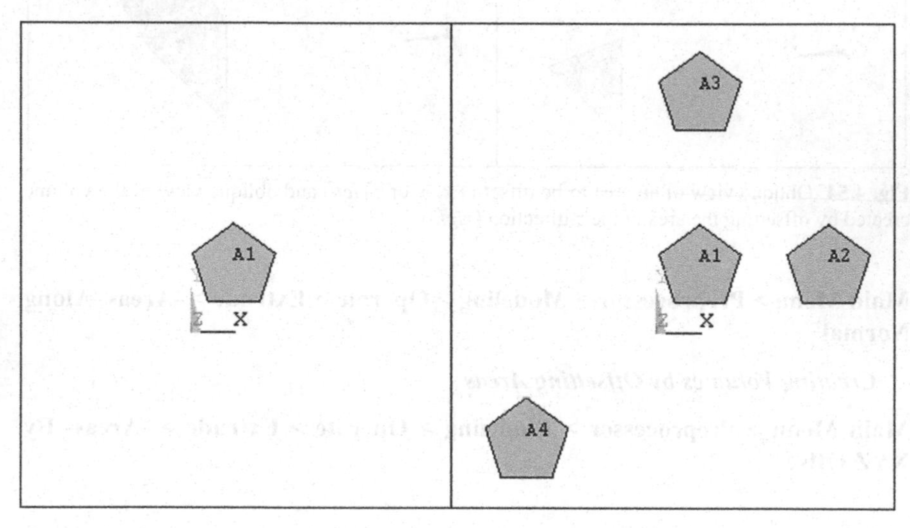

Fig. 4.56 Original area (*left*) and copies of the area created by offsets in x (A2), in y (A3), and in both x and y (A4) (*right*)

Main Menu > Preprocessor > Modeling > Move/Modify > -Keypoints- Single KP

Main Menu > Preprocessor > Modeling > Move/Modify > Lines

Main Menu > Preprocessor > Modeling > Move/Modify > -Areas- Areas

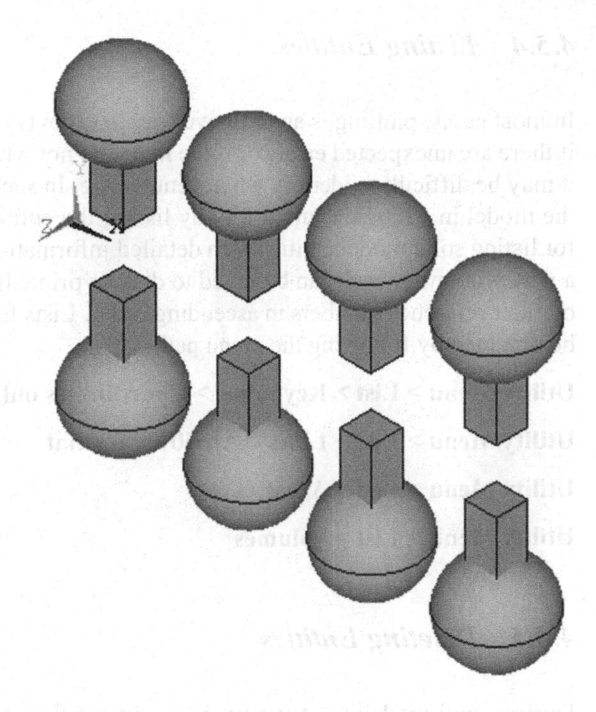

Fig. 4.57 The first volume is copied three times with 0.4 unit offset in the x-direction and then reflected with respect to the x-z plane

Main Menu > Preprocessor > Modeling > Move/Modify > Volumes

The common menu paths for copying entities are specified as

Main Menu > Preprocessor > Modeling > Copy > Keypoints

Main Menu > Preprocessor > Modeling > Copy > Lines

Main Menu > Preprocessor > Modeling > Copy > Areas

Main Menu > Preprocessor > Modeling > Copy > Volumes

4.5.3 Keeping/Deleting Original Entities

During the performance of Boolean-type operations, there are input entities (e.g., original areas to be added or, when dividing a line with a volume, the original line and volume) and output entities. By default, ANSYS will delete the input entities, keeping only the output entity.However, the input entities can be "kept" through the menu path:

Main Menu > Preprocessor > Modeling > Operate > Booleans > Settings

This brings up a dialog box requesting the user to specify certain settings. The first setting option controls whether the input entities will be kept or deleted. Answering ***Yes*** instructs ANSYS to keep the input entities; otherwise, they are deleted.

4.5.4 Listing Entities

In most cases, plotting is an effective way to quickly examine the model. However, if there are unexpected errors or if the model is not what the user intended to create, it may be difficult to identify what went wrong. In such cases, the user can examine the model in a more accurate way by listing the entities. ANSYS provides options for listing solid model entities with detailed information. All of the lists are given in a new window (which can be saved to disk or printed) where the entities are sorted by their reference numbers in ascending order. Lists for the solid model entities can be obtained by following the menu paths:

Utility Menu > List > Keypoints > Coordinates only

Utility Menu > List > Lines > Attribute format

Utility Menu > List > Areas

Utility Menu > List > Volumes

4.5.5 Deleting Entities

During solid modeling, it is very common for the user to create unintended solid modeling entities. These extra entities might make the modeling phase confusing and may potentially cause errors.In order to eliminate this possibility, the user should "clean up" the model by deleting these entities. The hierarchy of the solid model entities is important in that the entity (or entities) must not be used for the definition of any higher-order entities in order to be deleted.For example, the existence of an area automatically implies that lines and keypoints are attached to this area.None of the lines can be deleted as long as the area exists. The area must first be deleted, then the lower-order entities can be deleted. Similarly, a KP cannot be deleted as long as the line(s) to which the KP is attached exist(s). Only after the line(s) is(are) deleted, can the KP can be deleted. Solid model entities can be deleted in two different methods:

1. Delete the entity without deleting the lower-order entities that are attached to it.
2. Delete the entity and all the lower-order entities that are attached to it. In this case, if some of the lower-order entities are associated with other entities, they will not be deleted.

The following menu paths are used for these two methods:

To Delete Entities Only

Main Menu > Preprocessor > Modeling > Delete > Keypoints

Main Menu > Preprocessor > Modeling > Delete > Lines Only

Main Menu > Preprocessor > Modeling > Delete > Areas Only

Main Menu > Preprocessor > Modeling > Delete > Volumes Only

To Delete Entities and Below

Main Menu > Preprocessor > Modeling > Delete > Lines and Below

Main Menu > Preprocessor > Modeling > Delete > Areas and Below

Main Menu > Preprocessor > Modeling > Delete > Volumes and Below

4.6 Viewing a Model

ANSYS provides a very robust graphic utility to view solid model entities, nodes, elements, material properties, boundary constraints, loads, and results. The graphics-related utilities are accessible through the *Plot* and *PlotCtrls* (stands for Plot Controls) submenus under the *Utility Menu*. All of the entities can be viewed through the *Plot* submenu. However, the *PlotCtrls* submenu (as its name indicates) provides many options that enhance the use of the plot utility for many different purposes, such as plotting the numbers associated with entities, plotting the entities in different colors,[1] and viewpoint and viewing angle adjustments.

For the sake of brevity, only the most frequently used items are explained in detail here.However, many more features are discussed in the examples.

4.6.1 Plotting: Pan, Zoom, and Rotate Functions

The *Pan-Zoom-Rotate* tool is a very effective function of ANSYS for manipulating the view by panning, zooming, and rotating the model. The following menu path is used to activate this function:

Utility Menu > PlotCtrls > Pan, Zoom, Rotate

The *Pan-Zoom-Rotate* window appears, as shown in Fig. 4.58. There are eight different fields in this window:

1. *Active Window Field*: The *Graphics Window* can be divided into "sub"-windows (up to 5) in ANSYS; only one of them can be the "active" window at any one time.This button identifies which window(s) are to be affected by the operations performed within the *Pan-Zoom-Rotate* window.
2. *Viewing Direction Field*: This group of buttons changes the viewpoint. Clicking on the ***Top*** button will redraw the model (or entities) as seen from the top. In ANSYS, "top" corresponds to the positive global Y-direction. Similarly, "front" and "right" will redraw the model as seen from the positive global Z-and

[1] Colors have not been used in the printed version of the figures. See the accompanying CD-ROM for color versions of the figures.

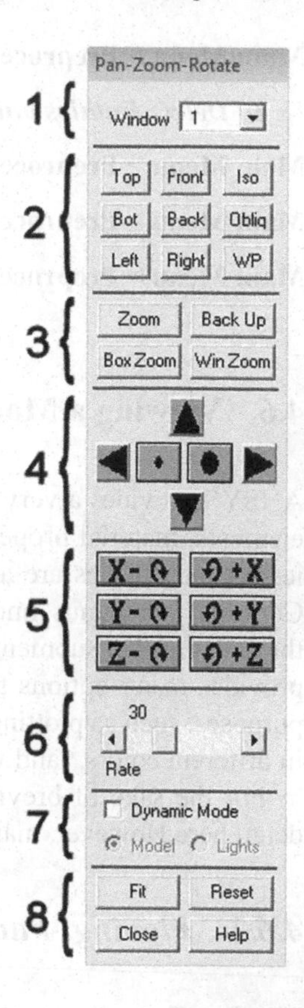

Fig. 4.58 The eight fields in the *Pan-Zoom-Rotate* window

X-directions, respectively. The ***Iso*** and ***Obliq*** buttons redraw the model as seen from a point that lies on a line that passes through the origin and (1, 1, 1) and (1, 2, 3), respectively. Finally, the ***WP*** button redraws the model by taking the positive Working Plane *z*-direction as the front of the model.

3. *Zoom Field*: Provides different zooming methods:

Zoom: Clicking on this button, followed by a single left-click, chooses the center of the region of interest. After the first click, moving the mouse toward and away from the center will display a moving square outline of the potential target region that the user would like to zoom in. Once decided, a second left-click will zoom in to the region indicated by the outline.

Box Zoom: This function works in a similar way. The user picks two corners of the zoom-in region. After picking the first corner by clicking the left mouse button, moving the mouse over the *Graphics Window* will show a moving outline of the potential zoom-in region. A second click will pick the second corner and ANSYS will redraw the zoom-in region.

Backup: Clicking on this button redraws the model in the previous viewing configuration.

Win Zoom*:** This button works like the ***Box Zoom button except that after picking the first point, ANSYS locks the aspect ratio of the potential zoom-in region at the same values as the aspect ratios of the active window (the redraw of the zoom-in region will fit perfectly in the window).

4. *Pan/Zoom Field*: The arrow buttons pan the model in the indicated directions and the dots zoom in and out. A small dot indicates zooming out and a large dot indicates zooming in. The *Sliding Rate Control Bar*, explained below, dictates the rate at which pan and zoom actions operate.
5. *Rotate Field*: These six buttons rotate the model about "Screen" *x*-, *y*-, and *z*-directions. The "screen origin" is the center of the active window. The positive "Screen" *x*-direction starts from the center of the window and extends to the right.Likewise, the positive "Screen" *y*- and *z*-directions start at the center of the window and extend to, respectively, the top and out (of the monitor).
6. *Rate Control Field*: The Sliding Bar controls the rate of pan, zoom, and rotate that is performed in the active window. The range is from 1 to 100 (rate 1 pans/zooms at a smaller rate than rate 100 would).
7. *Dynamic Mode Field*: By clicking on this radio button, the user toggles on/off the option to pan and rotate dynamically. When the *Dynamic Mode* is active, the mouse pointer changes shape when it is over the *Graphics Window*. Pressing the left mouse button (without releasing) and moving around in the *Graphics Window* pans the model. Similarly, the right mouse button is used for rotating the model dynamically.
8. *Action Field*: Includes four action buttons:

***Fit*:** Fits the whole model in the active window.
***Reset*:** Restores the default orientation and size for viewing (front view).
***Close*:** Closes the *Pan-Zoom-Rotate* window.
***Help*:** Brings the help page for *Pan-Zoom-Rotate* window.

4.6.2 Plotting/Listing Entities

The following menu paths are used to plot and list the solid model (keypoints, lines, areas, and volumes) and mesh (nodes and elements) entities:

Utility Menu > Plot > Keypoints > Keypoints

Utility Menu > Plot > Lines

Utility Menu > Plot > Areas

Utility Menu > Plot > Volumes

Utility Menu > List > Volumes

Utility Menu > Plot > Nodes

Utility Menu > List > Nodes

Utility Menu > Plot > Elements

Utility Menu > List > Elements > Nodes + Attributes

The resulting plots are displayed in the *Graphics Window* and can be examined using the *Pan-Zoom-Rotate* window discussed in the previous section.

4.6.3 Numbers in the Graphics Window

Whenever an entity is being created, ANSYS either asks for a reference number or assigns the lowest available number for that type of entity. Therefore, every entity differs from the other entities of the same type by this reference number. When plotting these entities in the *Graphics Window*, by default, ANSYS will not show the entity numbers. Often times, it is important for the user to see the numbers printed when plotting entities. This can be done using the following menu path:

Utility Menu > PlotCtrls > Numbering

which brings up the *Plot Numbering Controls* dialog box, as shown in Fig. 4.59. The entity numbers for keypoints, lines, areas, volumes, and nodes can simply be turned on by placing a checkmark in the corresponding boxes. Element numbers

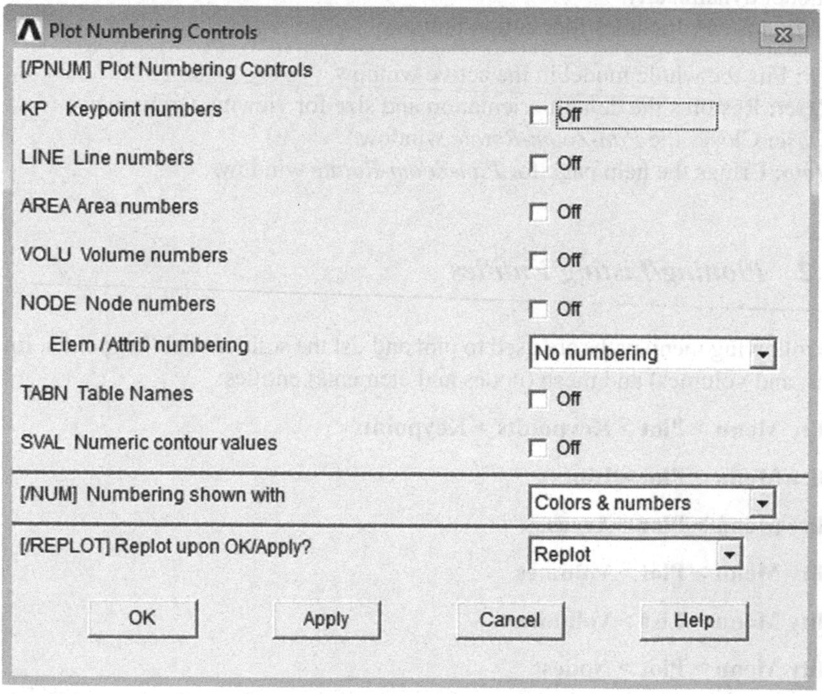

Fig. 4.59 *Plot numbering controls* dialog box

can be turned on using the ***Elem/Attrib numbering*** pull-down menu. Instead of the element numbers, the user can display element attribute numbers (element type, real constant, and material) using the same option.Also, colors may be assigned to each entity number for more convenient viewing.The ***[/NUM] Numbering shown with*** pull-down menu in this dialog box allows the user to plot numbers with or without color assignments, as well as to plot using colors only (without numbers).

4.7 Meshing

As mentioned previously (Sect. 4.3), the mesh of the geometry under consideration may be generated directly, i.e., generation of nodes and elements, one at a time. However, this may prove to be a challenging task. Almost always *Solid Modeling* constitutes a part of the finite element analysis. Thus, the sole purpose of *Solid Modeling* is to create the mesh of the geometry, as conveniently and efficiently as possible. Once the *Solid Model* is completed, the user is ready to perform meshing. Regardless of whether a *Solid Model* is generated or not, the meshing can be performed only after the specification of element type(s). ANSYS offers several convenient options to assist in meshing. These include *Automatic Meshing*, *Smart Sizing*, and *Mapped Meshing*. In the following subsections, topics related to meshing are discussed in more detail.

4.7.1 Automatic Meshing

One of the most powerful features of ANSYS is automatic mesh generation. ANSYS meshes the solid model entities upon execution of an “appropriate” single command. With automatic meshing, the user can still provide specific preferences for mesh density and shape. If no preferences are specified by the user, ANSYS uses the default preferences. The following menu paths are used for automatic mesh generation after solid model generation:

Mesh Using Line Elements
This option is used for models utilizing one-dimensional elements, such as trusses and beams. It requires existing lines. The following menu path is used to mesh lines:

Main Menu > Preprocessor > Meshing > Mesh > Lines

This brings up a *Pick Menu* asking the user to either enter the line number(s) through the *text field* or pick line(s) from the *Graphics Window*. When all the lines are input (picked), hitting ***OK*** in the *Pick Menu* generates the mesh.

Mesh Using Area Elements
This option is used for models utilizing 2-D elements, and it requires existing areas. The following menu path is used to mesh areas:

Main Menu > Preprocessor > Meshing > Mesh > Areas > Mapped > 3 or 4 Sided

Main Menu > Preprocessor > Meshing > Mesh > Areas > Free

Meshing can be accomplished through either the *Mapped* or *Free Meshing* methods.If free meshing is chosen, the second menu path is used, bringing up a *Pick Menu* asking the user to either enter the area number(s) through the *text field* or pick area(s) from the *Graphics Window*. When all the areas are input (picked), hitting ***OK*** in the *Pick Menu* generates the mesh. The *Mapped meshing* option is discussed in a later subsection.

Mesh Using Volume Elements
This option is used for models using 3-D elements, and it requires existing volumes.

Main Menu > Preprocessor > Meshing > Mesh > Volumes > Mapped > 4–6 Sided

Main Menu > Preprocessor > Meshing > Mesh > Volumes > Free

This brings up a *Pick Menu* asking the user to either enter the volume number(s) through the *text field* or pick volumes(s) from the *Graphics Window*. When all the volumes are input (picked), hitting ***OK*** in the *Pick Menu* will generate the mesh.

ANSYS allows the user to control the mesh density of the domains defined by solid model entities. The desired mesh density can be achieved by:

- Defining a target element edge size on the domain boundaries.
- Defining a default number of element edges on all lines.
- Defining the number of element edges on specific lines.
- Using *smart sizing*.
- Using *mapped meshing*.

These methods are discussed in detail in the following subsections.

4.7.1.1 Specifying Mesh Density Globally

There are two approaches for enforcing the mesh density globally. The first one involves specification of the element edge size; ANSYS attempts to generate a mesh with all elements having edge sizes as close as possible to the specified value. The second possibility is to specify a fixed number of elements along all the lines within the solid model. The following menu path is used for specifying the mesh density globally:

Main Menu > Preprocessor > Meshing > Size Cntrls > ManualSize > Global > Size

This brings up the *Global Element Sizes* dialog box with two input parameters: ***SIZE*** and ***NDIV.SIZE*** denotes the target element edge length, and ***NDIV*** is the target number of elements along the lines. If ***SIZE*** is specified, ***NDIV*** is ignored. The following example explains these concepts. Consider a square area with sides

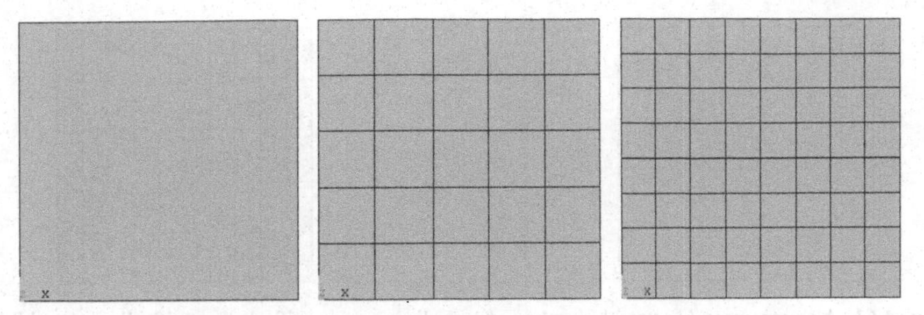

Fig. 4.60 A square area (*left*) meshed with global element size (*SIZE*) specified (*middle*) and number of line divisions (*NDIV*) specified globally (*right*)

5 units long, as shown in Fig. 4.60 (left). If the global element size is specified as 1 (***SIZE*** = 1), the mesh shown in Fig. 4.60 (middle) is generated with each element having an edge size of 1 unit. If the user chooses to the specify the number of elements along lines instead of element sizes, then ***SIZE*** is left untouched (zero), and ***NDIV*** is set to a specific value, say 8. As a result of this operation, the mesh shown in Fig. 4.60 (right) is generated, with 8 elements along each line.

Specification of mesh density globally works well when the geometry of the problem is regular, with aspect ratio close to one. When domains of irregular shapes are considered, applying the same meshing targets to lines of different sizes results in meshes with high aspect ratios, leading to potentially erroneous results. Therefore, the techniques explained in the following subsections are more desirable.

4.7.1.2 Specifying Number of Element Edges on Specific Lines

When the geometry of the problem is irregular, i.e., not basic shapes such as triangles and rectangles, specifying the number of element edges along specific lines may be a good way to avert possible meshing problems. This strategy also helps to refine the mesh around regions where it may be crucial for accuracy. Similarly, certain regions in the geometry may not be critical, and keeping the mesh around these regions may help reduce the computational cost without losing accuracy. The number of element edges on specific lines can be specified using the following menu path:

Main Menu > Preprocessor > Meshing > Size Cntrls > ManualSize > Lines > Picked Lines

which brings up the *Pick Menu* for line picking. After the user picks the lines and clicks on ***OK***, the *Element Sizes on Picked Lines* dialog box appears. The second parameter, ***NDIV***, dictates how many elements will be placed along the picked lines. The third parameter, ***SPACE***, which stands for spacing ratio, is important when a mesh graded (biased) toward a direction is desired. The default value for ***SPACE*** is 1 (no bias, uniform spacing). If it is positive, the spacing is biased from one end of

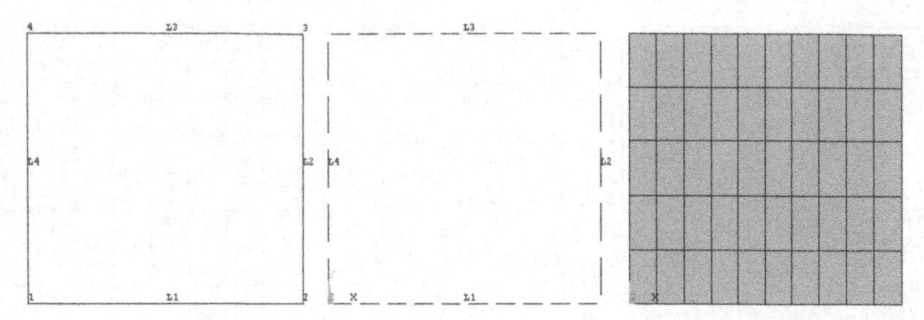

Fig. 4.61 A square area (*left*) with number of line divisions specified at specific lines (*middle*) and the resulting mesh (*right*)

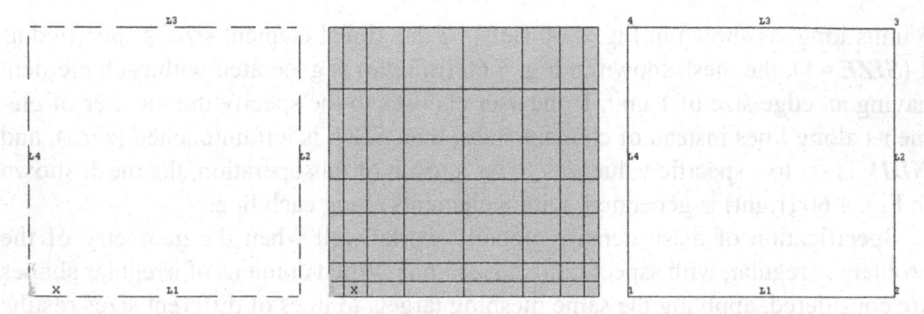

Fig. 4.62 Biased line divisions (*left*), resulting mesh (*middle*), and lines with keypoint numbers plotted (*right*)

the line to the other end.If it is negative, then the bias is from the center toward the ends.Its magnitude defines the ratio of the largest division size to the smallest.These concepts are explained in the following examples. Consider the square area used for the example in the previous subsection, shown in Fig. 4.61 (left) with line numbers. Using the menu path above, ***NDIV*** is specified to be 5 for lines 2 and 4, and 10 for lines 1 and 3. After this operation, lines are plotted with the specified divisions clearly visible [Fig. 4.61 (middle)].Meshing of the area produces the one shown in Fig. 4.61 (right). The same example is considered, this time with the specification of spacing ratios.The goal is to have a mesh graded from coarse at the center to fine at the edges in the x-direction, and from coarse at the top to fine at the bottom (in the y-direction). The same number of divisions is used for all the lines with a value of 8 (***NDIV***=8). Using the menu path given above, spacing ratios (***SPACE***) for lines 1, 2, 3, and 4 are specified as –4, 4, –4, and 0.25, respectively. The line plot after this operation is shown in Fig. 4.62 (left), and the corresponding mesh is given in Fig. 4.62 (middle). It is worth noting the reason why the parameter ***SPACE*** is different for lines 2 and 4. Figure 4.62 (right) shows the line plot with the keypoint numbers. Line 2 is defined from keypoint 2 to keypoint 3 (from bottom to top) whereas line 4 is defined from keypoint 4 to keypoint 1 (from top to bottom). When ***SPACE*** is positive and greater than 1, its value defines the ratio of the division length at the end of the line to the length of the division at the beginning of the line.

4.7.1.3 Smart Sizing

Instead of specifying the number of line divisions or element edge sizes, one can use the ANSYS "smart sizing" feature, where the mesh density is specified in a cumulative sense. In this method, the user specifies a level of refinement that ranges from **1** to **10**; the smaller the number, the finer the mesh. To use smart sizing, follow the menu path given below:

Main Menu > Preprocessor > Meshing > Size Cntrls > SmartSize > Basic

This brings up a dialog box with a pull-down menu asking the user to choose a level of refinement. Selecting the level, followed by hitting ***OK***, activates the smart sizing. Now, the user is ready to mesh the solid model entities. As this option uses meshing options involving advanced geometry, there is no easy way to explain how it works.Therefore, it is suggested that the user experiment with it, and build a knowledge base that will be helpful later on.

4.7.1.4 Mapped Meshing

Another very commonly used (by experienced ANSYS users) meshing method is *Mapped Meshing*. The mapped meshing concept is valid only in two- and three-dimensional problems (no line elements). The solid model entities (areas and volumes) meshed with this option use quadrilateral area elements or hexahedral (brick) volume elements.

The reason why mapped meshing is desirable is that it generates regular, thus computationally well-behaving, meshes. Not every area or volume can be mapped meshed. The areas or volumes to be mapped meshed must be "regular." This regularity is governed by two properties of the solid model entity: the number of sides (lines for areas and areas for volumes) and number of divisions on opposite sides (opposite sides must have an equal number of divisions). For areas, the acceptable number of sides is 3 or 4. If the area has 3 sides (defined by 3 lines), then the number of divisions in all 3 lines must be equal and even. If 4 lines define the area, as stated before, the lines on opposite sides must have the same number of divisions. These considerations are similar for mapped meshing of volumes. The number of areas that define the volume must be either 4 (tetrahedron), 5 (prism), or 6 (hexahedron). The number of divisions on opposite sides must be equal. If 4 or 5 areas define the volume, the number of divisions on the triangular areas must be equal and even.

To use mapped meshing, follow the menu path given below:

Main Menu > Preprocessor > Meshing > Mesh > Areas > Mapped > 3 or 4 Sided

Main Menu > Preprocessor > Meshing > Mesh > Volumes > Mapped > 4 to 6 Sided

which brings up a *Pick Menu* for area picking.After the areas are picked and the ***OK*** button is pressed, the mesh is generated. Figure 4.63 (left) shows a triangular area

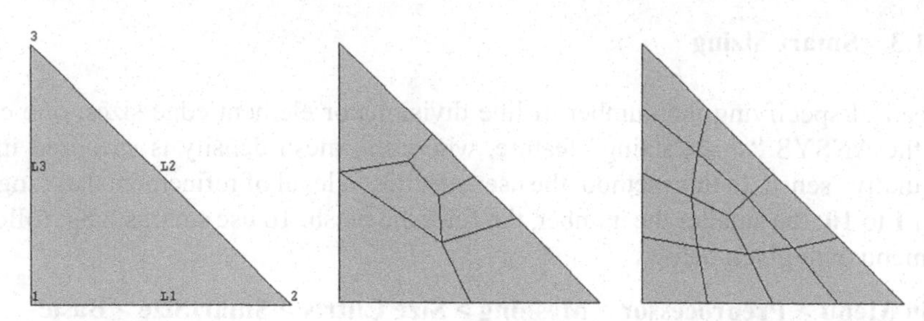

Fig. 4.63 Triangular area (*left*) and corresponding free (*middle*) and mapped (*right*) meshes

with corresponding free and mapped meshes given in Fig. 4.63 (middle) and 4.63 (right), respectively. It is clear from these figures that the mapped mesh delivers elements with controlled and desirable aspect ratios.

If the areas or volumes do not have the required number of sides that are given above, there might still be a way to "mapped mesh" these entities. For this, the user looks for sides that could be considered as a single side when combined. This way the number of sides can be reduced to the required numbers. This operation is performed through "concatenating" lines (for meshing of areas with more than 4 sides) and areas (for meshing of volumes with more than 6 sides). Line and area concatenations are performed using the following menu paths:

Main Menu > Preprocessor > Meshing > Mesh > Areas > Mapped > Concatenate > Lines

Main Menu > Preprocessor > Meshing > Mesh > Volumes > Mapped > Concatenate > Areas

Main Menu > Preprocessor > Meshing > Mesh > Volumes > Mapped > Concatenate > Lines

which brings up a *Pick Menu* for lines to be concatenated. The concatenation is explained through the following example. Consider the irregular area, shown in Fig. 4.64 (top), with line numbers plotted. Free meshing of this area produces the mesh given in Fig. 4.64 (middle), with elements having large aspect ratios. As observed from Fig. 4.64 (top), the area to be meshed is enclosed by 7 lines (sides). When mapped meshing areas, the maximum number of sides is 4. Therefore, if the user wants to mapped mesh this area, line concatenations must be performed. For this purpose, lines 1, 2, and 3 are concatenated to produce a new line (line 8). Also, lines 4 and 5 are concatenated, producing line 9. With these concatenations, the number of sides defining the area is reduced from 7 to 4 and mapped meshing is possible. After specifying the number of divisions on lines 6 and 7 as 3, mapped meshing is performed by using the menu path given above, producing the mesh given in Fig. 4.64 (bottom), with elements having acceptable aspect ratios.

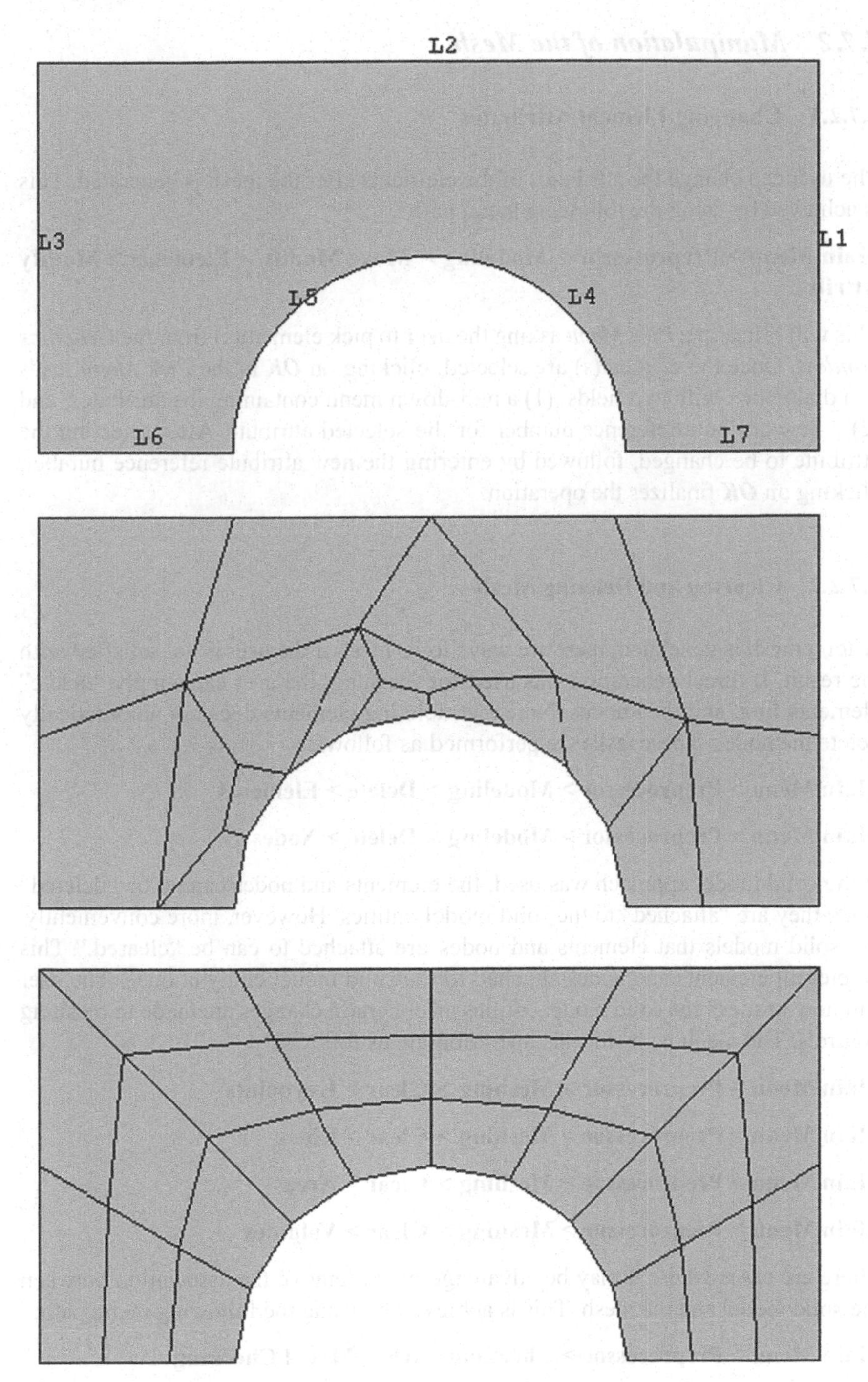

Fig. 4.64 An irregular area (*top*) meshed using free meshing (*middle*) and meshed with mapped meshing after line concatenation (*bottom*)

4.7.2 Manipulation of the Mesh

4.7.2.1 Changing Element Attributes

The user can change the attributes of the elements after the mesh is generated. This is achieved by using the following menu path:

Main Menu > Preprocessor > Modeling > Move/Modify > Elements > Modify Attrib

This will bring up a *Pick Menu* asking the user to pick element(s) from the *Graphics Window*. Once the element(s) are selected, clicking on ***OK*** in the *Pick Menu* leads to a dialog box with two fields: (1) a pull-down menu containing the attributes, and (2) a new attribute reference number for the selected attribute. After selecting the attribute to be changed, followed by entering the new attribute reference number, clicking on ***OK*** finalizes the operation.

4.7.2.2 Clearing and Deleting Mesh

After a mesh is generated, there are ways to re-mesh if the user is not satisfied with the result. If direct generation was used for meshing, the user can simply "delete" elements first, and then nodes. Note that deleting elements does not automatically delete the nodes. These tasks are performed as follows:

Main Menu > Preprocessor > Modeling > Delete > Elements

Main Menu > Preprocessor > Modeling > Delete > Nodes

If the solid model approach was used, the elements and nodes cannot be "deleted" since they are "attached" to the solid model entities. However, more conveniently, the solid models that elements and nodes are attached to can be "cleared." This deletes all elements and nodes attached to the solid model entity at once. The user can now re-mesh the solid model entities after certain changes are made in meshing controls. The menu paths for this operation are as follows:

Main Menu > Preprocessor > Meshing > Clear > Keypoints

Main Menu > Preprocessor > Meshing > Clear > Lines

Main Menu > Preprocessor > Meshing > Clear > Areas

Main Menu > Preprocessor > Meshing > Clear > Volumes

There are cases where it may be advantageous to remove the association between the solid model and the mesh. This is achieved by using the following menu path:

Main Menu > Preprocessor > Checking Ctrls > Model Checking

which brings up a dialog box with a pull-down menu.Selecting the item ***Detach*** in the pull-down menu and clicking on ***OK*** removes the association between the solid model and the mesh.

4.7.2.3 Numbering Controls

When dealing with complicated geometries, the Boolean operations explained in Sect. 4.4 are used regularly. These operations often generate new entities while removing existing ones, which creates gaps in the entity numbering. For example, when an area (say, area 1) is subtracted from another one (area 2), the resulting area is given the smallest available area number (in this case, area 3). Immediately after the creation of this area (area 3), ANSYS internally deletes the input areas (areas 1 and 2). Similarly, all the keypoints and lines associated with the new area are given new numbers while the ones associated with the old areas are removed. Similar considerations apply to nodes and elements. ANSYS provides the user with the option of "compressing" the entity numbers, which is performed as follows:

Main Menu > Preprocessor > Numbering Ctrls > Compress Numbers

which leads to a dialog box with a pull-down menu. After the entity label is selected from this menu, clicking ***OK*** finalizes this operation.

Another important concept in solid modeling and meshing is the possible existence of duplicate entities. This usually occurs when the user creates new entities by copying existing ones or by reflection about a plane. If the old and new entities occupy the same space and if the material is supposed to be continuous along the line (or plane) where duplicate items lie, then they must be merged. Although it may not be apparent in the *Graphics Window*, the existence of duplicate entities compromises the continuity of the mesh, and thus may lead to invalid solutions. The following menu path is used to merge entities:

Main Menu > Preprocessor > Numbering Ctrls > Merge Items

This brings up a dialog box in which the first field is a pull-down menu for the label of entities to be merged.

4.8 Selecting and Components

In the case of a three-dimensional finite element model, graphical picking may become a tedious and, at times, frustrating experience. In such cases, the selection tool provided by ANSYS is highly useful. This tool is efficient as it utilizes concepts from logics. The selected entity can then be saved in the ANSYS database as a "component." Thus, the next time this group of entities needs to be selected, selection of the component is sufficient. Selecting operations are discussed first, and the components next.

4.8.1 Selecting Operations

In ANSYS, entities are stored in separate sets (e.g., a set of areas, a set of volumes, etc.). Initially all of the full sets are active, until a selection operation is performed.

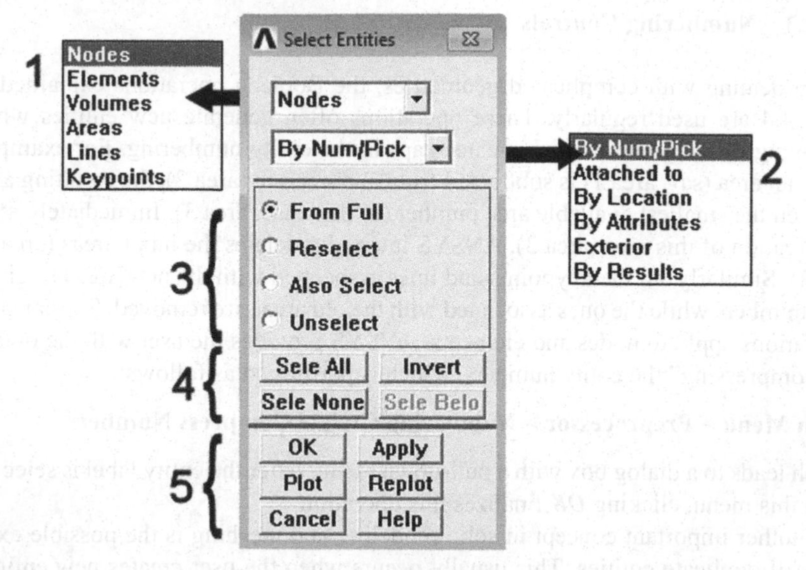

Fig. 4.65 *Select entities* dialog box (by number and picking)

When individual entities in a full set are selected (subset), they become active. Entity sets are independent of each other, i.e., selecting a group of lines does not cause any change in the selection status of the sets of keypoints or areas. Selections can be made based on several criteria, as explained below.

Selections can be performed by using the following menu path:

Utility Menu > Select > Entities

which brings up the *Select Entities* dialog box, as shown in Fig. 4.65. This dialog box has five distinct fields:

1. *Entity Field*: The entity to be selected is chosen using this pull-down menu.
2. *Criterion Field*: The entity chosen in the *Entity Field* is selected based on the criterion chosen in this pull-down menu. The following criteria are possible:

By Num/Pick: Clicking ***OK*** after choosing this criterion starts the *Pick Menu* and the entities are selected by picking.

Attached to: As discussed previously, the entities in ANSYS are associated with each other. For example, a line is composed of at least two keypoints; an area is made up of at least three lines; etc. Thus, keypoints and lines, and lines and areas, are mutually attached (the list may be extended).When this criterion is chosen, another field appears in the *Select Entities* dialog box, listing the possibilities for attachment.If ***Areas*** in the *Entity Field* and ***Attached to*** in the *Criterion Field* are chosen, the new field lists ***Lines*** and ***Volumes*** as possible attachments; choosing ***Volumes*** and clicking on ***OK*** results in the selection of volumes that are attached to the currently selected areas.

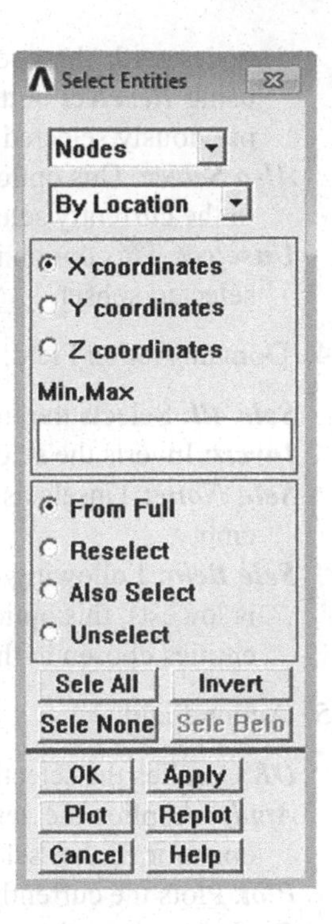

Fig. 4.66 *Select entities* dialog box (by location)

By Location: Selects entities based on their location. Upon choosing this criterion, a new field appears in the *Select Entities* dialog box with radio-buttons for *x*-, *y*-, and *z*-coordinates and a text field for the minimum and maximum values for the coordinate (shown in Fig. 4.66). For example, in order to select the nodes located between $y=2$ and $y=5$, the radio-button for the *y*-coordinate is activated and the expression "***2, 5***" (without the quotation marks) is entered in the text field.

By Attributes: This criterion is used for selecting entities based on their attributes (element type, material, real constant, etc.).

Exterior: Using this criterion, entities along the outer boundaries of the model are selected.

By Results: If a solution is obtained, then entities (only nodes and elements) can be selected based on result values.

3. *Domain Field*: This field determines the domain of the entity set with which the criterion is applied, as explained below:

From Full: Selection is made from the full set of entities regardless of the selection status of the particular entity set.

Reselect: This option is used to refine the selection.It is used to select entities from a previously selected subset. For example, if the goal is to select all the nodes having coordinates $x=2$ and $y=3$ (both at the same time), then the nodes with the coordinate $x=2$ are first selected ***From Full*** set and, then, using ***Reselect*** button, the nodes with coordinate $y=3$ are selected from the previously selected subset of nodes with coordinate $x=2$.

Also Select: This option is used to expand the selection. It is used to add entities to the currently selected subset, based on a different criterion.

Unselect: This option is used to deactivate (unselect) a group of entities from the selected subset.

4. Domain Action Field:

Sele All: Selects the full set of a specific entity.

Invert: Inverts the selected set; active entities become inactive and vice versa.

Sele None: Unselects the full set of a specific entity; the active set becomes empty.

Sele Belo: Following the hierarchy of entities (i.e., volume is highest and node is lowest), this option selects the lower entities attached to the selected set of entities chosen in the Entity field.

5. Action Field:

OK: Applies the selection operation and closes the Select Entities dialog box.

Apply: Applies the selection operation; the Select Entities dialog box remains open for further selections.

Plot: Plots the currently selected set of a specific entity.

Replot: Updates the plot.

Cancel: Closes the Select Entities dialog box without applying the selection operation.

Help: Displays the help pages related to selection operations.

In order to select "everything" (reset all entities to their full sets), the following menu path is used:

Utility Menu > Select > Everything

4.8.2 Components

Groups of selected entities can be saved in an ANSYS database for easy retrieval. These groups are called components, and they can only contain entities of the same kind. The main advantage of defining components is to avoid multiple selection operations every time the user needs to select the same group of entities. The following menu path is used for defining components:

Utility Menu > Select > Comp/Assembly > Create Component

which is followed by a dialog box requesting the name to be given to the component and the type of entity to include in the component. Upon clicking ***OK***, the component is created using the currently selected subset of the entity type chosen. The following menu path is used when a component has to be selected:

Utility Menu > Select > Comp/Assembly > Select Comp/Assembly

Listing and deletion of components is performed by using the following menu paths:

Utility Menu > Select > Comp/Assembly > List Comp/Assembly

Utility Menu > Select > Comp/Assembly > Delete Comp/Assembly

Chapter 5
ANSYS Solution and Postprocessing

5.1 Overview

A typical ANSYS session, regardless of the discipline, involves the following steps:

1. Model Generation

   – Specify *jobname* (this step is optional but recommended).
   – Enter *Preprocessor.*
   – Define element types and options.
   – Define real constant for the element types (if the element type(s) require real constants).
   – Define material properties.
   – Create the model:

     - Build solid model (using either top-down or bottom-up approach).
     - Define meshing controls.
     - Create the mesh.

   – Exit the Preprocessor.

2. Boundary/Initial Conditions and Solution

   – Enter Solution Processor.
   – Define analysis type and analysis options.
   – Specify boundary/initial conditions:

     - Degree of freedom constraints.
     - Nodal force loads.
     - Surface loads.
     - Body loads.
     - Inertia loads.
     - Initial conditions (if the analysis type is transient).

   – Save database (this step is not required but is recommended).

The online version of this book (doi: 10.1007/978-1-4939-1007-6_5) contains supplementary material, which is available to authorized users

E. Madenci, I. Guven, *The Finite Element Method and Applications in Engineering Using ANSYS®*, DOI 10.1007/978-1-4899-7550-8_5

– Initiate solution.
– Exit the *Solution Processor.*

3. Review Results

– Enter the appropriate Postprocessor (*General Postprocessor* or *Time History Postprocessor*).
– Display results.
– List results.

The first step involves operations concerning the *ANSYS Preprocessor* and was covered in detail in Chap. 4. The operations pertaining to the solution and post-processing of the results are discussed in detail in this chapter. At the end, specific steps are demonstrated by considering a one-dimensional transient heat transfer problem.

5.2 Solution

After preprocessing, the model generation, including meshing, is complete. The user is ready to begin the solution phase of the ANSYS session. First, the analysis type is specified from among the three main types:

- Static.
- Transient (time-dependent).
- Submodeling and substructuring (discussed in Sects. 11.3 and 11.4).

If the problem under consideration falls into the *Structural Analysis* discipline, then there are additional analysis types, such as *modal*, *harmonic*, *spectrum*, and *eigenvalue buckling*. There are two main deciding factors in choosing the analysis type:

Loading conditions: If the boundary conditions change as a function of time ***or*** there are initial conditions, then the analysis type is *Transient*. However, if the analysis discipline is structural ***and*** if the loading is a sinusoidal function of time, then the analysis type is *Harmonic*. Similarly, if the loading is a seismic spectrum, the analysis type is *Spectrum*.

Results of interest: If the analysis discipline is structural ***and*** if the results of interest are the natural structural frequencies, then the analysis type is *Modal*. Similarly, if the interest is in determining the load at which the structure looses stability (buckles), then the analysis type is *eigenvalue buckling*.

The analysis type is specified by using the following menu path:

Main Menu > Solution > Analysis Type > New Analysis

This brings up the dialog box shown in Fig. 5.1. The user selects a particular analysis type by clicking on the corresponding radio-button and clicks ***OK***. The common

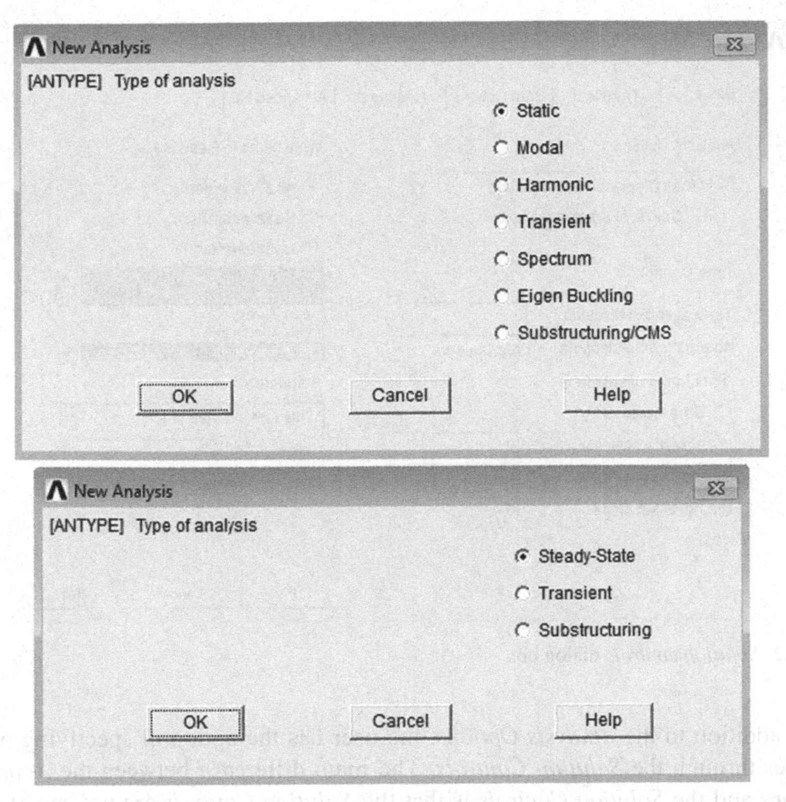

Fig. 5.1 Dialog boxes for selecting the type of analysis for structural (*top*) and thermal (*bottom*) disciplines

solution operations used in almost every ANSYS session are discussed in the following subsections.

5.2.1 Analysis Options/Solution Controls

ANSYS allows the user to select certain options during the solution phase. They are specified through either *Analysis Options* or *Solution Controls*. The *Analysis Options*, specific to the *Analysis Type*, permit the user to select the method of solution and related details; this step requires familiarity with the *Analysis Type*. *Analysis Options* can be specified by the following the menu path:

Main Menu > Solution > Analysis Type > Analysis Options

Because the *Analysis Options* are specific to the particular problems under consideration, related discussions are covered in various example problems throughout this book.

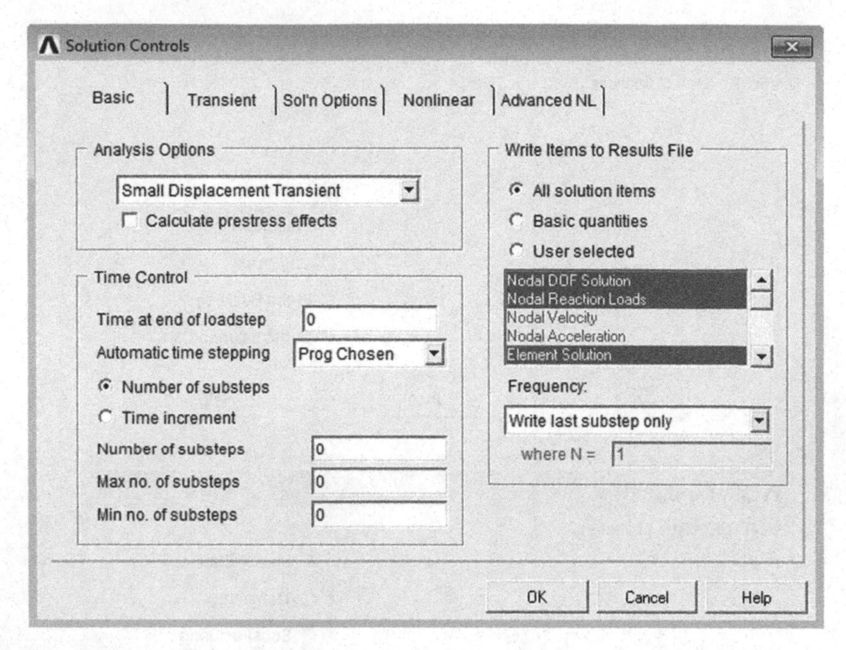

Fig. 5.2 *Solution controls* dialog box

In addition to the *Analysis Options*, the user has the option of specifying preferences through the *Solution Controls*. The main difference between the *Analysis Options* and the *Solution Controls* is that the *Solution Controls* are not specific to the *Analysis Type*. The same set of options in the *Solution Controls* can be used in a structural or thermal analysis. The *Solution Controls* dialog box can be activated by using the following menu path:

Main Menu > Solution > Analysis Type > Sol'n Controls

As shown in Fig. 5.2, the *Solution Controls* dialog box has five different tabs:

Basic: Involves selection of options specific to the analysis type, time-domain-related parameters, and results items to be written to the *Results File*.

Transient: This option provides control over the way the loading is applied (stepped or ramped over time), the damping coefficients, and the time integration parameters.

Sol'n Options: The equation solver is chosen under this option. Also, if the current analysis is a *Restart* from a previous analysis or is intended to be "restarted" later, this option controls the number of restart files to write and the frequency at which they are written.

Nonlinear: Involves nonlinear options, specification of the maximum number of equilibrium iterations, and the limits on physical values used to perform bisection when performing nonlinear structural analyses, such as plasticity deformation, creep, etc.

Advanced NL: This option is used to specify what the software should do when convergence is not achieved during a nonlinear analysis.

Within the *Solution Controls*, all of the options have default values that the user is **not** required to specify while performing the analysis. However, if the analysis fails to produce convergence, manual specification of these options may improve the chances of convergence. As part of the example problems solved throughout this book, the *Solution Controls* items are manually specified for several problems in Chaps. 8 and 9.

5.2.2 Boundary Conditions

In a well-posed mathematical problem, the conditions along the entire boundary must be known. These conditions are referred to as the *boundary conditions*, and they can be specified in three different ways:

Type I: Specification of the primary variable (degree of freedom).
Type II: Specification of variables related to the derivative of the primary variable.
Type III: Specification of a linear combination of the primary variable and its derivative.

In a *Structural* problem, the primary variables are the displacement components (see Sect. 2.2.1.3). When *Type I* boundary conditions are used, the displacement constraints are specified along a segment of the boundary. If tractions are specified along the boundary, the boundary conditions fall under *Type II* because the tractions are related to the derivatives of the displacement components. A special case of the traction boundary conditions is the point load (also called the *Force/Moment* load). When the structure is subjected to tractions over a rather small area of the boundary, it is reasonable to idealize this condition as a concentrated load applied at a point. While conducting an analysis with **BEAM** or **SHELL** element types, moment loads can also be applied. Displacements, pressures (normal tractions), forces, and moments can be specified using the following menu paths:

Main Menu > Solution > Define Loads > Apply > Structural > Displacement
Main Menu > Solution > Define Loads > Apply > Structural > Force/Moment
Main Menu > Solution > Define Loads > Apply > Structural > Pressure

In a *Thermal* problem, the temperature is the primary variable. Similar to *Structural* problems, *Type I* boundary conditions correspond to the specification of the primary variable, i.e., temperature over a portion of the boundary. Specified heat flux conditions fall under *Type II* boundary conditions. Finally, convective conditions correspond to *Type III* boundary conditions. Temperature, heat flux, and convective conditions along the boundaries can be specified using the following menu paths:

Main Menu > Solution > Define Loads > Apply > Thermal > Temperature
Main Menu > Solution > Define Loads > Apply > Thermal > Heat Flux
Main Menu > Solution > Define Loads > Apply > Thermal > Heat Flow
Main Menu > Solution > Define Loads > Apply > Thermal > Convection

Each of the boundary conditions discussed in this section can be applied on *Nodes* or on appropriate solid model entities such as *Keypoints*, *Lines*, or *Areas*. If they are applied on the solid model entities, ANSYS transfers them to the nodes when the solution is initiated. Although the paths for specification of boundary conditions are shown under the *Solution Processor*, it is possible to apply them under the *Preprocessor*.

5.2.3 Initial Conditions

A well-posed transient problem requires the specification of initial conditions. For *Structural* problems, initial conditions may involve components of displacement, rotation, velocity, or acceleration. In a *Thermal* problem, initial conditions are typically the temperature distribution within the domain. Initial conditions can be specified only when the *Analysis Type* is selected as *Transient*; if the *Analysis Type* is selected as *Static*, the specification of initial conditions does not appear as an option in the menus. Initial conditions can be specified using the following menu paths:

Main Menu > Solution > Define Loads > Apply > Initial Condit'n

5.2.4 Body Loads

Body loads can be generated internally or externally as the result of a physical field acting on the body. They act within the domain expressed volumetrically. Gravity, inertia loads, and temperature change represent body loads in a *Structural* problem. They can be specified using one of the following menu paths:

Main Menu > Solution > Define Loads > Apply > Structural > Temperature
Main Menu > Solution > Define Loads > Apply > Structural > Inertia > Angular Velocity
Main Menu > Solution > Define Loads > Apply > Structural > Inertia > Angular Accel
Main Menu > Solution > Define Loads > Apply > Structural > Inertia > Coriolis Effects
Main Menu > Solution > Define Loads > Apply > Structural > Inertia > Gravity

Heat generation within the domain is also represented as a body load for *Thermal* problems and can be specified using the following menu path:

Main Menu > Solution > Define Loads > Apply > Thermal > Heat Generat

5.2.5 *Solution in Single and Multiple Load Steps*

After completing the finite element mesh and specifying the loading conditions (boundary, initial, and body loads), the solution can be initiated using the following menu path:

Main Menu > Solution > Solve > Current LS

However, there are cases in which the loads are time-dependent, and the solution is achieved in multiple steps. Different load steps must be used if the loading on the structure changes abruptly. The use of load steps also becomes necessary if the response of the structure at specific points in time is desired. ANSYS accommodates the application of time-dependent loads through the use of multiple *Load Steps*. Time-dependent loading is commonly encountered in analyses involving the determination of dynamic response and viscoplastic- and creep-type material behaviors. Simulation of manufacturing processes also involves time-dependent thermal loading. Figure 5.3 illustrates different profiles of impact loading as a function of time. The solid lines designate the actual loading while the dashed lines denote the loading profiles as specified in ANSYS. The solid circles indicate the times at which a load step starts or ends. As observed in Fig. 5.3, ANSYS permits the user to specify either step or ramped loading. In all of the cases, the last load step is necessary in order to capture the response of the structure at times after the load is removed. The multiple load steps are also necessary in modeling a viscoplastic material subjected to thermal cycling, as shown in Fig. 5.4.

The following steps are used in order to use multiple load step solution method:

1. **Apply the initial conditions as explained in Sect. 5.2.3.**
2. **Apply the boundary conditions appropriate for the first load step.**
3. **Specify time-related parameters**: This is performed by using the following menu path:

Main Menu > Solution > Load Step Opts > Time/Frequenc > Time—Time Step

This brings up the *Time and Time Step Options* dialog box (Fig. 5.5) Enter the *time at end of load step* (***TIME***) and the *time step size* (***DELTIM***), which is optional. Choose between whether the loads are applied in a stepped or ramped manner (***KBC***). If the *Automatic Time Stepping* is ***OFF***, then the user must specify the *time step size*. If the *time step size* is not specified and the *Automatic Time Stepping* is set as ***Prog Chosen*** (stands for program chosen), then ANSYS turns the *Automatic Time Stepping* ***ON***. If the *time step size* is specified and the *Automatic Time Stepping* is ***ON***, ANSYS starts the solution with the specified *time step size* and modifies it based on the convergence.

4. **Write Load Step file:** Load steps are written to *Load Step Files* by using the following menu path:

Main Menu >Solution > Load Step Opts > Write LS File

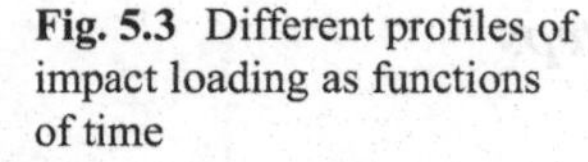

Fig. 5.3 Different profiles of impact loading as functions of time

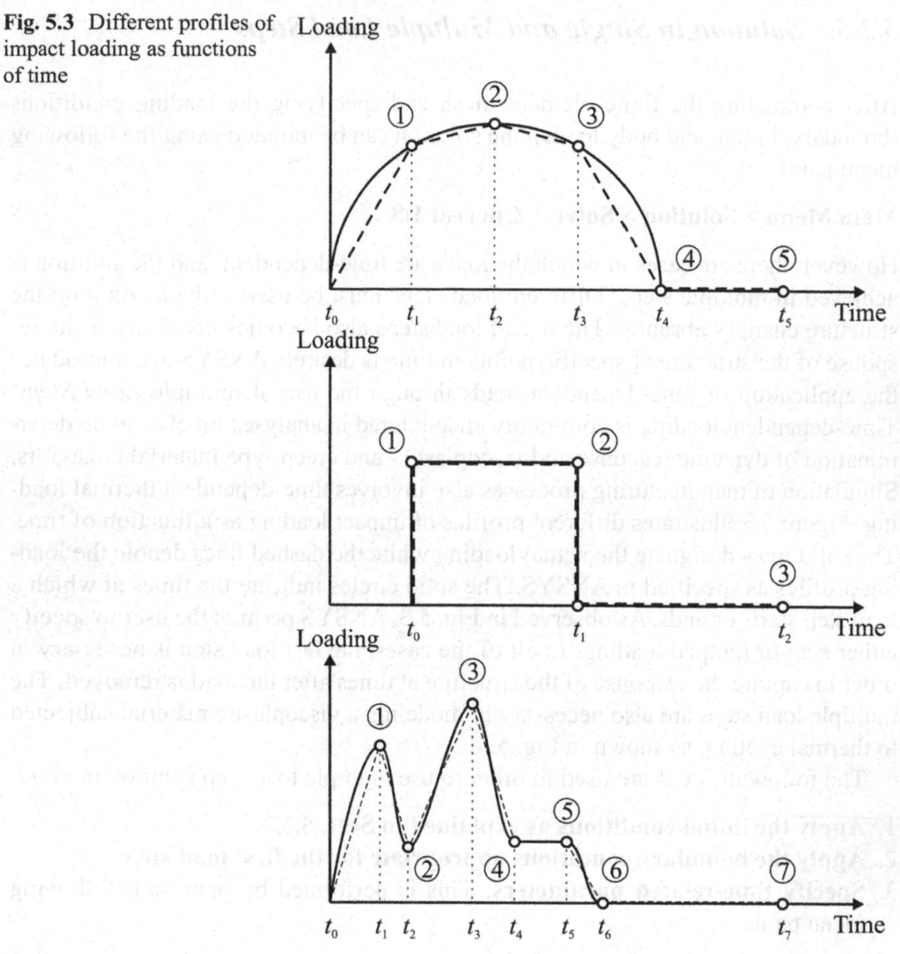

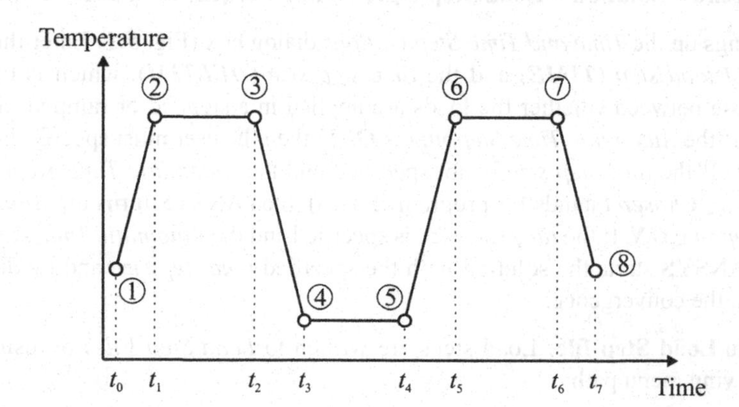

Fig. 5.4 Cyclic thermal loading

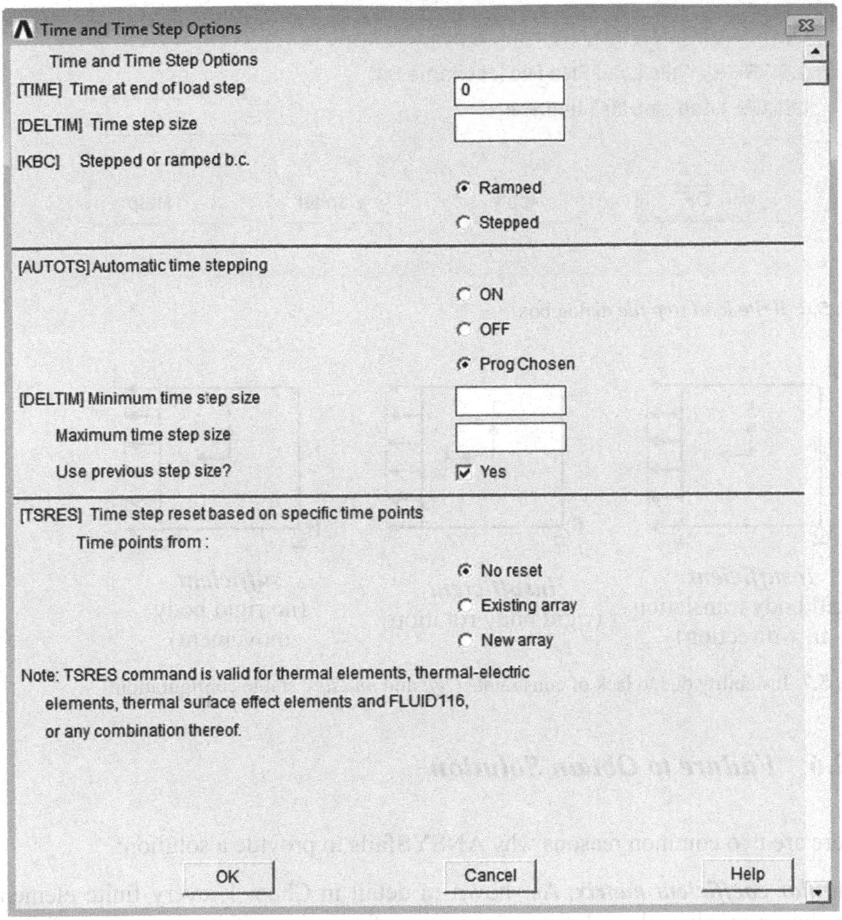

Time and Time Step Options
Time and Time Step Options
[TIME] Time at end of load step 0
[DELTIM] Time step size
[KBC] Stepped or ramped b.c.
Ramped
Stepped
[AUTOTS] Automatic time stepping
ON
OFF
Prog Chosen
[DELTIM] Minimum time step size
Maximum time step size
Use previous step size? Yes
[TSRES] Time step reset based on specific time points
Time points from :
No reset
Existing array
New array
Note: TSRES command is valid for thermal elements, thermal-electric elements, thermal surface effect elements and FLUID116, or any combination thereof.
OK Cancel Help

Fig. 5.5 *Time and time step options* dialog box

This brings up the *Write Load Step* File dialog box (Fig. 5.6). Enter the load step file number (***LSNUM***) and hit ***OK***. This file is stored in the *Working Directory* and contains all the solution options, the time, and time-related parameters, as well as the boundary conditions.

5. **Repeat steps 2–4 for the remainder of the load steps.**
6. **Initiate solution from *Load Step files*:** Once all of the *load step files* are written, the solution is initiated by using the following menu path:

Main Menu > Solution > Solve > From LS Files

which brings up the *Solve Load Step Files* dialog box. Enter the starting and ending load step file numbers (***LSMIN*** and ***LSMAX***) and hit ***OK***.

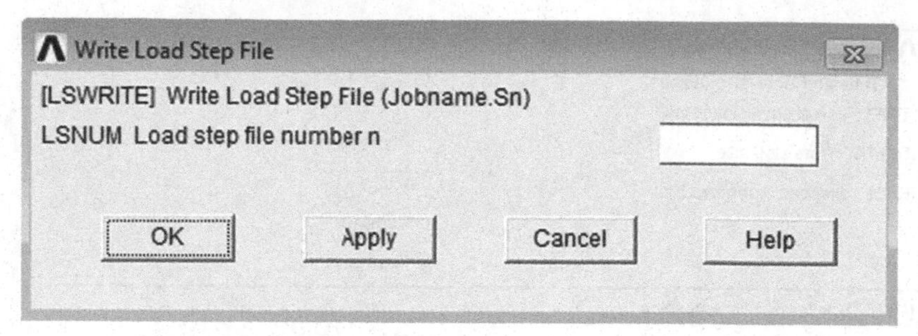

Fig. 5.6 *Write load step file* dialog box

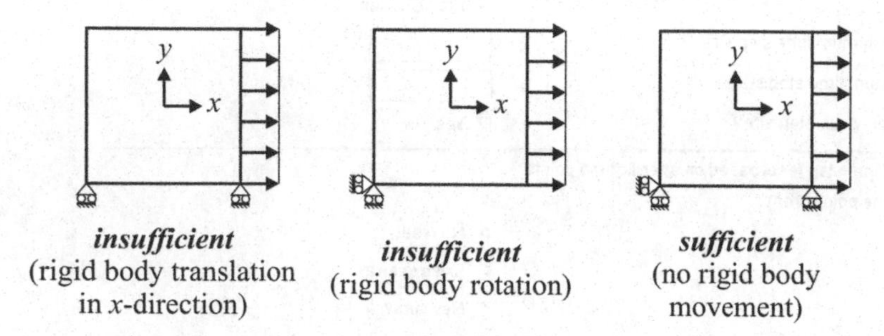

Fig. 5.7 Instability due to lack of constraints (*left* and *middle*); stable configuration

5.2.6 Failure to Obtain Solution

There are two common reasons why ANSYSfails to provide a solution:

Singular coefficient matrix: As shown in detail in Chap. 1, every finite element solution involves the solution of a system of equations with a known coefficient matrix (stiffness), an unknown degree of freedom vector, and a known right-hand-side (force) vector. If the coefficient matrix is singular, the solution fails. The most common reasons why the coefficient matrix becomes singular are as follows:

1. Instability in the structure due to lack of constraints in static structural analyses. This leads to rigid-body translations and rotations, which makes the stiffness matrix singular. As an example of this phenomenon, Fig. 5.7 shows three distinct constraint configurations applied on the same 2-D square structure subjected to a distributed tensile load in the x-direction. The first configuration involves two constraints, both suppressing displacements in the y-direction, along the bottom surface of the structure. Because there are no constraints suppressing displacements in the x-direction, the structure is free to move in the x-direction under the applied load, thus leading to a singular stiffness matrix. In the second configuration, displacements in both the x- and y-directions are suppressed at the

same point (bottom left corner). Although this configuration prevents rigid-body translations, it fails to prevent rigid-body rotation around the corner node where displacement constraints are applied, causing the stiffness matrix to become singular. Finally, in the third configuration, two corners are constrained in the y-direction, with one of them also constrained in the x-direction. This is a stable configuration, preventing all possible rigid-body movements, leading to a nonsingular stiffness matrix and thus a successful unique solution.

- Material properties that are physically impossible may make the coefficient matrix singular. Examples include zero or negative Young's modulus, thermal conductivity, density, or specific heat.
- There are structural elements within the ANSYS element library that carry loads only along their line of direction (**SPAR** elements simulating truss structures). Stability concerns of *Statics* apply to structures made up of these elements, and the user must make sure that the structure is stable.

2. ***Failed convergence***: In finite element analyses, problems involving nonlinearity are solved through iterations. As described in Chap. 2, these nonlinearities arise through the material behavior (plasticity, creep, viscoelasticity, viscoplasticity, etc.) or geometric configuration (large deformations) of the structure. The "correct" solution is approached in small *steps*, referred to as convergence iterations. If the problem is time-dependent, then the small *steps* are taken in the time domain. If the problem is not dependent on time (e.g., plasticity), these small steps are taken in the application of the loads. At the end of each iteration, ANSYS checks whether the solution satisfies a *convergence criterion* "built-in" for different analysis types. If the criterion is not satisfied, the last step is repeated with a smaller step size. This is repeated until the convergence criterion is satisfied. However, there are limits on the number of convergence iterations and, if a converged solution is not achieved within those limits, ANSYS terminates the solution process. Because each nonlinear analysis type is different, there is no straightforward answer as to what to do to improve the chances of a successful convergence. However, several nonlinear problems are considered in Chap. 10 that may give the reader some ideas on convergence considerations.

5.3 Postprocessing

After a solution is obtained in an ANSYS session, the user can review the results in either the *General Postprocessor* or the *Time History Postprocessor*. If the problem is static (or steady state), then the *General Postprocessor* is the only postprocessor where the results can be reviewed. However, if the problem is dependent on time (transient), both processors are useful for distinctly different tasks. The postprocessors and common postprocessing operations are discussed briefly in the following (5.3.1–5.3.6).

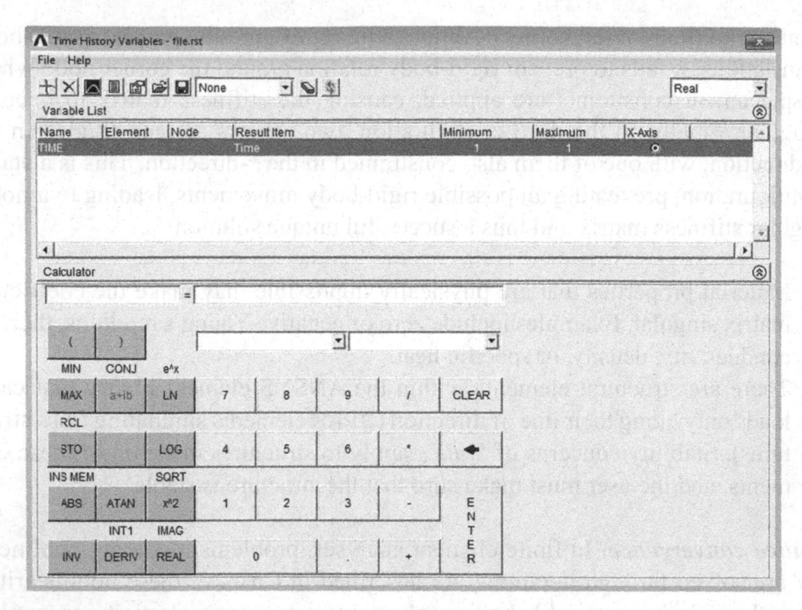

Fig. 5.8 *Time history variables* dialog box

5.3.1 General Postprocessor

In the *General Postprocessor*, the results of a solution at a specific time (if the problem is time dependent) are reviewed. Available options for review include graphical displays and a listing of results. It is also possible to perform sorting and mathematical operations on the results.

5.3.2 Time History Postprocessor

When the problem under consideration is time dependent, the time variation of the results at specific locations (nodes) are reviewed under the *Time History Postprocessor*. Upon entering this postprocessor, the *Time History Variables* dialog box appears (Fig. 5.8). This dialog box has three distinct areas: ***Toolbar***, ***Variables***, and ***Calculator***. The first four buttons (from the left) in the ***Toolbar*** are the most commonly used ones:

Add Data Button: This button is used to define new variables, such as displacements, temperatures, etc., at specific nodes.
Delete Data Button: Used for deleting defined variables.
Graph Data: Using this button, the user can plot the time variation of variables.
List Data: Similar to plotting, this button is used for listing the results as functions of time.

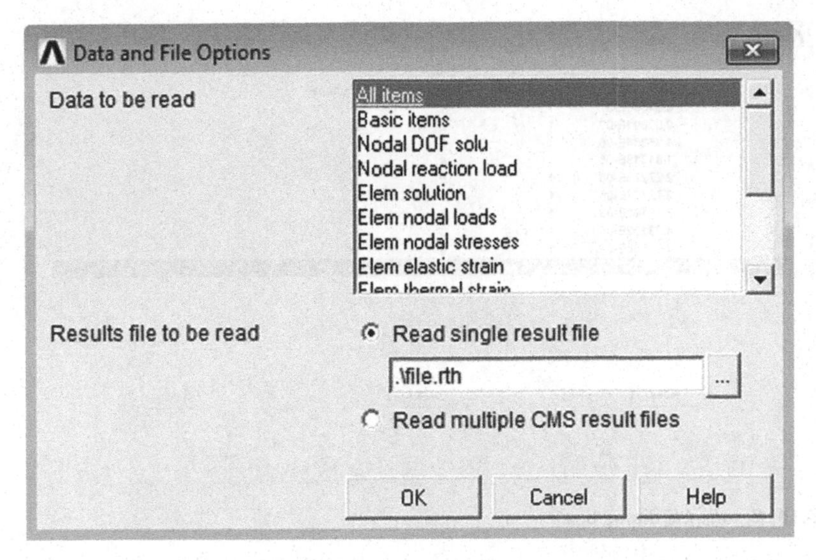

Fig. 5.9 *Data and file options* dialog box

As the new variables are defined, they appear in the ***Variables*** area. By default, ***TIME*** is the first variable and cannot be removed. In addition to the name of the variable, the ***Variables*** area includes useful information about the variable, such as its element or node number, what result item it corresponds to, and the range of its values.

The last item (located at the right-most side) is the ***X-Axis*** button, which enables the user to select which variable to display on the x-axis in the graphical representations.

An example problem (time-dependent heat transfer) demonstrating the use of the *Time History Postprocessor* is given in Sect. 5.4.

5.3.3 Read Results

The results, obtained through the *Solution Processor*, are saved in results files (***jobname.rst*** for structural, ***jobname.rth*** for thermal, and ***jobname.rfl*** for fluids problems), which are stored in the working directory. In order to review the results, the user needs to guide ANSYS so that the correct results file is selected. This is done by using the following menu path:

Main Menu > General Postproc > Data & File Opts

This brings up the *Data and File Options* dialog box (Fig. 5.9). The results file is selected by clicking on the browse button (button with three dots). After selecting the correct file, click on ***OK***.

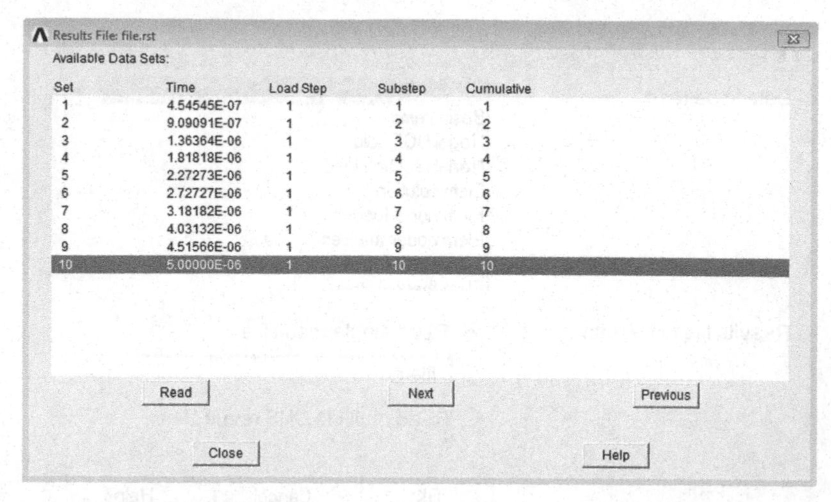

Fig. 5.10 *Results file* dialog box

If the solution does ***not*** involve multiple substeps and load steps, then there is only one results "set" the user can review. However, when the solution involves multiple substeps and load steps, there are many results sets and the user should select the correct (intended) one. The results sets can be selected using the following options:

First Set: Results related to the first available set are read into the database using the following menu path:

Main Menu > General Postproc > Read Results > First Set

Next Set: Results related to the set available immediately after the current set are read into the database using the following menu path:

Main Menu > General Postproc > Read Results > Next Set

Previous Set: Results related to the set available immediately before the current set are read into the database using the following menu path:

Main Menu > General Postproc > Read Results > Previous Set

Last Set: Results related to the last available set are read into the database using the following menu path:

Main Menu > General Postproc > Read Results > Last Set

Read By Picking: The following menu path is used for this option:

Main Menu > General Postproc > Read Results > By Pick

which brings up a dialog box (Fig. 5.10) listing the available results sets. The user selects the desired results set and clicks on ***Read*** and ***Close*** for the results to be read into the database.

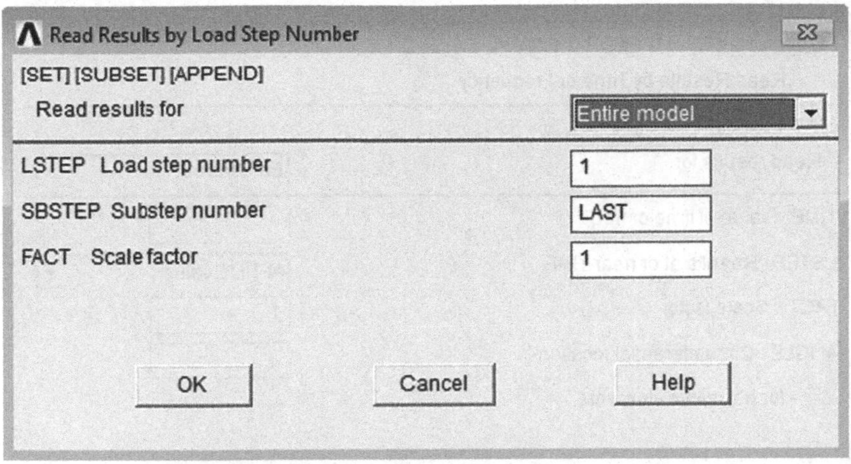

Fig. 5.11 *Read results by load step number* dialog box

Read By Load Step Number: Results related to a specific load step and substep are read into the database using the following menu path:

Main Menu > General Postproc > Read Results > By Load Step

which brings up the *Read Results by Load Step Number* dialog box (Fig. 5.11) in which the user specifies the load step number (***LSTEP***) and substep (***SBSTEP***) number within that load step and clicks on ***OK*** for the results to be read into the database.

Read by Time: Results related to a specific time (or frequency) value are read into the database using the following menu path:

Main Menu > General Postproc > Read Results > By Time/Freq

which brings up the *Read Results by Time or Frequency* dialog box (Fig. 5.12) in which the user specifies the value of time (or frequency) (***TIME***) and clicks on ***OK*** for the results to be read into the database.

5.3.4 Plot Results

After the desired results set is read into the database, the result quantities can be reviewed through graphics displays. The types of graphics displays include deformed shapes (structural analysis), contour plots, vector displays (thermal), and path plots.

In structural analyses, the deformed shape resulting from the applied loads and boundary conditions is displayed using the following menu path:

Main Menu > General Postproc > Plot Results > Deformed Shape

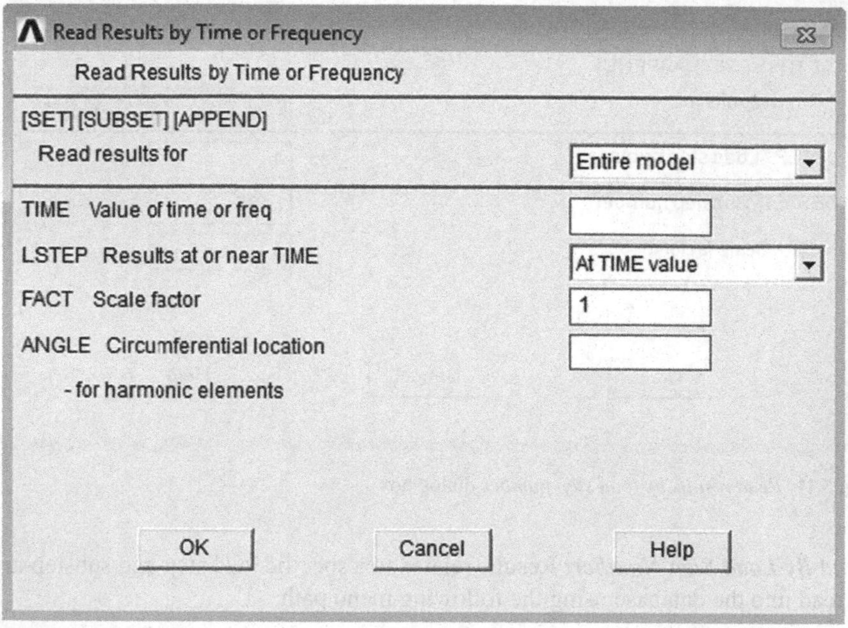

Fig. 5.12 *Read results by time or frequency* dialog box

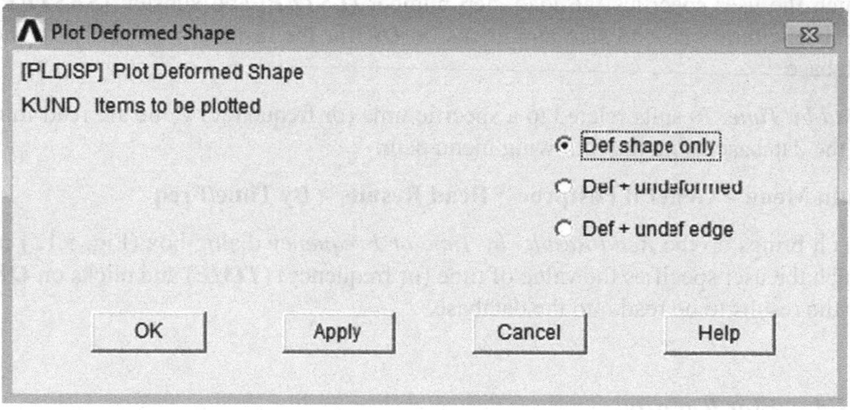

Fig. 5.13 *Plot deformed shape* dialog box

which brings up the *Plot Deformed Shape* dialog box (Fig. 5.13). The user is offered three distinct display modes:

- Display deformed shape only.
- Display deformed and undeformed shapes together.
- Display deformed shape with the outer boundary (edge) of the undeformed shape.

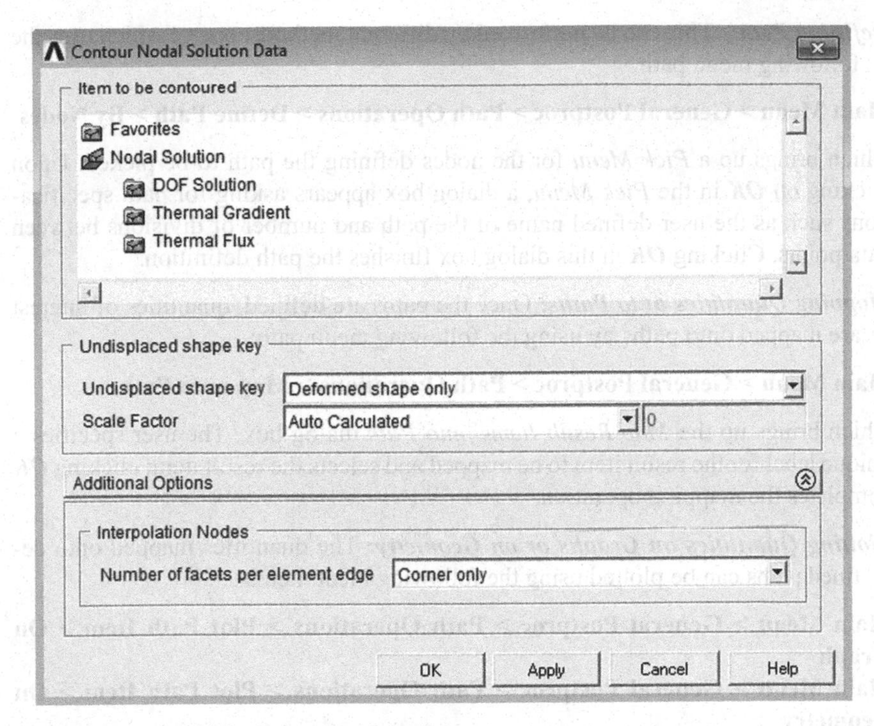

Fig. 5.14 *Contour nodal solution data* dialog box

After the user makes a choice and clicks on ***OK***, the deformed shape appears in the *Graphics Window*.

Contour plots are obtained using one of the following menu paths:

Main Menu > General Postproc > Plot Results > Contour Plot > Nodal Solu
Main Menu > General Postproc > Plot Results > Contour Plot > Element Solu

which brings up the *Contour Nodal* (*Element*) *Solution Data* dialog box (Fig. 5.14). In this dialog box, both the degree of freedom (DOF) solution (displacements, temperatures, etc.) and derived quantities (stresses, strains, fluxes, etc.) are available for plotting. Once the user makes the selection, upon clicking ***OK***, the contour plot appears in the *Graphics Window*.

Vector plots are obtained using the following menu path:

Main Menu > General Postproc > Plot Results > Vector Plot >Predefined

which brings up the *Vector Plot of Predefined Vectors* dialog box. Similar to the contour plots, this dialog box has two fields identifying the quantity to be plotted. Once the user makes a selection, upon clicking ***OK***, the vector plot appears in the *Graphics Window*.

In the ANSYS *General Postprocessor*, it is possible to obtain line plots along a path. Utilizing path plots involves:

Defining Paths: This can be performed by different methods, one of which uses the following menu path:

Main Menu > General Postproc > Path Operations > Define Path > By Nodes

which brings up a *Pick Menu* for the nodes defining the path to be picked. Upon clicking on ***OK*** in the *Pick Menu*, a dialog box appears asking for path specifications such as the user-defined name of the path and number of divisions between data points. Clicking ***OK*** in this dialog box finishes the path definition.

Mapping Quantities onto Paths: Once the paths are defined, quantities of interest are mapped onto paths by using the following menu path:

Main Menu > General Postproc > Path Operations > Map onto Path

which brings up the *Map Result Items onto Path* dialog box. The user specifies a unique label for the result item to be mapped and selects the result item; clicking ***OK*** completes the mapping operation.

Plotting Quantities on Graphs or on Geometry: The quantities mapped onto defined paths can be plotted using the following menu items:

Main Menu > General Postproc > Path Operations > Plot Path Item > On Graph
Main Menu > General Postproc > Path Operations > Plot Path Item > On Geometry

which brings up the *Plot of Path Items on Graph* (*Geometry*) dialog box. The user selects the path item to be plotted from the list of defined path items and clicks on ***OK***; the plot appears in the *Graphics Window*.

The operations related to path plots are demonstrated through an example problem in Sect. 5.4.

5.3.5 Element Tables

In ANSYS, each element type possesses numerous output quantities available upon completion of the solution. Although several of these quantities are offered by their names under the postprocessors, some are not directly accessible, and the user needs to take additional steps in order to access them. One important purpose of using *Element Tables* is to access these result items. Another important role of *Element Tables* is that they enable the user to perform arithmetic operations involving several result items. Element tables are defined by using the following menu path:

Main Menu > General Postproc > Element Table > Define Table

which brings up the *Element Table Data* dialog box (Fig. 5.15). In order to add new items to the element table, the user needs to click on the ***Add*** button, which brings up the *Define Additional Element Table Items* dialog box (Fig. 5.16). After selecting

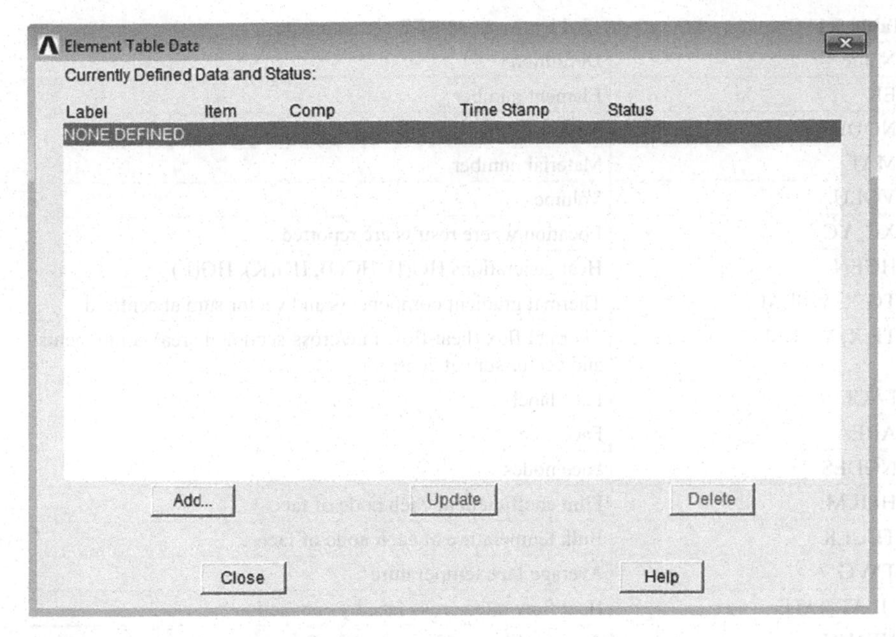

Fig. 5.15 *Element table data* dialog box

Define Additional Element Table Items
[ETABLE] Define Additional Element Table Items
Lab User label for item
Item,Comp Results data item
DOF solution
Flux & gradient
Nodal force data
Energy
Error estimation
Geometry
By sequence num
Temperature TEMP
Temperature TEMP
(For "By sequence num", enter sequence no. in Selection box. See Table 4.xx-3 in Elements Manual for seq. numbers.)
OK Apply Cancel Help

Fig. 5.16 *Define additional element table items* dialog box

the result quantity and specifying a user label for it, clicking on ***OK*** completes the element table item definition.

In order to explain the usage of element tables, an example based on the **PLANE55** element type is considered. In Table 5.1, the output quantities provided by the element type **PLANE55** are given. Table 5.2 lists the results quantities accessible through the element tables. For example, the heat flow rate per unit area across the element faces caused by input heat flux, denoted by ***HFLXAVG*** in Tables 5.1

Table 5.1 Output quantities provided by the **PLANE55** element type

Name	Definition
EL	Element number
NODES	Nodes: I, J, K, L
MAT	Material number
VOLU	Volume
XC, YC	Location where results are reported
HGEN	Heat generations HG(I), HG(J), HG(K), HG(L)
TG:X, Y, SUM	Thermal gradient components and vector sum at centroid
TF:X, Y, SUM	Thermal flux (heat flow rate/cross-sectional area) components and vector sum at centroid
FACE	Face label
AREA	Face area
NODES	Face nodes
HFILM	Film coefficient at each node of face
TBULK	Bulk temperature at each node of face
TAVG	Average face temperature
HEAT RATE	Heat flow rate across face by convection
HFAVG	Average film coefficient of the face
TBAVG	Average face bulk temperature
HFLXAVG	Heat flow rate per unit area across face caused by input heat flux
HEAT RATE/AREA	Heat flow rate per unit area across face by convection
HFLUX	Heat flux at each node of face

Table 5.2 Quantities obtained via the element table

Output quantity name	Element table input				
	Item	FC1	FC2	FC3	FC4
AREA	NMISC	1	7	13	19
HFAVG	NMISC	2	8	14	20
TAVG	NMISC	3	9	15	21
TBAVG	NMISC	4	10	16	22
HEAT RATE	NMISC	5	11	17	23
HFLXAVG	NMISC	6	12	18	24

and 5.2, is available for definition in the element tables. In Table 5.2, the matching item for this quantity is given as ***NMISC***, along with numbers 6, 12, 18, and 24 corresponding to different faces of the element. To store ***HFLXAVG*** at the 4th face of each element, in the *Define Additional Element Table Items* dialog box, ***By sequence num*** from the left list and ***NMISC*** from the right list must be selected. Entering 24 in the text field underneath the right list ensures that the ***HFLXAVG*** at the 4th face of each element will be stored in the element table.

Element tables are plotted and listed using the following menu paths:

Main Menu > General Postproc > Element Table > Plot Elem Table
Main Menu > General Postproc > Element Table > List Elem Table

It is also possible to perform arithmetic operations within each column or between the columns of the element table. Examples of such operations include: finding absolute values, finding the sum of each element table item, adding and multiplying element table items, etc.

5.3.6 List Results

Results of an ANSYS solution can be reviewed through lists. Although there are numerous different options for listing the results under postprocessors, only two of them are discussed in this section: nodal and element solutions. In order to list results computed at the nodes, the following menu path is used:

Main Menu > General Postproc > List Results > Nodal Solu

which brings up the *List Nodal Solution* dialog box. Once the user makes a selection as to what result quantities are to be reviewed and clicks on ***OK***, the list appearsw in a separate window. Similar to nodal solution listings, the element results are listed by using the following menu path:

Main Menu > General Postproc > List Results > Element Solu

The usage of this option is similar to the nodal solution lists.

5.4 Example: One-dimensional Transient Heat Transfer

Consider the one-dimensional transient heat transfer problem shown in Fig. 5.17. The problem is time dependent; therefore, in addition to thermal conductivity, the specific heat and the density of the material are taken into account. The governing equation for this problem is written as

$$\rho c \frac{\partial T}{\partial t} = \kappa \frac{\partial^2 T}{\partial x^2} \qquad 0 \le x \le l \tag{5.1}$$

with the boundary and initial conditions

$$\begin{aligned} T(x=0,t) &= T_a = 100 \\ T(x=l,t) &= T_b = 0 \\ T(x,t=0) &= f(x) = 0 \end{aligned} \tag{5.2}$$

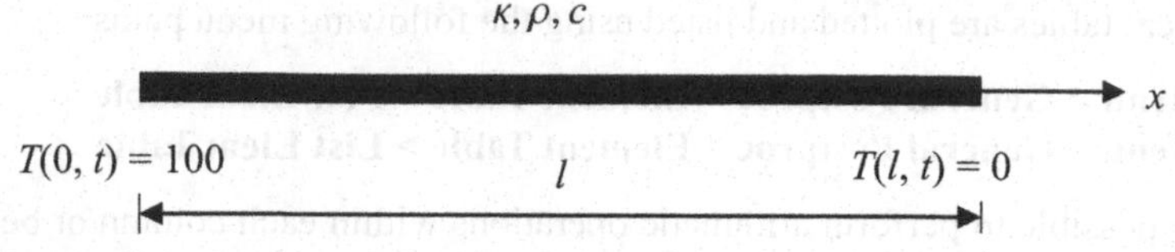

Fig. 5.17 One-dimensional transient heat transfer problem

The analytical solution for this problem is given by (Carslaw and Jaeger 1959, pp. 99–100):

$$T(x,t) = T_a + (T_b - T_a)\frac{x}{l} + \frac{2}{\pi}\sum_{n=1}^{\infty}\frac{T_b\cos n\pi - T_a}{n}\sin\frac{n\pi x}{l}e^{-\alpha n^2\pi^2 t/l^2}$$
$$+\frac{2}{l}\sum_{n=1}^{\infty}\sin\frac{n\pi x}{l}e^{-\alpha n^2\pi^2 t/l^2}\int_0^l f(x')\sin\frac{n\pi x'}{l}dx' \quad (5.3)$$

where $\alpha = \kappa/(\rho c)$. Substituting $T_a = 100$, $T_b = 0$, $f(x) = 0$, $l = 2$, $\kappa = 1$, $\rho = 10$, and $c = 3$, Eq. (5.3) yields

$$T(x,t) = 100 - 100\frac{x}{2} + \frac{2}{\pi}\sum_{n=1}^{\infty}\left(-\frac{100}{n}\right)\sin\frac{n\pi x}{2}e^{-n^2\pi^2 t/120} \quad (5.4)$$

When computing the exact solution using this equation, the number of terms, *n*, is truncated at 40 for satisfactory convergence.

Subjected to the boundary conditions indicated in Fig. 5.18 in the time range $0 \le t \le 5$, this problem is solved by using two-dimensional **PLANE55** elements in ANSYS.

The model is generated by using 4 element divisions along the vertical boundaries and 20 element divisions along the horizontal boundaries. The temperature variations along the midline at times t=0.1, 0.5, and 5 are obtained by ANSYS and the exact solution is plotted.

Model Generation

- Specify the ***jobname*** as ***1d_dif*** using the following menu path:

Utility Menu > File > Change Jobname

- In the dialog box, type ***1d_dif*** in the ***[/FILNAM] Enter new jobname*** text field; click on the check box for ***New log and error files*** to show ***Yes***; click on ***OK***.

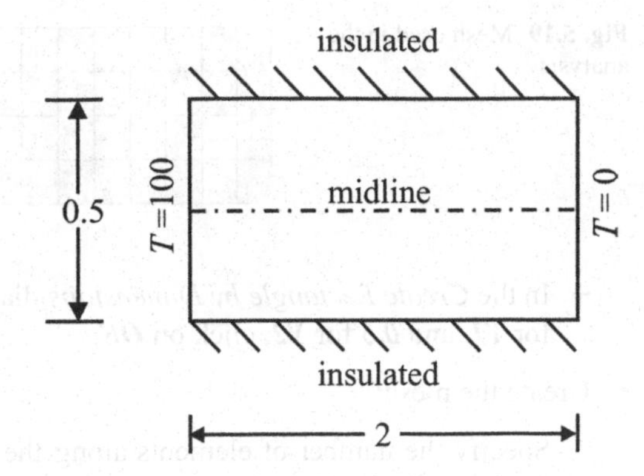

Fig. 5.18 Two-dimensional representation of the 1-D transient heat transfer problem and corresponding boundary conditions

- Define the element type (**ET** command) using the following menu path:

Main Menu > Preprocessor > Element Type > Add/Edit/Delete

  - Click on ***Add***.
  - Select ***Solid*** immediately below ***Thermal Mass*** on the left list and ***Quad 4node 55*** on the right list; click on ***OK***.
  - Click on ***Close***.

- Specify material properties (**MP** command) using the following menu path:

Main Menu > Preprocessor > Material Props > Material Models

  - In the *Define Material Model Behavior* dialog box, in the right window, successively left-click on ***Thermal***, ***Conductivity***, and, finally, ***Isotropic***, which brings up another dialog box.
  - ***Enter 1 for KXX; click on OK.***
  - In the *Define Material Model Behavior* dialog box, in the right window, left-click on ***Specific Heat***, which brings up another dialog box.
  - Enter ***3*** for ***C***; click on ***OK***.
  - In the *Define Material Model Behavior* dialog box, in the right window, left-click on ***Density***, which brings up another dialog box.
  - Enter ***10*** for ***DENS***; click on ***OK***.
  - Close the *Define Material Model Behavior* dialog box by using the following menu path:

Material > Exit

- Create the solid model:

  - Create a rectangular area using the following menu path:

Main Menu > Preprocessor > Modeling > Create > Areas > Rectangle > By Dimensions

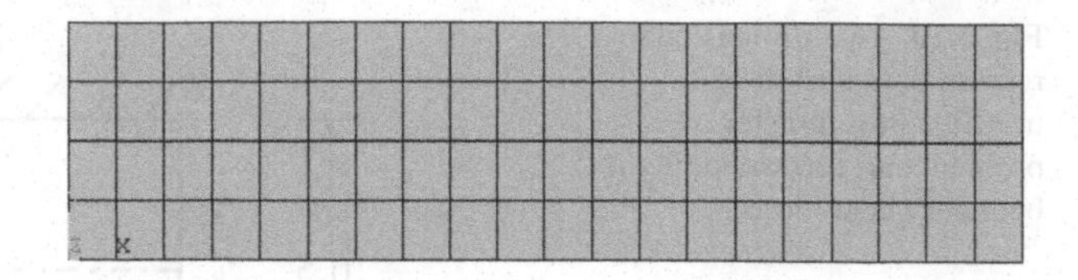

Fig. 5.19 Mesh used in the analysis

- In the *Create Rectangle by Dimensions* dialog box, type ***0*** for ***X1***, ***2*** for ***X2***, ***0*** for ***Y1***, and ***0.5*** for ***Y2***; click on ***OK***.

- Create the mesh:
  - Specify the number of elements along the vertical boundaries using the following menu path:

Main Menu > Preprocessor > Meshing > Size Cntrls > ManualSize > Lines > Picked Lines

- Pick the two vertical lines; click on OK.
- *Element Sizes on Lines* dialog box appears; type ***4*** in the text field corresponding to ***NDIV*** (the second text field), and uncheck the first check box; click on ***OK***.
- Specify the number of elements along the horizontal boundaries using the following menu path:

Main Menu > Preprocessor > Meshing > Size Cntrls > ManualSize > Lines > Picked Lines

- Pick the two horizontal lines; click on OK.
- *Element Sizes on Lines dialog box reappears; type 20 in the text field corresponding to NDIV (the second text field), and uncheck the first check box; click on OK.*
- Create the mesh using the following menu path:

Main Menu > Preprocessor > Meshing > Mesh > Areas > Mapped > 3 or 4 sided

- In the *Pick Menu*, click on ***Pick All***.
- Figure 5.19 shows the mesh.

- Save the model using the following menu path:

Utility Menu > File > Save as Jobname.db

The model is saved under the name ***1d_dif.db*** in the ***working directory***.

Solution

- Specify the analysis type as transient using the following menu path:

Main Menu > Solution > Analysis Type > New Analysis

- Click on ***Transient***; click on ***OK***.
- A new dialog box appears; click on ***OK***.

- Specify temperature boundary conditions along the vertical boundaries using the following menu path:

Main Menu > Solution > Define Loads > Apply > Thermal > Temperature > On Nodes

- Pick Menu appears; click on the Box radio-button and draw a rectangle around the nodes along the left vertical boundary; click on OK.
- *Apply TEMP on Nodes* dialog box appears; highlight ***TEMP***, enter ***100*** for ***VALUE Load TEMP value;*** click on ***Apply***.
- *Pick Menu* reappears; click on the ***Box*** radio-button and draw a rectangle around the nodes along the right vertical boundary; click on ***OK***.
- *Apply TEMP on Nodes* dialog box reappears; highlight ***TEMP***, enter ***0*** for ***VALUE Load TEMP value;*** click on ***OK***.

- Specify initial conditions within the domain using the following menu path:

Main Menu > Solution > Define Loads > Apply > Initial Condit'n > Define

- *Pick Menu* appears; click on ***Pick All***.
- *Define Initial Conditions* dialog box appears; select ***TEMP*** on the ***Lab*** pull-down menu; enter ***0*** in the ***VALUE*** text field; click on ***OK***.

- Specify time parameters using the following menu path:

Main Menu > Solution > Load Step Opts > Time/Frequenc > Time—Time Step

- Time and Time Step Options dialog box appears.
- As shown in Fig. 5.20, enter **5** in the **[TIME] Time at end of load step** text field and **5/100** in the **[DELTIM] Time step size** text field, and click on the **Stepped** radio-button for **[KBC]**; click on **OK**.

- Specify output controls using the following menu path:

Main Menu > Solution > Load Step Opts > Output Ctrls > DB/Results File

- Controls for Database and Results File Writing dialog box appears.
- As shown in Fig. 5.21, click on the **Every substep** radio-button for **FREQ File write frequency**; click on **OK**.

- Obtain solution using the following menu path:

Main Menu > Solution > Solve > Current LS

- Confirmation Window appears along with Status Report Window.
- Review status; if **OK**, close the Status Report Window and click on **OK** in the Confirmation Window.
- Wait until ANSYS responds with ***Solution is done!***

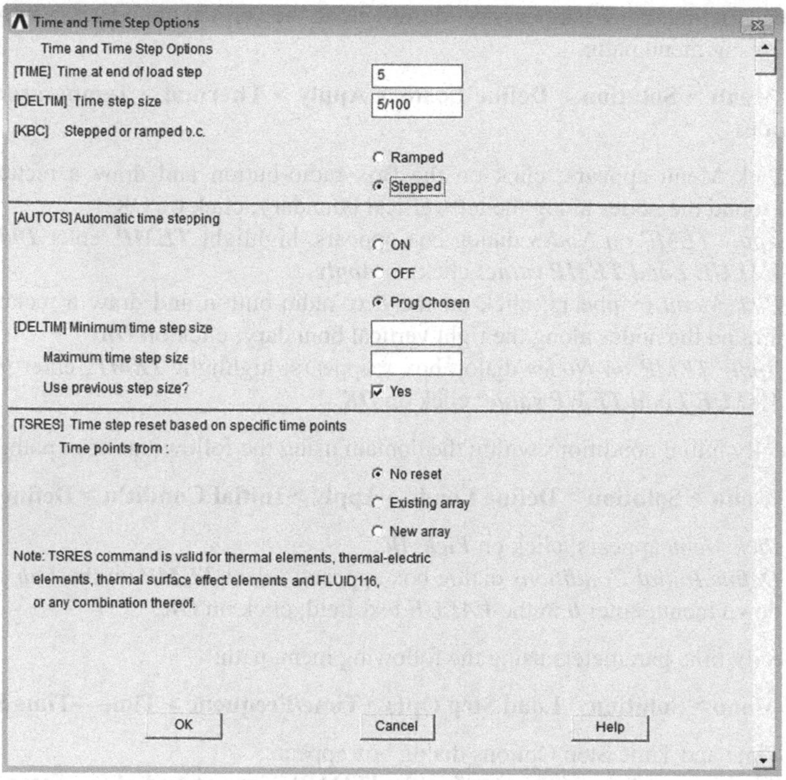

Time and Time Step Options

Time and Time Step Options
[TIME] Time at end of load step 5
[DELTIM] Time step size 5/100
[KBC] Stepped or ramped b.c.
Ramped
Stepped
[AUTOTS] Automatic time stepping
ON
OFF
Prog Chosen
[DELTIM] Minimum time step size
Maximum time step size
Use previous step size? Yes
[TSRES] Time step reset based on specific time points
Time points from :
No reset
Existing array
New array
Note: TSRES command is valid for thermal elements, thermal-electric elements, thermal surface effect elements and FLUID116, or any combination thereof.
OK Cancel Help

Fig. 5.20 *Time and time step options* dialog box used in the analysis

General Postprocessing

- Review results at the end of first substep using the following menu paths:

Main Menu > General Postproc > Read Results > First Set Main Menu > General Postproc > Plot Results > Contour Plot > Nodal Solu

  - Contour Nodal Solution Data dialog box appears; click on **DOF Solution** and **Nodal Temperature**; click on **OK**.
  - Fig. 5.22 shows the contour plot of the temperature distribution at $t = 0.05$ as it appears in the Graphics Window.

- Review results at the "next" substep using the following menu paths:

Main Menu > General Postproc > Read Results > Next SetMain Menu > General Postproc > Plot Results > Contour Plot > Nodal Solu

  - Contour Nodal Solution Data dialog box appears; click on **DOF Solution** and **Nodal Temperature**; click on **OK**.
  - Figure 5.23 shows the contour plot of temperature distribution at $t = 0.1$ as it appears in the Graphics Window.

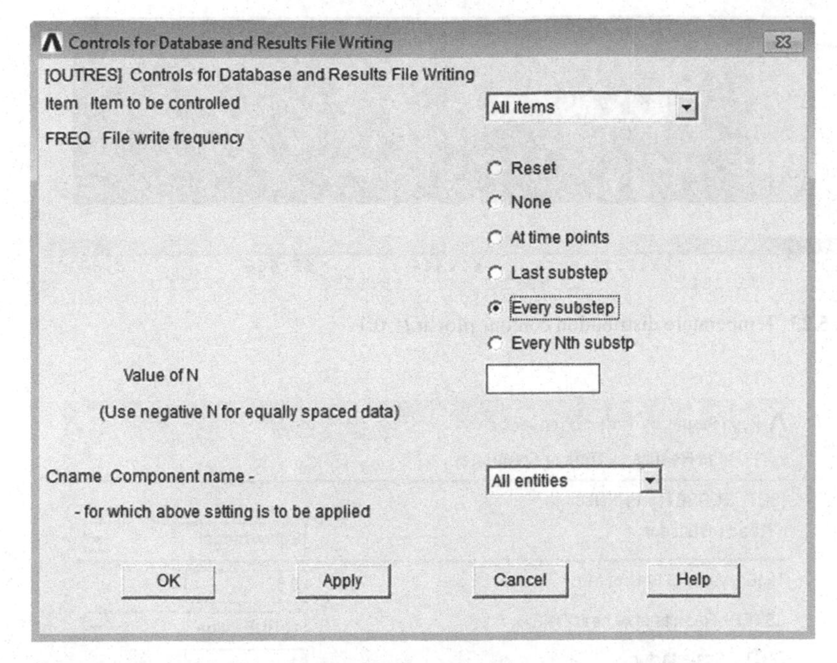

Fig. 5.21 *Output controls* dialog box used in the analysis

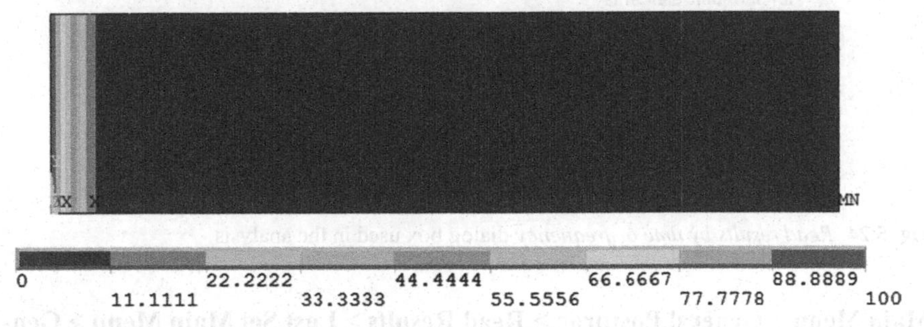

Fig. 5.22 Temperature distribution contour plot at $t = 0.05$

- Review results at $t = 0.5$ using the following menu path:

Main Menu > General Postproc > Read Results > By Time/Freq

  - As shown in Fig. 5.24, Read Results by Time or Frequency dialog box appears; enter 0.5 for TIME Value of time or freq; click on OK.
  - View the temperature contours, as shown in Fig. 5.25.

- Review results at the “last” substep using the following menu paths:

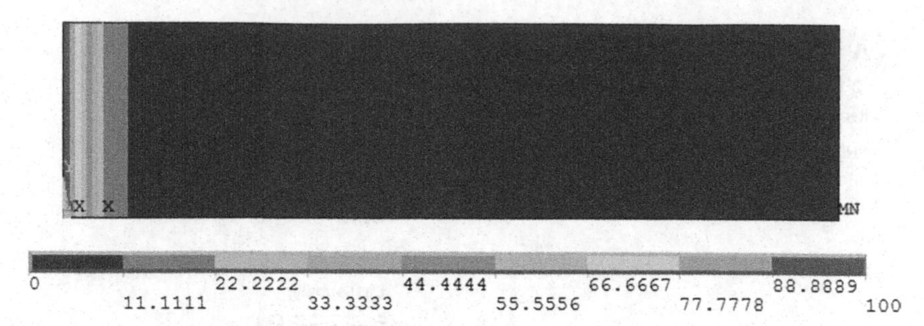

Fig. 5.23 Temperature distribution contour plot at $t=0.1$

Read Results by Time or Frequency

Read Results by Time or Frequency

[SET] [SUBSET] [APPEND]

Read results for — Entire model

TIME Value of time or freq — 0.5

LSTEP Results at or near TIME — At TIME value

FACT Scale factor — 1

ANGLE Circumferential location

- for harmonic elements

OK Cancel Help

Fig. 5.24 *Read results by time or frequency* dialog box used in the analysis

Main Menu > General Postproc > Read Results > Last Set Main Menu > General Postproc > Plot Results > Contour Plot > Nodal Solu

- Contour Nodal Solution Data dialog box appears; click on ***DOF Solution*** and ***Nodal Temperature***; click on **OK**.
- Figure 5.26 shows the contour plot of the temperature distribution at $t=5$ as it appears in the Graphics Window.

- Review thermal flux vector plot at $t=5$ using the following menu path:

Main Menu > General Postproc > Plot Results > Vector Plot > Predefined

- *Vector Plot of Predefined Vectors* dialog box appears; select ***Thermal flux TF*** on the right list and click on ***OK***.

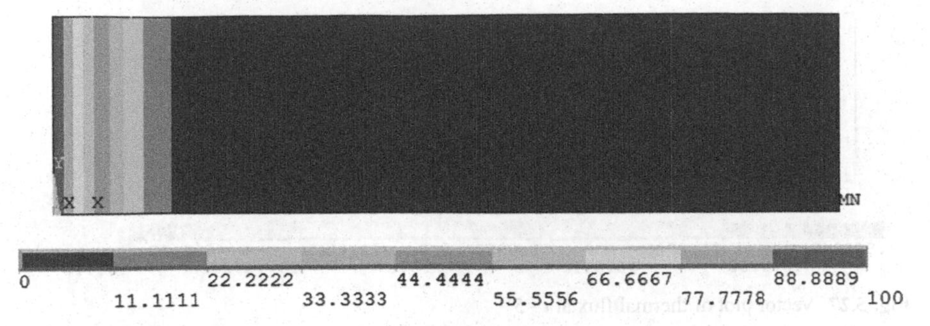

Fig. 5.25 Temperature distribution contour plot at $t=0.5$

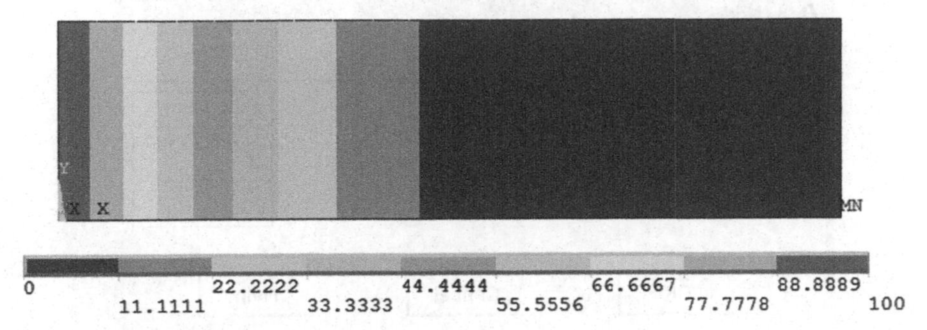

Fig. 5.26 Temperature distribution contour plot at $t=5$

- Figure 5.27 shows the vector plot of the thermal flux as it appears in the *Graphics Window*.

• Review results by path plots:

- Plot elements using the following menu path:

Utility Menu > Plot > Elements

- Turn node numbering on using the following menu path:

Utility Menu > PlotCtrls > Numbering

- *Plot Numbering Controls* dialog box appears; click on the check box for ***NODE Node numbers*** to show ***On***; click on ***OK***.
- Define the path using the following menu path:

Main Menu > General Postproc > Path Operations > Define Path > By Nodes

- ***Pick Menu appears; pick nodes 47 (x=0, y=0.25) and 24 (x=2, y=0.25); click on OK.***
- *Define Path By Nodes* dialog box appears, as shown in Fig. 5.28; enter a unique name (e.g., ***y025***) identifying the path; click on ***OK***.

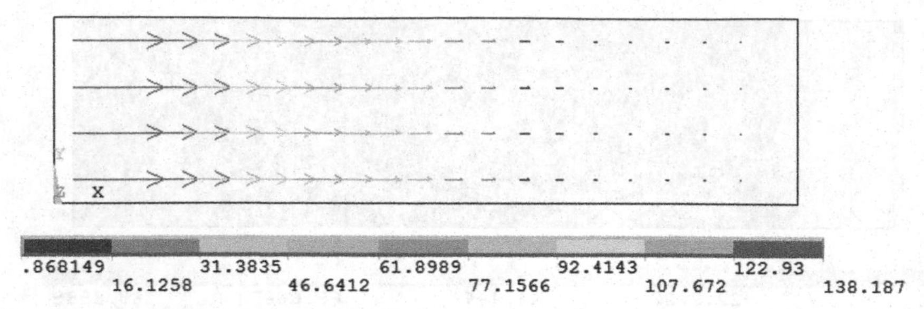

Fig. 5.27 Vector plot of thermal flux at $t=5$

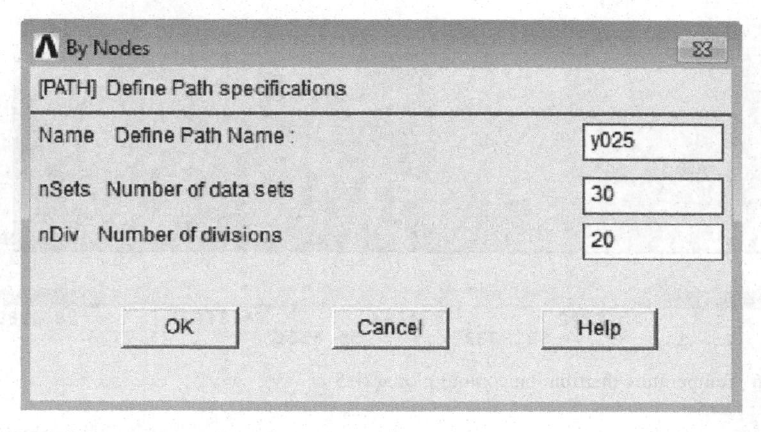

Fig. 5.28 Defining path specifications

- Close the *Path Status Information Window.*
- Turn node numbering off using the following menu path:

Utility Menu > PlotCtrls > Numbering

- In the *Plot Numbering Controls* dialog box, click on the check box for ***NODE Node numbers*** to show ***Off***; click on ***OK***.
- Plot the path on geometry using the following menu path:

Main Menu > General Postproc > Path Operations > Plot Paths

- Figure 5.29 shows the result of this action.

• Map the temperature results onto the defined path using the following menu path:

Main Menu > General Postproc > Path Operations > Map onto Path

- *Map Result Items onto Path* dialog box appears, as shown in Fig. 5.30; enter a unique name for the result item (e.g., ***t025***, note that this is different than the

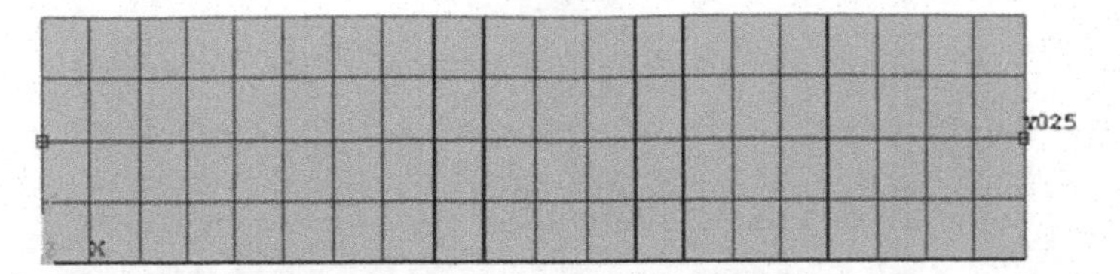

Fig. 5.29 Geometry plot of path `Y025`

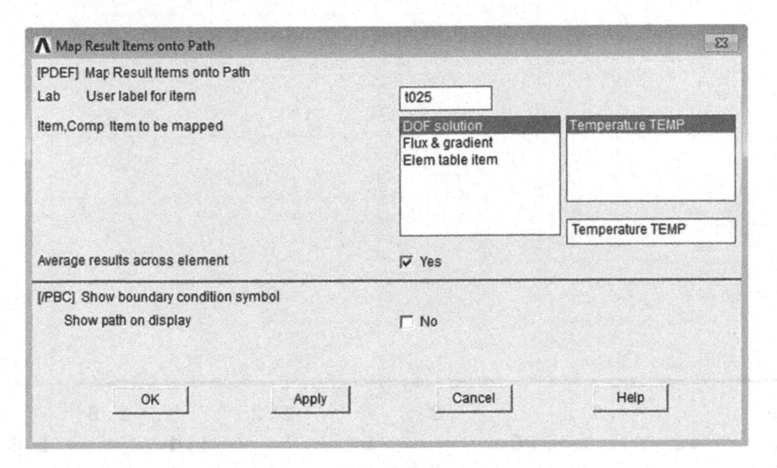

Fig. 5.30 *Map result items onto path* dialog box used for line plot of temperature along the path `Y025`

name given for the path); select ***DOF solution*** from the left list and ***Temperature TEMP*** from the right list for the item; click on ***OK***.

- Path plot the temperature results using the following menu path:

Main Menu > General Postproc > Path Operations > Plot Path Item > On Graph

- *Plot of Path Items on Graph* dialog box appears; select ***T025*** from the list; click on ***OK***.
- Observe the temperature variation along the path ***y025*** as it appears in the *Graphics Window*, as shown in Fig. 5.31.
- Map the flux results onto the defined path using the following menu path:

Main Menu > General Postproc > Path Operations > Map onto Path

- *Map Result Items onto Path* dialog box appears, as shown in Fig. 5.32; enter a unique name for the result item (e.g., ***q025***); select ***Flux & gradient*** from the left list and ***Thermal flux TFX*** from the right list for the item; click on ***OK***.
- Path plot the flux results using the following menu path:

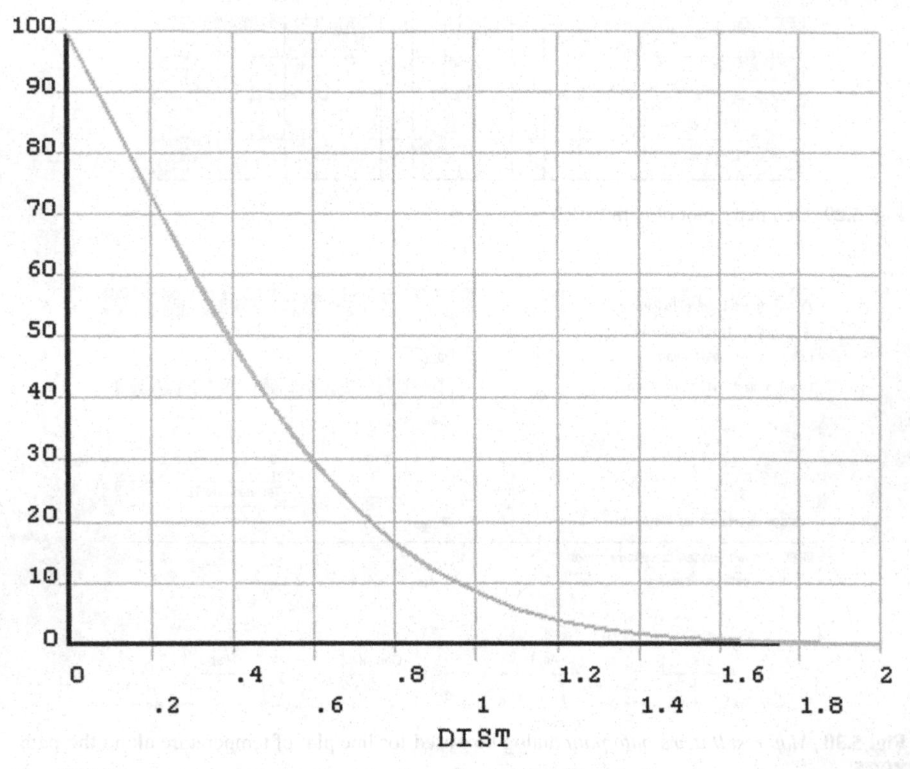

Fig. 5.31 Temperature variation line plot along the path **Y025** at $t=5$

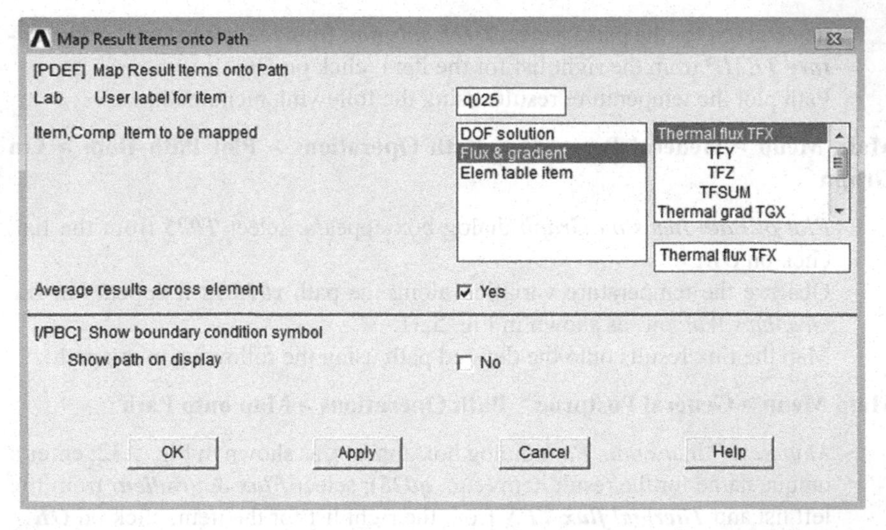

Fig. 5.32 *Map result items onto path* dialog box used for line plot of flux along the path **Y025**

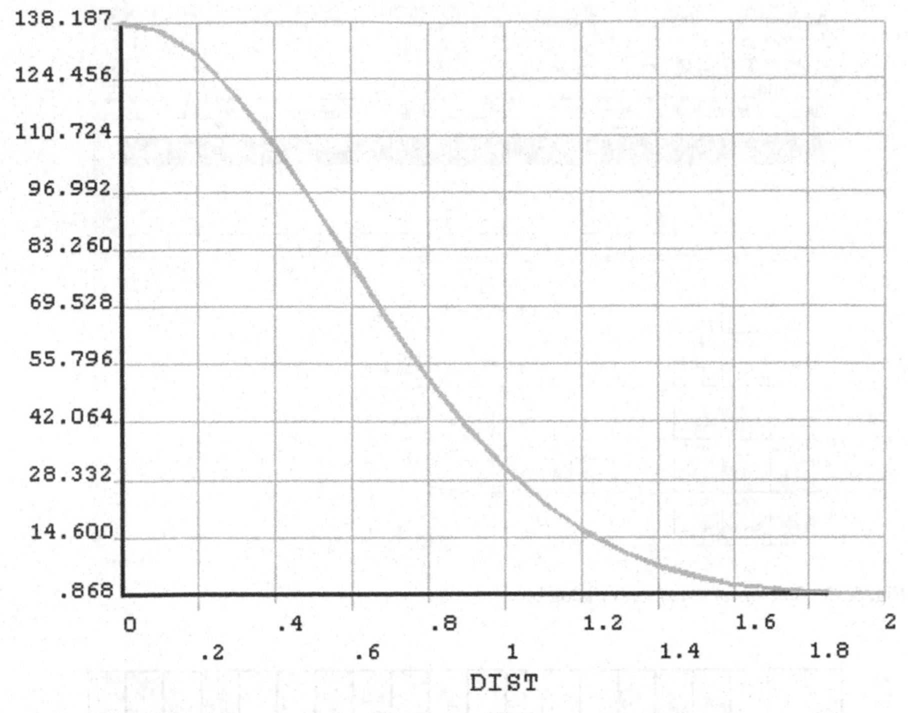

Fig. 5.33 Flux variation line plot along the path **Y025** at $t=5$

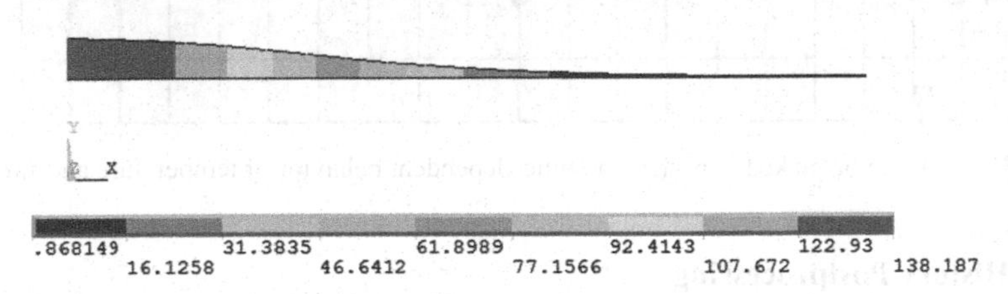

Fig. 5.34 Path plot on geometry for the variation of flux

Main Menu > General Postproc > Path Operations > Plot Path Item > On Graph

- *Plot of Path Items on Graph* dialog box appears; unselect ***T025*** and select ***Q025*** from the list; click on ***OK***.
- Observe the flux variation along the path ***y025*** as it appears in the *Graphics Window*, as shown in Fig. 5.33.
- Finally, plot the flux on actual geometry using the following menu path:

Main Menu > General Postproc > Path Operations > Plot Path Item > On Geometry

- *Plot of Path Items on Geometry* dialog box appears; select ***Q025*** from the list; click on ***OK***.
- Fig. 5.34 shows the path plot of thermal flux as it appears in the *Graphics Window*.

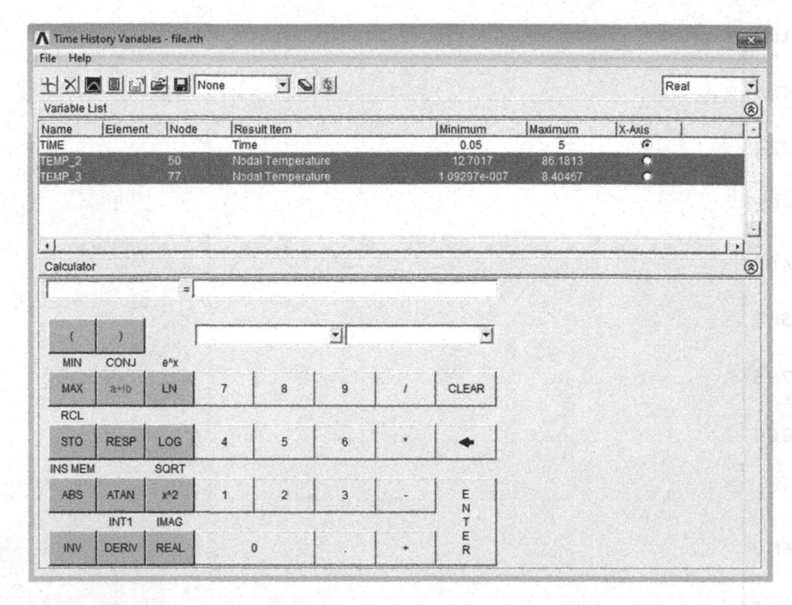

Fig. 5.35 *Time history variables* dialog box

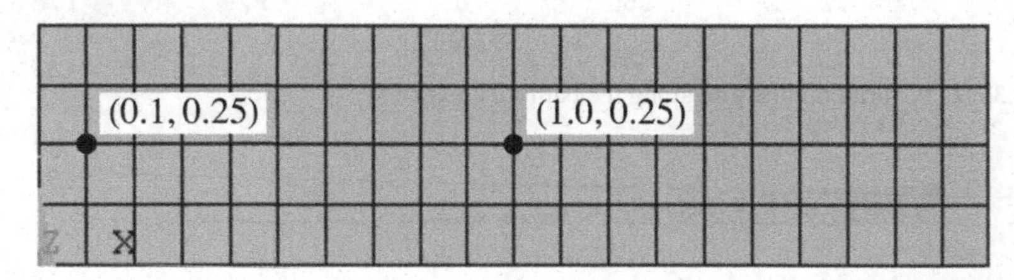

Fig. 5.36 Nodes to be picked to review the time-dependent behavior of temperature and flux

Time History Postprocessing

- Review time-dependent behavior of temperature at nodes located at $(0.1, 0.25)$ and $(1.0, 0.25)$ using the following menu path:

Main Menu > TimeHist Postpro

  - *Time History Variables* dialog box appears (Fig. 5.35).
  - Click on the button with the green plus sign at the top-left to define a variable.
  - Add Time History Variable dialog box appears.
  - Successively click on the items Nodal Solution, DOF Solution, and Nodal Temperature; click on ***OK***.
  - Pick Menu appears; pick the node located at $x = 0.1, y = 0.25$ (as indicated in Fig. 5.36); click on ***OK***.
  - Note the new variable ***TEMP_2*** in the Time History Variables dialog box.

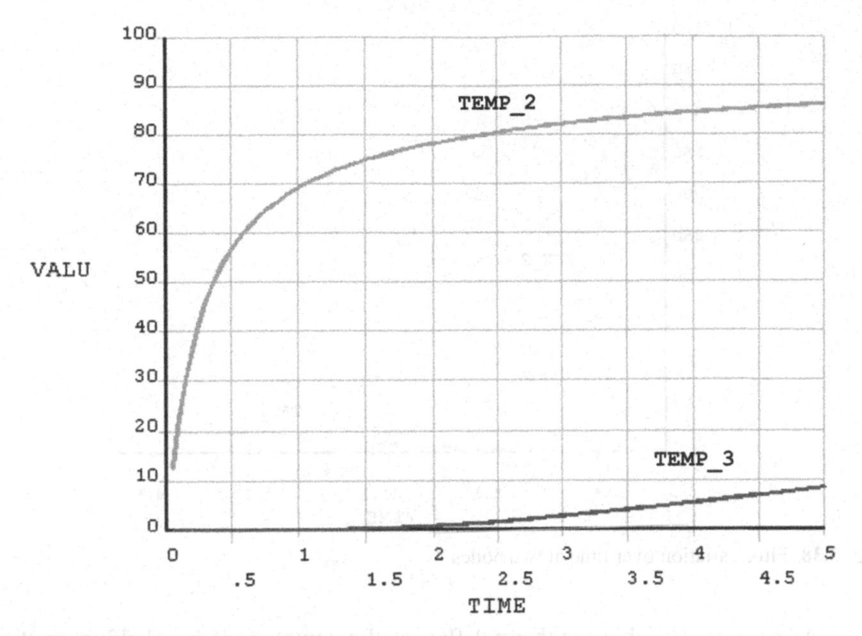

Fig. 5.37 Temperature variation over time at two nodes

- Add a new variable for temperature at the center node by clicking on the button with the green plus sign and successively clicking on the items ***Nodal Solution***, ***DOF Solution***, and ***Nodal Temperature***; click on ***OK***.
- Pick the node located at $x=1, y=0.25$ (as indicated in Fig. 5.36); click on ***OK***.
- Note the new variable ***TEMP_3*** in the *Time History Variables* dialog box.
- Highlight the rows ***TEMP_2*** and ***TEMP_3*** from the list (by pressing ***Ctrl*** on the keyboard and clicking on the rows with the left mouse button); click on the third from the left button to plot the time variation of these temperatures.
- The plot appears in the *Graphics Window*, as shown in Fig. 5.37.

- Review time-dependent behavior of thermal flux at nodes located at $(0.1, 0.25)$ and $(1.0, 0.25)$.
  - In the *Time History Variables* dialog box, highlight the rows ***TEMP_2*** and ***TEMP_3*** from the list and click on the second from the left button (button with a red cross) to delete the temperature variables.
  - In order to add thermal flux variables, click on the button with the green plus sign at the top-left to define a variable.
  - *Add Time History Variable* dialog box appears.
  - Successively click on the items ***Nodal Solution***, ***Thermal Flux***, and ***X-Component of thermal flux***; click on ***OK***.
  - The *Pick Menu* appears; pick the node located at ($x=0.1, y=0.25$) (as indicated in Fig. 5.36); and click on ***OK***.
  - Note the new variable ***TFX_2*** in the *Time History Variables* dialog box.

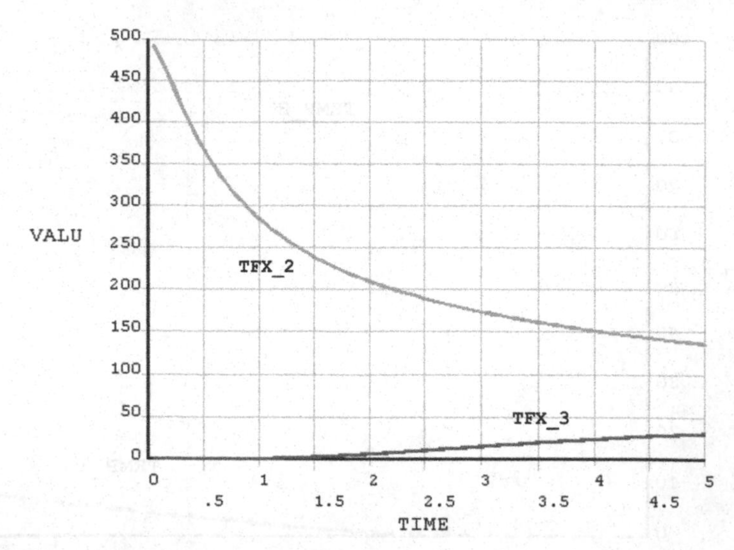

Fig. 5.38 Flux variation over time at two nodes

- Add a new variable for thermal flux at the center node by clicking on the button with the green plus sign and successively clicking on the items ***Nodal Solution***, ***Thermal Flux***, and ***X-Component of thermal flux***; click on ***OK***.
- Pick the node located at ($x = 1, y = 0.25$) (as indicated in Fig. 5.36); click on ***OK***.
- Note the new variable ***TFX_3*** in the *Time History Variables* dialog box.
- Highlight the rows ***TFX_2*** and ***TFX_3*** from the list (by pressing ***Ctrl*** on the keyboard and clicking on the rows with the left mouse button); click on the third from the left button to plot the time variation of these thermal fluxes.
- The plot appears in the *Graphics Window*, as shown in Fig. 5.38.
- Close *Time History Variables* dialog box.

Table 5.3 lists the temperature values along the midline ($y = 0.25$) obtained by ANSYS (columns 2–4) and the analytical solution given by Eq. (5.4) (columns 5–7) at times t=0.1, 0.5, and 5. The analytical solution is obtained by using $n = 40$ in the series. Three separate ANSYS solutions are obtained with the final times t=0.1, 0.5, and 5, each utilizing 100 equal time steps. Figure 5.39 shows a graphical comparison of the analytical and ANSYS solutions.

Table 5.3 Temperature values along the midline ($y=0.25$) obtained by ANSYS and Eq. (5.4) ($t=0.1, 0.5, 5.0$)

x	ANSYS			EXACT ($n=40$)		
	t=0.1	t=0.5	t=5.0	t=0.1	t=0.5	t=5.0
0.00	100.0000	100.0000	100.0000	100.0000	100.0000	100.0000
0.10	24.2230	56.7890	88.1810	22.0671	58.3882	86.2490
0.20	3.7816	26.6910	72.7800	1.4306	27.3322	72.9034
0.30	0.4497	10.7250	60.1760	0.0239	10.0348	60.3332
0.40	0.0443	3.7954	48.6770	9.84E–05	2.8460	48.8422
0.50	0.0038	1.2112	38.5010	2.09E–06	0.6170	38.6475
0.60	0.0003	0.3550	29.7610	1.88E–06	0.1015	29.8698
0.70	2.07E–05	0.0969	22.4760	1.73E–06	0.0126	22.5346
0.80	1.36E–06	0.0249	16.5800	1.56E–06	0.0012	16.5857
0.90	8.47E–08	0.0061	11.9450	1.40E–05	8.24E–05	11.9033
1.00	5.00E–09	0.0014	8.4047	1.24E–06	4.32E–06	8.3264
1.10	2.83E–10	0.0003	5.7753	1.09E–06	1.69E–07	5.6746
1.20	1.54E–11	6.94E–05	3.8761	9.43E–07	4.94E–09	3.7666
1.30	8.14E–13	1.46E–05	2.5411	8.08E–07	1.08E–10	2.4340
1.40	4.17E–14	3.00E–06	1.6271	6.80E–07	1.74E–12	1.5307
1.50	2.08E–15	6.00E–07	1.0169	5.58E–07	0.00E+00	0.9360
1.60	1.03E–16	1.17E–07	0.6187	4.41E–07	0.00E+00	0.5551
1.70	4.85E–18	2.25E–08	0.3628	3.27E–07	0.00E+00	0.3167
1.80	2.27E–19	4.24E–09	0.1981	2.17E–07	0.00E+00	0.1684
1.90	1.04E–20	7.61E–10	0.0868	1.08E–07	−1.69E–14	0.0723
2.00	0	0	0	−8.39E–15	−4.03E–15	−3.85E–15

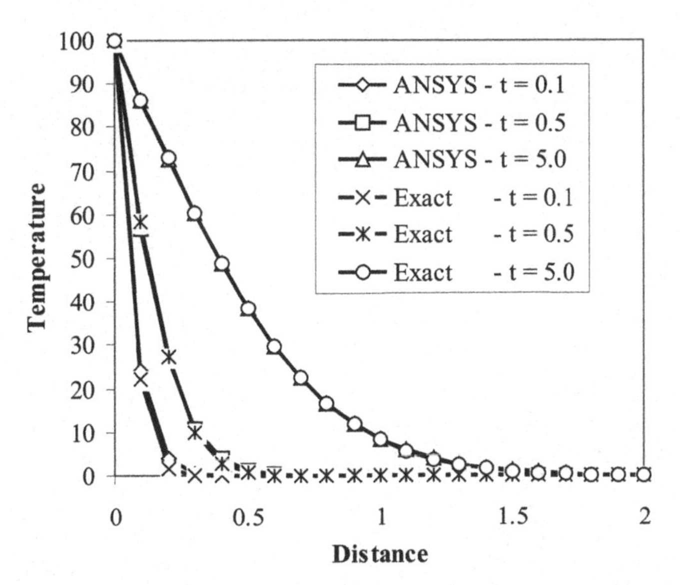

Fig. 5.39 Graphical comparison of ANSYS and analytical solutions

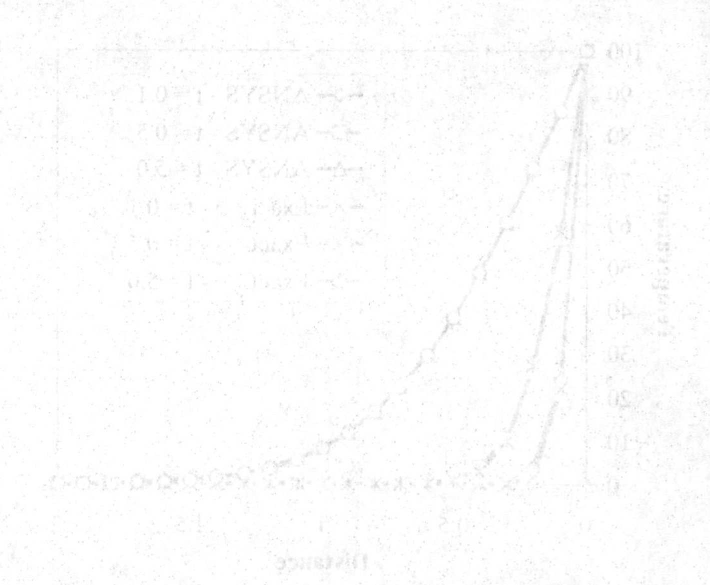

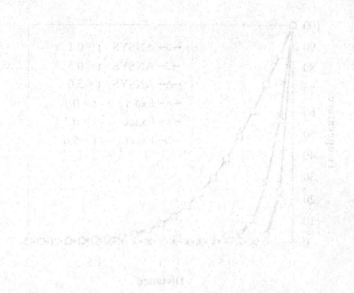

Chapter 6
Finite Element Equations

Finite element equations capture the characteristics of the field equations. Their derivation is based on either the governing differential equation or the global energy balance of the physical problem. The approach involving the governing differential equation is referred to as the *method of weighted residuals* or *Galerkin's method.* The approach utilizing the global energy balance is referred to as the *variational method* or *Rayleigh-Ritz method.*

6.1 Method of Weighted Residuals

The method of weighted residuals involves the approximation of the functional behavior of the dependent variable in the governing differential equation (Finlayson 1972). When substituted into the governing differential equation, the approximate form of the dependent variable leads to an error called the "residual." This residual error is required to vanish in a weighted average sense over the domain. If the weighting functions are chosen to be the same as the element shape (interpolation) functions used in the element approximation functions, the method of weighted residuals is referred to as Galerkin's method.

The governing differential equation for the physical problem in domain D described in Fig. 6.1 can be expressed in the form

$$L(\phi) - f = 0 \tag{6.1}$$

where ϕ is a dependent variable and f is a known forcing function. The ordinary or partial differential operator, L whose order is specified by $p,$ can be linear or nonlinear. The boundary conditions are given by

$$B_j(\phi) = g_j \text{ on } C_1 \tag{6.2}$$

The online version of this book (doi: 10.1007/978-1-4939-1007-6_6) contains supplementary material, which is available to authorized users

E. Madenci, I. Guven, *The Finite Element Method and Applications in Engineering Using ANSYS®*, DOI 10.1007/978-1-4899-7550-8_6

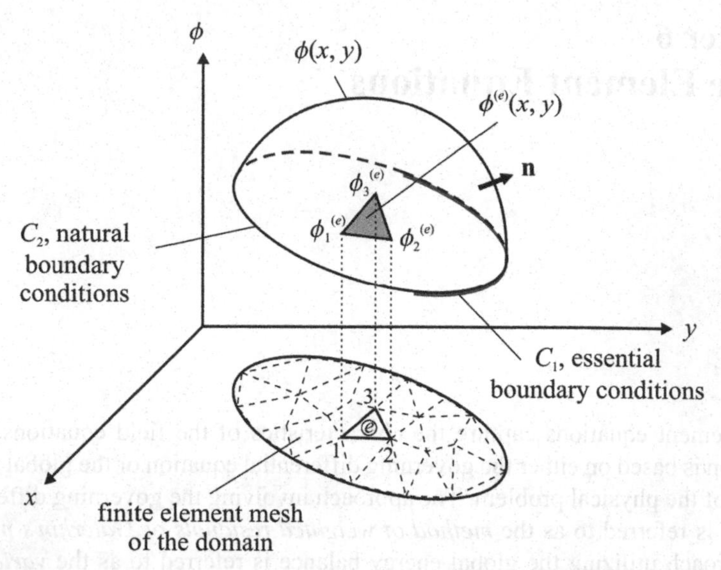

Fig. 6.1 Variation of the dependent (field) variable over a two-dimensional domain under specified boundary conditions

and

$$E_j(\phi) = h_j \text{ on } C_2 \tag{6.3}$$

in which B_j and E_j are operators, with $j = 1,2,3,\ldots,p$. The known functions g_j and h_j prescribe the boundary conditions on the dependent vari-able and its derivatives, respectively. The conditions on the dependent variable over C_1 are referred to as *essential* or *forced* boundary conditions, and the ones involving the derivatives of the dependent variable over C_2 are referred to as *natural* boundary conditions.

The method of weighted residuals requires that

$$\int_D \left[L(\tilde{\phi}) - f \right] W_k \, dD = 0, \quad \text{with} \quad k = 1,2,3,\ldots,n \tag{6.4}$$

where W_k are the weighting functions approximating the dependent variable as

$$\phi \approx \tilde{\phi} = \sum_{k=1}^{n} \alpha_k W_k \tag{6.5}$$

while satisfying the essential boundary conditions on C_1. The unknown coefficients, α_k, are determined by solving for the resulting system of algebraic equations.

Since the governing differential equation is valid for the entire domain, D, partitioning the domain into subdomains or elements, $D^{(e)}$, and applying Galerkin's method with weighting functions $W_k = N_k^{(e)}$ over the element domain results in

$$\sum_{e=1}^{E} \int_{D^{(e)}} \mathbf{N}^{(e)} \left(L\left(\tilde{\phi}^{(e)}\right) - f \right) dD = 0 \tag{6.6}$$

in which E is the number of elements and the superscript "e" denotes a specific element whose domain is $D^{(e)}$. The approximation to the dependent variable within the element can be expressed as

$$\tilde{\phi}^{(e)} = \sum_{i=1}^{n} N_i^{(e)} \phi_i^{(e)} \tag{6.7}$$

or

$$\tilde{\phi}^{(e)} = \mathbf{N}^{(e)T} \boldsymbol{\varphi}^{(e)} \tag{6.8}$$

where

$$\mathbf{N}^{(e)T} = \left\{ N_1^{(e)} \quad N_2^{(e)} \quad N_3^{(e)} \quad \ldots \quad N_n^{(e)} \right\} \tag{6.9}$$

and

$$\boldsymbol{\varphi}^{(e)T} = \left\{ \phi_1^{(e)} \quad \phi_2^{(e)} \quad \phi_3^{(e)} \quad \ldots \quad \phi_n^{(e)} \right\} \tag{6.10}$$

with n representing the number of nodes associated with element e. The nodal unknowns and shape functions are denoted by $\phi_i^{(e)}$ and $N_i^{(e)}$, with $i = 1, 2, .., n$, respectively. The shape functions need not satisfy the boundary conditions; however, they satisfy the inter-element continuity conditions necessary for assembly of the element equations. The essential boundary conditions are imposed after assembling the global matrix. The natural boundary conditions are not imposed directly. However, their influence emerges in the derivation of the element equations.

The required order of the element continuity is equal to one less than the highest derivative of the dependent variable appearing in the integrand. This requirement is relaxed by applying integration by parts in the minimization procedure of the residual error in Galerkin's method.

6.1.1 Example: One-Dimensional Differential Equation with Line Elements

The application of Galerkin's method is introduced by considering the ordinary differential equation given by

$$\frac{d^2\phi(x)}{dx^2} + \phi(x) - f(x) = 0 \tag{6.11}$$

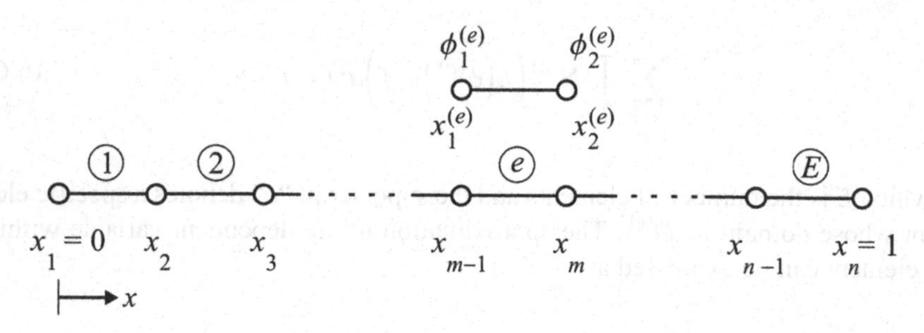

Fig. 6.2 Domain of the one-dimensional differential equation, discretized into E elements

in domain D defined by $0 \le x \le 1$. The known forcing function is given by

$$f(x) = -x \tag{6.12}$$

The boundary conditions, identified as the essential type, are $\phi(0) = 0$ and $\phi(1) = 0$. As shown in Fig. 6.2, the domain can be discretized with E linear line elements, each having two nodes ($n = 2$). There are a total of N nodes, and global coordinates of each node in domain D are specified by x_i, with $i = 1, 2, \ldots, N$. Nodal values of the dependent variable associated with element e are specified at its first and second nodes by $\phi_1^{(e)}$ and $\phi_2^{(e)}$, respectively.

The linear approximation function for the dependent variable in element e can be expressed in the form

$$\tilde{\phi}^{(e)} = N_1^{(e)}\phi_1^{(e)} + N_2^{(e)}\phi_2^{(e)} \tag{6.13}$$

or

$$\tilde{\phi}^{(e)} = \mathbf{N}^{(e)T}\boldsymbol{\varphi}^{(e)} \tag{6.14}$$

where

$$\mathbf{N}^{(e)T} = \left\{ N_1^{(e)} \quad N_2^{(e)} \right\} \text{ and } \boldsymbol{\varphi}^{(e)T} = \left\{ \phi_1^{(e)} \quad \phi_2^{(e)} \right\} \tag{6.15}$$

in which the shape functions are given by

$$N_1^{(e)} = \frac{x_2^{(e)} - x}{x_2^{(e)} - x_1^{(e)}} \text{ and } N_2^{(e)} = \frac{x - x_1^{(e)}}{x_2^{(e)} - x_1^{(e)}} \tag{6.16}$$

They are the same as the length coordinates given by Eq. (3.9). Applying Galerkin's method by Eq. (6.6) leads to

$$\sum_{e=1}^{E} \int_{x_1^{(e)}}^{x_2^{(e)}} \mathbf{N}^{(e)} \left(\frac{d^2\tilde{\phi}^{(e)}(x)}{dx^2} + \tilde{\phi}^{(e)}(x) - f(x) \right) dx = 0 \tag{6.17}$$

Integrating the first term in the integral by parts results in

$$\sum_{e=1}^{E}\left[\mathbf{N}^{(e)} \frac{d\tilde{\phi}^{(e)}(x)}{dx}\Bigg|_{x_1^{(e)}}^{x_2^{(e)}} - \int_{x_1^{(e)}}^{x_2^{(e)}} \frac{d\mathbf{N}^{(e)}}{dx}\frac{d\tilde{\phi}^{(e)}}{dx}dx \right. \\ \left. + \int_{x_1^{(e)}}^{x_2^{(e)}} \mathbf{N}^{(e)}\tilde{\phi}^{(e)}(x)dx - \int_{x_1^{(e)}}^{x_2^{(e)}} \mathbf{N}^{(e)} f(x)dx \right] = 0 \tag{6.18}$$

Substituting for the element approximation function ($\tilde{\phi}^{(e)} = \mathbf{N}^{(e)T}\boldsymbol{\varphi}^{(e)}$) yields

$$\sum_{e=1}^{E}\mathbf{k}^{(e)}\boldsymbol{\varphi}^{(e)} = \sum_{e=1}^{E}\mathbf{f}^{(e)} \tag{6.19}$$

where

$$\mathbf{k}^{(e)} = -\int_{x_1^{(e)}}^{x_2^{(e)}} \frac{d\mathbf{N}^{(e)}}{dx}\frac{d\mathbf{N}^{(e)T}}{dx}dx + \int_{x_1^{(e)}}^{x_2^{(e)}} \mathbf{N}^{(e)}\mathbf{N}^{(e)T}dx \tag{6.20}$$

and

$$\mathbf{f}^{(e)} = \int_{x_1^{(e)}}^{x_2^{(e)}} \mathbf{N}^{(e)} f(x)dx - \left[\mathbf{N}^{(e)}\frac{d\tilde{\phi}^{(e)}(x)}{dx} \right]_{x_1^{(e)}}^{x_2^{(e)}} \tag{6.21}$$

After substituting for the shape functions and their derivatives, as well as the forcing function, the expressions for the element characteristic matrix, $\mathbf{k}^{(e)}$, and the right-hand-side vector, $\mathbf{f}^{(e)}$, become

$$\mathbf{k}^{(e)} = -\frac{1}{\left(x_2^{(e)} - x_1^{(e)}\right)^2}\int_{x_1^{(e)}}^{x_2^{(e)}} \begin{bmatrix} 1 & -1 \\ -1 & 1 \end{bmatrix} dx \\ + \int_{x_1^{(e)}}^{x_2^{(e)}} \begin{bmatrix} N_1^{(e)}N_1^{(e)} & N_1^{(e)}N_2^{(e)} \\ N_2^{(e)}N_1^{(e)} & N_2^{(e)}N_2^{(e)} \end{bmatrix} dx \tag{6.22}$$

$$\mathbf{f}^{(e)} = -\int\limits_{x_1^{(e)}}^{x_2^{(e)}} \begin{Bmatrix} N_1^{(e)} \\ N_2^{(e)} \end{Bmatrix} x\,dx - \left[\begin{Bmatrix} N_1^{(e)} \\ N_2^{(e)} \end{Bmatrix} \frac{d\tilde{\phi}^{(e)}(x)}{dx} \right]_{x_1^{(e)}}^{x_2^{(e)}} \tag{6.23}$$

Evaluation of these integrals leads to the final form of the element characteristic matrix, $\mathbf{k}^{(e)}$, and the right-hand-side vector, $\mathbf{f}^{(e)}$

$$\mathbf{k}^{(e)} = -\frac{1}{\left(x_2^{(e)} - x_1^{(e)}\right)} \begin{bmatrix} 1 & -1 \\ -1 & 1 \end{bmatrix} + \frac{\left(x_2^{(e)} - x_1^{(e)}\right)}{6} \begin{bmatrix} 2 & 1 \\ 1 & 2 \end{bmatrix} \tag{6.24}$$

and

$$\mathbf{f}^{(e)} = -\frac{1}{6}\left(x_2^{(e)} - x_1^{(e)}\right) \begin{Bmatrix} 2x_1^{(e)} + x_2^{(e)} \\ x_1^{(e)} + 2x_2^{(e)} \end{Bmatrix} - \left[\frac{d\tilde{\phi}^{(e)}\left(x_2^{(e)}\right)}{dx} \begin{Bmatrix} N_1^{(e)}\left(x_2^{(e)}\right) \\ N_2^{(e)}\left(x_2^{(e)}\right) \end{Bmatrix} - \frac{d\tilde{\phi}^{(e)}\left(x_1^{(e)}\right)}{dx} \begin{Bmatrix} N_1^{(e)}\left(x_1^{(e)}\right) \\ N_2^{(e)}\left(x_1^{(e)}\right) \end{Bmatrix} \right] \tag{6.25}$$

or

$$\mathbf{f}^{(e)} = -\frac{1}{6}\left(x_2^{(e)} - x_1^{(e)}\right) \begin{Bmatrix} 2x_1^{(e)} + x_2^{(e)} \\ x_1^{(e)} + 2x_2^{(e)} \end{Bmatrix} - \begin{Bmatrix} -\dfrac{d\tilde{\phi}^{(e)}}{dx}\left(x_1^{(e)}\right) \\ \dfrac{d\tilde{\phi}^{(e)}}{dx}\left(x_2^{(e)}\right) \end{Bmatrix} \tag{6.26}$$

The local and global nodes for the domain discretized with three elements, $E = 3$, and four nodes, $N = 4$, are numbered as shown in Table 6.1.

With the appropriate value of the nodal coordinates from Eq. (6.24) and (6.26), the element characteristic matrices and vectors are calculated as

$$\begin{array}{cc} \boxed{1} & \boxed{2} \end{array}$$

$$\mathbf{k}^{(1)} = \frac{1}{18} \begin{bmatrix} 52 & -55 \\ -55 & 52 \end{bmatrix} \begin{matrix} \boxed{1} \\ \boxed{2} \end{matrix} \tag{6.27}$$

$$\begin{array}{cc} \boxed{2} & \boxed{3} \end{array}$$

$$\mathbf{k}^{(2)} = \frac{1}{18} \begin{bmatrix} 52 & -55 \\ -55 & 52 \end{bmatrix} \begin{matrix} \boxed{2} \\ \boxed{3} \end{matrix} \tag{6.28}$$

Table 6.1 Element connectivity and nodal coordinates

Element Number (e)	Node 1	Node 2	$x_1^{(e)}$	$x_2^{(e)}$
1	1	2	0	1/3
2	2	3	1/3	2/3
3	3	4	2/3	1

$$\begin{matrix} & \boxed{3} & \boxed{4} & \\ \mathbf{k}^{(3)} = \frac{1}{18} & \begin{bmatrix} 52 & -55 \\ -55 & 52 \end{bmatrix} & & \begin{matrix}\boxed{3}\\ \boxed{4}\end{matrix} \end{matrix} \tag{6.29}$$

$$\mathbf{f}^{(1)} = \frac{1}{54}\begin{Bmatrix} 1 \\ 2 \end{Bmatrix} + \begin{Bmatrix} -\dfrac{d\tilde{\phi}^{(1)}(0)}{dx} \\ \dfrac{d\tilde{\phi}^{(1)}(1/3)}{dx} \end{Bmatrix} \begin{matrix}\boxed{1}\\ \boxed{2}\end{matrix} \tag{6.30}$$

$$\mathbf{f}^{(2)} = \frac{1}{54}\begin{Bmatrix} 4 \\ 5 \end{Bmatrix} + \begin{Bmatrix} -\dfrac{d\tilde{\phi}^{(2)}(1/3)}{dx} \\ \dfrac{d\tilde{\phi}^{(2)}(2/3)}{dx} \end{Bmatrix} \begin{matrix}\boxed{2}\\ \boxed{3}\end{matrix} \tag{6.31}$$

$$\mathbf{f}^{(3)} = \frac{1}{54}\begin{Bmatrix} 7 \\ 8 \end{Bmatrix} + \begin{Bmatrix} -\dfrac{d\tilde{\phi}^{(3)}(2/3)}{dx} \\ \dfrac{d\tilde{\phi}^{(3)}(1)}{dx} \end{Bmatrix} \begin{matrix}\boxed{3}\\ \boxed{4}\end{matrix} \tag{6.32}$$

As reflected by the element connectivity in Table 6.1, the boxed numbers indicate the rows and columns of the global matrix, **K**, and global right-hand-side vector, **F**, to which the individual coefficients are added. The global coefficient matrix, **K**, and the global right-hand-side vector, **F**, are obtained from the "expanded" element coefficient matrices, $\mathbf{k}^{(e)}$, and the element right-hand-side vectors, $\mathbf{f}^{(e)}$, by summation in the form

$$\mathbf{K} = \sum_{e=1}^{E} \mathbf{k}^{(e)} \quad \text{and} \quad \mathbf{F} = \sum_{e=1}^{E} \mathbf{f}^{(e)} \tag{6.33}$$

The "expanded" element matrices are the same size as the global matrix but have rows and columns of zeros corresponding to the nodes not associated with element (e). Specifically, the expanded form of the element stiffness and load vector becomes

$$\mathbf{k}^{(1)} = \frac{1}{18}\begin{array}{c} \begin{array}{cccc} \boxed{1} & \boxed{2} & \boxed{3} & \boxed{4} \end{array} \\ \begin{bmatrix} 52 & -55 & 0 & 0 \\ -55 & 52 & 0 & 0 \\ 0 & 0 & 0 & 0 \\ 0 & 0 & 0 & 0 \end{bmatrix} \end{array} \begin{array}{c} \\ \boxed{1} \\ \boxed{2} \\ \boxed{3} \\ \boxed{4} \end{array} \tag{6.34}$$

$$\mathbf{f}^{(1)} = \frac{1}{54}\begin{Bmatrix} 1 \\ 2 \\ 0 \\ 0 \end{Bmatrix} + \begin{Bmatrix} -\dfrac{d\tilde{\phi}^{(1)}(0)}{dx} \\ \dfrac{d\tilde{\phi}^{(1)}(1/3)}{dx} \\ 0 \\ 0 \end{Bmatrix} \begin{array}{c} \boxed{1} \\ \boxed{2} \\ \boxed{3} \\ \boxed{4} \end{array} \tag{6.35}$$

$$\mathbf{k}^{(2)} = \frac{1}{18}\begin{array}{c} \begin{array}{cccc} \boxed{1} & \boxed{2} & \boxed{3} & \boxed{4} \end{array} \\ \begin{bmatrix} 0 & 0 & 0 & 0 \\ 0 & 52 & -55 & 0 \\ 0 & -55 & 52 & 0 \\ 0 & 0 & 0 & 0 \end{bmatrix} \end{array} \begin{array}{c} \\ \boxed{1} \\ \boxed{2} \\ \boxed{3} \\ \boxed{4} \end{array} \tag{6.36}$$

$$\mathbf{f}^{(2)} = \frac{1}{54}\begin{Bmatrix} 0 \\ 4 \\ 5 \\ 0 \end{Bmatrix} + \begin{Bmatrix} 0 \\ -\dfrac{d\tilde{\phi}^{(2)}(1/3)}{dx} \\ \dfrac{d\tilde{\phi}^{(2)}(2/3)}{dx} \\ 0 \end{Bmatrix} \begin{array}{c} \boxed{1} \\ \boxed{2} \\ \boxed{3} \\ \boxed{4} \end{array} \tag{6.37}$$

$$\mathbf{k}^{(3)} = \frac{1}{18}\begin{array}{c} \begin{array}{cccc} \boxed{1} & \boxed{2} & \boxed{3} & \boxed{4} \end{array} \\ \begin{bmatrix} 0 & 0 & 0 & 0 \\ 0 & 0 & 0 & 0 \\ 0 & 0 & 52 & -55 \\ 0 & 0 & -55 & 52 \end{bmatrix} \end{array} \begin{array}{c} \\ \boxed{1} \\ \boxed{2} \\ \boxed{3} \\ \boxed{4} \end{array} \tag{6.38}$$

$$
\mathbf{f}^{(3)} = \frac{1}{54}\begin{Bmatrix} 0 \\ 0 \\ 7 \\ 8 \end{Bmatrix} + \begin{Bmatrix} 0 \\ 0 \\ -\dfrac{d\tilde{\phi}^{(3)}(2/3)}{dx} \\ \dfrac{d\tilde{\phi}^{(3)}(1)}{dx} \end{Bmatrix} \begin{matrix} \boxed{1} \\ \boxed{2} \\ \boxed{3} \\ \boxed{4} \end{matrix} \tag{6.39}
$$

In accordance with Eq. (6.33) and (6.19), the assembly of the element characteristic matrices and vectors results in the global equilibrium equations

$$
\frac{1}{18}\begin{bmatrix} 52 & -55 & 0 & 0 \\ -55 & 52+52 & -55 & 0 \\ 0 & -55 & 52+52 & -55 \\ 0 & 0 & -55 & 52 \end{bmatrix} \begin{Bmatrix} \phi_1 = \phi_1^{(1)} \\ \phi_2 = \phi_2^{(1)} = \phi_1^{(2)} \\ \phi_3 = \phi_2^{(2)} = \phi_1^{(3)} \\ \phi_4 = \phi_2^{(3)} \end{Bmatrix} = \frac{1}{54}\begin{Bmatrix} 1 \\ 2+4 \\ 5+7 \\ 8 \end{Bmatrix}
$$

$$
+\begin{Bmatrix} -\dfrac{d\tilde{\phi}^{(1)}(0)}{dx} \\ \cancelto{0}{\dfrac{d\tilde{\phi}^{(1)}(1/3)}{dx} - \dfrac{d\tilde{\phi}^{(2)}(1/3)}{dx}} \\ \cancelto{0}{\dfrac{d\tilde{\phi}^{(2)}(2/3)}{dx} - \dfrac{d\tilde{\phi}^{(3)}(2/3)}{dx}} \\ \dfrac{d\tilde{\phi}^{(3)}(1)}{dx} \end{Bmatrix} \tag{6.40}
$$

or

$$
\frac{1}{18}\begin{bmatrix} 52 & -55 & 0 & 0 \\ -55 & 104 & -55 & 0 \\ 0 & -55 & 104 & -55 \\ 0 & 0 & -55 & 52 \end{bmatrix} \begin{Bmatrix} \phi_1 \\ \phi_2 \\ \phi_3 \\ \phi_4 \end{Bmatrix} = \frac{1}{54}\begin{Bmatrix} 1 \\ 6 \\ 12 \\ 8 \end{Bmatrix} + \begin{Bmatrix} -\dfrac{d\tilde{\phi}^{(1)}(0)}{dx} \\ 0 \\ 0 \\ \dfrac{d\tilde{\phi}^{(3)}(1)}{dx} \end{Bmatrix} \tag{6.41}
$$

or

$$
\mathbf{K\Phi} = \mathbf{F} \tag{6.42}
$$

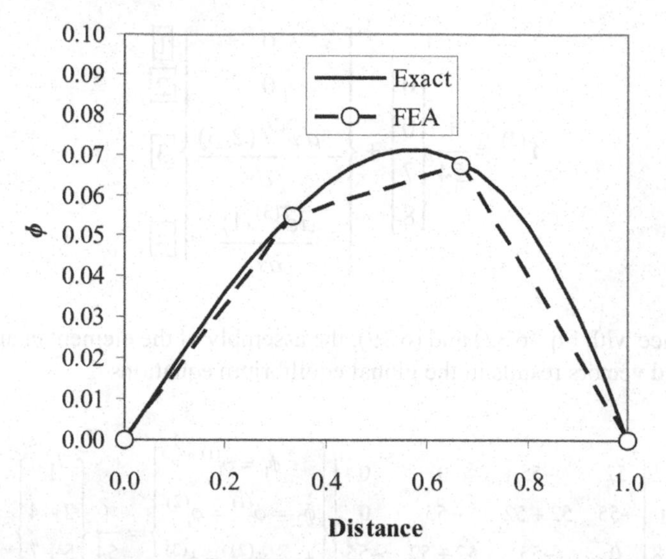

Fig. 6.3 Comparison of the exact and FEA (approximate) solutions to the 1D differential equation

After imposing the essential boundary conditions, $\phi_1 = 0$ and $\phi_4 = 0$, the global system of equations is reduced by deleting the row and column corresponding to ϕ_1 and ϕ_4, leading to

$$\frac{1}{18}\begin{bmatrix} 104 & -55 \\ -55 & 104 \end{bmatrix}\begin{Bmatrix} \phi_2 \\ \phi_3 \end{Bmatrix} = \frac{1}{54}\begin{Bmatrix} 6 \\ 12 \end{Bmatrix} \tag{6.43}$$

Its solution yields

$$\begin{Bmatrix} \phi_2 \\ \phi_3 \end{Bmatrix} = \begin{Bmatrix} 0.05493 \\ 0.06751 \end{Bmatrix} \tag{6.44}$$

The exact solution to the differential equation given by

$$\phi(x) = \frac{\sin(x)}{\sin(1)} - x \tag{6.45}$$

provides

$$\begin{Bmatrix} \phi_2 \\ \phi_3 \end{Bmatrix} = \begin{Bmatrix} 0.0555 \\ 0.0682 \end{Bmatrix} \tag{6.46}$$

The exact and FEM calculations of ϕ along the x-axis are shown in Fig. 6.3.

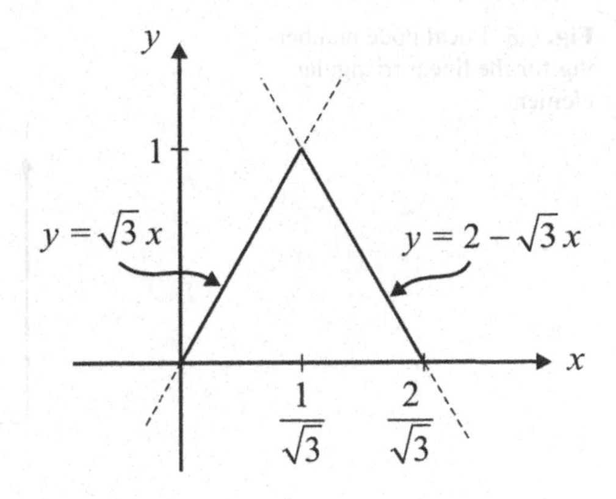

Fig. 6.4 The equilateral triangular domain

6.1.2 *Example: Two-Dimensional Differential Equation with Linear Triangular Elements*

6.1.2.1 Galerkin's Method

The application of Galerkin's method in solving two-dimensional problems with linear triangular elements is demonstrated by considering the partial differential equation given by

$$\frac{\partial^2 \phi(x,y)}{\partial x^2} + \frac{\partial^2 \phi(x,y)}{\partial y^2} + A = 0 \tag{6.47}$$

in domain D, defined by the intersection of $y = 0, y = 2 - \sqrt{3}x$, and $y = \sqrt{3}x$ (as shown in Fig. 6.4), where $A = 1$.

The boundary conditions are specified as

$$-n_y \frac{\partial \phi(x, y=0)}{\partial y} = [\phi(x, y=0) - (B=1)] \text{ for } 0 \leq x \leq 2/\sqrt{3} \tag{6.48}$$

$$\phi(x, y=\sqrt{3}x) = 0 \quad \text{for} \quad 0 \leq x \leq 1/\sqrt{3} \tag{6.49}$$

$$\phi(x, y=2-\sqrt{3}x) = 0 \text{ for } 1/\sqrt{3} \leq x \leq 2/\sqrt{3} \tag{6.50}$$

When independent of time, these equations provide the temperature field, $\phi(x, y)$, due to heat conduction in a domain having a heat generation of A with one of its boundaries subjected to a convective heat transfer. The thermal conductivity and the film (surface) heat transfer coefficient are equal to unity, and the temperature of the surrounding medium is B.

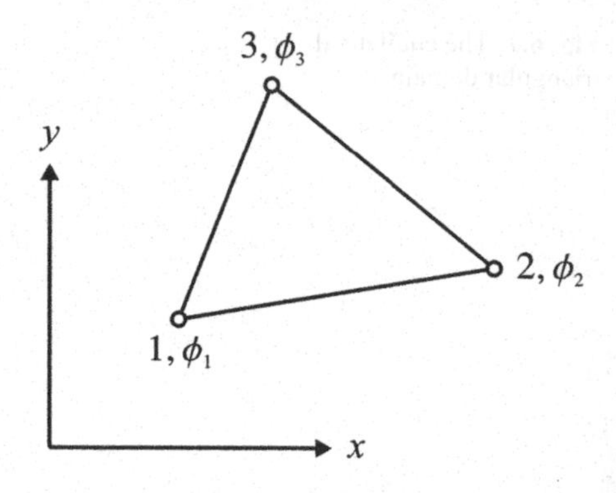

Fig. 6.5 Local node numbering for the linear triangular element

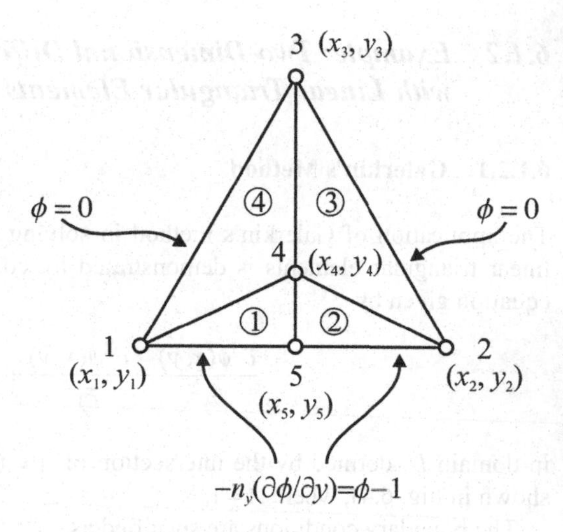

Fig. 6.6 Finite element discretization of the domain

The triangular domain can be discretized into four linear triangular elements, each having three nodes identified as 1, 2, and 3 (local node numbering), as illustrated in Fig. 6.5.

As shown in Fig. 6.6, the global coordinates of each node in domain D are specified by (x_i, y_i), with $i = 1, 2, 3, 4$, and 5. These coordinates are presented in Table 6.2.

The nodal values of the dependent variable associated with the global coordinates are denoted by ϕ_i ($i = 1, 2, 3, 4$, and 5). As shown in Fig. 6.5, the nodal values of the dependent variable associated with element e are specified at its first, second, and third nodes by $\phi_1^{(e)}$, $\phi_2^{(e)}$, and $\phi_3^{(e)}$, respectively.

The linear element approximation function for the dependent field variable in a triangular element "e" is written as

$$\tilde{\phi}^{(e)} = N_1^{(e)}\phi_1^{(e)} + N_2^{(e)}\phi_2^{(e)} + N_3^{(e)}\phi_3^{(e)} \tag{6.51}$$

Table 6.2 Nodal coordinates

Global node number	Nodal coordinates	Nodal unknowns
1	$(0,0)$	ϕ_1
2	$(2/\sqrt{3},0)$	ϕ_2
3	$(1/\sqrt{3},1)$	ϕ_3
4	$(1/\sqrt{3},1/3)$	ϕ_4
5	$(1/\sqrt{3},0)$	ϕ_5

or

$$\tilde{\phi}^{(e)} = \mathbf{N}^{(e)T}\boldsymbol{\varphi}^{(e)} \tag{6.52}$$

As derived in Chap. 3, the element shape functions in Eq. (3.17) are taken as

$$\begin{Bmatrix} N_1^{(e)} \\ N_2^{(e)} \\ N_3^{(e)} \end{Bmatrix} = \frac{1}{2\Delta^{(e)}} \begin{bmatrix} (x_2y_3 - x_3y_2) & y_{23} & x_{32} \\ (x_3y_1 - x_1y_3) & y_{31} & x_{13} \\ (x_1y_2 - x_2y_1) & y_{12} & x_{21} \end{bmatrix} \begin{Bmatrix} 1 \\ x \\ y \end{Bmatrix} \tag{6.53}$$

where $x_{mn} = x_m - x_n$, $y_{mn} = y_m - y_n$, and $\Delta^{(e)}$ is the area of the element computed by

$$2\Delta^{(e)} = \begin{vmatrix} 1 & 1 & 1 \\ x_1 & x_2 & x_3 \\ y_1 & y_2 & y_3 \end{vmatrix} \tag{6.54}$$

Applying Eq. (6.6), Galerkin's method, leads to

$$\sum_{e=1}^{E} \int_{D^{(e)}} \mathbf{N}^{(e)} \left(\frac{\partial^2 \tilde{\phi}^{(e)}(x,y)}{\partial x^2} + \frac{\partial^2 \tilde{\phi}^{(e)}(x,y)}{\partial y^2} + A \right) dx\,dy = 0 \tag{6.55}$$

Since the element approximation function is C^0 continuous, the second-order derivatives in the integrand must be reduced by one so that the inter-element continuity is achieved during the assembly of the global matrix. This reduction is achieved by observing that

$$\mathbf{N}^{(e)} \frac{\partial^2 \tilde{\phi}^{(e)}}{\partial x^2}(x,y) = \frac{\partial}{\partial x}\left(\mathbf{N}^{(e)} \frac{\partial \tilde{\phi}^{(e)}}{\partial x}(x,y) \right) - \frac{\partial \mathbf{N}^{(e)}}{\partial x} \frac{\partial \tilde{\phi}^{(e)}}{\partial x}(x,y) \tag{6.56}$$

and

$$\mathbf{N}^{(e)}\frac{\partial^2\tilde{\phi}^{(e)}}{\partial y^2}(x,y)=\frac{\partial}{\partial y}\left(\mathbf{N}^{(e)}\frac{\partial\tilde{\phi}^{(e)}}{\partial y}(x,y)\right)-\frac{\partial\mathbf{N}^{(e)}}{\partial y}\frac{\partial\tilde{\phi}^{(e)}}{\partial y}(x,y) \tag{6.57}$$

Their substitution into the integrand in Eq. (6.55) and rearrangement of the terms result in

$$\sum_{e=1}^{E}\left\{\int_{D^{(e)}}\left[\frac{\partial}{\partial x}\left(\mathbf{N}^{(e)}\frac{\partial\tilde{\phi}^{(e)}}{\partial x}\right)+\frac{\partial}{\partial y}\left(\mathbf{N}^{(e)}\frac{\partial\tilde{\phi}^{(e)}}{\partial y}\right)\right]dx\,dy\right.$$
$$\left.+\int_{D^{(e)}}\left[-\frac{\partial\mathbf{N}^{(e)}}{\partial x}\frac{\partial\tilde{\phi}^{(e)}}{\partial x}-\frac{\partial\mathbf{N}^{(e)}}{\partial y}\frac{\partial\tilde{\phi}^{(e)}}{\partial y}+\mathbf{N}^{(e)}A\right]dx\,dy\right\}=0 \tag{6.58}$$

Applying the divergence theorem to the first integral renders the domain integral to the boundary integral, and it yields

$$\sum_{e=1}^{E}\left\{\oint_{C^{(e)}}\left[\left(\mathbf{N}^{(e)}\frac{\partial\tilde{\phi}^{(e)}}{\partial x}\right)n_x^{(e)}+\left(\mathbf{N}^{(e)}\frac{\partial\tilde{\phi}^{(e)}}{\partial y}\right)n_y^{(e)}\right]ds\right.$$
$$\left.+\int_{D^{(e)}}\left[-\frac{\partial\mathbf{N}^{(e)}}{\partial x}\frac{\partial\tilde{\phi}^{(e)}}{\partial x}-\frac{\partial\mathbf{N}^{(e)}}{\partial y}\frac{\partial\tilde{\phi}^{(e)}}{\partial y}+\mathbf{N}^{(e)}A\right]dx\,dy\right\}=0 \tag{6.59}$$

where $n_x^{(e)}$ and $n_y^{(e)}$ are, respectively, the x- and y-components of the outward normal vector along the closed boundary defining the area of the element, $C^{(e)}$.

Substituting for the element approximation function yields

$$\sum_{e=1}^{E}\left\{\oint_{C^{(e)}}\mathbf{N}^{(e)}\left[\frac{\partial\tilde{\phi}^{(e)}}{\partial x}n_x^{(e)}+\frac{\partial\tilde{\phi}^{(e)}}{\partial y}n_y^{(e)}\right]ds\right.$$
$$-\int_{D^{(e)}}\left[\frac{\partial\mathbf{N}^{(e)}}{\partial x}\frac{\partial\mathbf{N}^{(e)T}}{\partial x}+\frac{\partial\mathbf{N}^{(e)}}{\partial y}\frac{\partial\mathbf{N}^{(e)T}}{\partial y}\right]dx\,dy\boldsymbol{\varphi}^{(e)} \tag{6.60}$$
$$\left.+\int_{D^{(e)}}\mathbf{N}^{(e)}A\,dx\,dy\right\}=0$$

This equation can be recast in matrix form as

$$\sum_{e=1}^{E}(-\mathbf{k}^{(e)}\boldsymbol{\varphi}^{(e)}+\mathbf{f}^{(e)}+\mathbf{Q}^{(e)})=0 \tag{6.61}$$

where

$$\mathbf{k}^{(e)}=\int_{D^{(e)}}\left[\frac{\partial \mathbf{N}^{(e)}}{\partial x}\frac{\partial \mathbf{N}^{(e)T}}{\partial x}+\frac{\partial \mathbf{N}^{(e)}}{\partial y}\frac{\partial \mathbf{N}^{(e)T}}{\partial y}\right]dx\,dy \tag{6.62}$$

$$\mathbf{f}^{(e)}=\int_{D^{(e)}}\mathbf{N}^{(e)}A\,dx\,dy \tag{6.63}$$

$$\mathbf{Q}^{(e)}=\oint_{C^{(e)}}\mathbf{N}^{(e)}\left[\frac{\partial\tilde{\phi}^{(e)}}{\partial x}n_x^{(e)}+\frac{\partial\tilde{\phi}^{(e)}}{\partial y}n_y^{(e)}\right]ds \tag{6.64}$$

in which $\mathbf{k}^{(e)}$ is the element characteristic matrix, $\mathbf{f}^{(e)}$ is the element right-hand-side vector, and $\mathbf{Q}^{(e)}$ is often referred to as the inter-element vector that includes the derivative terms along the boundary of the element. The boundary integral around each element is evaluated in a counterclockwise direction, i.e., this boundary integral is the sum of three integrals taken along each side of the element.

Depending on whether the element has an exterior boundary or not, the inter-element vector is divided into two parts

$$\mathbf{Q}^{(e)}=\mathbf{Q}_e^{(e)}+\mathbf{Q}_i^{(e)} \tag{6.65}$$

in which $\mathbf{Q}_e^{(e)}$ represents the contribution of the derivative terms specified along the external boundary of the element $C_e^{(e)}$, and $\mathbf{Q}_i^{(e)}$ represents the contribution from the internal boundaries of the element shared with other adjacent elements. Because each of the boundary integrals is evaluated in a counterclockwise direction, the contributions coming from the vector $\mathbf{Q}_i^{(e)}$ vanish when the global system of equations are assembled, thus no further discussion is necessary. However, in the case of specified derivative boundary conditions, the contribution coming from $\mathbf{Q}_e^{(e)}$ must be included.

In view of the boundary conditions given by Eq. (6.48) and the discretization of the domain, the 1–5 side of element 1 and the 5–2 side of element 2 are subjected to derivative boundary conditions.

With $n_x^{(1)}=n_x^{(2)}=0$ and $n_y^{(1)}=n_y^{(2)}=-1$, the contribution of the derivative boundary conditions appearing in Eq. (6.64) leads to the inter-element vectors as

$$\mathbf{Q}_e^{(1)}=\oint_{C_{1-5}^{(1)}}\mathbf{N}^{(1)}\left[B-\phi_C\right]ds \quad\text{and}\quad \mathbf{Q}_e^{(2)}=\oint_{C_{5-2}^{(2)}}\mathbf{N}^{(2)}\left[B-\phi_C\right]ds \tag{6.66}$$

where ϕ_C is the *unknown* value of the field variable on the external boundary of the element C_e, along which the derivative boundary condition is specified.

Approximating the unknown field variable, ϕ_C, by $\tilde{\phi}^{(e)} = \mathbf{N}^{(e)T}\boldsymbol{\varphi}^{(e)}$ in these equations leads to

$$\mathbf{Q}_e^{(1)} = \oint_{C_{1-5}^{(1)}} \mathbf{N}^{(1)}\left[B - \mathbf{N}^{(1)T}\boldsymbol{\varphi}^{(1)}\right]ds \tag{6.67a}$$

and

$$\mathbf{Q}_e^{(2)} = \oint_{C_{5-2}^{(2)}} \mathbf{N}^{(2)}\left[B - \mathbf{N}^{(2)T}\boldsymbol{\varphi}^{(2)}\right]ds \tag{6.67b}$$

which can be rewritten as

$$\mathbf{Q}_e^{(1)} = \oint_{C_{1-5}^{(1)}} \mathbf{N}^{(1)}Bds - \left\{\oint_{C_{1-5}^{(1)}} \mathbf{N}^{(1)}\mathbf{N}^{(1)T}ds\right\}\boldsymbol{\varphi}^{(1)} \text{ or } \mathbf{Q}_e^{(1)} = \mathbf{g}^{(1)} - \mathbf{h}^{(1)}\boldsymbol{\varphi}^{(1)} \tag{6.68}$$

and

$$\mathbf{Q}_e^{(2)} = \oint_{C_{5-2}^{(2)}} \mathbf{N}^{(2)}Bds - \left\{\oint_{C_{5-2}^{(2)}} \mathbf{N}^{(2)}\mathbf{N}^{(2)T}Bds\right\}\boldsymbol{\varphi}^{(2)} \tag{6.69a}$$

or

$$\mathbf{Q}_e^{(2)} = \mathbf{g}^{(2)} - \mathbf{h}^{(2)}\boldsymbol{\varphi}^{(2)} \tag{6.69b}$$

where

$$\mathbf{h}^{(1)} = \oint_{C_{1-5}^{(1)}} \mathbf{N}^{(1)}\mathbf{N}^{(1)T}ds \text{ and } \mathbf{h}^{(2)} = \oint_{C_{2-5}^{(2)}} \mathbf{N}^{(2)}\mathbf{N}^{(2)T}ds \tag{6.70}$$

and

$$\mathbf{g}^{(1)} = \oint_{C_{1-5}^{(1)}} \mathbf{N}^{(1)}Bds \text{ and } \mathbf{g}^{(2)} = \oint_{C_{2-5}^{(2)}} \mathbf{N}^{(2)}Bds \tag{6.71}$$

With this representation of the inter-element vector, the element equilibrium equations given by Eq. (6.61) can be rewritten in their final form as

$$\begin{aligned}
\left(\mathbf{k}^{(1)} + \mathbf{h}^{(1)}\right)\boldsymbol{\varphi}^{(1)} &= \mathbf{f}^{(1)} + \mathbf{g}(1) \\
\left(\mathbf{k}^{(2)} + \mathbf{h}^{(2)}\right)\boldsymbol{\varphi}^{(2)} &= \mathbf{f}^{(2)} + \mathbf{g}^{(2)} \\
\mathbf{k}^{(3)}\boldsymbol{\varphi}^{(3)} &= \mathbf{f}^{(3)} \\
\mathbf{k}^{(4)}\boldsymbol{\varphi}^{(4)} &= \mathbf{f}^{(4)}
\end{aligned} \tag{6.72}$$

With the derivatives of the shape functions obtained as

$$\begin{Bmatrix} \dfrac{\partial N_1^{(e)}}{\partial x} \\ \dfrac{\partial N_2^{(e)}}{\partial x} \\ \dfrac{\partial N_3^{(e)}}{\partial x} \end{Bmatrix} = \frac{1}{2\Delta^{(e)}} \begin{Bmatrix} y_{23} \\ y_{31} \\ y_{12} \end{Bmatrix} \quad \text{and} \quad \begin{Bmatrix} \dfrac{\partial N_1^{(e)}}{\partial y} \\ \dfrac{\partial N_2^{(e)}}{\partial y} \\ \dfrac{\partial N_3^{(e)}}{\partial y} \end{Bmatrix} = \frac{1}{2\Delta^{(e)}} \begin{Bmatrix} x_{32} \\ x_{13} \\ x_{21} \end{Bmatrix} \tag{6.73}$$

the evaluation of the area integrals in Eq. (6.62) and (6.63) by using Eq. (3.19) leads to the final form of the element coefficient matrix, $\mathbf{k}^{(e)}$, and right-hand-side vector, $\mathbf{f}^{(e)}$

$$\mathbf{k}^{(e)} = \frac{1}{4\Delta^{(e)}} \begin{bmatrix} x_{32}^2 + y_{23}^2 & x_{32}x_{13} + y_{23}y_{31} & x_{32}x_{21} + y_{23}y_{12} \\ x_{32}x_{13} + y_{23}y_{31} & x_{13}^2 + y_{31}^2 & x_{13}x_{21} + y_{31}y_{12} \\ x_{32}x_{21} + y_{23}y_{12} & x_{13}x_{21} + y_{31}y_{12} & x_{21}^2 + y_{12}^2 \end{bmatrix} \tag{6.74}$$

and

$$\mathbf{f}^{(e)} = \frac{A\Delta^{(e)}}{3} \begin{Bmatrix} 1 \\ 1 \\ 1 \end{Bmatrix} \tag{6.75}$$

Their numerical evaluation results in

$$\mathbf{k}^{(1)} = \frac{\sqrt{3}}{6} \begin{bmatrix} 1 & -1 & 0 \\ -1 & 4 & -3 \\ 0 & -3 & 3 \end{bmatrix} \quad \text{and} \quad \mathbf{f}^{(1)} = \frac{1}{18\sqrt{3}} \begin{Bmatrix} 1 \\ 1 \\ 1 \end{Bmatrix} \tag{6.76}$$

$$\mathbf{k}^{(2)} = \frac{\sqrt{3}}{6} \begin{bmatrix} 4 & -1 & -3 \\ -1 & 1 & 0 \\ -3 & 0 & 3 \end{bmatrix} \quad \text{and} \quad \mathbf{f}^{(2)} = \frac{1}{18\sqrt{3}} \begin{Bmatrix} 1 \\ 1 \\ 1 \end{Bmatrix} \tag{6.77}$$

$$\mathbf{k}^{(3)} = \frac{\sqrt{3}}{12} \begin{bmatrix} 4 & 2 & -6 \\ 2 & 4 & -6 \\ -6 & -6 & 12 \end{bmatrix} \quad \text{and} \quad \mathbf{f}^{(3)} = \frac{1}{9\sqrt{3}} \begin{Bmatrix} 1 \\ 1 \\ 1 \end{Bmatrix} \tag{6.78}$$

$$\mathbf{k}^{(4)} = \frac{\sqrt{3}}{12}\begin{bmatrix} 4 & 2 & -6 \\ 2 & 4 & -6 \\ -6 & -6 & 12 \end{bmatrix} \quad \text{and} \quad \mathbf{f}^{(4)} = \frac{1}{9\sqrt{3}}\begin{Bmatrix} 1 \\ 1 \\ 1 \end{Bmatrix} \tag{6.79}$$

in which the area of each element is computed as

$$\Delta^{(1)} = \frac{1}{2}\begin{vmatrix} 1 & 1 & 1 \\ 0 & 1/\sqrt{3} & 1/\sqrt{3} \\ 0 & 0 & 1/3 \end{vmatrix} = \frac{1}{6\sqrt{3}} \tag{6.80}$$

$$\Delta^{(2)} = \frac{1}{2}\begin{vmatrix} 1 & 1 & 1 \\ 1/\sqrt{3} & 2/\sqrt{3} & 1/\sqrt{3} \\ 0 & 0 & 1/3 \end{vmatrix} = \frac{1}{6\sqrt{3}} \tag{6.81}$$

$$\Delta^{(3)} = \frac{1}{2}\begin{vmatrix} 1 & 1 & 1 \\ 2/\sqrt{3} & 1/\sqrt{3} & 1/\sqrt{3} \\ 0 & 1 & 1/3 \end{vmatrix} = \frac{1}{3\sqrt{3}} \tag{6.82}$$

$$\Delta^{(4)} = \frac{1}{2}\begin{vmatrix} 1 & 1 & 1 \\ 1/\sqrt{3} & 0 & 1/\sqrt{3} \\ 1 & 0 & 1/3 \end{vmatrix} = \frac{1}{3\sqrt{3}} \tag{6.83}$$

Associated with the inter-element vector, the boundary integrals in Eq. (6.70) and (6.71) are rewritten as

$$\mathbf{h}^{(1)} = \oint_{C_{1-5}^{(1)}} \begin{bmatrix} N_1^{(1)}N_1^{(1)} & N_1^{(1)}N_2^{(1)} & N_1^{(1)}N_3^{(1)} \\ N_2^{(1)}N_1^{(1)} & N_2^{(1)}N_2^{(1)} & N_2^{(1)}N_3^{(1)} \\ N_3^{(1)}N_1^{(1)} & N_3^{(1)}N_2^{(1)} & N_3^{(1)}N_3^{(1)} \end{bmatrix} ds \tag{6.84a}$$

$$\mathbf{g}^{(1)} = \oint_{C_{1-5}^{(1)}} \begin{Bmatrix} N_1^{(1)} \\ N_2^{(1)} \\ N_3^{(1)} \end{Bmatrix} ds \tag{6.84b}$$

and

$$\mathbf{h}^{(2)} = \oint_{C_{5-2}^{(2)}} \begin{bmatrix} N_1^{(2)}N_1^{(2)} & N_1^{(2)}N_2^{(2)} & N_1^{(2)}N_3^{(2)} \\ N_2^{(2)}N_1^{(2)} & N_2^{(2)}N_2^{(2)} & N_2^{(2)}N_3^{(2)} \\ N_3^{(2)}N_1^{(2)} & N_3^{(2)}N_2^{(2)} & N_3^{(2)}N_3^{(2)} \end{bmatrix} ds \tag{6.85a}$$

$$\mathbf{g}^{(2)} = \oint_{C_{5-2}^{(2)}} \begin{Bmatrix} N_1^{(2)} \\ N_2^{(2)} \\ N_3^{(2)} \end{Bmatrix} ds \tag{6.85b}$$

in which $N_3^{(1)}$ and $N_3^{(2)}$ are zero along side 1–5 (with length L_{1-5}) and along side 5–2 (with length L_{2-5}), respectively. The remaining shape functions $N_1^{(1)}$, $N_2^{(1)}$, $N_1^{(2)}$, and $N_2^{(2)}$ reduce to a one-dimensional form as

$$N_1^{(1)} = \frac{L_{1-5} - s}{L_{1-5}} \quad \text{and} \quad N_2^{(1)} = \frac{s}{L_{1-5}} \tag{6.86}$$

$$N_1^{(2)} = \frac{L_{5-2} - s}{L_{5-2}} \quad \text{and} \quad N_2^{(2)} = \frac{s}{L_{5-2}} \tag{6.87}$$

in which s is the local coordinate in the range of $(0 \le s \le L_{1-5})$ along side 1–5 and $(0 \le s \le L_{5-2})$ along side 5–2, $L_{1-5} = 1/\sqrt{3}$, and $L_{5-2} = 1/\sqrt{3}$. With these shape functions, the evaluation of $\mathbf{h}^{(1)}$, $\mathbf{g}^{(1)}$, $\mathbf{h}^{(2)}$, and $\mathbf{g}^{(2)}$ leads to

$$\mathbf{h}^{(1)} = \frac{1}{6\sqrt{3}} \begin{bmatrix} 2 & 1 & 0 \\ 1 & 2 & 0 \\ 0 & 0 & 0 \end{bmatrix} \quad \text{and} \quad \mathbf{g}^{(1)} = \frac{1}{2\sqrt{3}} \begin{Bmatrix} 1 \\ 1 \\ 0 \end{Bmatrix} \tag{6.88}$$

and

$$\mathbf{h}^{(2)} = \frac{1}{6\sqrt{3}} \begin{bmatrix} 2 & 1 & 0 \\ 1 & 2 & 0 \\ 0 & 0 & 0 \end{bmatrix} \quad \text{and} \quad \mathbf{g}^{(2)} = \frac{1}{2\sqrt{3}} \begin{Bmatrix} 1 \\ 1 \\ 0 \end{Bmatrix} \tag{6.89}$$

Considering the correspondence between the local and global node numbering presented in Table 6.3, the element characteristic matrices and vectors can be rewritten as

Table 6.3 Element connectivity

Element number (e)	Node 1	Node 2	Node 3
1	1	5	4
2	5	2	4
3	2	3	4
4	3	1	4

Element 1:

$$\begin{matrix} \boxed{1} & \boxed{5} & \boxed{4} \end{matrix}$$
$$\begin{bmatrix} k_{11}^{(1)}+h_{11}^{(1)} & k_{12}^{(1)}+h_{12}^{(1)} & k_{13}^{(1)}+h_{13}^{(1)} \\ k_{21}^{(1)}+h_{21}^{(1)} & k_{22}^{(1)}+h_{22}^{(1)} & k_{23}^{(1)}+h_{23}^{(1)} \\ k_{31}^{(1)}+h_{31}^{(1)} & k_{32}^{(1)}+h_{32}^{(1)} & k_{33}^{(1)}+h_{33}^{(1)} \end{bmatrix} \begin{Bmatrix} \phi_1^{(1)} \\ \phi_2^{(1)} \\ \phi_3^{(1)} \end{Bmatrix} = \begin{Bmatrix} f_1^{(1)}+g_1^{(1)} \\ f_2^{(1)}+g_2^{(1)} \\ f_3^{(1)}+g_3^{(1)} \end{Bmatrix} \begin{matrix} \boxed{1} \\ \boxed{5} \\ \boxed{4} \end{matrix} \tag{6.90}$$

Element 2:

$$\begin{matrix} \boxed{5} & \boxed{2} & \boxed{4} \end{matrix}$$
$$\begin{bmatrix} k_{11}^{(2)}+h_{11}^{(2)} & k_{12}^{(2)}+h_{12}^{(2)} & k_{13}^{(2)}+h_{13}^{(2)} \\ k_{21}^{(2)}+h_{21}^{(2)} & k_{22}^{(2)}+h_{22}^{(2)} & k_{23}^{(2)}+h_{23}^{(2)} \\ k_{31}^{(2)}+h_{31}^{(2)} & k_{32}^{(2)}+h_{32}^{(2)} & k_{33}^{(2)}+h_{33}^{(2)} \end{bmatrix} \begin{Bmatrix} \phi_1^{(2)} \\ \phi_2^{(2)} \\ \phi_3^{(2)} \end{Bmatrix} = \begin{Bmatrix} f_1^{(2)}+g_1^{(2)} \\ f_2^{(2)}+g_2^{(2)} \\ f_3^{(2)}+g_3^{(2)} \end{Bmatrix} \begin{matrix} \boxed{5} \\ \boxed{2} \\ \boxed{4} \end{matrix} \tag{6.91}$$

Element 3:

$$\begin{matrix} \boxed{2} & \boxed{3} & \boxed{4} \end{matrix}$$
$$\begin{bmatrix} k_{11}^{(3)} & k_{12}^{(3)} & k_{13}^{(3)} \\ k_{21}^{(3)} & k_{22}^{(3)} & k_{23}^{(3)} \\ k_{31}^{(3)} & k_{32}^{(3)} & k_{33}^{(3)} \end{bmatrix} \begin{Bmatrix} \phi_1^{(3)} \\ \phi_2^{(3)} \\ \phi_3^{(3)} \end{Bmatrix} = \begin{Bmatrix} f_1^{(3)} \\ f_2^{(3)} \\ f_3^{(3)} \end{Bmatrix} \begin{matrix} \boxed{2} \\ \boxed{3} \\ \boxed{4} \end{matrix} \tag{6.92}$$

Element 4:

$$\begin{matrix} \boxed{3} & \boxed{1} & \boxed{4} \end{matrix}$$
$$\begin{bmatrix} k_{11}^{(4)} & k_{12}^{(4)} & k_{13}^{(4)} \\ k_{21}^{(4)} & k_{22}^{(4)} & k_{23}^{(4)} \\ k_{31}^{(4)} & k_{32}^{(4)} & k_{33}^{(4)} \end{bmatrix} \begin{Bmatrix} \phi_1^{(4)} \\ \phi_2^{(4)} \\ \phi_3^{(4)} \end{Bmatrix} = \begin{Bmatrix} f_1^{(4)} \\ f_2^{(4)} \\ f_3^{(4)} \end{Bmatrix} \begin{matrix} \boxed{3} \\ \boxed{1} \\ \boxed{4} \end{matrix} \tag{6.93}$$

In the assembly of the element characteristic matrices and vectors, the boxed numbers indicate the rows and columns of the global matrix, **K**, and global right-hand-side vector, **F**, to which the individual coefficients are added, resulting in

$$\mathbf{K}\mathbf{\Phi} = \mathbf{F} \tag{6.94}$$

where

$$\mathbf{K} = \begin{bmatrix} k_{11}^{(1)} + h_{11}^{(1)} + k_{22}^{(4)} & 0 & k_{21}^{(4)} & k_{13}^{(1)} + h_{13}^{(1)} + k_{23}^{(4)} & k_{12}^{(1)} + h_{12}^{(1)} \\ 0 & k_{22}^{(2)} + h_{22}^{(2)} + k_{11}^{(3)} & k_{12}^{(3)} & k_{23}^{(2)} + h_{23}^{(2)} + k_{13}^{(3)} & k_{21}^{(2)} + h_{21}^{(2)} \\ k_{12}^{(4)} & k_{21}^{(3)} & k_{22}^{(3)} + k_{11}^{(4)} & k_{23}^{(3)} + k_{13}^{(4)} & 0 \\ k_{31}^{(1)} + h_{31}^{(1)} + k_{32}^{(4)} & k_{32}^{(2)} + h_{32}^{(2)} + k_{31}^{(3)} & k_{32}^{(3)} + k_{31}^{(4)} & k_{33}^{(1)} + h_{33}^{(1)} + k_{33}^{(2)} + h_{33}^{(2)} + k_{33}^{(3)} + k_{33}^{(4)} & k_{32}^{(1)} + h_{32}^{(1)} + k_{31}^{(2)} + h_{31}^{(2)} \\ k_{21}^{(1)} + h_{21}^{(1)} & k_{12}^{(2)} + h_{12}^{(2)} & 0 & k_{23}^{(1)} + h_{23}^{(1)} + k_{13}^{(2)} + h_{13}^{(2)} & k_{22}^{(1)} + h_{22}^{(1)} + k_{11}^{(2)} + h_{11}^{(2)} \end{bmatrix} \tag{6.95a}$$

$$\mathbf{F} = \begin{Bmatrix} f_1^{(1)} + g_1^{(1)} + f_2^{(4)} \\ f_2^{(2)} + g_2^{(2)} + f_1^{(3)} \\ f_2^{(3)} + f_1^{(4)} \\ f_3^{(1)} + g_3^{(1)} + f_3^{(2)} + g_3^{(2)} + f_3^{(3)} + f_3^{(4)} \\ f_2^{(1)} + g_2^{(1)} + f_1^{(2)} + g_1^{(2)} \end{Bmatrix} \tag{6.95b}$$

$$\mathbf{\Phi} = \begin{Bmatrix} \phi_1 \\ \phi_2 \\ \phi_3 \\ \phi_4 \\ \phi_5 \end{Bmatrix} \tag{6.95c}$$

After imposing the essential boundary conditions, the global system of equations are reduced by deleting the rows and columns corresponding to ϕ_1, ϕ_2, and ϕ_3, leading to

$$\begin{bmatrix} k_{33}^{(1)} + h_{33}^{(1)} + k_{33}^{(2)} + h_{33}^{(2)} + k_{33}^{(3)} + k_{33}^{(4)} & k_{32}^{(1)} + h_{32}^{(1)} + k_{31}^{(2)} + h_{31}^{(2)} \\ k_{23}^{(1)} + h_{23}^{(1)} + k_{13}^{(2)} - h_{13}^{(2)} & k_{22}^{(1)} + h_{22}^{(1)} + k_{11}^{(2)} + h_{11}^{(2)} \end{bmatrix} \begin{Bmatrix} \phi_4 \\ \phi_5 \end{Bmatrix} = \begin{Bmatrix} f_3^{(1)} + g_3^{(1)} + f_3^{(2)} + g_3^{(2)} + f_3^{(3)} + f_3^{(4)} \\ f_2^{(1)} + g_2^{(1)} + f_1^{(2)} + g_1^{(2)} \end{Bmatrix} \tag{6.96}$$

With the explicit values of the coefficients, the nodal unknowns, ϕ_4 and ϕ_5, are determined as

$$\phi_4 = \frac{4}{27} = 0.14815 \tag{6.97a}$$

$$\phi_5 = \frac{1}{3} = 0.33333 \tag{6.97b}$$

The expressions for $\mathbf{h}^{(1)}$ and $\mathbf{h}^{(2)}$ in Eq. (6.70) are derived based on a formulation consistent with the derivation of the element coefficient matrices, $\mathbf{k}^{(e)}$. An alternative to the consistent formulation is the use of lumped diagonal matrices and expressing $\mathbf{h}^{(1)}$ and $\mathbf{h}^{(2)}$ in the form

$$\mathbf{h}^{(1)} = \oint_{C_{1-5}^{(1)}} \begin{bmatrix} N_1^{(1)} & 0 & 0 \\ 0 & N_2^{(1)} & 0 \\ 0 & 0 & N_3^{(1)} \end{bmatrix} ds = \frac{1}{6\sqrt{3}} \begin{bmatrix} 3 & 0 & 0 \\ 0 & 3 & 0 \\ 0 & 0 & 0 \end{bmatrix} \tag{6.98}$$

and

$$\mathbf{h}^{(2)} = \oint_{C_{5-2}^{(2)}} \begin{bmatrix} N_1^{(2)} & 0 & 0 \\ 0 & N_2^{(2)} & 0 \\ 0 & 0 & N_3^{(2)} \end{bmatrix} ds = \frac{1}{6\sqrt{3}} \begin{bmatrix} 3 & 0 & 0 \\ 0 & 3 & 0 \\ 0 & 0 & 0 \end{bmatrix} \tag{6.99}$$

Replacing the components of $\mathbf{h}^{(1)}$ and $\mathbf{h}^{(2)}$ in Eq. (6.96) with the values obtained in Eq. (6.98) and (6.99), the nodal unknowns ϕ_4 and ϕ_5 are determined as

$$\phi_4 = \frac{5}{36} = 0.13889 \tag{6.100a}$$

$$\phi_5 = \frac{11}{36} = 0.30556 \tag{6.100b}$$

Note that the discrepancy in the value of ϕ_4 and ϕ_5 obtained from the two methods is due to the small number of elements in the discretization of the domain.

6.1.2.2 ANSYS Solution

The governing equations for a steady-state heat transfer, described by Eq. (6.47) through (6.50), also can be solved using ANSYS. The solution procedure is outlined as follows:

Model Generation

- Specify the element type (**ET** command) using the following menu path:

Main Menu > Preprocessor > Element Type > Add/Edit/Delete

  - Click on ***Add***.
  - Select ***Solid*** immediately below ***Thermal Mass*** from the left list and ***Quad 4node 55*** from the right list; click on ***OK***.
  - Click on ***Close***.

- Specify material properties (**MP** command) using the following menu path:

Main Menu > Preprocessor > Material Props > Material Models

  - In the *Define Material Model Behavior* dialog box, in the right window, successively left-click on ***Thermal***, ***Conductivity***, and, finally, ***Isotropic***, which brings up another dialog box.
  - Enter ***1*** for ***KXX***, and click on ***OK***.
  - Close the *Define Material Model Behavior* dialog box by using the following menu path:

Material > Exit

- Create nodes (**N** command) using the following menu path:

Main Menu > Preprocessor > Modeling > Create > Nodes > In Active CS

  - A total of 5 nodes are created (Table 6.2).
  - Referring to Table 6.2, enter x- and y-coordinates of node 1, and click on ***Apply***. This action keeps the *Create Nodes in Active Coordinate System* dialog box open. If the ***Node number*** field is left blank, then ANSYS assigns the lowest available node number to the node that is being created.
  - Repeat the same procedure for the nodes 2 through 5.
  - After entering the x- and y-coordinates of node 5, click on ***OK*** (instead of ***Apply***).
  - The nodes should appear in the *Graphics Window*, as shown in Fig. 6.7.

- Create elements (**E** command) using the following menu path:

Main Menu > Preprocessor > Modeling > Create > Elements > Auto Numbered > Thru Nodes

  - *Pick Menu* appears; refer to Fig. 6.8 to create elements by picking ***three*** nodes at a time and clicking on ***Apply*** in between.
  - Observe the elements created after clicking on ***Apply*** in the *Pick Menu*.

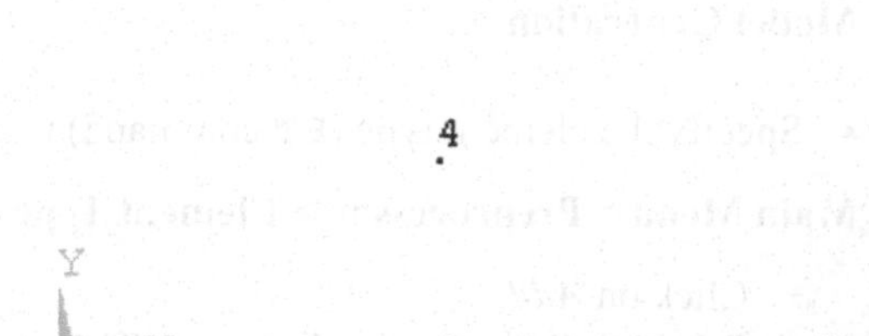

Fig. 6.7 Generation of nodes

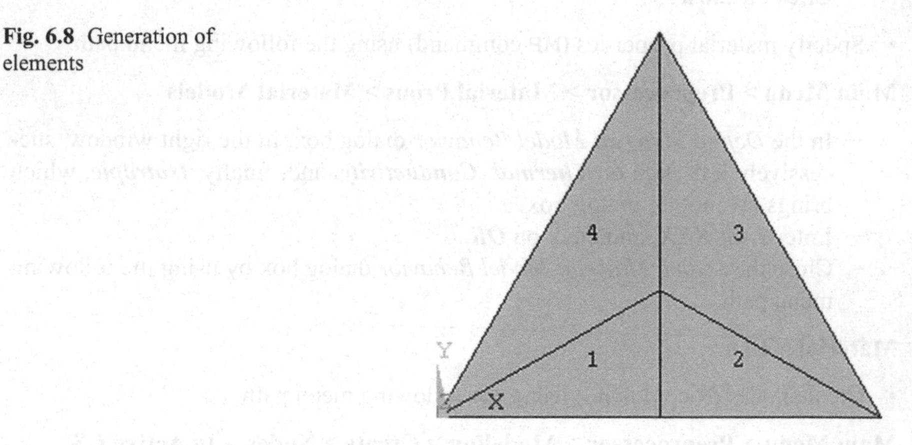

Fig. 6.8 Generation of elements

- Repeat until the last element is created.
- Click on ***OK*** when the last element is created.

• Review elements:

- Turn on element numbering using the following menu path:

Utility Menu > PlotCtrls > Numbering

- Select ***Element numbers*** from the first pull-down menu; click on ***OK***.
- Plot elements (**`EPLOT`** command) using the following menu path:

Utility Menu > Plot > Elements

- Figure 6.8 shows the outcome of this action as it appears in the *Graphics Window*.
- Turn off element numbering and turn on node numbering using the following menu path:

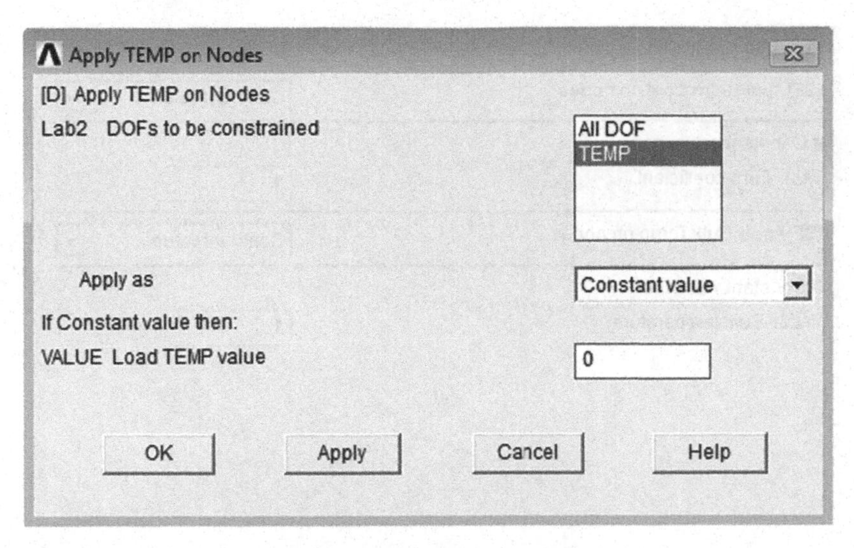

Fig. 6.9 Application of temperature boundary conditions on nodes

Utility Menu > PlotCtrls > Numbering

- Place a *checkmark* by clicking on the empty box next to ***NODE Node numbers***.
- Select ***No numbering*** from the first pull-down menu.
- Click on ***OK***.
- Plot nodes (**NPLOT** command) using the following menu path:

Utility Menu > Plot > Nodes

- Figure 6.7 shows the outcome of this action as it appears in the *Graphics Window*.

Solution

- Apply temperature boundary conditions (Dcommand) using the following menu path:

Main Menu > Solution > Define Loads > Apply > Thermal > Temperature > On Nodes

- *Pick Menu* appears; pick nodes 1, 2, and 3 (Fig. 6.7); click on ***OK*** on *Pick Menu*.
- Highlight ***TEMP*** and enter **0** for ***VALUE***; click on ***OK*** (Fig. 6.9).
- Apply convection boundary conditions (**SF** command) using the following menu path:

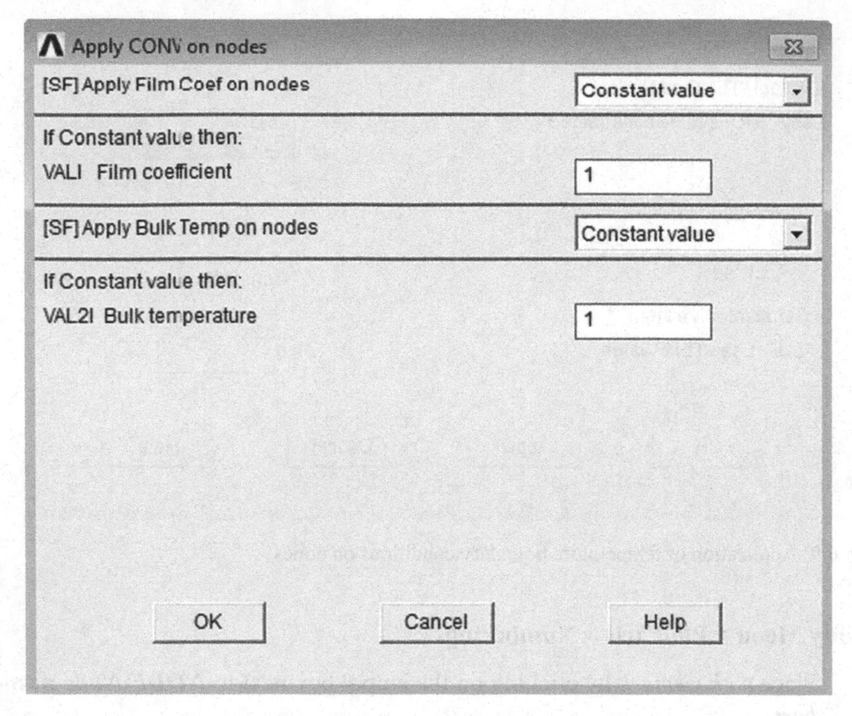

Fig. 6.10 Application of convection boundary conditions on nodes

Main Menu > Solution > Define Loads > Apply > Thermal > Convection > On Nodes

- *Pick Menu* appears; pick nodes 1, 2 and 5 along the boundary (Fig. 6.7); click on ***OK*** on *Pick Menu.*
- Enter ***1*** for both ***VALI Film coefficient*** and ***VAL2I Bulk temperature***; click on ***OK*** (Fig. 6.10).

- Apply body load on elements (**BFE** command) using the following menu path:

Main Menu > Solution > Define Loads > Apply > Thermal > Heat Generat > On Elements

- *Pick Menu* appears; click on ***Pick All.***
- Enter ***1*** for ***VAL1*** leave other fields untouched, as shown in Fig. 6.11.
- Click on ***OK***.

- Obtain solution (**SOLVE** command) using the following menu path:

Main Menu > Solution > Solve >Current LS

- *Confirmation Window* appears along with *Status Report Window.*
- Review status. If OK, close the *Status Report Window* and click on ***OK*** in *Confirmation Window.*
- Wait until ANSYS responds with ***Solution is done!***

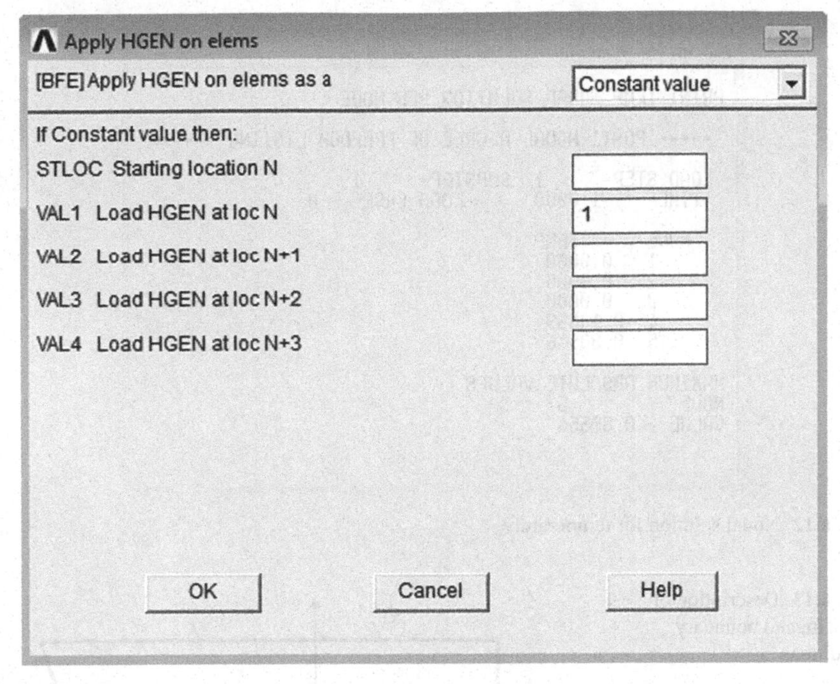

Fig. 6.11 Application of heat generation condition on elements

Postprocessing

- Review temperature values (**PRNSOL** command) using the following menu path:

Main Menu > General Postproc > List Results > Nodal Solution

- Click on ***DOF Solution*** and ***Nodal Temperature***; click on ***OK***.
- The list appears in a new window, as shown in Fig. 6.12.

6.1.3 Example: Two-Dimensional Differential Equation with Linear Quadrilateral Elements

6.1.3.1 Galerkin's Method

In solving two-dimensional problems with quadrilateral isoparametric elements, Galerkin's method is demonstrated by considering the partial differential equation given by

$$\frac{\partial^2\phi(x,y)}{\partial x^2}+\frac{\partial^2\phi(x,y)}{\partial y^2}-A=0 \tag{6.101}$$

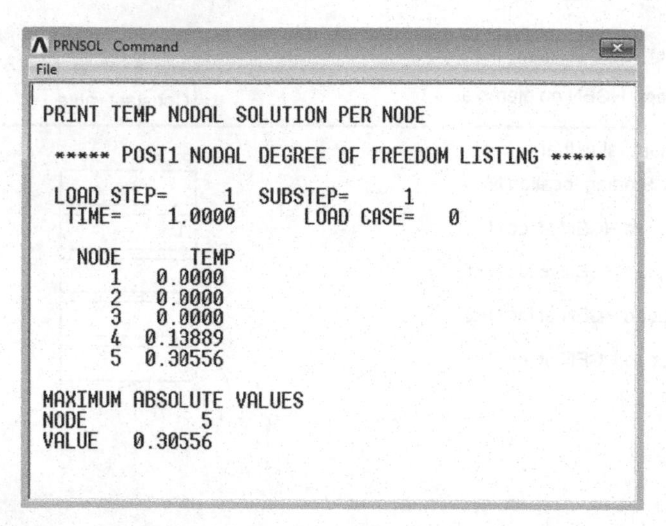

Fig. 6.12 Nodal solution for temperature

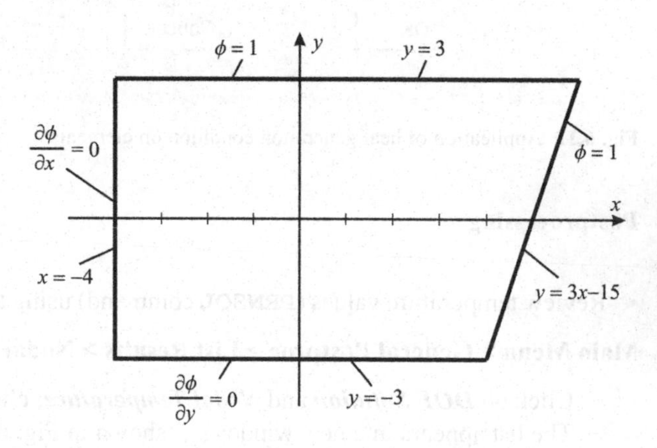

Fig. 6.13 Description of domain, and boundary conditions

in domain D defined by the intersection of $y = -3$, $x = -4$, $y = 3$, and $y = 3x - 15$. The constant, A, is known. As shown in Fig. 6.13, the flux vanishes along the boundary of the domain specified by $y = -3$ and $x = -4$, and along the remaining part of the boundary specified by $y = 3$, and $y = 3x - 15$, the dependent variable, $\phi(x, y)$, has a value of unity. These boundary conditions are expressed as

$$\phi(x, y) = 1 \quad \text{for} \quad 4 \le x \le 6, y = 3x - 15 \tag{6.102}$$

$$\frac{\partial}{\partial x}\phi(x, y = -3) = 0 \quad \text{for} \quad -4 \le x \le 4 \tag{6.103}$$

$$\frac{\partial}{\partial x}\phi(x = -4, y) = 0 \quad \text{for} \quad -3 \le y \le 3 \tag{6.104}$$

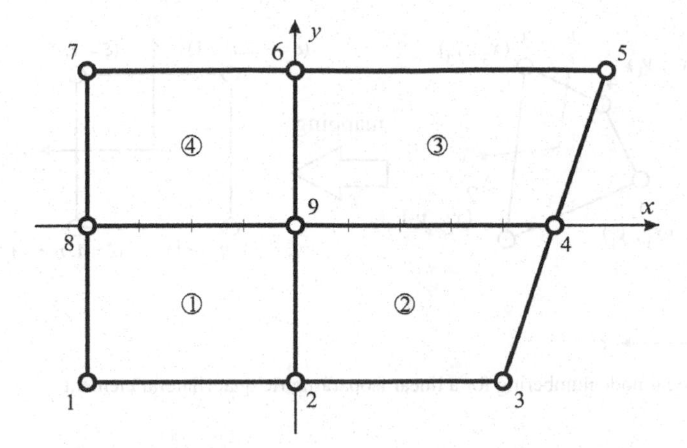

Fig. 6.14 FEM discretization of the domain into four quadrilaterals

$$\phi(x, y=3)=1 \quad \text{for} \quad -4 \le x \le 6 \tag{6.105}$$

The domain is discretized with four linear quadrilateral isoparametric elements, each having four nodes identified as 1, 2, 3, and 4, shown in Fig. 6.14. The nodal values of the dependent variable associated with element e are specified at its first, second, third, and fourth nodes by $\phi_1^{(e)}$, $\phi_2^{(e)}$, $\phi_3^{(e)}$, and $\phi_4^{(e)}$, respectively. The discretization of the domain with global node numbering is shown in Fig. 6.14. The global coordinates of the nodal values of the dependent variable denoted by ϕ_i $(i=1,2,\ldots,9)$ are presented in Table 6.4.

The linear element approximation function for the dependent field variable in a quadrilateral isoparametric element "e" is written as

$$\tilde{\phi}^{(e)} = N_1^{(e)}\phi_1^{(e)} + N_2^{(e)}\phi_2^{(e)} + N_3^{(e)}\phi_3^{(e)} + N_4^{(e)}\phi_4^{(e)} \tag{6.106}$$

Table 6.4 Nodal coordinates

Global node number	Nodal coordinates	Nodal variables
1	$(-4,-3)$	ϕ_1
2	$(0,-3)$	ϕ_2
3	$(4,-3)$	ϕ_3
4	$(5,0)$	ϕ_4
5	$(6,3)$	ϕ_5
6	$(0,3)$	ϕ_6
7	$(-4,3)$	ϕ_7
8	$(-4,0)$	ϕ_8
9	$(0,0)$	ϕ_9

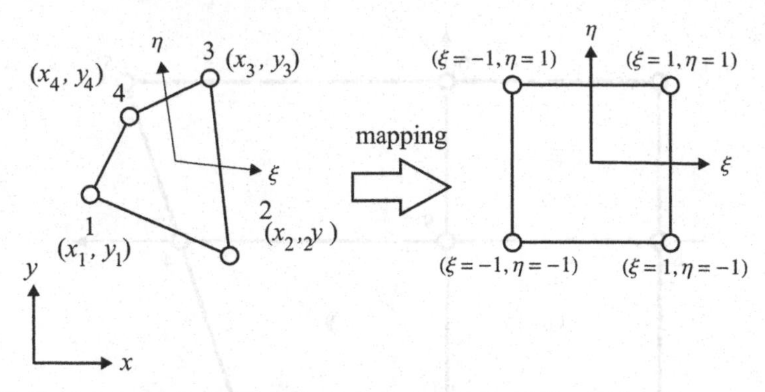

Fig. 6.15 Local node numbering for a linear isoparametric quadrilateral element

or

$$\tilde{\phi}^{(e)} = \mathbf{N}^{(e)T} \boldsymbol{\varphi}^{(e)} \tag{6.107}$$

where

$$\mathbf{N}^{(e)} = \begin{Bmatrix} N_1^{(e)} \\ N_2^{(e)} \\ N_3^{(e)} \\ N_4^{(e)} \end{Bmatrix} \quad \text{and} \quad \boldsymbol{\varphi}^{(e)} = \begin{Bmatrix} \phi_1^{(e)} \\ \phi_2^{(e)} \\ \phi_3^{(e)} \\ \phi_4^{(e)} \end{Bmatrix} \tag{6.108}$$

in which the shape functions $N_1^{(e)}$, $N_2^{(e)}$, $N_3^{(e)}$, and $N_4^{(e)}$ are expressed in terms of the centroidal or natural coordinates, (ξ, η), shown in Fig. 6.15. For a linear (straight-sided) quadrilateral illustrated in Fig. 6.15, they are of the form

$$N_i^{(e)} = \frac{1}{4}(1+\xi\xi_i)(1+\eta\eta_i) \quad \text{with} \quad i = 1,2,3,4 \tag{6.109}$$

where ξ_i and η_i represent the coordinates of the corner nodes in the natural coordinate system, $(\xi_1 = -1, \eta_1 = -1)$, $(\xi_2 = 1, \eta_2 = -1)$, $(\xi_3 = 1, \eta_3 = 1)$, and $(\xi_4 = -1, \eta_4 = 1)$.

Applying Eq. (6.6), Galerkin's method, leads to

$$\sum_{e=1}^{E} \int_{D^{(e)}} \mathbf{N}^{(e)} \left(\frac{\partial^2 \tilde{\phi}^{(e)}(x,y)}{\partial x^2} + \frac{\partial^2 \tilde{\phi}^{(e)}(x,y)}{\partial y^2} - A \right) dx\,dy = 0 \tag{6.110}$$

Since the element approximation function is C^0 continuous, the second-order derivatives in the integrand must be reduced by one so that inter-element continuity is achieved during the assembly of the global matrix. This reduction is achieved by observing that

$$\mathbf{N}^{(e)} \frac{\partial^2 \tilde{\phi}^{(e)}}{\partial x^2}(x,y) = \frac{\partial}{\partial x}\left(\mathbf{N}^{(e)} \frac{\partial \tilde{\phi}^{(e)}}{\partial x}(x,y) \right) - \frac{\partial \mathbf{N}^{(e)}}{\partial x} \frac{\partial \tilde{\phi}^{(e)}}{\partial x}(x,y) \tag{6.111}$$

and

$$\mathbf{N}^{(e)} \frac{\partial^2 \tilde{\phi}^{(e)}}{\partial y^2}(x,y) = \frac{\partial}{\partial y}\left(\mathbf{N}^{(e)} \frac{\partial \tilde{\phi}^{(e)}}{\partial y}(x,y) \right) - \frac{\partial \mathbf{N}^{(e)}}{\partial y} \frac{\partial \tilde{\phi}^{(e)}}{\partial y}(x,y) \tag{6.112}$$

Their substitution into the integrand in Eq. (6.110) and rearrangement of the terms result in

$$\sum_{e=1}^{E} \left\{ \int_{D^{(e)}} \left[\frac{\partial}{\partial x}\left(\mathbf{N}^{(e)} \frac{\partial \tilde{\phi}^{(e)}}{\partial x} \right) + \frac{\partial}{\partial y}\left(\mathbf{N}^{(e)} \frac{\partial \tilde{\phi}^{(e)}}{\partial y} \right) \right] dx\,dy + \int_{D^{(e)}} \left[-\frac{\partial \mathbf{N}^{(e)}}{\partial x} \frac{\partial \tilde{\phi}^{(e)}}{\partial x} - \frac{\partial \mathbf{N}^{(e)}}{\partial y} \frac{\partial \tilde{\phi}^{(e)}}{\partial y} + \mathbf{N}^{(e)} A \right] dx\,dy \right\} = 0 \tag{6.113}$$

Applying the divergence theorem to the first integral renders the domain integral to the boundary integral, and it yields

$$\sum_{e=1}^{E} \left\{ \oint_{C^{(e)}} \left[\left(\mathbf{N}^{(e)} \frac{\partial \tilde{\phi}^{(e)}}{\partial x} \right) n_x^{(e)} + \left(\mathbf{N}^{(e)} \frac{\partial \tilde{\phi}^{(e)}}{\partial y} \right) n_y^{(e)} \right] ds + \int_{D^{(e)}} \left[-\frac{\partial \mathbf{N}^{(e)}}{\partial x} \frac{\partial \tilde{\phi}^{(e)}}{\partial x} - \frac{\partial \mathbf{N}^{(e)}}{\partial y} \frac{\partial \tilde{\phi}^{(e)}}{\partial y} + \mathbf{N}^{(e)} A \right] dx\,dy \right\} = 0 \tag{6.114}$$

where $n_x^{(e)}$ and $n_y^{(e)}$ are, respectively, the x- and y-components of the outward normal vector along the closed boundary defining the area of the element $C^{(e)}$.

Substituting for the element approximation function yields

$$\sum_{e=1}^{E}\left\{\oint_{C^{(e)}} \mathbf{N}^{(e)}\left[\frac{\partial\tilde{\phi}^{(e)}}{\partial x}n_x^{(e)}+\frac{\partial\tilde{\phi}^{(e)}}{\partial y}n_y^{(e)}\right]ds\right.$$

$$+\int_{D^{(e)}}\left[-\frac{\partial\mathbf{N}^{(e)}}{\partial x}\frac{\partial\mathbf{N}^{(e)T}}{\partial x}-\frac{\partial\mathbf{N}^{(e)}}{\partial y}\frac{\partial\mathbf{N}^{(e)T}}{\partial y}\right]dx\,dy\boldsymbol{\varphi}^{(e)} \tag{6.115}$$

$$\left.-\int_{D^{(e)}}\mathbf{N}^{(e)}A\,dx\,dy\right\}=0$$

This equation can be recast in matrix form as

$$\sum_{e=1}^{E}\left(\mathbf{k}^{(e)}\boldsymbol{\varphi}^{(e)}-\mathbf{f}^{(e)}+\mathbf{Q}^{(e)}\right)=0 \tag{6.116}$$

where

$$\mathbf{k}^{(e)}=-\int_{D^{(e)}}\left[\frac{\partial\mathbf{N}^{(e)}}{\partial x}\frac{\partial\mathbf{N}^{(e)T}}{\partial x}+\frac{\partial\mathbf{N}^{(e)}}{\partial y}\frac{\partial\mathbf{N}^{(e)T}}{\partial y}\right]dx\,dy \tag{6.117}$$

$$\mathbf{f}^{(e)}=\int_{D^{(e)}}A\mathbf{N}^{(e)}dx\,dy \tag{6.118}$$

$$\mathbf{Q}^{(e)}=\oint_{C^{(e)}}\mathbf{N}^{(e)}\left[\frac{\partial\tilde{\phi}^{(e)}}{\partial x}n_x^{(e)}+\frac{\partial\tilde{\phi}^{(e)}}{\partial y}n_y^{(e)}\right]ds \tag{6.119}$$

in which $\mathbf{k}^{(e)}$ is the element characteristic matrix, $\mathbf{f}^{(e)}$ is the element right-hand-side vector, and $\mathbf{Q}^{(e)}$ is often referred to as the inter-element vector that includes the derivative terms along the boundary of the element. The boundary integral around each element is evaluated in a counterclockwise direction, i.e., this boundary integral is the sum of four integrals taken along each side of the element.

Because the specified derivatives have zero values along the element boundaries, the inter-element vector, $\mathbf{Q}^{(e)}$ vanishes, i.e., $\mathbf{Q}^{(e)}=0$, thus reducing the element equilibrium equations to

$$\sum_{e=1}^{E}\left(\mathbf{k}^{(e)}\boldsymbol{\varphi}^{(e)}-\mathbf{f}^{(e)}\right)=0 \tag{6.120}$$

The integrals contributing to the characteristic element matrix, $\mathbf{k}^{(e)}$, and the right-hand-side vector, $\mathbf{f}^{(e)}$, are evaluated over a square region in the natural coordinate system after an appropriate coordinate transformation given by

$$x = \sum_{i=1}^{4} N_i^{(e)}(\xi,\eta)x_i^{(e)} \quad \text{and} \quad y = \sum_{i=1}^{4} N_i^{(e)}(\xi,\eta)y_i^{(e)} \tag{6.121}$$

Application of the chain rule of differentiation yields

$$\begin{Bmatrix} \dfrac{\partial N_i^{(e)}}{\partial \xi} \\ \dfrac{\partial N_i^{(e)}}{\partial \eta} \end{Bmatrix} = \begin{bmatrix} \dfrac{\partial x}{\partial \xi} & \dfrac{\partial y}{\partial \xi} \\ \dfrac{\partial x}{\partial \eta} & \dfrac{\partial y}{\partial \eta} \end{bmatrix} \begin{Bmatrix} \dfrac{\partial N_i^{(e)}}{\partial x} \\ \dfrac{\partial N_i^{(e)}}{\partial y} \end{Bmatrix} \quad \text{with} \quad i = 1,2,3,4 \tag{6.122}$$

or

$$\begin{Bmatrix} \dfrac{\partial}{\partial \xi} \\ \dfrac{\partial}{\partial \eta} \end{Bmatrix} N_i^{(e)} = \mathbf{J} \begin{Bmatrix} \dfrac{\partial}{\partial x} \\ \dfrac{\partial}{\partial y} \end{Bmatrix} N_i^{(e)} \tag{6.123}$$

where $\mathbf{J}$ is called the Jacobian matrix. It can be expressed as

$$\mathbf{J} = \begin{bmatrix} J_{11} & J_{12} \\ J_{21} & J_{22} \end{bmatrix} \tag{6.124}$$

in which

$$J_{11} = \frac{\partial x}{\partial \xi} = \frac{1}{4}\left\{-(1-\eta)x_1^{(e)} + (1-\eta)x_2^{(e)} + (1+\eta)x_3^{(e)} - (1+\eta)x_4^{(e)}\right\} \tag{6.125}$$

$$J_{12} = \frac{\partial y}{\partial \xi} = \frac{1}{4}\left\{-(1-\eta)y_1^{(e)} + (1-\eta)y_2^{(e)} + (1+\eta)y_3^{(e)} - (1+\eta)y_4^{(e)}\right\} \tag{6.126}$$

$$J_{21} = \frac{\partial x}{\partial \eta} = \frac{1}{4}\left\{-(1-\xi)x_1^{(e)} - (1+\xi)x_2^{(e)} + (1+\xi)x_3^{(e)} + (1-\xi)x_4^{(e)}\right\} \tag{6.127}$$

$$J_{22} = \frac{\partial y}{\partial \eta} = \frac{1}{4}\left\{-(1-\xi)y_1^{(e)} - (1+\xi)y_2^{(e)} + (1+\xi)y_3^{(e)} + (1-\xi)y_4^{(e)}\right\} \tag{6.128}$$

Also, the Jacobian can be rewritten in the form

$$\mathbf{J}=\begin{bmatrix}\dfrac{\partial N_1^{(e)}}{\partial \xi} & \dfrac{\partial N_2^{(e)}}{\partial \xi} & \dfrac{\partial N_3^{(e)}}{\partial \xi} & \dfrac{\partial N_4^{(e)}}{\partial \xi}\\ \dfrac{\partial N_1^{(e)}}{\partial \eta} & \dfrac{\partial N_2^{(e)}}{\partial \eta} & \dfrac{\partial N_3^{(e)}}{\partial \eta} & \dfrac{\partial N_4^{(e)}}{\partial \eta}\end{bmatrix}\begin{bmatrix}x_1^{(e)} & y_1^{(e)}\\ x_2^{(e)} & y_2^{(e)}\\ x_3^{(e)} & y_3^{(e)}\\ x_4^{(e)} & y_4^{(e)}\end{bmatrix} \tag{6.129}$$

or

$$\mathbf{J}=\frac{1}{4}\begin{bmatrix}-(1-\eta) & (1-\eta) & (1+\eta) & -(1+\eta)\\ -(1-\xi) & -(1+\xi) & (1+\xi) & (1-\xi)\end{bmatrix}\begin{bmatrix}x_1^{(e)} & y_1^{(e)}\\ x_2^{(e)} & y_2^{(e)}\\ x_3^{(e)} & y_3^{(e)}\\ x_4^{(e)} & y_4^{(e)}\end{bmatrix} \tag{6.130}$$

Because the transformation between the natural and global coordinates has a one-to-one correspondence, the inverse of the Jacobian exists, and it can be expressed as

$$\mathbf{J}^{-1}=\frac{1}{|\mathbf{J}|}\begin{bmatrix}J_{22} & -J_{12}\\ -J_{21} & J_{11}\end{bmatrix} \tag{6.131}$$

When the element is degenerated into a triangle by increasing an internal angle to 180°, $\mathbf{J}$ is singular at that corner. The inverse of the Jacobian matrix permits the expression for the derivatives in terms of global coordinates

$$\begin{Bmatrix}\dfrac{\partial N_i^{(e)}}{\partial x}\\ \dfrac{\partial N_i^{(e)}}{\partial y}\end{Bmatrix}=\mathbf{J}^{-1}\begin{Bmatrix}\dfrac{\partial N_i^{(e)}}{\partial \xi}\\ \dfrac{\partial N_i^{(e)}}{\partial \eta}\end{Bmatrix} \tag{6.132}$$

Defining the element shape matrix $\mathbf{B}^{(e)}$ as

$$\mathbf{B}^{(e)}=\begin{bmatrix}\dfrac{\partial N_1^{(e)}}{\partial x} & \dfrac{\partial N_2^{(e)}}{\partial x} & \dfrac{\partial N_3^{(e)}}{\partial x} & \dfrac{\partial N_4^{(e)}}{\partial x}\\ \dfrac{\partial N_1^{(e)}}{\partial y} & \dfrac{\partial N_2^{(e)}}{\partial y} & \dfrac{\partial N_3^{(e)}}{\partial y} & \dfrac{\partial N_4^{(e)}}{\partial y}\end{bmatrix}=\begin{Bmatrix}\dfrac{\partial}{\partial x}\\ \dfrac{\partial}{\partial y}\end{Bmatrix}\mathbf{N}^{(e)^T} \tag{6.133}$$

permits the element matrix $\mathbf{k}^{(e)}$ be written in the form

$$\mathbf{k}^{(e)}=-\int_{D^{(e)}}\mathbf{B}^{(e)T}\mathbf{B}^{(e)}dx\,dy=-\int_{-1}^{1}\int_{-1}^{1}\mathbf{B}^{(e)T}\mathbf{B}^{(e)}|\mathbf{J}|d\xi\,d\eta \tag{6.134}$$

A similar operation is performed for evaluation of $\mathbf{f}^{(e)}$

$$\mathbf{f}^{(e)} = A \int_{D^{(e)}} \mathbf{N}^{(e)} dx\, dy = A \int_{-1}^{1}\int_{-1}^{1} \mathbf{N}^{(e)} |\mathbf{J}|\, d\xi\, d\eta \tag{6.135}$$

Due to the difficulty of obtaining an analytical expression for the determinant and inverse of the Jacobian matrix, these integrals are evaluated numerically by the Gaussian integration technique described in detail in Sec. 3.6.

Prior to the calculation of the element characteristic matrices, their Jacobian matrices are obtained for each element using Eq. (6.130) as

$$\mathbf{J}^{(1)} = \frac{1}{4}\begin{bmatrix} -(1-\eta) & (1-\eta) & (1+\eta) & -(1+\eta) \\ -(1-\xi) & -(1+\xi) & (1+\xi) & (1-\xi) \end{bmatrix}\begin{bmatrix} -4 & -3 \\ 0 & -3 \\ 0 & 0 \\ -4 & 0 \end{bmatrix} \tag{6.136}$$

$$= \frac{1}{4}\begin{bmatrix} 8 & 0 \\ 0 & 6 \end{bmatrix} \quad \text{with } \left|\mathbf{J}^{(1)}\right| = 3$$

$$\mathbf{J}^{(2)} = \frac{1}{4}\begin{bmatrix} -(1-\eta) & (1-\eta) & (1+\eta) & -(1+\eta) \\ -(1-\xi) & -(1+\xi) & (1+\xi) & (1-\xi) \end{bmatrix}\begin{bmatrix} 0 & -3 \\ 4 & -3 \\ 5 & 0 \\ 0 & 0 \end{bmatrix} \tag{6.137}$$

$$= \frac{1}{4}\begin{bmatrix} 9+\eta & 0 \\ 1+\xi & 6 \end{bmatrix} \quad \text{with } \left|\mathbf{J}^{(2)}\right| = \frac{3}{8}(9+\eta)$$

$$\mathbf{J}^{(3)} = \frac{1}{4}\begin{bmatrix} -(1-\eta) & (1-\eta) & (1+\eta) & -(1+\eta) \\ -(1-\xi) & -(1+\xi) & (1+\xi) & (1-\xi) \end{bmatrix}\begin{bmatrix} 0 & 0 \\ 5 & 0 \\ 6 & 3 \\ 0 & 3 \end{bmatrix} \tag{6.138}$$

$$= \frac{1}{4}\begin{bmatrix} 11+\eta & 0 \\ 1+\xi & 6 \end{bmatrix} \quad \text{with } \left|\mathbf{J}^{(3)}\right| = \frac{3}{8}(11+\eta)$$

$$\mathbf{J}^{(4)} = \frac{1}{4}\begin{bmatrix} -(1-\eta) & (1-\eta) & (1+\eta) & -(1+\eta) \\ -(1-\xi) & -(1+\xi) & (1+\xi) & (1-\xi) \end{bmatrix}\begin{bmatrix} -4 & 0 \\ 0 & 0 \\ 0 & 3 \\ -4 & 3 \end{bmatrix} \tag{6.139}$$

$$= \frac{1}{4}\begin{bmatrix} 8 & 0 \\ 0 & 6 \end{bmatrix} \quad \text{with } \left|\mathbf{J}^{(4)}\right| = 3$$

The inverse of the Jacobian matrices are obtained as

$$\left[\mathbf{J}^{(1)}\right]^{-1} = 4\begin{bmatrix} \frac{1}{8} & 0 \\ 0 & \frac{1}{6} \end{bmatrix} \tag{6.140}$$

$$\left[\mathbf{J}^{(2)}\right]^{-1} = \frac{4}{6(9+\eta)}\begin{bmatrix} 6 & 0 \\ -(1+\xi) & 9+\eta \end{bmatrix} \tag{6.141}$$

$$\left[\mathbf{J}^{(3)}\right]^{-1} = \frac{4}{6(11+\eta)}\begin{bmatrix} 6 & 0 \\ -(1+\xi) & 11+\eta \end{bmatrix} \tag{6.142}$$

$$\left[\mathbf{J}^{(4)}\right]^{-1} = 4\begin{bmatrix} \frac{1}{8} & 0 \\ 0 & \frac{1}{6} \end{bmatrix} \tag{6.143}$$

The element shape matrices $\mathbf{B}^{(e)}$ are obtained as

$$\mathbf{B}^{(1)} = \begin{bmatrix} -\frac{1}{8}(1-\eta) & \frac{1}{8}(1-\eta) & \frac{1}{8}(1+\eta) & -\frac{1}{8}(1+\eta) \\ -\frac{1}{6}(1-\xi) & -\frac{1}{6}(1+\xi) & \frac{1}{6}(1+\xi) & \frac{1}{6}(1-\xi) \end{bmatrix} \tag{6.144}$$

$$\mathbf{B}^{(2)} = \frac{1}{9+\eta}\begin{bmatrix} -(1-\eta) & (1-\eta) & (1+\eta) & -(1+\eta) \\ -\frac{1}{3}(4-5\xi+\eta) & -\frac{5}{3}(1+\xi) & \frac{4}{3}(1+\xi) & \frac{1}{3}(5-4\xi+\eta) \end{bmatrix} \tag{6.145}$$

$$\mathbf{B}^{(3)} = \frac{1}{11+\eta}\begin{bmatrix} -(1-\eta) & (1-\eta) & (1+\eta) & -(1+\eta) \\ -\frac{1}{3}(5-6\xi+\eta) & -2(1+\xi) & \frac{5}{3}(1+\xi) & \frac{1}{3}(6-5\xi+\eta) \end{bmatrix} \tag{6.146}$$

$$\mathbf{B}^{(4)} = \begin{bmatrix} -\frac{1}{8}(1-\eta) & \frac{1}{8}(1-\eta) & \frac{1}{8}(1+\eta) & -\frac{1}{8}(1+\eta) \\ -\frac{1}{6}(1-\xi) & -\frac{1}{6}(1+\xi) & \frac{1}{6}(1+\xi) & \frac{1}{6}(1-\xi) \end{bmatrix} \tag{6.147}$$

Numerical evaluation of the element characteristic matrices results in

$$\mathbf{k}^{(1)} = \begin{bmatrix} \frac{25}{36} & -\frac{1}{36} & -\frac{25}{72} & -\frac{23}{72} \\ -\frac{1}{36} & \frac{25}{36} & -\frac{23}{72} & -\frac{25}{72} \\ -\frac{25}{72} & -\frac{23}{72} & \frac{25}{36} & -\frac{1}{36} \\ -\frac{23}{72} & -\frac{25}{72} & -\frac{1}{36} & \frac{25}{36} \end{bmatrix} \tag{6.148}$$

$$\mathbf{k}^{(2)} = \begin{bmatrix} 0.688943 & -0.0222762 & -0.282179 & -0.384488 \\ -0.0222762 & 0.85561 & -0.384488 & -0.448846 \\ -0.282179 & -0.384488 & 0.60759 & 0.0590766 \\ -0.384488 & -0.448846 & 0.0590766 & 0.774257 \end{bmatrix} \tag{6.149}$$

$$\mathbf{k}^{(3)} = \begin{bmatrix} 0.753348 & 0.0799856 & -0.316655 & -0.516679 \\ 0.0799856 & 0.920014 & -0.516679 & -0.483321 \\ -0.316655 & -0.516679 & 0.680566 & 0.152768 \\ -0.516679 & -0.483321 & 0.152768 & 0.847232 \end{bmatrix} \tag{6.150}$$

$$\mathbf{k}^{(4)} = \begin{bmatrix} \frac{25}{36} & -\frac{1}{36} & -\frac{25}{72} & -\frac{23}{72} \\ -\frac{1}{36} & \frac{25}{36} & -\frac{23}{72} & -\frac{25}{72} \\ -\frac{25}{72} & -\frac{23}{72} & \frac{25}{36} & -\frac{1}{36} \\ -\frac{23}{72} & -\frac{25}{72} & -\frac{1}{36} & \frac{25}{36} \end{bmatrix} \tag{6.151}$$

Similarly, the right-hand-side vectors are calculated as

$$\mathbf{f}^{(1)} = A\begin{Bmatrix} 3 \\ 3 \\ 3 \\ 3 \end{Bmatrix}, \quad \mathbf{f}^{(2)} = A\begin{Bmatrix} 3.25 \\ 3.25 \\ 3.5 \\ 3.5 \end{Bmatrix}, \quad \mathbf{f}^{(3)} = A\begin{Bmatrix} 4 \\ 4 \\ 4.25 \\ 4.25 \end{Bmatrix}, \quad \mathbf{f}^{(4)} = A\begin{Bmatrix} 3 \\ 3 \\ 3 \\ 3 \end{Bmatrix}, \tag{6.152}$$

The element definitions (or connectivity of elements), as shown in Fig. 6.14, are presented in Table 6.5.

Table 6.5 Element connectivity

Element number (e)	Node 1	Node 2	Node 3	Node 4
1	1	2	9	8
2	2	3	4	9
3	9	4	5	6
4	8	9	6	7

Considering the correspondence between the local and global node numbering as shown in Table 6.5, the element equations can be rewritten as

$$\text{Element 1:}\quad \begin{matrix} \boxed{1} & \boxed{2} & \boxed{9} & \boxed{8} \\ \end{matrix}$$
$$\begin{bmatrix} k_{11}^{(1)} & k_{12}^{(1)} & k_{13}^{(1)} & k_{14}^{(1)} \\ k_{21}^{(1)} & k_{22}^{(1)} & k_{23}^{(1)} & k_{24}^{(1)} \\ k_{31}^{(1)} & k_{32}^{(1)} & k_{33}^{(1)} & k_{34}^{(1)} \\ k_{41}^{(1)} & k_{42}^{(1)} & k_{43}^{(1)} & k_{44}^{(1)} \end{bmatrix} \begin{Bmatrix} \phi_1^{(1)} \\ \phi_2^{(1)} \\ \phi_3^{(1)} \\ \phi_4^{(1)} \end{Bmatrix} = \begin{Bmatrix} f_1^{(1)} \\ f_2^{(1)} \\ f_3^{(1)} \\ f_4^{(1)} \end{Bmatrix} \begin{matrix} \boxed{1} \\ \boxed{2} \\ \boxed{9} \\ \boxed{8} \end{matrix} \tag{6.153}$$

$$\text{Element 2:}\quad \begin{matrix} \boxed{2} & \boxed{3} & \boxed{4} & \boxed{9} \\ \end{matrix}$$
$$\begin{bmatrix} k_{11}^{(2)} & k_{12}^{(2)} & k_{13}^{(2)} & k_{14}^{(2)} \\ k_{21}^{(2)} & k_{22}^{(2)} & k_{23}^{(2)} & k_{24}^{(2)} \\ k_{31}^{(2)} & k_{32}^{(2)} & k_{33}^{(2)} & k_{34}^{(2)} \\ k_{41}^{(2)} & k_{42}^{(2)} & k_{43}^{(2)} & k_{44}^{(2)} \end{bmatrix} \begin{Bmatrix} \phi_1^{(2)} \\ \phi_2^{(2)} \\ \phi_3^{(2)} \\ \phi_4^{(2)} \end{Bmatrix} = \begin{Bmatrix} f_1^{(2)} \\ f_2^{(2)} \\ f_3^{(2)} \\ f_4^{(2)} \end{Bmatrix} \begin{matrix} \boxed{2} \\ \boxed{3} \\ \boxed{4} \\ \boxed{9} \end{matrix} \tag{6.154}$$

$$\text{Element 3:}\quad \begin{matrix} \boxed{9} & \boxed{4} & \boxed{5} & \boxed{6} \\ \end{matrix}$$
$$\begin{bmatrix} k_{11}^{(3)} & k_{12}^{(3)} & k_{13}^{(3)} & k_{14}^{(3)} \\ k_{21}^{(3)} & k_{22}^{(3)} & k_{23}^{(3)} & k_{24}^{(3)} \\ k_{31}^{(3)} & k_{32}^{(3)} & k_{33}^{(3)} & k_{34}^{(3)} \\ k_{41}^{(3)} & k_{42}^{(3)} & k_{43}^{(3)} & k_{44}^{(3)} \end{bmatrix} \begin{Bmatrix} \phi_1^{(3)} \\ \phi_2^{(3)} \\ \phi_3^{(3)} \\ \phi_4^{(3)} \end{Bmatrix} = \begin{Bmatrix} f_1^{(3)} \\ f_2^{(3)} \\ f_3^{(3)} \\ f_4^{(3)} \end{Bmatrix} \begin{matrix} \boxed{9} \\ \boxed{4} \\ \boxed{5} \\ \boxed{6} \end{matrix} \tag{6.155}$$

$$\text{Element 4:}\quad \begin{matrix} \boxed{8} & \boxed{9} & \boxed{6} & \boxed{7} \\ \end{matrix}$$
$$\begin{bmatrix} k_{11}^{(4)} & k_{12}^{(4)} & k_{13}^{(4)} & k_{14}^{(4)} \\ k_{21}^{(4)} & k_{22}^{(4)} & k_{23}^{(4)} & k_{24}^{(4)} \\ k_{31}^{(4)} & k_{32}^{(4)} & k_{33}^{(4)} & k_{34}^{(4)} \\ k_{41}^{(4)} & k_{42}^{(4)} & k_{43}^{(4)} & k_{44}^{(4)} \end{bmatrix} \begin{Bmatrix} \phi_1^{(4)} \\ \phi_2^{(4)} \\ \phi_3^{(4)} \\ \phi_4^{(4)} \end{Bmatrix} = \begin{Bmatrix} f_1^{(4)} \\ f_2^{(4)} \\ f_3^{(4)} \\ f_4^{(4)} \end{Bmatrix} \begin{matrix} \boxed{8} \\ \boxed{9} \\ \boxed{6} \\ \boxed{7} \end{matrix} \tag{6.156}$$

In the assembly of the element characteristic matrices and vectors, the boxed numbers indicate the rows and columns of the global matrix, $\mathbf{K}$, and global right-hand-side vector, $\mathbf{F}$, to which the individual coefficients are added, resulting in

$$\left[\begin{array}{ccccccccc}
k_{11}^{(1)} & k_{12}^{(1)} & 0 & 0 & 0 & 0 & 0 & k_{14}^{(1)} & k_{13}^{(1)} \\
k_{21}^{(1)} & k_{22}^{(1)}+k_{11}^{(2)} & k_{12}^{(2)} & k_{13}^{(2)} & 0 & 0 & 0 & k_{24}^{(1)} & k_{23}^{(1)}+k_{14}^{(2)} \\
0 & k_{21}^{(2)} & k_{22}^{(2)} & k_{23}^{(2)} & 0 & 0 & 0 & 0 & k_{24}^{(2)} \\
0 & k_{31}^{(2)} & k_{32}^{(2)} & k_{33}^{(2)}+k_{22}^{(3)} & k_{23}^{(3)} & k_{24}^{(3)} & 0 & 0 & k_{34}^{(2)}+k_{21}^{(3)} \\
0 & 0 & 0 & k_{32}^{(3)} & k_{33}^{(3)} & k_{34}^{(3)} & 0 & 0 & k_{31}^{(3)} \\
0 & 0 & 0 & k_{42}^{(3)} & k_{43}^{(3)} & k_{44}^{(3)}+k_{33}^{(4)} & k_{34}^{(4)} & k_{31}^{(4)} & k_{41}^{(3)}+k_{32}^{(4)} \\
0 & 0 & 0 & 0 & 0 & k_{43}^{(4)} & k_{44}^{(4)} & k_{41}^{(4)} & k_{42}^{(4)} \\
k_{41}^{(1)} & k_{42}^{(1)} & 0 & 0 & 0 & k_{13}^{(4)} & k_{14}^{(4)} & k_{44}^{(1)}+k_{11}^{(4)} & k_{43}^{(1)}+k_{12}^{(4)} \\
k_{31}^{(1)} & k_{32}^{(1)}+k_{41}^{(2)} & k_{42}^{(2)} & k_{43}^{(2)}+k_{12}^{(3)} & k_{13}^{(3)} & k_{14}^{(3)}+k_{23}^{(4)} & k_{24}^{(4)} & k_{34}^{(1)}+k_{21}^{(4)} & k_{33}^{(1)}+k_{44}^{(2)}+k_{11}^{(3)}+k_{22}^{(4)}
\end{array}\right]
\left\{\begin{array}{c}\phi_1\\ \phi_2\\ \phi_3\\ \phi_4\\ \phi_5\\ \phi_6\\ \phi_7\\ \phi_8\\ \phi_9\end{array}\right\}
=\left\{\begin{array}{c}
f_1^{(1)} \\
f_2^{(1)}+f_1^{(2)} \\
f_2^{(2)} \\
f_3^{(2)}+f_2^{(3)} \\
f_3^{(3)} \\
f_4^{(3)}+f_3^{(4)} \\
f_4^{(4)} \\
f_4^{(1)}+f_1^{(4)} \\
f_3^{(1)}+f_4^{(2)}+f_1^{(3)}+f_2^{(4)}
\end{array}\right\} \tag{6.157}$$

or

$$\mathbf{K\Phi}=\mathbf{F} \tag{6.158}$$

the global stiffness matrix and right-hand-side vector are numerically evaluated as

$$\mathbf{K} = \begin{bmatrix} 0.694444 & -0.0277778 & 0 & 0 & 0 & 0 & 0 & -0.319444 & -0.347222 \\ -0.0277778 & 1.38339 & -0.0222762 & -0.282179 & 0 & 0 & 0 & -0.347222 & -0.703932 \\ 0 & -0.0222762 & 0.85561 & -0.384488 & 0 & 0 & 0 & 0 & -0.448846 \\ 0 & -0.282179 & -0.384488 & 1.5276 & -0.516679 & -0.483321 & 0 & 0 & 0.139062 \\ 0 & 0 & 0 & -0.516679 & 0.680566 & 0.152768 & 0 & 0 & -0.316655 \\ 0 & 0 & 0 & -0.483321 & 0.152768 & 1.54168 & -0.0277778 & -0.347222 & -0.836123 \\ 0 & 0 & 0 & 0 & 0 & -0.0277778 & 0.694444 & -0.319444 & -0.347222 \\ -0.319444 & -0.347222 & 0 & 0 & 0 & -0.347222 & -0.319444 & 1.38889 & -0.0555556 \\ -0.347222 & -0.703932 & -0.448846 & 0.139062 & -0.316655 & -0.836123 & -0.347222 & -0.0555556 & 2.91649 \end{bmatrix} \tag{6.159}$$

and

$$\mathbf{F} = \begin{Bmatrix} 3 \\ 6.25 \\ 3.25 \\ 7.5 \\ 4.25 \\ 7.25 \\ 3 \\ 6 \\ 13.5 \end{Bmatrix} \tag{6.160}$$

After imposing the essential boundary conditions, i.e., $\phi_3 = \phi_4 = \phi_5 = \phi_6 = \phi_7 = 1$, the global system of equations is reduced by deleting the rows and columns corresponding to ϕ_3, ϕ_4, ϕ_5, ϕ_6, and ϕ_7, leading to

$$\begin{bmatrix} k_{11}^{(1)} & k_{12}^{(1)} & k_{14}^{(1)} & k_{13}^{(1)} \\ k_{21}^{(1)} & k_{22}^{(1)}+k_{11}^{(2)} & k_{24}^{(1)} & k_{23}^{(1)}+k_{14}^{(2)} \\ k_{41}^{(1)} & k_{42}^{(1)} & k_{44}^{(1)}+k_{11}^{(4)} & k_{43}^{(1)}+k_{12}^{(4)} \\ k_{31}^{(1)} & k_{32}^{(1)}+k_{41}^{(2)} & k_{34}^{(1)}+k_{21}^{(4)} & k_{33}^{(1)}+k_{44}^{(2)}+k_{11}^{(3)}+k_{22}^{(4)} \end{bmatrix} \begin{Bmatrix} \phi_1 \\ \phi_2 \\ \phi_8 \\ \phi_9 \end{Bmatrix}$$

$$= \begin{Bmatrix} f_1^{(1)} \\ f_2^{(1)}+f_1^{(2)}-k_{12}^{(2)}-k_{13}^{(2)} \\ f_4^{(1)}+f_1^{(4)}-k_{13}^{(4)}-k_{14}^{(4)} \\ f_3^{(1)}+f_4^{(2)}+f_1^{(3)}+f_2^{(4)}-k_{42}^{(2)}-\left(k_{43}^{(2)}+k_{12}^{(3)}\right) \\ -k_{13}^{(3)}-\left(k_{14}^{(3)}+k_{23}^{(4)}\right)-k_{24}^{(4)} \end{Bmatrix} \qquad (6.161)$$

which is numerically evaluated as

$$\mathbf{K} = \begin{bmatrix} 0.694444 & -0.0277778 & -0.319444 & -0.347222 \\ -0.0277778 & 1.38339 & -0.347222 & -0.703932 \\ -0.319444 & -0.347222 & 1.38889 & -0.0555556 \\ -0.347222 & -0.703932 & -0.0555556 & 2.91649 \end{bmatrix} \qquad (6.162)$$

and

$$\mathbf{F} = \begin{Bmatrix} 3 \\ 6.55446 \\ 6.66667 \\ 15.3098 \end{Bmatrix} \qquad (6.163)$$

Finally, the solution of the reduced global system yields

$$\begin{Bmatrix} \phi_1 \\ \phi_2 \\ \phi_8 \\ \phi_9 \end{Bmatrix} = \begin{Bmatrix} 15.8119 \\ 13.5401 \\ 12.2471 \\ 10.6332 \end{Bmatrix} \qquad (6.164)$$

6.1.3.2 ANSYS Solution

The governing equations for a steady-state heat transfer, described by Eq. (6.101) through (6.105), also can be solved using ANSYS. The solution procedure is outlined as follows:

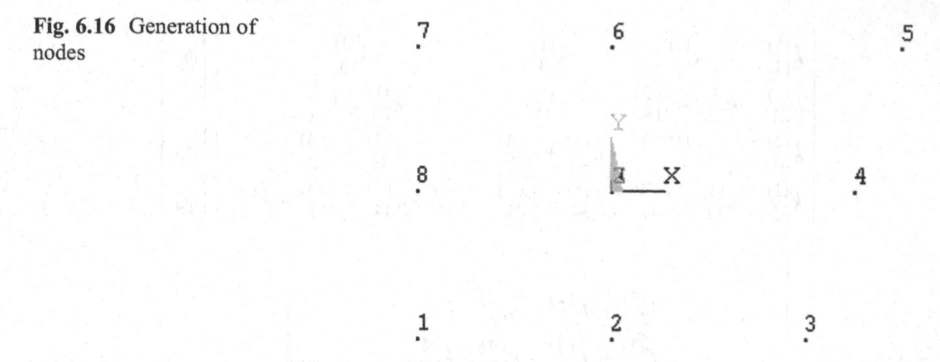

Fig. 6.16 Generation of nodes

Model Generation

- Specify the element type (**ET** command) using the following menu path:

Main Menu > Preprocessor > Element Type > Add/Edit/Delete

- Click on ***Add***.
- Select ***Solid*** immediately below ***Thermal Mass*** from the left list and ***Quad 4node 55*** from the right list; click on ***OK***.
- Click on ***Close***.

- Specify material properties (**MP** command) using the following menu path:

Main Menu > Preprocessor > Material Props > Material Models

- In the Define Material Model Behavior dialog box, in the right window, successively left-click on Thermal, Conductivity, and, finally, Isotropic, which brings up another dialog box.
- Enter ***1*** for ***KXX***, and click on ***OK***.
- Close the Define Material Model Behavior dialog box by using the following menu path:

Material > Exit

- Create nodes (**N** command) using the following menu path:

Main Menu > Preprocessor > Modeling > Create > Nodes > In Active CS

- A total of 9 nodes will be created (Table 6.4).
- Referring to Table 6.4, enter x- and y-coordinates of node 1, and Click on Apply. This action will keep the Create Nodes in Active Coordinate System dialog box open. If the Node number field is left blank, then ANSYS will assign the lowest available node number to the node that is being created.
- Repeat the same procedure for the nodes 2 through 9.
- After entering the x- and y-coordinates of node 9, click on OK (instead of App***ly***).
- The nodes should appear in the *Graphics Window*, as shown in Fig. 6.16.

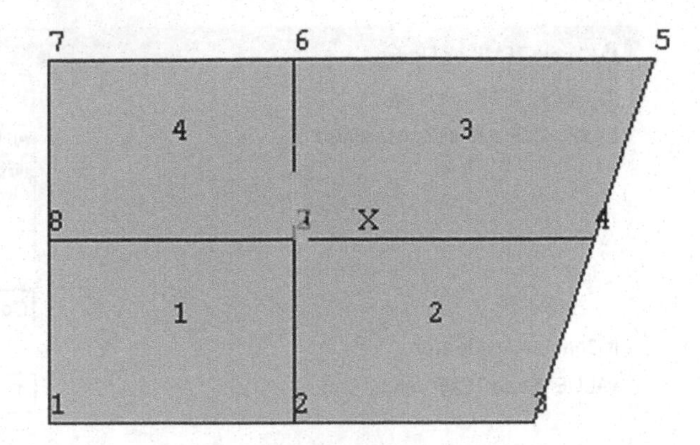

Fig. 6.17 Generation of elements

- Create elements (**E** command) using the following menu path:

Main Menu > Preprocessor > Modeling > Create > Elements > Auto Numbered > Thru Nodes

- *Pick Menu* appears; refer to Fig. 6.17 to create elements by picking ***four*** nodes at a time and clicking on ***Apply*** in between.
- Observe the elements created after clicking on ***Apply*** in the *Pick Menu*.
- Repeat until the last element is created.
- Click on ***OK*** when the last element is created.

- Review elements:

- Turn on element numbering using the following menu path:

Utility Menu > PlotCtrls > Numbering

- Select ***Element numbers*** from the first pull-down menu; click on ***OK***.
- Plot elements (**EPLOT** command) using the following menu path:

Utility Menu > Plot > lements

- Figure 6.17 shows the outcome of this action as it appears in the *Graphics Window*.
- Turn off element numbering and turn on node numbering using the following menu path:

Utility Menu > PlotCtrls > Numbering

- Place a *checkmark* by clicking on the empty box next to ***NODE Node numbers***.
- Select ***No numbering*** from the first pull-down menu.
- Click on ***OK***.
- Plot nodes (**NPLOT** command) using the following menu path:

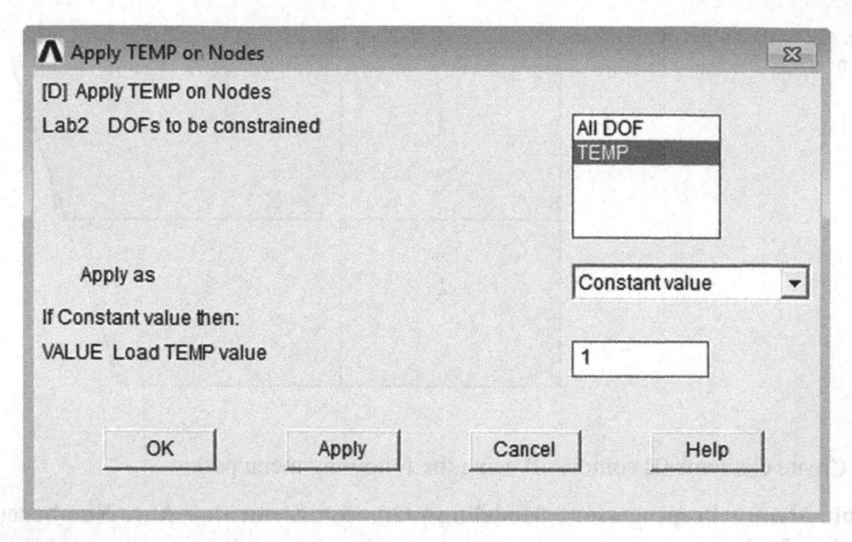

Fig. 6.18 Application of temperature boundary conditions on nodes

Utility Menu > Plot > Nodes

– Figure 6.16 shows the outcome of this action as it appears in the *Graphics Window*.

Solution

- Apply temperature boundary conditions (`D` command) using the following menu path:

Main Menu > Solution > Define Loads > Apply > Thermal > Temperature > On Nodes

– *Pick Menu* appears; pick nodes 3 through 7 along the boundary (Fig. 6.16) and click on ***OK*** on *Pick Menu*.
– Highlight ***TEMP*** and enter ***1*** for ***VALUE***; click on ***OK*** (Fig. 6.18).

- Apply body load on elements (`BFE` command) using the following menu path:

Main Menu > Solution > Define Loads > Apply > Thermal > Heat Generat > On Elements

– *Pick Menu* appears; click on ***Pick All***.
– Enter ***1*** for ***VAL1*** (leave other fields untouched, as shown in Fig. 6.19).
– Click on ***OK***.

- Obtain solution (`SOLVE` command) using the following menu path:

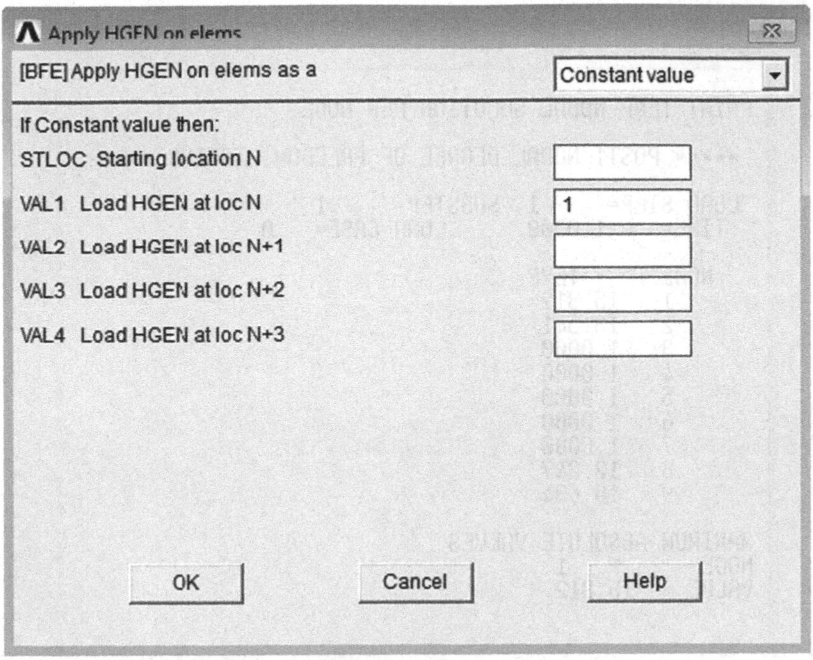

Apply HGEN on elems

[BFE] Apply HGEN on elems as a — Constant value

If Constant value then:

STLOC Starting location N

VAL1 Load HGEN at loc N — 1

VAL2 Load HGEN at loc N+1

VAL3 Load HGEN at loc N+2

VAL4 Load HGEN at loc N+3

OK Cancel Help

Fig. 6.19 Application of heat generation condition on elements

Main Menu > Solution > Solve > Current LS

- *Confirmation Window* appears along with *Status Report Window*.
- Review status/ If OK, close the *Status Report Window* and click on ***OK*** in the *Confirmation Window*.
- Wait until ANSYS responds with ***Solution is done!***

Postprocessing

- Review temperature values (**PRNSOL** command) using the following menu path:

Main Menu > General Postproc > List Results > Nodal Solution

- Click on ***DOF Solution*** and ***Nodal Temperature***; click on ***OK***.
- The list will appear in a new window, as shown in Fig. 6.20.

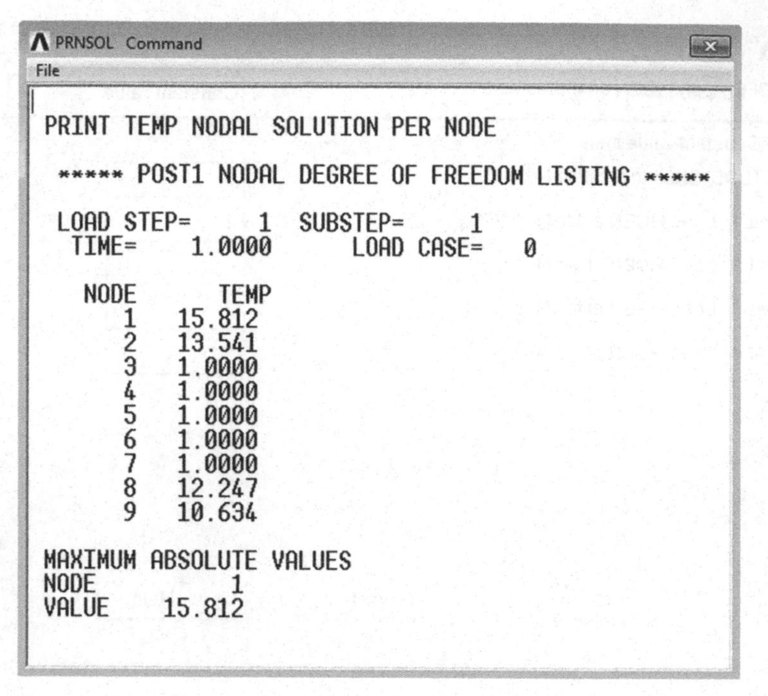

Fig. 6.20 Nodal solution for temperature

6.2 Principle of Minimum Potential Energy

Galerkin's method is not always suitable for all structural problems because of difficulties in mathematically describing the structural geometry and/or the boundary conditions. An alternative to Galerkin's method is the principle of minimum potential energy (Washizu 1982; Dym and Shames 1973).

The energy method involves determination of the stationary values of the global energy. This requires the approximation of the functional behavior of the dependent variable so that the global energy becomes stationary. The stationary value can be a maximum, a minimum or a neutral point. With an understanding of variational calculus, the minimum stationary value leading to stable equilibrium (Fig. 6.21) is obtained by requiring the first variation of the global energy to vanish.

Avoiding the details of variational calculus, the concepts of differential calculus can be used to perform the minimization of the global energy. In solid mechanics, this is known as the principle of minimum potential energy, which states that among all compatible displacement fields satisfying the boundary conditions (kinematically admissible), the correct displacement field satisfying the equilibrium equations is the one that renders the potential energy an absolute minimum. A solution satisfying both equilibrium equations and boundary conditions is, of course, "exact"; however, such solutions are difficult, if not impossible, to construct for complex problems. Therefore, approximate solutions are obtained by assuming kinematically admissible displacement fields with unknown coefficients. The values of

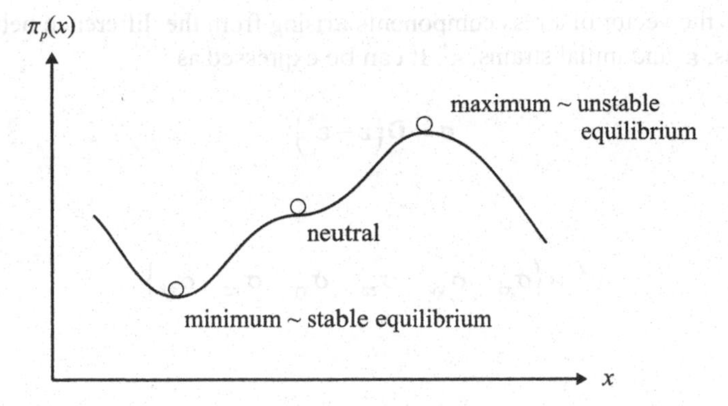

Fig. 6.21 Schematics of stable, neutral, and unstable equilibrium points of the global energy

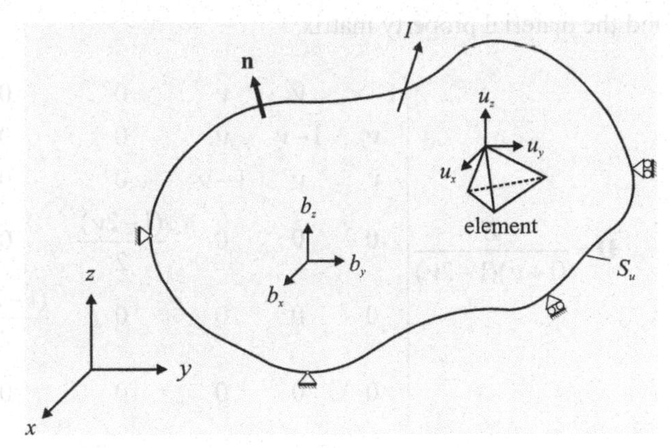

Fig. 6.22 A 3D body with displacement constraints, body and concentrated forces, and surface tractions

these coefficients are determined in such a way that the total potential energy of the system is a minimum.

The principle of virtual work is applicable for any material behavior, whereas the principle of minimum potential energy is applicable only for elastic materials. However, both principles yield the same element equations for elastic materials.

The total potential energy of the structural system shown in Fig. 6.22 is defined as

$$\pi_p = W + \Omega \tag{6.165}$$

in which W is the strain energy and Ω is the potential energy arising from the presence of body forces, surface tractions, and the initial residual stresses. Strain energy is the capacity of the internal forces (or stresses) to do work through strains in the structure.

For a linear elastic material, the strain energy of the deformed structure is given by

$$W = \frac{1}{2}\int_V \left(\boldsymbol{\varepsilon} - \boldsymbol{\varepsilon}^*\right)^T \boldsymbol{\sigma}\, dV \tag{6.166}$$

where **A** is the vector of stress components arising from the difference between the total strains, $\boldsymbol{\varepsilon}$, and initial strains, $\boldsymbol{\varepsilon}^*$. It can be expressed as

$$\boldsymbol{\sigma} = \mathbf{D}\left(\boldsymbol{\varepsilon} - \boldsymbol{\varepsilon}^*\right) \tag{6.167}$$

in which

$$\boldsymbol{\sigma}^T = \left\{\sigma_{xx} \quad \sigma_{yy} \quad \sigma_{zz} \quad \sigma_{xy} \quad \sigma_{yz} \quad \sigma_{zx}\right\} \tag{6.168}$$

and

$$\boldsymbol{\varepsilon}^T = \left\{\varepsilon_{xx} \quad \varepsilon_{yy} \quad \varepsilon_{zz} \quad \gamma_{xy} \quad \gamma_{yz} \quad \gamma_{zx}\right\} \tag{6.169}$$

and the material property matrix

$$\mathbf{D} = \frac{E}{(1+\nu)(1-2\nu)} \begin{bmatrix} 1-\nu & \nu & \nu & 0 & 0 & 0 \\ \nu & 1-\nu & \nu & 0 & 0 & 0 \\ \nu & \nu & 1-\nu & 0 & 0 & 0 \\ 0 & 0 & 0 & \dfrac{(1-2\nu)}{2} & 0 & 0 \\ 0 & 0 & 0 & 0 & \dfrac{(1-2\nu)}{2} & 0 \\ 0 & 0 & 0 & 0 & 0 & \dfrac{(1-2\nu)}{2} \end{bmatrix} \tag{6.170}$$

where σ_{ij} and ε_{ij} represent the stress and strain components, with $i, j = x, y, z$ being the Cartesian coordinates. The elastic modulus and Poisson's ratio are denoted by E and ν, respectively. In the presence of temperature change, the initial strains can be expressed as

$$\boldsymbol{\varepsilon}^{*T} = \{\alpha\Delta T \quad \alpha\Delta T \quad \alpha\Delta T \quad 0 \quad 0 \quad 0\} \tag{6.171}$$

where α is the coefficient of thermal expansion and ΔT is the temperature change with respect to a reference state.

The potential energy arising from the presence of body forces, **b**, surface tractions, **T**, and the initial residual stresses, $\boldsymbol{\sigma}^*$, is given by

$$\Omega = -\int_V \mathbf{u}^T\mathbf{b}dV - \int_{S_\sigma} \mathbf{u}^T\mathbf{T}dS + \int_V \boldsymbol{\varepsilon}^T\boldsymbol{\sigma}^* dV \tag{6.172}$$

with

$$\mathbf{b}^T = \left\{b_x \quad b_y \quad b_z\right\} \tag{6.173}$$

$$\mathbf{T}^T = \left\{ T_x \;\; T_y \;\; T_z \right\} \tag{6.174}$$

$$\mathbf{u}^T = \left\{ u_x \;\; u_y \;\; u_z \right\} \tag{6.175}$$

in which b_x, b_y, and b_z are the components of body force (in units of force per unit volume), and T_x, T_y, and T_z represent the components of the applied traction vector (in units of force per unit area) over the surface defined by S_σ. The entire surface of the body having a volume of V is defined by S, with segments S_u and S_σ subjected to displacement and traction conditions, respectively. The displacement components are given by u_x, u_y, and u_z in the x-, y-, and z-directions, respectively. Also, included in the expression for the total potential is the initial residual stresses denoted by $\boldsymbol{\sigma}^*$. The initial stresses could be measured, but their prediction without full knowledge of the material's history is impossible.

After partitioning the entire domain occupied by volume V into E number of elements with volume V^e, the total potential energy of the system can be rewritten as

$$\pi_p(u_x,u_y,u_z) = \sum_{e=1}^{E} \pi_p^{(e)}(u_x,u_y,u_z) \tag{6.176}$$

in which

$$\begin{aligned} \pi_p^{(e)} = \frac{1}{2} \int_{V^{(e)}} \boldsymbol{\varepsilon}^T \mathbf{D} \boldsymbol{\varepsilon} dV - \int_{V^{(e)}} \boldsymbol{\varepsilon}^T \mathbf{D} \boldsymbol{\varepsilon}^* dV + \frac{1}{2} \int_{V^{(e)}} \boldsymbol{\varepsilon}^{*T} \mathbf{D} \boldsymbol{\varepsilon}^* dV \\ - \int_{V^{(e)}} \mathbf{u}^T \mathbf{b} \, dV - \int_{S_\sigma^{(e)}} \mathbf{u}^T \mathbf{T} dS + \int_{V^{(e)}} \boldsymbol{\varepsilon}^T \boldsymbol{\sigma}^* dV \end{aligned} \tag{6.177}$$

where the superscript e denotes a specific element.

Based on kinematical considerations, the components of the total strain vector, $\boldsymbol{\varepsilon}$, in terms of the displacement components are expressed as

$$\begin{Bmatrix} \varepsilon_{xx} \\ \varepsilon_{yy} \\ \varepsilon_{zz} \\ \gamma_{xy} \\ \gamma_{yz} \\ \gamma_{zx} \end{Bmatrix} = \begin{bmatrix} \frac{\partial}{\partial x} & 0 & 0 \\ 0 & \frac{\partial}{\partial y} & 0 \\ 0 & 0 & \frac{\partial}{\partial z} \\ \frac{\partial}{\partial y} & \frac{\partial}{\partial x} & 0 \\ 0 & \frac{\partial}{\partial z} & \frac{\partial}{\partial y} \\ \frac{\partial}{\partial z} & 0 & \frac{\partial}{\partial x} \end{bmatrix} \begin{Bmatrix} u_x \\ u_y \\ u_z \end{Bmatrix} \quad \text{or} \quad \boldsymbol{\varepsilon} = \mathbf{L}\mathbf{u} \tag{6.178}$$

in which **L** is the differential operator matrix.

The finite element process seeks a minimum in the potential energy based on the approximate form of the dependent variables (displacement components) within each element. The greater the number of degrees of freedom associated with the element (usually means increasing the number of nodes), the more closely the solution will approximate the true equilibrium position. Within each element, the approximation to the displacement components can be expressed as

$$u_x^{(e)} \approx \tilde{u}_x^{(e)} = \sum_{r=1}^{n} N_r^{(e)} u_{x_r}^{(e)}$$
$$u_y^{(e)} \approx \tilde{u}_y^{(e)} = \sum_{r=1}^{n} N_r^{(e)} u_{y_r}^{(e)} \tag{6.179}$$
$$u_z^{(e)} \approx \tilde{u}_z^{(e)} = \sum_{r=1}^{n} N_r^{(e)} u_{z_r}^{(e)}$$

with n representing the number of nodes associated with element e. The nodal unknowns and shape functions are denoted by $u_{x_r}^{(e)}$, $u_{y_r}^{(e)}$, $u_{z_r}^{(e)}$, and $N_r^{(e)}$, respectively. In matrix form, the approximate displacement components can be expressed as

$$\tilde{\mathbf{u}}^{(e)} = \mathbf{N}^{(e)T}\mathbf{U}^{(e)} \tag{6.180}$$

in which

$$\tilde{\mathbf{u}}^{(e)T} = \left\{ \tilde{u}_x^{(e)} \quad \tilde{u}_y^{(e)} \quad \tilde{u}_z^{(e)} \right\} \tag{6.181}$$

$$\mathbf{N}^{(e)T} = \begin{bmatrix} N_1 & 0 & 0 & N_2 & 0 & 0 & . & . & . & N_n & 0 & 0 \\ 0 & N_1 & 0 & 0 & N_2 & 0 & . & . & . & 0 & N_n & 0 \\ 0 & 0 & N_1 & 0 & 0 & N_2 & . & . & . & 0 & 0 & N_n \end{bmatrix}_{3\times 3n} \tag{6.182}$$

$$\mathbf{U}^{(e)T} = \left\{ u_{x_1}^{(e)} \quad u_{y_1}^{(e)} \quad u_{z_1}^{(e)} \quad u_{x_2}^{(e)} \quad u_{y_2}^{(e)} \quad u_{z_2}^{(e)} \quad \cdots \quad u_{x_n}^{(e)} \quad u_{y_n}^{(e)} \quad u_{z_n}^{(e)} \right\} \tag{6.183}$$

With the approximate form of the displacement components, the strain components within each element can be expressed as

$$\boldsymbol{\varepsilon} \approx \mathbf{B}^{(e)}\mathbf{U}^{(e)} \tag{6.184}$$

where

$$\mathbf{B}^{(e)} = \mathbf{L}\mathbf{N}^{(e)T} \tag{6.185}$$

leading to the expression for the total potential in terms of element nodal displacements, $\mathbf{U}^{(e)}$

$$\pi_p^{(e)} = \frac{1}{2}\mathbf{U}^{(e)T}\mathbf{k}^{(e)}\mathbf{U}^{(e)} - \mathbf{U}^{(e)T}\mathbf{p}^{(e)} + \frac{1}{2}\int_{V^{(e)}} \boldsymbol{\varepsilon}^{*T}\mathbf{D}\boldsymbol{\varepsilon}^{*}dV \tag{6.186}$$

in which the element stiffness matrix, $\mathbf{k}^{(e)}$, and the element force vector, $\mathbf{p}^{(e)}$, are defined as

$$\mathbf{k}^{(e)} = \int_{V^{(e)}} \mathbf{B}^{(e)T}\mathbf{D}\mathbf{B}^{(e)}dV \tag{6.187}$$

and

$$\mathbf{p}^{(e)} = \mathbf{p}_{\mathbf{b}}^{(e)} + \mathbf{p}_{\mathbf{T}}^{(e)} + \mathbf{p}_{\varepsilon^*}^{(e)} - \mathbf{p}_{\sigma^*}^{(e)} \tag{6.188}$$

with $\mathbf{p}_{\mathbf{b}}^{(e)}$, $\mathbf{p}_{\mathbf{T}}^{(e)}$, $\mathbf{p}_{\varepsilon^*}^{(e)}$, and $\mathbf{p}_{\sigma^*}^{(e)}$ representing the element load vectors due to body forces, surface tractions (forces), initial strains, and initial stresses, respectively, defined by

$$\begin{aligned}
\mathbf{p}_{\mathbf{b}}^{(e)} &= \int_{V^{(e)}} \mathbf{N}^{(e)}\mathbf{b}\,dV \\
\mathbf{p}_{\mathbf{T}}^{(e)} &= \int_{S_\sigma^{(e)}} \mathbf{N}^{(e)}\mathbf{T}\,dS \\
\mathbf{p}_{\varepsilon^*}^{(e)} &= \int_{V^{(e)}} \mathbf{B}^{(e)T}\mathbf{D}\boldsymbol{\varepsilon}^{*}dV \\
\mathbf{p}_{\sigma^*}^{(e)} &= \int_{V^{(e)}} \mathbf{B}^{(e)T}\boldsymbol{\sigma}^{*}dV
\end{aligned} \tag{6.189}$$

Evaluation of these integrals results in the statically equivalent nodal forces in the elements affected by the body force, the surface tractions, and the initial strains and initial stresses. In the presence of external concentrated forces acting on various nodes, the potential energy is modified as

$$\begin{aligned}
\pi_p = &\frac{1}{2}\mathbf{U}^T\left\{\sum_{e=1}^{E}\mathbf{k}^{(e)}\right\}\mathbf{U} - \mathbf{U}^T\left\{\sum_{e=1}^{E}\left(\mathbf{p}_{\mathbf{b}}^{(e)} + \mathbf{p}_{\mathbf{T}}^{(e)} + \mathbf{p}_{\varepsilon^*}^{(e)} - \mathbf{p}_{\sigma^*}^{(e)}\right) - \mathbf{P}_{\mathbf{c}}\right\} \\
&+ \frac{1}{2}\sum_{e=1}^{E}\int_{V^{(e)}} \boldsymbol{\varepsilon}^{*T}\mathbf{D}\boldsymbol{\varepsilon}^{*}dV
\end{aligned} \tag{6.190}$$

where $\mathbf{P}_{\mathbf{c}}$ is the vector of nodal forces and $\mathbf{U}$ represents the vector of nodal displacements for the entire structure. Note that each component of the element nodal displacement vector, $\mathbf{U}^{(e)}$, appears in the global (system) nodal displacement vector, $\mathbf{U}$. Therefore, the element nodal displacement vector $\mathbf{U}^{(e)}$ can be replaced by $\mathbf{U}$ with the appropriate enlargement of the element matrices and vectors in the expression for the potential energy by adding the required number of zero elements and rearranging. The summation in the expression for the potential energy implies the expansion of the element matrices to the size of the global (system) matrix while collecting the overlapping terms.

Minimization of the total potential energy requires that

$$\left\{\frac{\partial \pi_p}{\partial \mathbf{U}}\right\} = 0 \tag{6.191}$$

leading to the system (global) equilibrium equations in the form

$$\mathbf{KU} = \mathbf{P} \tag{6.192}$$

in which $\mathbf{K}$ and $\mathbf{P}$ are the assembled (global) stiffness matrix and the assembled (global) nodal load vector, respectively, defined by

$$\mathbf{K} = \sum_{e=1}^{E} \mathbf{k}^{(e)} \tag{6.193}$$

and

$$\mathbf{P} = \sum_{e=1}^{E} (\mathbf{p}_{\mathbf{b}}^{(e)} + \mathbf{p}_{\mathbf{T}}^{(e)} + \mathbf{p}_{\varepsilon^*}^{(e)} - \mathbf{p}_{\sigma^*}^{(e)}) - \mathbf{P}_{\mathbf{c}} \tag{6.194}$$

This global equilibrium equation cannot be solved unless boundary constraints are imposed to suppress the rigid-body motion. Otherwise, the global stiffness matrix becomes singular.

After obtaining the solution to the nodal displacements of the system equilibrium equations, the stresses within the element can be determined from

$$\boldsymbol{\sigma} = \mathbf{DB}^{(e)}\mathbf{U}^{(e)} - \mathbf{D}\boldsymbol{\varepsilon}^* + \boldsymbol{\sigma}^* \tag{6.195}$$

The global stiffness matrix and the load vector require the evaluation of the integrals associated with the element stiffness matrix and the element nodal load vector.

6.2.1 Example: One-Dimensional Analysis with Line Elements

The application of this approach is demonstrated by computing the displacements and strains in a rod constructed of three concentric sections of different materials. As shown in Fig. 6.23, the rod has a uniform cross section and is subjected to a

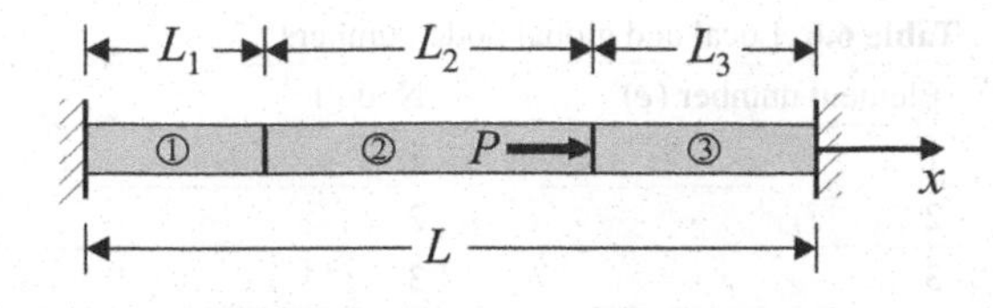

Fig. 6.23 A rod constrained at both ends, subjected to a concentrated force

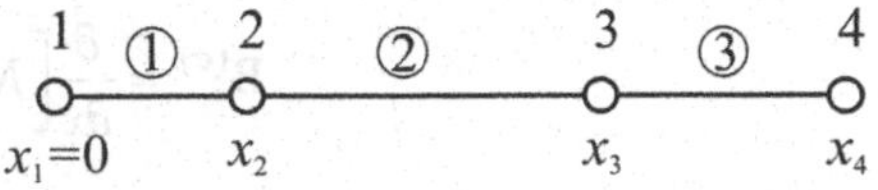

1 ① 2 ② 3 ③ 4

$x_1 = 0$ x_2 x_3 x_4

Fig. 6.24 Finite element discretization of the rod with three elements

concentrated horizontal load, P, at the second joint, and the boundary conditions are specified as $u_x(x=0)=0$ and $u_x(x=L)=0$.

The domain is discretized with 3 linear line elements having two nodes, as shown in Fig. 6.24. The global coordinates of each node in domain D are specified by x_i, with $i=1,2,3,4$. The nodal values of the dependent variable associated with element e are specified at its first and second nodes by $u_{x_i}^{(e)}$ and $u_{x_j}^{(e)}$, respectively.

For the domain discretized with three elements and four nodes, the local and global nodes are numbered as shown in Table 6.6.

Within each element, the approximation to the displacement component can be expressed as

$$u_x^{(e)} \approx \tilde{u}_x^{(e)} = \sum_{r=1}^{2} N_r^{(e)} u_{x_r}^{(e)} \tag{6.196}$$

The nodal unknowns and shape functions are denoted by $u_{x_r}^{(e)}$ and $N_r^{(e)}$, respectively. In matrix form, the approximate displacement components can be expressed as

$$\tilde{\mathbf{u}}^{(e)} = \mathbf{N}^{(e)T} \mathbf{U}^{(e)} \tag{6.197}$$

with

$$\mathbf{N}^{(e)T} = \left\{ N_1^{(e)} \quad N_2^{(e)} \right\} \quad \text{and} \quad \mathbf{U}^{(e)} = \begin{Bmatrix} u_{x_1}^{(e)} \\ u_{x_2}^{(e)} \end{Bmatrix} \tag{6.198}$$

in which the shape functions are

$$N_1^{(e)} = \frac{x_2^{(e)} - x}{x_2^{(e)} - x_1^{(e)}} \quad \text{and} \quad N_2^{(e)} = \frac{x - x_1^{(e)}}{x_2^{(e)} - x_1^{(e)}} \tag{6.199}$$

With the approximate form of the displacement components and $\mathbf{L} = \partial / \partial x$, the shape matrix can be obtained from

Table 6.6 Local and global node numbers

Element number (e)	Node 1	Node 2
1	1	2
2	2	3
3	3	4

$$\mathbf{B}^{(e)} = \frac{\partial}{\partial x}\begin{bmatrix} N_1^{(e)} & N_2^{(e)} \end{bmatrix} \tag{6.200}$$

For a constant cross section, $A^{(e)}$, and elastic modulus, $E^{(e)}$, in each element, the element stiffness matrix is

$$\begin{aligned}\mathbf{k}^{(e)} &= \int_{V^{(e)}} \mathbf{B}^{(e)T}\mathbf{D}\mathbf{B}^{(e)}\,dV \\ &= A^{(e)}\int_{x_1^{(e)}}^{x_2^{(e)}} \frac{\partial}{\partial x}\begin{bmatrix} N_1^{(e)} \\ N_2^{(e)} \end{bmatrix} E^{(e)}\frac{\partial}{\partial x}\begin{bmatrix} N_1^{(e)} & N_2^{(e)} \end{bmatrix} dx\end{aligned} \tag{6.201}$$

Substituting for the shape functions, the element stiffness matrix becomes

$$\mathbf{k}^{(e)} = -\frac{A^{(e)}E^{(e)}}{\left(x_2^{(e)} - x_1^{(e)}\right)^2}\int_{x_1^{(e)}}^{x_2^{(e)}} \begin{bmatrix} 1 & -1 \\ -1 & 1 \end{bmatrix} dx \tag{6.202}$$

Integration along the element length results in

$$\begin{aligned}\mathbf{k}^{(e)} &= -\frac{A^{(e)}E^{(e)}}{\left(x_2^{(e)} - x_1^{(e)}\right)}\begin{bmatrix} 1 & -1 \\ -1 & 1 \end{bmatrix} = -\frac{A^{(e)}E^{(e)}}{L^{(e)}}\begin{bmatrix} 1 & -1 \\ -1 & 1 \end{bmatrix} \\ &= -\alpha^{(e)}\begin{bmatrix} 1 & -1 \\ -1 & 1 \end{bmatrix}\end{aligned} \tag{6.203}$$

in which $L^{(e)} = (x_2^{(e)} - x_1^{(e)})$and $\alpha^{(e)} = A^{(e)}E^{(e)} / L^{(e)}$. The element stiffness matrices are computed as

$$\mathbf{k}^{(1)} = \begin{bmatrix} \alpha^{(1)} & -\alpha^{(1)} \\ -\alpha^{(1)} & \alpha^{(1)} \end{bmatrix} \begin{matrix} \boxed{1} \\ \boxed{2} \end{matrix} \quad \text{(columns: } \boxed{1}\ \boxed{2}\text{)} \tag{6.204}$$

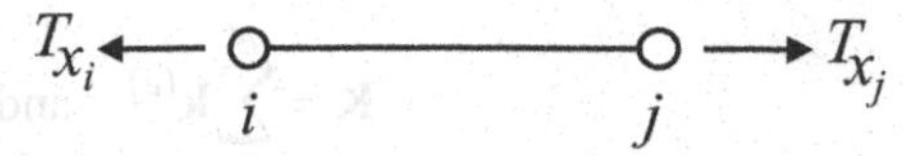

Fig. 6.25 A typical linear line element with two nodes

$$\begin{array}{c} \qquad\quad \boxed{2} \qquad\quad \boxed{3} \\ \mathbf{k}^{(2)} = \begin{bmatrix} \alpha^{(2)} & -\alpha^{(2)} \\ -\alpha^{(2)} & \alpha^{(2)} \end{bmatrix} \begin{matrix} \boxed{2} \\ \boxed{3} \end{matrix} \end{array} \tag{6.205}$$

$$\begin{array}{c} \qquad\quad \boxed{3} \qquad\quad \boxed{4} \\ \mathbf{k}^{(3)} = \begin{bmatrix} \alpha^{(3)} & -\alpha^{(3)} \\ -\alpha^{(3)} & \alpha^{(3)} \end{bmatrix} \begin{matrix} \boxed{3} \\ \boxed{4} \end{matrix} \end{array} \tag{6.206}$$

The element load vector, $\mathbf{p}_{\mathbf{T}}^{(e)}$, due to the unknown nodal forces, T_{x_i} and T_{x_j} at nodes i and j, respectively (Fig. 6.25), can be obtained from

$$\begin{aligned} \mathbf{p}_{\mathbf{T}}^{(e)} = \int_{S_{\sigma}^{(e)}} \mathbf{N}^{(e)\mathrm{T}} \mathbf{T}\, dS &= \begin{Bmatrix} N_1^{(e)}\left(x = x_1^{(e)}\right) \\ N_2^{(e)}\left(x = x_1^{(e)}\right) \end{Bmatrix} T_{x_1} \\ &+ \begin{Bmatrix} N_1^{(e)}\left(x = x_2^{(e)}\right) \\ N_2^{(e)}\left(x = x_2^{(e)}\right) \end{Bmatrix} T_{x_2} \end{aligned} \tag{6.207}$$

Evaluating the shape functions results in a load vector of the form

$$\begin{aligned} \mathbf{p}_{\mathbf{T}}^{(e)} &= \begin{Bmatrix} -1 \\ 0 \end{Bmatrix} T_{x_1} + \begin{Bmatrix} 0 \\ 1 \end{Bmatrix} T_{x_2} \\ &= \begin{Bmatrix} -T_{x_1} \\ 0 \end{Bmatrix} + \begin{Bmatrix} 0 \\ T_{x_2} \end{Bmatrix} = \begin{Bmatrix} -T_{x_1} \\ T_{x_2} \end{Bmatrix} \end{aligned} \tag{6.208}$$

Associated with each element, the load vectors become

$$\mathbf{p}_{\mathbf{T}}^{(1)} = \begin{Bmatrix} -T_{x_1} \\ T_{x_2} \end{Bmatrix}, \quad \mathbf{p}_{\mathbf{T}}^{(2)} = \begin{Bmatrix} -T_{x_2} \\ T_{x_3} + P \end{Bmatrix}, \quad \mathbf{p}_{\mathbf{T}}^{(3)} = \begin{Bmatrix} -T_{x_3} \\ T_{x_4} \end{Bmatrix} \tag{6.209}$$

The global coefficient matrix, $\mathbf{K}$, and the load vector, $\mathbf{P}_{\mathbf{T}}$, are obtained from the "expanded" element coefficient matrices, $\mathbf{k}^{(e)}$, and the element load vectors, $\mathbf{p}_{\mathbf{T}}^{(e)}$, by summation in the form

$$\mathbf{K} = \sum_{e=1}^{E} \mathbf{k}^{(e)} \quad \text{and} \quad \mathbf{P_T} = \sum_{e=1}^{E} \mathbf{p}_{\mathbf{T}}^{(e)} \tag{6.210}$$

The "expanded" element matrices are the same size as the global matrix but have rows and columns of zeros corresponding to the nodes not associated with element (e). Specifically, the expanded form of the element stiffness and load vector becomes

$$\begin{array}{c} \quad\ \boxed{1} \qquad \boxed{2} \quad \boxed{3}\ \boxed{4} \\ \mathbf{k}^{(1)} = \begin{bmatrix} \alpha^{(1)} & -\alpha^{(1)} & 0 & 0 \\ -\alpha^{(1)} & \alpha^{(1)} & 0 & 0 \\ 0 & 0 & 0 & 0 \\ 0 & 0 & 0 & 0 \end{bmatrix} \begin{matrix} \boxed{1} \\ \boxed{2} \\ \boxed{3} \\ \boxed{4} \end{matrix} \end{array} ; \quad \begin{array}{c} \boxed{1} \\ \mathbf{p}_{\mathbf{T}}^{(1)} = \begin{bmatrix} -T_{x_1} \\ T_{x_2} \\ 0 \\ 0 \end{bmatrix} \begin{matrix} \boxed{1} \\ \boxed{2} \\ \boxed{3} \\ \boxed{4} \end{matrix} \end{array} \tag{6.211}$$

$$\begin{array}{c} \quad\ \boxed{1} \quad \boxed{2} \quad \boxed{3} \quad \boxed{4} \\ \mathbf{k}^{(2)} = \begin{bmatrix} 0 & 0 & 0 & 0 \\ 0 & \alpha^{(2)} & -\alpha^{(2)} & 0 \\ 0 & -\alpha^{(2)} & \alpha^{(2)} & 0 \\ 0 & 0 & 0 & 0 \end{bmatrix} \begin{matrix} \boxed{1} \\ \boxed{2} \\ \boxed{3} \\ \boxed{4} \end{matrix} \end{array} ; \quad \begin{array}{c} \boxed{1} \\ \mathbf{p}_{\mathbf{T}}^{(2)} = \begin{bmatrix} 0 \\ -T_{x_2} \\ T_{x_3} + P \\ 0 \end{bmatrix} \begin{matrix} \boxed{1} \\ \boxed{2} \\ \boxed{3} \\ \boxed{4} \end{matrix} \end{array} \tag{6.212}$$

$$\begin{array}{c} \quad\ \boxed{1}\ \boxed{2} \quad \boxed{3} \qquad \boxed{4} \\ \mathbf{k}^{(3)} = \begin{bmatrix} 0 & 0 & 0 & 0 \\ 0 & 0 & 0 & 0 \\ 0 & 0 & \alpha^{(3)} & -\alpha^{(3)} \\ 0 & 0 & -\alpha^{(3)} & \alpha^{(3)} \end{bmatrix} \begin{matrix} \boxed{1} \\ \boxed{2} \\ \boxed{3} \\ \boxed{4} \end{matrix} \end{array} ; \quad \begin{array}{c} \boxed{1} \\ \mathbf{p}_{\mathbf{T}}^{(3)} = \begin{bmatrix} 0 \\ 0 \\ -T_{x_3} \\ T_{x_4} \end{bmatrix} \begin{matrix} \boxed{1} \\ \boxed{2} \\ \boxed{3} \\ \boxed{4} \end{matrix} \end{array} \tag{6.213}$$

In accordance with Eq. (6.210) and (6.192), the global equilibrium equations can be written as

$$\begin{bmatrix} \alpha^{(1)} & -\alpha^{(1)} & 0 & 0 \\ -\alpha^{(1)} & \left(\alpha^{(1)} + \alpha^{(2)}\right) & -\alpha^{(2)} & 0 \\ 0 & -\alpha^{(2)} & \left(\alpha^{(2)} + \alpha^{(3)}\right) & -\alpha^{(3)} \\ 0 & 0 & -\alpha^{(3)} & \alpha^{((3))} \end{bmatrix} \begin{Bmatrix} u_{x_1} \\ u_{x_2} \\ u_{x_3} \\ u_{x_4} \end{Bmatrix} = \begin{Bmatrix} -T_{x_1} \\ T_{x_2} - T_{x_2} \\ T_{x_3} + P - T_{x_3} \\ T_{x_4} \end{Bmatrix} \tag{6.214}$$

Enforcing the boundary conditions of $u_{x_1} = u_{x_4} = 0$ leads to

$$\begin{bmatrix} \alpha^{(1)} & -\alpha^{(1)} & 0 & 0 \\ -\alpha^{(1)} & \left(\alpha^{(1)}+\alpha^{(2)}\right) & -\alpha^{(2)} & 0 \\ 0 & -\alpha^{(2)} & \left(\alpha^{(2)}+\alpha^{(3)}\right) & -\alpha^{(3)} \\ 0 & 0 & -\alpha^{(3)} & \alpha^{(3)} \end{bmatrix} \begin{Bmatrix} 0 \\ u_{x_2} \\ u_{x_3} \\ 0 \end{Bmatrix} = \begin{Bmatrix} -T_{x_1} \\ 0 \\ P \\ T_{x_4} \end{Bmatrix} \tag{6.215}$$

This system of equations can be partitioned in the form

$$\begin{bmatrix} -\alpha^{(1)} & \left(\alpha^{(1)}+\alpha^{(2)}\right) & -\alpha^{(2)} & 0 \\ 0 & -\alpha^{(2)} & \left(\alpha^{(2)}+\alpha^{(3)}\right) & 0 \end{bmatrix} \begin{Bmatrix} 0 \\ u_{x2} \\ u_{x3} \\ 0 \end{Bmatrix} = \begin{Bmatrix} 0 \\ P \end{Bmatrix} \tag{6.216a}$$

or

$$\begin{bmatrix} \left(\alpha^{(1)}+\alpha^{(2)}\right) & -\alpha^{(2)} \\ -\alpha^{(2)} & \left(\alpha^{(2)}+\alpha^{(3)}\right) \end{bmatrix} \begin{Bmatrix} u_{x2} \\ u_{x3} \end{Bmatrix} = \begin{Bmatrix} 0 \\ P \end{Bmatrix} \tag{6.216b}$$

and

$$\begin{bmatrix} \alpha^{(1)} & -\alpha^{(1)} & 0 & 0 \\ 0 & 0 & -\alpha^{(3)} & \alpha^{(3)} \end{bmatrix} \begin{Bmatrix} 0 \\ u_{x_2} \\ u_{x_3} \\ 0 \end{Bmatrix} = \begin{Bmatrix} -T_{x_1} \\ T_{x_4} \end{Bmatrix} \tag{6.217a}$$

or

$$T_{x_1} = \alpha^{(1)} u_{x_2} \text{ and } T_{x_4} = -\alpha^{(3)} u_{x_3} \tag{6.217b}$$

Solution to nodal displacements results in

$$u_{x_2} = \frac{\alpha^{(2)}}{\left(\alpha^{(1)}\alpha^{(2)} + \alpha^{(1)}\alpha^{(3)} + \alpha^{(2)}\alpha^{(3)}\right)} P \tag{6.218}$$

$$u_{x_3} = \frac{\alpha^{(1)} + \alpha^{(2)}}{\left(\alpha^{(1)}\alpha^{(2)} + \alpha^{(1)}\alpha^{(3)} + \alpha^{(2)}\alpha^{(3)}\right)} P \tag{6.219}$$

With these nodal displacements, the reaction forces are computed as

$$T_{x_1} = \frac{\alpha^{(1)}\alpha^{(2)}}{\left(\alpha^{(1)}\alpha^{(2)} + \alpha^{(1)}\alpha^{(3)} + \alpha^{(2)}\alpha^{(3)}\right)} P \tag{6.220}$$

$$T_{x_4} = -\frac{\alpha^{(3)}\left(\alpha^{(1)} + \alpha^{(2)}\right)}{\left(\alpha^{(1)}\alpha^{(2)} + \alpha^{(1)}\alpha^{(3)} + \alpha^{(2)}\alpha^{(3)}\right)} P \tag{6.221}$$

Finally, the strains are computed as

$$\varepsilon_{xx}^{(1)} = \frac{1}{L^{(1)}}(u_{x_2} - u_{x_1}) = \frac{\alpha^{(2)}}{\left(\alpha^{(1)}\alpha^{(2)} + \alpha^{(1)}\alpha^{(3)} + \alpha^{(2)}\alpha^{(3)}\right)L^{(1)}} P \tag{6.222}$$

$$\varepsilon_{xx}^{(2)} = \frac{1}{L^{(2)}}(u_{x_3} - u_{x_2}) = \frac{\alpha^{(1)}}{\left(\alpha^{(1)}\alpha^{(2)} + \alpha^{(1)}\alpha^{(3)} + \alpha^{(2)}\alpha^{(3)}\right)L^{(2)}} P \tag{6.223}$$

$$\varepsilon_{xx}^{(3)} = \frac{1}{L^{(3)}}(u_{x_4} - u_{x_3}) = -\frac{\left(\alpha^{(1)} + \alpha^{(2)}\right)}{\left(\alpha^{(1)}\alpha^{(2)} + \alpha^{(1)}\alpha^{(3)} + \alpha^{(2)}\alpha^{(3)}\right)L^{(3)}} P \tag{6.224}$$

6.2.2 *Two-Dimensional Structural Analysis*

The three-dimensional analysis of either "thin" or "long" components subjected to in-plane external loading conditions can be reduced to a two-dimensional analysis under certain assumptions referred to as "plane stress" and "plane strain" conditions.

6.2.2.1 Plane Stress Conditions

A state of plane stress exists for thin components subjected only to in-plane external loading, i.e., no lateral loads (Fig. 6.26). Due to a small thickness-to-characteristic length ratio and in-plane external loading only, there is no out-of-plane displacement component, u_z, and the shear strain components associated with the thickness direction, γ_{xz} and γ_{yz}, are very small and assumed to be zero. Therefore, the stress components, σ_{zz}, σ_{xz}, and σ_{yz}, associated with the thickness direction vanish. Under these assumptions, the displacement, $\mathbf{u}$, stress, $\mathbf{A}$, strain, $\boldsymbol{\varepsilon}$, and traction, $\mathbf{T}$, vectors, and material property matrix, $\mathbf{D}$, reduce to

$$\begin{aligned}
\mathbf{u}^T &= \{u_x \quad u_y\} \\
\boldsymbol{\sigma}^T &= \{\sigma_{xx} \quad \sigma_{yy} \quad \sigma_{xy}\} \\
\boldsymbol{\varepsilon}^T &= \{\varepsilon_{xx} \quad \varepsilon_{yy} \quad \gamma_{xy}\} \\
\mathbf{T}^T &= \{T_x \quad T_y\}
\end{aligned} \tag{6.225}$$

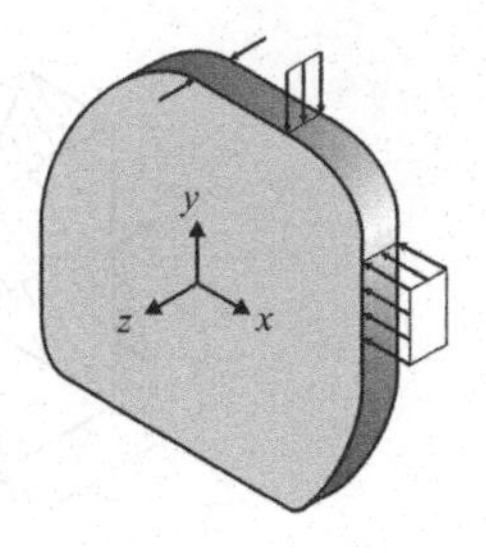

Fig. 6.26 Thin body with in-plane loading; suitable for plane stress idealization

and

$$\mathbf{D} = \frac{E}{1-\nu^2}\begin{bmatrix} 1 & \nu & 0 \\ \nu & 1 & 0 \\ 0 & 0 & \frac{(1-\nu)}{2} \end{bmatrix} \tag{6.226}$$

with

$$\varepsilon_{zz} = -\frac{\nu}{E}\left(\sigma_{xx} + \sigma_{yy}\right) \tag{6.227}$$

The initial strains arising from ΔT, the temperature change with respect to the reference state, can be expressed as

$$\boldsymbol{\varepsilon}^{*T} = \left[\alpha\Delta T \quad \alpha\Delta T \quad 0\right] \tag{6.228}$$

6.2.2.2 Plane Strain Conditions

A state of plane strain exists for a cylindrical component that is either "long" or fully constrained in the length direction under the action of only uniform lateral external loads (two examples are shown in Fig. 6.27). Because the ends of the cylindrical component are prevented from deforming in the thickness direction, it is assumed that the displacement component u_z vanishes at every cross section of the body. The uniform loading and cross-sectional geometry eliminates any variation in the length direction, leading to $\partial() / \partial z = 0$. Also, planes perpendicular to the thickness direction before deformation remain perpendicular to the thickness direction after deformation. These assumptions result in zero transverse shear strains, $\gamma_{xz} = \gamma_{yz} = 0$. Under these assumptions, the displacement, **u**, stress, **A**, strain, ε, and traction, **T**, vectors, and material property matrix, **D,** reduce to

$$\begin{aligned} \mathbf{u}^T &= \left\{u_x \quad u_y\right\} \\ \boldsymbol{\sigma}^T &= \left\{\sigma_{xx} \quad \sigma_{yy} \quad \sigma_{xy}\right\} \\ \boldsymbol{\varepsilon}^T &= \left\{\varepsilon_{xx} \quad \varepsilon_{yy} \quad \gamma_{xy}\right\} \\ \mathbf{T}^T &= \left\{T_x \quad T_y\right\} \end{aligned} \tag{6.229}$$

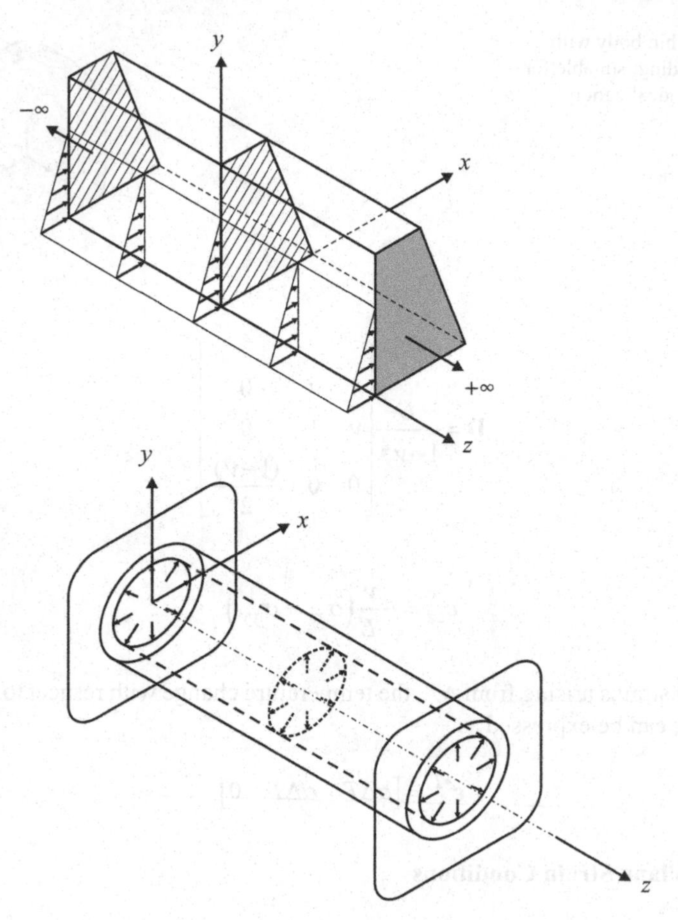

Fig. 6.27 Long bodies with in-plane loading; suitable for plane strain idealization

and

$$\mathbf{D} = \frac{E}{(1+\nu)(1-2\nu)}\begin{bmatrix} 1-\nu & \nu & 0 \\ \nu & 1-\nu & 0 \\ 0 & 0 & \frac{(1-2\nu)}{2} \end{bmatrix} \tag{6.230}$$

The initial strain vector due to this temperature change can be expressed as

$$\boldsymbol{\varepsilon}^{*T} = \left[(1+\nu)\alpha\Delta T \quad (1+\nu)\alpha\Delta T \quad 0\right] \tag{6.231}$$

where ΔT is the temperature change with respect to a reference state.

The material property matrices for both plane stress and strain conditions have the same form, and it is convenient to present it in the form

$$\mathbf{D} = \begin{bmatrix} D_1 & D_1 D_2 & 0 \\ D_1 D_2 & D_1 & 0 \\ 0 & 0 & D_{12} \end{bmatrix} \tag{6.232}$$

where

$$D_{12} = \frac{D_1(1-D_2)}{2} \tag{6.233}$$

with $D_1 = E/(1-\nu)^2$ and $D_2 = \nu$ for plane stress, and $D_1 = E(1-\nu)/\ (1+\nu)(1-2\nu)$ and $D_2 = \nu/(1-\nu)$ for plane strain.

6.2.2.3 Finite Element Equations with Linear Triangular Elements

The displacement components u_x and u_y within a triangular element can be approximated as

$$\begin{aligned} u_x^{(e)} &= \tilde{u}_x^{(e)} = N_1^{(e)} u_{x_1}^{(e)} + N_2^{(e)} u_{x_2}^{(e)} + N_3^{(e)} u_{x_3}^{(e)} \\ u_y^{(e)} &= \tilde{u}_y^{(e)} = N_1^{(e)} u_{y_1}^{(e)} + N_2^{(e)} u_{y_2}^{(e)} + N_3^{(e)} u_{y_3}^{(e)} \end{aligned} \tag{6.234}$$

in which $N_1^{(e)}$, $N_2^{(e)}$, and $N_3^{(e)}$ are the linear shape functions and $(u_{x_1}^{(e)}, u_{y_1}^{(e)})$, $(u_{x_2}^{(e)}, u_{y_2}^{(e)})$, and $(u_{x_3}^{(e)}, u_{y_3}^{(e)})$ are the nodal unknowns (degrees of freedom) associated with first, second, and third nodes, respectively. An example of a triangular element with its nodal degrees of freedom and local nodal numbering is shown in Fig. 6.28. In matrix form, the approximate displacement components become

$$\tilde{\mathbf{u}}^{(e)} = \mathbf{N}^{(e)T} \mathbf{U}^{(e)} \tag{6.235}$$

in which

$$\tilde{\mathbf{u}}^{(e)T} = \left\{ \tilde{u}_x^{(e)} \quad \tilde{u}_y^{(e)} \right\} \tag{6.236}$$

and

$$\mathbf{N}^{(e)T} = \begin{bmatrix} N_1 & 0 & N_2 & 0 & N_3 & 0 \\ 0 & N_1 & 0 & N_2 & 0 & N_3 \end{bmatrix} \tag{6.237}$$

and

$$\mathbf{U}^{(e)T} = \left\{ u_{x_1}^{(e)} \quad u_{y_1}^{(e)} \quad u_{x_2}^{(e)} \quad u_{y_2}^{(e)} \quad u_{x_3}^{(e)} \quad u_{y_3}^{(e)} \right\} \tag{6.238}$$

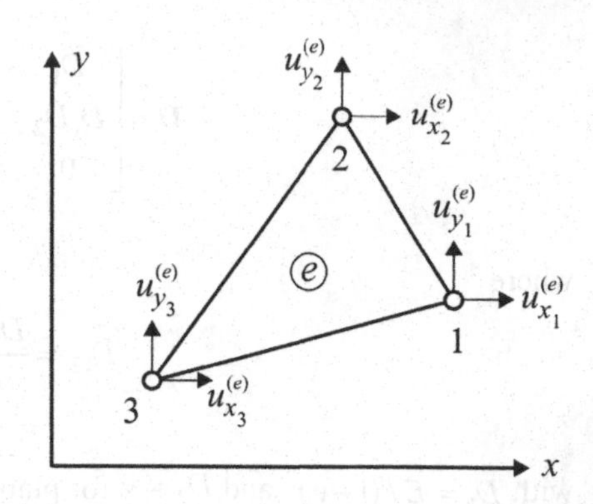

Fig. 6.28 Typical linear triangular element with nodal degrees of freedom

The element shape matrix, $\mathbf{B}^{(e)}$, becomes

$$\mathbf{B}^{(e)} = \begin{bmatrix} \frac{\partial N_1}{\partial x} & 0 & \frac{\partial N_2}{\partial x} & 0 & \frac{\partial N_3}{\partial x} & 0 \\ 0 & \frac{\partial N_1}{\partial y} & 0 & \frac{\partial N_2}{\partial y} & 0 & \frac{\partial N_3}{\partial y} \\ \frac{\partial N_1}{\partial y} & \frac{\partial N_1}{\partial x} & \frac{\partial N_2}{\partial y} & \frac{\partial N_2}{\partial x} & \frac{\partial N_3}{\partial y} & \frac{\partial N_3}{\partial x} \end{bmatrix} \tag{6.239}$$

Substituting for the derivatives of the shape functions, this matrix simplifies to

$$\mathbf{B}^{(e)} = \frac{1}{2\Delta^{(e)}} \begin{bmatrix} y_{23}^{(e)} & 0 & y_{31}^{(e)} & 0 & y_{12}^{(e)} & 0 \\ 0 & x_{32}^{(e)} & 0 & x_{13}^{(e)} & 0 & x_{21}^{(e)} \\ x_{32}^{(e)} & y_{23}^{(e)} & x_{13}^{(e)} & y_{31}^{(e)} & x_{21}^{(e)} & y_{12}^{(e)} \end{bmatrix} \tag{6.240}$$

Both the element shape and material property matrices are independent of the spatial coordinates, x and y, thus leading to the evaluation of the element stiffness matrix, $\mathbf{k}^{(e)}$, as

$$\mathbf{k}^{(e)} = \mathbf{B}^{(e)T}\mathbf{D}\mathbf{B}^{(e)}V^{(e)} \tag{6.241}$$

where $V^{(e)} = t\Delta^{(e)}$, with element area $\Delta^{(e)}$ and constant thickness t. The evaluation of the load vectors, $\mathbf{p}_{\mathbf{b}}^{(e)}$ and $\mathbf{p}_{\mathbf{T}}^{(e)}$, arising from the body forces and surface tractions (forces), respectively, involve integrals of the form

$$\int dx\,dy, \int x dx\,dy, \int y\,dx\,dy \tag{6.242}$$

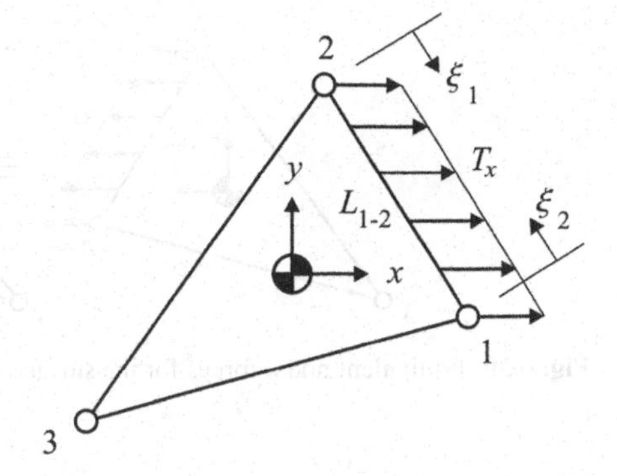

Fig. 6.29 Surface force along side 1–2 of the triangular element

By choosing the centroid of the triangle as the origin of the (x, y) coordinate system, the integrals involving either x or y in the integrand vanish. The load vector arising from the body forces can be obtained from

$$\mathbf{p}_{\mathbf{b}}^{(e)} = \int_{V^{(e)}} \begin{bmatrix} N_1 & 0 \\ 0 & N_1 \\ N_2 & 0 \\ 0 & N_2 \\ N_3 & 0 \\ 0 & N_3 \end{bmatrix} \begin{Bmatrix} b_x \\ b_y \end{Bmatrix} dV = \int_{V^{(e)}} \begin{Bmatrix} N_1 b_x \\ N_1 b_y \\ N_2 b_x \\ N_2 b_y \\ N_3 b_x \\ N_3 b_y \end{Bmatrix} dV \tag{6.243}$$

reducing to

$$\mathbf{p}_{\mathbf{b}}^{(e)T} = \frac{t\Delta^{(e)}}{3} \begin{bmatrix} b_x & b_y & b_x & b_y & b_x & b_y \end{bmatrix} \tag{6.244}$$

in which b_x and b_y are the components of the body force vector.

The evaluation of the element load vector due to the applied traction forces (distributed loads as shown in Fig. 6.29) requires their explicit variation along the edges of the element. For an element of constant thickness subjected to uniform load of T_x in the x-direction along its 1–2 edge, the vector $\mathbf{p}_{\mathbf{T}}^{(e)}$ can be written as

$$\mathbf{p}_{\mathbf{T}}^{(e)} = t \int_{L_{1-2}} \begin{bmatrix} N_1 & 0 \\ 0 & N_1 \\ N_2 & 0 \\ 0 & N_2 \\ 0 & 0 \\ 0 & 0 \end{bmatrix} \begin{Bmatrix} T_x \\ 0 \end{Bmatrix} dl = t \int_{L_{1-2}} \begin{Bmatrix} N_1 T_x \\ 0 \\ N_2 T_x \\ 0 \\ 0 \\ 0 \end{Bmatrix} dl \tag{6.245}$$

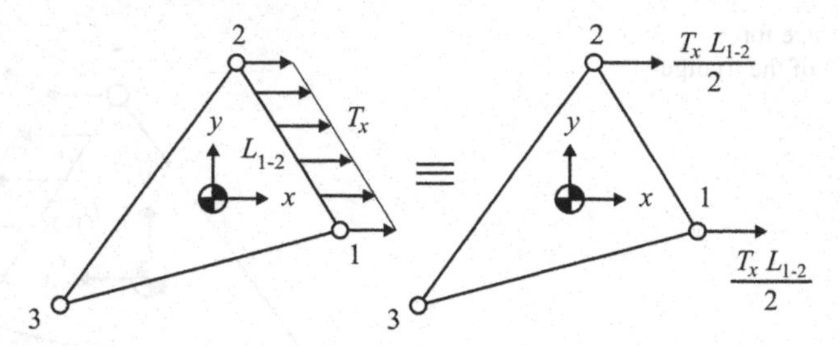

Fig. 6.30 Equivalent nodal forces for the surface force along side 1–2 of the triangular element

in which $N_3 = 0$ along the 1–2 edge and L_{1-2} is the length of the 1–2 edge. Since N_1 and N_2 vary linearly along the 1–2 edge, they can be expressed in terms of the natural coordinates, ξ_1 and ξ_2, as derived in Chap. 3

$$N_1 = \xi_1 = \frac{x_2^{(e)} - x}{x_2^{(e)} - x_1^{(e)}} \quad \text{and} \quad N_2 = \xi_2 = \frac{x - x_1^{(e)}}{x_2^{(e)} - x_1^{(e)}} \tag{6.246}$$

The integrals in the expression for $\mathbf{p}_{\mathbf{T}}^{(e)}$ are evaluated as

$$\begin{aligned} \int_{L_{1-2}} N_1 \, T_x dl &= \int_0^1 \xi_1 \, T_x L_{1-2} d\xi_1 = \frac{T_x L_{1-2}}{2} \\ \int_{L_{1-2}} N_2 \, T_x dl &= \int_0^1 \xi_2 \, T_x L_{1-2} d\xi_2 = \frac{T_x L_{1-2}}{2} \end{aligned} \tag{6.247}$$

Thus, the load vector, $\mathbf{p}_{\mathbf{T}}^{(e)}$, takes the form

$$\mathbf{p}_{\mathbf{T}}^{(e)T} = t \frac{T_x L_{1-2}}{2} [1 \quad 0 \quad 1 \quad 0 \quad 0 \quad 0] \tag{6.248}$$

as illustrated in Fig. 6.30.

Note that this result corresponds to equivalent point forces acting at the first and second nodes. The element load vectors arising from the initial strains and stresses can be written as

$$\begin{aligned} \mathbf{p}_{\varepsilon^*}^{(e)} &= \mathbf{B}^{(e)T} \mathbf{D} \boldsymbol{\varepsilon}^* \, V^{(e)} \\ \mathbf{p}_{\sigma^*}^{(e)} &= \mathbf{B}^{(e)T} \boldsymbol{\sigma}^* \, V^{(e)} \end{aligned} \tag{6.249}$$

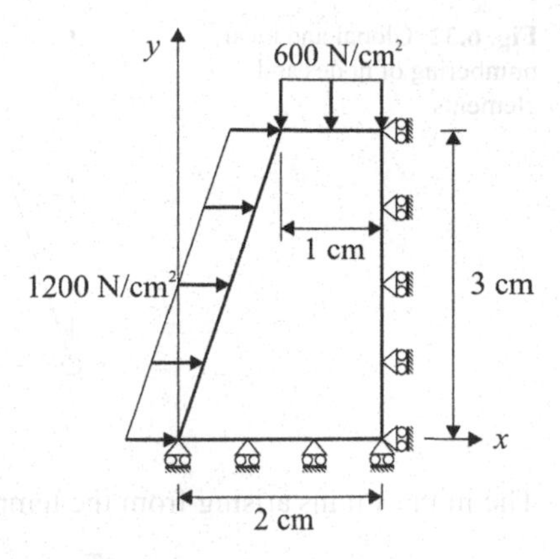

Fig. 6.31 Geometry and loading of the problem

6.2.2.4 Example of a Plane Stress Analysis with Linear Triangular Elements

Derivation of a System of Equations and Its Solution

Using linear triangular elements, determine the nodal displacements and the element stresses in a thin plate subjected to displacement constraints and surface tractions as shown in Fig. 6.31. Also, the plate is exposed to a temperature change of 10 °C from the reference temperature. The plate thickness is 0.5 cm and the Young's modulus, E, and the Poisson's ratio, ν, are $15 \times 10^6\,\text{N/cm}^2$ and 0.25, respectively. The coefficient of thermal expansion is 6×10^{-6} / °C.

In order to illustrate the solution method, the plate is discretized into two triangular elements, as shown in Fig. 6.32.

The global coordinates of each node are specified by (x_p, y_p), with $p = 1,2,3,4$, and are presented in Table 6.7.

The global unknown nodal displacement vector is given by

$$\mathbf{U}^T = \left\{ u_{x_1} \quad u_{y_1} \quad u_{x_2} \quad u_{y_2} \quad u_{x_3} \quad u_{y_3} \quad u_{x_4} \quad u_{y_4} \right\} \tag{6.250}$$

Considering the correspondence between the local and global node numbering schemes, the elements are defined (connected) as shown in Table 6.8.

The areas of each element are calculated to be

$$\Delta^{(1)} = 3\ \text{cm}^2 \quad \text{and} \quad \Delta^{(2)} = 3/2\ \text{cm}^2 \tag{6.251}$$

Under plane stress assumptions, the material property matrix becomes

$$\mathbf{D} = 10^6 \begin{bmatrix} 16 & 4 & 0 \\ 4 & 16 & 0 \\ 0 & 0 & 6 \end{bmatrix} \text{N/cm}^2 \tag{6.252}$$

Fig. 6.32 Global and local numbering of nodes and elements

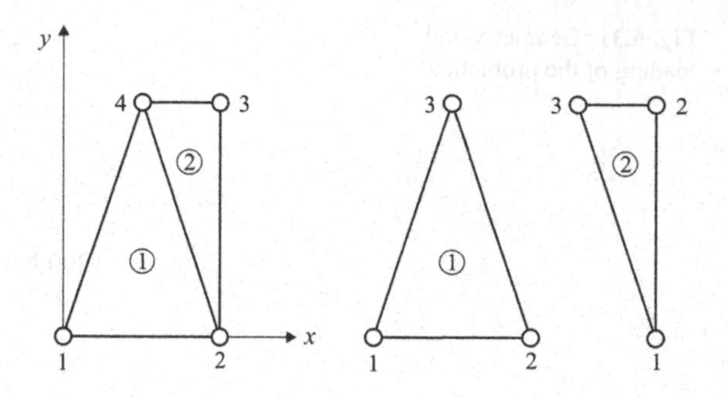

The initial strains arising from the temperature change is written as

$$\boldsymbol{\varepsilon}^{*T} = 10^{-6}\begin{bmatrix}60 & 60 & 0\end{bmatrix} \tag{6.253}$$

The element load vectors arising from the applied tractions are

$$\mathbf{p}_{\mathbf{T}}^{(1)T} = t\frac{T_x L_{1-4}}{2}\begin{bmatrix}1 & 0 & 0 & 0 & 1 & 0\end{bmatrix} \tag{6.254}$$

$$\mathbf{p}_{\mathbf{T}}^{(2)T} = t\frac{T_y L_{3-4}}{2}\begin{bmatrix}0 & 0 & 0 & 1 & 0 & 1\end{bmatrix} \tag{6.255}$$

With the specified values of the thickness and the distributed loads, these element load vectors become

$$\mathbf{p}_{\mathbf{T}}^{(1)T} = 300\sqrt{10}\begin{bmatrix}1 & 0 & 0 & 0 & 1 & 0\end{bmatrix}N \tag{6.256}$$

and

$$\mathbf{p}_{\mathbf{T}}^{(2)T} = -150\begin{bmatrix}0 & 0 & 0 & 1 & 0 & 1\end{bmatrix}N \tag{6.257}$$

For the first element, $e = 1$, the components of the element shape matrix $\mathbf{B}^{(1)}$ are computed as

$$\begin{aligned}
y_{23}^{(1)} &= y_2^{(1)} - y_3^{(1)} = y_2 - y_4 = -3\,, & x_{32}^{(1)} &= x_3^{(1)} - x_2^{(1)} = x_4 - x_2 = -1\\
y_{31}^{(1)} &= y_3^{(1)} - y_1^{(1)} = y_4 - y_1 = 3\,, & x_{13}^{(1)} &= x_1^{(1)} - x_3^{(1)} = x_1 - x_4 = -1\\
y_{12}^{(1)} &= y_1^{(1)} - y_2^{(1)} = y_1 - y_2 = 0\,, & x_{21}^{(1)} &= x_2^{(1)} - x_1^{(1)} = x_2 - x_1 = 2
\end{aligned} \tag{6.258}$$

leading to

$$\mathbf{B}^{(1)} = \frac{1}{6}\begin{bmatrix}-3 & 0 & 3 & 0 & 0 & 0\\ 0 & -1 & 0 & -1 & 0 & 2\\ -1 & -3 & -1 & 3 & 2 & 0\end{bmatrix} \tag{6.259}$$

Table 6.7 Global nodal coordinates

Global node number	Nodal coordinates	Nodal unknowns
1	(0,0)	u_{x_1}, u_{y_1}
2	(2,0)	u_{x_2}, u_{y_2}
3	(2,3)	u_{x_3}, u_{y_3}
4	(1,3)	u_{x_4}, u_{y_4}

Table 6.8 Element connectivity

Element Number (e)	Node 1	Node 2	Node 3
1	1	2	4
2	2	3	4

For the second element, $e = 2$, the components of the element shape matrix $\mathbf{B}^{(2)}$ are computed as

$$\begin{aligned} y_{23}^{(2)} &= y_2^{(2)} - y_3^{(2)} = y_3 - y_4 = 0\,, & x_{32}^{(2)} &= x_3^{(2)} - x_2^{(2)} = x_4 - x_3 = -1 \\ y_{31}^{(2)} &= y_3^{(2)} - y_1^{(2)} = y_4 - y_2 = 3\,, & x_{13}^{(2)} &= x_1^{(2)} - x_3^{(2)} = x_2 - x_4 = 1 \\ y_{12}^{(2)} &= y_1^{(2)} - y_2^{(2)} = y_2 - y_3 = -3\,, & x_{21}^{(2)} &= x_2^{(2)} - x_1^{(2)} = x_3 - x_2 = 0 \end{aligned} \tag{6.260}$$

leading to

$$\mathbf{B}^{(2)} = \frac{1}{3}\begin{bmatrix} 0 & 0 & 3 & 0 & -3 & 0 \\ 0 & -1 & 0 & 1 & 0 & 0 \\ -1 & 0 & 1 & 3 & 0 & -3 \end{bmatrix} \tag{6.261}$$

The evaluation of the stiffness matrices, $\mathbf{k}^{(1)}$ and $\mathbf{k}^{(2)}$, requires the products of $\mathbf{B}^{(1)T}\mathbf{D}$ and $\mathbf{B}^{(2)T}\mathbf{D}$. Also, these products appear in the evaluation of the element load vectors arising from the temperature change. Therefore,

$$\mathbf{B}^{(1)T}\mathbf{D} = \frac{10^6}{6}\begin{bmatrix} -48 & -12 & -6 \\ -4 & -16 & -18 \\ 48 & 12 & -6 \\ -4 & -16 & 18 \\ 0 & 0 & 12 \\ 8 & 32 & 0 \end{bmatrix} \tag{6.262}$$

$$\mathbf{B}^{(2)T}\mathbf{D} = \frac{10^6}{3}\begin{bmatrix} 0 & 0 & -6 \\ -4 & -16 & 0 \\ 48 & 12 & 6 \\ 4 & 16 & 18 \\ -48 & -12 & 0 \\ 0 & 0 & -18 \end{bmatrix} \tag{6.263}$$

The element stiffness matrices become

$$\begin{array}{ccc} \boxed{i=1} & \boxed{j=2} & \boxed{k=4} \end{array}$$

$$\mathbf{k}^{(1)} = \frac{10^6}{12}\begin{bmatrix} 75 & 15 & -69 & -3 & -6 & -12 \\ 15 & 35 & 3 & -19 & -18 & -16 \\ -69 & 3 & 75 & -15 & -6 & 12 \\ -3 & -19 & -15 & 35 & 18 & -16 \\ -6 & -18 & -6 & 18 & 12 & 0 \\ -12 & -16 & 12 & -16 & 0 & 32 \end{bmatrix} \tag{6.264}$$

and

$$\begin{array}{ccc} \boxed{i=2} & \boxed{j=3} & \boxed{k=4} \end{array}$$

$$\mathbf{k}^{(2)} = \frac{10^6}{12}\begin{bmatrix} 6 & 0 & -6 & -18 & 0 & 18 \\ 0 & 16 & -12 & -16 & 12 & 0 \\ -6 & -12 & 150 & 30 & -144 & -18 \\ -18 & -16 & 30 & 70 & -12 & -54 \\ 0 & 12 & -144 & -12 & 144 & 0 \\ 18 & 0 & -18 & -54 & 0 & 54 \end{bmatrix} \tag{6.265}$$

The boxed numbers above each column pair indicate the nodal order of degrees of freedom in each element stiffness matrix.

The thermal load vectors associated with each element are obtained as

$$\mathbf{p}_{\varepsilon^*}^{(1)} = \begin{Bmatrix} -900 \\ -300 \\ 900 \\ -300 \\ 0 \\ 600 \end{Bmatrix} N \quad \text{and} \quad \mathbf{p}_{\varepsilon^*}^{(2)} = \begin{Bmatrix} 0 \\ -300 \\ 900 \\ 300 \\ -900 \\ 0 \end{Bmatrix} N \tag{6.266}$$

Rewriting the element stiffness matrices and the load vectors, in the expanded order and rearranged form according to the increasing nodal degrees of freedom of the global stiffness matrix, **K** yields

Associated with the first element:

$$\mathbf{k}^{(1)} = \frac{10^6}{12}\begin{array}{c} \begin{array}{cccccccc} \multicolumn{2}{c}{\boxed{1}} & \multicolumn{2}{c}{\boxed{2}} & \multicolumn{2}{c}{\boxed{3}} & \multicolumn{2}{c}{\boxed{4}} \end{array} \\ \begin{bmatrix} 75 & 15 & -69 & -3 & 0 & 0 & -6 & -12 \\ 15 & 35 & 3 & -19 & 0 & 0 & -18 & -16 \\ -69 & 3 & 75 & -15 & 0 & 0 & -6 & 12 \\ -3 & -19 & -15 & 35 & 0 & 0 & 18 & -16 \\ 0 & 0 & 0 & 0 & 0 & 0 & 0 & 0 \\ 0 & 0 & 0 & 0 & 0 & 0 & 0 & 0 \\ -6 & -18 & -6 & 18 & 0 & 0 & 12 & 0 \\ -12 & -16 & 12 & -16 & 0 & 0 & 0 & 32 \end{bmatrix} \end{array} \tag{6.267}$$

$$\mathbf{p}_{\mathbf{T}}^{(1)} = 300\sqrt{10}\begin{Bmatrix} 1 \\ 0 \\ 0 \\ 0 \\ 0 \\ 0 \\ 1 \\ 0 \end{Bmatrix} N \quad \text{and} \quad \mathbf{p}_{\varepsilon^*}^{(1)} = \begin{Bmatrix} -900 \\ -300 \\ 900 \\ -300 \\ 0 \\ 0 \\ 0 \\ 600 \end{Bmatrix} N \tag{6.268}$$

Associated with the second element:

$$\mathbf{k}^{(2)} = \frac{10^6}{12}\begin{array}{c} \begin{array}{cccccccc} \multicolumn{2}{c}{\boxed{1}} & \multicolumn{2}{c}{\boxed{2}} & \multicolumn{2}{c}{\boxed{3}} & \multicolumn{2}{c}{\boxed{4}} \end{array} \\ \begin{bmatrix} 0 & 0 & 0 & 0 & 0 & 0 & 0 & 0 \\ 0 & 0 & 0 & 0 & 0 & 0 & 0 & 0 \\ 0 & 0 & 6 & 0 & -6 & -18 & 0 & 18 \\ 0 & 0 & 0 & 16 & -12 & -16 & 12 & 0 \\ 0 & 0 & -6 & -12 & 150 & 30 & -144 & -18 \\ 0 & 0 & -18 & -16 & 30 & 70 & -12 & -54 \\ 0 & 0 & 0 & 12 & -144 & -12 & 144 & 0 \\ 0 & 0 & 18 & 0 & -18 & -54 & 0 & 54 \end{bmatrix} \end{array} \tag{6.269}$$

$$\mathbf{p}_{\mathbf{T}}^{(2)} = -150\begin{Bmatrix} 0 \\ 0 \\ 0 \\ 0 \\ 0 \\ 1 \\ 0 \\ 1 \end{Bmatrix} N \quad \text{and} \quad \mathbf{p}_{\varepsilon^*}^{(2)} = \begin{Bmatrix} 0 \\ 0 \\ 0 \\ -300 \\ 900 \\ 300 \\ -900 \\ 0 \end{Bmatrix} N \tag{6.270}$$

Summation of the element stiffness matrices

$$\mathbf{K} = \sum_{e=1}^{E} \mathbf{k}^{(e)} \tag{6.271}$$

and load vectors

$$\mathbf{P} = \sum_{e=1}^{E=2} \left(\mathbf{p}_{\mathbf{T}}^{(e)} + \mathbf{p}_{\varepsilon^*}^{(e)} \right) \tag{6.272}$$

results in the global stiffness matrix and the global load vector as

$$\mathbf{K} = \frac{10^6}{12} \begin{bmatrix} 75 & 15 & -69 & -3 & 0 & 0 & -6 & -12 \\ 15 & 35 & 3 & -19 & 0 & 0 & -18 & -16 \\ -69 & 3 & (75+6) & -15 & -6 & -18 & -6 & (12+18) \\ -3 & -19 & -15 & (35+6) & -12 & -16 & (18+12) & -16 \\ 0 & 0 & -6 & -12 & 150 & 30 & -144 & -18 \\ 0 & 0 & -18 & -16 & 30 & 64 & -12 & -48 \\ -6 & -18 & -6 & (18+12) & -144 & -12 & (12+144) & 0 \\ -12 & -16 & (12+18) & -16 & -18 & -48 & 0 & (32+48) \end{bmatrix} \tag{6.273}$$

and

$$\mathbf{P} = \begin{Bmatrix} (300\sqrt{10} - 900) \\ -300 \\ 900 \\ (-300 - 300) \\ 900 \\ (-150 + 300) \\ (300\sqrt{10} - 900) \\ -150 + 600 \end{Bmatrix} N \tag{6.274}$$

The final form of the global system of equations becomes

$$\frac{10^6}{12}\begin{bmatrix} 75 & 15 & -69 & -3 & 0 & 0 & -6 & -12 \\ 15 & 35 & 3 & -19 & 0 & 0 & -18 & -16 \\ -69 & 3 & (75+6) & -15 & -6 & -18 & -6 & (12+18) \\ -3 & -19 & -15 & (35+6) & -12 & -16 & (18+12) & -16 \\ 0 & 0 & -6 & -12 & 150 & 30 & -144 & -18 \\ 0 & 0 & -18 & -16 & 30 & 70 & -12 & -54 \\ -6 & -18 & -6 & (18+12) & -144 & -12 & (12+144) & 0 \\ -12 & -16 & (12+18) & -16 & -18 & -54 & 0 & (32+54) \end{bmatrix} \times \begin{Bmatrix} u_{x_1} \\ u_{y_1} \\ u_{x_2} \\ u_{y_2} \\ u_{x_3} \\ u_{y_3} \\ u_{x_4} \\ u_{y_4} \end{Bmatrix} = \begin{Bmatrix} (300\sqrt{10}-900) \\ -300 \\ 900 \\ (-300-300) \\ 900 \\ (-150+300) \\ (300\sqrt{10}-900) \\ -150+600 \end{Bmatrix} \tag{6.275}$$

Applying the prescribed values of the displacement components leads to

$$\frac{10^6}{12}\begin{bmatrix} 75 & 15 & -69 & -3 & 0 & 0 & -6 & -12 \\ 15 & 35 & 3 & -19 & 0 & 0 & -18 & -16 \\ -69 & 3 & (75+6) & -15 & -6 & -18 & -6 & (12+18) \\ -3 & -19 & -15 & (35+6) & -12 & -16 & (18+12) & -16 \\ 0 & 0 & -6 & -12 & 150 & 30 & -144 & -18 \\ 0 & 0 & -18 & -16 & 30 & 70 & -12 & -54 \\ -6 & -18 & -6 & (18+12) & -144 & -12 & (12+144) & 0 \\ -12 & -16 & (12+18) & -16 & -18 & -54 & 0 & (32+54) \end{bmatrix} \times \begin{Bmatrix} u_{x_1} \\ 0 \\ 0 \\ 0 \\ 0 \\ u_{y_3} \\ u_{x_4} \\ u_{y_4} \end{Bmatrix} = \begin{Bmatrix} (300\sqrt{10}-900) \\ -300 \\ 900 \\ (-300-300) \\ 900 \\ (-150+300) \\ (300\sqrt{10}-900) \\ -150+600 \end{Bmatrix} \tag{6.276}$$

Eliminating the rows and columns corresponding to zero displacement components simplifies the global system of equations to

$$\frac{10^6}{12}\begin{bmatrix} 75 & 0 & -6 & -12 \\ 0 & 70 & -12 & -54 \\ -6 & -12 & (12+144) & 0 \\ -12 & -54 & 0 & (32+54) \end{bmatrix}\begin{Bmatrix} u_{x_1} \\ u_{y_3} \\ u_{x_4} \\ u_{y_4} \end{Bmatrix} = \begin{Bmatrix} (300\sqrt{10}-900) \\ (-150+300) \\ (300\sqrt{10}-900) \\ (-150+600) \end{Bmatrix} \quad (6.277)$$

The solution to this system of equations results in the values for the unknown displacement components as

$$\begin{Bmatrix} u_{x_1} \\ u_{y_3} \\ u_{x_4} \\ u_{y_4} \end{Bmatrix} = \begin{Bmatrix} 0.0000357839 \\ 0.000157003 \\ 0.0000171983 \\ 0.000166367 \end{Bmatrix} \text{cm} \quad (6.278)$$

6.2.2.5 ANSYS Solution

The nodal displacements of the plate subjected to uniform temperature can also be obtained using ANSYS. The solution procedure is outlined as follows:

Model Generation

Specify the element type (**ET** command) using the following menu path:

Main Menu > Preprocessor > Element Type > Add/Edit/Delete

- Click on ***Add***.
- Select ***Solid*** immediately below ***Structural Mass*** from the left list and ***Quad 4node 182*** from the right list; click on ***OK***.
- Click on ***Options***.
- In order to specify the 2-D idealization as plane stress with thickness, in the newly appeared dialog box pull down the menu for ***Element behavior K3*** and select ***Plane strs w/thk***; click on ***OK*** (Fig. 6.33).
- Click on ***Close***.

- Specify real constants (**R** command) using the following menu path:

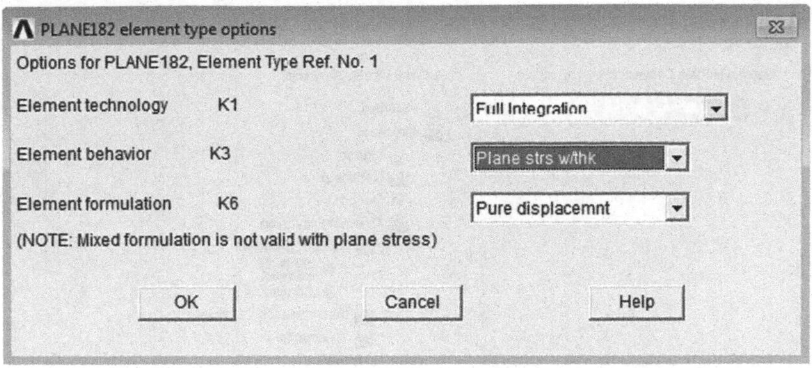

PLANE182 element type options

Options for PLANE182, Element Type Ref. No. 1

Element technology K1 Full Integration

Element behavior K3 Plane strs w/thk

Element formulation K6 Pure displacemnt

(NOTE: Mixed formulation is not valid with plane stress)

OK Cancel Help

Fig. 6.33 Specification of element options

Main Menu > Preprocessor > Real Constants > Add/Edit/Delete

- Click on ***Add***.
- Click on ***OK***.
- Enter ***5e−3*** for ***Thickness THK***; click on ***OK***.
- Click on ***Close***.

- Specify material properties (`MP` command) using the following menu path:

Main Menu > Preprocessor > Material Props > Material Models

- In the *Define Material Model Behavior* dialog box, in the right window, successively left-click on ***Structural***, ***Linear***, ***Elastic***, and, finally, ***Isotropic***, which will bring another dialog box.
- Enter ***150e9*** for ***EX***, and ***0.25*** for ***PRXY***; click on ***OK***.
- In the *Define Material Model Behavior* dialog box, in the right window, under ***Structural*** find ***Thermal Expansion, Secant Coefficient***, and ***Isotropic***, which will bring another dialog box (Fig. 6.34).
- Enter ***6e−6*** for ***APLX***; click on ***OK***.
- Close the *Define Material Model Behavior* dialog box by using the following menu path:

Material > Exit

- Create nodes (`N` command) using the following menu path:

Main Menu > Preprocessor > Modeling > Create > Nodes > In Active CS

- A total of 4 nodes will be created (Table 6.7).
- Referring to Table 6.7, enter x- and y-coordinates of node 1 (be sure to convert the coordinates to meters), and Click on ***Apply***. This action will keep the *Create Nodes in Active Coordinate System* dialog box open. If the ***Node number*** field is left blank, then ANSYS will assign the lowest available node number to the node that is being created.

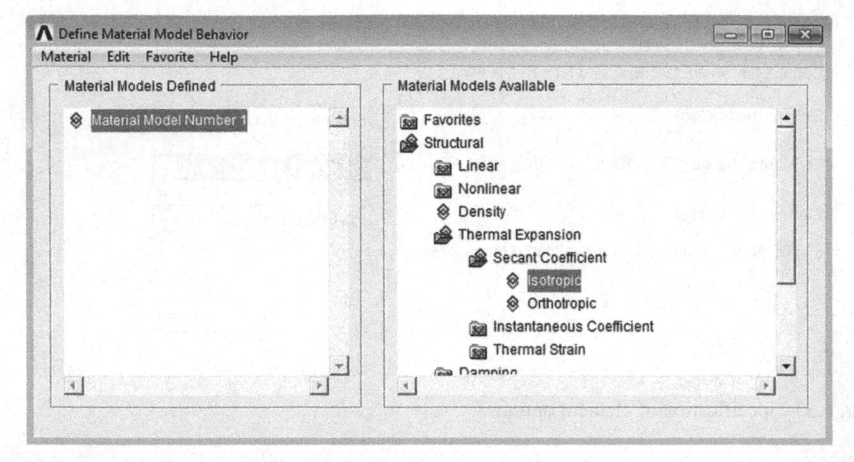

Fig. 6.34 Specification of material behavior

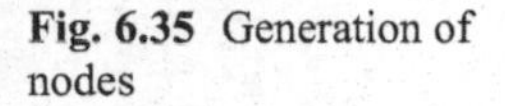

Fig. 6.35 Generation of nodes

- Repeat the same procedure for the nodes 2 through 4.
- After entering the *x*- and *y*-coordinates of node 4, click on ***OK*** (instead of ***Apply***).
- The nodes should appear in the *Graphics Window*, as shown in Fig. 6.35.

• Create elements (**E** command) using the following menu path:

Main Menu > Preprocessor > Modeling > Create > Elements > Auto Numbered > Thru Nodes

- *Pick Menu* appears; refer to Fig. 6.36 to create elements by picking ***three*** nodes at a time and clicking on ***Apply*** in between.
- Observe the elements created after clicking on ***Apply*** in the *Pick Menu.*
- Repeat until the last element is created.
- Click on ***OK*** when the last element is created.

• Review elements:

- Turn on element numbering using the following menu path:

Utility Menu > PlotCtrls > Numbering

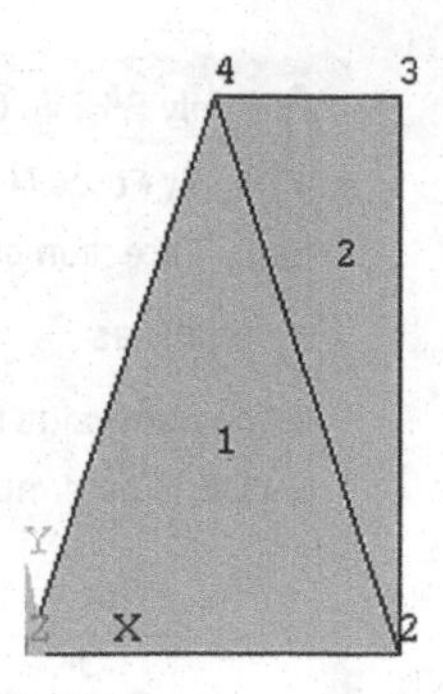

Fig. 6.36 Generation of elements

- Select ***Element numbers*** from the first pull-down menu; click on ***OK***.
- Plot elements (**EPLOT** command) using the following menu path:

Utility Menu > Plot > Elements

- Figure 6.36 shows the outcome of this action as it appears in the *Graphics Window*.
- Turn off element numbering and turn on node numbering using the following menu path:

Utility Menu > PlotCtrls > Numbering

- Place a *checkmark* by clicking on the empty box next to ***NODE Node numbers***.
- Select ***No numbering*** from the first pull-down menu.
- Click on ***OK***.
- Plot nodes (**NPLOT** command) using the following menu path:

Utility Menu > Plot > Nodes

- Figure 6.35 shows the outcome of this action as it appears in the *Graphics Window*.

Solution

- Apply displacement boundary conditions (**D** command) using the following menu path:

Main Menu > Solution > Define Loads > Apply > Structural > Displacement > On Nodes

- *Pick Menu* appears; pick nodes 1 and 2 along the bottom horizontal boundary (Fig. 6.35) and click on ***OK*** on *Pick Menu*.
- Highlight ***UY*** and enter ***0*** for ***VALUE***; click on ***Apply***.
- ***Pick Menu*** reappears; pick nodes 2 and 3 along the right vertical boundary (Fig. 6.35) and click on ***OK*** on *Pick Menu*.
- Highlight ***UX*** and remove the highlight on ***UY***; enter ***0*** for ***VALUE***; click on ***OK***.

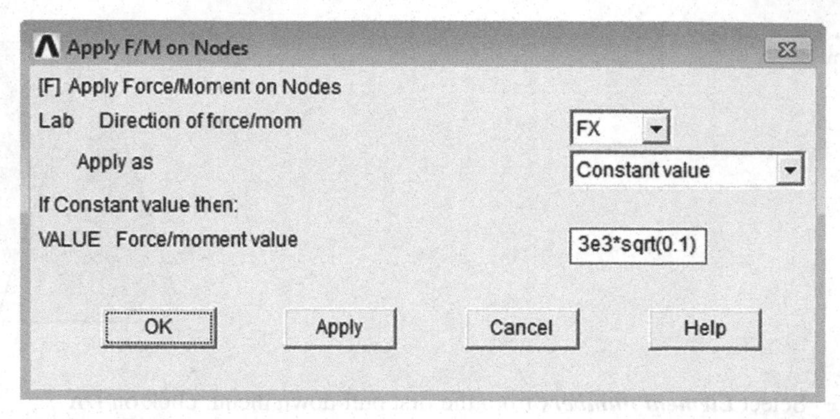

Fig. 6.37 Application of external loads

- Apply force boundary conditions on nodes (**F** command) using the following menu path:

Main Menu > Solution > Define Loads > Apply > Structural > Force/Moment > On Nodes

  - *Pick Menu* appears; pick nodes 1 and 4 along the slanted boundary; click on ***OK***.
  - Enter ***3e3*sqrt(0.1)*** for ***VALUE*** (Fig. 6.37).
  - Click on ***Apply***.
  - *Pick Menu* reappears; pick nodes 4 and 3 along the top horizontal boundary; click on ***OK***.
  - Pull down the menu for ***Direction of force/mom*** and select ***FY***; Enter ***−150*** for ***VALUE***; click on ***OK***.

- Apply thermal load (**TUNIF** command) using the following menu path:

Main Menu > Solution > Define Loads > Apply > Structural > Temperature > Uniform Temp

  - *Uniform Temperature* dialog box appears; Enter ***10*** for ***Uniform temperature***.
  - Click on ***OK***.

- Obtain solution (**SOLVE** command) using the following menu path:

Main Menu > Solution > Solve > Current LS

  - *Confirmation Window* appears along with *Status Report Window*.
  - Review status. If OK, close the *Status Report Window* and click on ***OK*** in *Confirmation Window*.
  - Wait until ANSYS responds with ***Solution is done!***

Postprocessing

- Review deformed shape (**PLDISP** command) using the following menu path:

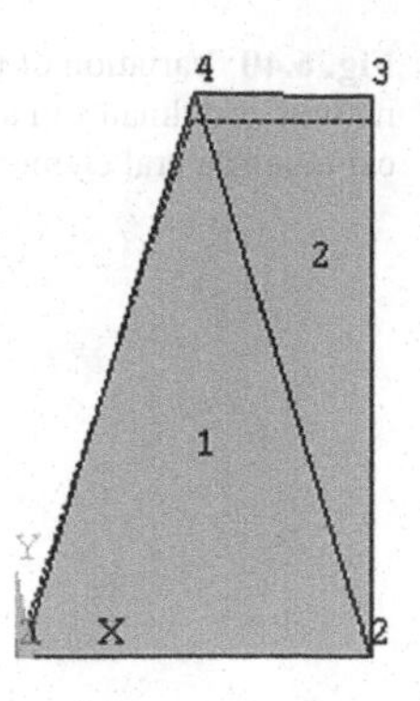

Fig. 6.38 Deformed configuration

Main Menu > General Postproc > Plot Results > Deformed Shape

- In the *Plot Deformed Shape* dialog box, choose the radio-button for ***Def+undef edge***; click on ***OK***.
- The deformed shape will appear in the *Graphics Window*, as shown in Fig. 6.38.

- Review displacement values (**PRNSOL** command) using the following menu path:

Main Menu > General Postproc > List Results > Nodal Solution

- Under ***Nodal Solution***, click on ***DOF Solution*** and ***Displacement vector sum***; click on ***OK***.
- The list will appear in a new window, as shown in Fig. 6.39.

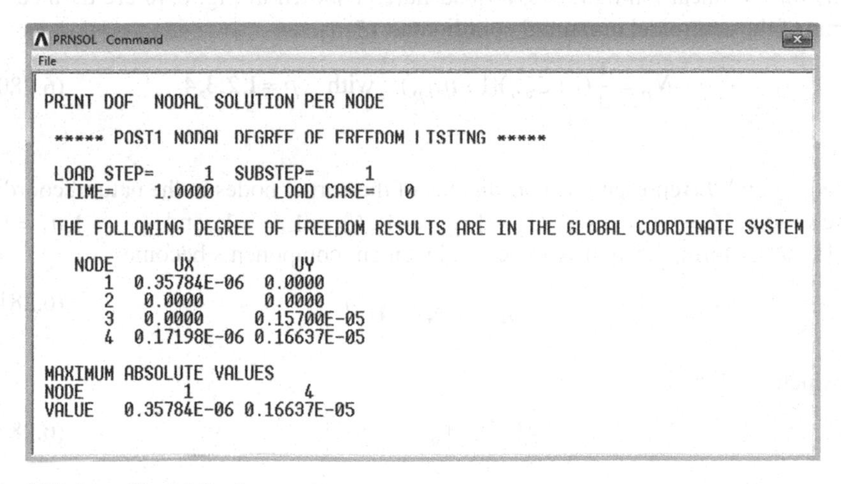
PRNSOL Command
File

```
PRINT DOF   NODAL SOLUTION PER NODE

 ***** POST1 NODAL DEGREE OF FREEDOM LISTING *****

 LOAD STEP=     1  SUBSTEP=     1
  TIME=   1.0000      LOAD CASE=  0

 THE FOLLOWING DEGREE OF FREEDOM RESULTS ARE IN THE GLOBAL COORDINATE SYSTEM

    NODE       UX          UY
       1   0.35784E-06  0.0000
       2    0.0000      0.0000
       3    0.0000      0.15700E-05
       4   0.17198E-06  0.16637E-05

MAXIMUM ABSOLUTE VALUES
NODE           1           4
VALUE    0.35784E-06 0.16637E-05
```

Fig. 6.39 List of nodal displacements

Fig. 6.40 Variation of the natural coordinates in a typical quadrilateral element

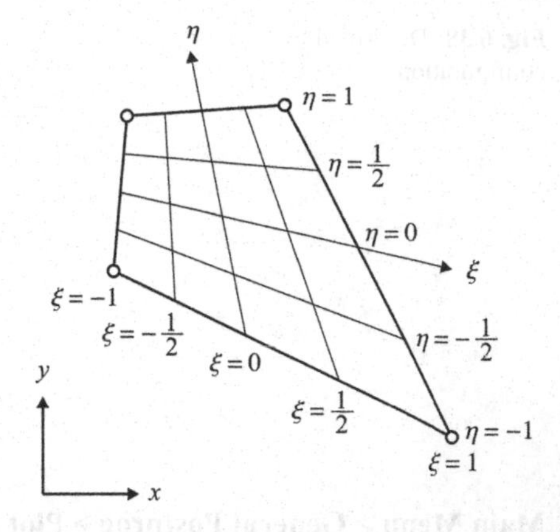

Finite Element Equations with Linear Quadrilateral Isoparametric Elements

The displacement components u_x and u_y within a quadrilateral element can be approximated as

$$\begin{aligned} u_x^{(e)} &= \tilde{u}_x^{(e)} = N_1^{(e)} u_{x_1}^{(e)} + N_2^{(e)} u_{x_2}^{(e)} + N_3^{(e)} u_{x_3}^{(e)} + N_4^{(e)} u_{x_4}^{(e)} \\ u_y^{(e)} &= \tilde{u}_y^{(e)} = N_1^{(e)} u_{y_1}^{(e)} + N_2^{(e)} u_{y_2}^{(e)} + N_3^{(e)} u_{y_3}^{(e)} + N_4^{(e)} u_{y_4}^{(e)} \end{aligned} \tag{6.279}$$

in which $N_1^{(e)}$, $N_2^{(e)}$, $N_3^{(e)}$, and $N_4^{(e)}$ are the linear shape functions and $(u_{x_1}^{(e)}, u_{y_1}^{(e)})$, $(u_{x_2}^{(e)}, u_{y_2}^{(e)})$, $(u_{x_3}^{(e)}, u_{y_3}^{(e)})$, and $(u_{x_4}^{(e)}, u_{y_4}^{(e)})$ are the nodal unknowns (degrees of freedom) associated with first, second, third, and fourth nodes, respectively. The shape functions for the linear (straight-sided) quadrilateral shown in Fig. 6.40 are defined in terms of the centroidal or natural coordinates, (ξ, η), as

$$N_p = \frac{1}{4}(1 + \xi\xi_p)(1 + \eta\eta_p) \quad \text{with} \quad p = 1,2,3,4 \tag{6.280}$$

where ξ_p and η_p represent the coordinates of the corner nodes in the natural coordinate system, $(\xi_1 = -1, \eta_1 = -1)$, $(\xi_2 = 1, \eta_2 = -1)$, $(\xi_3 = 1, \eta_3 = 1)$, and $(\xi_4 = -1, \eta_4 = 1)$.

In matrix form, the approximate displacement components become

$$\tilde{\mathbf{u}}^{(e)} = \mathbf{N}^{(e)T} \mathbf{U}^{(e)} \tag{6.281}$$

in which

$$\tilde{\mathbf{u}}^{(e)T} = \left\{ \tilde{u}_x^{(e)} \quad \tilde{u}_y^{(e)} \right\} \tag{6.282}$$

and

$$\mathbf{N}^{(e)T} = \begin{bmatrix} N_1 & 0 & N_2 & 0 & N_3 & 0 & N_4 & 0 \\ 0 & N_1 & 0 & N_2 & 0 & N_3 & 0 & N_4 \end{bmatrix} \tag{6.283}$$

and

$$\mathbf{U}^{(e)T} = \left\{ u_{x_1}^{(e)} \quad u_{y_1}^{(e)} \quad u_{x_2}^{(e)} \quad u_{y_2}^{(e)} \quad u_{x_3}^{(e)} \quad u_{y_3}^{(e)} \quad u_{x_4}^{(e)} \quad u_{y_4}^{(e)} \right\} \tag{6.284}$$

The element shape matrix $\mathbf{B}^{(e)}$ can be expressed as

$$\mathbf{B}^{(e)} = \mathbf{L}\mathbf{N}^{(e)T} \tag{6.285}$$

in which the differential operator matrix is

$$\mathbf{L} = \begin{bmatrix} \dfrac{\partial}{\partial x} & 0 \\ 0 & \dfrac{\partial}{\partial y} \\ \dfrac{\partial}{\partial x} & \dfrac{\partial}{\partial y} \end{bmatrix} \tag{6.286}$$

The element shape matrix can be rewritten as

$$\mathbf{B}^{(e)} = \begin{bmatrix} \dfrac{\partial N_1}{\partial x} & 0 & \dfrac{\partial N_2}{\partial x} & 0 & \dfrac{\partial N_3}{\partial x} & 0 & \dfrac{\partial N_4}{\partial x} & 0 \\ 0 & \dfrac{\partial N_1}{\partial y} & 0 & \dfrac{\partial N_2}{\partial y} & 0 & \dfrac{\partial N_3}{\partial y} & 0 & \dfrac{\partial N_4}{\partial y} \\ \dfrac{\partial N_1}{\partial y} & \dfrac{\partial N_1}{\partial x} & \dfrac{\partial N_2}{\partial y} & \dfrac{\partial N_2}{\partial x} & \dfrac{\partial N_3}{\partial y} & \dfrac{\partial N_3}{\partial x} & \dfrac{\partial N_4}{\partial y} & \dfrac{\partial N_4}{\partial x} \end{bmatrix} \tag{6.287}$$

However, the shape functions are defined in terms of the centroidal or natural coordinates, (ξ,η). Therefore, they cannot be differentiated directly with respect to the x- and y-coordinates. In order to overcome this difficulty, the global coordinates are expressed in terms of the shape functions in the form

$$x = \sum_{p=1}^{4} N_p(\xi,\eta) x_p \quad \text{and} \quad y = \sum_{p=1}^{4} N_p(\xi,\eta) y_p \tag{6.288}$$

With this transformation utilizing the same shape functions as those used for the displacement components, the concept of *isoparametric* element emerges, and the element is referred to as an *isoparametric* element.

The derivatives of the shape functions can be obtained as

$$\begin{aligned} \frac{\partial N_p}{\partial x} &= \frac{\partial N_p}{\partial \xi}\frac{\partial \xi}{\partial x} + \frac{\partial N_p}{\partial \eta}\frac{\partial \eta}{\partial x} \\ \frac{\partial N_p}{\partial y} &= \frac{\partial N_p}{\partial \xi}\frac{\partial \xi}{\partial y} + \frac{\partial N_p}{\partial \eta}\frac{\partial \eta}{\partial y} \end{aligned} \quad \text{with} \quad p = 1,2,3,4 \tag{6.289}$$

Fig. 6.41 Internal angle exceeding 180°

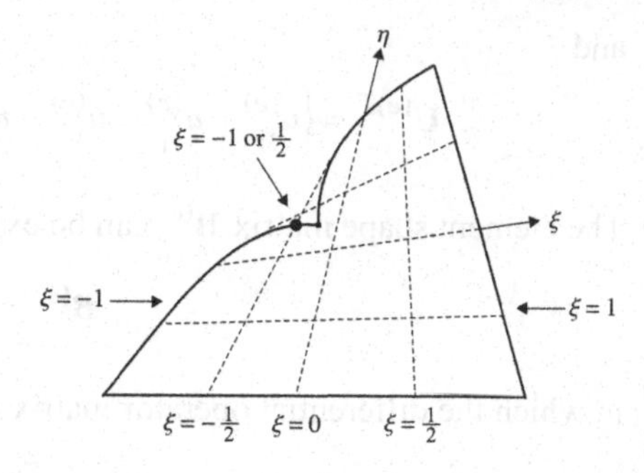

Application of the chain rule of differentiation yields

$$\begin{aligned}\frac{\partial N_p}{\partial \xi} &= \frac{\partial N_p}{\partial x}\frac{\partial x}{\partial \xi} + \frac{\partial N_p}{\partial y}\frac{\partial y}{\partial \xi} \\ \frac{\partial N_p}{\partial \eta} &= \frac{\partial N_p}{\partial x}\frac{\partial x}{\partial \eta} + \frac{\partial N_p}{\partial y}\frac{\partial y}{\partial \eta}\end{aligned} \quad \text{with} \quad p = 1,2,3,4 \tag{6.290}$$

In matrix form, it can be expressed as

$$\begin{Bmatrix} \dfrac{\partial}{\partial \xi} \\ \dfrac{\partial}{\partial \eta} \end{Bmatrix} = \begin{bmatrix} \dfrac{\partial x}{\partial \xi} & \dfrac{\partial y}{\partial \xi} \\ \dfrac{\partial x}{\partial \eta} & \dfrac{\partial y}{\partial \eta} \end{bmatrix} \begin{Bmatrix} \dfrac{\partial}{\partial x} \\ \dfrac{\partial}{\partial y} \end{Bmatrix} \quad \text{or} \quad \begin{Bmatrix} \dfrac{\partial}{\partial \xi} \\ \dfrac{\partial}{\partial \eta} \end{Bmatrix} = \mathbf{J} \begin{Bmatrix} \dfrac{\partial}{\partial x} \\ \dfrac{\partial}{\partial y} \end{Bmatrix} \tag{6.291}$$

where **J** is called the Jacobian matrix, whose inverse does not exist if there is excessive distortion of the element leading to the intersection of lines of constant ξ and η inside or on the element boundaries, as illustrated in Fig. 6.41. If the quadrilateral element is degenerated into a triangle by increasing an internal angle to 180°, then **J** is singular at that corner. It is possible to obtain the element stiffness because **J** is still unique at the Gaussian integration points. However, the stresses at that corner are indeterminate. A similar situation occurs when two adjacent corner nodes are made coincident to produce a triangular element. Therefore, any internal angle of each corner node should be less than 180°,and there is a loss of accuracy as the internal angle approaches 180°.

In the absence of excessive distortion, the transformation between the natural and global coordinates has a one-to-one correspondence and $\mathbf{J}^{-1}$ inverse exists. It can be expressed as

$$\mathbf{J}^{-1} = \frac{1}{|\mathbf{J}|}\begin{bmatrix} \dfrac{\partial y}{\partial \eta} & -\dfrac{\partial y}{\partial \xi} \\ -\dfrac{\partial x}{\partial \eta} & \dfrac{\partial x}{\partial \xi} \end{bmatrix} \tag{6.292}$$

where the determinant of the Jacobian matrix is

$$|\mathbf{J}| = \frac{\partial x}{\partial \xi}\frac{\partial y}{\partial \eta} - \frac{\partial x}{\partial \eta}\frac{\partial y}{\partial \xi} \tag{6.293}$$

in which

$$\begin{aligned}
\frac{\partial x}{\partial \xi} &= \sum_{p=1}^{4} \frac{\partial N_p}{\partial \xi} x_p = \frac{1}{4}\{-(1-\eta)x_1 + (1-\eta)x_2 + (1+\eta)x_3 - (1+\eta)x_4\} \\
\frac{\partial y}{\partial \xi} &= \sum_{p=1}^{4} \frac{\partial N_p}{\partial \xi} y_p = \frac{1}{4}\{-(1-\eta)y_1 + (1-\eta)y_2 + (1+\eta)y_3 - (1+\eta)y_4\} \\
\frac{\partial x}{\partial \eta} &= \sum_{p=1}^{4} \frac{\partial N_p}{\partial \eta} x_p = \frac{1}{4}\{-(1-\xi)x_1 - (1+\xi)x_2 + (1+\xi)x_3 + (1-\xi)x_4\} \\
\frac{\partial y}{\partial \eta} &= \sum_{p=1}^{4} \frac{\partial N_p}{\partial \eta} y_p = \frac{1}{4}\{-(1-\xi)y_1 - (1+\xi)y_2 + (1+\xi)y_3 + (1-\xi)y_4\}
\end{aligned} \tag{6.294}$$

Substituting for the derivatives and rearranging the terms permit the Jacobian to be rewritten in the form

$$\mathbf{J} = \begin{bmatrix} \frac{\partial N_1}{\partial \xi} & \frac{\partial N_2}{\partial \xi} & \frac{\partial N_3}{\partial \xi} & \frac{\partial N_4}{\partial \xi} \\ \frac{\partial N_1}{\partial \eta} & \frac{\partial N_2}{\partial \eta} & \frac{\partial N_3}{\partial \eta} & \frac{\partial N_4}{\partial \eta} \end{bmatrix} \begin{bmatrix} x_1 & y_1 \\ x_2 & y_2 \\ x_3 & y_3 \\ x_4 & y_4 \end{bmatrix} \tag{6.295}$$

or

$$\mathbf{J} = \frac{1}{4}\begin{bmatrix} -(1-\eta) & (1-\eta) & (1+\eta) & -(1+\eta) \\ -(1-\xi) & -(1+\xi) & (1+\xi) & (1-\xi) \end{bmatrix} \begin{bmatrix} x_1 & y_1 \\ x_2 & y_2 \\ x_3 & y_3 \\ x_4 & y_4 \end{bmatrix} \tag{6.296}$$

Its determinant can be expressed in the form

$$|\mathbf{J}| = \frac{1}{8}\begin{bmatrix} x_1 & x_2 & x_3 & x_4 \end{bmatrix} \begin{bmatrix} 0 & 1-\eta & -\xi+\eta & -1+\xi \\ -1+\eta & 0 & 1+\xi & -\xi-\eta \\ \xi-\eta & -1-\xi & 0 & 1+\eta \\ 1-\xi & \xi+\eta & -1-\eta & 0 \end{bmatrix} \begin{Bmatrix} y_1 \\ y_2 \\ y_3 \\ y_4 \end{Bmatrix} \tag{6.297}$$

In a concise form, the determinant can be also rewritten as

$$|\mathbf{J}| = \frac{1}{8}[(x_{31}y_{42} - x_{42}y_{31}) + \xi(x_{12}y_{23} - x_{23}y_{12}) + \eta(x_{41}y_{32} - x_{32}y_{41})] \quad (6.298)$$

where

$$x_{ij} = x_i - x_j \quad \text{and} \quad y_{ij} = y_i - y_j \quad (6.299)$$

Determination of the inverse of the Jacobian matrix permits the expression for the derivatives of the natural coordinates in terms of the global coordinates, x and y

$$\begin{Bmatrix} \dfrac{\partial \xi}{\partial x} \\ \dfrac{\partial \xi}{\partial y} \end{Bmatrix} = \frac{1}{|\mathbf{J}|} \begin{bmatrix} \dfrac{\partial y}{\partial \eta} & -\dfrac{\partial y}{\partial \xi} \\ -\dfrac{\partial x}{\partial \eta} & \dfrac{\partial x}{\partial \xi} \end{bmatrix} \begin{Bmatrix} \dfrac{\partial \xi}{\partial \xi} \\ \dfrac{\partial \xi}{\partial \eta} \end{Bmatrix} \quad (6.300a)$$

and

$$\begin{Bmatrix} \dfrac{\partial \eta}{\partial x} \\ \dfrac{\partial \eta}{\partial y} \end{Bmatrix} = \frac{1}{|\mathbf{J}|} \begin{bmatrix} \dfrac{\partial y}{\partial \eta} & -\dfrac{\partial y}{\partial \xi} \\ -\dfrac{\partial x}{\partial \eta} & \dfrac{\partial x}{\partial \xi} \end{bmatrix} \begin{Bmatrix} \dfrac{\partial \eta}{\partial \xi} \\ \dfrac{\partial \eta}{\partial \eta} \end{Bmatrix} \quad (6.300b)$$

By substituting for the derivatives of the global coordinates in terms of the natural coordinates, these expressions can be rewritten as

$$\frac{\partial \xi}{\partial x} = \frac{1}{|\mathbf{J}|} \sum_{p=1}^{4} \frac{\partial N_p}{\partial \eta} y_p \quad \text{and} \quad \frac{\partial \xi}{\partial y} = -\frac{1}{|\mathbf{J}|} \sum_{p=1}^{4} \frac{\partial N_p}{\partial \eta} x_p \quad (6.301)$$

$$\frac{\partial \eta}{\partial x} = -\frac{1}{|\mathbf{J}|} \sum_{p=1}^{4} \frac{\partial N_p}{\partial \xi} y_p \quad \text{and} \quad \frac{\partial \eta}{\partial y} = \frac{1}{|\mathbf{J}|} \sum_{p=1}^{4} \frac{\partial N_p}{\partial \xi} x_p$$

Finally, the derivatives in the shape matrix becomes

$$\left.\begin{aligned} \frac{\partial N_p}{\partial x} &= \frac{1}{|\mathbf{J}|} \left\{ \frac{\partial N_p}{\partial \xi} \sum_{q=1}^{4} \frac{\partial N_q}{\partial \eta} y_q - \frac{\partial N_p}{\partial \eta} \sum_{q=1}^{4} \frac{\partial N_q}{\partial \xi} y_q \right\} \\ \frac{\partial N_p}{\partial y} &= \frac{1}{|\mathbf{J}|} \left\{ -\frac{\partial N_p}{\partial \xi} \sum_{q=1}^{4} \frac{\partial N_q}{\partial \eta} x_q + \frac{\partial N_p}{\partial \eta} \sum_{q=1}^{4} \frac{\partial N_q}{\partial \xi} x_q \right\} \end{aligned}\right. \quad \text{with} \quad p = 1,2,3,4 \quad (6.302)$$

These explicit expressions for the derivatives appearing in the element shape matrix permit the determination of the element stiffness matrix, $\mathbf{k}^{(e)}$, defined as

$$\mathbf{k}^{(e)} = \int_{V^{(e)}} \mathbf{B}^{(e)T} \mathbf{D} \mathbf{B}^{(e)} \, dV \tag{6.303}$$

in which $V^{(e)} = t\Delta^{(e)}$, with $\Delta^{(e)}$ and t representing the element area and constant element thickness. It can be rewritten in the form

$$\mathbf{k}^{(e)} = t \int_{\Delta^{(e)}} \mathbf{B}^{(e)T} \mathbf{D} \mathbf{B}^{(e)} \, dA \tag{6.304}$$

The material property matrix $\mathbf{D}$ is usually independent of the spatial coordinates, x and y, while the element shape matrix $\mathbf{B}^{(e)}$ requires differentiation of the shape functions with respect to x and y. In order to overcome this difficulty, the integrals are evaluated over a square region in the natural coordinate system, with the transformation of coordinates given by

$$x = \sum_{p=1}^{4} N_p(\xi,\eta) x_p \quad \text{and} \quad y = \sum_{p=1}^{4} N_p(\xi,\eta) y_p \tag{6.305}$$

With this transformation and utilizing the following relation

$$\iint_A dx\, dy = \int_{-1}^{1} \int_{-1}^{1} |\mathbf{J}| \, d\xi \, d\eta \tag{6.306}$$

the element stiffness matrix, $\mathbf{k}^{(e)}$, can be rewritten as

$$\mathbf{k}^{(e)} = t \int_{-1}^{1} \int_{-1}^{1} \mathbf{B}^{(e)T} \, \mathbf{D} \mathbf{B}^{(e)} |\mathbf{J}| \, d\xi \, d\eta \tag{6.307}$$

Due to the difficulty of obtaining analytical expression for the determinant and inverse of the Jacobian matrix, these integrals are evaluated numerically by the Gaussian integration technique. The element stiffness matrix can be evaluated numerically as

$$\mathbf{k}^{(e)} = t \sum_{p=1}^{P} \sum_{q=1}^{Q} w_p w_q \mathbf{B}^{(e)}(\xi_p,\eta_q)^T \, \mathbf{D} \mathbf{B}^{(e)}(\xi_p,\eta_q) \left| \mathbf{J}(\xi_p,\eta_q) \right| \tag{6.308}$$

in which w_p and w_q are the weights and ξ_p and η_q are the integration points of the Gaussian integration technique explained in Sec. 3.6. For this quadrilateral isoparametric element, $P = 2$ and $Q = 2$ are sufficient for accurate integration.

For an element of constant thickness subjected to a uniform load of T_x and T_y in the x- and y-directions, respectively, along its 1–2 edge, the vector $\mathbf{p}_{\mathbf{T}}^{(e)}$, arising from tractions can be written as

$$\mathbf{p}_{\mathbf{T}}^{(e)} = t\int_{L_{1-2}} \begin{bmatrix} N_1 & 0 \\ 0 & N_1 \\ N_2 & 0 \\ 0 & N_2 \\ N_3 & 0 \\ 0 & N_3 \\ N_4 & 0 \\ 0 & N_4 \end{bmatrix} \begin{Bmatrix} T_x \\ T_y \end{Bmatrix} dl = t\int_{L_{1-2}} \begin{Bmatrix} N_1\,T_x \\ N_1\,T_y \\ N_2\,T_x \\ N_2\,T_y \\ N_3\,T_x \\ N_3 T_y \\ N_4\,T_x \\ N_4\,T_y \end{Bmatrix} dl \tag{6.309}$$

Referring to Fig. 6.40, along the 1–2 edge whose length is L_{1-2}, the coordinate η has a constant value of -1 and ξ varies between -1 and 1, leading to

$$\mathbf{p}_{\mathbf{T}}^{(e)} = t\frac{L_{1-2}}{2}\int_{-1}^{1} \begin{Bmatrix} N_1\,T_x \\ N_1\,T_y \\ N_2\,T_x \\ N_2\,T_y \\ N_3\,T_x \\ N_3 T_y \\ N_4\,T_x \\ N_4\,T_y \end{Bmatrix} d\xi \tag{6.310}$$

Along $\xi = -1$ to 1 and $\eta = -1$,

$$\begin{aligned} N_1 &= \frac{1}{4}(1-\xi)(1-\eta) = \frac{1}{2}(1-\xi) \\ N_2 &= \frac{1}{4}(1+\xi)(1-\eta) = \frac{1}{2}(1+\xi) \\ N_3 &= \frac{1}{4}(1+\xi)(1+\eta) = 0 \\ N_4 &= \frac{1}{4}(1-\xi)(1+\eta) = 0 \end{aligned} \tag{6.311}$$

The integrals in the expression for $\mathbf{p}_{\mathbf{T}}^{(e)}$ are evaluated as

$$t\frac{L_{1-2}}{2}\int_{-1}^{1} N_1\, T_x d\xi = t\frac{L_{1-2}}{4}\int_{-1}^{1}(1-\xi)T_x d\xi = t\frac{L_{1-2}}{2}T_x \tag{6.312}$$

and

$$t\frac{L_{1-2}}{2}\int_{-1}^{1} N_2\, T_y d\xi = t\frac{L_{1-2}}{4}\int_{-1}^{1}(1+\xi)T_y d\xi = t\frac{L_{1-2}}{2}T_y \tag{6.313}$$

Thus, the load vector, $\mathbf{p}_{\mathbf{T}}^{(e)}$, takes the form

$$\mathbf{p}_{\mathbf{T}}^{(e)T} = t\frac{L_{1-2}}{2}\begin{bmatrix} T_x & T_y & T_x & T_y & 0 & 0 & 0 & 0 \end{bmatrix} \tag{6.314}$$

Note that this result implies that the applied load is distributed equally at the first and second nodes of the 1–2 edge. This is a result of the linear variation of the shape function along the edges.

As carried out in the derivation of the element stiffness matrix, the load vectors due to body forces, initial strains, and initial stresses can be rewritten as

$$\mathbf{p}_{\mathbf{b}}^{(e)} = t\int_{-1}^{1}\int_{-1}^{1}\mathbf{N}^{(e)}\mathbf{b}\left|\mathbf{J}\right| d\xi d\eta \tag{6.315}$$

$$\mathbf{p}_{\varepsilon^*}^{(e)} = t\int_{-1}^{1}\int_{-1}^{1}\mathbf{B}^{(e)T}\mathbf{D}\boldsymbol{\varepsilon}^*\left|\mathbf{J}\right| d\xi d\eta \tag{6.316}$$

$$\mathbf{p}_{\sigma^*}^{(e)} = t\int_{-1}^{1}\int_{-1}^{1}\mathbf{B}^{(e)T}\boldsymbol{\sigma}^*\left|\mathbf{J}\right| d\xi d\eta \tag{6.317}$$

Application of the Gaussian integration technique leads to the evaluation of these load vectors in the form

$$\mathbf{p}_{\mathbf{b}}^{(e)} = t\sum_{p=1}^{P}\sum_{q=1}^{Q} w_p w_q \mathbf{N}^{(e)}(\xi_p,\eta_q)\mathbf{b}\left|\mathbf{J}(\xi_p,\eta_q)\right| \tag{6.318}$$

$$\mathbf{p}_{\varepsilon^*}^{(e)} = t\sum_{p=1}^{P}\sum_{q=1}^{Q} w_p w_q \mathbf{B}^{(e)}(\xi_p,\eta_q)^T\mathbf{D}\boldsymbol{\varepsilon}^*\left|\mathbf{J}(\xi_p,\eta_q)\right| \tag{6.319}$$

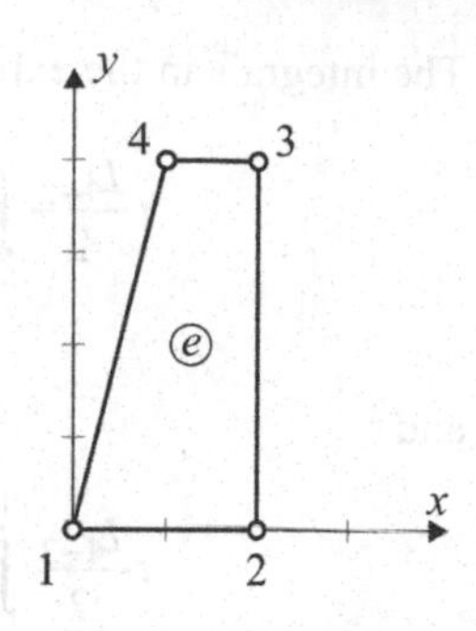

Fig. 6.42 Local numbering scheme of the FEM discretization with a quadrilateral element

$$\mathbf{p}_{\sigma^*}^{(e)} = t\sum_{p=1}^{P} \sum_{q=1}^{Q} w_p w_q \mathbf{B}^{(e)}(\xi_p,\eta_q)^T \boldsymbol{\sigma}^* \left|\mathbf{J}(\xi_p,\eta_q)\right| \tag{6.320}$$

in which w_p and w_q are the weights and ξ_p and η_q are the integration points of the Gaussian integration technique.

6.2.2.6 Example of a Plane Stress Analysis with Linear Quadrilateral Isoparametric Elements

Derivation of a System of Equations and Its Solution

The previous example discussed in Sec. 6.2.2.4 is reconsidered to compute the nodal displacements and the element stresses. In order to illustrate the finite element solution method, the plate is discretized into one quadrilateral isoparametric element, as shown in Fig. 6.42.

The global coordinates of each node are specified by (x_p, y_p), with $p = 1,2,3,4$, and are tabulated in Table 6.9.

The global unknown nodal displacement vector is given by

$$\mathbf{U}^T = \left\{u_{x_1} \quad u_{y_1} \quad u_{x_2} \quad u_{y_2} \quad u_{x_3} \quad u_{y_3} \quad u_{x_4} \quad u_{y_4}\right\} \tag{6.321}$$

Considering the correspondence between the local and global node numbering schemes the elements are defined in Table 6.10.

For this element, $e = 1$, the coefficients of the Jacobian matrix are determined from

$$\begin{aligned}\frac{\partial x}{\partial \xi} &= \frac{1}{4}\{-(1-\eta)(x_1 = 0) + (1-\eta)(x_2 = 2) + (1+\eta)(x_3 = 2) \\ &\quad -(1+\eta)(x_4 = 1)\} = \frac{1}{4}(3-\eta)\end{aligned} \tag{6.322a}$$

$$\begin{aligned}\frac{\partial y}{\partial \xi} &= \frac{1}{4}\{-(1-\eta)(y_1 = 0) + (1-\eta)(y_2 = 0) + (1+\eta)(y_3 = 3) \\ &\quad -(1+\eta)(y_4 = 3)\} = 0\end{aligned} \tag{6.322b}$$

Table 6.9 Global nodal coordinates

Global node number	Nodal coordinates	Nodal unknowns
1	$(x_1 = 0, y_1 = 0)$	u_{x_1}, u_{y_1}
2	$(x_2 = 2, y_2 = 0)$	u_{x_2}, u_{y_2}
3	$(x_3 = 2, y_3 = 3)$	u_{x_3}, u_{y_3}
4	$(x_4 = 1, y_4 = 3)$	u_{x_4}, u_{y_4}

Table 6.10 Element connectivity

Element number (e)	Node 1	Node 2	Node 3	Node 4
1	1	2	3	4

$$\begin{aligned}\frac{\partial x}{\partial \eta} &= \frac{1}{4}\{-(1-\xi)(x_1 = 0) - (1+\xi)(x_2 = 2) + (1+\xi)(x_3 = 2) \\ &\quad +(1-\xi)(x_4 = 1)\} = \frac{1}{4}(1-\xi)\end{aligned} \tag{6.322c}$$

$$\begin{aligned}\frac{\partial y}{\partial \eta} &= \frac{1}{4}\{-(1-\xi)(y_1 = 0) - (1+\xi)(y_2 = 0) + (1+\xi)(y_3 = 3) \\ &\quad +(1-\xi)(y_4 = 3) = \frac{6}{4}\end{aligned} \tag{6.322d}$$

leading to the Jacobian matrix given by

$$\mathbf{J} = \begin{bmatrix} \frac{1}{4}(3-\eta) & 0 \\ \frac{1}{4}(1-\xi) & \frac{6}{4} \end{bmatrix} \tag{6.323}$$

with its determinant

$$|\mathbf{J}| = \frac{3}{8}(3-\eta) \tag{6.324}$$

The inverse of the Jacobian matrix becomes

$$\mathbf{J}^{-1} = \begin{bmatrix} \dfrac{4}{(3-\eta)} & 0 \\ -\dfrac{2(1-\xi)}{3(3-\eta)} & \dfrac{2}{3} \end{bmatrix} \tag{6.325}$$

The determinant of the Jacobian matrix can be also determined from

$$|\mathbf{J}| = \frac{1}{8}[(x_{31}y_{42} - x_{42}y_{31}) + \xi(x_{12}y_{23} - x_{23}y_{12}) + \eta(x_{41}y_{32} - x_{32}y_{41})] \quad (6.326)$$

in which

$$\begin{aligned}
&x_{31} = x_3 - x_1 = 2 \quad &x_{43} = x_4 - x_3 = -1 \quad &x_{32} = x_3 - x_2 = 0 \\
&y_{31} = y_3 - y_1 = 3 \quad &y_{43} = y_4 - y_3 = 0 \quad &y_{32} = y_3 - y_2 = 3 \\
&y_{42} = y_4 - y_2 = 3 \quad &x_{21} = x_2 - x_1 = 2 \quad &x_{41} = x_4 - x_1 = 1 \\
&x_{42} = x_4 - x_2 = -1 \quad &y_{21} = y_2 - y_1 = 0 \quad &y_{41} = y_4 - y_1 = 3
\end{aligned} \quad (6.327)$$

Substituting for the following derivatives

$$\begin{aligned}
&\sum_{p=1}^{4} \frac{\partial N_p}{\partial \xi} x_p = \frac{\partial x}{\partial \xi} = \frac{1}{4}(3-\eta) \\
&\sum_{p=1}^{4} \frac{\partial N_p}{\partial \xi} y_p = \frac{\partial y}{\partial \xi} = 0 \\
&\sum_{p=1}^{4} \frac{\partial N_p}{\partial \eta} x_p = \frac{\partial x}{\partial \eta} = \frac{1}{4}(1-\xi) \\
&\sum_{p=1}^{4} \frac{\partial N_p}{\partial \eta} y_p = \frac{\partial y}{\partial \eta} = \frac{3}{2}
\end{aligned} \quad (6.328)$$

permits the derivatives of the shape functions as

$$\begin{aligned}
\frac{\partial N_p}{\partial x} &= \frac{8}{3(3+\eta)} \left\{ -\frac{6}{4} \frac{\partial N_p}{\partial \xi} \right\} = -\frac{4}{(3+\eta)} \frac{\partial N_p}{\partial \xi} \\
\frac{\partial N_p}{\partial y} &= \frac{2(1+\xi)}{3(3+\eta)} \frac{\partial N_p}{\partial \xi} - \frac{2}{3} \frac{\partial N_p}{\partial \eta}
\end{aligned} \quad \text{with} \quad p = 1,2,3,4 \quad (6.329)$$

Thus, the components of the element shape matrix, $\mathbf{B}^{(1)}$ are computed as

$$\frac{\partial N_1}{\partial x} = -\frac{(1-\eta)}{(3-\eta)}, \quad \frac{\partial N_2}{\partial x} = \frac{(1-\eta)}{(3-\eta)},$$
$$\frac{\partial N_3}{\partial x} = \frac{(1+\eta)}{(3-\eta)}, \quad \frac{\partial N_4}{\partial x} = -\frac{(1+\eta)}{(3-\eta)}$$
$$\frac{\partial N_1}{\partial y} = -\frac{1}{3}\frac{(1-\xi)}{(3-\eta)}, \quad \frac{\partial N_2}{\partial y} = -\frac{(2+\xi-\eta)}{3(3-\eta)}, \tag{6.330}$$
$$\frac{\partial N_3}{\partial y} = \frac{(1+2\xi-\eta)}{3(3-\eta)}, \quad \frac{\partial N_4}{\partial y} = \frac{2(1-\xi)}{3(3-\eta)}$$

$$\mathbf{B}^{(1)} = \begin{bmatrix} -\frac{(1-\eta)}{(3-\eta)} & 0 & \frac{(1-\eta)}{(3-\eta)} & 0 & \frac{(1+\eta)}{(3-\eta)} & 0 & -\frac{(1+\eta)}{(3-\eta)} & 0 \\ 0 & -\frac{1}{3}\frac{(1-\xi)}{(3-\eta)} & 0 & -\frac{(2+\xi-\eta)}{3(3-\eta)} & 0 & \frac{(1+2\xi-\eta)}{3(3-\eta)} & 0 & \frac{2(1-\xi)}{3(3-\eta)} \\ -\frac{1}{3}\frac{(1-\xi)}{(3-\eta)} & -\frac{(1-\eta)}{(3-\eta)} & -\frac{(2+\xi-\eta)}{3(3-\eta)} & \frac{(1-\eta)}{(3-\eta)} & \frac{(1+2\xi-\eta)}{3(3-\eta)} & \frac{(1+\eta)}{(3-\eta)} & \frac{2(1-\xi)}{3(3-\eta)} & -\frac{(1+\eta)}{(3-\eta)} \end{bmatrix} \tag{6.331}$$

Under plane stress assumptions, the material property matrix, **D** becomes

$$\mathbf{D} = 10^6 \begin{bmatrix} 16 & 4 & 0 \\ 4 & 16 & 0 \\ 0 & 0 & 6 \end{bmatrix} \text{N/cm}^2 \tag{6.332}$$

The element stiffness matrix, $\mathbf{k}^{(1)}$, is computed as

$$\mathbf{k}^{(1)} = 10^6 \begin{bmatrix} 4.8666 & 0.76713 & -4.3666 & 0.23287 & -2.7668 & -0.96574 & 2.2668 & -0.034264 \\ 0.76713 & 2.3545 & 0.73287 & -1.0211 & -0.96574 & -1.1244 & -0.53426 & -0.20891 \\ -4.3666 & 0.73287 & 5.3666 & -1.7329 & 2.2668 & -0.53426 & -3.2668 & 1.5343 \\ 0.23287 & -1.0211 & -1.7329 & 3.6878 & -0.034264 & -0.20891 & 1.5343 & -2.4578 \\ -2.7668 & -0.96574 & 2.2668 & -0.034264 & 6.9663 & 0.56853 & -6.4663 & 0.43147 \\ -0.96574 & -1.1244 & -0.53426 & -0.20891 & 0.56853 & 3.5845 & 0.93147 & -2.2512 \\ 2.2668 & -0.53426 & -3.2668 & 1.5343 & -6.4663 & 0.93147 & 7.4663 & -1.9315 \\ -0.034264 & -0.20891 & 1.5343 & -2.4578 & 0.43147 & -2.2512 & -1.9315 & 4.9178 \end{bmatrix} \quad (6.333)$$

The initial strains arising from the temperature change are included in the vector $\boldsymbol{\varepsilon}^*$ as

$$\boldsymbol{\varepsilon}^{*T} = 10^{-6}\begin{bmatrix} 60 & 60 & 0 \end{bmatrix} \quad (6.334)$$

The element load vectors, $\mathbf{p}_{\mathbf{T}\,1-4}^{(1)}$ and $\mathbf{p}_{\mathbf{T}\,3-4}^{(1)}$, arising from the applied tractions are

$$\mathbf{p}_{\mathbf{T}\,1-4}^{(1)T} = t\frac{T_x L_{1-4}}{2}\begin{bmatrix} 1 & 0 & 0 & 0 & 0 & 0 & 1 & 0 \end{bmatrix} \quad (6.335)$$

$$\mathbf{p}_{\mathbf{T}\,3-4}^{(1)T} = t\frac{T_y L_{3-4}}{2}\begin{bmatrix} 0 & 0 & 0 & 0 & 0 & 1 & 0 & 1 \end{bmatrix} \quad (6.336)$$

With the specified values of the thickness and the distributed loads, these element load vectors become

$$\mathbf{p}_{\mathbf{T}\,1-4}^{(1)T} = 300\sqrt{10}\begin{bmatrix} 1 & 0 & 0 & 0 & 0 & 0 & 1 & 0 \end{bmatrix}\mathrm{N} \quad (6.337)$$

$$\mathbf{p}_{\mathbf{T}\,3-4}^{(1)T} = -150\begin{bmatrix} 0 & 0 & 0 & 0 & 0 & 1 & 0 & 1 \end{bmatrix}\mathrm{N} \quad (6.338)$$

The element load vector from all the applied tractions is

$$\mathbf{p}_{\mathbf{T}\,1-4}^{(1)} + \mathbf{p}_{\mathbf{T}\,3-4}^{(1)} = \begin{Bmatrix} 300\sqrt{10} \\ 0 \\ 0 \\ 0 \\ 0 \\ -150 \\ 300\sqrt{10} \\ -150 \end{Bmatrix} \mathrm{N} \tag{6.339}$$

The thermal load vector of the element, $\mathbf{p}_{\varepsilon^*}^{(1)}$, is obtained as

$$\mathbf{p}_{\varepsilon^*}^{(1)} = \begin{Bmatrix} -900 \\ -300 \\ 900 \\ -600 \\ 900 \\ 300 \\ -900 \\ 600 \end{Bmatrix} \mathrm{N} \tag{6.340}$$

Thus, the total element load vector, **P** is

$$\mathbf{P} = \begin{Bmatrix} (300\sqrt{10} - 900) \\ -300 \\ 900 \\ (-300 - 300) \\ 900 \\ (-150 + 300) \\ (300\sqrt{10} - 900) \\ -150 + 600 \end{Bmatrix} \mathrm{N} \tag{6.341}$$

After applying the boundary conditions, the global stiffness matrix is reduced to

$$\mathbf{K} = 10^6 \begin{bmatrix} 4.8666 & -0.96574 & 2.2668 & -0.034264 \\ -0.96574 & 3.5845 & 0.93147 & -2.2512 \\ 2.2668 & 0.93147 & 7.4663 & -1.9315 \\ -0.034264 & -2.2512 & -1.9315 & 4.9178 \end{bmatrix} \tag{6.342}$$

and the reduced load vector is

$$\mathbf{P} = \begin{Bmatrix} (300\sqrt{10} - 900) \\ 150 \\ (300\sqrt{10} - 900) \\ 450 \end{Bmatrix} \text{N} \tag{6.343}$$

The solution is given by

$$\begin{Bmatrix} u_{x_1} \\ u_{y_3} \\ u_{x_4} \\ u_{y_4} \end{Bmatrix} = \begin{Bmatrix} 0.0000307806 \\ 0.000150801 \\ 0.0000222016 \\ 0.000169468 \end{Bmatrix} \text{cm} \tag{6.344}$$

ANSYS Solution

The nodal displacements of the plate subjected to uniform temperature can also be obtained using ANSYS. The solution procedure is outlined as follows:

Model Generation

- Specify the element type (`ET` command) using the following menu path:

Main Menu > Preprocessor > Element Type > Add/Edit/Delete

 - Click on ***Add***.
 - Select ***Solid*** immediately below ***Structural Mass*** from the left list and ***Quad 4node 182*** from the right list; click on ***OK***.
 - Click on ***Options***.
 - In order to specify the 2-D idealization as plane stress with thickness, in the newly appeared dialog box, pull down the menu for ***Element behavior K3*** and select ***Plane strs w/thk***; click on ***OK*** (Fig. 6.43).
 - Click on ***Close***.

- Specify real constants (`R` command) using the following menu path:

Main Menu > Preprocessor > Real Constants > Add/Edit/Delete

 - Click on ***Add***.
 - Click on ***OK***.
 - Enter ***5e−3*** for ***Thickness THK***; click on ***OK***.
 - Click on ***Close***.

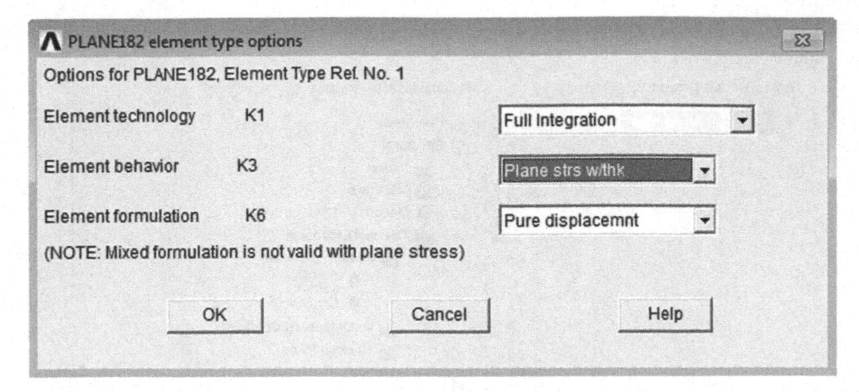

Fig. 6.43 Specification of element options

- Specify material properties (**MP** command) using the following menu path:

Main Menu > Preprocessor > Material Props > Material Models

- In the *Define Material Model Behavior* dialog box, in the right window, successively left-click on ***Structural***, ***Linear,*** ***Elastic***, and, finally, ***Isotropic***, which will bring another dialog box.
- Enter ***150e9*** for ***EX***, and ***0.25*** for ***PRXY***; click on ***OK***.
- In the *Define Material Model Behavior* dialog box, in the right window, under ***Structural***, find ***Thermal Expansion, Secant Coefficient***, and ***Isotropic***, which will bring another dialog box (Fig. 6.44).
- Enter ***6e−6*** for ***APLX***; click on ***OK***.
- Close the *Define Material Model Behavior* dialog box by using the following menu path:

Material >Exit

- Create nodes (**N** command) using the following menu path:

Main Menu > Preprocessor > Modeling > Create > Nodes > In Active CS

- A total of four nodes will be created (Table 6.7).
- Referring to Table 6.7, enter *x*- and *y*-coordinates of node 1 (be sure to convert the coordinates to meters), and Click on ***Apply***. This action will keep the *Create Nodes in Active Coordinate System* dialog box open. If the ***Node number*** field is left blank, then ANSYS will assign the lowest available node number to the node that is being created.
- Repeat the same procedure for the nodes 2 through 4.
- After entering the *x*- and *y*-coordinates of node 4, click on ***OK*** (instead of ***Apply***).
- The nodes should appear in the *Graphics Window*, as shown in Fig. 6.45.

- Create one element (**E** command) using the following menu path:

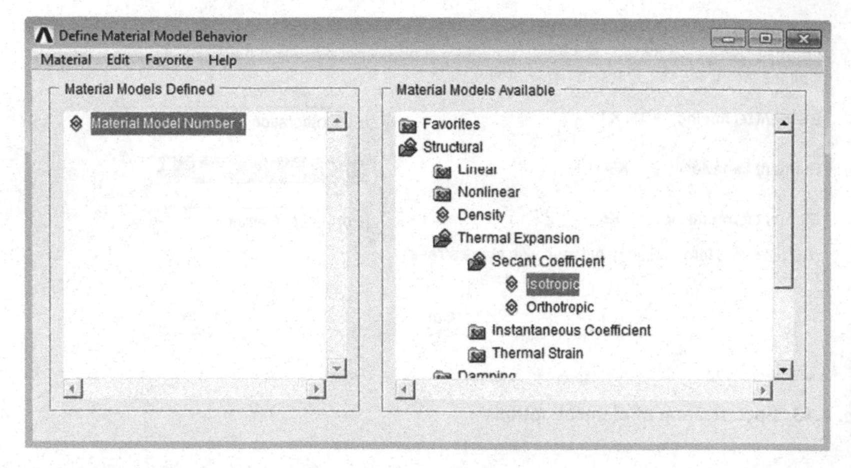

Fig. 6.44 Specification of material behavior

Fig. 6.45 Generation of nodes

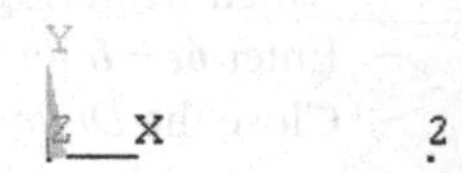

Main Menu > Preprocessor > Modeling > Create > Elements > Auto Numbered > Thru Nodes

- *Pick Menu* appears; pick ***four*** nodes in a clockwise (or counterclockwise) order.
- Click on ***OK***.

Solution

- Apply displacement boundary conditions (**D** command) using the following menu path:

Main Menu > Solution > Define Loads > Apply > Structural > Displacement > On Nodes

- *Pick Menu* appears; pick nodes 1 and 2 along the bottom horizontal boundary (Fig. 6.45) and click on ***OK*** on *Pick Menu*.
- Highlight ***UY*** and enter ***0*** for ***VALUE***; click on ***Apply***.
- *Pick Menu* reappears; pick nodes 2 and 3 along the right vertical boundary (Fig. 6.45); click on ***OK*** on *Pick Menu*.

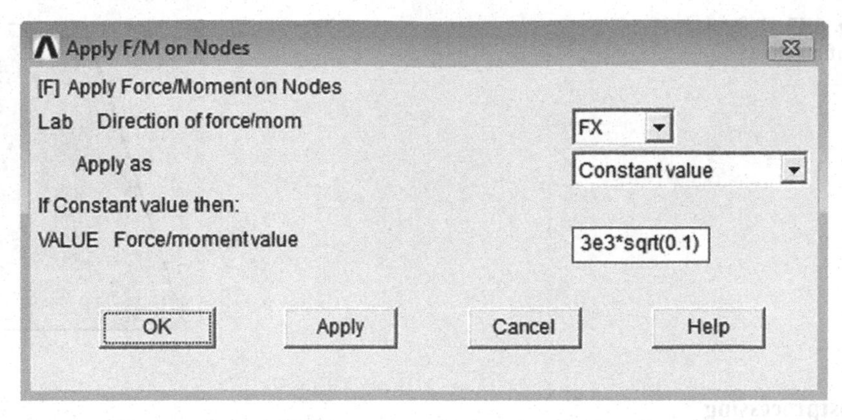

Fig. 6.46 Application of external loads

- Highlight ***UX*** and remove the highlight from ***UY***; Enter ***0*** for ***VALUE***; click on ***OK***.

- Apply force boundary conditions on nodes (**F** command) using the following menu path:

Main Menu > Solution > Define Loads > Apply > Structural > Force/Moment > On Nodes

- *Pick Menu* appears; pick nodes 1 and 4 along the slanted boundary; click on ***OK***.
- Enter ***3e3*sqrt(0.1)*** for ***VALUE*** (Fig. 6.46).
- Click on ***Apply***.
- *Pick Menu* reappears; pick nodes 4 and 3 along the top horizontal boundary; click on ***OK***.
- Pull down the menu for ***Direction of force/mom*** and select ***FY***; Enter ***−150*** for ***VALUE***; click on ***OK***.

- Apply thermal load (**TUNIF** command) using the following menu path:

Main Menu > Solution > Define Loads > Apply > Structural > Temperature > Uniform Temp

- *Uniform Temperature* dialog box appears; Enter ***10*** for ***Uniform temperature***.
- Click on ***OK***.

- Obtain solution (**SOLVE** command) using the following menu path:

Main Menu > Solution > Solve > Current LS

- *Confirmation Window* appears along with *Status Report Window*.
- Review status. If OK, close the Status *Report Window* and click on ***OK*** in *Confirmation Window*.
- Wait until ANSYS responds with ***Solution is done!***

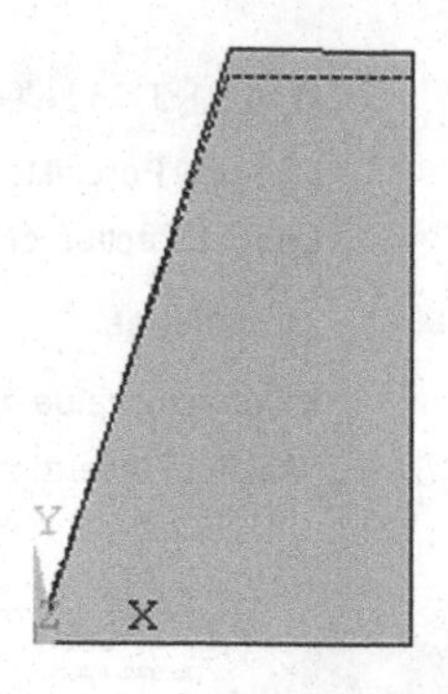

Fig. 6.47 Deformed configuration

Postprocessing

- Review deformed shape (**PLDISP** command) using the following menu path:

Main Menu > General Postproc > Plot Results > Deformed Shape

- In the *Plot Deformed Shape* dialog box, choose the radio-button for ***Def+undef edge***; click on ***OK***.
- The deformed shape will appear in the *Graphics Window*, as shown in Fig. 6.47.

- Review displacement values (**PRNSOL** command) using the following menu path:

Main Menu > General Postproc > List Results > Nodal Solution

- Click on ***DOF Solution*** and ***Displacement vector sum***; click on ***OK***.
- The list will appear in a new window, as shown in Fig. 6.48.

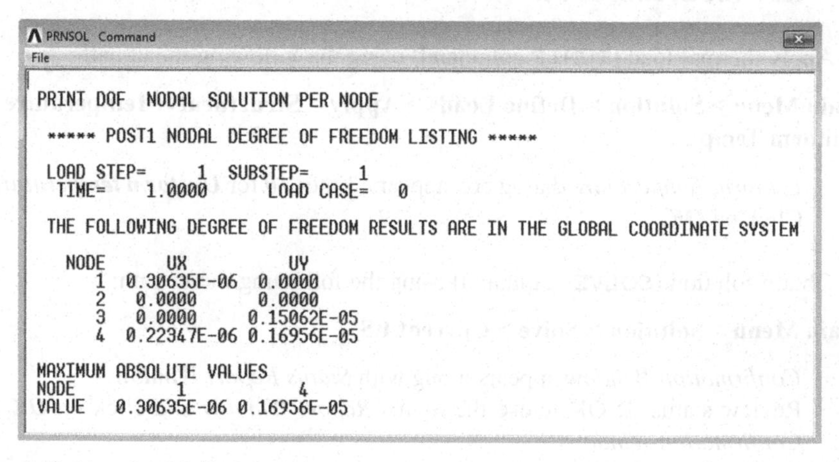

```
PRNSOL Command
File

PRINT DOF  NODAL SOLUTION PER NODE

 ***** POST1 NODAL DEGREE OF FREEDOM LISTING *****

 LOAD STEP=     1  SUBSTEP=     1
  TIME=    1.0000      LOAD CASE=   0

 THE FOLLOWING DEGREE OF FREEDOM RESULTS ARE IN THE GLOBAL COORDINATE SYSTEM

   NODE       UX           UY
      1  0.30635E-06  0.0000
      2   0.0000      0.0000
      3   0.0000     0.15062E-05
      4  0.22347E-06 0.16956E-05

 MAXIMUM ABSOLUTE VALUES
 NODE          1            4
 VALUE   0.30635E-06 0.16956E-05
```

Fig. 6.48 List of nodal displacements

6.3 Problems

6.1 Construct the finite element equations for the solution of the linear second-order ordinary differential equation given in the form

$$p(x)\frac{d^2u(x)}{dx^2}+\frac{dp(x)}{dx}\frac{du(x)}{dx}+q(x)u(x)=f(x)$$

subject to the conditions given as

$$u(x_0)=A,\ u(x_n)=B$$

by using the Galerkin technique within the realm of finite element method with linear interpolation functions.

6.2 By using a one-dimensional (line) C^1 continuous cubic element, derive the element coefficient matrix for the solution of the differential equation given as

$$\frac{d^2u(x)}{dx^2}=f(x)$$

Assume equally spaced nodal points.

6.3 By using quadratic interpolation functions, derive the element coefficient matrix for the solution of the differential equation given as

$$\frac{d^2u}{dx^2}=e^x$$

subject to the conditions

$$u(0)=1 \text{ and } \frac{du}{dx}(4)=0$$

Also, explicitly assemble both the global coefficient matrix and the right-hand vector for equally spaced nodal points located at $x=0$, 1, 2, 3, and 4.

6.4 Without giving any consideration to the boundary conditions, write down the contribution from the four elements, shown in Fig. 6.49, in the finite element formulation for the Poisson equation $\nabla^2\phi=C$. Denote all entries in the element coefficient matrices symbolically and write your answer in the form $[\mathbf{K}]\{\boldsymbol{\varphi}\}+\{\mathbf{F}\}=\{\mathbf{0}\}$.

6.5 In Problem 6.4, note that the interaction of the internal node 5 with all the adjacent elements is included in forming the equation arising from the field variable

Fig. 6.49 Four linear triangular elements forming a quadrilateral element

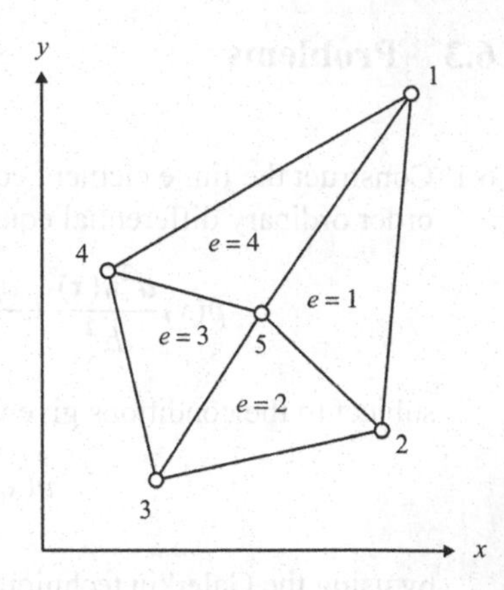

ϕ_5 associated with the 5th node. In the absence of external loads, the last row of the vector-matrix expression in the previous problem may be set directly equal to zero. Using the resulting equation, eliminate ϕ_5 from the remaining four rows of the vector-matrix expression to obtain the element coefficient matrix and the contribution to the right-hand-side vector of a *quadrilateral* element made up of four simpler triangular elements.

6.6 Suppose a collection of elements (part of some larger collection) has a total of n interior nodes and m exterior (or boundary) nodes. The contribution from this collection to the global finite element equations can be written as

$$[\mathbf{K}]^e\{\boldsymbol{\varphi}\}^e + \{\mathbf{f}\}^e$$

The contributions from the exterior nodes, ϕ_i^E $(i = 1, 2, \ldots, m)$, and the interior nodes, ϕ_i^I $(i = m+1, \ldots, n+m)$, may be partitioned as

$$\begin{bmatrix} \mathbf{K}^E & \mathbf{K}^* \\ \mathbf{K}^{*T} & \mathbf{K}^I \end{bmatrix} \begin{Bmatrix} \boldsymbol{\varphi}^E \\ \boldsymbol{\varphi}^I \end{Bmatrix} + \begin{Bmatrix} \mathbf{f}^E \\ \mathbf{f}^I \end{Bmatrix}$$

where $[\mathbf{K}^E]$ is an $m \times m$ submatrix, $[\mathbf{K}^I]$is an $n \times n$ submatrix, etc. Consideration of all of the contributions to the interior nodes results in

$$[\mathbf{K}^*]^T\{\boldsymbol{\varphi}^E\} + [\mathbf{K}^I]\{\boldsymbol{\varphi}^I\} + \{\mathbf{f}^I\} = \{\mathbf{0}\}$$

Proceeding from this point, eliminate the quantities φ_i^I from the remaining equations to express the contribution from this collection of elements in the form

$$[\mathbf{K}^R]^e\{\boldsymbol{\varphi}^E\} + \{\mathbf{f}^R\}$$

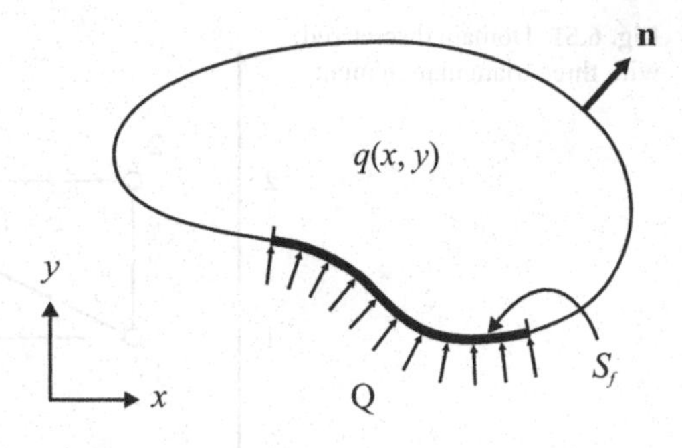

Fig. 6.50 Heat generation within the body and flux boundary condition along S_f

where $[\mathbf{K}^R]$ is an $m \times m$ matrix. This technique is called substructuring.

6.7 For two-dimensional heat transfer in an isotropic body, the governing equation is

$$\frac{\partial}{\partial x}\left(K\frac{\partial T}{\partial x}\right)+\frac{\partial}{\partial y}\left(K\frac{\partial T}{\partial y}\right)+q(x,y)=0$$

where T is temperature, K is thermal conductivity, and $q(x,y)$ is the heat generation rate over the domain. Suppose the heat flux out of some portion, S_f, of the boundary is specified to have a constant value, Q, as shown in Fig. 6.50. Then, the boundary condition over S_f becomes

$$K\left(\frac{\partial T}{\partial v}\right)+Q=K\left[\left(\frac{\partial T}{\partial x}\right)n_x+\left(\frac{\partial T}{\partial y}\right)n_y\right]+Q=0$$

where $\mathbf{n} =< n_x, n_y >$ is the unit normal vector to the boundary. Using the *Galerkin technique*, show in a general way how this boundary condition enters the right-hand-side vector.

6.8 Suppose that the heat flux is specified to be Q over the side 4–5 of the domain as shown in Fig. 6.51. Find *explicitly* the contribution of the interpolating function associated with node 4 to the right-hand-side vector in the system of equations derived in Problem 6.7:

a. for the case where element 3 is a linear triangular element.
b. for the case where element 3 is a quadratic triangular element with a mid-side node between nodes 4 and 5.

Hint: Use a local coordinate, s, directed along the side of the triangle from node 4 to node 5. Note that the interpolating function associated with node 4 is linear in s for linear interpolation and quadratic for quadratic interpolation.

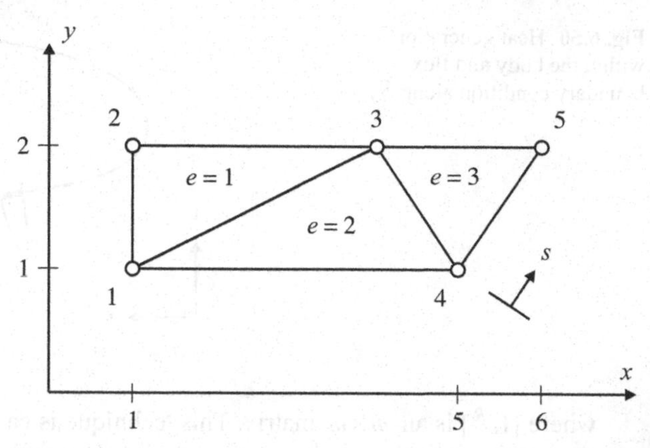

Fig. 6.51 Domain discretized with three triangular elements

6.9 Explicitly evaluate the element coefficient matrix for the problem

$$\frac{\partial^2 \psi}{\partial x^2} + \frac{\partial^2 \psi}{\partial y^2} = G$$

using 2×2 Gaussian integration for a 4-noded quadrilateral element whose nodal point locations are given by

Node No.	x	y
1	6.0	3.0
2	−4.0	3.0
3	−5.0	−3.0
4	4.0	−3.0

6.10 Using quadratic interpolation over a 6-noded triangle (shown in Fig. 6.52), derive explicit expressions for the entries K_{11}, K_{44}, and K_{15} in the element coefficient matrix for the Poisson equation

$$\frac{\partial^2 \phi}{\partial x^2} + \frac{\partial^2 \phi}{\partial y^2} = f(x, y)$$

6.11 Consider the 3-noded triangular element subjected to traction boundary conditions along the 2–3 side as shown in Fig. 6.53. Assuming plane stress idealization with thickness $t = 0.01$ m, $E = 200$ GPa, and $\nu = 0.25$, construct:

a. the stiffness matrix.
b. the equivalent nodal force vector.

6.12 Assume that the nodal displacement components of the triangular element considered in Problem 6.11 are as follows:

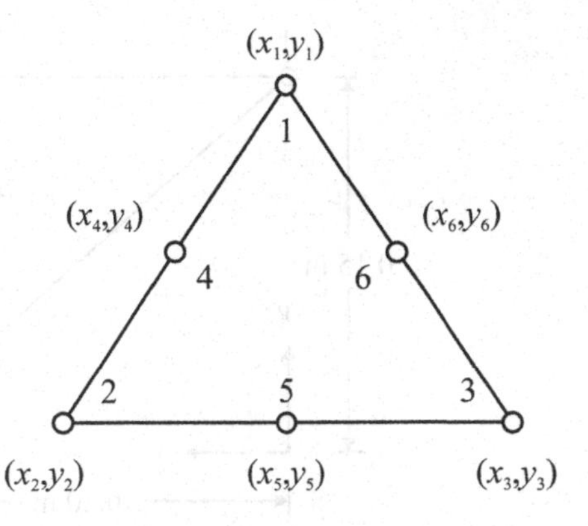

Fig. 6.52 A six-noded triangular element

$$u_1 = 0 \qquad v_1 = 0$$

$$u_2 = 3.30078 \times 10^{-4}\,\text{m} \qquad v_2 = 0$$

$$u_3 = 1.85937 \times 10^{-4}\,\text{m} \qquad v_3 = 4.6875 \times 10^{-6}\,\text{m}$$

Find the stress components (σ_{xx}, σ_{yy} and σ_{xy}).

6.13 Assuming that the triangular element considered in Problem 6.11 is subjected to gravitational acceleration in the negative y-direction with mass density $\rho = 7850\text{kg/m}^3$, find the equivalent nodal force vector.

6.14 Derive the equivalent nodal force vector for a 3-noded triangular element when it is subjected to a uniform temperature change of ΔT. The coefficient of thermal expansion of the material is α.

6.15 The equations governing the *time-dependent* motion of an elastic body are

$$\frac{\partial}{\partial x_j}[\sigma_{ij}] - \rho \frac{\partial^2 u_i}{\partial t^2} = 0$$

where ρ is the mass density of the body. The term $\rho \partial^2 u_i / \partial t^2$ may be interpreted as an "inertia" force, which is a special type of body force.

a. Identifying the inertia force as a body force with $F_i = -\rho \partial^2 u_i / \partial t^2$, derive the contribution from a single element to the global finite element formulation for the case of plane strain.

b. If no tractions are specified over the surface of the body, write down the general form of the global finite element equations. Assuming

$$\{\mathbf{u}\} = \{\bar{\mathbf{u}}\} e^{i\omega t}$$

write down an equation for ω, the natural frequencies of vibration.

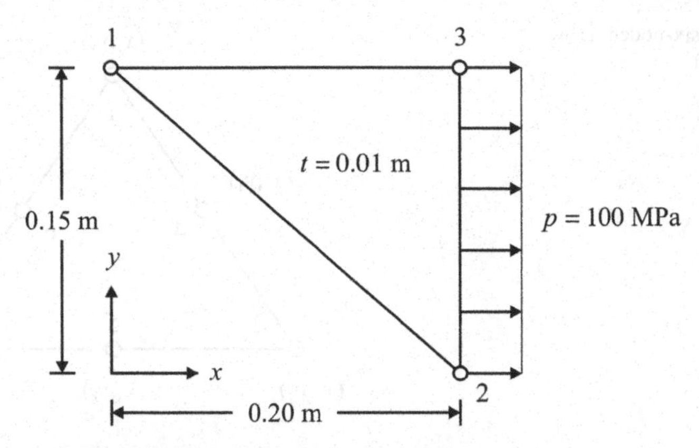

Fig. 6.53 Three-noded triangular element under uniform traction

6.16 A two-dimensional situation that is often of theoretical interest (although less seldom of practical interest) is that of *antiplane strain*, in which $u_1 = u_2 = 0$ and $u_3 = u_3(x_1, x_2)$. Hence, the only non-zero components of strain are ε_{13} and ε_{23} and those of stress are σ_{13} and σ_{23}, which are related by Hooke's law:

$$\sigma_{13} = \frac{E\varepsilon_{13}}{(1+v)}, \sigma_{23} = \frac{E\varepsilon_{23}}{(1+v)}$$

Find the element coefficient matrix for this problem for the linear triangle (3-noded) using the integration formulas for area coordinates given previously.

6.17 Newton's method is a familiar recursive technique for finding the roots of a transcendental equation. Suppose the roots of n transcendental equations, $\{g_i(a_j)\} = 0$, in n unknowns are to be found. Then, Newton's method can be generalized to

$$\{x_i\}^{(m+1)} = \{x_i\}^{(m)} - \left(\left[\frac{\partial g_i}{\partial x_j}\right]^{-1}\right)^{(m)} \{g_i\}^{(m)}$$

where

$$\left[\frac{\partial g_i}{\partial x_j}\right]^{(m)} = \begin{bmatrix} \frac{\partial g_1}{\partial a_1} & \frac{\partial g_1}{\partial a_2} & \cdots & \frac{\partial g_1}{\partial a_n} \\ \frac{\partial g_2}{\partial a_1} & \frac{\partial g_2}{\partial a_2} & \cdots & \frac{\partial g_2}{\partial a_n} \\ \vdots & \vdots & \ddots & \vdots \\ \frac{\partial g_n}{\partial a_1} & \frac{\partial g_n}{\partial a_2} & \cdots & \frac{\partial g_n}{\partial a_n} \end{bmatrix}^{(m)}$$

and $\{g_i\}^{(m)}$ and $[\partial g_i / \partial a_j]$ are evaluated at $\{a_i\}^{(m)}$.
The finite element equations resulting from the nonlinear two-point boundary value problem

$$\frac{d^2u}{dx^2} + g(u,x) = 0$$

have the form

$$[K_{ij}]\{a_i\} + \{f_i(a_j)\} = \{0\}(i = 1,2,\ldots,n)$$

where $\{a_i\}$ are the nodal values and $\{f_i(a_j)\}$ is some nonlinear function of the nodal values. Apply Newton's method to this problem to obtain a recursive formula for the nodal values. What is the major drawback of this approach?

Chapter 7
Use of Commands in ANSYS

The distinct differences between the two modes of ANSYS usage, i.e., the *Graphical User Interface* (*GUI*) and *Batch Mode*, are covered briefly in Chap. 2, and the most common operations within the *Preprocessor*, *Solution*, and *Postprocessors*, mainly using the *GUI*, are covered in Chap. 4 and 5. This chapter is devoted to using the *Batch Mode* of ANSYS, which is the method preferred by advanced ANSYS users.

As mentioned in Chap. 2, every action taken by the user within the ANSYS GUI platform has an equivalent ANSYS *command*. Using ANSYS through the *Batch Mode* involves text (ASCII) files with specific ANSYS *commands*. These commands, along with specific rules, form a special programming language, *ANSYS Parametric Design Language*, or *APDL*, which utilizes concepts and structures very similar to common scientific programming languages such as BASIC, FORTRAN, etc. Using the *APDL*, the user can create (a) an *Input File* to solve a specific problem and (b) *Macro File(s)* that act as special functions, accepting several arguments as input. In either case, each line consists of a single command, and the lines are executed sequentially.

The basic ANSYS commands, operators, and functions are discussed in the following sections. After solving a simple problem by using the *Batch Mode*, more advanced APDL features are covered. The *Batch Mode* command files for each example problem included in this book are given on the accompanying CD-ROM.

7.1 Basic ANSYS Commands

There are around 1500 ANSYS commands, each with a specific syntax and function. It is impractical (and perhaps impossible) for the user to learn the use of all of the commands. This apparent obstacle is overcome by using the ANSYS *Help System*, accessible from within the program, which is covered in Sect. 2.7. However, the solution of a typical problem often involves a limited number of commonly used commands. A selection of these common commands is presented in tabular form in

---

The online version of this book (doi: 10.1007/978-1-4939-1007-6_7) contains supplementary material, which is available to authorized users

E. Madenci, I. Guven, *The Finite Element Method and Applications in Engineering Using ANSYS®*, DOI 10.1007/978-1-4899-7550-8_7

Table 7.1 Session and database commands

Command	Description
/CLEAR	Clear the database (and memory)
/PREP7	Enter the *Preprocessor*
/SOLU	Enter the *Solution*
/POST1	Enter the *General Postprocessor*
/POST26	Enter the *Time History Postprocessor*
FINISH	Exit the current processor; go to *Begin* level
/EOF	Marks the end of file (stop reading)
/FILNAME	Specify *jobname*
HELP	Display help pages related to the command
SAVE	Save the database
RESUME	Resume from an existing database
KSEL, LSEL,ASEL, VSEL, NSEL, ESEL,CMSEL	Select keypoints, lines, areas, volumes, nodes, elements, and components
ALLSEL	Select all entities
CLOCAL, LOCAL	Define local coordinate systems
CSYS	Switch between coordinate systems
!	Start comment—ANSYS ignores the characters to the right of the exclamation mark

Table 7.2 APDL commands

Command	Description
*AFUN	Switch between degrees and radians to be used for angles
*GET	Store model or result information into parameters
*VWRITE	Write formatted output to external files
*DO,*ENDDO	Beginning and ending of do loops
*IF,*ELSE,*ELSEIF, *ENDIF	Commands related to IF-THEN-ELSE blocks
*SET	Define parameters

this section. Within the context of this book, they are grouped into the following six categories:

- Session and Database Commands (Table 7.1).
- APDL Commands (Table 7.2).
- Preprocessor Solid Model Generation Commands (Table 7.3).
- Preprocessor Meshing Commands (Table 7.4).
- Solution Commands (Table 7.5).
- General Postprocessor Commands (Table 7.6).

In Tables 7.1–7.6, the first column gives the command and the corresponding description is given in the second column. With the exception of some APDL commands, the commands can also be issued as a command line input in the *Input*

Table 7.3 Preprocessor solid model generation commands

Command	Description
Command	Description
BLC4	Create rectangular area or prism volume
CYL4	Create circular area or cylindrical volume
K, L, A, AL, V, VA	Create keypoints, lines, areas, and volumes
LARC	Create circular arc
SPLINE, BSPLIN	Create line through spline fit to keypoints
ADRAG	Create an area by dragging a line along a path
VRAG	Create a volume by dragging an area along a path
VEXT	Create a volume by extruding an area
AAD, VADD	Add areas and volumes
LGLUE, AGLUE, VGLUE	Glue lines, areas, and volumes
LOVLAP, AOVLAP, VOVLAP	Overlap lines, areas, and volumes
CM	Create components
KDELE, LDELE, ADELE, VDELE, CMDELE	Delete keypoints, lines, areas, volumes, and components
KPLOT, LPLOT, APLOT, VPLOT	Plot keypoints, lines, areas, and volumes in the *Graphics Window*
KLIST, LLIST, ALIST, VLIST, CMLIST	List keypoints, lines, areas, volumes, and components

Table 7.4 Preprocessor meshing commands

Command	Description
ET	Specify element type
R	Specify real constants
MP	Specify material properties
N	Create nodes
E	Create elements
TYPE	Specify default element type attribute number
REAL	Specify default real constant set attribute number
MAT	Specify default material property set attribute number
LMESH, AMESH,VMESH	Mesh the lines, areas, and volumes
LCLEAR, ACLEAR, VCLEAR	Clear the mesh from lines, areas, and volumes (deletes the nodes and elements attached to those entities)
LESIZE	Specify number of elements or element sizes along selected lines
MSHKEY	Specify whether to use mapped or free meshing
NDELE, EDELE	Delete nodes and elements
NPLOT, EPLOT	Plot nodes and elements in the *Graphics Window*
NLIST, ELIST	List nodes and elements

Table 7.5 Solution commands

Command	Description
SOLVE	Start solution for the current load step
LSSOLVE	Start solution from multiple load step files
D	Specify DOF constraints on nodes
F	Specify concentrated load boundary conditions on nodes
SF, SFE,SFL, SFA	Specify surface (distributed) loads on nodes, elements, lines, and areas
BF, BFE	Specify body loads on nodes and elements
TUNIF	Specify uniform thermal load on all nodes
IC	Specify initial conditions
LSREAD, LSWRITE	Read from and write to load step files

Table 7.6 General postprocessor commands

Command	Description
FILE	Specify the results file for the results to be read from
SET	Specify the load step and substep numbers to be loaded
PLDISP	Plot deformed shape
PLNSOL	Plot contours of nodal solution
PLESOL	Plot contours of element solution
PRNSOL	List nodal solution items
PRESOL	List element solution items

Field in the ANSYS GUI. It is worth noting that some ANSYS commands are valid only in a specific processor or BEGIN level while the remaining ones are valid at all times. Most of the ANSYS commands require arguments separated by commas. For example, the syntax for the **K** command (to create keypoints) given in Table 7.3 is

```
K, NPT, X, Y, Z
```

where **NPT** is the keypoint number and **X**, **Y**, and **Z** are the *x*-, *y*-, and z-coordinates of the node,

As explained in Sect. 2.7, the help page related to the use of this command can be retrieved by issuing the following command line input in the *Input Field* in ANSYS:

```
HELP,K
```

This command brings up detailed information about the arguments.

Table 7.7 List of operators within ANSYS

Operator	Description
+	Addition
−	Subtraction
*	Multiplication
/	Division
**	Exponentiation
<	Less-than comparison
>	Greater-than comparison
=	Equal to (used in defining parameters)

Table 7.8 Selected ANSYS functions

Function	Description
ABS(X)	Absolute value of *X*
EXP(X)	Exponential of *X*
LOG(X)	Natural logarithm of *X*
LOG10(X)	Base 10 logarithm of *X*
SQRT(X)	Square root of *X*
NINT(X)	Nearest integer to *X*
RAND(X, Y)	Random number within the range *X*–*Y*
SIN(X), COS(X), TAN(X)	Sine, cosine, and tangent of *X*
SINH(X), COSH(X), TANH(X)	Hyperbolic sine, hyperbolic cosine, and hyperbolic tangent of *X*
ASIN(X), ACOS(X), ATAN(X)	Inverse sine, inverse cosine, and inverse tangent of *X*

Tables 7.1–7.6 serve as an introduction to the ANSYS commands. However, it is highly recommended that the user read the help pages before usage.

7.1.1 Operators and Functions

In the ANSYS Parametric Design Language (APDL), several fundamental mathematical operations can be utilized through the use of common operators and functions. A complete list of operators is given in Table 7.7. Table 7.8 lists selected mathematical functions available within APDL. Section 7.1.2 provides several examples demonstrating the definition and use of parameters in APDL. These examples are also useful in understanding the way mathematical operators and functions are used in ANSYS.

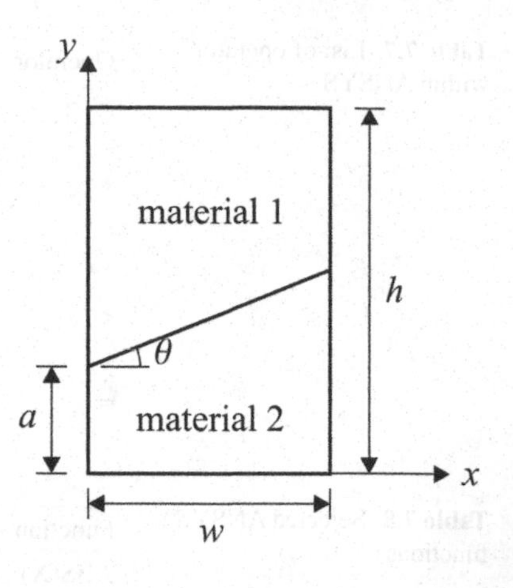

Fig. 7.1 A rectangular area consisting of two dissimilar materials

7.1.2 Defining Parameters

Parameters in APDL can be defined by using either the **`*SET`** command or the "equal to" sign (=). For example the parameter "**`USRPRM`**" can be defined to have the value 22 by either

```
*SET,USRPRM,22
```

or

```
USRPRM=22
```

The rules for naming of parameters are:

- The first character of a parameter name must be a letter.
- Within the parameter name, only letters, numbers, and the underscore character (_) are allowed.
- The maximum number of characters within a parameter name is 32.

The use of common mathematical operations and functions (Tables 7.7–7.8) in parameter definitions is illustrated in the example below. Similar input files for various examples considered in this book are also provided on the CD-ROM.

A rectangular area consisting of two dissimilar materials, shown in Fig. 7.1, has a width and height of w and h, respectively. The material interface starts on the left edge at point (0, a), with an inclination angle θ. Assuming the numerical values of $w=2$, $h=4$, $a=1$, and $\theta=30°$, the following APDL block creates the solid model shown in Fig. 7.2:

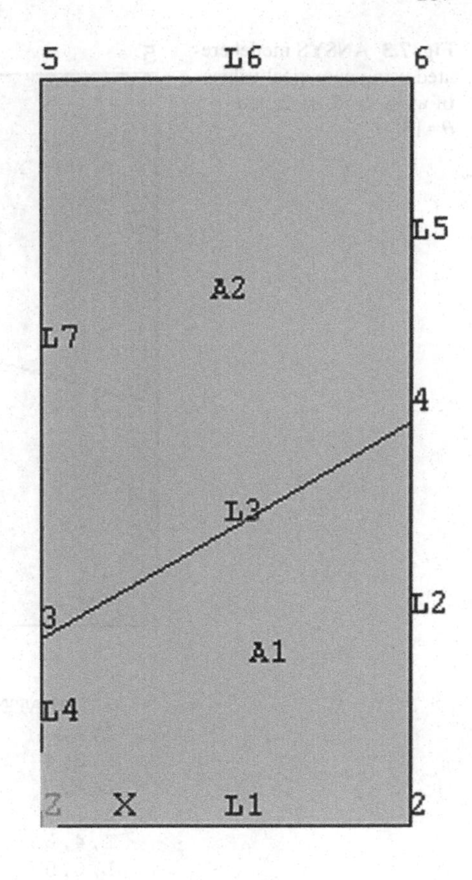

Fig. 7.2 ANSYS model created using of two dissimilar materials. using numerical values of $w=2$, $h=4$, $a=1$, and $\theta=30°$

```
/PREP7                          ! ENTER PREPROCESSOR

*AFUN,DEG                       ! SWITCH TO DEGREES
W=2                             ! WIDTH
H=4                             ! HEIGHT
A=1                             ! Y-COORDINATE OF MATERIAL
                                !INTERFACE AT LEFT EDGE
THETA=30                        ! INCLINATION ANGLE
B= W*TAN(THETA)

! CREATE KEYPOINTS
K,1,0,0
K,2,W,0
K,3,0,A
K,4,W,A+B
K,5,0,H
K,6,W,H
```

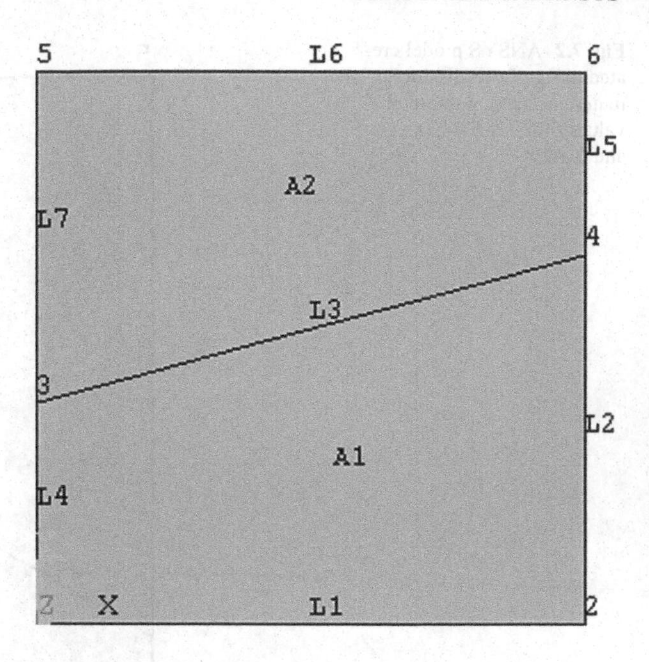

Fig. 7.3 ANSYS model created using numerical values of $w=5$, $h=5$, $a=2$, and $\theta=15°$

```
! CREATE LINES
L,1,2
L,2,4
L,4,3
L,1,3
L,4,6
L,6,5
L,3,5

! CREATE AREAS
AL,1,2,3,4
AL,3,5,6,7
```

Note that the distance b is calculated using a mathematical operator (*) and the function **TAN** to create *keypoint* 4. The input block can be saved as a text file, "*example.txt,*" in the *Working Directory* and read from within ANSYS using the following menu path:

Utility Menu > File > Read Input from

It is also possible to read input files by issuing the **/INPUT** command in the *Input Field* in ANSYS GUI as follows:

```
/INPUT,EXAMPLE,TXT
```

Convenience in using the *Batch Mode* is demonstrated by modifying the length and angle parameters defined in the previous example. Fig. 7.3 shows the solid model generated using $w=5$, $h=5$, $a=2$, and $\theta=15°$ and Fig. 7.4 shows the one using $w=1$, $h=5$, $a=2$, and $\theta=60°$.

Fig. 7.4 ANSYS model created using numerical values of $w=1$, $h=5$, $a=2$, and $\theta=60°$

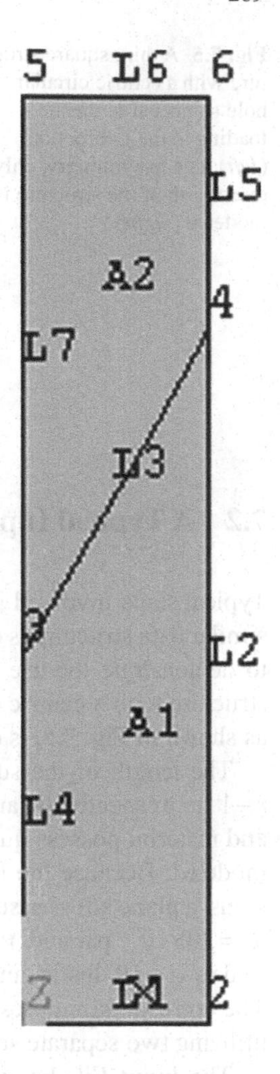

It is worth noting that if a parameter is redefined in the input file, the new value is ***not*** reflected in the entities or parameters defined previously. For example, *keypoint* 4 is created using parameters *w*, *a*, and *b*. If the parameter *w* is redefined (from 2 to 5) after the creation of *keypoint* 4 as shown below,

the new value of *w* is not reflected in the definition of *keypoint* 4 and the *x*-coordinate of *keypoint* 4 remains as 2.

```
W=2
A=1
THETA=30
B= W*TAN(THETA)
K,4,W,A+B ! CREATE KEYPOINT 4
W=5
```

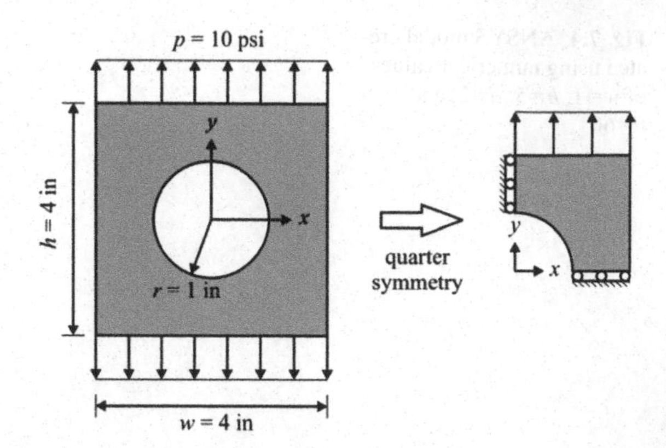

Fig. 7.5 A thin, square structure with a centric circular hole subjected to tensile loading in the *y*-direction (*left*); due to symmetry, only one-fourth of the structure is modeled (*right*)

7.2 A Typical Input File

Typical steps involved in solving an engineering problem are listed in Sect. 5.1. A similar data structure is observed in the *Input Files* used in the *Batch Mode*. In order to demonstrate the use of the *Batch Mode* for a complete analysis, a thin, square structure with a centric circular hole subjected to tensile loading in the *y*-direction, as shown in Fig. 7.5, is considered.

The length of the square and the radius of the circular hole are $w=4$ in. and $r=1$ in., respectively, and the thickness of the structure is $t=0.1$ in. The geometry and material possess quarter-symmetry, therefore only one-fourth of the domain is modeled. Because the thickness is significantly smaller than the in-plane dimensions, a plane stress assumption is used. The elastic modulus and Poisson's ratio are $E=30\times10^6$ psi and $\nu=0.30$, respectively. The distributed tensile load is specified as $q=10$ lbs/in, and it is applied in the form of pressure loading with $q=-10$. The corresponding pressure is input as $q=-10$. The analysis is demonstrated by utilizing two separate solid modeling approaches: *Bottom-up* and *Top-down*.

The *Input File* below uses the *Bottom-up* approach, which starts building the model with *keypoints*, then *line* from *keypoints,* and, finally, *areas* using *lines* (explanations are given along with the commands; the commands between the dashed lines correspond to the *Bottom-up* approach in solid modeling).

```
/FILNAM,BOT-UP               ! SPECIFY JOBNAME
/PREP7                       ! ENTER PREPROCESSOR
ET,1,182                     ! SELECT ELEMENT TYPE AS
                             ! PLANE182
KEYOPT,1,3,3                 ! SPECIFY PLANE STRESS WITH
                             ! THICKNESS
R,1,0.1                      ! SPECIFY REAL CONSTANT
                             ! (THICKNESS)
MP,EX,1,30E6                 ! SPECIFY ELASTIC MODULUS
MP,NUXY,1,0.3                ! SPECIFY POISSON'S RATIO
W=4                          ! SIDE LENGTH OF SQUARE
R=1                          ! HOLE RADIUS
P=10                         ! APPLIED SURFACE LOAD
```

```
K,1,0,0                 ! CREATE KEYPOINTS
K,2,R,0
K,3,W/2,0
K,4,0,R
K,5,0,W/2
K,6,W/2,W/2
L,2,3                   ! CREATE LINES
L,3,6
L,6,5
L,5,4
LARC,4,2,1,R            ! CREATE ARC
LESIZE,1,,,10           ! SPECIFY NUMBER OF
LESIZE,4,,,10           ! ELEMENTS ALONG LINES
LESIZE,2,,,15
LESIZE,3,,,15
LESIZE,5,,,30
AL,1,2,3,4,5            ! CREATE AREA
LCCAT,2,3               ! CONCATENATE LINES FOR MAPPED
                        ! MESHING
```

```
MSHKEY,1                ! USE MAPPED MESHING
AMESH,ALL               ! MESH AREA
/SOLU                   ! ENTER SOLUTION
NSEL,S,LOC,X,0          ! SELECT NODES AT X = 0
D,ALL,UX                ! SUPPRESS X-DISPLACEMENTS
                        ! AT SELECTED NODES
NSEL,S,LOC,Y,0          ! SELECT NODES AT Y = 0
D,ALL,UY                ! SUPPRESS Y-DISPLACEMENTS
                        ! AT SELECTED NODES
NSEL,S,LOC,Y,W/2        ! SELECT NODES AT Y = W/2
SF,ALL,PRES,-P          ! APPLY SURFACE LOADS
ALLSEL                  ! SELECT EVERYTHING
SOLVE                   ! SOLVE
/POST1                  ! ENTER POSTPROCESSOR
PLDISP,2                ! PLOT DEFORMED SHAPE
PLNSOL,S,Y              ! PLOT STRESS IN Y-DIR
/EOF                    ! MARK THE END OF FILE
```

Solid modeling using the *Top-down* approach to accomplish the same task is given below; the methods are interchangeable and the results are the same.

```
RECTNG,0,W/2,0,W/2        ! CREATE RECTANGLE
PCIRC,1                   ! CREATE CIRCLE
ASBA,1,2                  ! SUBTRACT CIRCLE FROM
                          ! RECTANGLE
LSEL,S,LOC,X,0            ! SELECT LINES AT X = 0
LSEL,A,LOC,Y,0            ! ADD LINES AT Y = 0 TO THE
                          ! SELECTED SET
LESIZE,ALL,,,10           ! SPECIFY NUMBER OF ELEMENTS
                          ! ALONG LINES
LSEL,S,LOC,X,W/2          ! SELECT LINES AT X = W/2
LSEL,A,LOC,Y,W/2          ! ADD LINES AT Y = W/2 TO
                          ! THE SELECTED SET
LESIZE,ALL,,,15           ! SPECIFY NUMBER OF ELEMENTS
                          ! ALONG LINES
LCCAT,ALL                 ! CONCATENATE SELECTED LINES
CSYS,1                    ! SWITCH TO GLOBAL CYLINDRICAL
                          ! COORDINATE SYSTEM
LSEL,S,LOC,X,R            ! SELECT LINES AT r = R
LESIZE,ALL,,,30           ! SPECIFY NUMBER OF ELEMENTS
                          ! ALONG LINES
CSYS,0                    ! SWITCH TO GLOBAL CARTESIAN
                          ! COORDINATE SYSTEM
ALLSEL                    ! SELECT EVERYTHING
```

The deformed shape of the structure and the contour variation of stresses in the *y*-direction after the solution are shown in Fig. 7.6 and 7.7, respectively.

7.3 Selecting Operations

Selecting operations play a key role when programming with APDL. The most commonly used ANSYS commands for selecting operations are given in Table 7.9.

The basic group of selection commands involves the ones that allow the user to select a subset of entities, i.e., KSEL, LSEL, ASEL, VSEL, NSEL, and ESEL. The syntax for these commands is as follows:

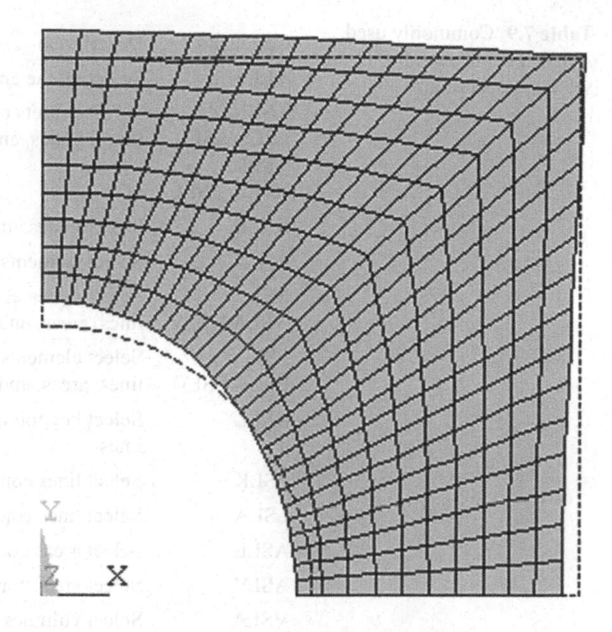

Fig. 7.6 Deformed shape of the structure

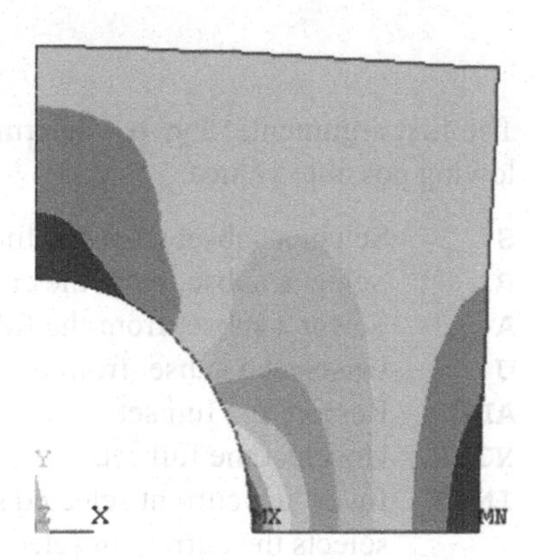

Fig. 7.7 Contour plot of normal stress in the *y*-direction

```
KSEL, Type, Item, Comp, VMIN, VMAX, VINC, KABS
LSEL, Type, Item, Comp, VMIN, VMAX, VINC, KSWP
ASEL, Type, Item, Comp, VMIN, VMAX, VINC, KSWP
VSEL, Type, Item, Comp, VMIN, VMAX, VINC, KSWP
NSEL, Type, Item, Comp, VMIN, VMAX, VINC, KABS
ESEL, Type, Item, Comp, VMIN, VMAX, VINC, KABS
```

Table 7.9 Commonly used ANSYS commands for selecting operations

Command	Description
ALLSEL	Select all the entities
KSEL, LSEL,ASEL, VSEL, NSEL, ESEL	Select subsets of keypoints, lines, areas, volumes, nodes, and elements
NSLE	Select nodes attached to the selected elements
ESLN	Select elements containing the selected nodes
NSL, NSLA,NSLV	Select nodes associated with the selected lines, areas, and volumes
ESL, ESLA,ESLV	Select elements associated with the selected lines, areas, and volumes
KSLL	Select keypoints contained in the selected lines
LSLK	Select lines containing the selected keypoints
LSLA	Select lines contained in the selected areas
ASLL	Select areas containing the selected lines
ASLV	Select areas contained in the selected volumes
VSLA	Select volumes containing the selected areas

The first argument, "**Type**," determines the specific type of selection with the following possible values:

S Select a subset from the full set.
R Select a subset from the current selected set.
A Select a subset from the full set and add it to the current selected set.
U Unselect a subset from the current selected set.
ALL Restore the full set.
NONE Unselect the full set.
INVE Invert the current selected set, which unselects the current selected set and selects the current unselected set.

Figure 7.8 graphically illustrates the concepts behind the argument **Type**. The following examples demonstrate the use of **Type**, along with the remaining arguments.

The argument **Item**, depending on the entity, may have several different meanings. The third through sixth arguments (**Comp**, **VMIN**, **VMAX**, and **VINC**) refer to the argument Item. The most commonly used Item arguments are:

- **Entity name**: **KP** for keypoints, **LINE** for lines, **AREA** for areas, **VOLU** for volumes,**NODE** for nodes, and **ELEM** for elements. In this case, the Comp field (stands for component) is left blank, and **VMIN**, **VMAX**, and **VINC** refer to the minimum and maximum values of the item range and value increment in range (if **VINC** is not specified, its default value is 1), respectively. For example, in order to select keypoints 21 through 30, the following statement is used:

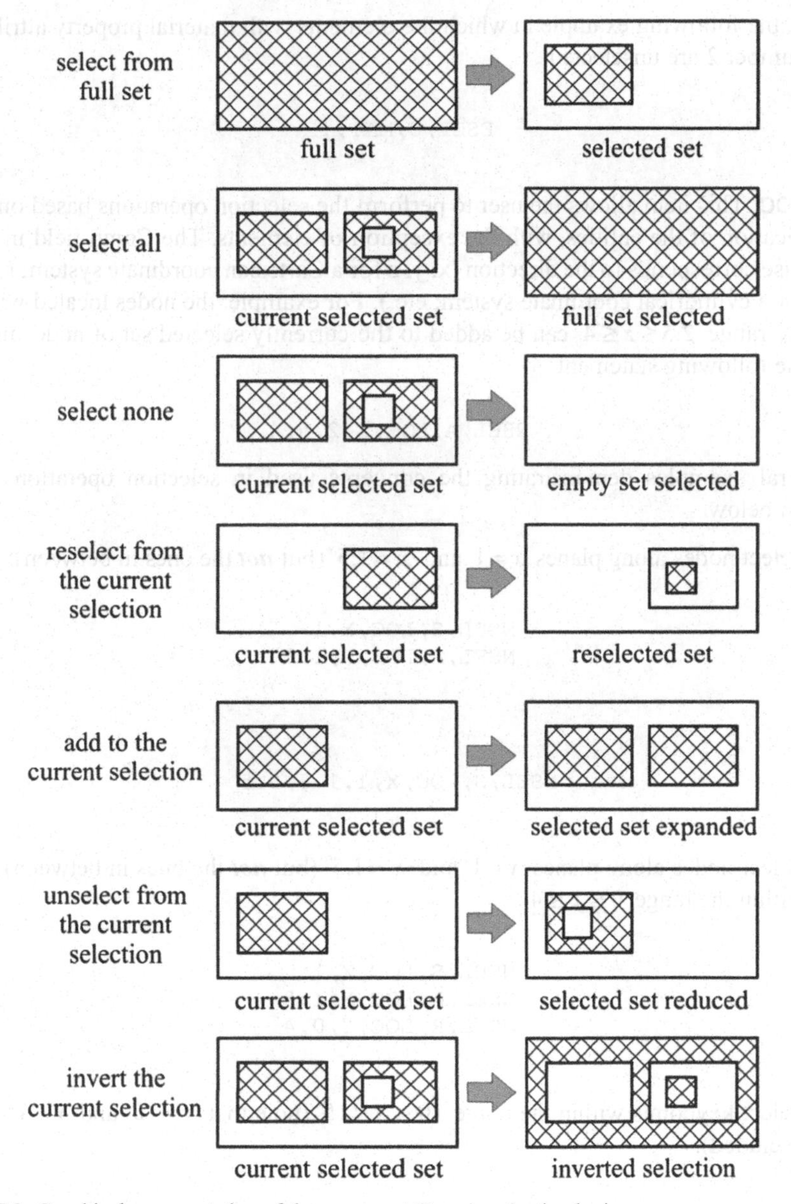

Fig. 7.8 Graphical representation of the argument Type in selection logic

```
KSEL,S,KP,,21,30
```

- **MAT**, **REAL**, **Type**: Selects the entities based on their association with material, real constant, and element type attributes, with the exception of nodes. Similar to the entity name, the Comp field is left blank. The use of this item is demonstrated in the following example in which the elements with material property attribute number 2 are unselected:

```
ESEL,U,MAT,,2
```

- **LOC**: This item allows the user to perform the selection operations based on the location of the entities, with the exception of elements. The Comp field in this case corresponds to the direction (x, y, z for a Cartesian coordinate system; r, θ, z for a cylindrical coordinate system; etc.). For example, the nodes located within the range $2.5 \le z \le 4$ can be added to the currently selected set of nodes using the following statement:

```
NSEL,A,LOC,Z,2.5,4
```

Several examples demonstrating the concepts used in selection operations are given below.

- Select nodes along planes $x = 1$ and $x = 1.5$ (but ***not*** the ones in between):

```
NSEL,S,LOC,X,1
NSEL,A,LOC,X,1.5
```

or

```
NSEL,S,LOC,X,1,1.5,0.5
```

- Select nodes along planes $x = 1$ and $x = 1.5$ (but ***not*** the ones in between) and within the range $0 \le y \le 4$:

```
NSEL,S,LOC,X,1
NSEL,A,LOC,X,1.5
NSEL,R,LOC,Y,0,4
```

- Select keypoints within the range $10 < x < 15$ (note that $x = 10$ and $x = 15$ are excluded):

```
TINY=1E-6 ! DEFINE SMALL NUMBER
KSEL,S,LOC,X,10+TINY,15-TINY
```

- Select keypoints with $x \le 10$ and $x > 15$:

```
TINY=1E-6
KSEL,S,LOC,X,10+TINY,15
KSEL,INVE ! INVERT SELECTION
```

- Select elements 1 through 101 with the increment 2:

```
ESEL,S,ELEM,,1,101,2
```

- Select elements with material attribute number 5 but without those whose real constant attribute number is 3:

```
ESEL,S,MAT,,5
ESEL,U,REAL,,3
```

The remaining commands included in Table 7.9 perform more specific tasks, mostly utilizing the association between the entities. For example **NSLE**, the command is used for performing selection operations on nodes associated with the currently selected set of elements. The command line input given in the following example unselects the nodes that are attached to the selected set of elements:

```
NSLE,U
```

Command **ESLN** selects elements attached to the currently selected set of nodes. Analogous to the other selection commands, the first argument is **Type**, which determines the type of selection. The value of the second argument, **EKEY**, determines which elements are to be selected:

If **EKEY** = 0, elements are selected if **any** of their nodes are in the selected node set.

If **EKEY** = 1, elements are selected only if **all** of their nodes are in the selected node set.

The following lines demonstrate the use of this command:

```
ESLN,S,0 ! CASE 1
ESLN,S,1 ! CASE 2
```

7.4 Extracting Information from ANSYS

Programming with the ANSYS Parametric Design Language often requires the extraction of data such as entity numbers and locations, geometric information, results, etc. Considering the fact that a typical FEA mesh consists of thousands of nodes and elements, the user does not usually have control over the entity numbering. Thus, the information that nodes exist at a specific location may be known without knowledge of their numbers. So, if the user is interested in extracting the variation of a certain solution item along a specific path, the aforementioned data extraction tasks must be performed. These tasks are achieved using the ***GET** command. The syntax of the ***GET** command is as follows:

```
*GET,Par,Entity,ENTNUM,Item1,IT1NUM,Item2,IT2NUM
```

The ***GET** command retrieves and subsequently stores data into parameters. The first argument, **Par** , is the parameter name given by the user. The help page for ***GET** the command provides a complete list of possible argument combinations, and it is highly recommended that the user refer to it. In order to explain the use of the ***GET** command, we consider the examples given below.

- Store the maximum and minimum node numbers in the currently selected node set in parameters *maxnod* and *minnod*:

```
*GET,maxnod,NODE,0,NUM,MAX
*GET,minnod,NODE,0,NUM,MIN
```

- Store the maximum and minimum element numbers in the currently selected element set in parameters *maxel* and *minel*:

```
*GET,maxel,ELEM,0,NUM,MAX
*GET,minel,ELEM,0,NUM,MIN
```

- Store the number of nodes and elements in the currently selected node and element sets in parameters *numnod* and *numel*:

```
*GET,numnod,NODE,0,COUNT
*GET,numel,ELEM,0,COUNT
```

- Store the x-, y-, and z-coordinates of the node numbered *maxnod* in parameters x1, y1, and z1:

```
*GET,x1,NODE,maxnod,LOC,X
*GET,y1,NODE,maxnod,LOC,Y
*GET,z1,NODE,maxnod,LOC,Z
```

- Store the x-, y-, and z-displacements of the node numbered *minnod* in parameters u2, v2, and w2:

```
*GET,u2,NODE,minnod,U,X
*GET,v2,NODE,minnod,U,Y
*GET,w2,NODE,minnod,U,Z
```

- Store the rotations of the node numbered *minnod* about the x-, y-, and z-axes in parameters r_x, r_y, and r_z:

```
*GET,r_x,NODE,minnod,ROT,X
*GET,r_y,NODE,minnod,ROT,Y
*GET,r_z,NODE,minnod,ROT,Z
```

- Store the shear stresses σ_{xy}, σ_{yz}, and σ_{xz} and von Mises stress σ_{eqv} at the node numbered *maxnod* in parameters *s_xy, s_yz, s_xz*, and *s_eqv*:

```
*GET,s_xy,NODE,maxnod,S,XY
*GET,s_yz,NODE,maxnod,S,YZ

*GET,s_xz,NODE,maxnod,S,XZ
*GET,s_eqv,NODE,maxnod,S,EQV
```

- Store the normal strains ε_{xx}, ε_{yy}, and ε_{zz} at the node numbered *minnod* in parameters *eps_xx, eps_yy*, and *eps_zz*:

```
*GET,eps_xx,NODE,minnod,EPEL,X
*GET,eps_yy,NODE,minnod,EPEL,Y
*GET,eps_zz,NODE,minnod,EPEL,Z
```

- Store the x-, y-, and z-coordinates of the centroid of the element numbered *maxel* in parameters *ce_x, ce_y*, and *ce_z*:

```
*GET,ce_x,ELEM,maxel,CENT,X
*GET,ce_y,ELEM,maxel,CENT,Y
*GET,ce_z,ELEM,maxel,CENT,Z
```

- Store the area of the element numbered *minel* in parameters *e_area*:

```
*GET,e_area,ELEM,minel,AREA
```

As an alternative to the syntax given above, one can use readily available ***GET** functions that are predefined in compact form. A few of these functions are listed in Table 7.10. For example, the *x*-, *y*-, and *z*-displacements of the node numbered *minnod* can be stored in parameters *u2, v2,* and *w2* by using the following:

```
u2=UX(minnod)
v2=UY(minnod)
w2=UZ(minnod)
```

Table 7.10 Selected compact ***GET** functions

Command	Description
NX(**n**), NY(**n**), NZ(**n**)	Retrieve *x*-, *y*-, and *z*-coordinates of node numbered **n**
NDNEXT(**n**)	Retrieve node number of the next selected node having a node number greater than node **n**
ELNEXT(**e**)	Retrieve element number of the next selected element having an element number greater than element **e**
UX(**n**), UY(**n**), UZ(**n**)	Retrieve *x*-, *y*-, and *z*-displacements of node numbered **n**
ROTX(**n**), ROTY(**n**), ROTZ(**n**)	Retrieve rotations about *x*-, *y*-, and *z*-axes of node numbered **n**
TEMP(**n**)	Retrieve temperature at node numbered **n**
PRES(**n**)	Retrieve pressure at node numbered **n**

7.5 Programming with ANSYS

The ANSYS Parametric Design Language contains features that are common to other scientific programming languages. These include looping (**DO** loops) and conditional branching (**IF** statements), as well as writing formatted output to text files (**/OUTPUT** and ***VWRITE** commands). These concepts are discussed in the following subsections.

7.5.1 DO Loops

Do loops are program blocks containing a series of commands executed repeatedly, once for each value of the *loop index*. The APDL commands ***DO** and ***ENDDO** define the beginning and ending of a do loop, respectively. The syntax for the ***DO** command is

```
*DO,Par,IVAL,FVAL,INC
```

in which **Par** is the loop index and **IVAL** and **FVAL** designate the initial and final values of **Par** to be incremented by **INC**. For example, the following input block is used to find the arithmetic average of *x*-displacements along the boundary defined by $x = -2.5$:

```
/POST1                        ! ENTER POSTPROCESSOR
NSEL,S,LOC,X,-2.5             ! SELECT NODES ALONG X = -2.5
*GET,numnod,NODE,0,COUNT      ! RETRIEVE NUMBER OF NODES
*GET,minnod,NODE,0,NUM,MIN ! RETRIEVE MINIMUM NODE NUMBER
sum=0                         ! INITIALIZE SUM OF
                              ! DISPLACEMENTS
curnod=minnod                 ! INITIALIZE CURRENT NODE
                              ! NUMBER
-----------------------------------------------------------
*DO,ii,1,numnod               ! BEGIN DO LOOP
*GET,cux,NODE,curnod,U,X      ! RETRIEVE X- DISPLACEMENT OF
                              ! THE CURRENT NODE
sum=sum+cux                   ! UPDATE SUMMATION
*GET,nextnod,NODE,curnod,NXTH    ! STORE NEXT HIGHER NODE
                                 ! NUMBER IN nextnod
curnod=nextnod                ! UPDATE CURRENT NODE
*ENDDO                        ! END DO LOOP
-----------------------------------------------------------
avgdisp=sum/numnod            ! CALCULATE ARITHMETIC AVERAGE
```

In the example above, **ii** is the loop index with the initial and final values of **1** and **numnod**, respectively. Before the do loop begins, the necessary information is obtained by using the ***GET** command (**numnod** for number of nodes **minnod** and for the minimum node number in the selected set of nodes). Also, two new parameters are defined:

- **sum** for the summation of displacements, which is updated within the do loop and finally divided by the number of nodes (**numnod**) to find the arithmetic average.
- **curnod** designating the node number of the "current node" within the do loop. Its initial value is set as the minimum node number in the selected set of nodes (**minnod**), and it is updated within the loop.

Additional do loops may be used within do loops. For example, the following input block creates 216 nodes, starting from the origin, with increments of 0.25, 0.1, and 0.5 in the *x*-, *y*-, and *z*-directions, respectively.

```
/PREP7       ! ENTER PREPROCESSOR
dx=0.25      ! DEFINE PARAMETER FOR INCREMENT IN X
dy=0.1       ! DEFINE PARAMETER FOR INCREMENT IN Y
dz=0.5       ! DEFINE PARAMETER FOR INCREMENT IN Z
-----------------------------------------------------------
*DO,i,1,6                            ! BEGIN DO LOOP IN i
*DO,j,1,6                            ! BEGIN DO LOOP IN j
*DO,k,1,6                            ! BEGIN DO LOOP IN k
N,,(i-1)*dx,(j-1)*dy,(k-1)*dz        ! CREATE NODE
*ENDDO                               ! END DO LOOP IN k
*ENDDO                               ! END DO LOOP IN j
*ENDDO                               ! END DO LOOP IN i
```

7.5.2 *IF Statements*

Conditional branching, which is the execution of an input block based on a condition, is accomplished by using the ***IF** command. The syntax for the ***IF** command is

```
*IF,VAL1,Oper1,VAL2,Base1,VAL3,Oper2,VAL4,Base2
```

in which **VAL1**, **VAL2**, **VAL3**, and **VAL4** are numerical values or parameters that are compared (**VAL1** is compared to **VAL2**, and **VAL3** is compared to **VAL4**). The types of these comparisons are dictated by operator arguments **Oper1** and **Oper2**. Finally, the arguments **Base1** and **Base2** specify the action to be taken based on the comparisons. Operator arguments **Oper1** and **Oper2** may take the following selected logical values:

EQ Equal to (**VAL1**=**VAL2**).
NE Not equal (**VAL1**≠**VAL2**).
LT Less than (**VAL1**<**VAL2**).
GT Greater than (**VAL1**>**VAL2**).
LE Less than or equal to (**VAL1**≤**VAL2**).
GE Greater than or equal to (**VAL1**≥**VAL2**).

Common logical values for action arguments **Base1** and **Base2** are:

AND True if both comparisons dictated by **Oper1** and **Oper2** are true.
OR True if either one of the comparisons dictated by **Oper1** and **Oper2** is true.
XOR True if either one but not both of the comparisons dictated by **Oper2** and is true.
THEN If the preceding logical comparison is true, continue to the next line, otherwise skip to one of the following commands (whichever appears first): ***ELSE**, ***ELSEIF**, or ***ENDIF**. This point is explained in further detail below.

In the event that the first action argument **Base1** has the logical value **THEN**, which is often the case, then the conditional branching has the form

```
*IF,VAL1,Oper1,VAL2,THEN
```

and implies that this is an **IF-THEN-ELSE** block and that it ***must*** be ended by a ***ENDIF** command. Between the ***IF** (marking the beginning of the block) and ***ENDIF** (marking the end of the block) commands, the user may use ***ELSEIF** and ***ELSE** commands. A typical **IF-THEN-ELSE** block has the following form:

```
*IF,VAL1,Oper1,VAL2,THEN      ! COMPARISON-1
...                           ! APDL-1 (ANY NUMBER OF APDL
                              ! COMMANDS)
*ELSEIF,VAL1,Oper1,VAL2       ! COMPARISON-2 (OPTIONAL)
...                           ! APDL-2 (ANY NUMBER OF APDL
!                             ! COMMANDS)
*ELSE                         ! COMPARISON-3 (OPTIONAL)
...                           ! APDL-3 ANY NUMBER OF APDL
                              ! COMMANDS
*ENDIF
```

There may be several ***ELSEIF** commands ***ELSEIF**. command usage is the same as for the ***IF** command (as far as the arguments are concerned) whereas the ***ELSE** command does ***not*** have any arguments. There can only be one ***ELSE** command, and it is the last **IF-THEN-ELSE** block command before the ***ENDIF** command. In the example above, note that:

If **COMPARISON-1** is true, then the input block **APDL-1** is executed and the input blocks **APDL-2** and **APDL-3** are ignored. If **COMPARISON-1** is false ***and*** if **COMPARISON-2** is true, then the input block **APDL-2** is executed and the input blocks **APDL-1** and **APDL-3** are ignored. Finally, if neither **COMPARISON-1** nor **COMPARISON-2** is true, then the input block **APDL-3** is executed and the input blocks **APDL-1** and **APDL-2** are ignored.

The arithmetic average of x-displacements along the boundary defined by $x=-2.5$ was evaluated in the example considered in Sect. 7.5.1. In order to demonstrate the use of **IF-THEN-ELSE** blocks, the example is modified by computing the arithmetic averages of positive and negative x-displacements separately, and the number of nodes with zero x-displacement along the boundary specified as $x=-2.5$. This task can be performed by the following input block:

```
/POST1                                  ! ENTER POSTPROCESSOR
NSEL,S,LOC,X,-2.5                       ! SELECT NODES ALONG
                                        ! X = -2.5
*GET,numnod,NODE,0,COUNT                ! RETRIEVE NUMBER OF NODES
*GET,minnod,NODE,0,NUM,MIN              ! RETRIEVE MINIMUM NODE
                                        ! NUMBER
sum_p=0      ! INITIALIZE SUM OF POSITIVE DISPLACEMENTS
sum_n=0      ! INITIALIZE SUM OF NEGATIVE DISPLACEMENTS
cnt_p=0      ! INITIALIZE # OF NODES WITH POSITIVE
             ! DISPLACEMENTS
cnt_n=0      ! INITIALIZE # OF NODES WITH NEGATIVE
             ! DISPLACEMENTS
cnt_z=0      !INITIALIZE # OF NODES WITH ZERO DISPLACEMENTS
curnod=minnod                           !INITIALIZE CURRENT NODE
                                        !NUMBER
*DO,ii,1,numnod                         !BEGIN DO LOOP
*GET,cux,NODE,curnod,U,X                !RETRIEVE X-DISPLACEMENT
                                        !OF THE CURRENT NODE
*IF,cux,GT,0,THEN          !BEGIN IF-THEN-ELSE BLOCK
sum_p=sum_p+cux            !UPDATE SUM FOR POSITIVE
                           !DISPLACEMENTS
cnt_p=cnt_p+1              !UPDATE NUMBER OF NODES
*ELSEIF,cux,LT,0           !cux IS NEGATIVE
sum_n=sum_n+cux            !UPDATE SUM FOR NEGATIVE
                           !DISPLACEMENTS
cnt_n=cnt_n+1              !UPDATE NUMBER OF NODES
*ELSE                      !cux IS ZERO
cnt_z=cnt_z+1              !UPDATE NUMBER OF NODES
*ENDIF                     !END IF-THEN-ELSE BLOCK
-----------------------------------------------------------
*GET,nextnod,NODE,curnod,NXTH           ! STORE NEXT HIGHER NODE
                                        ! NUMBER IN nextnod
curnod=nextnod                          ! UPDATE CURRENT NODE
*ENDDO                                  ! END DO LOOP
ave_d_p=sum_p/cnt_p                     ! CALCULATE AVERAGE
                                        ! POSITIVE
ave_d_n=sum_n/cnt_n                     ! CALCULATE AVERAGE
                                        ! NEGATIVE
```

The arithmetic averages of positive and negative *x*-displacements are stored in parameters **ave_d_p** and **ave_d_n**, respectively. Also, the number of nodes with zero *x*-displacement is stored in parameter **cnt_z**.

*7.5.3 /OUTPUT and *VWRITE Commands*

APDL offers the option of writing formatted output to text files through use of **/OUTPUT** and ***VWRITE** commands. The **/OUTPUT** command redirects the text output, normally written in the *Output Window*, to a text (ASCII) file whereas the

***VWRITE** command allows desired parameters to be written with FORTRAN (or C) formatting.

The syntax for the **/OUTPUT** command is

```
/OUTPUT,Fname,Ext, ,Loc
```

in which **Fname** and **Ext** are the file name and extension, respectively, and **Loc** decides whether to start writing from the top of this file or to append to it. If the field for **Loc** is left blank, then the output is written from the top of the file. If the value of **Loc** is specified as **APPEND**, then the output is appended. Once the desired data are written to the text file, the output should be redirected back to the *Output Window* using the same command with ***no*** arguments, i.e.,

```
/OUTPUT
```

The syntax for the ***VWRITE** command is

```
*VWRITE,Par1,Par2,...,Par19
```

in which Par1 through Par19 are the parameters to be written with formatting. As observed, up to 19 parameters can be written at a time. A FORTRAN or C format can be used and ***must*** be supplied in the ***next*** line. The FORTRAN format must be enclosed in parentheses and ***only*** *real* or *alphanumeric* formatting is allowed.

When appended to the input block given in Sect. 7.5.2, the commands in the following input block write the arithmetic averages of positive and negative x-displacements, and the number of nodes with zero x-displacement along the boundary of $x = -2.5$ to three parameters (**ave_d_p**, **ave_d_n**, and **cnt_Z**) in a text file named ***data.out***:

```
/OUTPUT,data,out               ! REDIRECT OUTPUT TO FILE
*VWRITE,ave_d_p,ave_d_n        ! WRITE PARAMETERS ON THE
                               ! SAME LINE


(E15.8,2X,E15.8)               ! FORMAT STATEMENT
/OUTPUT                        ! REDIRECT OUTPUT BACK TO
                               ! OUTPUT WINDOW
/OUTPUT,data,out,,APPEND       ! APPEND TO EXISTING FILE
*VWRITE,cnt_z                  ! WRITE THE PARAMETER
(F8.0)                         ! FORMAT STATEMENT
/OUTPUT                        ! REDIRECT OUTPUT BACK TO
                               ! OUTPUT WINDOW
```

In this particular example, **E15.8** in the format statement allocates 15 spaces for the parameter, 8 of which are used for the numbers after the decimal point. The **2X** enforces 2 blank spaces between the parameters. The parameters **ave_d_p** and **ave_d_n** are written in the following format:

```
±0.12345678E±00 ±0.12345678E±00
```

Similarly, the format statement **F8.0** allocates a total of 8 spaces for the parameter with no space for the numbers after the decimal point, and the parametercnt_z must be an integer.

7.6 Macro Files

A more advanced level of APDL use is the *Macro Files*, which are similar to *subroutines* in the FORTRAN programming language. *Macro Files* are saved in separate text files with the file extension **mac** (e.g. ***macro1.mac***) and written using the APDL. If they are saved in the *Working Directory*, they are automatically recognized by the ANSYS program. Otherwise, the user must declare their location using the **/PSEARCH** command. They are particularly useful for tasks that are repeated many times with different values of model variables such as geometry, material properties, boundary conditions, etc. A simple example on how *Macro Files* are used is given below.

The example under consideration involves the modeling of a spring that has a helix shape. The user needs to generate several models with different geometric properties as part of a design requirement. The coordinates of a point on the helix are given by the following set of parametric equations:

$$\begin{aligned} x &= a\cos(t) \\ y &= a\sin(t) \\ z &= bt \end{aligned} \tag{7.1}$$

in which a is the radius of the helix as it is projected onto the x-y plane, $2\pi b$ is the distance in the z-direction of one full turn, and t is the independent parameter. For this purpose, two *Macro Files* are written, with names ***HELIX1.MAC*** and ***HELIX2.MAC***, as given below:

```
! HELIX1.MAC
! MACRO FOR HELIX GENERATION
! ARG1  : COEFFICIENT A IN EQ. 7.1
! ARG2  : COEFFICIENT B IN EQ. 7.1
! ARG3  : NUMBER OF SEGMENTS TO BE USED FOR QUARTER CIRCLE
! ARG4  : NUMBER OF HELIX STEPS
/PREP7 ! ENTER PREPROCESSOR
PI=4*ATAN(1)            ! DEFINE PI
T=0                     ! INITIAL VALUE OF T
DT=2*PI/(4*ARG3-1)      ! INCREMENT OF T
*DO,I,1,4*ARG3
K,,ARG1*COS(T),ARG1*SIN(T),ARG2*T
T=T+DT
*ENDDO
HELIX2,ARG3             ! CALL MACRO HELIX2
LGEN,ARG4,ALL,,,,,2*PI*ARG2
/EOF                    ! MARK END OF FILE
```

```
! HELIX2.MAC
! CALLED BY HELIX1.MAC
! ARG1 : NUMBER OF SEGMENTS TO BE USED FOR QUARTER CIRCLE

KSEL,S,KP,,1,ARG1+1
BSPLIN,ALL
KSEL,S,KP,,ARG1+1,2*ARG1+1
BSPLIN,ALL
KSEL,S,KP,,2*ARG1+1,3*ARG1+1
BSPLIN,ALL
KSEL,S,KP,,3*ARG1+1,4*ARG1+1
BSPLIN,ALL
ALLSEL
LGLUE,ALL

/EOF
```

As long as these files are located in the *Working Directory*, issuing the following command produces the helix shown in Fig. 7.9 (oblique view):

```
HELIX1,1,0.1,4,4
```

Note that the arguments are specified as $a=1$ and $b=0.1$, and four segments are used in creating a quarter circle. Finally, the geometry possesses a total of 4 helix steps. When the radius is modified to be $a=0.5$, and the number of helix steps is increased to 10 using

```
HELIX1,0.5,0.1,4,10
```

the geometry shown in Fig. 7.10 is obtained.

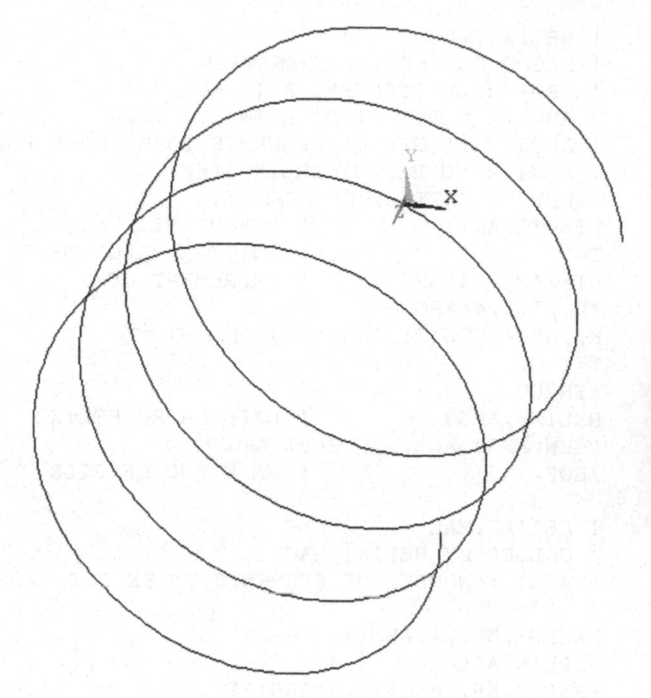

Fig. 7.9 Helix created upon execution of user-defined macro **HELIX1** with arguments ***1***, ***0.1***, ***4***, and ***4***

7.7 Useful Resources

There are three main resources for help in enhancing and accelerating knowledge of and skill in programming with APDL:

ANSYS Help System

Log File

ANSYS Verification Manual

The first topic is discussed in sufficient detail in Sect. 2.7. The following subsections briefly discuss the second and third topics.

7.7.1 Using the Log File for Programming

Every time ANSYS is used interactively (using GUI), a *Log File* is created in the *Working Directory* with the name ***jobname.log***. If *Jobname* is not specified, the default for the *Jobname* is ***file*** and the *Log File* is named ***file.log***. This file records every single action the user takes when using ANSYS, including the ones that are not directly related to the finite element method, such as graphics (zoom in/out, turning on/off entity numbering in the *Graphics Window*, etc.). Although it may appear to be a little "messy," it is extremely useful in learning which commands are used for certain actions when using GUI.

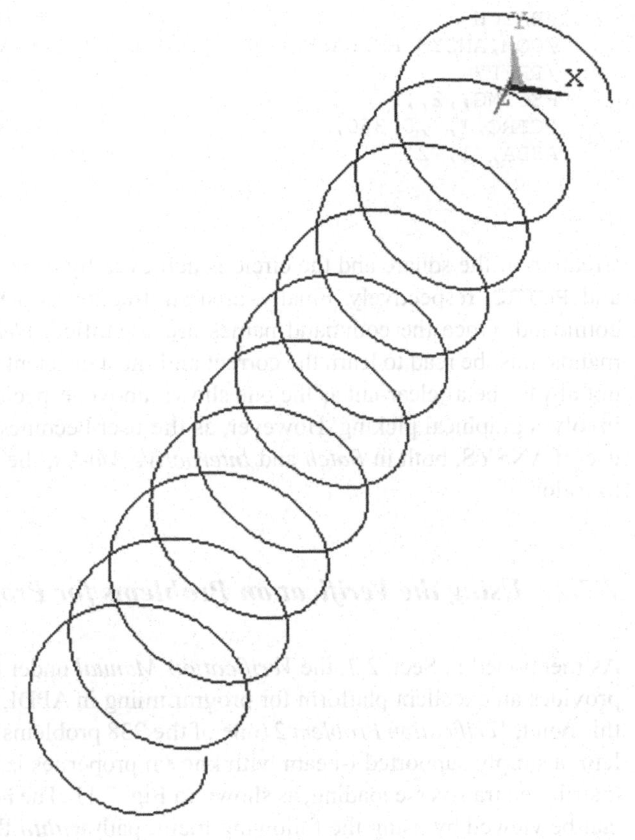

Fig. 7.10 Helix created upon execution of user-defined macro **HELIX1** with arguments *0.5*, *0.1*, *4*, and *10*

The following example illustrates how the user can utilize the *Log File* for learning certain commands. Suppose the domain is the same as the one considered in Sect. 7.2 and the user is required to write an input file to create the solid model. The model generation includes:

Creation of a square with side length of 2.
Creation of a circle with radius 1.
Subtraction of the circle from the square.

The details on how to perform these simple operations in GUI are left out as they should be clear based on the coverage in Chap. 4 and several examples throughout this book. Once the user performs these operations using the GUI, the *Log File* can be viewed from within ANSYS using

Utility Menu > List > Files > Log File

which should appear in a separate *Output Window* with contents as given below:

```
/BATCH
/COM,ANSYS RELEASE 8.0 UP20030930 14:49:33 06/16/2004
/PREP7
RECTNG,,2,,2,
PCIRC,1, ,0,360,
ASBA, 1, 2
```

Creation of the square and the circle is achieved by using the commands **RECTNG** and **PCIRC**, respectively. Finally, area subtraction is performed using the **ASBA** command. Once the command names are identified, *Help Pages* for these commands must be read to learn the correct and most efficient usage. The *Log File* may not always be as clear-cut as the one shown above, especially when the GUI action involves graphical picking. However, as the user becomes more accustomed to the use of ANSYS, both in *Batch* and *Interactive Modes*, the *Log File* becomes easier to follow.

7.7.2 Using the Verification Problems for Programming

As mentioned in Sect. 2.7, the *Verification Manual* under the ANSYS *Help System* provides an excellent platform for programming in APDL. In order to demonstrate this point, *Verification Problem 2* (one of the 238 problems) is selected. In this problem, a simply supported I-beam with known properties is subjected to a uniformly distributed transverse loading, as shown in Fig. 7.11. The help page for this problem can be viewed by using the following menu path ***within*** the ANSYS *Help System*, which appears on the left side of the *Help Window* (heading shown in Fig. 7.12):

Mechanical APDL > Verification Test Case Descriptions > VM2

The problem description, a sketch of the problem and corresponding FEA representation, the reference from which the problem is taken, and analysis assumptions are found at the top of the page. Included further down is a table (see Fig. 7.13) showing the results obtained by both analytical methods and the ANSYS software. As observed in Fig. 7.12, there is a hyperlink to the text file **vm2.dat**, which includes the input commands for the solution of this problem using the *Batch Mode*. Upon clicking on this hyperlink, the file appears as partially shown in Fig. 7.14. The user can go through this file line by line, referring frequently to the help pages of unfamiliar commands in order to learn the correct usage of commands.

Another important benefit from the *Verification Manual* is that one can learn how to solve problems with certain properties. The *Verification Test Case Descriptions* help page, accessed through the menu path

Mechanical APDL > Verification Test Case Descriptions

is a good place to start. For example, if the problem at hand involves materials with viscoelastic behavior, it would be a good idea to scan the test case descriptions to

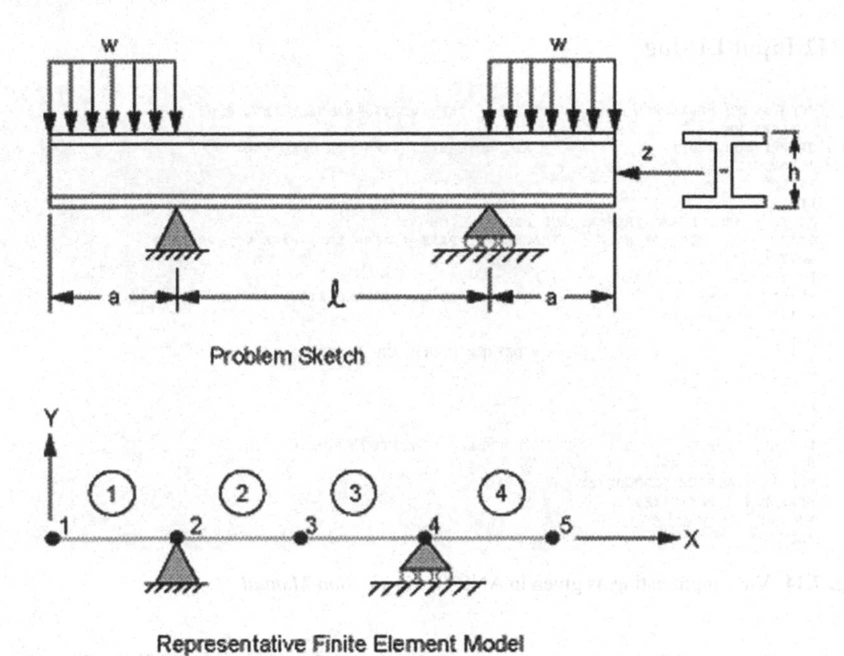

Fig. 7.11 Graphical description of *Verification Problem 2* as given in ANSYS *Verification Manual*

VM2

Beam Stresses and Deflections

Overview

Reference:	S. Timoshenko, *Strength of Material, Part I, Elementary Theory and Problems*, 3rd Edition, D. Van Nostrand Co., Inc., New York, NY, 1955, pg. 98, problem 4
Analysis Type (s):	Static Analysis (ANTYPE = 0)
Element Type (s):	3-D 2 Node Beam (BEAM188)
Input Listing:	vm2.dat

Fig. 7.12 Heading of the help page for *Verification Problem 2* as given in ANSYS *Verification Manual*

Results Comparison

	Target	ANSYS	Ratio
Stress, psi	-11400.000	-11440.746	1.004
Deflection, in	0.182	0.182	1.003

Fig. 7.13 Comparison of results as given in ANSYS *Verification Manual*

VM2 Input Listing

```
/COM,ANSYS MEDIA REL. 8.0 (9-17-2003) REF. VERIF. MANUAL: REL. 8.0
/VERIFY,VM2
JPGPRF,500,100,1              ! MACRO TO SET PREFS FOR JPEG PLOTS
/SHOW,JPEG
/PREP7
MP,PRXY,,0.3
/TITLE, VM2, BEAM STRESSES AND DEFLECTIONS
C***        STR. OF MATL., TIMOSHENKO, PART 1, 3RD ED., PAGE 98, PROB. 4
ANTYPE,STATIC
ET,1,BEAM3
KEYOPT,1,9,9                  ! OUTPUT AT 9 INTERMEDIATE LOCATIONS
R,1,50.65,7892,30
MP,EX,1,30E6
N,1                           ! DEFINE NODES AND ELEMENTS
N,5,480
FILL
E,1,2
EGEN,4,1,1
D,2,UX,,,,,UY                 ! BOUNDARY CONDITIONS AND LOADING
D,4,UY
SFBEAM,1,1,PRES,(10000/12)
SFBEAM,4,1,PRES,(1E4/12)
FINISH
/SOLU
```

Fig. 7.14 VM2 input listing as given in ANSYS *Verification Manual*

find a solved problem with such materials. A quick glance at the list of test case descriptions reveals that *Verification Problem 200* involves a viscoelastic material and that the user may benefit from examining this file in order to see how the problem is treated before moving on to the problem at hand, which is likely to be more complicated.

Chapter 8
Linear Structural Analysis

A linear analysis is conducted if a structure is expected to exhibit linear behavior. The deformation and load-carrying capability can be determined by employing one of the analysis types available in ANSYS, static or dynamic, depending on the nature of the applied loading. If the applied loading is determined as part of the solution for structural stability, a buckling analysis is conducted. If the structure is subjected to thermal loading, the analysis is referred to as thermomechanical.

8.1 Static Analysis

The behavior of structures under static loading can be analyzed by employing different types of elements within ANSYS. The nature of the structure dictates the type of elements utilized in the analysis. Discrete or framed structures are suitable for modeling with rod- and beam-type elements. However, the modeling of continuous structures usually requires a three-dimensional model with solid elements.

Under certain types of loading and geometric conditions, the three-dimensional type of analysis can be idealized as a two-dimensional analysis. If the component is subjected to in-plane loading only and its thickness is small with respect to the other length dimensions, it is idealized as a plane stress condition. If the component with a uniform cross section is long in the depth direction and is subjected to a uniform loading along the depth direction, it is idealized as a plane strain condition. If the component has a circular cross section and is subjected to uniform and concentric loading, it possesses axisymmetry. If thin structural components are subjected to lateral loading, the plate and shell elements are suitable for analysis.

8.1.1 Trusses

A truss is a structure that is made of straight structural members capable of carrying loads only in their own direction, i.e., no shear forces, no moments. Thus, each

The online version of this book (doi: 10.1007/978-1-4939-1007-6_8) contains supplementary material, which is available to authorized users

E. Madenci, I. Guven, *The Finite Element Method and Applications in Engineering Using ANSYS®*, DOI 10.1007/978-1-4899-7550-8_8

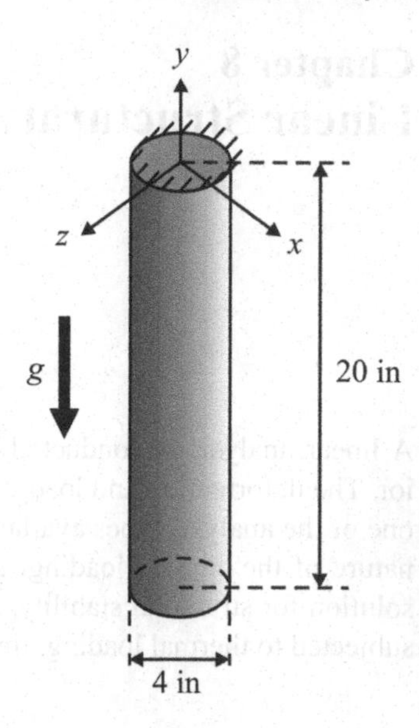

Fig. 8.1 Schematic of a bar deformed due to its own weight

member is under either axial tension or axial compression. These members are connected to each other by means of joints. It is assumed that loads can only be applied at the joints. `LINK180` is the element for modeling truss structures. The degrees of freedom at each node for truss elements are the displacement components u_x, u_y, and u_z. However, the vector sum of the deformations (elongation or contraction, *not* the displacements) is aligned with the direction of the element. Two example problems are given to demonstrate the usage of truss elements within ANSYS.

8.1.1.1 Elongation of a Bar Under its Own Weight Using Truss Elements

Consider a steel bar of uniform cross section whose upper end is supported such that it is fixed from translational motion. The mass density, elastic modulus, and Poisson's ratio of steel are $\rho = 0.284 \text{lb/in}^3$, $E = 30\times10^6$ psi, and $\nu = 0.3$, respectively. The radius and length of the bar are assumed to be $r = 2\text{in}$ and $l = 20\text{in}$, respectively, and the gravitational acceleration is $g = 386.2205\text{in/sec}^2$. The goal is to find the elongation of the bar at the lower end due to its own weight. The positive y-direction is the opposite direction of the gravitational acceleration, as shown in Fig. 8.1.

This problem can be solved using two-dimensional truss, two-dimensional axisymmetric plane, or three-dimensional elements. Since, we are interested in the elongation only, two-dimensional truss elements (`LINK180`) are used to obtain the solution.

Model Generation

- Specify the element type (`ET` command) using the following menu path:

Main Menu > Preprocessor > Element Type > Add/Edit/Delete

- Click on ***Add***.
- Select ***Link*** immediately below ***Structural Mass*** from the left list and ***3D finit stn 180*** from the right list; click on ***OK***.
- Click on ***Close***.

• Specify real constants (**R** command) using the following menu path:

Main Menu > Preprocessor > Real Constants > Add/Edit/Delete

- Click on ***Add***.
- Highlight ***Type 1 Link 180***; click on ***OK***.
- Enter ***12.5664*** (calculated based on radius, $r = 2$ in) for ***AREA***; click on ***OK***.
- Click on ***Close***.

• Specify material properties for the bar (**MP** command) using the following menu path:

Main Menu > Preprocessor > Material Props > Material Models

- In the *Define Material Model Behavior* dialog box, in the right window, successively left-click on ***Structural*** and ***Density***, which will bring up another dialog box.
- Enter ***0.284*** for ***DENS***; click on ***OK***.
- In order to specify the elastic modulus and Poisson's ratio, in the *Define Material Model Behavior* dialog box, in the right window, successively left-click on ***Structural***, ***Linear***, ***Elastic***, and, finally, ***Isotropic***, which will bring up another dialog box.
- Enter ***30e6*** for ***EX*** and ***0.3*** for ***PRXY***; click on ***OK***.
- Close the *Define Material Model Behavior* dialog box by using the following menu path:

Material > Exit

• Create keypoints (**K** command) using the following menu path:

Main Menu > Preprocessor > Modeling > Create > Keypoints > In Active CS

- A total of 2 keypoints will be created.
- Enter (x, y) coordinates of keypoint 1 as (0, 0); click on ***Apply***.
  This action will keep the *Create Keypoints in Active Coordinate System* dialog box open. If the ***NPT Keypoint number*** field is left blank, then ANSYS assigns the lowest available keypoint number to the keypoint that is being created.
- Repeat the same procedure for keypoint 2 using (0,−20) for the (x, y) coordinates.
- Click on ***OK*** (instead of ***Apply***).

• Create a line (**L** command) using the following menu path:

Main Menu > Preprocessor > Modeling > Create > Lines > Lines > Straight Line

 – *Pick Menu* appears; first pick keypoint 1, then keypoint 2; click on ***OK***.

- Specify the number of divisions on the line (**LESIZE** command) using the following menu path:

Main Menu > Preprocessor > Meshing > Size Cntrls > ManualSize > Lines > Picked Lines

 – *Pick Menu* appears; pick the line; click on ***OK***.
 – *Element Sizes on Picked Lines* dialog box appears; enter 20 for ***NDIV***; click on ***OK***.

- Create the mesh (**LMESH** command) using the following menu path:

Main Menu > Preprocessor > Meshing > Mesh > Lines

 – *Pick Menu* appears; pick the line; click on ***OK***.

- Review elements.

 – Turn on element numbering using the following menu path:

Utility Menu > PlotCtrls > Numbering

 – Select ***Element numbers*** from the first pull-down menu.
 – Plot elements (**EPLOT** command) using the following menu path:

Utility Menu > Plot > Elements

 – Turn off element numbering and turn on node numbering using the following menu path:

Utility Menu > PlotCtrls > Numbering

 – Place a *checkmark* by clicking on the empty box next to ***NODE Node numbers***.
 – Select ***No numbering*** from the first pull-down menu.
 – Click on ***OK***.
 – Plot nodes (**NPLOT** command) using the following menu path:

Utility Menu > Plot > Nodes

Solution

- Apply displacement constraints (**D** command) using the following menu path:

Main Menu > Solution > Define Loads > Apply > Structural > Displacement > On Nodes

 – *Pick Menu* appears; pick node 1 (upper end); click on ***OK*** in the *Pick Menu*.
 – Highlight ***All DOF***; click on ***OK***.

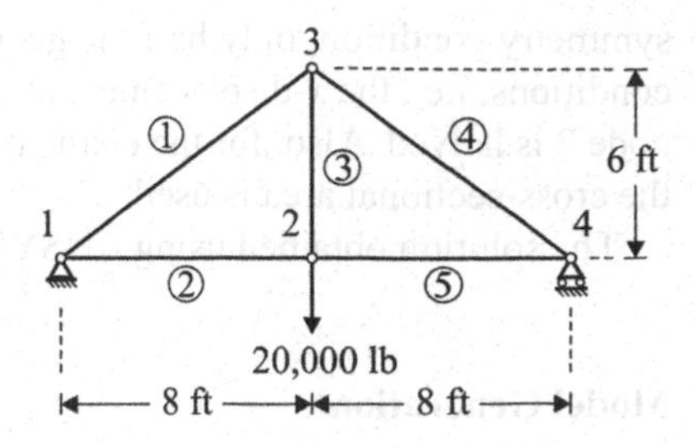

Fig. 8.2 Schematic of the truss structure with symmetry

- Apply gravitational acceleration (**ACEL** command) using the following menu path:

Main Menu > Solution > Define Loads > Apply > Structural > Inertia > Gravity > Global

  - *Apply (Gravitational) Acceleration* dialog box appears.
  - Enter ***386.2205*** for ***ACELY***; click on ***OK***.

- Obtain the solution (**SOLVE** command) using the following menu path:

Main Menu > Solution > Solve > Current LS

  - *Confirmation Window* appears along with *Status Report Window*.
  - Review status; if OK, close the *Status Report Window*; click on ***OK*** in the *Confirmation Window*.
  - Wait until ANSYS responds with ***Solution is done!***

Postprocessing

- Review displacement values (**PRNSOL** command) using the following menu path:

Main Menu > General Postproc > List Results > Nodal Solution

  - Click on ***Nodal Solution, DOF Solution***, and ***Y-component of displacement***; click on OK.
  - The list appears. Note that the value for the y-displacement at node 2 (lower end) is listed as ***−0.73124E–03*** (in inches).

8.1.1.2 Analysis of a Truss Structure with Symmetry

Consider the steel truss structure shown in Fig. 8.2, which possesses symmetry with respect to the ordinate. Node and element numbers are also shown in this figure. Element 3 has a cross-sectional area of $A = 20\ \text{in}^2$, while the other elements have $A = 10\ \text{in}^2$. The elastic modulus for all of the elements is $E = 30 \times 10^6$ psi. The goal is to find the displacements at the nodes and the stresses in the elements. Due to the symmetry condition, only half the geometry is modeled with appropriate boundary conditions, i.e., the x-displacement at nodes 2 and 3 is zero and the applied force at node 2 is halved. Also, for the element located along the symmetry line, one half of the cross-sectional area is used.

The solution obtained using ANSYS is as follows:

Model Generation

- Specify the element type (**ET** command) using the following menu path:

Main Menu > Preprocessor > Element Type > Add/Edit/Delete

- Click on ***Add***.
- Select ***Link*** immediately below ***Structural Mass*** from the left list and ***3D finit stn 180*** from the right list; click on ***OK***.
- Click on Close.

- Specify real constants (**R** command) using the following menu path:

Main Menu > Preprocessor > Real Constants > Add/Edit/Delete

- Click on ***Add***.
- Highlight ***Type 1 Link 180***; click on ***OK***.
- Enter ***10*** for ***AREA***; click on ***OK***.
- Click on ***Close***.

- Specify material properties for the bar (**MP** command) using the following menu path:

Main Menu > Preprocessor > Material Props > Material Models

- In the *Define Material Model Behavior* dialog box, in the right window, successively left-click on ***Structural***, ***Linear***, ***Elastic***, and, finally, ***Isotropic***, which will bring up another dialog box.
- Enter ***30e6*** for ***EX*** and ***0*** for ***PRXY***; click on ***OK***. Click on ***OK*** in the subsequent warning message.
- Close the *Define Material Model Behavior* dialog box by using the following menu path:

Material > Exit

- Create nodes (**N** command) using the following menu path:

Main Menu > Preprocessor > Modeling > Create > Nodes > In Active CS

- A total of 3 nodes will be created.
- Enter the (x, y) coordinates of node 1 as (0, 0); click on ***Apply***.
- Repeat the same procedure for nodes 2 and 3 using (96, 0) and (96, 72), respectively, for the (x, y) coordinates.
- After entering the coordinates for node 3, click on ***OK*** (instead of ***Apply***).

- A total of 3 elements will be created. Element 1 is defined by nodes 1 and 3 [1–3]. Similarly, elements 2 and 3 are defined by nodes [1–2] and [2–3], respectively. Create elements (`E` command) using the following menu path:

Main Menu > Preprocessor > Modeling > Create > Elements > Auto Numbered > Thru Nodes

  – *Pick Menu* appears; create elements by picking two nodes at a time and clicking on ***Apply*** in between.
  – Observe the elements created after clicking on ***Apply*** in the *Pick Menu.*
  – Repeat until element 3 is created; click on ***OK***.

Solution

- Apply displacement constraints (`D` command) using the following menu path:

Main Menu > Solution > Define Loads > Apply > Structural > Displacement > On Nodes

  – *Pick Menu* appears; pick node 1; click on ***OK*** in the *Pick Menu.*
  – Highlight ***UY***; click on ***Apply***.
  – *Pick Menu* reappears; pick nodes 2 and 3; click on ***OK*** in the *Pick Menu.*
  – Click on ***UY*** to remove the highlight then click on ***UX*** to highlight.
  – Click on ***OK***.

- Apply force boundary conditions (`F` command) using the following menu path:

Main Menu > Solution > Define Loads > Apply > Structural > Force/Moment > On Nodes

  – *Pick Menu* appears; pick node 2; click on ***OK*** in the *Pick Menu.*
  – Select ***FY*** from pull-down menu and enter ***−10000*** for ***Force/moment value***; click on ***OK***.

- Obtain the solution (`SOLVE` command) using the following menu path:

Main Menu > Solution > Solve > Current LS

  – *Confirmation Window* appears along with *Status Report Window.*
  – Review status, if OK, close the *Status Report Window*; click on ***OK*** in the *Confirmation Window.*
  – Wait until ANSYS responds with ***Solution is done!***

Postprocessing

- Review displacement values (`PRNSOL` command) using the following menu path:

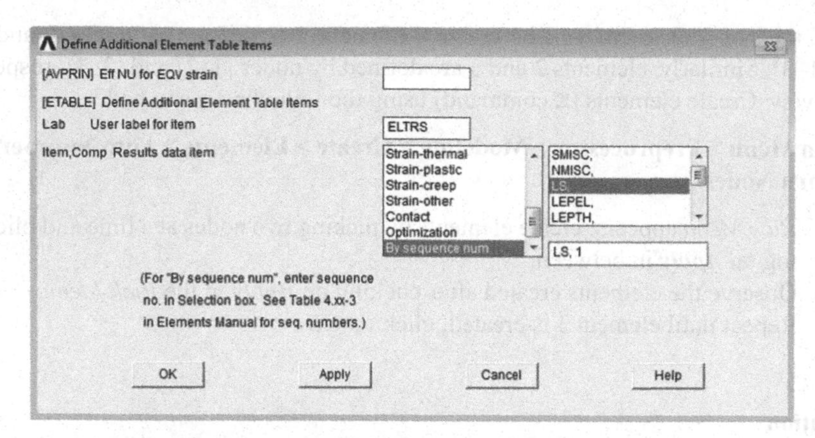

Fig. 8.3 Dialog box for retrieving element results based on sequence numbers

Main Menu > General Postproc > List Results > Nodal Solution

- *List Nodal Solution* dialog box appears. Click on ***DOF Solution*** and ***Y-component of displacement***; click on ***OK***.
- The list appears with the values for the y-displacement at nodes 2 and 3 as $-$***0.19200E–01*** and $-$***0.16800E–01***, respectively.

- Review element stress values (**ETABLE** command) using the following menu path:

Main Menu > General Postproc > Element Table > Define Table

- *Element Table Data* dialog box appears. Click on ***Add***, which brings up the *Define Additional Element Table Items* dialog box. Enter a label (***Lab***) for element stresses, say ***ELSTRS***. Scroll down in the left list; click on ***By Sequence num***; click on ***LS*** in the right list. Finally, enter ***LS,1*** in the last text field, as shown in Fig. 8.3; click on ***OK***.
- Note that the element table ***ELSTRS*** is now listed in the *Element Table Data* dialog box; click on ***Close***.
- List the element table (**PRETAB** command) using the following menu path:

Main Menu > General Postproc > Element Table > List Elem Table

- In the *List Element Table Data* dialog box, highlight ***ELSTRS***; click on ***OK***.
- The list appears with stresses in elements 1, 2, and 3 as $-$***1666.7***, ***1333.3***, and ***1000***, respectively.

8.1.2 Beams

A beam is a structural member capable of carrying axial, shear, and moment loads. Unlike truss members, loads can be applied anywhere along the beam geometry.

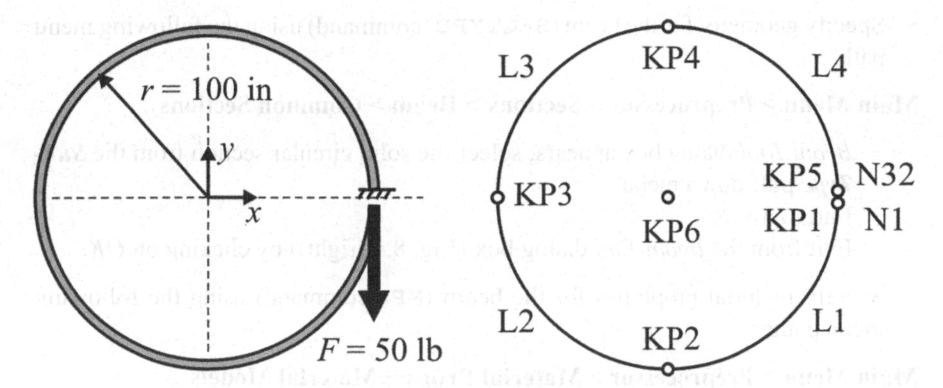

Fig. 8.4 Schematic of a circular steel ring and the corresponding solid model (*left*) and *Beam Tool* dialog box (*right*)

ANSYS provides several element types for modeling beams. The most commonly used one is **BEAM188** for two- and three-dimensional analyses, respectively. At each node, both displacements and rotations are the degrees of freedoms for structural beam elements (u_x, u_y, and θ_z for 2-D; u_x, u_y, u_z, θ_x, θ_y, and θ_z for 3-D). Two example problems are considered in this section for the demonstration of the usage of beam elements within ANSYS.

8.1.2.1 Analysis of a Slit Ring

A circular steel ring with a slit, as shown in Fig. 8.4, is subjected to a 50-lb vertical force acting in the negative *y*-direction at the termination point while translations and rotations are constrained in every direction. The ring has a solid circular cross section with radius 1 in. The structure is modeled using beam elements with cross-sectional area $A = \pi$, elastic modulus $E = 30\times10^6$, Poisson's ratio $\nu = 0.3$, and moment of inertia $I_{zz} = \pi/4$. The goal is to find the displacements at the nodes and the moment diagram. The solid model used in the ANSYS solution is also shown in Fig. 8.4 (left), with the keypoint and line numbers indicated.

Model Generation

- Specify the element type (**ET** command) using the following menu path:

Main Menu > Preprocessor > Element Type > Add/Edit/Delete

- Click on ***Add***.
- Select ***Beam*** immediately below ***Structural Mass*** from the left list and ***2node 188*** from the right list; click on ***OK***.
- Click on ***Close***.

- Specify geometry for the beam (`SECTYPE` command) using the following menu path:

Main Menu > Preprocessor > Sections > Beam > Common Sections

  - *Beam Tool* dialog box appears; select the solid circular section from the ***Sub-Type*** pull-down menu.
  - Enter ***1*** for ***R***.
  - Exit from the *Beam Tool* dialog box (Fig. 8.4 (right)) by clicking on ***OK***.

- Specify material properties for the beam (`MP` command) using the following menu path:

Main Menu > Preprocessor > Material Props > Material Models

  - In the *Define Material Model Behavior* dialog box, in the right window, successively left-click on ***Structural***, ***Linear***, ***Elastic***, and, finally, ***Isotropic***, which will bring up another dialog box.
  - Enter ***30e6*** for ***EX*** and ***0.3*** for ***PRXY***; click on ***OK***.
  - Close the *Define Material Model Behavior* dialog box by using the following menu path:

Material > Exit

- Create keypoints (`K` command) using the following menu path:

Main Menu > Preprocessor > Modeling > Create > Keypoints > In Active CS

  - A total of 6 keypoints will be created.
  - Enter the (x, y) coordinates of keypoint 1 as (***100***, ***0***); click on ***Apply***. This action will keep the *Create Keypoints in Active Coordinate System* dialog box open. If the ***NPT Keypoint number*** field is left blank, then ANSYS assigns the lowest available keypoint number to the keypoint that is being created.
  - Referring to Fig. 8.4, repeat the same procedure for keypoints 2, 3, 4, 5, and 6 using (***0***, ***−100***), (***−100***, ***0***), (***0***, ***100***), (***100***, ***0***), and (***0***, ***0***), respectively, for the (x, y) coordinates.
  - After generating keypoint 6, click on ***OK*** (instead of ***Apply***).
  - Note that keypoints 1 and 5 are coincident. This is intentional, so the slit can be modeled properly.

- Create arcs (`LARC` command) using the following menu path:

Main Menu > Preprocessor > Modeling > Create > Lines > Arcs > By End KPs & Rad

  - A total of 4 lines (arcs) will be created.
  - *Pick Menu* appears; pick keypoints 1 and 2 (end points of the arc); click on ***OK*** in the *Pick Menu*.
  - Pick keypoint 6 (center of the arc); click on ***OK*** in the *Pick Menu*.

- *Arc by End KPs & Radius* dialog box appears; enter ***100*** for ***RAD Radius of the arc***.
- Click on ***Apply***; line 1 is created.
- Repeat this procedure for lines 2, 3, and 4 using keypoint pairs (2, 3), (3, 4), and (4, 5), respectively. All lines use keypoint 6 as the center and 100 as the radius.

• Specify the number of divisions on all lines (`LESIZE` command) using the following menu path:

Main Menu > Preprocessor > Meshing > Size Cntrls > ManualSize > Lines > All Lines

- *Element Sizes on All Selected Lines* dialog box appears; enter ***10*** for ***NDIV***; click on ***OK***.

• Create the mesh (`LMESH` command) using the following menu path:

Main Menu > Preprocessor > Meshing > Mesh > Lines

- *Pick Menu* appears; click on ***Pick All***.

Solution

• Apply displacement constraints (`D` command) using the following menu path:

Main Menu > Solution > Define Loads > Apply > Structural > Displacement > On Nodes

- *Pick Menu* appears; pick one of the nodes at $x=100$ and $y=0$. There are two nodes at this location: nodes 1 and 32. When picking, ANSYS asks the user which one of the nodes is to be picked. Click on the ***Next*** button in this *Warning Window* so that it shows ***Node 32***; click on ***OK*** in the *Pick Menu*.
- *Apply U, Rot on Nodes* dialog box appears; highlight ***All DOF***; click on ***OK***.

• Apply force boundary conditions (`F` command) using the following menu path:

Main Menu > Solution > Define Loads > Apply > Structural > Force/Moment > On Nodes

- *Pick Menu* appears; this time pick node 1 (instead of node 32); click on ***OK*** in the *Pick Menu*.
- Select ***FY*** from the pull-down menu and enter ***−50*** for ***Force/moment value***; click on ***OK***.

• Obtain the solution (`SOLVE` command) using the following menu path:

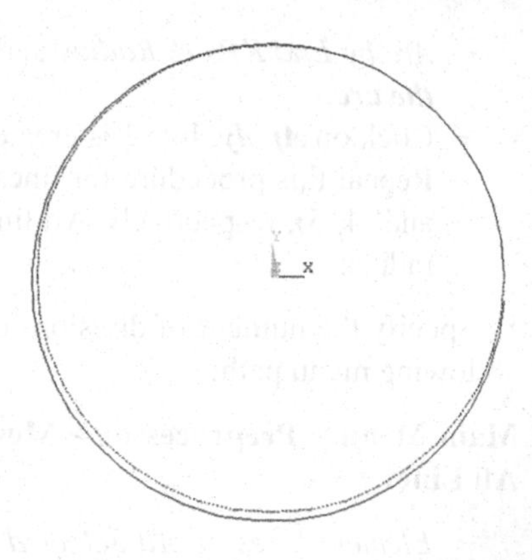

Fig. 8.5 Deformed shape of the steel ring under applied boundary conditions

Main Menu > Solution > Solve > Current LS

- *Confirmation Window* appears along with *Status Report Window*.
- Review status; if OK, close the *Status Report Window*; click on ***OK*** in the *Confirmation Window*.
- Wait until ANSYS responds with ***Solution is done!***

Postprocessing

- Review the deformed shape (**PLDISP** command) using the following menu path:

Main Menu > General PostProc > Plot Results > Deformed Shape

- Select ***Def + undeformed***; click on ***OK***.
- The deformed shape is shown in Fig. 8.5 as it appears in the *Graphics Window*.

- Store bending moment values in the element table (**ETABLE** command) using the following menu path:

Main Menu > General Postproc > Element Table > Define Table

- *Element Table Data* dialog box appears; click on ***Add***.
- *Define Additional Element Table Items* dialog box appears. Enter a label name, say MZI, in the User label for item text field. In the left list, scroll down to select ***By sequence number*** and select ***SMISC*** in the right list. Finally, type ***SMISC,3*** in the last text field; click on ***Apply*** (Fig. 8.6).
- Repeat this procedure for ***MZJ*** using ***SMISC, 16.*** When done, click on ***OK*** (instead of ***Apply***).
- Note that ***SMIS3*** now appears in the list in the *Element Table Data* dialog box. Exit from the *Element Table Data* dialog box by clicking on ***Close***.
- Plot the moment diagram (**PLLS** command) using the following menu path:

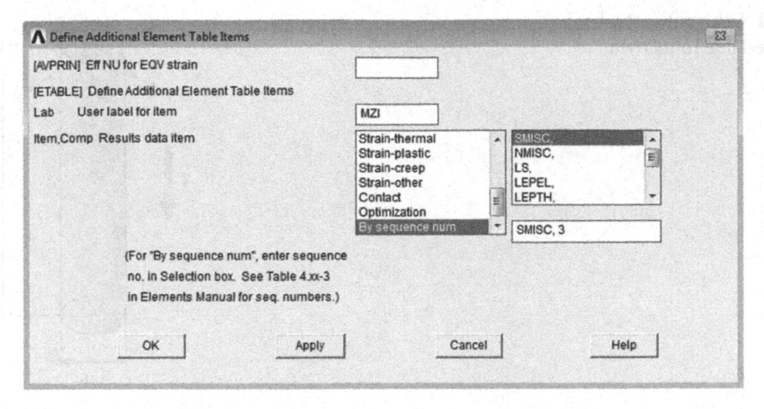

Fig. 8.6 *Define Additional Element Table Items* dialog box for extracting nodal moment values

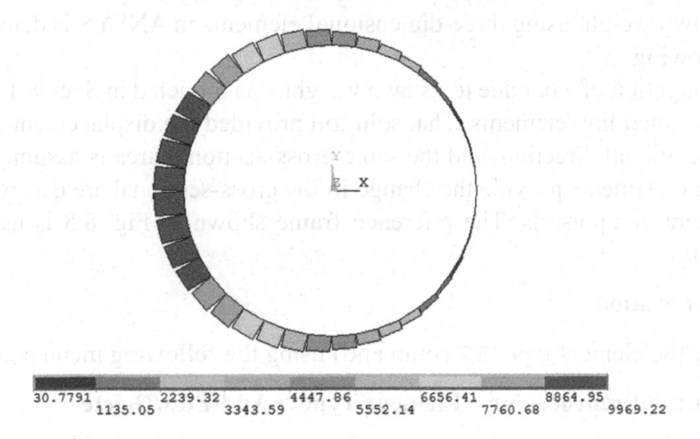

Fig. 8.7 Moment diagram of the steel ring

Main Menu > General Postproc > Plot Results > Contour Plot > Line Elem Res

- *Plot Line-Element Results* dialog box appears; click on ***OK***.
- Figure 8.7 shows the resulting moment diagram as displayed in the *Graphics Window*.

8.1.3 Three-Dimensional Problems

Almost all engineering problems are three-dimensional (3-D) by nature. However, depending on the specific geometry, loading conditions, and quantities of interest, it is common to approach the problem with the idealization of a lower dimensionality. If a representative idealization cannot be utilized, then a three-dimensional model must be created. The most commonly used three-dimensional structural element is, **`SOLID185`** which is an 8-noded brick element. The degrees of freedom at each

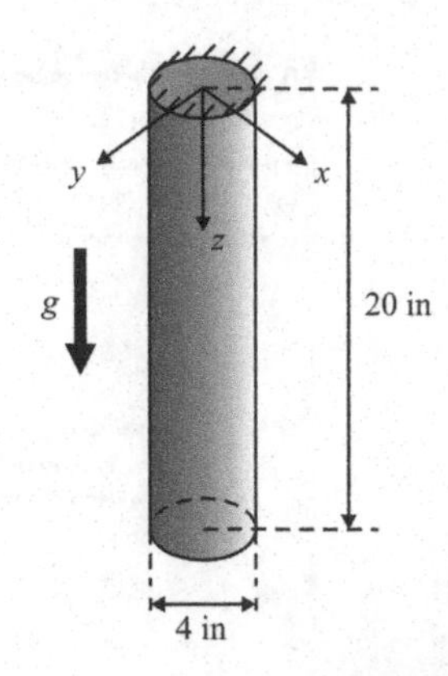

Fig. 8.8 Schematic of a bar deformed due to its own weight

node for 3-D problems are u_x, u_y, and u_z. Determining the deformation of a bar under its own weight using three-dimensional elements in ANSYS is demonstrated in the following.

The elongation of a bar due to its own weight was modeled in Sect. 8.1.1.1 using two-dimensional link elements. That solution provided the displacement of the bar in the longitudinal direction, and the same cross-sectional area is assumed. Three-dimensional elements provide the change in the cross-sectional area, as well as the displacement components. The reference frame shown in Fig. 8.8 is used in the 3-D solution.

Model Generation

- Specify the element type (**ET** command) using the following menu path:

Main Menu > Preprocessor > Element Type > Add/Edit/Delete

  - Click on ***Add***.
  - Select ***Solid*** immediately below ***Structural Mass*** from the left list and ***Brick 8node 185*** from the right list; click on ***OK***.
  - Click on ***Close***.

- Specify material properties for the bar (**MP** command) using the following menu path:

Main Menu > Preprocessor > Material Props > Material Models

  - In the *Define Material Model Behavior* dialog box, in the right window, successively left-click on ***Structural*** and ***Density***, which will bring up another dialog box.
  - Enter ***0.2839605*** for ***DENS***; click on ***OK***.
  - In order to specify the elastic modulus and Poisson's ratio, in the *Define Material Model Behavior* dialog box, in the right window, successively left-click on ***Structural***, ***Linear***, ***Elastic***, and, finally, ***Isotropic***, which will bring up another dialog box.

- Enter ***30e6*** for ***EX*** and ***0.3*** for ***PRXY***; click on ***OK***.
- Close the *Define Material Model Behavior* dialog box by using the following menu path:

Material > Exit

- Create a volume (**CYLIND** command) using the following menu path:

Main Menu > Preprocessor > Modeling > Create > Volumes > Cylinder > By Dimensions

- *Create Cylinder by Dimensions* dialog box appears. Enter ***2*** for ***RAD2***, ***20*** for ***Z2***, and ***90*** for ***THETA2***; click on ***OK***.

- Create additional volumes by reflection (**VSYMM** command) using the following menu path:

Main Menu > Preprocessor > Modeling > Reflect > Volumes

- *Pick Menu* appears; click on ***Pick All*** button, which brings up the *Reflect Volumes* dialog box.
- Click on the ***Y-Z Plane X*** radio-button; click on ***Apply***.
- *Pick Menu* reappears; click on ***Pick All*** button and in the *Reflect Volumes* dialog box click on the ***X-Z Plane Y*** radio-button; click on ***OK***.

- Glue the volumes (**VGLUE** command) using the following menu path:

Main Menu > Preprocessor > Modeling > Operate > Booleans > Glue > Volumes

- *Pick Menu* appears; click on ***Pick All*** button.

- Specify the global element size (**ESIZE** command) using the following menu path:

Main Menu > Preprocessor > Meshing > Size Cntrls > ManualSize > Global > Size

- *Global Element Sizes* dialog box appears; enter ***1*** for ***SIZE***; click on ***OK***.

- Create the mesh (**VMESH** command) using the following menu path:

Main Menu > Preprocessor > Meshing > Mesh > Volumes > Mapped > 4 to 6 sided

- *Pick Menu* appears; click on ***Pick All***.

Solution

- Apply displacement constraints (**D** command) using the following menu path:

Main Menu > Solution > Define Loads > Apply > Structural > Displacement > On Nodes

- *Pick Menu* appears; pick all the nodes at $z=0$ (use different viewpoints if necessary); click on ***OK*** in the *Pick Menu*.
- Highlight ***All DOF***; click on ***OK***.

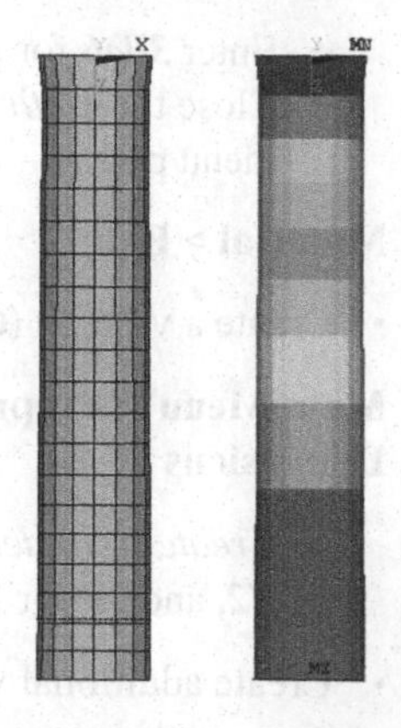

Fig. 8.9 Deformed shape (*left*) and contour plot of the *z*-displacement (*right*) of the bar due to its own weight

- Apply gravitational acceleration (**ACEL** command) using the following menu path:

Main Menu > Solution > Define Loads > Apply > Structural > Inertia > Gravity > Global

  - *Apply (Gravitational) Acceleration* dialog box appears.
  - Enter ***−386.2205*** for *ACELZ*; click on ***OK***.

- Obtain the solution (**SOLVE** command) using the following menu path:

Main Menu > Solution > Solve > Current LS

  - Confirmation Window appears along with Status Report Window.
  - Review status, if OK, close the Status Report Window; click on ***OK*** in the Confirmation Window.
  - Wait until ANSYS responds with **Solution is done!**

Postprocessing

- Review the deformed shape (**PLDISP** command) using the following menu path:

Main Menu > General PostProc > Plot Results > Deformed Shape

  - Select ***Def + undef edge***; click on ***OK***.
  - The deformed shape is shown in Fig. 8.9 as it appears in the *Graphics Window*.

- Review *z*-displacement contours (**PLNSOL** command) using the following menu path:

Main Menu > General PostProc > Plot Results > Contour Plot > Nodal Solu

  - Contour Nodal Solution Data dialog box appears. Click on **Nodal Solution**, **DOF Solution**, and then **Z-component of displacement**; click on **OK**.
  - The contour plot is shown in Fig. 8.9 as it appears in the Graphics Window.

- Review displacement values (**PRNSOL** command) using the following menu path:

Main Menu > General Postproc > List Results > Nodal Solution

- Click on ***Nodal Solution, DOF Solution***, and then ***Z-component of displacement***; click on ***OK***.
- The list appears in a separate window. It is a long list of z-displacements.
- At the bottom of the window maximum displacement value is printed as ***0.72386E–03***.

8.1.4 Two-Dimensional Idealizations

As mentioned in Sect. 6.2.2, the reduction of the dimensionality of a problem from three to two through an idealization may reduce the computational cost significantly. There are three distinct two-dimensional idealizations: plane stress, plane strain, and axisymmetry.

Plane stress and strain idealizations are discussed in Sects. 6.2.2.1 and 6.2.2.2, respectively. Therefore, the descriptions given in the following subsections are brief.

8.1.4.1 Plane Stress

In a structural problem, if one of the dimensions is much smaller than the in-plane dimensions, and if the structure is subjected to only in-plane loads along the boundary, then the plane stress idealization is valid. It reduces the computational cost significantly without a loss of accuracy in the quantities of interest. Plane stress idealization is demonstrated by considering a plate with a circular hole and a composite plate under axial tension.

Analysis of a Plate with a Circular Hole

A square plate (9×9 in^2) with a circular hole (radius $r=0.25$ in) is subjected to uniformly distributed tensile loading (1000 psi) in the vertical direction along its top surface while being fixed along the bottom surface (Fig. 8.10). The plate is stiffened by means of increased thickness, from 0.063 to 0.12 in. Plane stress idealization is used in the ANSYS solution, as the plate is thin and there are no lateral loads. The material properties are given as elastic modulus $E=10\times0^6$ psi and Poisson's ratio $\nu=0.25$. The goal is to obtain the displacement and stress fields resulting from the applied boundary conditions.

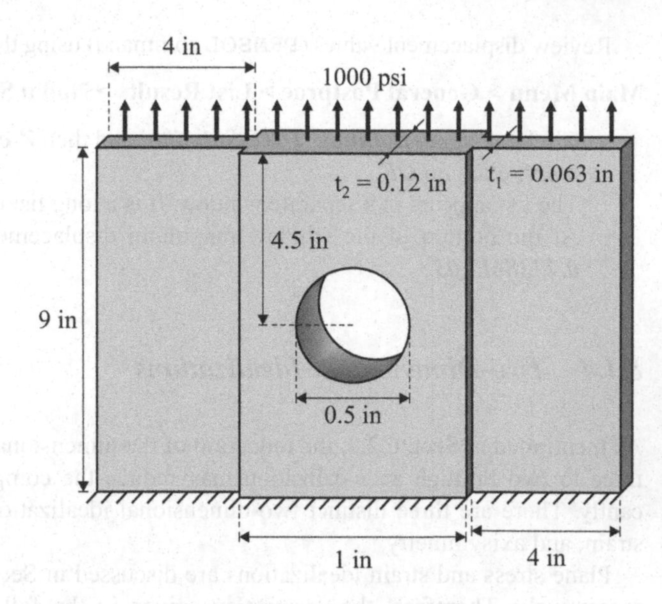

Fig. 8.10 Geometry and loading of the plate with a circular hole

Model Generation

- Define the element type (`ET` command) using the following menu path:

Main Menu > Preprocessor > Element Type > Add/Edit/Delete

- Click on ***Add***.
- Select ***Solid*** immediately below ***Structural Mass*** in the left list and ***Quad 4 Node 182*** in the right list; click on ***OK***.
- Click on ***Options***.
- *PLANE182 element type options* dialog box appears; select the ***Plane strs w/ thk*** item from the pull-down menu corresponding to ***Element behavior K3***.
- Click on ***OK***; click on ***Close***.

- Specify the thickness information using real constants (`R` command) using the following menu path:

Main Menu > Preprocessor > Real Constants > Add/Edit/Delete

- *Real Constants* dialog box appears; click on ***Add***. Click on ***OK***; *Real Constants Set Number 1 for PLANE182* dialog box appears.
- Type ***0.063*** in the ***Thickness THK*** text field; click on ***Apply***.
- Change the ***Real Constant Set No.*** from ***1*** to ***2*** and modify the ***Thickness THK*** text field to be ***0.12***; click on ***OK***.
- Exit from the *Real Constants* dialog box by clicking on ***Close***.

- Specify material properties (`MP` command) using the following menu path:

Main Menu > Preprocessor > Material Props > Material Models

- *Define Material Model Behavior* dialog box appears. In the right window, successively left-click on ***Structural***, ***Linear***, ***Elastic***, and, finally, ***Isotropic***, which brings up another dialog box.
- Enter ***10e6*** for ***EX*** and ***0.25*** for ***PRXY***; click on ***OK***.
- Close the *Define Material Model Behavior* dialog box by using the following menu path:

Material > Exit

- Create a square area (**RECTNG** command) using the following menu path:

Main Menu > Preprocessor > Modeling > Create > Areas > Rectangle > By Dimensions

- In the *Create Rectangle by Dimensions* dialog box, enter ***0*** and ***0.5*** for ***X1*** and ***X2*** and ***0*** and ***0.5*** for ***Y1*** and ***Y2***; click on ***OK***.

- Create a circular area for the hole geometry (**PCIRC** command) using the following menu path:

Main Menu > Preprocessor > Modeling > Create > Areas > Circle > By Dimensions

- In the Create Circle by Dimensions dialog box, type **0.25** for **Outer radius**; click on **OK**.

- Subtract the circle from the rectangle (**ASBA** command) using the following menu path:

Main Menu > Preprocessor > Modeling > Operate > Booleans > Subtract > Areas

- *Pick Menu* appears; pick the rectangle; click on ***OK***; pick the circle; click on ***OK***.

- Create additional rectangular areas (**RECTNG** command) using the following menu path:

Main Menu > Preprocessor > Modeling > Create > Areas > Rectangle > By Dimensions

- In the *Create Rectangle by Dimensions* dialog box, enter ***0.5*** and ***4.5*** for ***X1*** and ***X2*** and ***0*** and ***0.5*** for ***Y1*** and ***Y2***; click on ***Apply***.
- Now, enter ***0.5*** and ***4.5*** for ***X1*** and ***X2*** and ***0.5*** and ***4.5*** for ***Y1*** and ***Y2***; click on ***Apply***.
- Finally, enter ***0*** and ***0.5*** for ***X1*** and ***X2*** and ***0.5*** and ***4.5*** for ***Y1*** and ***Y2***; click on ***OK***.

- Glue the areas (**AGLUE** command) using the following menu path:

Fig. 8.11 Areas after gluing operation

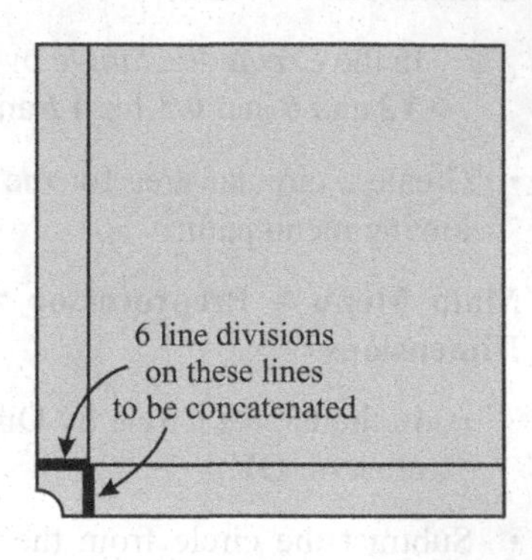

Fig. 8.12 Number of divisions on identified lines

Main Menu > Preprocessor > Modeling > Operate > Booleans > Glue > Areas

- *Pick Menu* appears; click on ***Pick All*** button.
- The areas appear in the *Graphics Window*, as shown in Fig. 8.11.

- Specify the global element size (**`ESIZE`** command) using the following menu path:

Main Menu > Preprocessor > Meshing > Size Cntrls > ManualSize > Global > Size

- *Global Element Sizes* dialog box appears; enter ***0.1*** for ***SIZE***; click on ***OK***.

- Specify the number of divisions on selected lines (**`LESIZE`** command) using the following menu path:

Main Menu > Preprocessor > Meshing > Size Cntrls > ManualSize > Lines > Picked Lines

- *Pick Menu* appears; pick the two lines identified in Fig. 8.12; click on ***OK***.
- *Element Sizes on Picked Lines* dialog box appears; enter 6 for ***NDIV*** and remove the checkmark next to ***KYNDIV SIZE, NDIV can be changed*** so that it shows ***No***; click on ***OK***.

- Concatenate lines (**`LCCAT`** command) using the following menu path:

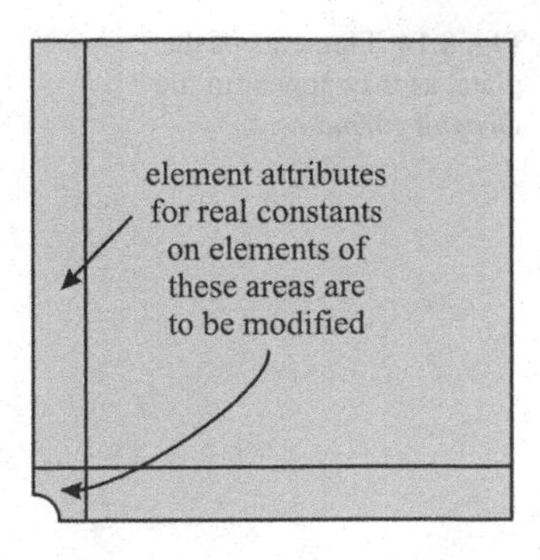

Fig. 8.13 Element attributes on elements attached to identified areas

Main Menu > Preprocessor > Meshing > Concatenate > Lines

- *Pick Menu* appears; pick the two lines identified in Fig. 8.12; click on ***OK***.

• Create the mesh (**AMESH** command) using the following menu path:

Main Menu > Preprocessor > Meshing > Mesh > Areas > Mapped > 3 or 4 sided

- *Pick Menu* appears; click on ***Pick All***.
- Modify the real constant set attribute of the elements corresponding to the thicker portion of the plate (**EMODIF** command) using the following menu path:

Main Menu > Preprocessor > Modeling > Move/Modify > Elements > Modify Attrib

- *Pick Menu* appears; pick the elements corresponding to the areas indicated in Fig. 8.13 (click on the ***Box*** radio-button in the *Pick Menu* and draw a rectangle in the *Graphics Window* to pick the elements). Clicking on ***OK*** brings up the *Modify Elem Attributes* dialog box.
- Select ***Real const REAL*** from the pull-down menu and enter ***2*** in the ***I1 New attribute number*** field; click on ***OK***.

• Create two successive reflective symmetric meshes (**ARSYM** command) using the following menu path:

Main Menu > Preprocessor > Modeling > Reflect > Areas

- *Pick Menu* appears; click on ***Pick All***.
- *Reflect Areas* dialog box appears; click on the ***Y-Z plane X*** radio-button; click on ***Apply***.
- *A Warning Window* appears; click on ***OK***.
- *Pick Menu* reappears; click on ***Pick All***.
- *Reflect Areas* dialog box reappears; click on the ***X-Z plane Y*** radio-button; click on ***OK***.
- Plot elements (**EPLOT** command) using the following menu path:

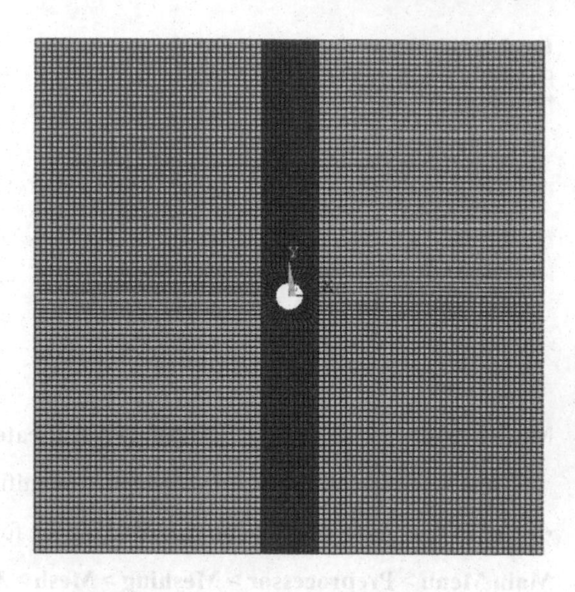

Fig. 8.14 Elements of the plate, as they appear in the *Graphics Window*

Utility Menu > Plot > Elements

– Although it is not apparent through visual inspection, there are duplicate entities (keypoints, lines, and nodes) along the symmetry lines, thus there is no continuity. Therefore, merge duplicate entities using the following menu path:

Main Menu > Preprocessor > Numbering Ctrls > Merge Items

– In the dialog box, select ***All*** from the first pull-down menu; click on ***OK***.
– Plot elements with different colors based on their real constant numbers using the following menu path:

Utility Menu > PlotCtrls > Numbering

– *Plot Numbering Controls* dialog box appears. Select ***Real const num*** from the first pull-down menu (corresponding to ***Elem/Attrib numbering***) and select ***Colors only*** from the second pull-down menu (corresponding to ***[/NUM] Numbering shown with***); click on ***OK***.
– Plot elements (**EPLOT** command) using the following menu path:

Utility Menu > Plot > Elements

– Figure 8.14 shows the corresponding element plot with different colors[1] based on material numbers.

[1] Colors have not been used in the printed version of the figures. See the accompanying CD-ROM for color versions of the figures.

Solution

- Apply displacement constraints (`D` command) using the following menu path:

Main Menu > Solution > Define Loads > Apply > Structural > Displacement > On Nodes

  - *Pick Menu* appears; pick the nodes along the bottom surface of the plate (click on the ***Box*** radio-button in the *Pick Menu* and draw a rectangle in the *Graphics Window* to pick the nodes); click on ***OK*** in the *Pick Menu.*
  - Highlight both ***UX*** and ***UY***; click on ***OK***.

- Apply surface force (pressure) boundary conditions (`SF` command) using the following menu path:

Main Menu > Solution > Define Loads > Apply > Structural > Pressure > On Nodes

  - *Pick Menu* appears; pick the nodes along the top surface of the plate; click on ***OK*** in the *Pick Menu.*
  - Type ***−1000*** (negative 1000) for ***VALUE Load PRES value***; click on ***OK***.
  - Pressure, by definition, acts normal toward the body along the surface. The direction of action in reference to the global coordinate system does not affect whether it is positive or negative. The only factor that dictates the sign is whether it acts toward or away from the body. Therefore, in order to apply the tensile loading, it is necessary to apply negative pressure.

- Obtain the solution (`SOLVE` command) using the following menu path:

Main Menu > Solution > Solve > Current LS

  - *Confirmation Window* appears along with *Status Report Window.*
  - Review status; if OK, close the *Status Report Window*; click on ***OK*** in the *Confirmation Window.*
  - Wait until ANSYS responds with ***Solution is done!***

Postprocessing

- Review the normal stress contour plots in the *x*- and *y*-directions (`PLNSOL` command) using the following menu path:

Main Menu > General Postproc > Plot Results > Contour Plot > Nodal Solu

  - Click on ***Nodal Solution***, ***Stress,*** and ***X-component of stress***; click on ***OK***.
  - The contour plot of σ_{xx} appears in the *Graphics Window*, as shown in Fig. 8.15.
  - The Contour plot of σ_{yy} is obtained similarly by selecting the ***Y-component of stress*** from the list and clicking on ***OK*** (shown in Fig. 8.16).

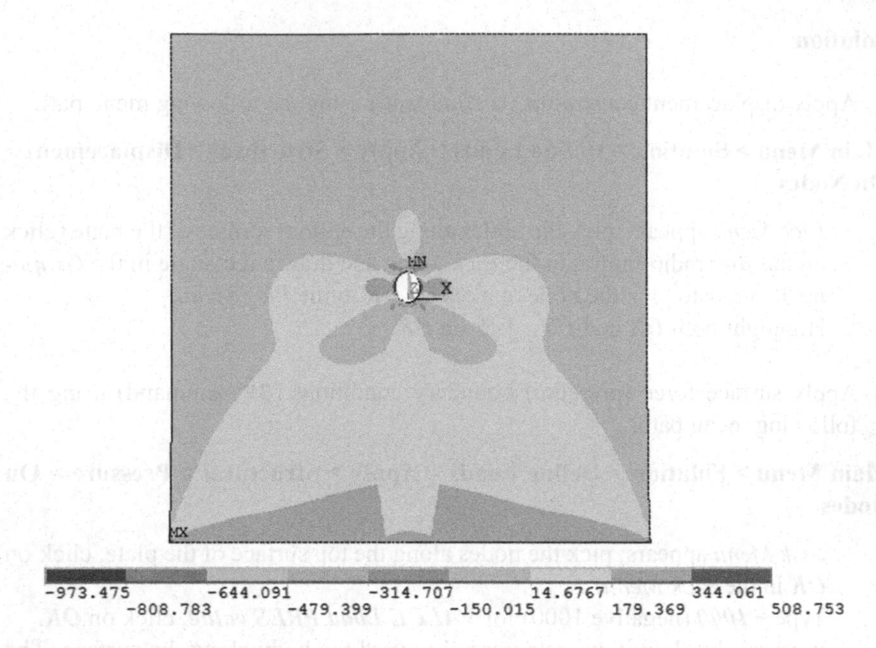

Fig. 8.15 Contour plot of σ_{xx}

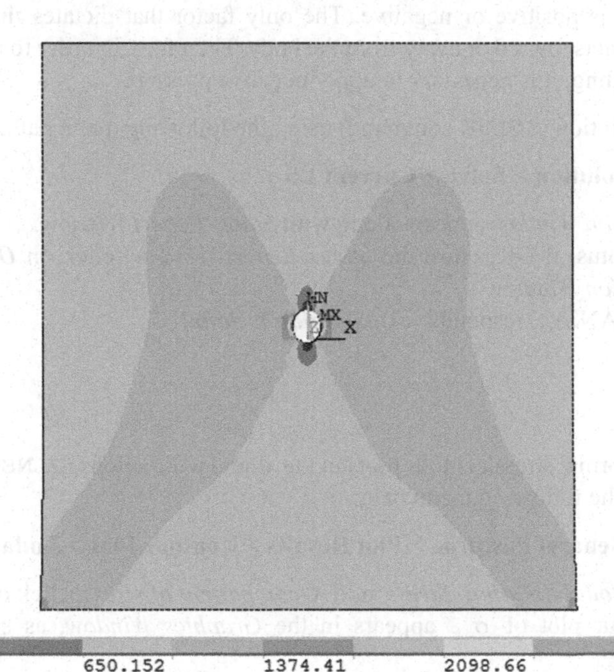

Fig. 8.16 Contour plot of σ_{yy}

- Review the variation of stresses along a path by means of a line plot. This operation requires the path to be defined first, followed by mapping the solution items of interest onto the path and, finally, obtaining the plot. The path that is defined in this case lies along the positive x-axis, starting from the left boundary of the hole and ending at the left boundary of the plate. Define the path (**PPATH** command) using the following menu path:

Main Menu > General Postproc > Path Operations > Define Path > By Nodes

- *Pick Menu* appears; pick the nodes with (x, y) coordinates (0.25, 0) and (4.5, 0); click on ***OK***.
- *By Nodes* dialog box appears; enter a name describing the path, say ***hrz***, in the ***Define Path Name*** text field; click on ***OK***.
- Close the *PATH Command Status Window*.
- Map results onto path (**PDEF** command) using the following menu path:

Main Menu > General Postproc > Path Operations > Map onto Path

- *Map Result Items onto Path* dialog box appears; select ***Stress*** from the left list and ***Y-direction SY*** from the right list; click on ***OK***.
- Obtain line plot of σ_{yy} along the path (**PLPATH** command) using the following menu path:

Main Menu > General Postproc > Path Operations > Plot Path Item > On Graph

- *Plot of Path Items on Graph* dialog box appears; select ***SY***; click on ***OK***.
- Figure 8.17 shows the line plots of σ_{xx} and σ_{yy} along the defined path.

Composite Plate Under Axial Tension

A fiber-reinforced square plate, shown in Fig. 8.18, is subjected to a uniform stress field of 20 ksi along the top and bottom boundaries. The sides of the plate are 10 in long, and the fibers are oriented at a 45° angle to the global Cartesian coordinate system. Material properties are specified as $E_1 = 10\times10^3$ ksi , $E_2 = 30\times10^3$ ksi , $G_{12} = 15\times10^3$ ksi , and $\nu_{12} = 0.1$. The goal is to find the displaced shape.

Model Generation

- Define the element type (**ET** command) using the following menu path:

Main Menu > Preprocessor > Element Type > Add/Edit/Delete

- Click on ***Add***.
- Select ***Solid*** immediately below ***Structural Mass*** in the left list and ***Quad 4 Node 182*** in the right list; click on ***OK***.
- Click on ***Close***.

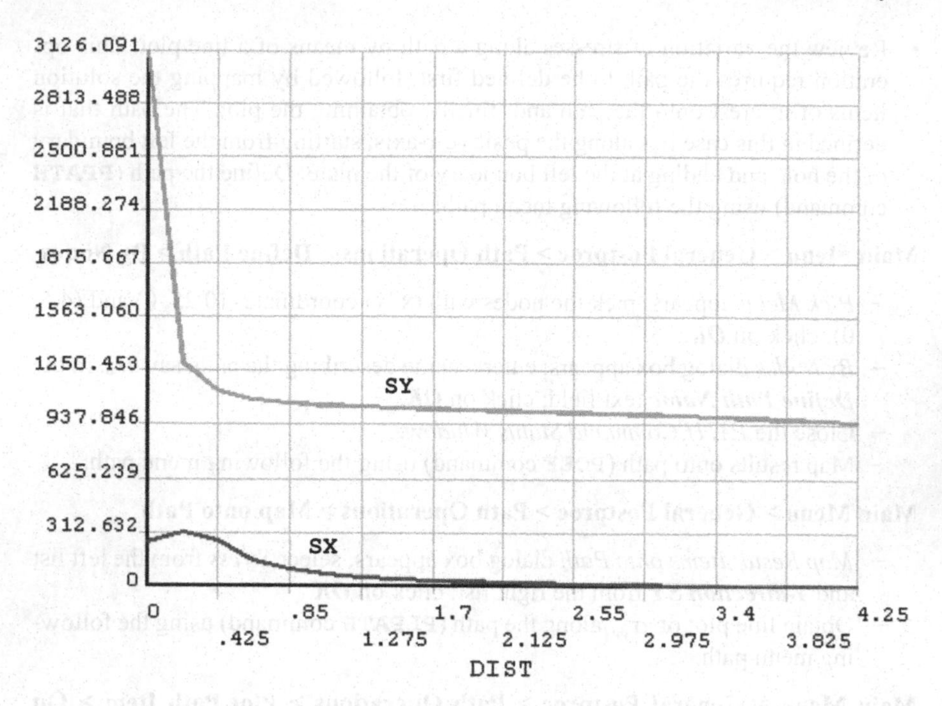

Fig. 8.17 Line plots of σ_{xx} and σ_{yy} along the defined path

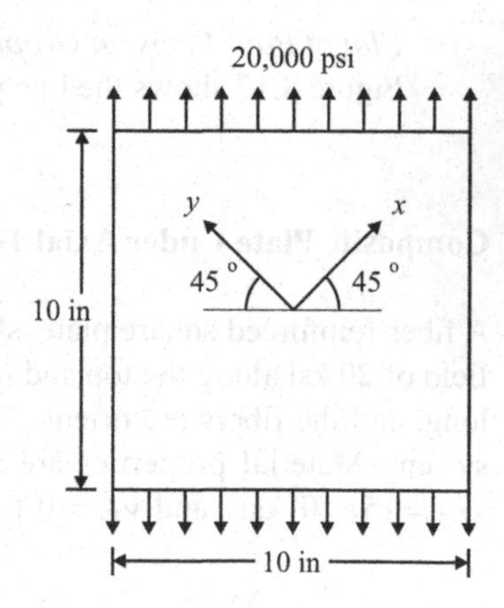

Fig. 8.18 Schematic of the composite plate, fiber orientation, and loading

- Specify material properties by typing the following four commands in the *Input Field* (at the end of each command, hit the Enter key to execute):

MP, EX,1,10E6
MP, EY,1,30E6
MP, PRXY,1,0.1
MP, GXY,1,15E6

- Create keypoints (**K** command) using the following menu path:

Main Menu > Preprocessor > Modeling > Create > Keypoints > In Active CS

  - A total of 4 keypoints will be created.
  - Enter (*x*, *y*) coordinates of keypoint 1 as (***−5, −5***); click on ***Apply***.
    This action will keep the *Create Keypoints in Active Coordinate System* dialog box open. If the ***NPT Keypoint number*** field is left blank, then ANSYS assigns the lowest available keypoint number to the keypoint that is being created.
  - Repeat the same procedure for the keypoints 2, 3, and 4 using (***5, −5***), (***5, 5***), and (***−5, 5***), respectively, for the (*x*, *y*) coordinates.
  - Once keypoint 4 is created, click on ***OK*** (instead of ***Apply***).

- Create the area through keypoints (**A** command) using the following menu path:

Main Menu > Preprocessor > Modeling > Create > Areas > Arbitrary > Through KPs

  - *Pick Menu* appears; pick keypoints 1 through 4 (in sequence); click on ***OK***.

- Material properties refer to the fiber directions. However, the global Cartesian coordinates and the fiber directions are at an angle of 45°. Therefore, the *element coordinate system* needs to be aligned with the fiber orientation. For this purpose, create a local coordinate system (**CLOCAL** command) using the following menu path:

Utility Menu > WorkPlane > Local Coordinate Systems > Create Local CS > At Specified Loc

  - *Pick Menu* appears; type ***0, 0, 0*** in the text field in the *Pick Menu*; click on ***OK***,
  - A dialog box appears; type ***45*** in the ***THXY Rotation about local Z*** text field; click on ***OK***; local coordinate system 11 is created.
  - Align the element coordinate system with local coordinate system 11 (**ESYS** command) using the following menu path:

Main Menu > Preprocessor > Meshing > Mesh Attributes > Default Attribs

  - *Meshing Attributes* dialog box appears. Select ***11*** from the ***ESYS Element coordinate sys*** pull-down menu; click on ***OK***.
  - Switch the active coordinate system to global Cartesian using the following menu path:

Utility Menu > Work Plane > Change Active CS to > Global Cartesian

- Specify the number of divisions on all lines (**LESIZE** command) using the following menu path:

Main Menu > Preprocessor > Meshing > Size Cntrls > ManualSize > Lines > All Lines

  - *Element Sizes on All Selected Lines* dialog box appears; enter ***20*** for **NDIV**; click on ***OK***.

- Mesh the square (**AMESH** command) using the following menu path:

Main Menu > Preprocessor > Meshing > Mesh > Areas > Mapped > 3 or 4 sided

  - *Pick Menu* appears; click on ***Pick All***.

Solution

- Apply displacement constraints (**D** command) using the following menu path:

Main Menu > Solution > Define Loads > Apply > Structural > Displacement > On Nodes

  - *Pick Menu* appears; pick the center node, i.e., $x = 0$ and $y = 0$; click on ***OK*** in the *Pick Menu*.
  - Highlight both ***UX*** and ***UY***; click on ***Apply***.
  - *Pick Menu* reappears; pick the right-side center node, i.e., $x = 5$ and $y = 0$; click on ***OK*** in the *Pick Menu*.
  - Remove the highlight on ***UX***, leaving ***UY*** highlighted; click on ***OK***.

- Apply surface force (pressure) boundary conditions (**SF** command) using the following menu path:

Main Menu > Solution > Define Loads > Apply > Structural > Pressure > On Nodes

  - *Pick Menu* appears; pick the nodes along the top and bottom surfaces of the plate; click on ***OK*** in the *Pick Menu*.
  - Type ***−20000*** (negative 20000) for ***VALUE Load PRES value***; click on ***OK***.
  - Pressure, by definition, acts normal toward the body along the surface. The direction of action in reference to the global coordinate system does not affect whether it is positive or negative. The only factor that dictates the sign is whether it acts toward or away from the body. Therefore, in order to apply the tensile loading, it is necessary to apply negative pressure.

- Obtain the solution (**SOLVE** command) using the following menu path:

Main Menu > Solution > Solve > Current LS

  - *Confirmation Window* appears along with *Status Report Window*.
  - Review status, if OK, close the *Status Report Window*; click on ***OK*** in the *Confirmation Window*.
  - Wait until ANSYS responds with ***Solution is done!***

Postprocessing

- Review the deformed shape (**PLDISP** command) using the following menu path:

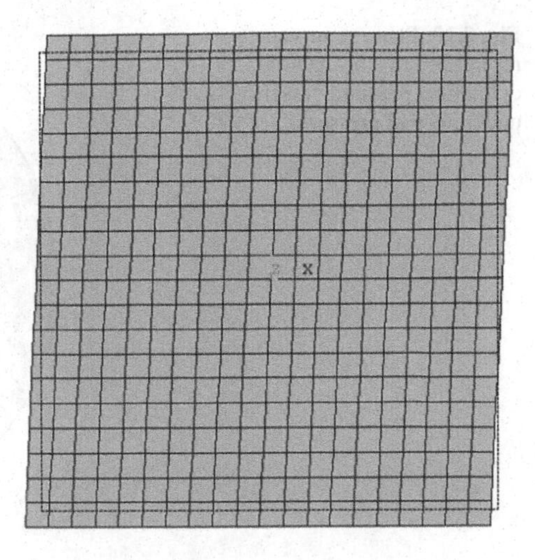

Fig. 8.19 Deformed shape of the composite plate

Main Menu > General PostProc > Plot Results > Deformed Shape

- Select ***Def + undef edge***; click on ***OK***.
- The deformed shape is shown in Fig. 8.19 as it appears in the *Graphics Window*.

• Review the x-displacement at the top-right and the y-displacement at the top-left nodes (**PRNSOL** command) using the following menu path:

Main Menu > General Postproc > List Results > Nodal Solution

- Click on ***Nodal Solution, DOF Solution***, and then ***Displacement vector sum***; click on ***OK***.
- The list appears. The x-displacement at the top-right node (node 22) is given as ***0.45E–2***, and the y-displacement at the top-left node (node 42) is given as ***0.45E–2***.
- In ANSYS, results can also be listed (or displayed) in different coordinate systems. By default, the *Results Coordinate System* is aligned with the Global Cartesian. Align the *Results Coordinate System* with local coordinate system 11 defined earlier using the following menu path:

Main Menu > General Postproc > Options for Outp

- *Options for Output* dialog box appears. Select ***Local system*** from the first pull-down menu and enter ***11*** for ***Local system reference no***; click on ***OK***.
- Now, review the nodal displacements one more time using the following menu path:

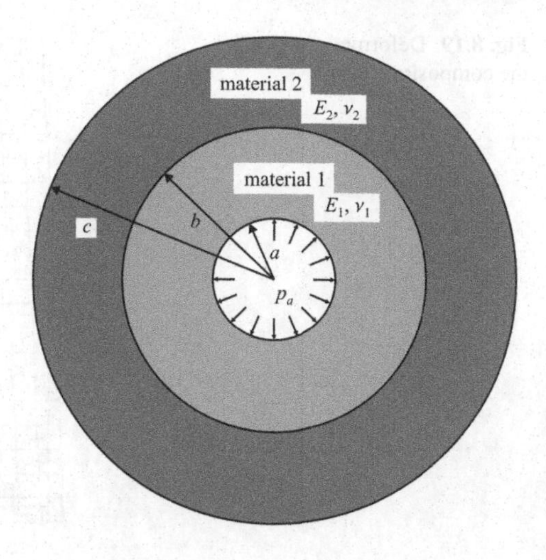

Fig. 8.20 Plane strain representation of a bi-material cylindrical pressure vessel under internal pressure

Main Menu > General Postproc > List Results > Nodal Solution

– When the results are transformed to the local coordinate system, the *x*-displacement at the top-right node becomes ***0.6364E–2***, and the *y*-displacement at the top-left node becomes ***0.16499E–2***. Corresponding analytical solution values are ***0.6364E–2*** and ***0.16495E–2***, producing negligible error values.

8.1.4.2 Plane Strain

In a structural problem, if one of the dimensions is significantly longer than the other dimensions defining a uniform cross-sectional area, and if the structure is subjected to only uniform lateral loads, then plane strain idealization is valid. Similar to plane stress idealization, because the number of nodes and elements in the model is reduced drastically, utilization of plane strain idealization leads to significant savings in computational cost without loss of accuracy in the quantities of interest. Stresses in a bi-material cylindrical pressure vessel are used to demonstrate plane strain idealization.

A bi-material cylinder is subjected to internal pressure, p_a, as shown in Fig. 8.20. The radius of the hollow portion is *a*, and the thicknesses of the inner and outer cylinders are $(b-a)$ and $(c-a)$, respectively. Perfect contact with no slipping is assumed along the interface, implying displacement continuity. Elastic properties of the inner and outer cylinders are (E_1, ν_1) and (E_2, ν_2), respectively. The goal is to compute the stress field. The problem is solved with ANSYS using $E_2/E_1 = 0.5$, $\nu_a = \nu_b = 0.33$, $b/a = 2$, and $c/a = 4$, with $a = 1$, $p_a = 1$, and $E_1 = 2$.

Model Generation

- Define the element type (**ET** command) using the following menu path:

Main Menu > Preprocessor > Element Type > Add/Edit/Delete

  - Click on ***Add***.
  - Select ***Solid*** immediately below ***Structural Mass*** in the left list and ***Quad 4 Node 182*** in the right list; click on ***OK***.
  - Click on ***Options***.
  - *PLANE182 element type options* dialog box appears; select ***Plane strain*** item from the pull-down menu corresponding to ***Element behavior K3***.
  - Click on ***OK***; click on ***Close***.

- Specify material properties (**MP** command) using the following menu path:

Main Menu > Preprocessor > Material Props > Material Models

  - The inner and outer cylinder will have material reference number 1 and 2, respectively. *Define Material Model Behavior* dialog box appears. In the right window, successively left-click on ***Structural***, ***Linear***, ***Elastic***, and, finally, ***Isotropic***, which brings up another dialog box.
  - Enter ***2*** for ***EX*** and ***0.33*** for ***PRXY***; click on ***OK***.
  - Add new material model using the following menu path:

Material > New Model

  - Click on ***OK*** in the new dialog box.
  - In the right window, successively left-click on ***Structural***, ***Linear***, ***Elastic***, and, finally, ***Isotropic***; Enter ***1*** for ***EX*** and ***0.33*** for ***PRXY***; click on ***OK***.
  - When finished, close the *Define Material Model Behavior* dialog box by using the following menu path:

Material > Exit

- Create partial hollow circular areas (**PCIRC** command) using the following menu path:

Main Menu > Preprocessor > Modeling > Create > Areas > Circle > By Dimensions

  - In the *Create Circle by Dimensions* dialog box, type **2** for ***Outer radius***, ***1*** for ***Inner radius***, ***0*** for ***Theta1***, and ***90*** for ***Theta2***; click on ***Apply***.
  - Now, type ***4*** for ***Outer radius***, ***2*** for ***Inner radius***, ***0*** for ***Theta1***, and ***90*** for ***Theta2***; click on ***OK***.

- Glue the areas (**AGLUE** command) using the following menu path:

Main Menu > Preprocessor > Modeling > Operate > Booleans > Glue > Areas

  - *Pick Menu* appears; click on ***Pick All*** button.

- Create the mesh. Since the problem involves two dissimilar materials, the inner circle (material 1) will be meshed first. Then the *default material attribute* will be changed to material 2 for the outer circle. Specify global element size (**ESIZE** command) using the following menu path:

Main Menu > Preprocessor > Meshing > Size Cntrls > ManualSize > Global > Size

- *Global Element Sizes* dialog box appears; enter ***0.1*** for ***SIZE***; click on ***OK***.
- Mesh the inner circle (**AMESH** command) using the following menu path:

Main Menu > Preprocessor > Meshing > Mesh > Areas > Mapped > 3 or 4 sided

- *Pick Menu* appears; pick the inner circle; click on ***OK***.
- Change the *default* material *attribute* to 2 (**MAT** command) using the following menu path:

Main Menu > Preprocessor > Meshing > Mesh Attributes > Default Attribs

- *Meshing Attributes* dialog box appears. Select *2* from the second pull-down menu; click on ***OK***.
- Mesh the outer circle (**AMESH** command) using the following menu path:

Main Menu > Preprocessor > Meshing > Mesh > Areas > Mapped > 3 or 4 sided

- *Pick Menu* appears; pick the outer circle; click on ***OK***.

Solution

- Apply displacement constraints (**D** command) using the following menu path:

Main Menu > Solution > Define Loads > Apply > Structural > Displacement > On Nodes

- *Pick Menu* appears; pick the nodes along $x=0$ (coincident with y-axis); click on ***OK*** in the *Pick Menu*.
- Highlight ***UX***; click on ***Apply***.
- *Pick Menu* reappears; pick the nodes along $y=0$ (coincident with x-axis); click on ***OK*** in the *Pick Menu*.
- Highlight ***UY*** and remove the highlight on ***UX***; click on ***OK***.

- Apply surface force (pressure) boundary conditions along the inner circular boundary. Since the boundary is circular, it is convenient to first switch to *Cylindrical Coordinates* and then select the nodes.

- Switch to *Cylindrical Coordinates* (**CSYS** command) using the following menu path:

Utility Menu > WorkPlane > Change Active CS to > Global Cylindrical

- Select nodes along the circular boundary (**NSEL** command) by using the following menu path:

Utility Menu > Select > Entities

- *Select Entities* dialog box appears; choose ***By Location*** in the second pull-down menu and type ***1*** in the ***Min, Max*** text field; click on ***OK***. Because the active coordinate system is cylindrical, any reference to the x-coordinate is treated as a reference to the r-coordinate by ANSYS.
- Now, apply pressure boundary conditions (**`SF`** command) by using the following menu path:

Main Menu > Solution > Define Loads > Apply > Structural > Pressure > On Nodes

- *Pick Menu* appears; click on ***Pick All***.
- Type ***1*** for ***VALUE Load PRES value***; click on ***OK***.
- Select everything (**`ALLSEL`** command) using the following menu path:

Utility Menu > Select > Everything

- Switch back to *Cartesian Coordinates* (**`CSYS`** command) using the following menu path:

Utility Menu > WorkPlane > Change Active CS to > Global Cartesian

- Obtain the solution (**`SOLVE`** command) using the following menu path:

Main Menu > Solution > Solve > Current LS

- *Confirmation Window* appears along with *Status Report Window*.
- Review status; if OK, close the *Status Report Window*; click on ***OK*** in the *Confirmation Window*.
- Wait until ANSYS responds with ***Solution is done!***

Postprocessing

- Obtain contour plots for σ_{xx} and σ_{yy} (**`PLNSOL`** command) using the following menu path:

Main Menu > General Postproc > Plot Results > Contour Plot > Nodal Solu

- Click on ***Stress*** and ***X-component of stress***; click on ***OK***.
- The contour plot appears in the *Graphics Window*, as shown in Fig. 8.21 (left).
- Repeat the same procedure for σ_{yy}, which produces the contour plot given in Fig. 8.21 (right)

- Since the problem possesses a circular geometry, it is often more useful to examine the stresses in cylindrical coordinates. For this purpose, change the results coordinate system to the global cylindrical system (**`RSYS`** command) using the following menu path:

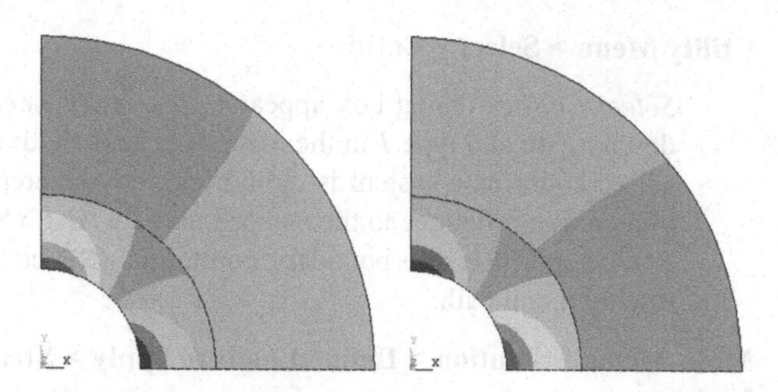

Fig. 8.21 Normal stresses in Cartesian coordinates: in x-direction, σ_{xx} (*left*), and in y-direction, σ_{yy} (*right*)

Main Menu > General Postproc > Options for Outp

- *Options for Output* dialog box appears. Select ***Global cylindric*** from the first pull-down; click on ***OK***.
- Now, obtain contour plots for σ_{rr} and $\sigma_{\theta\theta}$ (**PLNSOL** command) using the following menu path:

Main Menu > General Postproc > Plot Results > Contour Plot > Nodal Solu

- Select ***X-component of stress*** from the list; click on ***OK***.
- The contour plot appears in the *Graphics Window*, as shown in Fig. 8.22 (left).
- Repeat the same procedure for $\sigma_{\theta\theta}$, which produces the contour plot given in Fig. 8.22 (right).

• Review the variation of stresses along a path by means of a line plot. Define the path (**PPATH** command) using the following menu path:

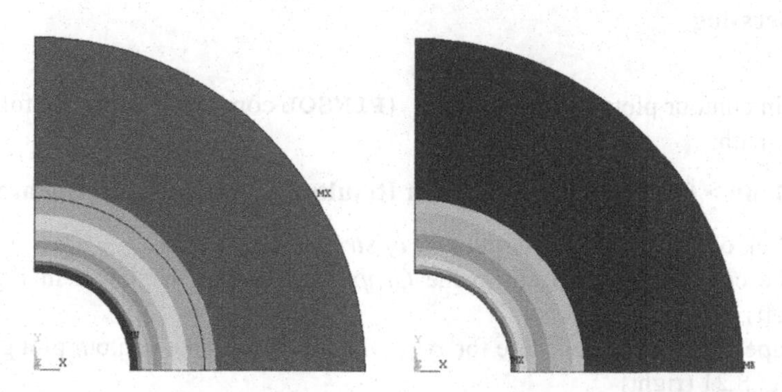

Fig. 8.22 Normal stresses in cylindrical coordinates: in r-direction, σ_{rr} (*left*), and in θ-direction, $\sigma_{\theta\theta}$ (*right*)

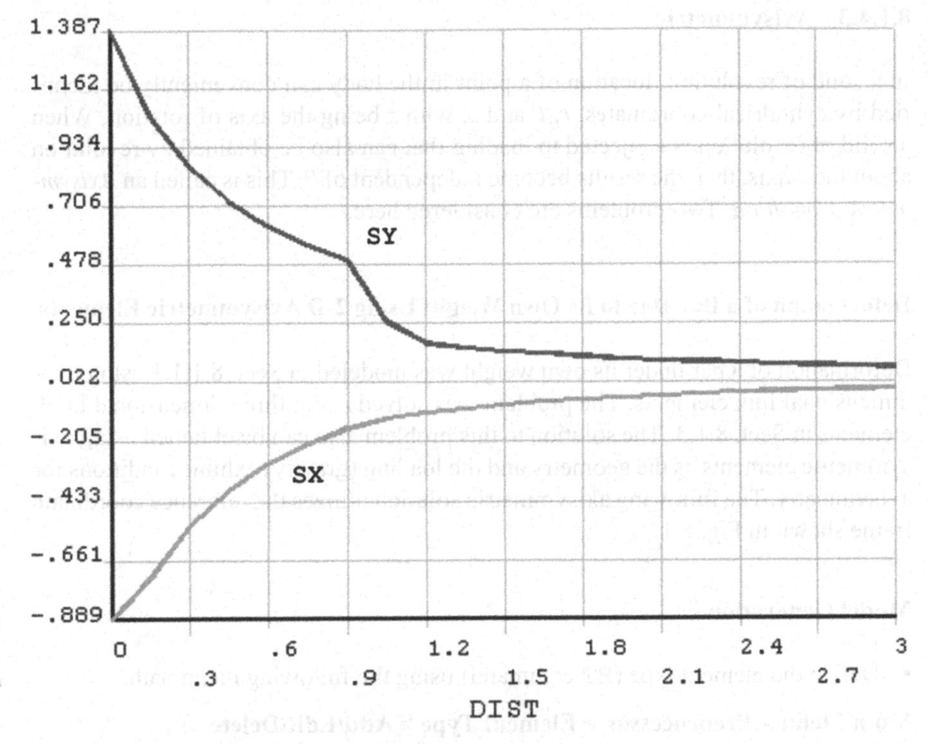

Fig. 8.23 Radial and hoop stresses along $y=0$

Main Menu > General Postproc > Path Operations > Define Path > By Nodes

- *Pick Menu* appears; pick the nodes with (x, y) coordinates (1, 0) and (4, 0); click on ***OK***. The path lies along the positive x-axis, starting from the boundary of the hole and ending at the left boundary of the structure.
- *By Nodes* dialog box appears; enter a name describing the path, say ***hrz***, in the ***Define Path Name*** text field; click on ***OK***.
- Close the *PATH Command Status Window*.
- Map results onto path (**PDEF** command) using the following menu path:

Main Menu > General Postproc > Path Operations > Map onto Path

- *Map Result Items onto Path* dialog box appears; select ***Stress*** from the left list and ***Y-direction SY*** from the right list; click on ***OK***.
- Obtain line plot of $\sigma_{\theta\theta}$ along the path (**PLPATH** command) using the following menu path:

Main Menu > General Postproc > Path Operations > Plot Path Item > On Graph

- *Plot of Path Items on Graph* dialog box appears; select ***SY***; click on ***OK***.
- Figure 8.23 shows the line plots of σ_{rr} and $\sigma_{\theta\theta}$ along the defined path.

8.1.4.3 Axisymmetric

In a solid of revolution, location of a point in the body can conveniently be identified by cylindrical coordinates, r, θ and z, with z being the axis of rotation. When a solid of revolution is subjected to loading that can also be obtained by revolution about the z-axis, then the results become independent of θ. This is called an *Axisymmetric Condition.* Two problems are considered here.

Deformation of a Bar Due to its Own Weight Using 2-D Axisymmetric Elements

Deformation of a bar under its own weight was modeled in Sect. 8.1.1.1 using two-dimensional link elements. The problem was solved using three-dimensional brick elements in Sect. 8.1.3. The solution to this problem also can be obtained using axisymmetric elements as the geometry and the loading (gravity) exhibit conditions for axisymmetry. The following axisymmetric solution utilizes the reference coordinate frame shown in Fig. 8.1.

Model Generation

- Define the element type (`ET` command) using the following menu path:

Main Menu > Preprocessor > Element Type > Add/Edit/Delete

- Click on ***Add***.
- Select ***Solid*** immediately below ***Structural Mass*** in the left list and ***Quad 4 Node 182*** in the right list; click on ***OK***.
- Click on ***Options***.
- *PLANE182 element type options* dialog box appears; select the ***Axisymmetric*** item from the pull-down menu corresponding to ***Element behavior K3***.
- Click on ***OK***; click on ***Close***.

- Specify material properties for the bar (`MP` command) using the following menu path:

Main Menu > Preprocessor > Material Props > Material Models

- In the *Define Material Model Behavior* dialog box, in the right window, successively left-click on ***Structural*** and ***Density***, which will bring up another dialog box.
- Enter ***0.2839605*** for ***DENS***; click on ***OK***.
- In order to specify the elastic modulus and Poisson's ratio, in the *Define Material Model Behavior* dialog box, in the right window, successively left-click on ***Structural***, ***Linear***, ***Elastic***, and, finally, ***Isotropic***, which will bring up another dialog box.
- Enter ***30e6*** for ***EX*** and ***0.3*** for ***PRXY***; click on ***OK***.

– Close the *Define Material Model Behavior* dialog box by using the following menu path:

Material > Exit

- Create the rectangle defining the axisymmetric cross section (**RECTNG** command) using the following menu path:

Main Menu > Preprocessor > Modeling > Create > Areas > Rectangle > By Dimensions

– In the *Create Rectangle by Dimensions* dialog box, enter ***0*** and ***2*** for ***X1*** and ***X2*** and ***0*** and ***−20*** for ***Y1*** and ***Y2***; click on ***OK***.

- Specify the global element size (**ESIZE** command) using the following menu path:

Main Menu > Preprocessor > Meshing > Size Cntrls > ManualSize > Global > Size

– *Global Element Sizes* dialog box appears; enter ***0.2*** for ***SIZE***; click on ***OK***.

- Create the mesh (**AMESH** command) using the following menu path:

Main Menu > Preprocessor > Meshing > Mesh > Areas > Free

– *Pick Menu* appears; click on ***Pick All***.

Solution

- Apply displacement constraints (**D** command) using the following menu path:

Main Menu > Solution > Define Loads > Apply > Structural > Displacement > On Nodes

– *Pick Menu* appears; pick the nodes at $y = 0$; click on ***OK*** in the *Pick Menu*.
– Highlight ***All DOF***; click on ***OK***.

- Apply gravitational acceleration (**ACEL** command) using the following menu path:

Main Menu > Solution > Define Loads > Apply > Structural > Inertia > Gravity > Global

– *Apply (Gravitational) Acceleration* dialog box appears.
– Enter ***386.2205*** for ***ACELY***; click on ***OK***.

- Obtain the solution (**SOLVE** command) using the following menu path:

Main Menu > Solution > Solve > Current LS

– *Confirmation Window* appears along with *Status Report Window*.
– Review status; if OK, close the *Status Report Window*; click on ***OK*** in the *Confirmation Window*.
– Wait until ANSYS responds with ***Solution is done!***

Postprocessing

- Review y-displacement contours (**PLNSOL** command) using the following menu path:

Main Menu > General PostProc > Plot Results > Contour Plot > Nodal Solu

  - Click on ***DOF Solution*** and ***Y-component of displacement***; click on ***OK***.
  - The contour plot is shown in Fig. 8.24 as it appears in the *Graphics Window*.

- Review displacement values (**PRNSOL** command) using the following menu path:

Main Menu > General Postproc > List Results > Nodal Solution

  - Click on ***DOF Solution*** and ***Y-component of displacement***; click on ***OK***.
  - The list appears in a separate window. It is a long list of z-displacements. At the bottom of the window, the maximum displacement value is printed as ***−0.72469E−03***.

Analysis of a Circular Plate Pushed Down by a Piston Head

An aluminum circular plate with a diameter of 40 in is pushed down by a steel piston head, as shown in Fig. 8.25. The piston head has two sections with diameters 20 and 2 in. The elastic modulus and Poisson's ratio for the aluminum plate are given as $E_{al} = 10 \times 10^6$ psi and $\nu_{al} = 0.35$, respectively, whereas the corresponding properties for steel are $E_{st} = 30 \times 10^6$ psi and $\nu_{st} = 0.3$. The aluminum plate is clamped along the boundary (all degrees of freedom constrained). The goal is to obtain the displacement and stress fields when the piston is pushed down (at the top) by an amount of 0.1 in. This problem possesses the conditions necessary for axisymmetry to be employed. Following is the solution utilizing axisymmetric elements in ANSYS.

Model Generation

- Define the element type (**ET** command) using the following menu path:

Main Menu > Preprocessor > Element Type > Add/Edit/Delete

  - Click on ***Add.***
  - Select ***Solid*** immediately below ***Structural Mass*** in the left list and ***Quad 4 Node 182*** in the right list; click on ***OK***.
  - Click on ***Options.***
  - *PLANE182 element type options* dialog box appears; select ***Axisymmetric*** item from the pull-down menu corresponding to ***Element behavior K3***.

Fig. 8.24 Contour plot of z-displacement of a bar elongated due to its own weight

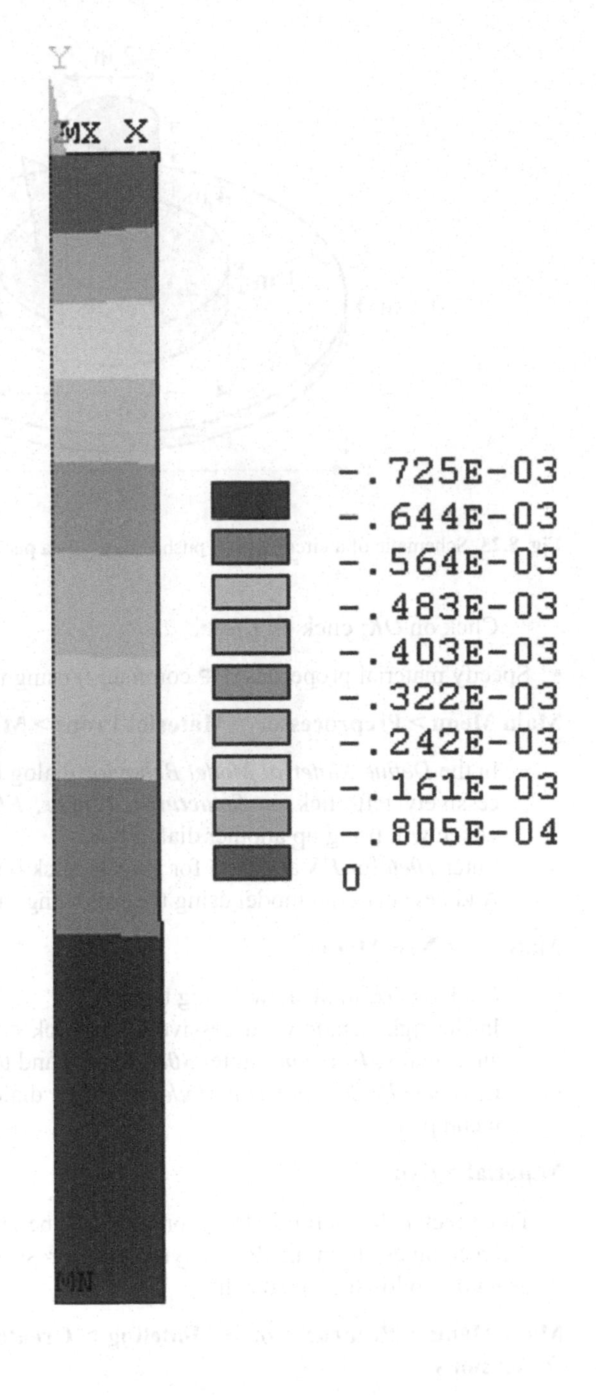

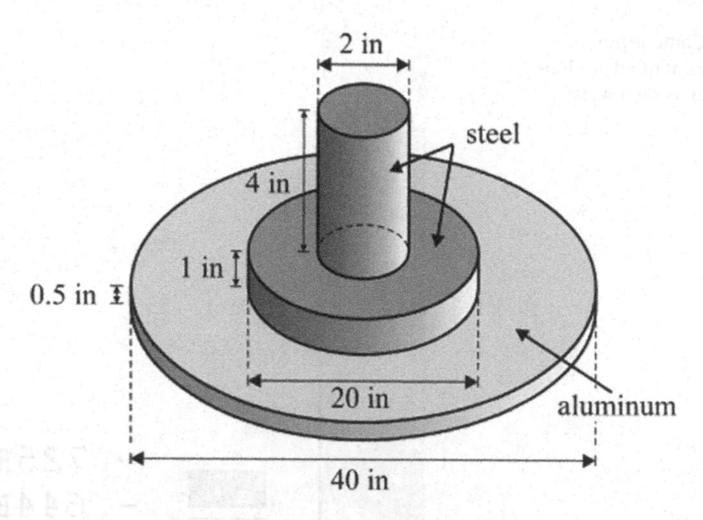

Fig. 8.25 Schematic of a circular plate pushed down by a piston head

- Click on ***OK***; click on ***Close***.

- Specify material properties (**`MP`** command) using the following menu path:

Main Menu > Preprocessor > Material Props > Material Models

- In the *Define Material Model Behavior* dialog box, in the right window, successively left-click on ***Structural***, ***Linear***, ***Elastic***, and, finally, ***Isotropic***, which will bring up another dialog box.
- Enter ***10e6*** for ***EX*** and ***0.35*** for ***PRXY***; click on ***OK***.
- Add new material model using the following menu path:

Material > New Model

- Click on ***OK*** in the new dialog box.
- In the right window, successively left-click on ***Structural***, ***Linear***, ***Elastic***, and, finally, ***Isotropic***; Enter ***30e6*** for ***EX*** and ***0.3*** for ***PRXY***; click on ***OK***.
- Close the *Define Material Model Behavior* dialog box by using the following menu path:

Material > Exit

- Three rectangles defining the geometry will be created and overlapped. Create the rectangles defining the axisymmetric cross section (**`RECTNG`** command) using the following menu path:

Main Menu > Preprocessor > Modeling > Create > Areas > Rectangle > By Dimensions

- In the *Create Rectangle by Dimensions* dialog box, enter ***0*** and ***20*** for ***X1*** and ***X2*** and ***0*** and ***0.5*** for ***Y1*** and ***Y2***; click on ***Apply***.

- Now, enter ***0*** and ***10*** for ***X1*** and ***X2*** and ***0*** and ***1.5*** for ***Y1*** and ***Y2***; click on ***Apply***.
- Finally, enter ***0*** and ***1*** for ***X1*** and ***X2*** and ***0*** and ***5.5*** for ***Y1*** and ***Y2***; click on ***OK***.

- Overlap the rectangles (**AOVLAP** command) using the following menu path:

Main Menu > Preprocessor > Modeling > Operate > Booleans > Overlap > Areas

- *Pick Menu* appears, click on ***Pick All***.
- The overlapping operation produces six areas (started with three), sharing lines along the interfaces.

- Specify the global element size (**ESIZE** command) using the following menu path:

Main Menu > Preprocessor > Meshing > Size Cntrls > ManualSize > Global > Size

- *Global Element Sizes* dialog box appears; enter ***0.2*** for ***SIZE***; click on ***OK***.

- Create the mesh for the aluminum plate (**AMESH** command) using the following menu path:

Main Menu > Preprocessor > Meshing > Mesh > Areas > Mapped > 3 or 4 sided

- *Pick Menu* appears; pick the bottom row of rectangles (corresponding to the aluminum plate); click on ***OK*** in the *Pick Menu*.
- Plot the areas (**APLOT** command) using the following menu path:

Utility Menu > Plot > Areas

- Change default element attribute for material number from 1 to 2 (**MAT** command) using the following menu path:

Main Menu > Preprocessor > Meshing > Mesh Attributes > Default Attribs

- *Meshing Attributes* dialog box appears; select ***2*** from the ***[MAT] Material number*** pull-down menu; click on ***OK***.
- Create mesh for the steel piston (**AMESH** command) using the following menu path:

Main Menu > Preprocessor > Meshing > Mesh > Areas > Mapped > 3 or 4 sided

- *Pick Menu* appears; pick the rectangles corresponding to the steel piston; click on ***OK*** in the *Pick Menu*.
- Plot elements with different colors based on their material numbers using the following menu path:

Utility Menu > PlotCtrls > Numbering

- *Plot Numbering Controls* dialog box appears. Select ***Material numbers*** from the first pull-down menu (corresponding to ***Elem/Attrib numbering***) and select ***Colors only*** from the second pull-down menu (corresponding to

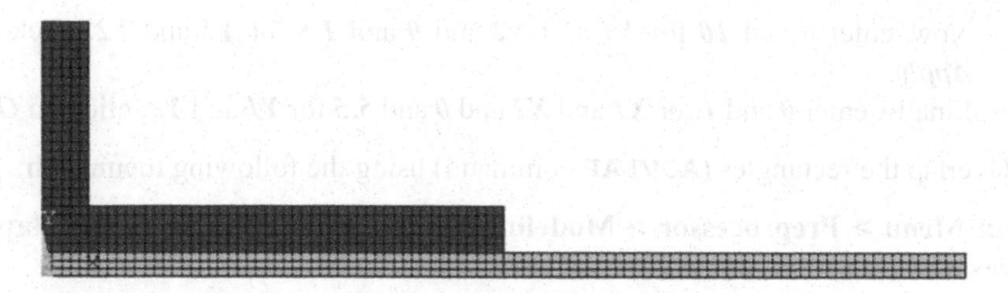

Fig. 8.26 Element plot with different colors based on material numbers

[/NUM] Numbering shown with); click on ***OK***. Figure 8.26 shows the corresponding element plot with different colors based on material numbers.

Solution

- Apply displacement constraints along the periphery of the aluminum plate (`D` command) using the following menu path:

Main Menu > Solution > Define Loads > Apply > Structural > Displacement > On Nodes

  - *Pick Menu* appears; pick the nodes along the right boundary ($x = 20$); click on ***OK*** in the *Pick Menu*.
  - Highlight ***All DOF***; click on ***OK***.

- Apply displacement constraints along the top surface of the steel piston (`D` command) using the following menu path:

Main Menu > Solution > Define Loads > Apply > Structural > Displacement > On Nodes

  - *Pick Menu* appears; pick the nodes along the top boundary ($y = 5.5$); click on ***OK*** in the *Pick Menu*.
  - *Remove* the highlight on ***All DOF*** and highlight ***UY***.
  - Enter ***−0.1*** in the text box for ***VALUE Displacement value***; click on ***OK***.

- Obtain the solution (`SOLVE` command) using the following menu path:

Main Menu > Solution > Solve > Current LS

  - *Confirmation Window* appears along with *Status Report Window*.
  - Review status, if OK, close the *Status Report Window*; click on ***OK*** in the *Confirmation Window*.
  - Wait until ANSYS responds with ***Solution is done!***

Postprocessing

- Review the deformed shape (`PLDISP` command) using the following menu path:

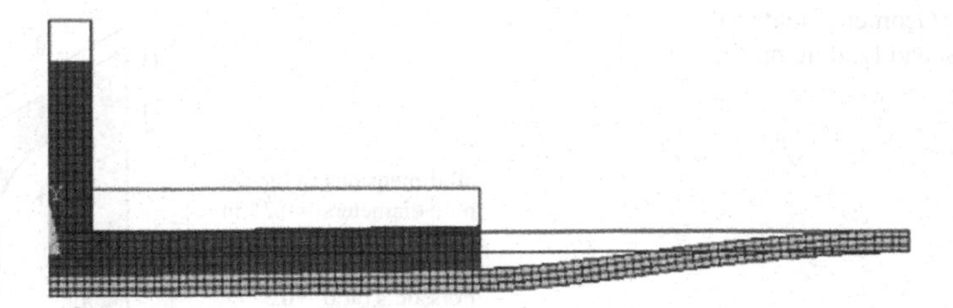

Fig. 8.27 Deformed shape with undeformed edge

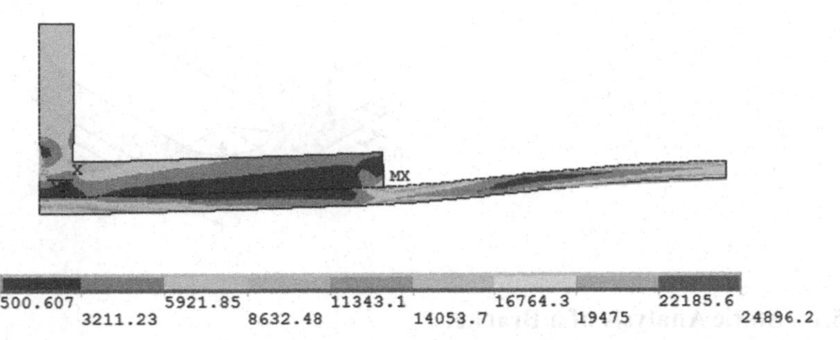

Fig. 8.28 Equivalent stress contours

Main Menu > General PostProc > Plot Results > Deformed Shape

- *Plot Deformed Shape* dialog box appears; select the ***Def + undef edge*** radio button; click on ***OK***.
- Corresponding deformed shape is shown in Fig. 8.27.

• Review the equivalent stress (von Mises) contour plot (**PLNSOL** command) using the following menu path:

Main Menu > General PostProc > Plot Results > Contour Plot > Nodal Solu

- *Contour Nodal Solution Data* dialog box appears. Click on ***Stress*** and scroll down to select ***von Mises stress.*** Click on ***OK***.
- Figure 8.28 shows the corresponding contour plot

8.1.5 Plates and Shells

Many engineering structures involve plates and shells where one dimension is much smaller than the other two. When these thin members are flat and only in-plane loads are applied, the problem can be solved using *Plane Stress* idealization. However, if they are curved and/or subjected to both in-plane and out-of-plane loads, it is necessary to solve the problem in 3-D using shell elements. At each node of the shell elements, both displacements and rotations are the degrees of freedom. Three problems are solved utilizing shell elements.

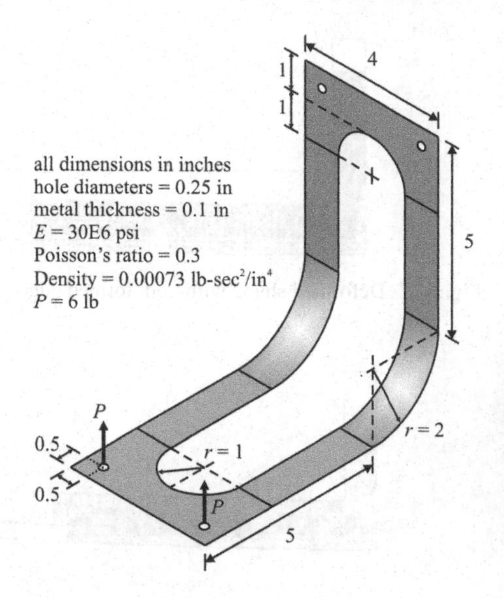

Fig. 8.29 Geometry, material properties, and loading on the bracket

8.1.5.1 Static Analysis of a Bracket

The bracket shown in Fig. 8.29 is clamped at the two top holes and is subjected to static vertical loading at the bottom two holes. Due to the symmetry in geometry, only one quarter of the structure is modeled at first. Once the top-left quarter is modeled and meshed, two symmetric reflection operations are utilized to create the rest of the bracket. The goal is to create the finite element model and obtain the static solution.

Model Generation

- Specify the ***jobname*** as ***bracket*** using the following menu path:

Utility Menu > File > Change Jobname

  - In the dialog box, type ***bracket*** in the ***[/FILNAM] Enter new jobname*** text field; click on the checkbox for ***New log and error files*** to show ***Yes***; click on ***OK***.

- Define the element type (**ET** command) using the following menu path:

Main Menu > Preprocessor > Element Type > Add/Edit/Delete

  - Click on ***Add***.
  - Select ***Shell*** immediately below ***Structural Mass*** in the left list and ***Elastic 4node 181*** in the right list; click on ***OK***.
  - Click on ***Close***.

- Specify material properties (**MP** command) using the following menu path:

Main Menu > Preprocessor > Material Props > Material Models

- In the *Define Material Model Behavior* dialog box, in the right window, successively left-click on ***Structural***, ***Linear***, ***Elastic***, and, finally, ***Isotropic***, which will bring up another dialog box.
- Enter ***30e6*** for ***EX*** and ***0.3*** for ***PRXY***; click on ***OK***.
- In the *Define Material Model Behavior* dialog box, in the right window, left-click on ***Density***, which will bring up another dialog box.
- Enter ***0.00073*** for ***DENS***; click on ***OK***.
- Close the *Define Material Model Behavior* dialog box by using the following menu path:

Material > Exit

- Specify the thickness for the shell (**SECTYPE** command) using the following menu path:

Main Menu > Preprocessor > Sections > Shell > Lay-up > Add/Edit

- *Create and Modify Shell Sections* dialog box appears; enter ***0.1*** for ***Thickness***.
- Exit from the *Create and Modify Shell Sections* dialog box by clicking on ***OK***.

- Create the solid model.

- Move *Working Plane origin* using the following menu path:

Utility Menu > WorkPlane > Offset WP by Increments

- *Offset WP* dialog box appears; type ***0, 3, −2*** in the ***X, Y, Z Offsets*** text field; click on ***OK***.
- Create a rectangular area using the following menu path:

Main Menu > Preprocessor > Modeling > Create > Areas > Rectangle > By Dimensions

- In the *Create Rectangle by Dimensions* dialog box, type ***−2*** for ***X1***, ***0*** for ***X2***, ***0*** for ***Y1***, and ***2*** for ***Y2***; click on ***OK***.
- Create a circular area using the following menu path:

Main Menu > Preprocessor > Modeling > Create > Areas > Circle > By Dimensions

- In the *Create Circle by Dimensions* dialog box, type ***1*** for ***Outer radius***, ***90*** for ***Theta1***, and ***180*** for ***Theta2***; click on ***OK***.
- Subtract the circle from the rectangle using the following menu path:

Main Menu > Preprocessor > Modeling > Operate > Booleans > Subtract > Areas

- *Pick Menu* appears; pick the rectangle; click on ***OK***; pick the circle; click on ***OK***.

- Move the *Working Plane* origin to the top-left hole center using the following menu path:

Utility Menu > WorkPlane > Offset WP by Increments

- *Offset WP* dialog box appears; type **−1.5, 1.5** in the ***X, Y, Z Offsets*** text field (because only *x*- and *y*-increments are entered, no move will be applied in *z*-direction); click on ***OK***.
- Create a circular area for the top-left hole using the following menu path:

Main Menu > Preprocessor > Modeling > Create > Areas > Circle > By Dimensions

- In the *Create Circle by Dimensions* dialog box, type ***0.25/2*** for ***Outer radius***, ***0*** for ***Theta1***, and ***360*** for ***Theta2***; click on ***OK***.
- Subtract the circle from the rest of the area using the following menu path:

Main Menu > Preprocessor > Modeling > Operate > Booleans > Subtract > Areas

- *Pick Menu* appears; pick the large area and click ***OK***; pick the circle; click on ***OK***.
- Move the *Working Plane* in order to create the additional rectangular area using the following menu path:

Utility Menu > WorkPlane > Offset WP by Increments

- *Offset WP* dialog box appears; type ***−0.5, 0.5*** in the ***X, Y, Z Offsets*** text field; click on ***OK***.
- Create additional rectangular area using the following menu path:

Main Menu > Preprocessor > Modeling > Create > Areas > Rectangle > By Dimensions

- In the *Create Rectangle by Dimensions* dialog box, type ***0*** for ***X1***, ***1*** for ***X2***, ***−2*** for ***Y1***, and ***−5*** for ***Y2***; click on ***OK***.
- In order to create a curved area, create keypoints that define the axis of rotation using the following menu path:

Main Menu > Preprocessor > Modeling > Create > Keypoints > In Active CS

- *Create Keypoints in Active Coordinate System* dialog box appears; type ***51*** for ***NPT Keypoint number*** and ***0*** in the ***X, Y, Z Location in active CS*** text fields; click on ***Apply***.
- In the same dialog box, type ***52*** for ***NPT Keypoint number*** and ***−0.5*** for *x* and ***0*** for *y* and *z* in the ***X, Y, Z Location in active CS*** text fields; click on ***OK***.
- Plot areas using the following menu path:

Fig. 8.30 Solid model of a quarter of the bracket

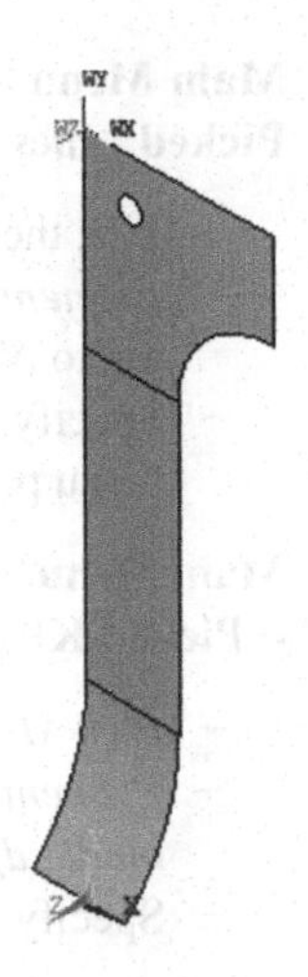

Utility Menu > Plot > Areas

– Create the curved area by sweeping the line at the bottom around an axis defined by the last two keypoints created using the following menu path:

Main Menu > Preprocessor > Modeling > Operate > Extrude > Lines > About Axis

– *Pick Menu* appears; the user is first asked to pick the line to be swept, and then to pick the keypoints defining the axis that the line to be swept about.
– Pick the horizontal line at the bottom; click on ***OK***; type ***51*** in the text field in the *Pick Menu* and hit ***Enter*** on the keyboard; type ***52*** followed by hitting ***Enter*** on the keyboard; click on ***OK***.
– ***Sweep Lines about Axis*** dialog box appears; type ***45*** for ***ARC Arc length in degrees***; click on ***OK***.
– Click on the ***Isometric View*** button.
– Figure 8.30 shows the result of this action.
– Although the areas created appear to be connected, ANSYS treats them as independent of each other (not connected). Therefore, the areas must be glued to each other. This is achieved by using the following menu path:

Main Menu > Preprocessor > Modeling > Operate > Booleans > Glue > Areas

– *Pick Menu* appears; click on ***Pick All***.

• Create the mesh.

– Specify the number of elements around the hole using the following menu path:

Main Menu > Preprocessor > Meshing > Size Cntrls > ManualSize > Lines > Picked Lines

- Pick the four circular segments defining the hole; click on ***OK***.
- *Element Sizes on Lines* dialog box appears; type ***2*** in the text field corresponding to ***NDIV*** (the second text field); uncheck the first checkbox; click on ***OK***.
- Specify mesh density in the vicinity of the top-left corner using the following menu path:

Main Menu > Preprocessor > Meshing > Size Cntrls > ManualSize > Keypoints > Picked KPs

- *Pick Menu* appears; pick the top-left keypoint; click on ***OK***.
- *Element Size at Picked Keypoints* dialog box appears; type ***0.3*** for ***SIZE Element edge length*** text field; click on ***OK***.
- Specify global mesh density using the following menu path:

Main Menu > Preprocessor > Meshing > Size Cntrls > ManualSize > Global > Size

- *Global Element Sizes* dialog box appears; type ***0.5*** for ***SIZE Element edge length*** text field; click on ***OK***.
- Create the mesh using the following menu path:

Main Menu > Preprocessor > Meshing > Mesh > Areas > Free

- In the *Pick Menu*, click on ***Pick All***.
- *A Warning Window* appears; click on ***OK***.

• Save the model using the following menu path:

Utility Menu > File > Save as Jobname.db

The model will be saved in the W*orking Directory* under the name ***bracket.db***.

• Create a reflective symmetric mesh using the following menu path:

Main Menu > Preprocessor > Modeling > Reflect > Areas

- *Pick Menu* appears; click on ***Pick All***.
- *Reflect Areas* dialog box appears; click on the ***Y-Z plane X*** radio-button; click on ***OK***.
- Although it is not apparent through visual inspection, there are duplicate entities (keypoints, lines and nodes) along the symmetry line, thus there is no continuity. Therefore, merge duplicate entities using the following menu path:

Main Menu > Preprocessor > Numbering Ctrls > Merge Items

- In the dialog box, select ***All*** from the first pull-down menu; click on ***OK***.

• Create a second reflective symmetric mesh.

- For this purpose, create a local coordinate system using the following menu path:

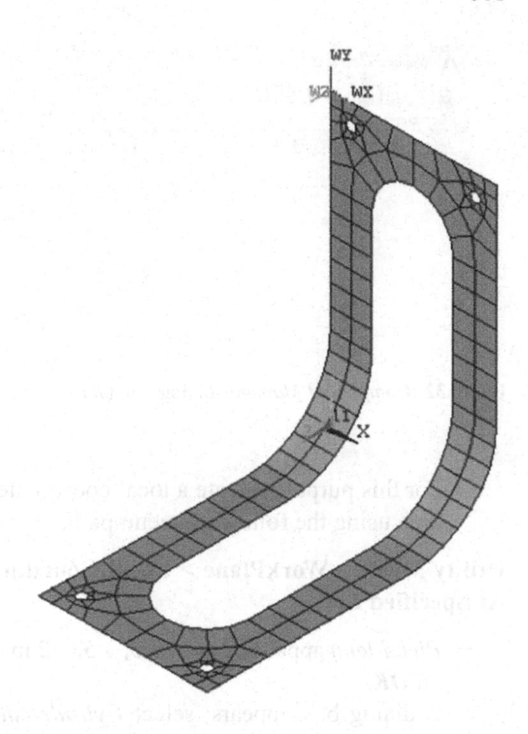

Fig. 8.31 Bracket after meshing and two reflection operations

Utility Menu > WorkPlane > Local Coordinate Systems > Create Local CS > At Specified Loc

- *Pick Menu* appears; type ***0, 0, 0*** in the text field in the *Pick Menu*; click on ***OK***.
- A dialog box appears; type ***−45*** in the ***THYZ Rotation about local X*** text field; click on ***OK***.
- Create a reflective symmetric mesh using the following menu path:

Main Menu > Preprocessor > Modeling > Reflect > Areas

- *Pick Menu* appears; click on ***Pick All***.
- *Reflect Areas* dialog box appears; click on the ***X-Z plane Y*** radio-button; click on ***OK***.
- Plot elements using the following menu path:

Utility Menu > Plot > Elements

- Figure 8.31 shows the ***isometric*** view of the mesh after the reflection.
- Merge duplicate entities using the following menu path:

Main Menu > Preprocessor > Numbering Ctrls > Merge Items

- In the dialog box, select ***All*** from the first pull-down menu; click on ***OK***.

- Define *components* for future use.

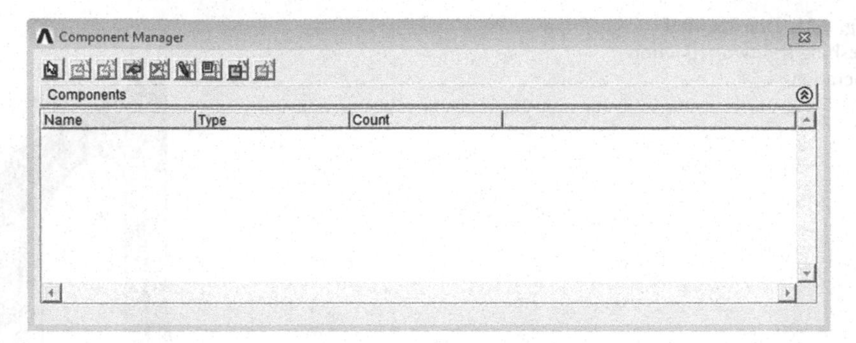

Fig. 8.32 *Component Manager* dialog box (*left-most button* is used for creating components)

- For this purpose, create a local coordinate system at the center of the top-left hole using the following menu path:

Utility Menu > WorkPlane > Local Coordinate Systems > Create Local CS > At Specified Loc

- *Pick Menu* appears; type ***−1.5, 4.5, −2*** in the text field in the *Pick Menu*; click on ***OK***.
- A dialog box appears; select ***Cylindrical 1*** in the ***KCS Type of coordinate system*** pull-down menu.
- ***Delete −45*** in the ***THYZ Rotation about local X*** text field; click on ***OK***.
- Select nodes along the top-left hole by using the following menu path:

Utility Menu > Select > Entities

- *Select Entities* dialog box appears; choose ***By Location*** in the second pull-down menu and type ***0.25/2*** in the ***Min, Max*** text field; click on ***OK***. Because the active coordinate system is cylindrical, any reference to the x-coordinate will be treated as a reference to the r-coordinate by ANSYS.
- Create the component by using the following menu path:

Utility Menu > Select > Component Manager

- *Component Manager* dialog box appears (Fig. 8.32); click on the first button on the left (***Create Component*** button).
- *Create Component* dialog box appears; click on the ***Nodes*** radio-button and name the component by typing ***TL_BOLT*** (stands for top-left bolt) in the text field (Fig. 8.33); click on ***OK***.
- Close the ***Component Manager***.
- Create components for top-right, bottom-left, and bottom-right bolts in the same manner. The origin of the local ***cylindrical*** coordinates for each of these are given as

  TR_BOLT: ***1.5, 4.5, −2***
  BL_BOLT: ***−1.5, −2, 4.5*** and use ***−90*** for the ***THYZ Rotation about local X***

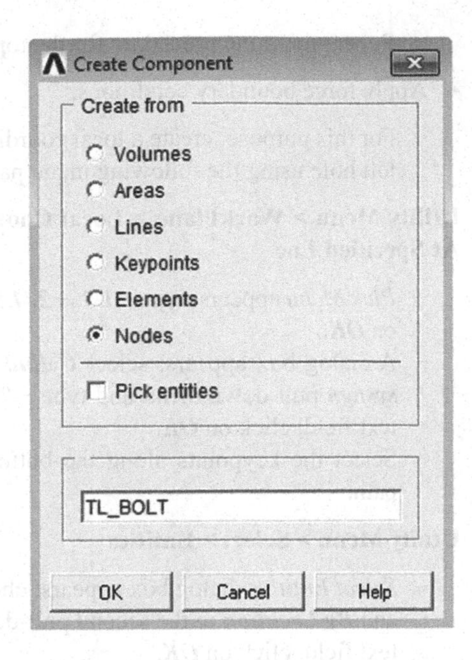

Fig. 8.33 Dialog box for creating components

BR_BOLT: ***1.5, −2, 4.5*** and use ***−90*** for the ***THYZ Rotation about local X***

- Save the model using the following menu path:

Utility Menu > File > Save as Jobname.db

Solution

- Constrain displacement and rotation degrees of freedom along the top-left and -right holes. For this purpose, first select the components created earlier for these holes (***TL_BOLT*** and ***TR_BOLT***) using the following menu path:

Utility Menu > Select > Comp/Assembly > Select Comp/Assembly

  - A dialog box appears; click on the ***by component name*** radio-button; click on ***OK***.
  - A new dialog box with the components listed appears; highlight ***TL_BOLT***; click on ***OK***. This action selects the nodes along the top-left hole.
  - Specify the displacement boundary conditions using the following menu path:

Main Menu > Solution > Define Loads > Apply > Structural > Displacement > On Nodes

  - *Pick Menu* appears; click on ***Pick All***.
  - In the new dialog box, highlight ***All DOF***; click on ***OK***.

- Repeat the same procedure for the top-right hole (***TR_BOLT***).

- Apply force boundary conditions.
  - For this purpose, create a local coordinate system at the center of the bottom-left hole using the following menu path:

Utility Menu > WorkPlane > Local Coordinate Systems > Create Local CS > At Specified Loc

- *Pick Menu* appears; type ***−1.5, −2, 4.5*** in the text field in the *Pick Menu*; click on ***OK***.
- A dialog box appears; select ***Cylindrical 1*** in the ***KCS Type of coordinate system*** pull-down menu and type ***−90*** in the ***THYZ Rotation about local X*** text field; click on ***OK***.
- Select the keypoints along the bottom-left hole using the following menu path:

Utility Menu > Select > Entities

- *Select Entities* dialog box appears; choose ***Keypoints*** in first pull-down menu and ***By Location*** in the second pull-down menu; type ***0.25/2*** in the ***Min, Max*** text field; click on ***OK***.
- Apply forces on the keypoints using the menu path:

Main Menu > Solution > Define Loads > Apply > Structural > Force/Moment > On Keypoints

- *Pick Menu* appears; click on ***Pick All***.
- In the new dialog box, select ***FY*** from pull-down menu and enter ***6/4*** for ***Force/moment value***; click on ***OK***.
- A *Warning Window* appears, informing the user that boundary conditions applied to solid modeling entities overwrite those that may have already been applied to finite element entities (nodes and elements) directly.
- Close the *Warning Window*.
- Repeat the same procedure for the bottom-right hole (use ***1.5, −2, 4.5*** for the local coordinate system origin and ***−90*** for the ***THYZ Rotation about local X***).

- Select everything using the following menu path:

Utility Menu > Select > Everything

- Save the model using the following menu path:

Utility Menu > File > Save as Jobname.db

- Obtain the solution using the following menu path:

Main Menu > Solution > Solve > Current LS

- *Confirmation Window* appears along with *Status Report Window*.

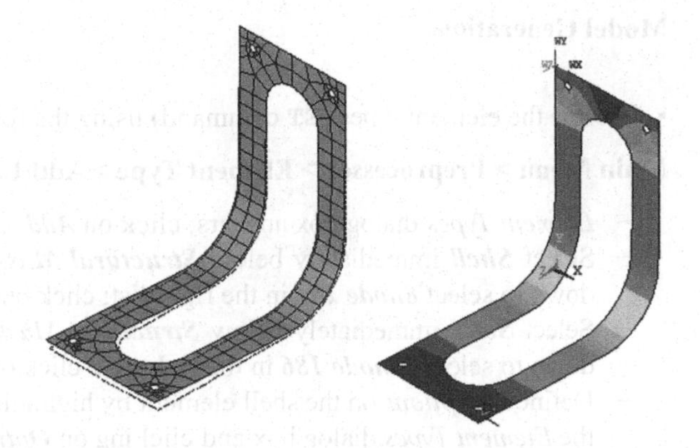

Fig. 8.34 Deformed shape (*left*), and contour plot of equivalent (von Mises) stresses (*right*)

- Review status; if OK, close the *Status Report Window*; click on ***OK*** in the *Confirmation Window.*
- If a dialog box with a message reminding the user about previous warnings issued appears, click on ***Yes***.
- Wait until ANSYS responds with ***Solution is done!***

Postprocessing

- Review the deformed shape using the following menu path:

Main Menu > General PostProc > Plot Results > Deformed Shape

- Select ***Def + undef edge***; click on ***OK***.
- The deformed shape is shown in Fig. 8.34 (left) as it appears in the *Graphics Window*.

- Review the stress contours using the following menu path:

Main Menu > General PostProc > Plot Results > Contour Plot > Nodal Solu

- Click on ***Stress***. Scroll down and click on ***von Mises stress***; click on ***OK***.

The equivalent stress contour plot is shown in Fig. 8.34 (right) as it appears in the *Graphics Window*.

8.1.5.2 Analysis of a Circular Plate Pushed Down by a Piston Head

The circular aluminum plate pushed down by a steel piston was analyzed in Sect. 8.1.4.3.2 by employing only axisymmetric elements. The geometry of the problem is shown in Fig. 8.25. The same problem is solved in this section by using a combination of shell and 3-D solid elements.

Model Generation

- Define the element types (`ET` command) using the following menu path:

Main Menu > Preprocessor > Element Type > Add/Edit/Delete

  - *Element Types* dialog box appears; click on ***Add***.
  - Select ***Shell*** immediately below ***Structural Mass*** in the left list and scroll down to select ***8node 281*** in the right list; click on ***Apply***.
  - Select ***Solid*** immediately below ***Structural Mass*** in the left list and scroll down to select ***20node 186*** in the right list; click on ***OK***.
  - Define *keyoptions* on the shell element by highlighting ***Type 1 SHELL281*** in the *Element Types* dialog box and clicking on ***Options***.
  - *SHELL281 element type options* dialog box appears; select ***All layers*** from the ***Storage of layer data K8*** pull-down menu.
  - Click on ***Close*** to exit from the *Element Types* dialog box.

- Specify material properties (`MP` command) using the following menu path:

Main Menu > Preprocessor > Material Props > Material Models

  - In the *Define Material Model Behavior* dialog box, in the right window, successively left-click on ***Structural***, ***Linear***, ***Elastic***, and, finally, ***Isotropic***, which brings up another dialog box.
  - Enter ***10e6*** for ***EX*** and ***0.35*** for ***PRXY***; click on ***OK***.
  - Add new material model using the following menu path:

Material > New Model

  - Click on ***OK***.
  - In the right window, successively left-click on ***Structural***, ***Linear***, ***Elastic***, and, finally, ***Isotropic***; Enter ***30e6*** for ***EX*** and ***0.3*** for ***PRXY***; click on ***OK***.
  - Close the *Define Material Model Behavior* dialog box by using the following menu path:

Material > Exit

- Specify the thickness for the shell (`SECTYPE` command) using the following menu path:

Main Menu > Preprocessor > Sections > Shell > Lay-up > Add/Edit

  - *Create and Modify Shell Sections* dialog box appears; enter ***0.5*** for ***Thickness.***
  - Select ***User-Input-Location*** from the ***Section Offset*** pull-down menu. Enter ***0.25*** in the text box for ***User Defined Value***.
  - Exit from the *Create and Modify Shell Sections* dialog box by clicking on ***OK***.

- Create quarter circular areas (`PCIRC` command) using the following menu path:

Main Menu > Preprocessor > Modeling > Create > Areas > Circle > By Dimensions

- In the *Create Circle by Dimensions* dialog box, type ***1*** for ***Outer radius***, ***0*** for ***Theta1***, and ***90*** for ***Theta2***; click on ***Apply***.
- Modify ***Outer radius*** to be ***10***; click on ***Apply***.
- Finally, modify ***Outer radius*** (one more time) to be ***20***; click on ***OK***.

- Overlap the areas (**AOVLAP** command) using the following menu path:

Main Menu > Preprocessor > Modeling > Operate > Booleans > Overlap > Areas

- *Pick Menu* appears; click on ***Pick All***.

- Create quarter cylindrical volumes (**CYLIND** command) using the following menu path:

Main Menu > Preprocessor > Modeling > Create > Volumes > Cylinder > By Dimensions

- In the Create *Cylinder by Dimensions* dialog box, type ***10*** for ***Outer radius***, ***0*** for ***Z1***, ***1*** for ***Z2***, ***0*** for ***Theta1***, and ***90*** for ***Theta2***; click on ***Apply***.
- Modify ***Outer radius*** to be ***1*** and ***Z2*** to be ***5***; click on ***OK***.

- Overlap the volumes (**VOVLAP** command) using the following menu path:

Main Menu > Preprocessor > Modeling > Operate > Booleans > Overlap > Volumes

- *Pick Menu* appears; click on ***Pick All***.

- Specify size controls for meshing. First, the number of divisions on specific lines will be specified, followed by specification of the global element size.
  - Select lines at $0.25 \le x \le 0.75$ and $0.25 \le y \le 0.75$ (**LSEL** command) using the following menu path:

Utility Menu > Select > Entities

- *Select Entities* dialog box appears; choose ***Lines*** from the first pull-down menu and ***By Location*** in the second pull-down menu. Type ***0.25,0.75*** in the ***Min, Max*** text field; click on ***Apply***.
- Now, select ***Y-coordinates*** and the ***Also Select*** radio-buttons without changing the text (***0.25, 0.75***) in the ***Min, Max*** text field; click on ***OK***.
- A total of 12 lines are selected. Specify the number of element divisions along the selected lines (**LESIZE** command) using the following menu path:

Main Menu > Preprocessor > Meshing > Size Cntrls > ManualSize > Lines > Picked Lines

- *Pick Menu* appears; click on ***Pick All***.

- *Element Sizes on Picked Lines* dialog box appears; enter ***4*** for ***NDIV***. Remove the checkmark next to ***KYNDIV SIZE, NDIV can be changed*** so that it shows ***No***; click on ***OK***.
- Specify global element size (**ESIZE** command) using the following menu path:

Main Menu > Preprocessor > Meshing > Size Cntrls > ManualSize > Global > Size

- *Global Element Sizes* dialog box appears; enter ***0.5*** for ***SIZE***; click on ***OK***.
- Select everything (**ALLSEL** command) using the following menu path:

Utility Menu > Select > Everything

- Generate the mesh in two stages: (i) the mesh associated with the aluminum plate is created using **SHELL281** elements followed by (ii) the generation of the mesh for the steel piston head using **SOLID186** elements.
  - Select areas attached to the volumes (**ASLV** command) using the following menu path:

Utility Menu > Select > Entities

- *Select Entities* dialog box appears; choose ***Areas*** from the first pull-down menu and ***Attached to*** in the second pull-down menu. Select ***Volumes*** and ***From Full*** radio-buttons; click on ***Apply***.
- Now, click on ***Invert*** button. This inverts the selection, i.e., selected areas are unselected and vice versa. At this point, areas associated with the aluminum plate are selected.
- Create the mesh for the aluminum plate (**AMESH** command) using the following menu path:

Main Menu > Preprocessor > Meshing > Mesh > Areas > Mapped > 3 or 4 sided

- *Pick Menu* appears; click on ***Pick All***.
- Select everything (**ALLSEL** command) using the following menu path:

Utility Menu > Select > Everything

- Change the *default element type attribute* to 2 (**TYPE** command) and *default material attribute* to 2 (**MAT** command) using the following menu path:

Main Menu > Preprocessor > Meshing > Mesh Attributes > Default Attribs

- *Meshing Attributes* dialog box appears. Select ***2 SOLID186*** from the first pull-down menu and select ***2*** from the second pull-down menu; click on ***OK***.
- Create the mesh for the steel piston head (**VMESH** command) using the following menu path:

Main Menu > Preprocessor > Meshing > Mesh > Volumes > Mapped > 4 to 6 sided

- *Pick Menu* appears; click on ***Pick All***.
- Obtain an isometric view of the mesh using the following menu path:

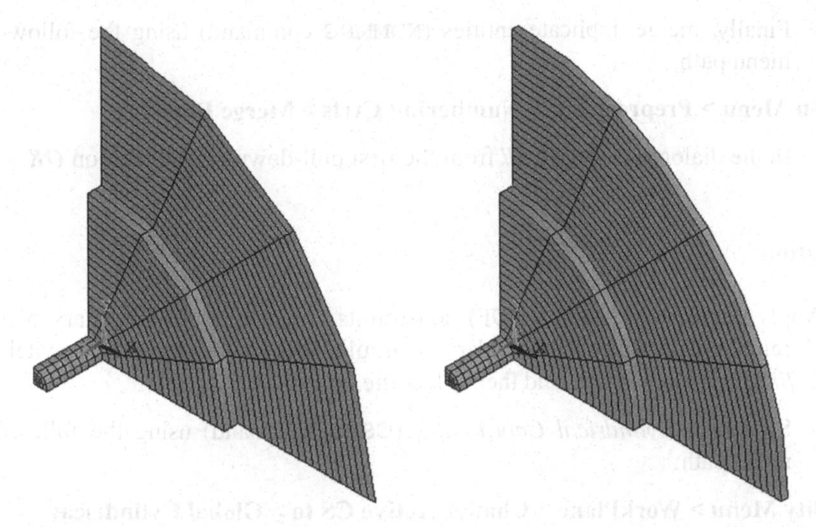

Fig. 8.35 Elements plotted without (*left*) and with (*right*) thickness information from real constants and curved surfaces

Utility Menu > PlotCtrls > Pan Zoom Rotate

- *Pan Zoom Rotate* window appears; click on ***Iso*** button.
- Plot elements with different colors based on their material numbers using the following menu path:

Utility Menu > PlotCtrls > Numbering

- *Plot Numbering Controls* dialog box appears. Select ***Material numbers*** from the first pull-down menu (corresponding to ***Elem/Attrib numbering***) and select ***Colors only*** from the second pull-down menu (corresponding to ***[/NUM] Numbering shown with***); click on ***OK***. Figure 8.35 (left) shows the corresponding element plot with different colors based on material numbers.
- Note in the element plot that the aluminum plate elements do not have a thickness. This is because the thickness information is stored in *real constants* and the **`SHELL281`** elements are plane elements. However, for visualization purposes, it is possible to plot plane elements with their thickness (**`/ESHAPE`** command) using the following menu path:

Utility Menu > PlotCtrls > Style > Size and Shape

- *Size and Shape* dialog box appears. Place a checkmark next *to **[/ESHAPE] Display of element shapes based on real constant descriptions*** so that it shows ***On***. In the same dialog box, select ***2 facets/edge** for **[/EFACET] Facets/element edge*** so that the elements with curved edges and surfaces are shown correctly (**`/EFACET`** command). Click on ***OK***.
- Figure 8.35 (right) shows the corresponding element plot with elements having curved edges/surfaces and thickness.

- Finally, merge duplicate entities (**NUMMRG** command) using the following menu path:

Main Menu > Preprocessor > Numbering Ctrls > Merge Items

- In the dialog box, select ***All*** from the first pull-down menu; click on ***OK***.

Solution

- Apply degree of freedom (DOF) constraints along the outer boundary of the aluminum plate. Since the boundary is circular, it is convenient to first switch to *Cylindrical Coordinates* and then select the nodes.
  - Switch to *Cylindrical Coordinates* (**CSYS** command) using the following menu path:

Utility Menu > WorkPlane > Change Active CS to > Global Cylindrical

- Select nodes along the circular boundary (**NSEL** command) by using the following menu path:

Utility Menu > Select > Entities

- *Select Entities* dialog box appears; choose ***Nodes*** in the first pull-down menu and ***By Location*** in the second pull-down menu. Click on ***X coordinate*** and ***From Full*** radio-buttons; type ***20*** in the ***Min, Max*** text field; click on ***OK***. Because the active coordinate system is cylindrical, any reference to the *x*-coordinate is treated as a reference to the *r*-coordinate by ANSYS.
- Now, apply DOF constraints (**D** command) by using the following menu path:

Main Menu > Solution > Define Loads > Apply > Structural > Displacement > On Nodes

- *Pick Menu* appears; click on ***Pick All***.
- In the new dialog box, highlight ***All DOF***; click on ***OK***.
- Select everything (**ALLSEL** command) using the following menu path:

Utility Menu > Select > Everything

- Switch back to *Cartesian Coordinates* (**CSYS** command) using the following menu path:

Utility Menu > WorkPlane > Change Active CS to > Global Cartesian

- Apply degree of freedom (DOF) constraints along the top surface of the steel piston.
  - Select nodes (**NSEL** command) by using the following menu path:

Utility Menu > Select > Entities

- *Select Entities* dialog box appears; choose ***By Location*** in the second pull-down menu; click on the ***Z coordinate*** and ***From Full*** radio-buttons; type ***5*** in the ***Min, Max*** text field; click on ***OK***.
- Apply DOF constraints (**D** command) by using the following menu path:

Main Menu > Solution > Define Loads > Apply > Structural > Displacement > On Nodes

- *Pick Menu* appears; click on ***Pick All***.
- In the new dialog box, remove the highlight on ***All DOF*** and highlight ***UZ***. Enter − ***0.1*** in the text box for ***VALUE Displacement value***; click on ***OK***.

• Apply symmetry conditions along the $x = 0$ and $y = 0$ planes for the entire structure.

- Select nodes (**NSEL** command) by using the following menu path:

Utility Menu > Select > Entities

- *Select Entities* dialog box appears; choose ***Nodes*** in the first pull-down menu and ***By Location*** in the second pull-down menu. Click on the ***X coordinate*** and ***From Full*** radio-buttons; type ***0*** in the ***Min, Max*** text field; click on ***OK***.
- Apply DOF constraints (**D** command) by using the following menu path:

Main Menu > Solution > Define Loads > Apply > Structural > Displacement > On Nodes

- *Pick Menu* appears; click on ***Pick All***.
- In the new dialog box, remove the highlight on ***UZ*** and highlight ***UX***. Enter ***0*** in the text box for ***VALUE Displacement value***; click on ***OK***.
- Select nodes (**NSEL** command) by using the following menu path:

Utility Menu > Select > Entities

- *Select Entities* dialog box appears; choose ***Nodes*** in the first pull-down menu and ***By Location*** in the second pull-down menu. Click on the ***Y coordinate*** and ***From Full*** radio-buttons; type ***0*** in the ***Min, Max*** text field; click on ***OK***.
- Apply DOF constraints (**D** command) by using the following menu path:

Main Menu > Solution > Define Loads > Apply > Structural > Displacement > On Nodes

- *Pick Menu* appears; click on ***Pick All***.
- In the new dialog box, remove the highlight on ***UX*** and highlight ***UY***; click on ***OK***.
- Select everything (**ALLSEL** command) using the following menu path:

Utility Menu > Select > Everything

• Obtain the solution using the following menu path:

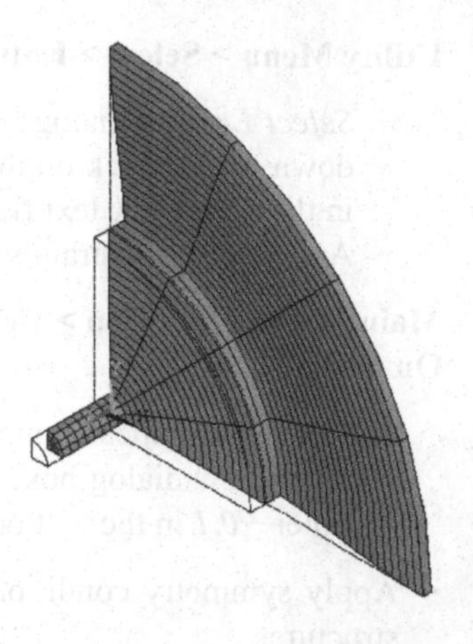

Fig. 8.36 Isometric view of the deformed shape

Main Menu > Solution > Solve > Current LS

- *Confirmation Window* appears along with *Status Report Window*.
- Review status; if OK, close the *Status Report Window*; click on ***OK*** in the *Confirmation Window*.
- Wait until ANSYS responds with ***Solution is done!***

Postprocessing

- Review the deformed shape using the following menu path:

Main Menu > General PostProc > Plot Results > Deformed Shape

- Select ***Def + undef edge***; click on ***OK***.
- The isometric view of the deformed shape is shown in Fig. 8.36 as it appears in the *Graphics Window*.

- Review the equivalent stress (von Mises) contour plot (`PLNSOL` command) using the following menu path:

Main Menu > General PostProc > Plot Results > Contour Plot > Nodal Solu

- *Contour Nodal Solution Data* dialog box appears. Click on ***Stress***. Scroll down and select ***von Mises stress***; click on ***OK***.
- Figure 8.37 shows the corresponding contour plot.

8.1.5.3 Analysis of an Axisymmetric Shell with Internal Pressure

Consider the pressure vessel shown in Fig. 8.38 with elastic properties $E = 10 \times 10^6$ psi and $\nu = 0.3$. Its radius changes, as shown in Fig. 8.38, while the

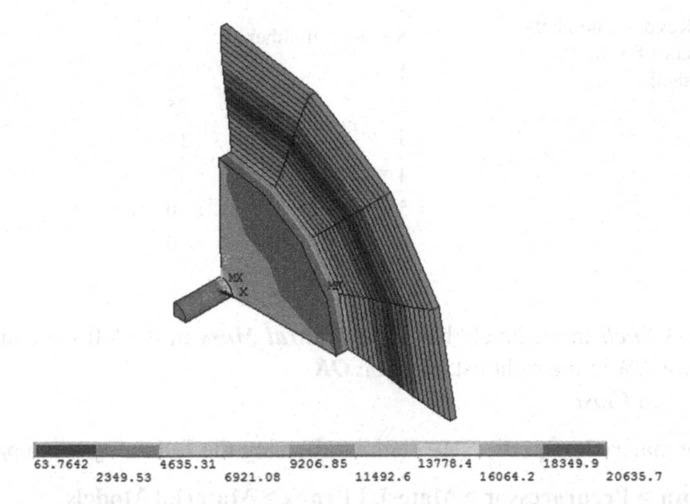

Fig. 8.37 Equivalent stress contours

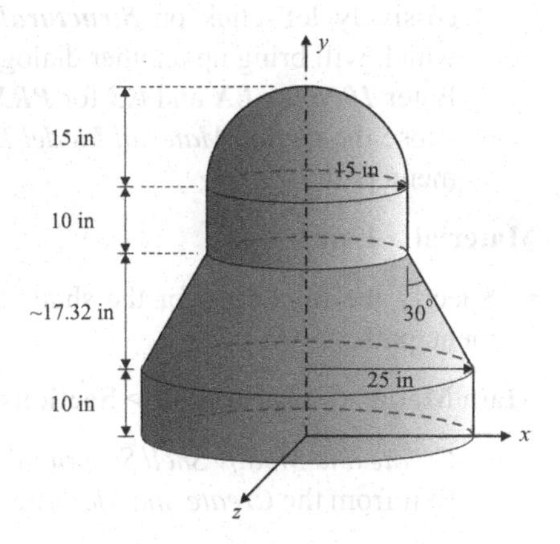

Fig. 8.38 Schematic of the axisymmetric shell with internal pressure

thickness remains constant, $t = 0.25$ in. The internal pressure is 300 psi. The goal is to find the meridional and hoop stresses in the shell. Examination of the geometry and loading reveals that the problem is axisymmetric. Therefore, axisymmetric shell elements within ANSYS are utilized in this section.

Model Generation

- Define the element type (**ET** command) using the following menu path:

Main Menu > Preprocessor > Element Type > Add/Edit/Delete

– Click on ***Add***.

Table 8.1 Keypoint numbers and coordinates for the axi-symmetric shell

Keypoint number	*x*	*y*
1	25	0
2	25	10
3	15	27.32
4	15	37.32
5	0	52.32
6	0	37.32

- Select ***Shell*** immediately below ***Structural Mass*** in the left list and ***Axisym 2node 208*** in the right list; click on ***OK***.
- Click on ***Close***.

• Specify material properties (**MP** command) using the following menu path:

Main Menu > Preprocessor > Material Props > Material Models

- In the *Define Material Model Behavior* dialog box, in the right window, successively left-click on ***Structural***, ***Linear***, ***Elastic***, and, finally, ***Isotropic***, which will bring up another dialog box.
- Enter ***10e6*** for ***EX*** and ***0.3*** for ***PRXY***; click on ***OK***.
- Close the *Define Material Model Behavior* dialog box by using the following menu path:

Material > Exit

• Specify the thickness for the shell (**SECTYPE** command) using the following menu path:

Main Menu > Preprocessor > Sections > Shell > Lay-up > Add/Edit

- *Create and Modify Shell Sections* dialog box appears; enter ***0.25*** for ***Thickness.***
- Exit from the *Create and Modify Shell Sections* dialog box by clicking on ***OK***.

• Create keypoints (**K** command) using the following menu path:

Main Menu > Preprocessor > Modeling > Create > Keypoints > In Active CS

- In the *Create Keypoints in Active Coordinate System* dialog box, type, ***25*** for ***X*** and ***0*** for ***Y***; click on ***Apply*** (keypoint 1 is created).
- Referring to the schematic of keypoints shown in Fig. 8.39 and tabulated in Table 8.1, repeat this procedure for keypoints 2 through 6.

• Create straight lines (**L** command) using the following menu path:

Main Menu > Preprocessor > Modeling > Create > Lines > Lines > Straight Line

- *Pick Menu* appears; pick keypoints 1 and 2; line 1 is created.

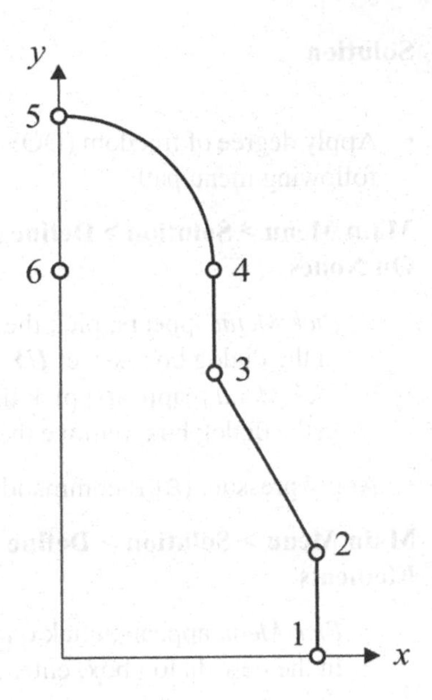

Fig. 8.39 Schematic of the keypoints used in the solid model

- Repeat this for lines 2 and 3 using keypoint pairs 2–3 and 3–4, respectively; click on ***OK*** in the *Pick Menu.*

- Create an arc (**LARC** command) using the following menu path:

Main Menu > Preprocessor > Modeling > Create > Lines > Arcs > By End KPs & Rad

- *Pick Menu* appears; pick keypoints 4 and 5 (end points of the arc); click on ***OK*** in the *Pick Menu.*
- Pick *keypoint* 6 (center of the arc); click on ***OK*** in the *Pick Menu.*
- *Arc by End KPs & Radius* dialog *box* appears; enter ***15*** for ***RAD Radius of the arc.***
- *Click* on ***OK***; line 4 is created.

- Specify the number of divisions on the lines (**LESIZE** command) using the following menu path:

Main Menu > Preprocessor > Meshing > Size Cntrls > ManualSize > Lines > All Lines

- *Element Sizes on Picked Lines* dialog box appears; enter ***20*** for ***NDIV***; click on ***OK***.

- Create the mesh (**LMESH** command) using the following menu path:

Main Menu > Preprocessor > Meshing > Mesh > Lines

- *Pick Menu* appears; click on ***Pick All***.

Solution

- Apply degree of freedom (DOF) constraints at end points (`D` command) using the following menu path:

Main Menu > Solution > Define Loads > Apply > Structural > Displacement > On Nodes

  - *Pick Menu* appears; pick the bottom node, i.e., $(x, y) = (25, 0)$; click on ***OK***.
  - In the dialog box, select ***UY***; click on ***Apply***.
  - *Pick Menu* reappears; pick the top node, i.e., $(x, y) = (0, 52.32)$; click on ***OK***.
  - In the dialog box, remove the highlight on ***UY*** and highlight ***UX***; click on ***OK***.

- Apply pressure (`SFE` command) using the following menu path:

Main Menu > Solution > Define Loads > Apply > Structural > Pressure > On Elements

  - *Pick Menu* appears; click on ***Pick All***.
  - In the new dialog box, enter ***300*** for ***Value Load PRES value***; click on ***OK***.

- Obtain the solution (`SOLVE` command) using the following menu path:

Main Menu > Solution > Solve > Current LS

  - *Confirmation Window* appears along with *Status Report Window*.
  - Review status; if OK, close the *Status Report Window*; click on ***OK*** in the *Confirmation Window*.
  - Wait until ANSYS responds with ***Solution is done!***

Postprocessing

- Review the deformed shape (`PLDISP` command) using the following menu path:

Main Menu > General PostProc > Plot Results > Deformed Shape

  - Select ***Def + undef edge***; click on ***OK***.
  - The deformed shape is shown in Fig. 8.40 (left) as it appears in the *Graphics Window*. It is clear from the figure that the bottom end of the conical section exhibits unexpected displacements/rotations. Problems with real applications (using realistic material properties, geometry, and loads) seldom produce displacements that can be visually detected. Therefore, ANSYS scales the displacements when displaying the deformed shape.
  - Change the displacement scaling (`/DSCALE` command) using the following menu path:

Utility Menu > PlotCtrls > Style > Displacement Scaling

  - This brings up the *Displacement Display Scaling* dialog box. Note the number ***25.253596413609*** in the ***User specified factor*** field. This means that the

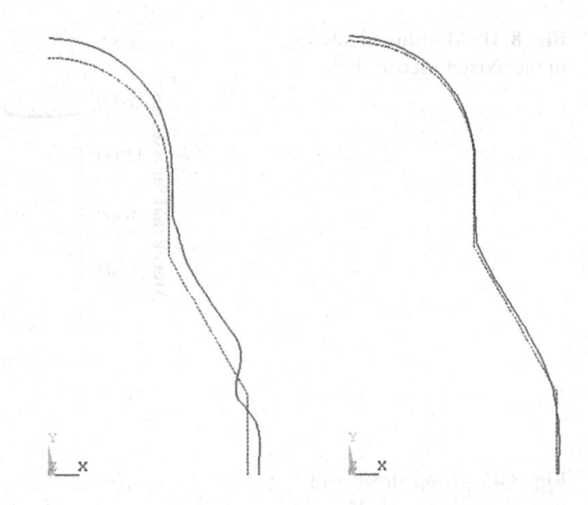

Fig. 8.40 Deformed shape with automatic scaling (*left*; amplified ~23 times) and manual scaling (*right*; amplified 5 times)

displacements are amplified by a factor of approximately 22, so they can be clearly viewed. In order to change this setting, click on the ***User specified*** radio-button; replace the existing scaling factor with the desired value. Figure 8.40 (right) shows the deformed shape amplified by a factor of 5.

- Store element stresses in the element table (**ETABLE** command) using the following menu path:

Main Menu > General PostProc > Element Table > Define Table

- *Element Table Data* dialog box appears; click on ***Add***.
- *Define Additional Element Table Items* dialog box appears; assign a user-defined label for the plate mid-plane meridional stresses, say ***STMR***, in the ***Lab User label for item*** field.
- In the left list, scroll down and select ***By sequence num***; in the right list, select ***SMISC***.
- Finally, in the last field, type ***SMISC,18***; click on ***Apply***. There are several quantities that are stored in sequences, i.e., ***EPEL, NL,*** or ***SMISC***. The information as to which quantity is stored under which sequence is given in *element help pages*. In this particular example, the help page for **SHELL208** contains tables explaining which quantities are stored under which sequence.
- Similarly, store plate mid-plane hoop stresses by assigning a user label, say ***STHP***, and entering ***SMISC,19***; click on ***OK***.
- Click on ***Close*** in the *Element Table Data* dialog box.

- View element table quantities (**PRETAB** command) using the following menu path:

Main Menu > General PostProc > Element Table > List Elem Table

- *List Element Table* dialog box appears; select ***STMR*** and ***STHP;*** click on ***OK***.
- Element numbers and requested element table items are displayed in columns in a separate window. Figures 8.41 and 8.42 show the meridional and hoop stresses, respectively (plotted outside ANSYS).

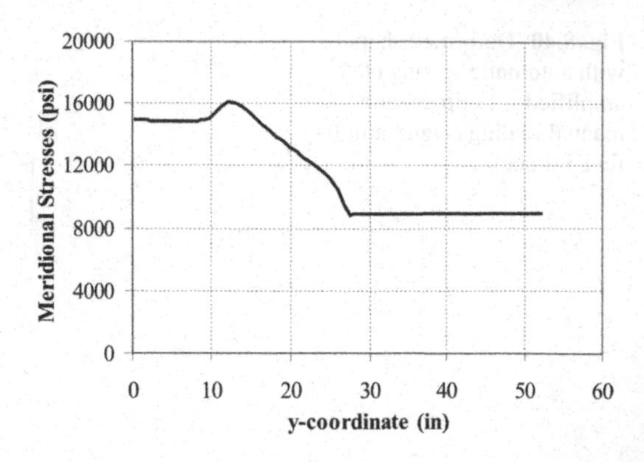

Fig. 8.41 Meridional stresses in the axisymmetric shell

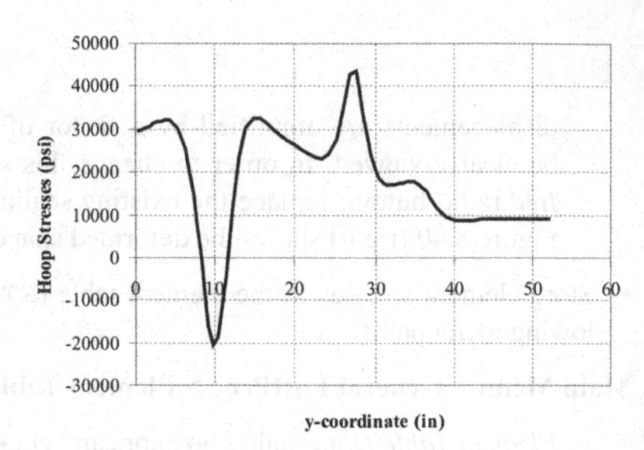

Fig. 8.42 Hoop stresses in the axisymmetric shell

8.1.5.4 Analysis of a Layered Composite Plate

A 10 in × 10 in square composite plate with a stacking sequence of [45°/0°/–45°/90°] is subjected to tensile loading of 100 MPa in the *y*-direction, as shown in Fig. 8.43. Unidirectional ply properties are $E_L = 161\text{GPa}$, $E_T = 9\text{GPa}$, $\nu_{LT} = 0.26$, and $G_{LT} = 6.1\text{GPa}$. The subscripts *L* and *T* designate longitudinal (fiber direction) and transverse (perpendicular to fiber direction), respectively. Each ply has a thickness of 0.16 mm. The goal is to find the displacement and stress fields in the plate.

Model Generation

- Define the element type (**ET** command) using the following menu path:

Main Menu > Preprocessor > Element Type > Add/Edit/Delete

– Click on ***Add***.
– Select ***Shell*** immediately below ***Structural Mass*** in the left list and ***Elastic 4node 181*** in the right list; click on ***OK***.

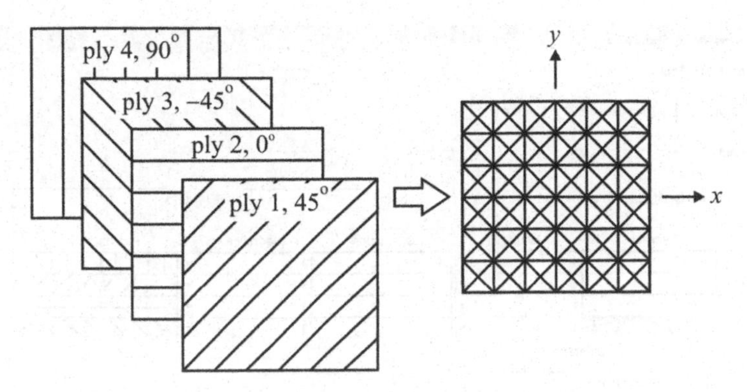

Fig. 8.43 Schematic of the layered composite plate

- Click on ***Close***.

- Specify material properties (**MP** command) using the following menu path:

Main Menu > Preprocessor > Material Props > Material Models

- In the *Define Material Model Behavior* dialog box, in the right window, successively left-click on ***Structural***, ***Linear***, ***Elastic***, and, finally, ***Orthotropic***, which will bring up another dialog box.
- Enter ***161e9*** for ***EX***, ***9e9*** for ***EY*** and ***EZ***, ***0.26*** for ***PRXY*** and ***PRXZ***, ***0.01*** for ***PRYZ***, ***6.1e9*** for ***GXY*** and ***GXZ***, and ***1e9*** for ***GYZ***; click on ***OK***.
- Close the *Define Material Model Behavior* dialog box by using the following menu path:

Material > Exit

- Specify layer information using sections (**SECTYPE** command) using the following menu path:

Main Menu > Preprocessor > Sections > Shell > Lay-up > Add/Edit

- Click on ***Add layer*** button until there are four rows to accommodate four layers.
- Dialog box for layer information appears; material number (***Material ID***), orientation angle (***Orientation***), and thickness (*Thickness*) information for each layer (ply) are entered in this dialog box. Enter the related quantities, as shown in Fig. 8.44.
- Exit from the *Create and Modify Shell Sections* dialog box by clicking on ***OK***.

- Create keypoints (**K** command) using the following menu path:

Main Menu > Preprocessor > Modeling > Create > Keypoints > In Active CS

- In the *Create Keypoints in Active Coordinate System* dialog box, type, ***−5*** for *X* and ***−5*** for *Y*; click on ***Apply*** (keypoint 1 is created).
- Repeat this procedure for keypoints 2 through 4 using the following data:

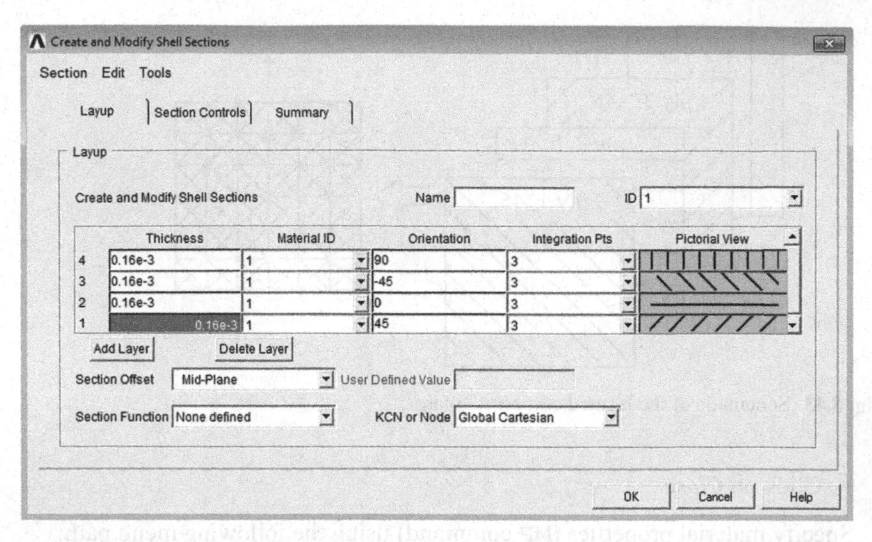

Fig. 8.44 Dialog box for entering layer information

$$x_2 = 5 \qquad y_2 = -5$$
$$x_3 = 5 \qquad y_3 = 5$$
$$x_4 = -5 \qquad y_4 = 5$$

– After creating keypoint 4, click on ***OK*** instead of ***Apply***.

- Create the area (**A** command) using the following menu path:

Main Menu > Preprocessor > Modeling > Create > Areas > Arbitrary > Through KPs

– *Pick Menu* appears; pick keypoints 1, 2, 3, and 4 in this order; click on ***OK*** in the *Pick Menu*.

- Specify the number of divisions on all lines (**LESIZE** command) using the following menu path:

Main Menu > Preprocessor > Meshing > Size Cntrls > ManualSize > Lines > All Lines

– *Element Sizes on All Selected Lines* dialog box appears; enter ***40*** for ***NDIV***; click on ***OK***.

- Create the mesh (**AMESH** command) using the following menu path:

Main Menu > Preprocessor > Meshing > Mesh > Areas > Mapped > 3 or 4 sided

– *Pick Menu* appears; click on ***Pick All***.

Solution

- Apply degree of freedom (DOF) constraints at the center node and right mid-node (`D` command) using the following menu path:

Main Menu > Solution > Define Loads > Apply > Structural > Displacement > On Nodes

  - *Pick Menu* appears; pick the center node, i.e., $(x, y)=(0, 0)$; click on ***OK***.
  - In the dialog box, select ***UX*** and ***UY***; click on ***Apply***.
  - *Pick Menu* reappears; pick the right mid-node, i.e., $(x, y)=(5, 0)$; click on ***OK***.
  - In the dialog box, remove the highlight on ***UX*** (leaving only ***UY*** highlighted); click on ***OK***.

- Constrain z degrees of freedom (DOF) in all nodes (`D` command) using the following menu path:

Main Menu > Solution > Define Loads > Apply > Structural > Displacement > On Nodes

  - *Pick Menu* appears; click on ***Pick All***.
  - In the dialog box, remove the highlight on ***UY*** and highlight ***UZ***; click on ***OK***.

- Plot nodes for clarity (`NPLOT` command) using the following menu path:

Utility Menu > Plot > Nodes

- Apply pressure (`SFE` command) using the following menu path:

Main Menu > Solution > Define Loads > Apply > Structural > Pressure > On Nodes

  - *Pick Menu* appears; using the ***Box*** radio-button in the *Pick Menu*, pick the nodes along $y=5$ and $y=-5$; click on ***OK***.
  - In the new dialog box, enter ***−100E6*4*.16E–3*** for ***Value Load PRES value***; click on ***OK***.

- Obtain the solution (`SOLVE` command) using the following menu path:

Main Menu > Solution > Solve > Current LS

  - *Confirmation Window* appears along with *Status Report Window*.
  - Review status; if OK, close the *Status Report Window*; click on ***OK*** in the *Confirmation Window*.
  - Wait until ANSYS responds with ***Solution is done!***

Postprocessing

- Review the deformed shape (`PLDISP` command) using the following menu path:

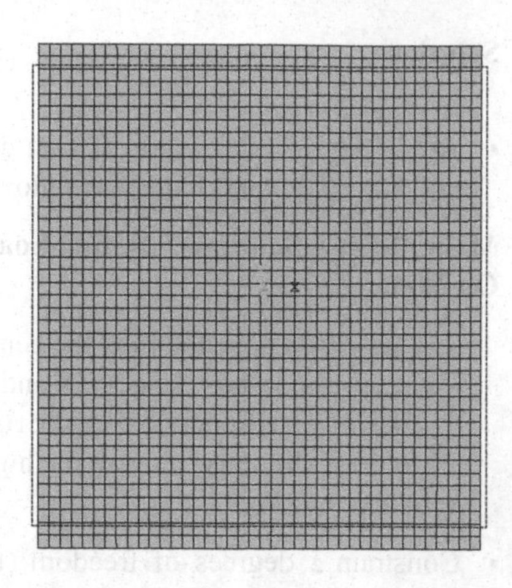

Fig. 8.45 Deformed shape of the composite plate under uni-axial tension in the y-direction

Main Menu > General PostProc > Plot Results > Deformed Shape

- Select ***Def + undef edge***; click on ***OK***.
- The deformed shape is shown in Fig. 8.45 as it appears in the *Graphics Window*.

• Obtain a contour plot of the y-displacement (u_y) (**PLNSOL** command) using the following menu path:

Main Menu > General Postproc > Plot Results > Contour Plot > Nodal Solu

- Select ***DOF Solution*** and ***Y-component of displacement***; click on ***OK.***
- The contour plot appears in the *Graphics Window*, as shown in Fig. 8.46.

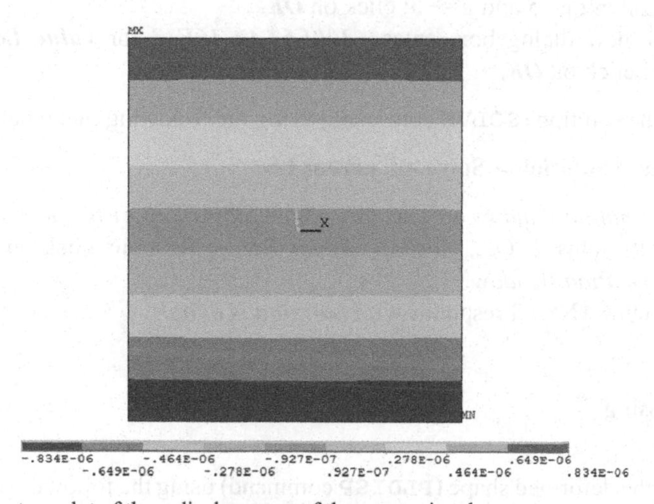

Fig. 8.46 Contour plot of the y-displacement of the composite plate

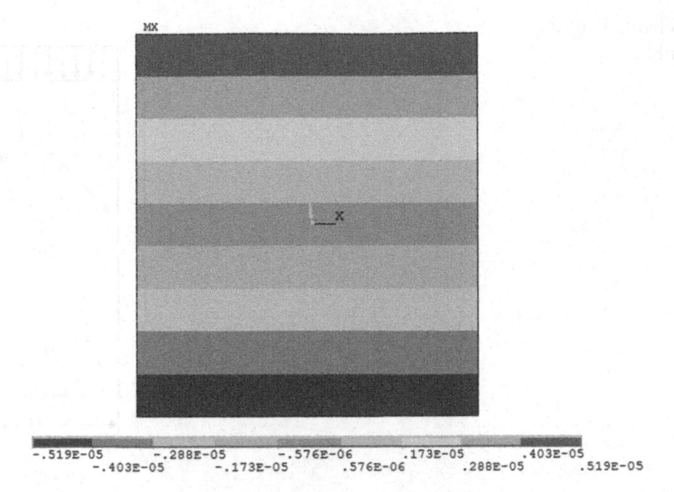

Fig. 8.47 Contour plot of the *y*-displacement of the composite plate utilizing two-dimensional plane stress idealization with orthotropic properties

The same problem can be solved following the procedure given in Sect. 8.1.4.1.2 using the following equivalent orthotropic material properties for the laminate:

$$E_x = E_y = 61.7\,\text{GPa}$$
$$\nu_{xy} = 0.3$$
$$G_{xy} = 23.8\,\text{GPa}$$

The contour plot for the *y*-displacement (u_y) obtained by using this two-dimensional approximation is shown in Fig. 8.47.

8.2 Linear Buckling Analysis

If the component is expected to exhibit structural instability, the search for the load that causes structural bifurcation is referred to as a "buckling load" analysis. Because the buckling load is not known a priori, the finite element equilibrium equations for this type of analysis involve the solution of homogeneous algebraic equations whose lowest eigenvalue corresponds to the buckling load, and the eigenvector represents the primary buckling mode.

There are two approaches in the ANSYS program for buckling analysis: (i) eigenvalue buckling (linear), and (ii) non-linear buckling. The first is considered here.

Eigenvalue buckling is used for calculating the theoretical buckling load of a linear elastic structure. Since it assumes the structure exhibits linearly elastic behavior, the predicted buckling loads are overestimated (unconservative). Steps involved in a typical *Eigenvalue Buckling* analysis are:

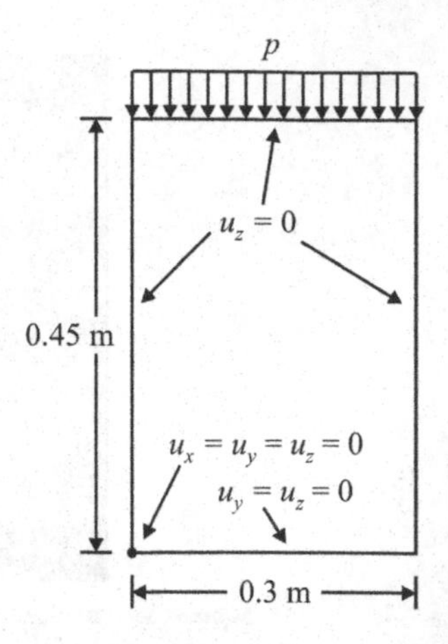

Fig. 8.48 Schematic of the rectangular plate

Build the model.
Obtain the static solution.
Obtain the *eigenvalue buckling* solution.
Expand the solution.
Review the results.

A static solution is needed to establish the stiffening of the structure under the applied load (*stress stiffening*). There are several buckling modes (theoretically, infinitely many!) in a structure. The first buckling mode is the one requiring the smallest load. The user specifies the number of buckling modes to be extracted. An *eigenvalue buckling* solution simply calculates the buckling loads for each of these modes. The solution is then *expanded* to include the deformation patterns in the structure (mode shapes) corresponding to the buckling loads. Results are reviewed in the *General Postprocessor*.

As an example, a rectangular plate is subjected to uniform compressive loading along its top edge while the bottom edge is constrained to move in the direction of loading, as shown in Fig. 8.48. The plate is 0.45 m long, 0.3 m wide, and 0.003 m thick. It is made of steel with elastic modulus $E=200$ GPa and Poisson's ratio $\nu=0.32$. The goal is to find the first four buckling modes and their corresponding buckling loads under given constraints and loading configuration.

Model Generation

- Define the element type (**ET** command) using the following menu path:

Main Menu > Preprocessor > Element Type > Add/Edit/Delete

- Click on ***Add***.
- Select ***Shell*** immediately below ***Structural Mass*** in the left list and ***Elastic 4node 181*** in the right list; click on ***OK***.
- Click on ***Close***.

- Specify material properties (**MP** command) using the following menu path:

Main Menu > Preprocessor > Material Props > Material Models

- In the *Define Material Model Behavior* dialog box, in the right window, successively left-click on ***Structural***, ***Linear***, ***Elastic***, and, finally, ***Isotropic***, which will bring up another dialog box.
- Enter ***200e9*** for ***EX*** and ***0.32*** for ***PRXY***; click on ***OK***.
- Close the *Define Material Model Behavior* dialog box by using the following menu path:

Material > Exit

- Specify the thickness for the shell (**SECTYPE** command) using the following menu path:

Main Menu > Preprocessor > Sections > Shell > Lay-up > Add/Edit

- *Create and Modify Shell Sections* dialog box appears; enter ***0.003*** for ***Thickness***.
- Exit from the *Create and Modify Shell Sections* dialog box by clicking on ***OK***.

- Create the solid model, a rectangular area in this case, using the following menu path:

Main Menu > Preprocessor > Modeling > Create > Areas > Rectangle > By Dimensions

- In the *Create Rectangle by Dimensions* dialog box, type ***0*** for ***X1***, ***0.3*** for ***X2***, ***0*** for ***Y1***, and ***0.45*** for ***Y2***; click on ***OK***.
- Turn line numbering on using the following menu path:

Utility Menu > PlotCtrls > Numbering

- Place a checkmark in the square box next to ***LINE Line numbers***; click on ***OK***.
- Plot lines using the menu path:

Utility Menu > Plot > Lines

- Specify the number of elements on selected lines for mapped meshing. On lines 1 and 3, use **15** divisions; on lines 2 and 4, use **25** divisions. Use the following menu path for this action:

Main Menu > Preprocessor > Meshing > Size Cntrls > ManualSize > Lines > Picked Lines

- Pick lines 1 and 3; click on ***OK***.
- *Element Sizes on Picked Lines* dialog box appears; type ***15*** in the text field corresponding to ***NDIV*** (the second text field), and uncheck the first checkbox; click on ***Apply***.
- Repeat this procedure for the next set of lines (2 and 4) with their corresponding number divisions as ***25***. After specifying the number of divisions for lines 2 and 4, click on ***OK*** in the *Element Sizes on Picked Lines* dialog box.

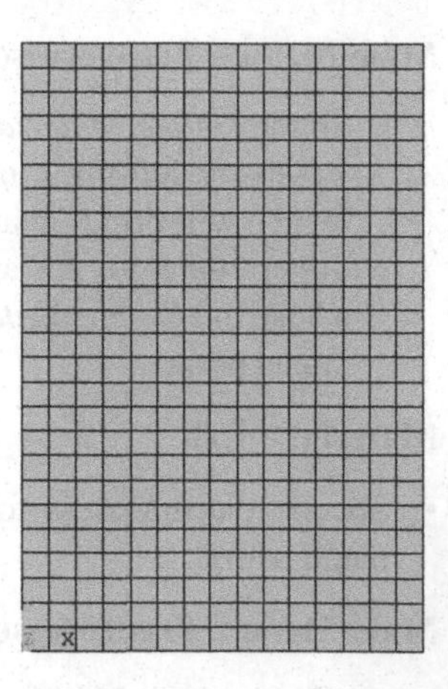

Fig. 8.49 Mesh of the rectangular

- Create the mesh using the following menu path:

Main Menu > Preprocessor > Meshing > Mesh > Areas > Mapped > 3 or 4 sided

  - Pick the area; click on ***OK***.
  - The mesh should appear in the *Graphics Window*, as shown in Fig. 8.49.

Solution

- Constrain the out-of-plane displacements (z-displacements) of the nodes along the entire boundary using the following menu path:

Main Menu > Solution > Define Loads > Apply > Structural > Displacement > On Nodes

  - In the *Pick Menu* click on the radio-button next to ***Box***. This enables the user to pick several nodes at a time by drawing an area in the *Graphics Window*. Move the mouse pointer to a location slightly left and above the top-left corner of the meshed area.
  - Click on the left mouse button (***without*** releasing it) and draw a rectangle that encloses only the nodes along the $x=0$ boundary; release the left button (Fig. 8.50). Observe that each selected node is identified by a small square.
  - Similarly, select nodes along all boundaries ($x=0.3$, $y=0$, and $y=0.45$); click on ***OK*** in the *Pick Menu*.
  - Highlight ***UZ***; click on ***OK***.

- Apply remaining displacement constraints in the same manner as in the previous step.
  - Constrain displacements in the y-direction along the $y=0$ boundary.
  - Constrain the displacement in the x-direction at the boundary point $(x, y)=(0,0)$.

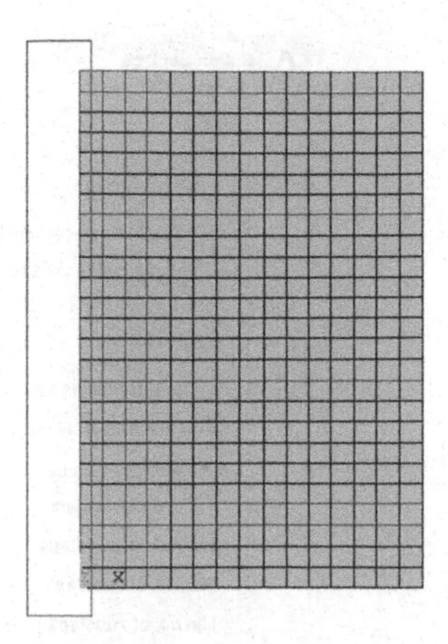

Fig. 8.50 Selecting nodes using plate. the *Box* option

- Apply the uniform load along the $y = 0.45$ boundary using the following menu path:

Main Menu > Solution > Define Loads > Apply > Structural > Pressure > On Nodes

  - Pick the nodes along the $y = 0.45$ boundary; click on ***OK***.
  - Type ***1***; click on ***OK***.
  - The eigenvalue buckling analysis calculates a scaling factor for the existing loads; therefore, if a unit load is applied, the scaling factor yields the buckling load.

- Turn on pre-stress effects using the following menu path:

Main Menu > Solution > Analysis Type > Sol'n Controls

  - A dialog box appears; place a checkmark in the box next to ***Calculate pre-stress effects***; click on ***OK*** (Fig. 8.51).

- Obtain the static solution (**`SOLVE`** command) using the following menu path:

Main Menu > Solution > Solve > Current LS

  - *Confirmation Window* appears along with *Status Report Window*.
  - Review status; if OK, close the *Status Report Window*; click on ***OK*** in the *Confirmation Window*.
  - Wait until ANSYS responds with ***Solution is done!***

- Exit *Solution Processor* using the following menu path:

Main Menu > Finish

- Re-enter the *Solution Processor* and change the analysis type to eigenvalue buckling using the following menu path:

Main Menu > Solution > Analysis Type > New Analysis

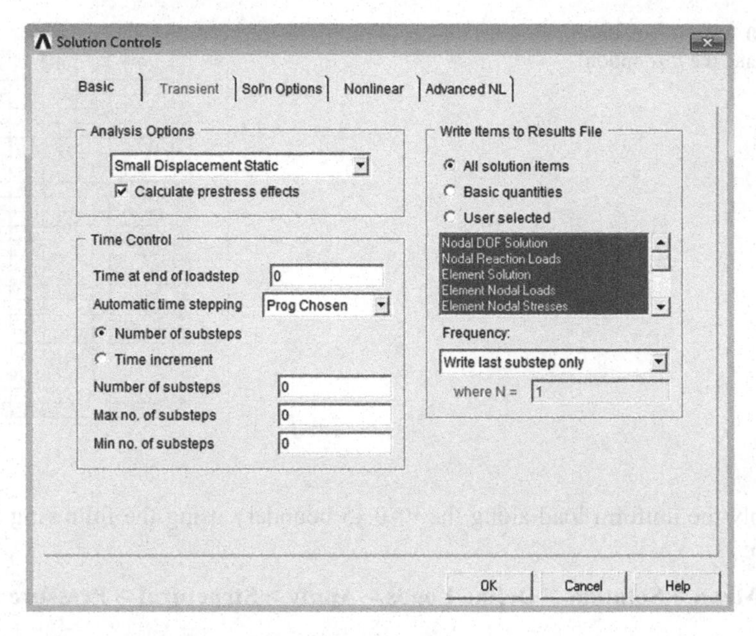

Fig. 8.51 *Solution Controls* dialog box (*Basic* tab shown)

- Click on ***Eigen Buckling***; click on ***OK***.

- Set analysis options using the following menu path:

Main Menu > Solution > Analysis Type > Analysis Options

- Type *4* in the text field next to ***NMODE No. of modes to extract***; click on ***OK***.
- In the new dialog box, click on ***OK***, leaving the settings at their default values.

- Instruct ANSYS to expand modes using the following menu path:

Main Menu > Solution > Load Step Opts > ExpansionPass > Single Expand > Expand Modes

- Type *4* in the text field next to ***NMODE No. of modes to extract***; click on ***OK***.

- Obtain the eigenvalue buckling solution (**SOLVE** command) using the following menu path:

Main Menu > Solution > Solve > Current LS

- *Confirmation Window* appears along with *Status Report Window*.
- Review status; if OK, close the *Status Report Window*; click on ***OK*** in the *Confirmation Window*.
- Wait until ANSYS responds with ***Solution is done!***

Postprocessing

- Review the buckling loads using the following menu path:

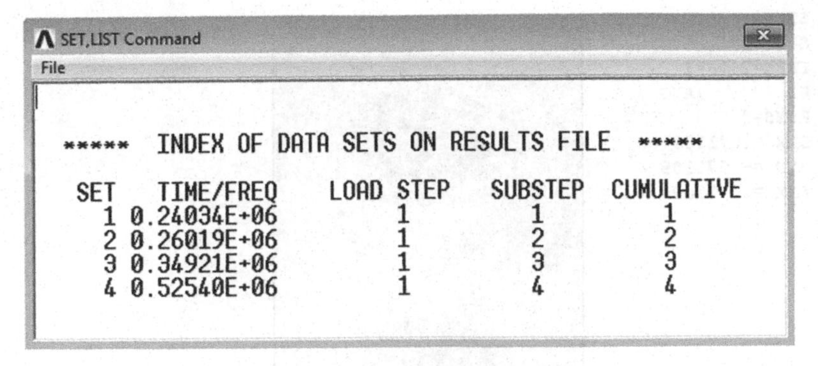
SET,LIST Command
File

***** INDEX OF DATA SETS ON RESULTS FILE *****

SET	TIME/FREQ	LOAD STEP	SUBSTEP	CUMULATIVE
1	0.24034E+06	1	1	1
2	0.26019E+06	1	2	2
3	0.34921E+06	1	3	3
4	0.52540E+06	1	4	4

Fig. 8.52 List of buckling loads for different buckling modes

Main Menu > General PostProc > Results Summary

- The list will appear in a new window, as shown in Fig. 8.52. The critical load for the first mode is given as 0.24034E+06 Pa (0.24034 MPa) in this list. This means that when the applied load p is increased to this value, the plate will buckle in the first mode.

• Review the buckling modes.

- Read the results for the first buckling load using the following menu path:

Main Menu > General PostProc > Read Results > First Set

- Obtain contour plot of the z-displacement using the following menu path:

Main Menu > General Postproc > Plot Results > Contour Plot > Nodal Solu

- Click on ***DOF Solution*** and ***Z-component of displacement***; click on ***OK***.
- The contour plot will appear in the *Graphics Window*, as shown in Fig. 8.53.
- Read the results for the second buckling load using the following menu path:

Main Menu > General PostProc > Read Results > Next Set

- Obtain contour plot of the z-displacement using the following menu path:

Main Menu > General Postproc > Plot Results > Contour Plot > Nodal Solu

- Click on ***DOF Solution*** and ***Z-component of displacement***; click on ***OK***.
- The contour plot will appear in the *Graphics Window*, as shown in Fig. 8.54.
- Repeat for modes 3 and 4 to obtain plots similar to those given in Fig. 8.55 and 8.56.

• Review thc buckling mode shapes.

- Read the results for the desired mode (as shown in previous step) and plot the deformed shape using the following menu path:

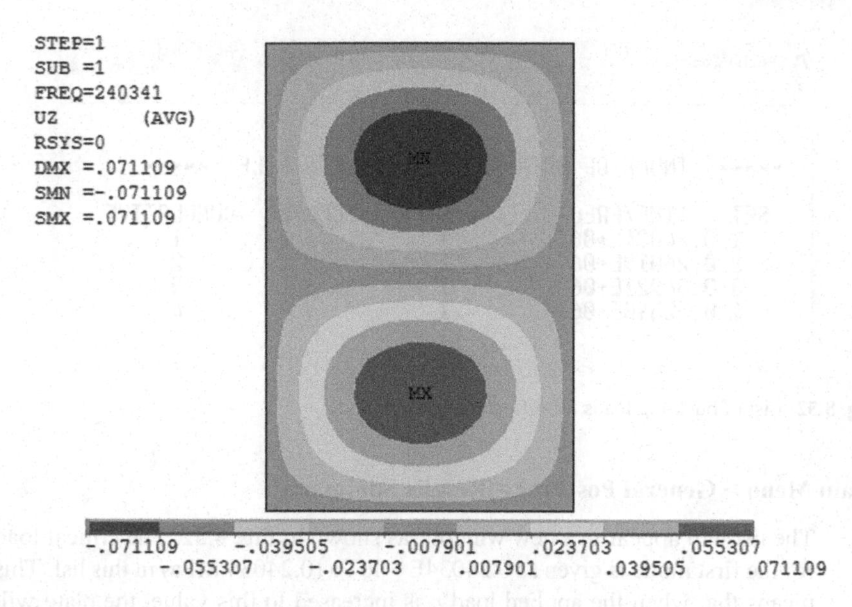

Fig. 8.53 Contour plot of u_z field (z-displacement) under first buckling mode

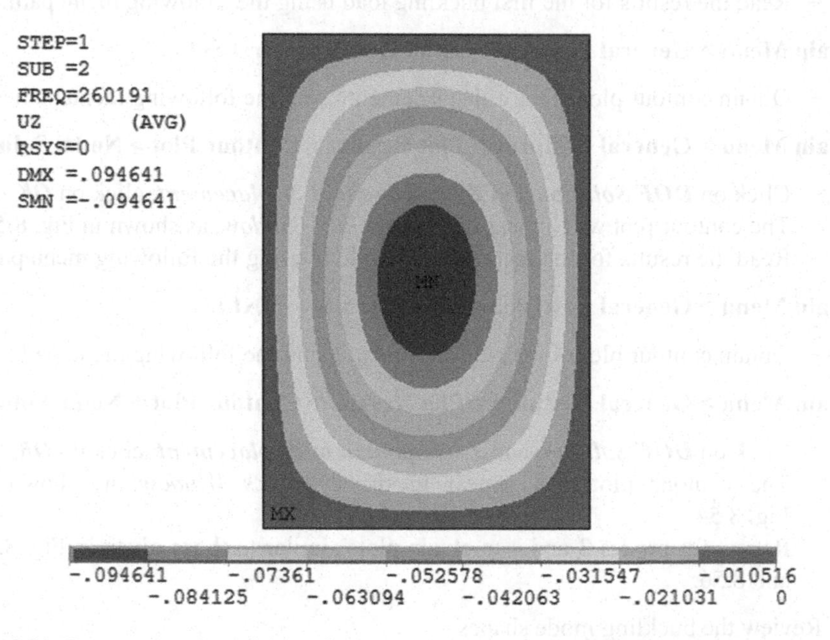

Fig. 8.54 Contour plot of u_z field (z-displacement) under second buckling mode

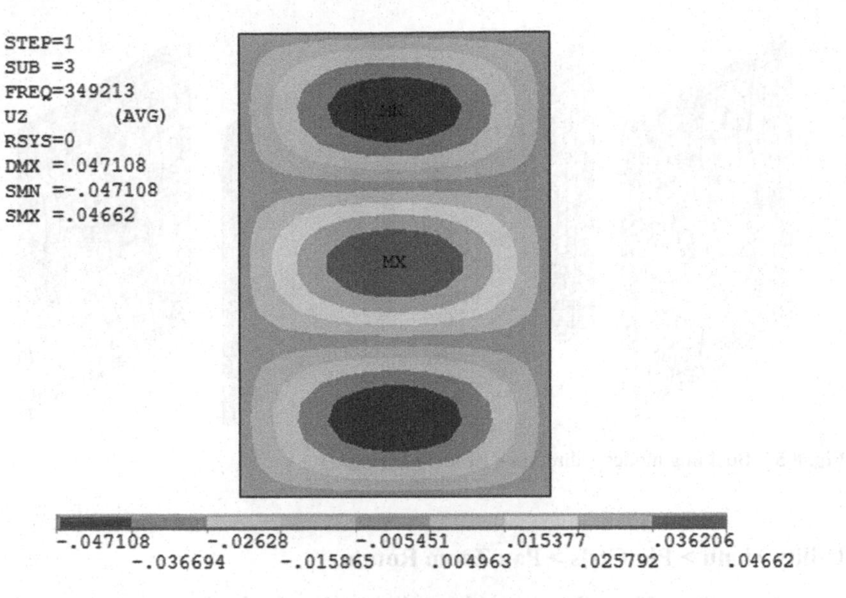

Fig. 8.55 Contour plot of u_z field (*z*-displacement) under third buckling mode

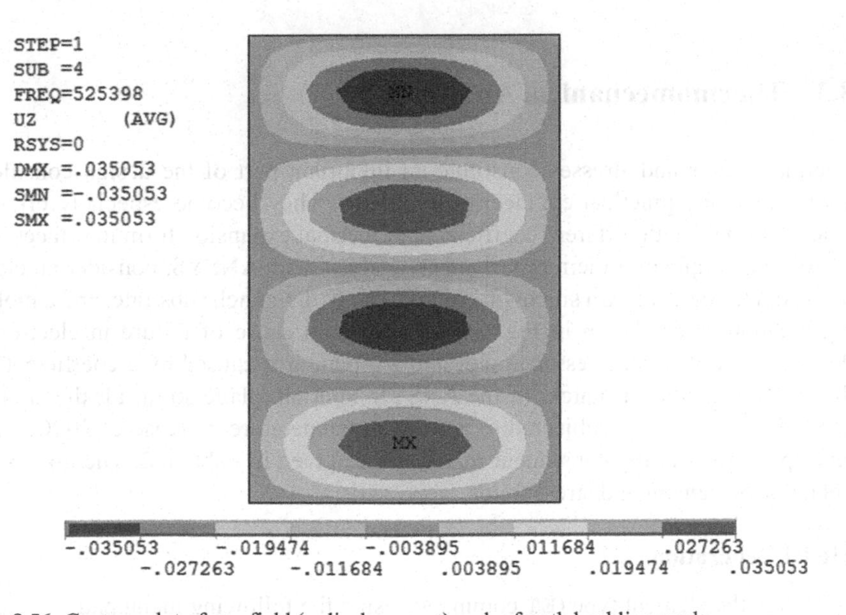

Fig. 8.56 Contour plot of u_z field (*z*-displacement) under fourth buckling mode

Main Menu > General Postproc > Plot Results > Deformed Shape

- Click on the ***Def shape only*** radio-button; click on ***OK***.
- Change the viewpoint to isometric using the menu path:

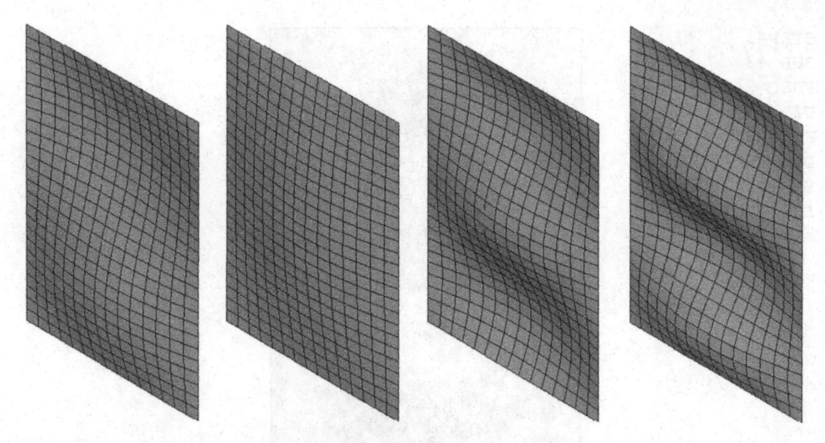

Fig. 8.57 Buckling modes 1 through 4 (from *left to right*)

Utility Menu > PlotCtrls > Pan Zoom Rotate

- In the ***Pan Zoom Rotate*** window, click on the ***Iso*** button.
- Figure 8.57 shows the first four mode shapes of the plate.

8.3 Thermomechanical Analysis

Thermal strains and stresses constitute an important part of the design considerations for many practical engineering problems. They become especially critical when materials with different coefficients of thermal expansion form interfaces.

As an example of a thermomechanical analysis with ANSYS, consider an electronic device containing a silicon die (chip), epoxy die-attach substrate, and a molding compound, as shown in Fig. 8.58. A common cause of failure in electronic devices is the thermal stresses at elevated temperatures caused by a coefficient of thermal expansion mismatch. In the ANSYS solution, plane strain idealization is utilized. The device is subjected to a uniform temperature increase of 30 °C. Material properties of the constituent materials are given in Table 8.2. The goal is to obtain displacement and stress fields.

Model Generation

- Define the element type (**ET** command) using the following menu path:

Main Menu > Preprocessor > Element Type > Add/Edit/Delete

- Click on ***Add***.
- Select ***Solid*** immediately below ***Structural Mass*** in the left list and ***Quad 4 Node 182*** in the right list; click on ***OK***.
- Click on ***Options***.

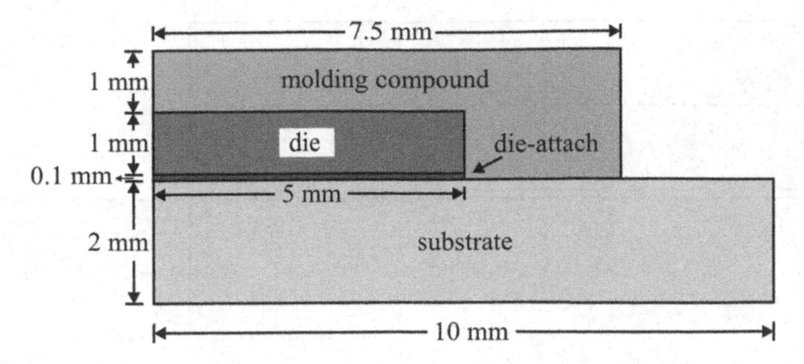

Fig. 8.58 Geometry of the electronic package

Table 8.2 Properties of the constituent materials in the electronic package

	E (GPa)	ν	α (10^{-6}/°C)	Material reference number
Substrate	22	0.39	18	1
Die-attach	7.4	0.4	52	2
Silicon	163	0.278	2.6	3
Molding compound	15	0.25	16	4

- *PLANE182 element type options* dialog box appears; select ***Plane strain*** item from the pull-down menu corresponding to ***Element behavior K3***.
- Click on ***OK***; click on ***Close***.

• Specify material properties (**MP** command) using the following menu path:

Main Menu > Preprocessor > Material Props > Material Models

- *Define Material Model Behavior* dialog box appears. In the right window, successively left-click on ***Structural***, ***Linear***, ***Elastic***, and, finally, ***Isotropic***, which brings up another dialog box.
- Referring to Table 8.2, enter ***22E9*** for ***EX*** and ***0.39*** for ***PRXY***; click on ***OK***.
- In the right list, successively left-click on ***Structural***, ***Thermal Expansion***, ***Secant Coefficient***, and, finally, ***Isotropic***, which brings up another dialog box.
- Enter ***18E–6*** for ***ALPX***; click on ***OK***.
- Add new material model using the following menu path:

Material > New Model

- Repeat the procedure for the remaining materials (2 through 4) referring to Table 8.2.
- When finished, close the *Define Material Model Behavior* dialog box by using the following menu path:

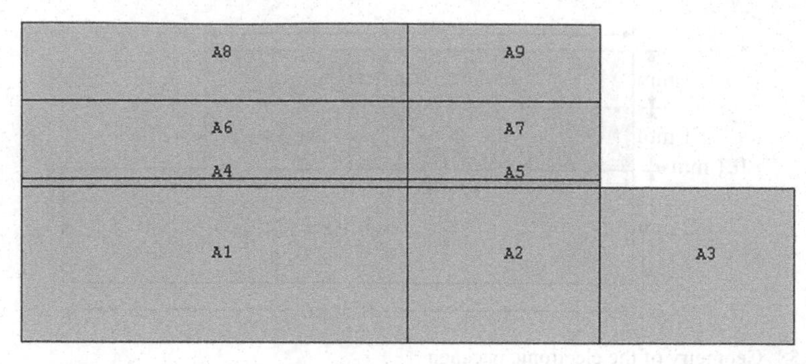

Fig. 8.59 Solid model of the electronic package

Material > Exit

- Create rectangles as identified in Fig. 8.59 (**RECTNG** command) using the following menu path:

Main Menu > Preprocessor > Modeling > Create > Areas > Rectangle > By Dimensions

- *Create Rectangle by Dimensions* dialog box appears. Referring to Table 8.3, enter ***0*** for ***X1***, ***5E−3*** for ***X2***, ***0*** for ***Y1***, and ***2E−3*** for ***Y2***; click on ***Apply***.
- Repeat the procedure for the remaining areas (2 through 9). When creating Area 9, click on ***OK*** after entering the coordinates.

- Glue the areas (**AGLUE** command) using the following menu path:

Main Menu > Preprocessor > Modeling > Operate > Booleans > Glue > Areas

- *Pick Menu* appears; click on ***Pick All*** button.

- Mesh the areas (**AMESH** command) using the following menu path:

Main Menu > Preprocessor > Meshing > Mesh > Areas > Mapped > 3 or 4 sided

- *Pick Menu* appears; click on ***Pick All***.
- At this point, all the elements have *Material Reference Number* 1. Attributes can be changed after the elements are created. For this purpose, areas are selected first. Then the elements that are attached to the selected areas are selected. Finally, elements are modified so they have the correct attributes. The correspondence between the areas and material numbers are given in Table 8.3. Select areas (**ASEL** command) using the following menu path:

Utility Menu > Select > Entities

- *Select Entities* dialog box appears; select ***Areas*** from the first pull-down menu and ***By Num/Pick*** from the second pull-down menu; click on ***OK***.
- *Pick Menu* appears; pick areas 5, 7, 8, and 9; click on ***OK***.
- Now, select the elements that are attached to the selected areas (**ESLA** command) using the following menu path:

Table 8.3 Coordinates defining the areas and the corresponding material reference numbers

Area number	*X1*	*X2*	*Y1*	*Y2*	Material reference number
	(mm)				
1	0	5	0	2	1
2	5	7.5	0	2	1
3	7.5	10	0	2	1
4	0	5	2	2.1	2
5	5	7.5	2	2.1	4
6	0	5	2.1	3.1	3
7	5	7.5	2.1	3.1	4
8	0	5	3.1	4.1	4
9	5	7.5	3.1	4.1	4

Utility Menu > Select > Entities

- *Select Entities* dialog box appears; select ***Elements*** from the first pull-down menu; select ***Attached to*** from the second pull-down menu. Click on the ***Areas*** radio-button; click on ***OK***.
- Modify the attributes of the selected set of elements (**EMODIF** command) using the following menu path:

Main Menu > Preprocessor > Modeling > Move/Modify > Elements > Modify Attrib

- *Pick Menu* appears; click on ***Pick All***, which brings up the *Modify Elem Attributes* dialog box.
- Select ***Material MAT*** from the pull-down menu and enter ***4*** in the ***I1 New attribute number*** field; click on ***OK***.
- Repeat this procedure for area 4 (material reference number 2) and area 6 (material reference number 3).
- When finished, select everything (**ALLSEL** command) using the following menu path:

Utility Menu > Select > Everything

Solution

- Apply displacement constraints (**D** command) using the following menu path:

Main Menu > Solution > Define Loads > Apply > Structural > Displacement > On Nodes

- *Pick Menu* appears; pick the nodes along $x = 0$ (y-axis); click on ***OK*** in the *Pick Menu*.
- Highlight ***UX***; click on ***Apply***.
- *Pick Menu* reappears; pick the bottom-left corner node ($x = 0$, $y = 0$); click on ***OK*** in the *Pick Menu*.

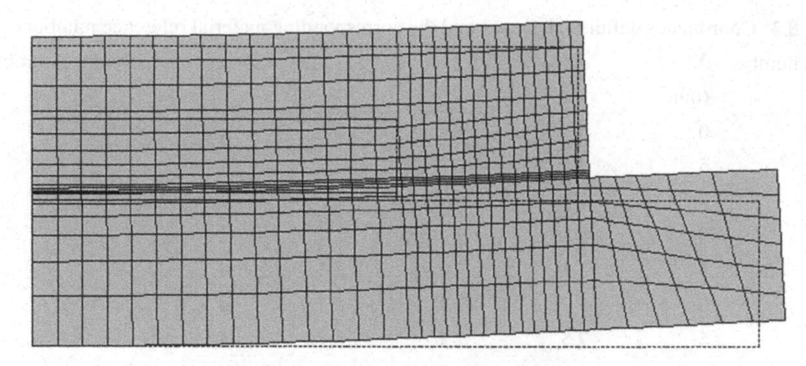

Fig. 8.60 Deformed shape of the electronic package under thermal load

- Remove highlight ***UY*** (leave the ***UX*** highlighted); click on ***OK***.

- Apply the thermal load (**TUNIF** command) using the following menu path:

Main Menu > Solution > Define Loads > Apply > Structural > Temperature > Uniform Temp

- *Uniform Temperature* dialog box appears; enter ***30*** for ***TUNIF***; click on ***OK***.

- Obtain the solution (**SOLVE** command) using the following menu path:

Main Menu > Solution > Solve > Current LS

- *Confirmation Window* appears along with *Status Report Window*.
- Review status; if OK, close the *Status Report Window*; click on ***OK*** in the *Confirmation Window*.
- Wait until ANSYS responds with ***Solution is done!***

Postprocessing

- Review the deformed shape (**PLDISP** command) using the following menu path:

Main Menu > General PostProc > Plot Results > Deformed Shape

- Select ***Def + undef edge***; click on ***OK***.
- The deformed shape is shown in Fig. 8.60 as it appears in the *Graphics Window*.

- Obtain the normal stress in the *y*-direction and shearing stress contour plots (**PLNSOL** command) using the following menu path:

Main Menu > General Postproc > Plot Results > Contour Plot > Nodal Solu

- In order to obtain the view of the normal stresses in the *y*-direction, select ***Stress*** and ***Y-component of stress***; click on ***OK***.
- The resulting contour plot, along with a zoomed-in view of the critical junction, is shown in Fig. 8.61.

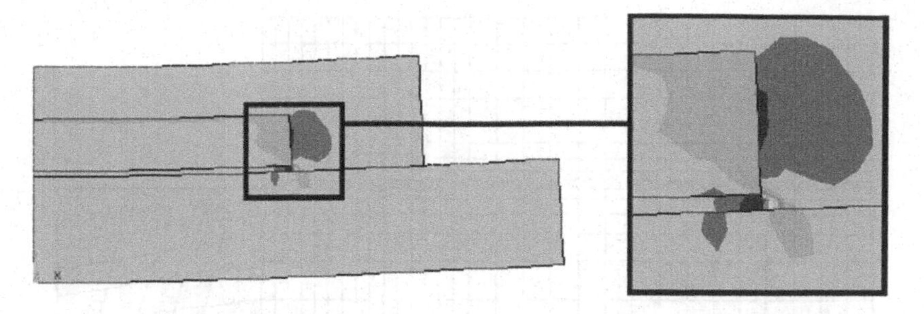

Fig. 8.61 Contour plot of the normal stress (σ_{yy}) in the y-direction: in the entire package (*left*) and in the vicinity of the die/die-attach interface (*right*)

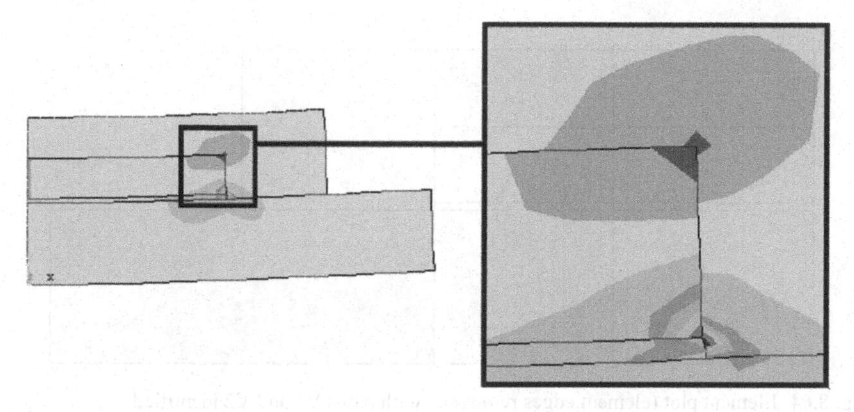

Fig. 8.62 Contour plot of the shear stress (σ_{xy}): in the entire package (*left*) and in the vicinity of the die/die-attach interface (*right*)

  - Similarly, in order to view the shear stresses, select ***Stress*** and ***XY Shear stress***; click on ***OK***.
  - The resulting contour plot, along with a zoomed-in view of the critical junction, is shown in Fig. 8.62.

- Plot elements (**`EPLOT`** command) using the following menu path:

Utility Menu > Plot > Elements

- Review variation of stresses along paths by means of line plots. Two paths are defined, both of which are vertical. The first path passes through the vertical cross section where the die and the die-attach terminate and form an interface with the molding compound ($x = 5mm.$). The second path is located approximately in the middle of the die and die-attach. Both paths are plotted in Fig. 8.63 and 8.64 (element edges are removed in Fig. 8.64 for clarity). Define the path (**`PPATH`** command) using the following menu path:

Main Menu > General Postproc > Path Operations > Define Path > By Nodes

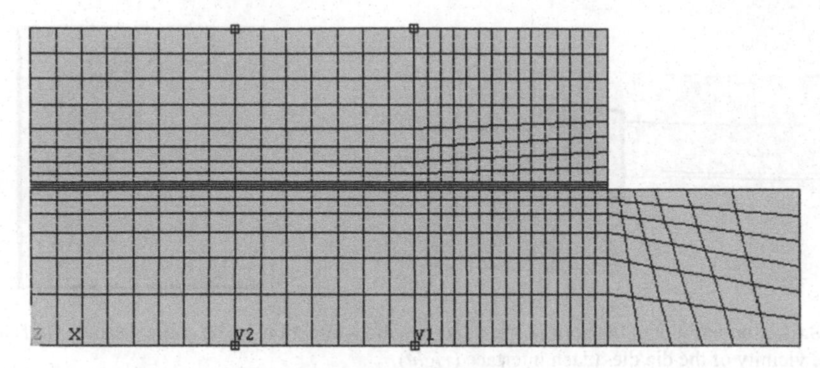

Fig. 8.63 Element plot with paths V1 and V2 identified

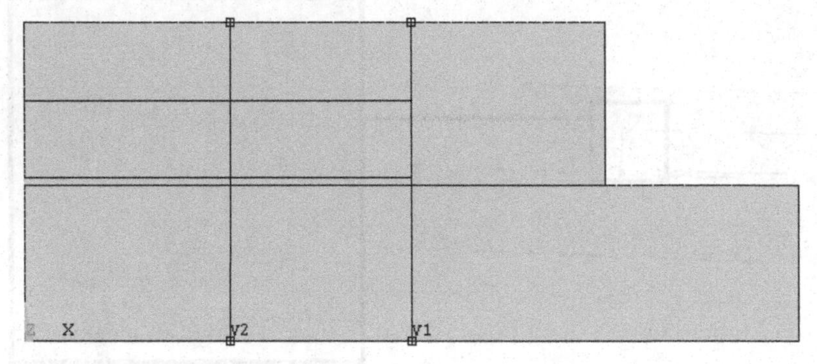

Fig. 8.64 Element plot (element edges removed) with paths V1 and V2 identified

- *Pick Menu* appears; pick the two nodes indicated with small squares, as shown in Fig. 8.63 (corresponding to the path V1); click on ***OK***.
- *By Nodes* dialog box appears; enter a name describing the path, say ***V1***, in the ***Define Path Name*** text field; click on ***OK***.
- Close the *PATH Command Status Window*.
- Define a second path as indicated in Fig. 8.63 and 8.64 (corresponding to the path V2); enter the name as ***V2***.
- When multiple paths are defined, only one path is active at a given time, and mapping of results is performed on the active path. Activate the path V1 (**PATH** command) using the following menu path:

Main Menu > General Postproc > Path Operations > Recall Path

- *Recall Path* dialog box appears. Select ***V1***; click on ***OK***.
- Map results onto path (**PDEF** command) using the following menu path:

Main Menu > General Postproc > Path Operations > Map onto Path

- *Map Result Items onto Path* dialog box appears; select ***Stress*** from the left list and ***Y-direction SY*** from the right list; click on ***Apply***.

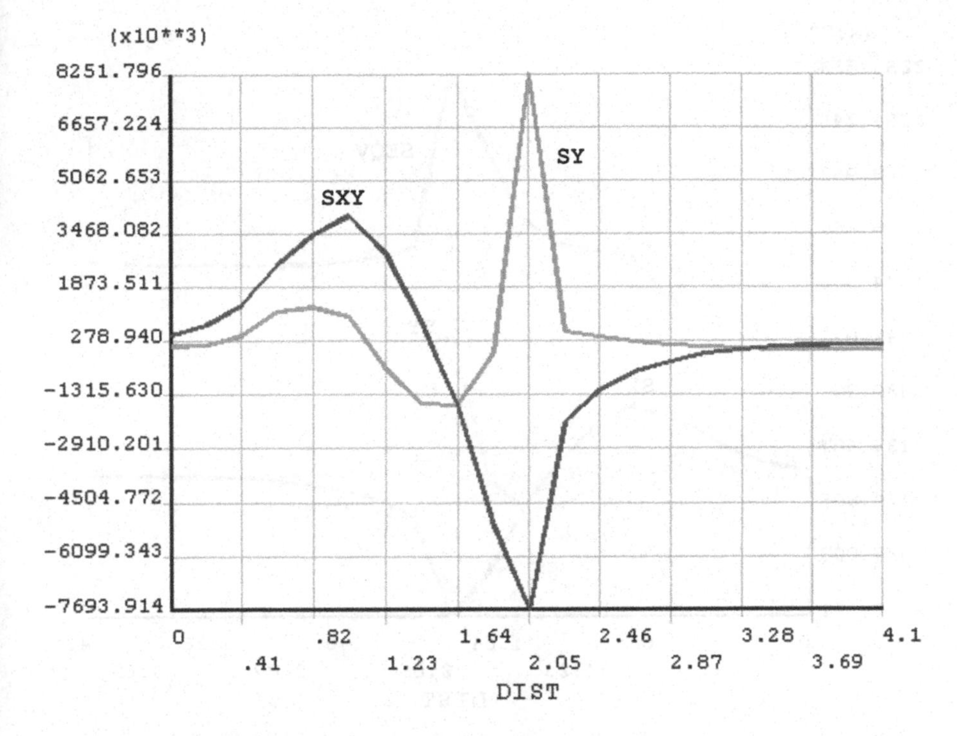

Fig. 8.65 Line plots of σ_{yy} and σ_{xy} along path V1

- *Map Result Items onto Path* dialog box remains active; select ***Stress*** from the left list and ***XY-shear SXY*** from the right list; click on ***Apply***.
- Now, select ***Stress*** from the left list and scroll down in the right list to select ***von Mises SEQV***; click on ***OK***.
- At this point, normal stress in the *y*-direction (σ_{yy}), *xy* shear stress (σ_{xy}), and equivalent stress (σ_{eqv}) values are mapped onto path V1. Obtain line plot of σ_{yy} and σ_{xy} along the path V1 (**PLPATH** command) using the following menu path:

Main Menu > General Postproc > Path Operations > Plot Path Item > On Graph

- *Plot of Path Items on Graph* dialog box appears; select ***SY*** and ***SXY***; click on ***OK***.
- Figure 8.65 shows the line plots of σ_{yy} and σ_{xy} along the defined path.
- Now, obtain line plot of σ_{eqv} in the same graph with σ_{yy} and σ_{xy} along path V1 (**PLPATH** command) using the following menu path:

Main Menu > General Postproc > Path Operations > Plot Path Item > On Graph

- *Plot of Path Items on Graph* dialog box appears; add ***SEQV*** to the existing selection (***SY*** and ***SXY***); click on ***OK***.

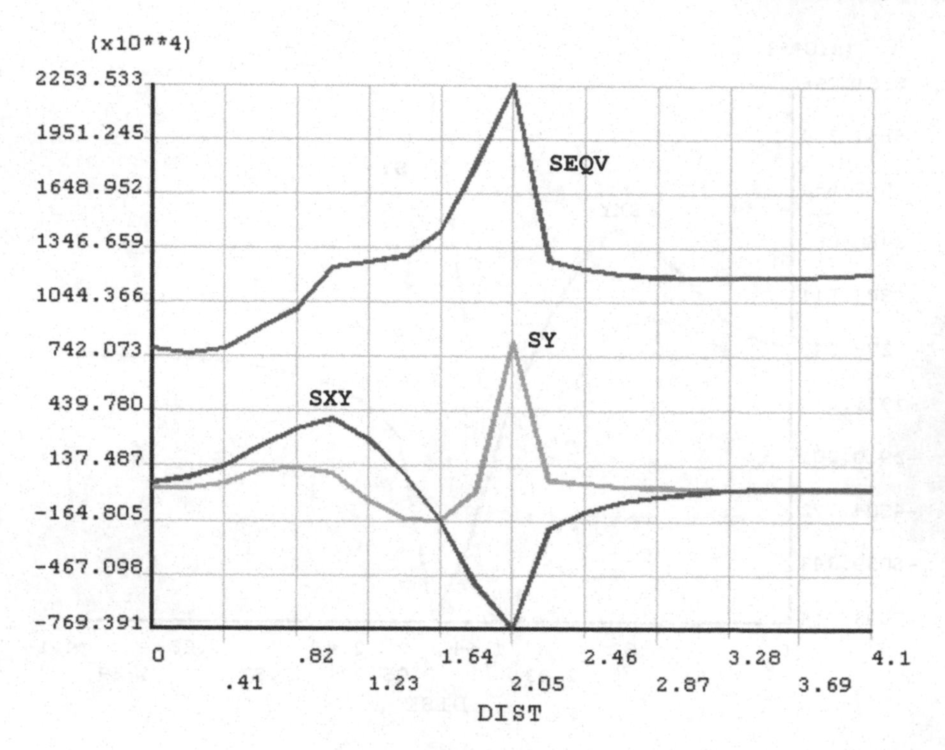

Fig. 8.66 Line plots of σ_{yy}, σ_{xy}, and σ_{eqv} along path V1

- Figure 8.66 shows the resulting line plot.
- Similar plots can be obtained for stresses along path V2. For this purpose, the user needs to activate path V2, followed by the mapping of quantities. Figure 8.67 shows the variation of σ_{yy}, σ_{xy}, and σ_{eqv} along the path.

8.4 Fracture Mechanics Analysis

Computation of fracture parameters, such as the stress intensity factors or energy release rate, using finite element analysis requires either a refined mesh around the crack tip or the use of "special elements" with embedded stress singularity near the crack tip. Although conceptually the stress intensity factors are obtained in a straightforward manner, finite element analyses with conventional elements near the crack tip always underestimate the sharply rising stress-displacement gradients.

Instead of trying to capture the well-known $1/\sqrt{r}$ singular behavior with smaller and smaller elements, Henshell and Shaw (1975) and Barsoum (1976, 1977) introduced a direct method by shifting the mid-side node of an 8-noded isoparametric quadrilateral element to the one-quarter point from the crack tip node. Relocating

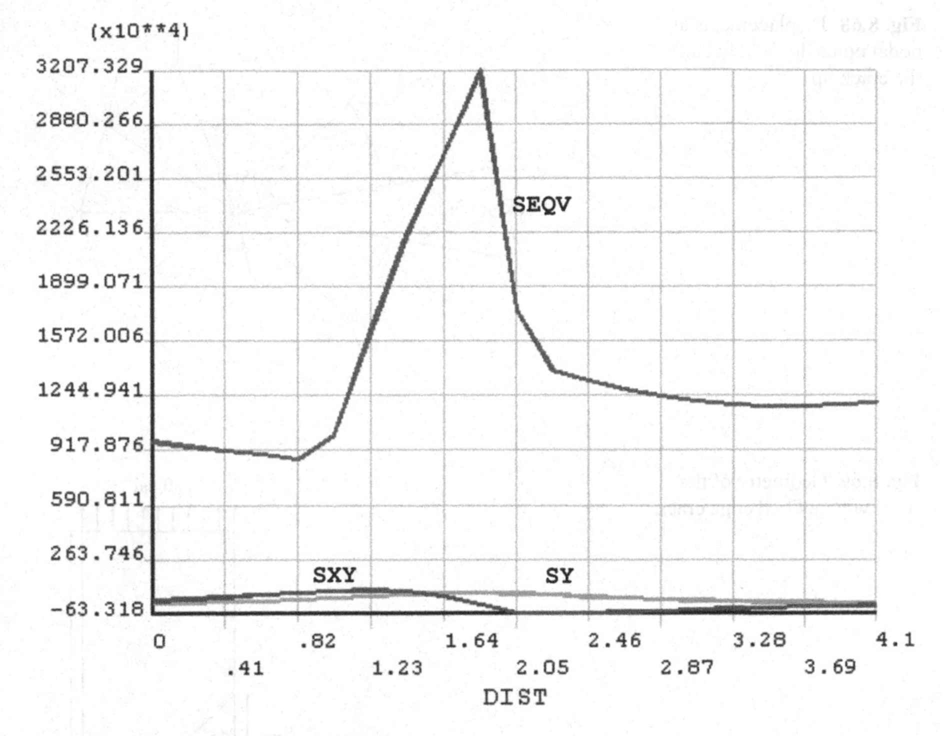

Fig. 8.67 Line plots of σ_{yy}, σ_{xy}, and σ_{eqv} along path V2

the mid-side nodes to the one-quarter point achieves the desired $1/\sqrt{r}$ singular behavior. In the case of linear elastic deformation, the elements **PLANE183** (2-D, 8-noded quadrilateral), and **SOLID186** (3-D, 20-noded brick) in ANSYS are used to obtain the well-established singular stress field by shifting the mid-side nodes one-quarter away from the crack tip.

Once an accurate stress field is obtained, fracture parameters (i.e., stress intensity factors, *J*-integral, and energy release rate) can be calculated within the ANSYS postprocessor.

As an extension of the node collapsing approach, Pu et al. (1978) showed that the stress intensity factors, K_I and K_{II} for opening and sliding modes, respectively, can be computed directly from the nodal displacements on opposite sides of the crack plane as

$$K_I = G\sqrt{\frac{2\pi}{r_0}}\,\frac{u_y(r_0,\theta=\pi)-u_y(r_0,\theta=-\pi)}{(\kappa+1)}$$

and

$$K_{II} = G\sqrt{\frac{2\pi}{r_0}}\,\frac{u_x(r_0,\theta=\pi)-u_x(r_0,\theta=-\pi)}{(\kappa+1)}$$

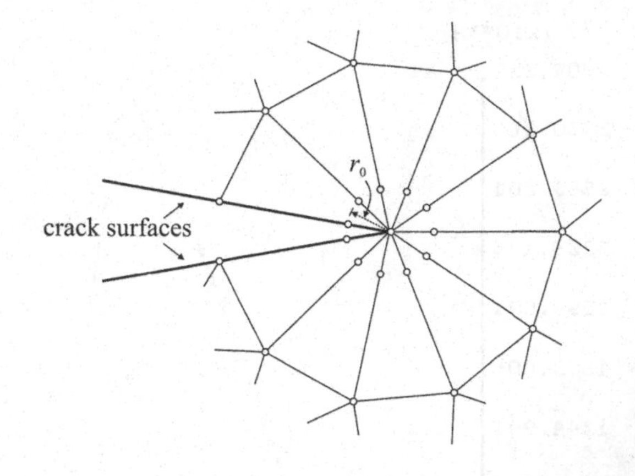

Fig. 8.68 Displacements at nodal points located behind the crack tip

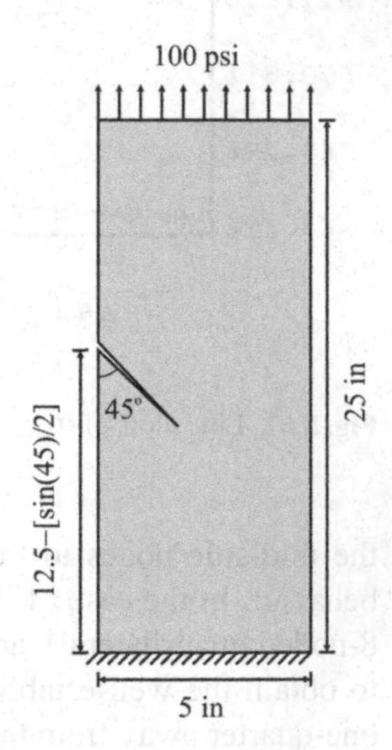

Fig. 8.69 Geometry of the strip with inclined edge crack

in which r_0, usually restricted to one or two percent of the crack length, is the distance from the crack tip to the first side-node behind the crack tip, as shown in Fig. 8.68. The shear modulus is G, and the parameters are $\kappa = 3-\nu/1+\nu$ and $\kappa = 3-4\nu$ for plane stress and strain idealizations, respectively.

Under plane strain assumptions, the computation of the stress intensity factors within ANSYS is demonstrated by considering a strip with an inclined edge crack, as shown in Fig. 8.69. The crack is 1 in long and has an inclination angle of 45°. The width and length of the strip are 5 and 25 in, respectively. The bottom surface of the strip is constrained in both directions while the top surface is subjected to a tensile load of 100 psi. Elastic modulus and Poisson's ratio of the strip are 30×10^6 psi and 0.3, respectively.

In the ANSYS solution, special meshing around the crack tip is utilized. It relocates the mid-side nodes to one-quarter away from the crack tip. Coordinates of the keypoints are listed in Table 8.4. Note that keypoint 2 is located at the crack tip, and

Table 8.4 Coordinates of the keypoints

Keypoint no.	x	y
1	0	$12.5+\frac{\sin(45)}{2}$
2	cos(45)	$12.5-\frac{\sin(45)}{2}$
3	0	$12.5+\frac{\sin(45)}{2}$
4	0	0
5	5	0
6	5	$12.5-\left\{[5-\cos(45)]\times\tan(45)+\frac{\sin(45)}{2}\right\}$
7	5	25
8	0	25

Table 8.5 Line-keypoint correspondence in the solid model

Line no.	Keypoint 1	Keypoint 2
1	4	5
2	5	6
3	6	2
4	2	1
5	1	4
6	6	7
7	7	8
8	8	3
9	3	2

keypoints 1 and 3 are coincident, each belonging to the opposite crack faces. Line numbers with their corresponding keypoints are listed in Table 8.5. The goal is to obtain stress intensity factors, as well as the displacement and stress fields.

Model Generation

- Define element type (**ET** command) using the following menu path:

Main Menu > Preprocessor > Element Type > Add/Edit/Delete

- Click on ***Add***.
- Select ***Solid*** immediately below ***Structural Mass*** from the left list and ***Quad 8node 183*** on the right list; click on ***OK***.
- Click on ***Options***.

– *PLANE183 element type options* dialog box appears; select ***Triangle*** from the pull-down menu corresponding to ***Element shape K1*** and ***Plane strain*** from the pull-down menu corresponding to ***Element behavior K3***.
– *Click* on ***OK***; click on ***Close***.

• Specify material properties (**MP** command) using the following menu path:

Main Menu > Preprocessor > Material Props > Material Models

– *Define Material Model Behavior* dialog box appears; in the right window, successively left-click on ***Structural***, ***Linear***, ***Elastic***, and, finally, ***Isotropic***, which brings up another dialog box.
– In the new dialog box, enter ***30E6*** for ***EX*** and ***0.3*** for ***PRXY***; click on ***OK***.
– Close the *Define Material Model Behavior* dialog box by using the following menu path:

Material > Exit

• Change the default angular unit to degrees (***AFUN** command) using the following menu path:

Utility Menu > Parameters > Angular Units

– *Angular Units for Parametric Functions* dialog box appears; select ***Degrees DEG*** from the pull-down menu; click on ***OK***.

• Create keypoints (**K** command) using the following menu path:

Main Menu > Preprocessor > Modeling > Create > Keypoints > In Active CS

– *Create Keypoints in Active Coordinate System* dialog box appears. Referring to Table 8.4, enter ***0*** and ***12.5+SIN(45)/2*** for ***X*** and ***Y***, leaving the text fields for ***NPT*** and ***Z*** blank. Click on ***Apply***.
– Repeat the procedure for the remaining keypoints (2 through 8). When creating keypoint 8, click on ***OK*** after entering the coordinates.

• Create lines (**L** command) using the following menu path:

Main Menu > Preprocessor > Modeling > Create > Lines > Lines > Straight Line

– *Pick Menu* appears, prompting the user to pick two keypoints forming the line. Referring to Table 8.5, pick the correct keypoints. When picking keypoint 1 or 3, ANSYS displays a warning message informing the user that there are two coincident keypoints at the particular location. By clicking on the ***Next*** button in this message, pick the correct keypoint.

• Turn line and keypoint numbering on (**/PNUM** command) using the following menu path:

Utility Menu > PlotCtrls > Numbering

- *Plot Numbering Controls* dialog box appears. Click on the boxes next to ***KP Keypoint numbers*** and ***LINE Line numbers*** (this places checkmarks), and select ***Numbers only*** from the ***[/NUM] Numbering shown*** with the pull-down menu. Click on ***OK***.
- Plot lines (**LPLOT** command) using the following menu path:

Utility Menu > Plot > Lines

- Figure 8.70 shows the line plot with both keypoint and line numbers printed. Observe that keypoints 1 and 3 and lines 4 and 9 are coincident.

• Two areas are created. The first area utilizes lines 1–5 while the second area is formed by lines 6, 7, 8, 9, and 3. Create areas using lines (**AL** command) using the following menu path:

Main Menu > Preprocessor > Modeling > Create > Areas > Arbitrary > By Lines

- *Pick Menu* appears, prompting the user to pick lines forming the area. Pick lines 1 through 5 (in this order) and click on ***OK*** in the *Pick Menu*. When picking line 4, ANSYS informs the user that there are two coincident lines at the picked location. Make sure to pick line 4.
- Repeat the same procedure for the second area by picking lines 6, 7, 8, 9, and 3. Similar to the previous case, make sure to pick line 9 (instead of line 4).

• For the stress intensity factor calculations, a local coordinate system aligned with the crack faces is needed. Create a local coordinate system using 3 keypoints (**CSKP** command) using the following menu path:

Utility Menu > WorkPlane > Local Coordinate Systems > Create Local CS > By 3 Keypoints

- *Pick Menu* appears; pick keypoints 2, 6, and 7 (in this order) and click on ***OK*** in the *Pick Menu*.
- *Create CS By 3 KPs* dialog box appears; click on ***OK***.
- The local coordinate system is now active. Activate the global Cartesian coordinate system (**CSYS** command) using the following menu path:

Utility Menu > WorkPlane > Change Active CS to > Global Cartesian

• Specify keypoint 2 to be the crack tip so that the elements around it have the singular stress capability (**KSCON** command) using the following menu path:

Main Menu > Preprocessor > Meshing > Size Cntrls > Concentrat KPs > Create

- *Pick Menu* appears; pick keypoint 2; click on ***OK*** in the *Pick Menu*.
- *Concentration Keypoint* dialog box appears; enter ***1/20*** for ***DELR Radius of 1st row of elems*** and ***6*** for ***NTHET No of elems around circumf.*** Select ***Skewed 1/4pt*** from the ***KCTIP midside node position*** pull-down menu and click on ***OK***.

• Specify mesh density around keypoints (**KESIZE** command) using the following menu path:

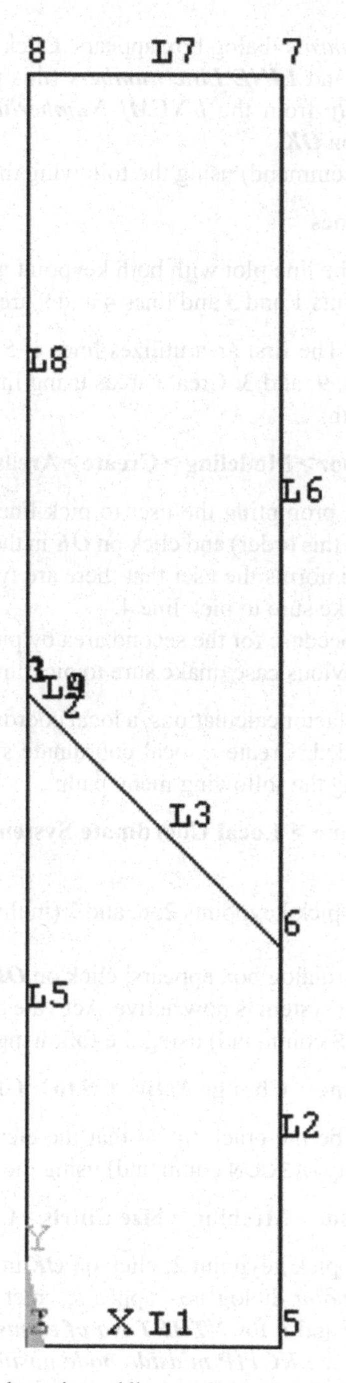

Fig. 8.70 Line plot with both keypoint and line numbers printed

Main Menu > Preprocessor > Meshing > Size Cntrls > ManualSize > Keypoints > All KPs

- *Element Size at All Keypoints* dialog box appears; enter ***5/3*** for ***SIZE Element edge length***; click on ***OK***.
- Specify mesh density around specific keypoints (**KESIZE** command) using the following menu path:

Main Menu > Preprocessor > Meshing > Size Cntrls > ManualSize > Keypoints > Picked KPs

- *Pick Menu* appears; pick keypoints 1 and 3 (since these keypoints are coincident, click on the location twice to pick both of them); click on ***OK*** in the *Pick Menu*.
- *Element Size at Picked Keypoints* dialog box appears; enter ***1/3*** for ***SIZE Element edge length***; click on ***OK***.
- Specify mesh density around crack tip (**KESIZE** command) using the following menu path:

Main Menu > Preprocessor > Meshing > Size Cntrls > ManualSize > Keypoints > Picked KPs

- *Pick Menu* appears; pick keypoint 2; click on ***OK*** in the *Pick Menu*.
- *Element Size at Picked Keypoints* dialog box appears; enter ***1/30*** for ***SIZE Element edge length***; click on ***OK***.

- Mesh the areas (**AMESH** command) using the following menu path:

Main Menu > Preprocessor > Meshing > Mesh > Areas > Free

- *Pick Menu* appears; click on ***Pick All***.
- Close the *Warning Window*.
- Zoom in around the crack tip and observe the mesh pattern around it (Fig. 8.71).

Solution

- Apply displacement constraints (**D** command) using the following menu path:

Main Menu > Solution > Define Loads > Apply > Structural > Displacement > On Nodes

- *Pick Menu* appears; pick the nodes along $y=0$ (x-axis); click on ***OK*** in *Pick Menu*.
- *Apply U, ROT on Nodes* dialog box appears; highlight ***UY***; click on ***Apply***.
- *Pick Menu* reappears; pick the bottom-left corner node ($x=0$, $y=0$); click on ***OK*** in *Pick Menu*.
- *Apply U, ROT on Nodes* dialog box reappears; highlight ***UX*** (leave the ***UY*** highlighted) and click on ***OK***.

- Apply surface load (**SF** command) using the following menu path:

Fig. 8.71 Mesh pattern around the crack tip

Main Menu > Solution > Define Loads > Apply > Structural > Pressure > On Nodes

- *Pick Menu* appears; pick the nodes along $y=25$; click on ***OK*** in *Pick Menu*.
- *Apply PRES on nodes* dialog box appears; enter ***−100*** for ***VALUE Load PRES value***; click on ***OK***.

• Obtain solution (**SOLVE** command) using the following menu path:

Main Menu > Solution > Solve > Current LS

- *Confirmation Window* appears along with *Status Report Window*.
- Review status; if OK, close the *Status Report Window* and click on ***OK*** in the *Confirmation Window*.
- If a dialog box with a message reminding the user about previous warnings issued appears, click on ***Yes***.
- Wait until ANSYS responds with ***Solution is done!***

Postprocessing

• Review deformed shape (**PLDISP** command) using the following menu path:

Main Menu > General PostProc > Plot Results > Deformed Shape

- Select ***Def shape only***; click on ***OK***.
- The deformed shape near the crack is shown in Fig. 8.72 as it appears in the *Graphics Window*.

• Activate local coordinate system 11 (**CSYS** command) using the following menu path:

Fig. 8.72 Deformed shape of the crack (*left*) and the deformed shape in the vicinity of the crack tip (*right*)

Utility Menu > WorkPlane > Change Active CS to>Specified Coord Sys

- Change Active CS to Specified CS dialog box appears; enter **11** for **KCN Coordinate system number**; click on **OK**.
- Enforce the use of the same coordinate system (11) for the results calculations and display (**RSYS** command) using the following menu path:

Main Menu > General Postproc > Options for Outp

- Options for Output dialog box appears; select **Local system** from the **[RSYS] Results coord system** pull-down menu and enter **11** for **Local system reference no.**; click on **OK**.

- In order to calculate stress intensity factors, a path along the crack faces in the vicinity of the crack tip is defined (**PPATH** command) using the following menu path:

Main Menu > General Postproc > Path Operations > Define Path > By Nodes

Pick Menu appears; a total of 5 nodes are needed for this operation. Crack tip node needs to be picked first, followed by the two nodes closest to the crack tip along the top crack face. Finally, the two nodes closest to the crack tip along the bottom crack face are picked. The crack tip node number is 18 in this particular problem. Node numbers for the two nodes closest to the crack tip along the top and bottom faces are 47 and 48, and 521 and 522, respectively. Before picking the nodes, it is recommended that the user zoom in around the crack tip and plot elements as shown in Fig. 8.73. Pick nodes 18, 47, 48, 522, and 521 (in this order) and click on ***OK***. The nodal locations to be picked are also shown in Fig. 8.73 (denoted by small squares).

- *By Nodes* dialog box appears; enter a path name (say ***crck***); click on ***OK***.
- Close the new information window.

- Calculate stress intensity factors (**KCALC** command) using the following menu path:

Fig. 8.73 Elements around the crack tip is zoomed in for picking operation

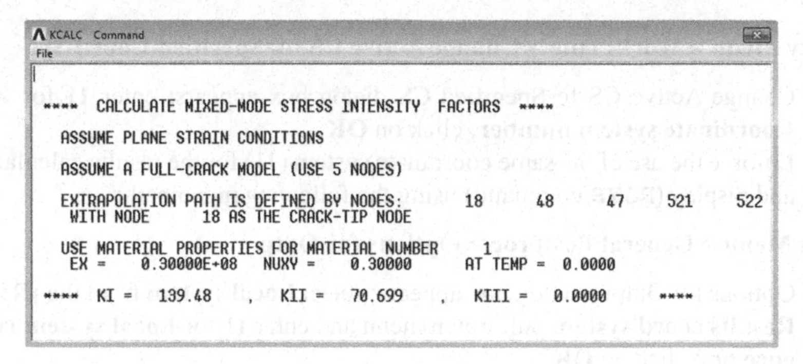

```
KCALC Command
File

**** CALCULATE MIXED-MODE STRESS INTENSITY FACTORS ****

ASSUME PLANE STRAIN CONDITIONS

ASSUME A FULL-CRACK MODEL (USE 5 NODES)

EXTRAPOLATION PATH IS DEFINED BY NODES:    18    48    47    521    522
 WITH NODE      18 AS THE CRACK-TIP NODE

USE MATERIAL PROPERTIES FOR MATERIAL NUMBER    1
 EX =    0.30000E+08   NUXY =    0.30000    AT TEMP =  0.0000

**** KI =   139.48    ,  KII =   70.699   ,  KIII =   0.0000    ****
```

Fig. 8.74 Stress intensity factor results

Main Menu > General PostProc > Nodal Calcs > Stress Int Factr

- *Stress Intensity Factor* dialog box appears; select ***Plane strain*** from the ***KPLAN Disp extrapolat based on*** pull-down menu and select ***Full-crack model*** from the ***KCSYM Model type*** pull-down menu. Click on ***OK***.
- Stress intensity factors are reported in a separate window (KI = 139.48, KII = 70.699), as shown in Fig. 8.74.

- Review normal and shear stresses ahead of the crack tip, in the direction of the crack. For this purpose, define a new path (**PPATH** command) using the following menu path:

Main Menu > General Postproc > Path Operations > Define Path > By Nodes

- *Pick Menu* appears; pick nodes 18 (crack tip) and 34, as shown in Fig. 8.75; click on ***OK*** in the *Pick Menu*.
- *By Nodes* dialog box appears; enter a path name (say ***strs***); click on ***OK***.

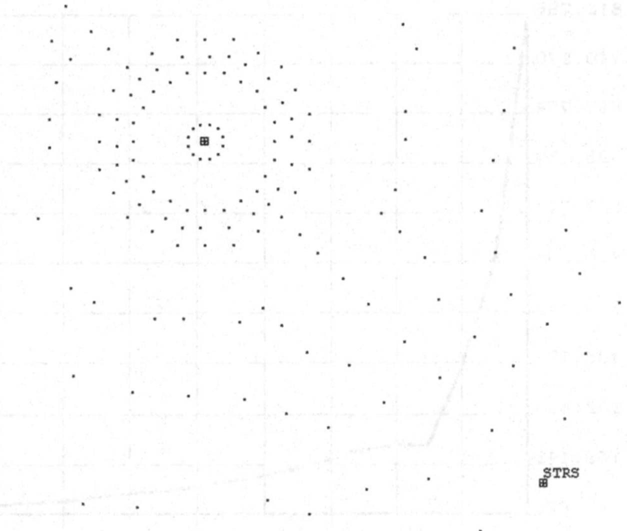

Fig. 8.75 Nodal plot showing the starting (crack tip, node 18) and ending (node 34) nodes for the path definition

- Close the new information window.
- Map stresses onto the path ***strs*** (**`PDEF`** command) using the following menu path:

Main Menu > General Postproc > Path Operations > Map onto Path

- *Map Result Items onto Path* dialog box appears; select ***Stress*** from the left list and ***Y-direction SY*** from the right list; click on ***Apply***.
- *Map Result Items onto Path* dialog box reappears; select ***Stress*** from the left list and ***XY-shear SXY*** from the right list and click on ***OK***.
- Plot stresses along the path on the graph (**`PLPATH`** command) using the following menu path:

Main Menu > General Postproc > Path Operations > Plot Path Item > On Graph

- *Plot of Path Items on Graph* dialog box appears; select ***SY*** from the list; click on ***OK***. The graph appears in the *Graphics Window*, as shown in Fig. 8.76. Similarly, shear stresses are plotted by selecting item ***SXY*** in the *Plot of Path Items on Graph* dialog box (shown in Fig. 8.77).

8.5 Dynamic Analysis

There are three commonly used dynamic analysis types in ANSYS: (i) modal analysis, (ii) harmonic analysis, and (iii) transient analysis.

The results related to these types of analyses can be reviewed in both postprocessors (*General Postprocessor* and *Time History Postprocessor*). The *General Postprocessor* is used to review results over the entire model at specific

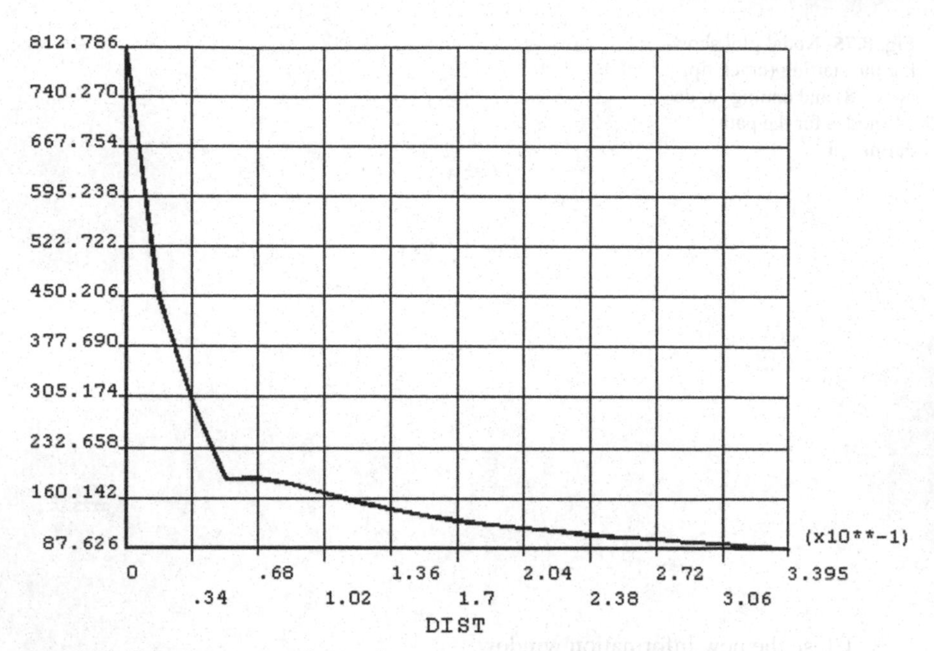

Fig. 8.76 Normal stress σ_y (in coordinate system 11) ahead of the crack tip

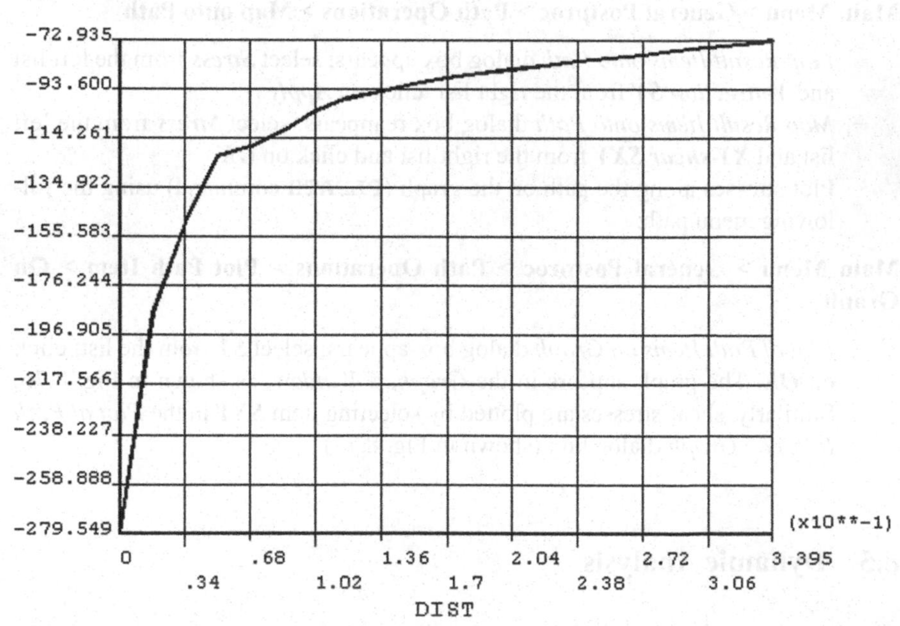

Fig. 8.77 Shear stress σ_{xy} (in coordinate system 11) ahead of the crack tip

times or frequencies while the *Time History Postprocessor* allows the user to review results at specific nodes in the model over the entire time or frequency range.

8.5.1 Modal Analysis

If the structural vibration is of concern in the absence of time-dependent external loads, a modal analysis is performed. Because the structural frequencies are not known a priori, the finite element equilibrium equations for this type of analysis involve the solution of homogeneous algebraic equations whose eigenvalues correspond to the frequencies, and the eigenvectors represent the vibration modes. The following steps are used in a typical modal analysis in ANSYS:

Build the model.
Apply loads and obtain the solution.
Expand the modes.
Review the results.

In certain cases, especially if the model has of a large number of degrees of freedom, it is advantageous to define *Master Degrees of Freedom* (MDOF). This procedure condenses the full matrices describing the structure into a smaller size, thus reducing the computational cost significantly. The only boundary conditions that are permissible in *modal analysis* are zero displacements. Any constraints/loads that are non-zero are ignored in the analysis. Once the modal analysis is complete, the solution is *expanded* to find results related to the complete structure—not just the MDOF. The results include natural frequencies, mode shapes, and corresponding parametric (relative) stress fields.

8.5.1.1 Modal Analysis of a Bracket

The bracket shown in Fig. 8.29 is clamped at the two top holes. The reduced method of modal analysis is used here, which requires master degrees of freedom. Master degrees of freedom are automatically selected by ANSYS. The reduced modal analysis is then expanded for the number of modes desired (in this case, four). The goal is to obtain the modal frequencies and corresponding mode shapes.

Model Generation

The finite element model of this bracket is created in a separate problem (Sect. 8.1.5.1). Therefore, the model will not be regenerated here; it will be generated interactively using an input file. It is worth noting that the components consisting of the nodes along the holes are defined in the input file.

- Create the model by reading the input file "***bracket.inp***" from the CD-ROM location ***Input_Files\ch08\bracket.inp*** using the following menu path:

Utility Menu >File > Read Input from

 – Browse for the ***bracket.inp*** file in the folder given above; click on ***OK***.
 – Wait until the model is generated.

Solution

- Specify the *Analysis Type* as *Modal*, using the following menu path (**ANTYPE** command), which will bring up the *New Analysis* dialog box:

Main Menu > Solution > Analysis Type > New Analysis

- Click on the ***Modal*** radio-button; click on ***OK***.
- Print 4 reduced mode shapes using the following menu path:

Main Menu > Solution > Analysis Type > Analysis Options

 – Click on the ***Reduced*** radio-button; click on ***OK***, which will bring up another dialog box.
 – Enter ***4*** for ***PRMODE***; click on ***OK***.

- Specify the master degrees of freedom using the following menu path:

Main Menu > Solution > Master DOFs > Program Selected

 – In the dialog box, type ***20*** for ***NTOT Total no. of master DOF***; click on the checkbox to show ***Yes*** for ***NRMDF Exclude rotational DOF***; click on ***OK***.

- Expand the first four modes using the following menu path:

Main Menu > Solution > Load Step Opts > ExpansionPass > Single Expand > Expand Modes

 – Enter ***4*** for ***NMODE***; click on ***OK***.

- Constrain displacement and rotation degrees of freedom along the top-left and -right holes. For this purpose, first select the components created earlier for these holes (***TL_BOLT*** and ***TR_BOLT***) using the following menu path:

Utility Menu > Select > Comp/Assembly > Select Comp/Assembly

 – A dialog box appears; click on the ***by component name*** radio-button; click on ***OK***.
 – A new dialog box with the components listed appears; highlight ***TL_BOLT***; click on ***OK***. This action selects the nodes along the top-left hole.
 – Specify the displacement boundary conditions using the following menu path:

Fig. 8.78 Listing of the mode frequencies

SET,LIST Command

File

***** INDEX OF DATA SETS ON RESULTS FILE *****

SET	TIME/FREQ	LOAD STEP	SUBSTEP	CUMULATIVE
1	22.797	1	1	1
2	40.998	1	2	2
3	70.657	1	3	3
4	185.39	1	4	4

Main Menu > Solution > Define Loads > Apply>Structural > Displacement>On Nodes

- *Pick Menu* appears; click on ***Pick All***.
- In the new dialog box, highlight ***All DOF***; click on ***OK***.
- Repeat the same procedure for the top-right hole (***TR_BOLT***).
- Select everything (**ALLSEL** command) using the following menu path:

Utility Menu > Select > Everything

- Obtain the solution using the following menu path:

Main Menu > Solution > Solve > Current LS

- Close the/*STATUS information window*.
- Click on ***OK*** to start the solution.
- If a dialog box with a message reminding the user about previous warnings issued appears, click on ***Yes***.
- Wait until ANSYS responds with ***Solution is done!***

Postprocessing

- Review the results using the following menu path:

Main Menu > General Postproc > Results Summary

- The mode frequencies appear in a separate window, as shown in Fig. 8.78.

- Plot the mode shapes.
  - Set the solution to ***First Set*** using the following menu path:

Main Menu > General Postproc > Read Results > First Set

- Plot the mode shape using the following menu path:

Main Menu > General Postproc > Plot Results > Deformed Shape

- Click on ***Def + undef edge*** radio-button; click on ***OK***.
- The mode shape is shown in Fig. 8.79 as it appears in the *Graphics Window*.

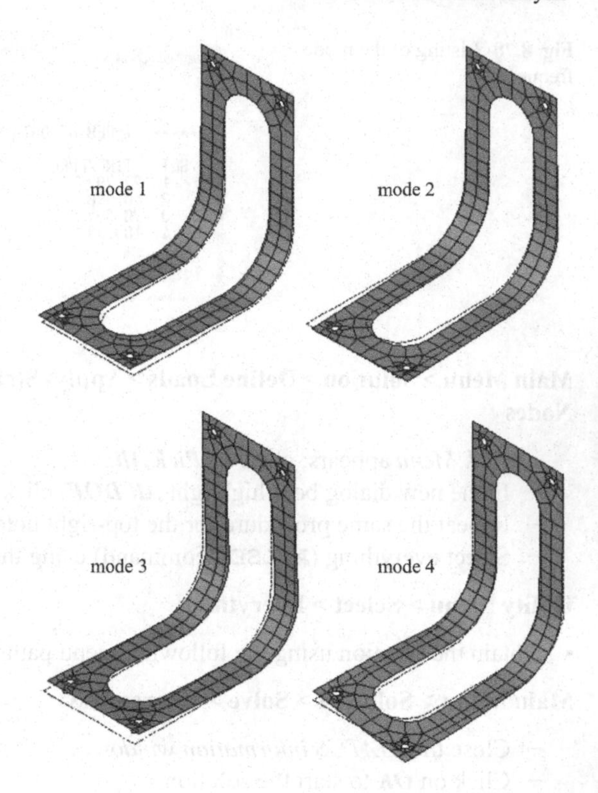

Fig. 8.79 Mode shapes 1 through 4

– Set the solution to ***Second Set*** using the following menu path:

Main Menu > General Postproc > Read Results > Next Set

– Plot the mode shape using the following menu path:

Main Menu > General Postproc > Plot Results > Deformed Shape

– Click on ***Def+ undef edge*** radio-button; click on ***OK***.
– The mode shape is shown in Fig. 8.79 as it appears in the *Graphics Window*.
– Repeat for the third and fourth modes.

8.5.1.2 Vibration of an Automobile Suspension

An automobile suspension system is simplified to consider only two major motions of the system: (i) up-and-down linear motion of the body, and (ii) pitching angular motion of the body.

The body is idealized as a lumped mass with weight, W, and radius of gyration, r, as shown in Fig. 8.80. The equivalent finite element model is depicted in Fig. 8.81. The numerical values of the geometric parameters used in Fig. 8.80 are as follows: $l_1 = 4.5$ ft, $l_2 = 5.5$ ft, and $r = 4$ ft; the gravitational acceleration is $g = 32.2$ ft/s^2.

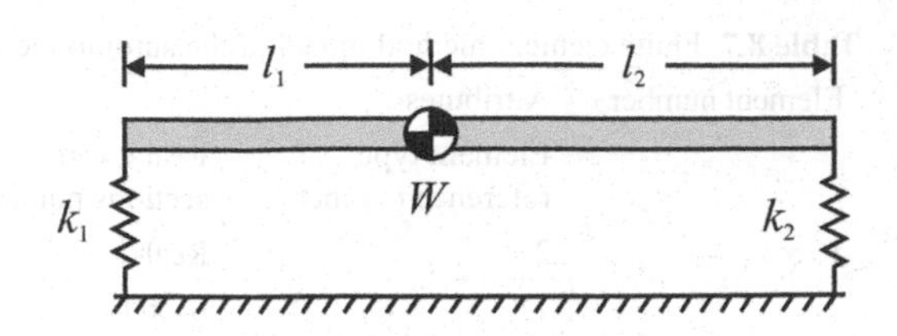

Fig. 8.80 Schematic of the automobile suspension system

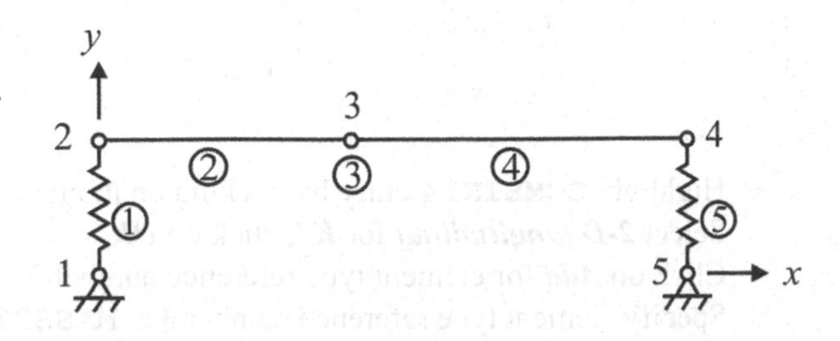

Fig. 8.81 Finite element method model of the automobile suspension system

Table 8.6 Node numbers and coordinates of the automobile suspension system model

Node number	x	y
1	0.0	0.0
2	0.0	1.0
3	4.5	1.0
4	10.0	1.0
5	10.0	0.0

The corresponding coupled frequencies, f_1 and f_2, are to be determined. The elastic modulus of the beam is $E=4\times10^9$ psf, the lumped weight is $W=3220$ lb, the spring constants are $k_1=2400$ lb/ft and $k_2=2600$ lb/ft. The nodal coordinates and element properties and connectivity are given in Tables 8.6 and 8.7, respectively.

Model Generation

- Define the element types (**`ET`** command) using the following menu path, which brings up the *Element Types* dialog box:

Main Menu > Preprocessor > Element Type > Add/Edit/Delete

  - Click on ***Add***.
  - Specify element type reference number 1 as **`BEAM188`** (***Beam*** in the left list immediately below ***Structural Mass***; ***2 node 188*** in the right list); click on ***OK***.
  - Click on ***Add*** for element type reference number 2.
  - Specify element type reference number 2 as **`COMBIN14`** (***Combination*** in the left list, ***Spring-damper 14*** in the right list); click on ***OK***.

Table 8.7 Finite element method model of the automobile suspension system

Element number	Attributes			Nodes	
	Element type reference number	Real const. & sections number	Material reference number	1	2
1	2	Real-1	1	1	2
2	1	Sect-1	1	2	3
3	3	Real-2	1	3	
4	1	Sect-2	1	3	4
5	2	Real-3	1	4	5

- Highlight **COMBIN14** entry by clicking on it; click on ***Options***.
- Select ***2-D longitudinal*** for ***K3***; click on ***OK***.
- Click on ***Add*** for element type reference number 3.
- Specify element type reference number 3 as **MASS21** (***Structural Mass*** in the left list, ***3D mass 21*** in the right list).
- Highlight **MASS21** entry by clicking on it; click on ***Options***.
- Select ***2-D w rot inert*** for ***K3***; click on ***OK***.
- Click on ***Close***.

• Specify the real constants (**R** command) using the following menu path:

Main Menu > Preprocessor > Real Constants > Add/Edit/Delete

- Click on ***Add***.
- Specify the left spring constant (k_1) as real constant set number 1:
  - Highlight ***Type 2 COMBIN14*** entry by clicking on it; click on ***OK***.
  - Enter ***2400*** for ***K***; click on ***OK***.
- Specify the mass properties as real constant set number 2:
  - Click on ***Add***.
  - Highlight ***Type 3 MASS21*** entry by clicking on it; click on ***OK***.
  - Enter ***100*** for ***MASS*** and ***1600*** for ***IZZ***; click on ***OK***.
- Specify the right spring constant (k_2) as real constant set number 3:
  - Highlight ***Type 2 COMBIN14*** entry by clicking on it; click on ***OK***.
  - Enter ***2600*** for ***K***; click on ***OK***.
- Click on ***Close***.

• Specify geometry for the beam (**SECTYPE** command) using the following menu path:

Main Menu > Preprocessor > Sections > Beam > Common Sections

- *Beam Tool* dialog box appears; enter ***1*** for ***B*** and ***1*** for ***H***; click on ***Apply***.
- Repeat the same procedure for ***ID 2***. When done, click on ***OK*** (instead of ***Apply***).

• Specify material properties for the beam (**MP** command) using the following menu path:

Main Menu > Preprocessor > Material Props > Material Models

- In the *Define Material Model Behavior* dialog box, in the right window, successively left-click on ***Structural***, ***Linear***, ***Elastic***, and, finally, ***Isotropic***, which brings up another dialog box.
- Enter ***4e9*** for ***EX***; click on ***OK***.
- Close the *Define Material Model Behavior* dialog box by using the following menu path:

Material > Exit

- Create the nodes (**N** command) using the following menu path, which brings up the *Create Nodes in Active Coordinate System* dialog box:

Main Menu > Preprocessor > Modeling > Create > Nodes > In Active CS

- A total of 5 nodes will be created.
- Referring to Table 8.6, enter the ***x***- and ***y***-coordinates of node 1; click on ***Apply.*** This action will keep the *Create Nodes in Active Coordinate System* dialog box open. If the ***Node number*** field is left blank, then ANSYS will assign the lowest available node number to the node that is being created.
- Repeat the same procedure for the nodes 2, 3 and 4.
- After creating node 5, click on ***OK*** (instead of ***Apply***).

- Create the elements (the element attributes and connectivity information are given in Table 8.7).
  - Referring to Table 8.7, element 1 is a spring element (*type 2*) with spring stiffness 2400 lb/ft (*real constant set number 1*), composed of nodes 1 and 2. By default, the *element attributes* are set to 1. Before element 1 is created, the element type attribute needs to be specified as 2. Perform this operation by using the following menu path, which will bring up the *Element Attributes* dialog box:

Main Menu > Preprocessor > Modeling > Create > Elements > Elem Attributes

- Select ***2 COMBIN14*** in the ***[TYPE] Element type number*** pull-down menu; click on ***OK.*** Any element that is created after this point will have these attributes (i.e., element type number 2, material number 1, and real constant set number 1).
- Create element 1 (**E** command) using the following menu path, which brings up a *Pick Menu*:

Main Menu > Preprocessor > Modeling > Create > Elements > Auto Numbered > Thru Nodes

- Pick (by clicking in the *Graphics Window*) nodes 1 and 2; click on ***OK*** in the *Pick Menu.*

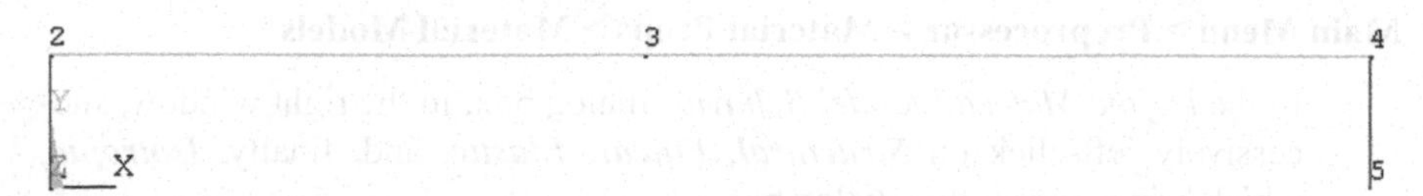

Fig. 8.82 Elements of the automobile suspension system model

```
ELIST Command
File

LIST ALL SELECTED ELEMENTS.  (LIST NODES)

   ELEM MAT TYP REL ESY SEC        NODES

      1   1   2   1   0   1      1   2
      2   1   1   1   0   1      2   3      0
      3   1   3   2   0   1      3
      4   1   1   2   0   2      3   4      0
      5   1   2   3   0   1      4   5
```

Fig. 8.83 Listing of the elements

- Create elements 2 through 5 in the same manner, each time setting the correct attributes as explained above.
- The elements should appear in the *Graphics Window*, as shown in Fig. 8.82.
- The commands equivalent to changing attributes are: **TYPE**, **MAT**, **REAL**, and **SECNUM**
- Obtain and review the list of elements by using the following menu path:

Utility Menu > List > Elements > Nodes + Attributes

Observe attributes (element type, real constants, material properties, and section numbers) and node numbers for each element (Fig. 8.83).

Solution

- Specify the *Analysis Type* as *Modal*, using the following menu path (**ANTYPE** command), which will bring up the *New Analysis* dialog box:

Main Menu > Solution > Analysis Type > New Analysis

- Click on the ***Modal*** radio-button; click on ***OK***.

- Expand the first mode (**MXPAND** command) using the following menu path:

Main Menu > Solution > Load Step Opts > ExpansionPass > Single Expand > Expand Modes

- Enter *1* for *NMODE*; click on ***OK***.

- Print two reduced mode shapes (`MODOPT` command) using the following menu path:

Main Menu > Solution > Analysis Type > Analysis Options

  - Click on the ***Reduced*** radio-button; click on ***OK***, which brings up another dialog box.
  - Enter *2* for ***PRMODE***; click on ***OK***.

- Specify the master degrees of freedom (`M` command) using the following menu path:

Main Menu > Solution > Master DOFs > User Selected>Define

  - After picking node 3, click on ***OK*** in the *Pick Menu*, which brings up the *Define Master DOFs* dialog box.
  - Select ***UY*** for ***Lab1*** and ***ROTZ*** for ***Lab2–6***; click on ***OK***.

- Specify boundary conditions (`D` command):
  - Constrain x- and y-displacements at nodes 1 and 5 using the following menu path:

Main Menu > Solution > Define Loads > Apply > Structural > Displacement > On Nodes

  - Pick nodes 1 and 5; click on ***OK*** in the *Pick Menu*, which brings up the *Apply U, ROT on Nodes* dialog box.
  - Highlight ***UX*** and ***UY***; click on ***OK***.
  - Constrain x-displacement at node 3 in the same manner.

- Instruct ANSYS to print the solution at every ***substep*** using the following menu path (`OUTPR` command):

Main Menu > Solution > Load Step Opts > Output Ctrls > Solu Printout

  - Select ***Nodal DOF solu*** from the first pull-down menu.
  - Click on the ***Every substep*** radio-button; click on ***OK***.

- Obtain the solution using the following menu path (`SOLVE` command):

Main Menu > Solution > Solve > Current LS

  - Close the*/STATUS information window*.
  - Click on ***OK*** to start the solution.
  - If a dialog box with a message reminding the user about previous warnings issued appears, click on ***Yes***.
  - Wait until ANSYS responds with ***Solution is done!***

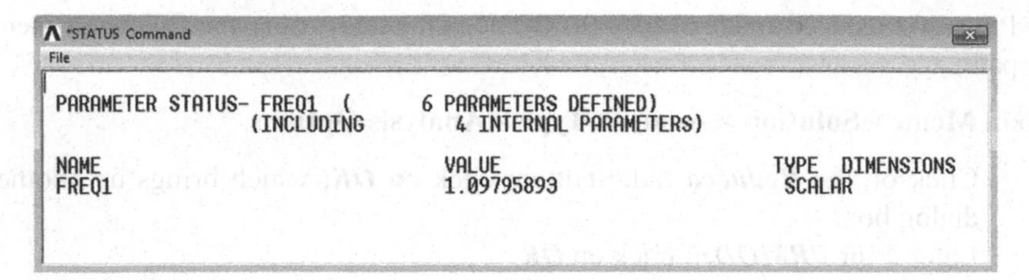

Fig. 8.84 First mode frequency for the automobile suspension system model

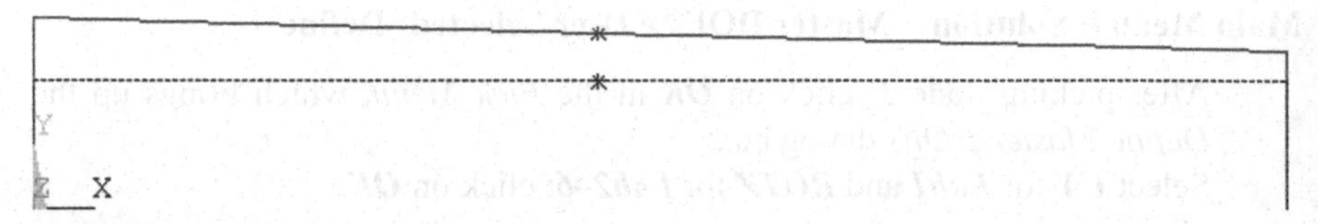

Fig. 8.85 First mode shape for the automobile suspension system model

Postprocessing

- Retrieve the coupled frequencies and store them in user-defined parameters using the following menu path:

Utility Menu > Parameters > Get Scalar Data

  - Highlight ***Results data*** in the left list, ***Modal results*** in the right list; click on ***OK***, which brings up another dialog box.
  - Enter ***freq1*** for ***Name of parameter to be defined*** field, and enter ***1*** for ***Mode number N***; click on ***OK***.
  - Repeat the procedure above to define parameter ***freq2*** corresponding to ***Mode 2***.

- List the parameters defined above using the following menu path:

Utility Menu > List > Other > Named Parameter

  - Highlight ***freq1*** or ***freq2*** from the list of parameters; click on ***OK***.
  - Figure 8.84 shows the outcome of this action for ***freq1***.

- Plot the first mode shape.
  - Set the solution to *First Set* using the following menu path:

Main Menu > General Postproc > Read Results > First Set

  - Plot the mode shape using the following menu path:

Main Menu > General Postproc > Plot Results > Deformed Shape

  - Click on ***Def + undeformed*** radio-button; click on ***OK***.
  - The mode shape is shown in Fig. 8.85 as it appears in the *Graphics Window*.

8.5.2 *Harmonic Analysis*

When a structure is subjected to cyclic loading, the resulting response is expected to be cyclic as well. ANSYS provides the user with the capability to solve this class of problems through the *Harmonic* analysis option. The restrictions on a harmonic analysis are:

All loads must be sinusoidal functions of time.
All loads must have the same frequency.
The structure must exhibit linearly elastic behavior (no geometric and material nonlinearities).

Sinusoidal loads are specified through the parameters *amplitude*, *phase angle*, and *forcing frequency range*. *Amplitude* is the peak value of the load, and *phase angle* is the time lag between multiple loads that are out of phase with each other. On the complex plane, it is the angle measured from the real axis. Finally, *forcing frequency range* is the frequency range of the harmonic load (in cycles/time).

8.5.2.1 Harmonic Analysis of a Bracket

A bracket, shown in Fig. 8.29, experiences a harmonic (cyclic) loading at two points of application. The forcing frequency varies from 0 to 400 Hz; however, the two loads are 120° out of phase, i.e., $F_1 = 1.5 \sin \omega t$ and $F_2 = 1.5 \sin(\omega - 120)t$. The amplitude of each load is 1.5 lb. The bracket is clamped at its two upper bolt holes. The structural damping ratio is 0.03. The mode superposition method is used to calculate the harmonic response. This method requires a modal analysis to be performed first. A reduced modal analysis is used, so the harmonic analysis results in a reduced solution, which is then expanded using *Expansion Pass*. Note that the phase angle input is crucial for the expansion pass. *Time History Postprocessing* of the reduced analysis reveals the appropriate phase for expansion.

Model Generation

The finite element model of this bracket was created in Sect. 8.1.5.1 (modeling of a bracket). Therefore, the model will not be regenerated here; it will be generated using the input file ***bracket.inp***. It is worth noting that the components consisting of the nodes along the holes are defined in the input file.

- Specify the *jobname* as ***br_harm*** using the following menu path:

Utility Menu > File > Change Jobname

- In the dialog box, type ***br_harm*** in the ***[/FILNAM] Enter new jobname*** text field; click on the checkbox for ***New log and error files*** to show ***Yes***; click on ***OK***.

- Create the model by reading the input file ***bracket.inp*** from the CD-ROM location/***Input_Files\ch08\bracket.inp*** using the following menu path:

Utility Menu > File > Read Input from

- Browse for the ***bracket.inp*** from the folder given above; click on ***OK***.
- Wait until the model is generated.

Solution

- Specify the *Analysis Type* as *Modal*, using the following menu path (**ANTYPE** command), which brings up the *New Analysis* dialog box:

Main Menu > Solution > Analysis Type > New Analysis

- Click on ***Modal*** radio-button; click on ***OK***.

- Specify the *reduced modal analysis* method using the following menu path:

Main Menu > Solution > Analysis Type > Analysis Options

- Click on the ***Reduced*** radio-button; click on ***OK***.
- A new dialog box appears; click on ***OK***.

- Constrain displacement and rotation degrees of freedom along the top-left and -right holes. For this purpose, first select the components created earlier for these holes (***TL_BOLT*** and ***TR_BOLT***) using the following menu path:

Utility Menu > Select > Comp/Assembly > Select Comp/Assembly

- A dialog box appears; click on the ***by component name*** radio-button; click on ***OK***.
- A new dialog box with the components listed appears; highlight ***TL_BOLT***; click on ***OK***. This action selects the nodes along the top-left hole.
- Specify the displacement boundary conditions using the following menu path:

Main Menu > Solution > Define Loads > Apply > Structural > Displacement > On Nodes

- *Pick Menu* appears; click on ***Pick All***.
- In the new dialog box, highlight ***All DOF***; click on ***OK***.
- Repeat the same procedure for the top-right hole (***TR_BOLT***).

- Specify the master degrees of freedom at the four nodes of interest (as shown in Fig. 8.86) and specify that an additional 40° of freedom are to be selected automatically by ANSYS.

- Select the bottom corner nodes using the following menu path:

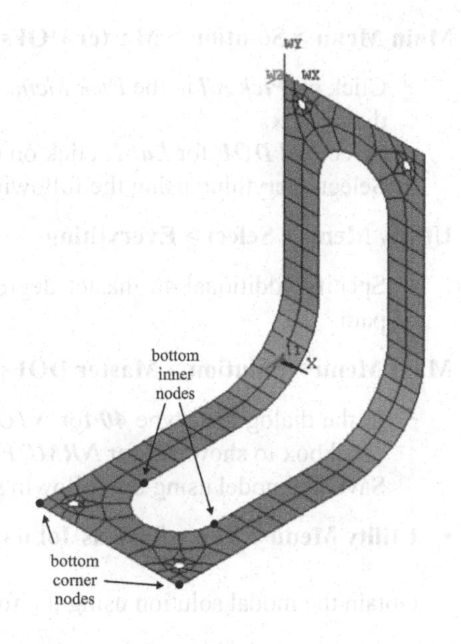

Fig. 8.86 Nodes at which master degrees of freedom are specified

Utility Menu > Select > Entities

- *Select Entities* dialog box appears; choose ***Nodes*** from the first pull-down menu and choose ***By Location*** in the second pull-down menu; click on the radio-button for ***Z coordinates*** and type ***5*** in the ***Min, Max*** text field; click on ***Apply***.
- In the same dialog box (*Select Entities* dialog box), click on the radio-button for ***X coordinates*** and type ***−1.99, 1.99*** in the ***Min, Max*** text field; click on the radio-button for ***Unselect;*** click on ***OK***.
- Specify the master degrees of freedom using the following menu path:

Main Menu > Solution > Master DOFs > User Selected>Define

- Click on ***Pick All*** in the *Pick Menu*, which brings up the *Define Master DOFs* dialog box.
- Select ***All DOF*** for ***Lab1***; click on ***OK***.
- Select bottom inner nodes using the following menu path:

Utility Menu > Select > Entities

- *Select Entities* dialog box appears; choose ***Nodes*** from the first pull-down menu and choose ***By Location*** in the second pull-down menu; click on the radio-button for ***X coordinates*** and type ***−1.01, 1.01*** in the ***Min, Max*** text field; click on the radio-button for ***From Full***; click on ***Apply***.
- In the same dialog box (*Select Entities* dialog box) click on the radio-button for ***Z coordinates*** and type ***3*** in the ***Min, Max*** text field; click on the radio-button for ***Reselect***; click on ***OK*** (the left node among the two is referred to as *bottom inner left node* in the postprocessing).
- Specify the master degrees of freedom using the following menu path:

Main Menu > Solution > Master DOFs > User Selected>Define

- Click on ***Pick All*** in the *Pick Menu*, which brings up the *Define Master DOFs* dialog box.
- Select ***All DOF*** for ***Lab1***; click on ***OK***.
- Select everything using the following menu path:

Utility Menu > Select > Everything

- Specify additional 40 master degrees of freedom using the following menu path:

Main Menu > Solution > Master DOFs > Program Selected

- In the dialog box, type ***40*** for ***NTOT Total no. of master DOF***; click on the checkbox to show ***Yes*** for ***NRMDF Exclude rotational DOF***; click on ***OK***.• Save the model using the following menu path:

- **Utility Menu > File > Save as Jobname.db**

- Obtain the modal solution using the following menu path:

Main Menu > Solution > Solve > Current LS

- Close the/*STATUS information window*.
- Click on ***OK*** to start the solution.
- If a dialog box with a message reminding the user about previous warnings issued appears, click on ***Yes***.
- Wait until ANSYS responds with ***Solution is done!***

- Exit the *Solution Processor* using the following menu path:

Main Menu > Finish

- Specify the *Analysis Type* as *Harmonic*, using the following menu path, which bring up the *New Analysis* dialog box:

Main Menu > Solution > Analysis Type > New Analysis

- Click on ***Harmonic*** radio-button; click on ***OK***.
- *A Warning Window* appears; click on ***OK***.

- Specify the *Mode Superposition* method using the following menu path:

Main Menu > Solution > Analysis Type > Analysis Options

- Select ***Mode superpos'n*** from the ***Solution method*** pull-down menu; click on ***OK***.
- *Mode Sup Harmonic Analysis* dialog box appears; click on radio-button for ***Cluster at modes***; click on ***OK***.

- Apply force at bottom-left and -right corner nodes.
  - Select bottom-left corner node using the following menu path:

Utility Menu > Select > Entities

- *Select Entities* dialog box appears; choose ***Nodes*** from the first pull-down menu and choose ***By Location*** in the second pull-down menu; click on the radio-button for ***Z coordinates*** and type ***5*** in the ***Min, Max*** text field; click on the ***From Full*** radio-button; click on ***Apply***.
- In the same dialog box (*Select Entities* dialog box) click on the radio-button for ***X coordinates*** and type ***−2*** in the ***Min, Max*** text field; click on the ***Reselect*** radio-button; click on ***OK***.
- Define scalar parameter using the following menu path:

Utility Menu > Parameters > Scalar Parameters

- In the text field within the dialog box, type ***AMPL=1.5***; click on ***Accept***. Close the *Scalar Parameters* dialog box by clicking on ***Close***.
- Apply the force on previously selected node (bottom-left corner) using the following menu path:

Main Menu > Solution > Define Loads > Apply > Structural > Force/Moment > On Nodes

- Click on ***Pick All*** in the *Pick Menu*.
- In the new dialog box, select ***FY*** from the first pull-down menu and type ***AMPL*** in the ***VALUE Real part of force/mom*** text field; click on ***OK***.
- Select bottom-right corner node using the following menu path:

Utility Menu > Select > Entities

- *Select Entities* dialog box appears; choose ***Nodes*** from the first pull-down menu and choose ***By Location*** in the second pull-down menu; click on the radio-button for ***Z coordinates*** and type ***5*** in the ***Min, Max*** text field; click on the ***From Full*** radio-button; click on ***Apply***.
- In the same dialog box (*Select Entities* dialog box) click on the radio-button for ***X coordinates*** and type ***2*** in the ***Min, Max*** text field; click on the ***Reselect*** radio-button; click on ***OK***.
- Select *degrees* as the unit for angles using the following menu path:

Utility Menu > Parameters > Angular Units

- In the dialog box, select ***Degrees*** for ***[*AFUN] Units for angular***; click on ***OK***.
- Define scalar parameter using the following menu path:

Utility Menu > Parameters > Scalar Parameters

- In the text field within the dialog box:
- Type ***PHASE=120***; click on ***Accept***.
- Type ***FR=AMPL * COS(PHASE)***; click on ***Accept***.
- Type ***FI=AMPL * SIN (PHASE)***; click on ***Accept***.
- Click on ***Close***.
- Apply the force on previously selected node (bottom-right corner) using the following menu path:

Main Menu > Solution > Define Loads > Apply > Structural > Force/Moment > On Nodes

- Click on ***Pick All*** in the *Pick Menu*.
- In the new dialog box, select ***FY*** from the pull-down menu and type ***FR*** in the ***VALUE Real part of force/mom*** text field and type ***FI*** in the ***VALUE2 Imag part of force/mom*** text field; click on ***OK***.
- Select everything using the following menu path:

Utility Menu > Select > Everything

- Specify the *harmonic frequencies* using the following menu path:

Main Menu > Solution > Load Step Opts > Time/Frequenc > Freq and Substeps

- Type ***0*** in the first text field and ***400*** in the second text field.
- Type ***40*** for ***NSUBST*** text field; select the ***Stepped*** radio-button; click on ***OK***.

- Specify the *damping* using the following menu path:

Main Menu > Solution > Load Step Opts > Time/Frequenc>Damping

- Type ***0.03*** in the ***DMPRAT*** text field (third text field); click on ***OK***.

- Save the model using the following menu path:

Utility Menu > File > Save as Jobname.db

- Obtain the harmonic solution using the following menu path:

Main Menu > Solution > Solve > Current LS

- Close the/*STATUS information window*.
- Click on ***OK*** to start the solution.
- If a dialog box with a message reminding the user about previous warnings issued appears, click on ***Yes***.
- Wait until ANSYS responds with ***Solution is done!***

Postprocessing

- Enter the *Time History Postprocessor* using the following menu path:

Main Menu > TimeHist Postpro

- *Time History Variables* dialog box appears. Click on the button with the green plus sign at the top-left to define a variable.
- *Add Time History Variable* dialog box appears.
- Successively click on ***Nodal Solution***, ***DOF Solution***, and ***Y-component of displacement***; in the text field, type ***UYRCNR***; click on ***OK***.
- *Pick Menu* appears; pick the bottom-right corner node; click on ***OK***.
- Note the new variable ***UYRCNR*** in *Time History Variables* dialog box.

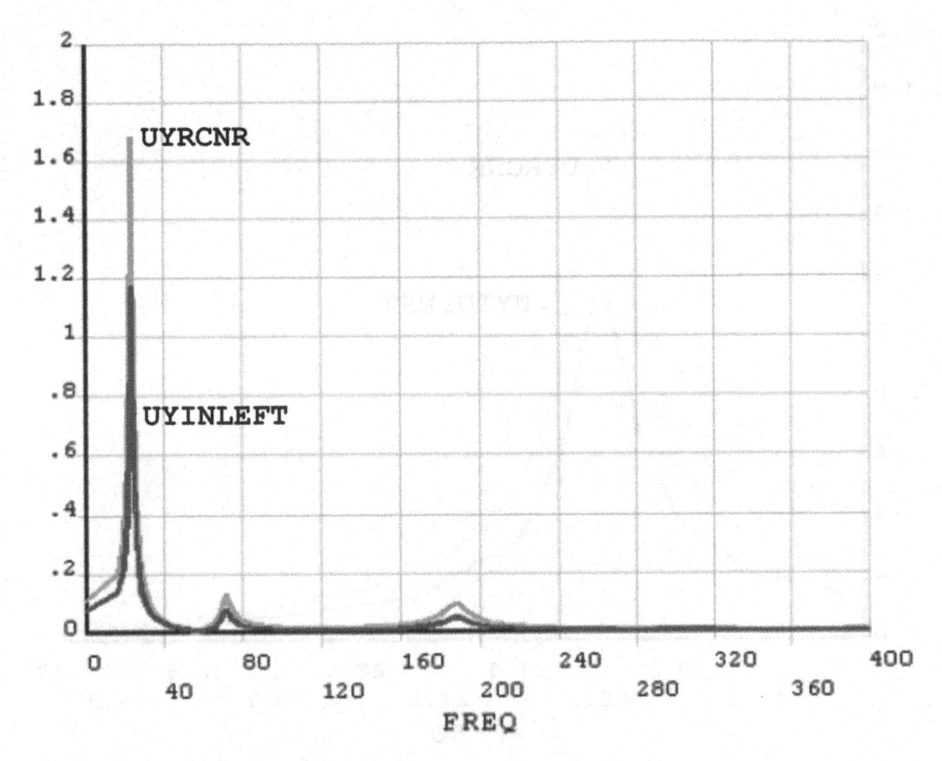

Fig. 8.87 Amplitudes of the displacements at two locations as functions of frequency under harmonic loading

- Add new variable for displacement in y-direction at the bottom inner-left node by clicking on the button with the green plus sign and successively clicking on ***Nodal Solution***, ***DOF Solution***, and ***Y-component of displacement***; in the text field, type ***UYINLEFT***; click on ***OK***.
- Pick the bottom inner-left node; click on ***OK***.
- Note the new variable ***UYINLEFT*** in *Time History Variables* dialog box.
- Highlight the rows ***UYRCNR*** and ***UYINLEFT*** from the list (by pressing ***Ctrl*** on the keyboard and clicking on the rows with left mouse
- button); click on the third from left button to plot the variation of these displacements.
- The plot appears in the *Graphics Window*, as shown in Fig. 8.87.
- Change the frequency axis (x-axis) limits in order to take a closer look at the first peak using the following menu path:

Utility Menu > PlotCtrls > Style > Graphs > Modify Axes

- In the dialog box, under ***[/XRANGE]***, select the radio button for ***Specified range*** and type ***16*** and ***37*** in the ***XMIN, XMAX Specified X range*** text fields; click on ***OK***.
- Obtain the plot with the new range using the following menu path:

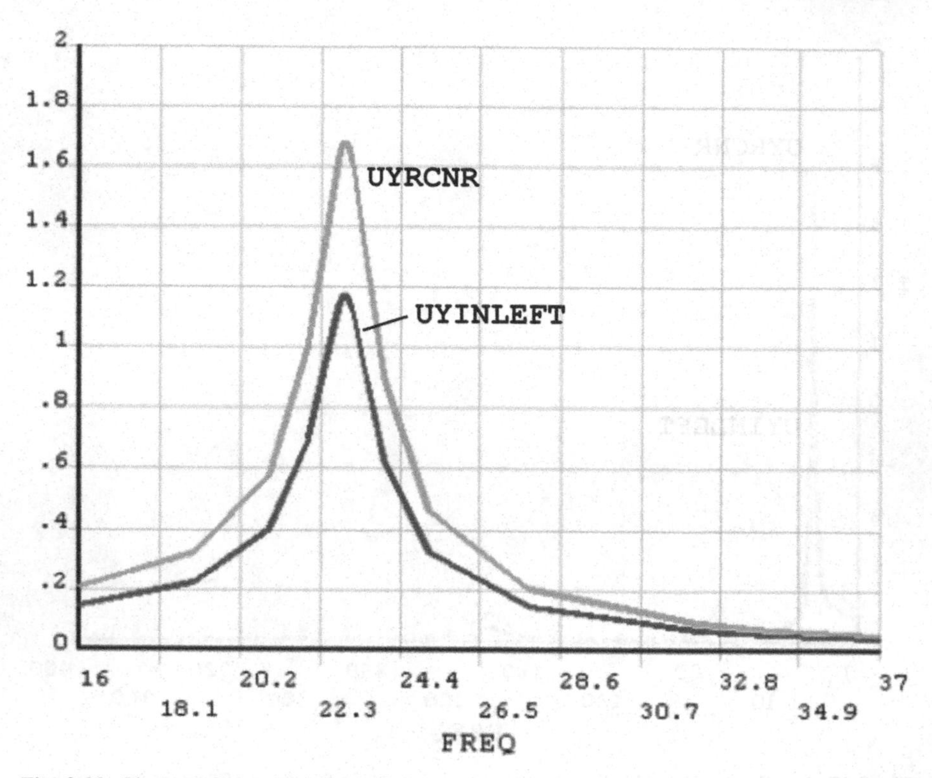

Fig. 8.88 Close-up view of amplitudes of the displacements at two locations as functions of frequency under harmonic loading

Utility Menu > Plot > Replot

– The plot appears in the *Graphics Window*, as shown in Fig. 8.88.
– Reset the frequency axis (x-axis) limits using the following menu path:

Utility Menu > PlotCtrls > Style > Graphs > Modify Axes

– In the dialog box, under ***[/XRANGE]***, select the radio button for ***Auto calculated***; click on ***OK***.
– In order to plot the phase angles, change the settings of the time history graphing using the following menu path:

Main Menu > TimeHist Postpro > Settings > Graph

– Select ***Phase Angle*** from the pull-down menu for ***[PLCPLX] Complex variable***; click on ***OK***.
– Click on the third from left button in *Time History Variables* dialog box to plot the variation of these phase angles.
– The plot appears in the *Graphics Window*, as shown in Fig. 8.89.

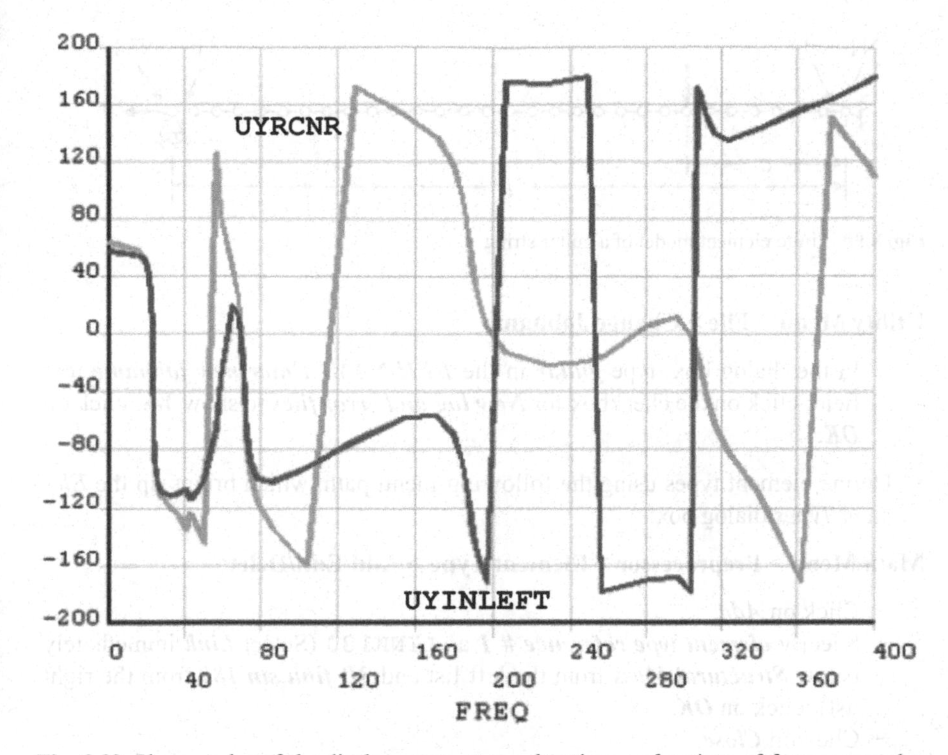

Fig. 8.89 Phase angles of the displacements at two locations as functions of frequency under harmonic loading

8.5.2.2 Harmonic Analysis of a Guitar String

A stainless-steel guitar string of length $l=710$ mm and diameter $d=0.254$ mm is stretched between two rigid supports by a tensioning force $F_1=84$ N applied at the right end (refer to Fig. 8.90). The string is struck at a location $c=165$ mm from the left end with a force $F_2=1$ N. The elastic modulus and density of the string are $E=190$ GPa and $\rho=7920$ kg/m^3, respectively. The fundamental frequency f_1 is determined using truss elements (**LINK180**), and a *Mode Superposition Harmonic Response Analysis* is performed. In order to perform the harmonic analysis, static and modal analyses are first performed.

Model Generation

- Specify the *jobname* as ***guitar*** using the following menu path:

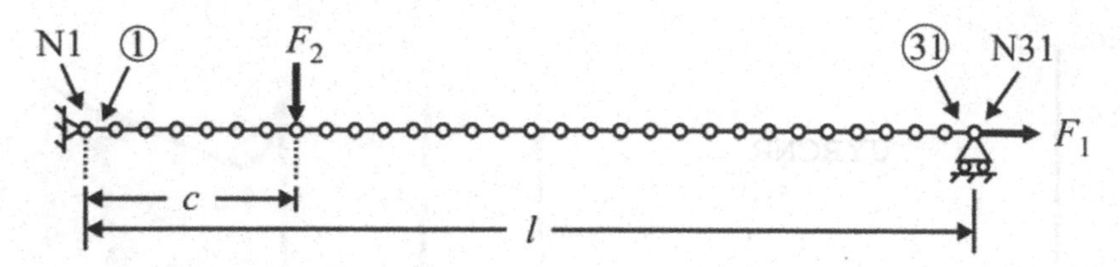

Fig. 8.90 Finite element model of a guitar string

Utility Menu > File > Change Jobname

- In the dialog box, type ***guitar*** in the ***[/FILNAM] Enter new jobname*** text field; click on the checkbox for ***New log and error files*** to show ***Yes***; click on ***OK***.

- Define element types using the following menu path, which brings up the *Element Types* dialog box:

Main Menu > Preprocessor > Element Type > Add/Edit/Delete

- Click on ***Add***.
- Specify ***element type reference # 1*** as `LINK180` (Select ***Link*** immediately below ***Structural Mass*** from the left list and ***3D finit stn 180*** from the right list); click on ***OK***.
- Click on ***Close***.

- Specify real constants using the following menu path:

Main Menu > Preprocessor > Real Constants > Add/Edit/Delete

- Click on ***Add***.
- Specify the cross-sectional area as real constant set number 1:
- Highlight ***Type 1 LINK180*** entry by clicking on it; click on ***OK***.
- Enter ***50671E−12*** for ***Cross-sectional area AREA***; click on ***OK***.
- Click on ***Close***.

- Specify material properties for the string using the following menu path:

Main Menu > Preprocessor > Material Props > Material Models

- In the *Define Material Model Behavior* dialog box, in the right window, successively left-click on ***Structural***, ***Linear***, ***Elastic***, and, finally, ***Isotropic***, which will bring up another dialog box.
- Enter ***190e9*** for ***EX***; click on ***OK***.
- In the *Define Material Model Behavior* dialog box, in the right window, left-click on ***Density***, which will bring up another dialog box.
- Enter ***7920*** for ***DENS***; click on ***OK***.
- Close the *Define Material Model Behavior* dialog box by using the following menu path:

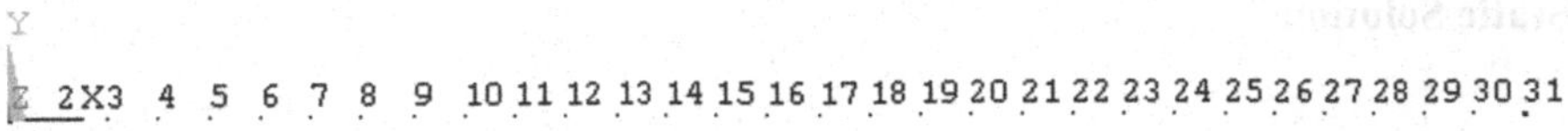

Y

Z 2X3 4 5 6 7 8 9 10 11 12 13 14 15 16 17 18 19 20 21 22 23 24 25 26 27 28 29 30 31

Fig. 8.91 Nodes of the guitar string finite element model

Y

Z 2X3 4 5 6 7 8 9 10 11 12 13 14 15 16 17 18 19 20 21 22 23 24 25 26 27 28 29 30 31

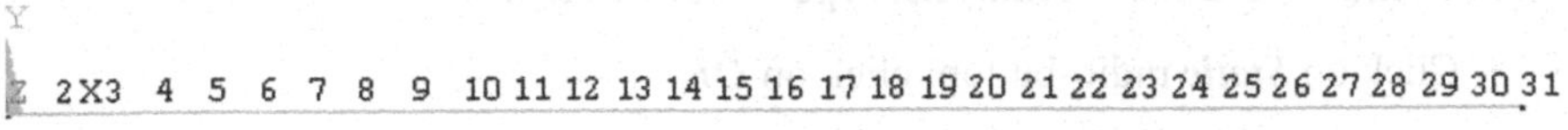

Fig. 8.92 Elements of the guitar string finite element model

Material > Exit

- Create nodes using the following menu path, which brings up the *Create Nodes in Active Coordinate System* dialog box:

Main Menu > Preprocessor > Modeling > Create > Nodes > In Active CS

 - Type ***1*** for ***NODE Node number***; type ***0*** for *x*- and *y*-coordinates; click on ***Apply***.
 - In the same dialog box, type ***31*** for ***NODE Node number***; type ***0.71*** for *x*-coordinate and ***0*** for *y*-coordinate; click on ***OK***.
 - Fill between nodes using the following menu path:

Main Menu > Preprocessor > Modeling > Create > Nodes > Fill Between Nds

 - *Pick Menu* appears; pick the two nodes; click on ***OK***.
 - A new dialog box appears; click on ***OK***.
 - The nodes should appear in the *Graphics Window*, as shown in Fig. 8.91.

- Create elements using the following menu path:

Main Menu > Preprocessor > Modeling > Create > Elements > Auto Numbered > Thru Nodes

 - *Pick Menu* appears; pick the first two nodes (nodes 1 and 2); click on ***OK***.
 - Create the rest of the elements by copying element 1 using the following menu path:

Main Menu > Preprocessor > Modeling > Copy > Elements > Auto Numbered

 - *Pick Menu* appears; pick the element; click on ***OK***.
 - A new dialog box appears; type ***30*** for ***ITIME Total number of copies***; click on ***OK***.
 - The elements should appear in the *Graphics Window*, as shown in Fig. 8.92.

Static Solution

- Specify the *Analysis Type* as *Static*, using the following menu path, which will bring up the *New Analysis* dialog box:

Main Menu > Solution > Analysis Type > New Analysis

  – Click on ***Static*** radio-button; click on ***OK***.

- Specify boundary conditions.

  – Constrain *x*- and *y*-displacements at node 1, and *y*-displacement at the remaining nodes using the following menu path:

Main Menu > Solution > Define Loads > Apply > Structural > Displacement > On Nodes

  – Pick node 1; click on ***OK***.
  – Highlight ***All DOF***; click on ***Apply***.
  – Click on the radio-button for ***Box*** in the *Pick Menu*; select all the nodes except node 1; click on ***OK***.
  – Highlight ***UY***, ***remove*** the highlight on ***All DOF*** by clicking on it; click on ***OK***.

- Apply the axial force using the following menu path:

Main Menu > Solution > Define Loads > Apply > Structural > Force/Moment > On Nodes

  – Pick the last node (node 31); click on ***OK***.
  – In the new dialog box, select ***FX*** from the pull-down menu and type ***84*** for ***VALUE Force/moment value***; click on ***OK***.

- Specify this analysis to be a *prestressed* analysis using the following menu path:

Main Menu > Solution > Analysis Type > Sol'n Controls

  – A dialog box with tabs appears. By default, the ***Basic*** tab is active.
  – On the left side, under ***Analysis Options***, put a checkmark for ***Calculate prestress effects***.
  – On the right side, under ***Write Items to Results File***, click on the radio-button for ***All solution items*** and select ***Write every substep*** from the ***Frequency*** pull-down menu.
  – Click on ***OK***.

- Save the model using the following menu path:

Utility Menu > File > Save as Jobname.db
The model will be saved in the *Working Directory* under the name ***guitar.db***.

- Obtain the solution using the following menu path:

Main Menu > Solution > Solve > Current LS

- Close the/*STATUS information window*.
- Click on ***OK*** to start the solution.
- Wait until ANSYS responds with ***Solution is done!***

- Exit the *Solution Processor* using the following menu path:

Main Menu > Finish

Modal Solution

- Specify the *Analysis Type* as *Modal*, using the following menu path, which will bring up the *New Analysis* dialog box:

Main Menu > Solution > Analysis Type > New Analysis

- Click on ***Modal*** radio-button; click on ***OK***.
- *A Warning Window* appears; click on ***OK***.

- Specify *analysis options* using the following menu path:

Main Menu > Solution > Analysis Type > Analysis Options

- Click on the ***Block Lanczos*** radio-button; type ***6*** for ***No. of modes to extract*** (first text field).
- Place a checkmark for ***[PSTRES] Incl prestress effects***; click on ***OK***, which brings up another dialog box.
- Click on ***OK***.

- Modify boundary conditions.

- Delete constraints in the *y*-direction at nodes 2 through 30 (leaving the displacement constraints at nodes 1 and 31 intact) using the following menu path:

Main Menu > Solution > Define Loads > Delete > Structural > Displacement > On Nodes

- Click on the radio-button for ***Box*** in the *Pick Menu*; select all the nodes except nodes 1 and 31; click on ***OK***.
- Select ***UY*** from the pull-down menu; click on ***OK***.

- Save the model using the following menu path:

Utility Menu > File > Save as Jobname.db
The model will be saved in the *Working Directory* under the name ***guitar.db***.

- Obtain the solution using the following menu path:

Main Menu > Solution > Solve > Current LS

- Close the/*STATUS information window*.
- Click on ***OK*** to start the solution.
- Wait until ANSYS responds with ***Solution is done!***

• Exit the *Solution Processor* using the following menu path:

Main Menu > Finish

Harmonic Solution

• Specify the *Analysis Type* as *Harmonic*, using the following menu path, which will bring up the *New Analysis* dialog box:

Main Menu > Solution > Analysis Type > New Analysis

- Click on ***Harmonic*** radio-button; click on ***OK***.
- *A Warning Window* appears; click on ***OK***.

• Specify the *mode superposition* method using the following menu path:

Main Menu > Solution > Analysis Type > Analysis Options

- Select ***Mode superpos'n*** from the ***[HROPT] Solution method*** pull-down menu.
- Select ***Amplitud + phase*** from the ***[HROUT] DOF printout format*** pull-down menu; click on ***OK***.
- Click on ***OK*** in the new dialog box.

• Modify the boundary conditions.

- Delete the axial force using the following menu path:

Main Menu > Solution > Define Loads > Delete > Structural > Force/Moment > On Nodes

- Pick the last node (node 31); click on ***OK***.
- In the new dialog box, select ***FX*** from the pull-down menu; click on ***OK***.
- Apply the vertical force using the following menu path:

Main Menu > Solution > Define Loads > Apply > Structural > Force/Moment > On Nodes

- Pick node 8; click on ***OK***.
- In the new dialog box, select ***FY*** from the pull-down menu and type ***−1*** for ***VALUE Real part of force/mom***; click on ***OK***.

• Specify the *harmonic frequencies* and enforce step loading using the following menu path:

Main Menu > Solution > Load Step Opts > Time/Frequenc > Freq and Substeps

- Type ***0*** in the first text field and ***2000*** in the second text field for ***HARFRQ***.
- Type ***250*** for ***NSUBST*** text field; select the ***Stepped*** radio-button; click on ***OK***.

- Save the model using the following menu path:

Utility Menu>File>Save as Jobname.db
The model will be saved in the W*orking Directory* under the name ***guitar.db***.

- Obtain the solution using the following menu path:

Main Menu > Solution > Solve > Current LS

- Close the/*STATUS information window*.
- Click on ***OK*** to start the solution.
- Wait until ANSYS responds with ***Solution is done!***

- Exit the *Solution Processor* using the following menu path:

Main Menu > Finish

Postprocessing

- Enter the *Time History Postprocessor* using the following menu path:

Main Menu > TimeHist Postpro

- *Time History* Variables dialog box appears. Click on the button with the green plus sign at the top-left to define a variable.
- *Add Time History Variable* dialog box appears.
- Successively click on the items ***Nodal Solution***, ***DOF Solution***, and ***Y-component of displacement***; click on ***OK***.
- *Pick Menu* appears; pick the midpoint node (node 16); click on ***OK***.
- Note the new variable ***UY_2*** in *Time History Variables* dialog box.
- Highlight the row ***UY_2*** from the list; click on the third from left button to plot the variation of these displacements.
- The plot appears in the *Graphics Window*, as shown in Fig. 8.93.

8.5.3 Transient Analysis

Practical engineering problems having a non-cyclic transient loading can be solved using the *Transient* analysis in ANSYS. Typically, the following steps are used for a *Transient* analysis:

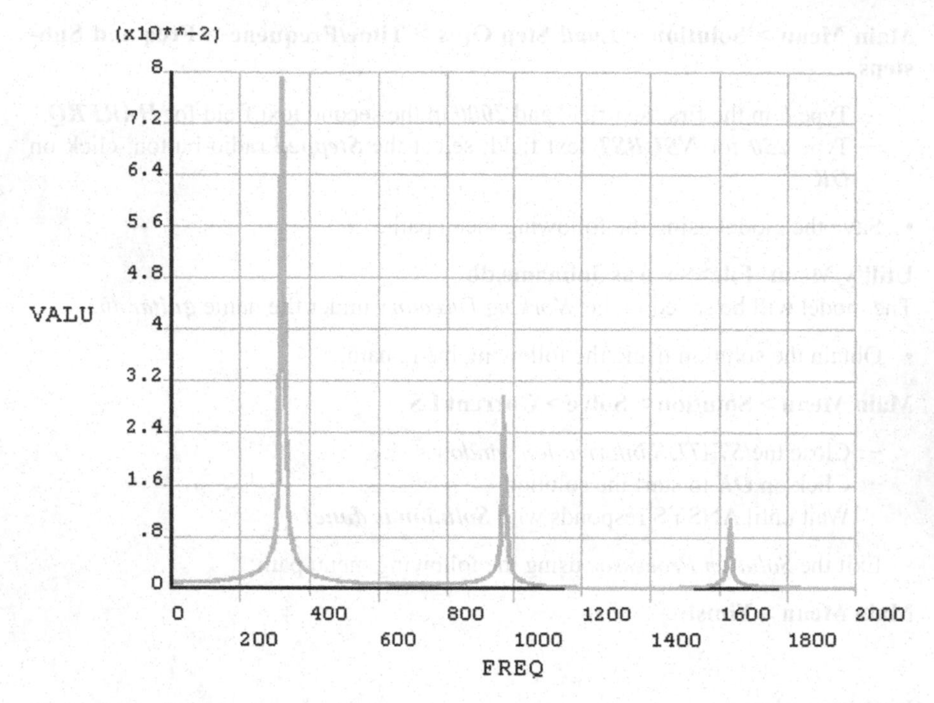

Fig. 8.93 Amplitudes of the displacement at the midpoint node as a function of frequency under harmonic loading

Build the model.
Specify initial conditions.
Specify *Solution Controls.*
Apply loads.
Write *Load Step File.*
Apply/change loads and Solution Controls for the next *Load Step(s)* and write *Load Step File(s).*
Obtain solution from *Load Step Files.*
Review the Results.

Solution Controls are specified using the *Solution Controls* dialog box, which is accessed through the following menu path:

Main Menu > Solution > Analysis Type > Sol'n Controls

There are five tabs in the *Solution Controls* dialog box. The first two tabs, ***Basic*** and ***Transient***, are sufficient for most transient analyses. The ***Basic*** tab contains options to specify nonlinear geometry (large deformation), time at the end of the load step, load step and substep numbers and sizes, automatic time stepping, and the amount and frequency of results data to be saved. The ***Transient*** tab involves options related to time integration and damping specifications.

Fig. 8.94 Time-dependent loading on the bracket

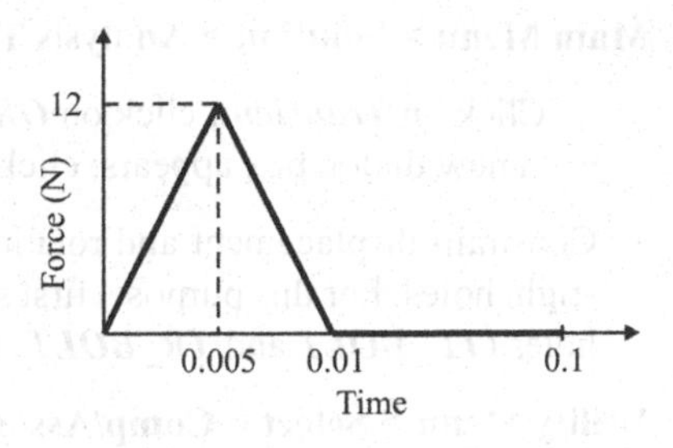

8.5.3.1 Dynamic Analysis of a Bracket

The bracket shown in Fig. 8.29 is clamped at the two top holes and subjected to an impact load at the bottom-left hole. The load is 12 lb, in the positive *z*-direction, and is applied for 0.01 s. Three load steps with ramped loading are used; as shown in Fig. 8.94, the times at the end of the first, second, and third load steps are 0.005, 0.01, and 0.1 s, respectively, with a time step size of 1E–4 s. The objective is to obtain the time-dependent response of the bracket.

Model Generation

The finite element model of this bracket was created in Sect. 8.1.5.1 (modeling of a bracket). Therefore, the model will not be regenerated here; it will be generated using an input file. It is worth noting that the components consisting of the nodes along the holes are defined in the input file.

- Create the model by reading the input file ***bracket.inp*** from the CD-ROM location/***Input_Files\ch08\bracket.inp*** using the following menu path:

Utility Menu > File > Read Input from

- Browse for the ***bracket.inp*** from the folder given above; click on ***OK***.
- Wait until the model is generated.

Solution

There are three distinct loading regiments: (1) from zero loading to maximum loading, (2) from maximum loading to complete removal of the load, and (3) time period after load removal during which the transient response of the bracket is sought. Each of these distinct periods will be defined as a *Load Step*. Each load step will be written to a load step file prior to solution. Once all the load step files are written, the solution will be obtained using these files. Displacement constraints do not change throughout the analysis; they will be specified only once at the beginning.

- Specify the *Analysis Type* as *transient* using the following menu path:

Main Menu > Solution > Analysis Type > New Analysis

- Click on ***Transient***; click on ***OK***.
- A new dialog box appears; click on ***OK***.

• Constrain displacement and rotation degrees of freedom along the top-left and -right holes. For this purpose, first select the components created earlier for these holes (***TL_BOLT*** and ***TR_BOLT***) using the following menu path:

Utility Menu > Select > Comp/Assembly > Select Comp/Assembly

- A dialog box appears; click on the ***by component name*** radio-button; click on ***OK***.
- A new dialog box with the components listed appears; highlight ***TL_BOLT***; click on ***OK***. This action selects the nodes along the top-left hole.
- Specify the displacement boundary conditions using the following menu path:

Main Menu > Solution > Define Loads > Apply > Structural > Displacement > On Nodes

- *Pick Menu* appears; click on ***Pick All***.
- In the new dialog box, highlight ***All DOF***; click on ***OK***.
- Repeat the same procedure for the top-right hole (***TR_BOLT***).

• Apply the load at the bottom-left hole.
  - For this purpose, create a local coordinate system at the center of the bottom-left hole using the following menu path:

Utility Menu > WorkPlane > Local Coordinate Systems > Create Local CS > At Specified Loc

- *Pick Menu* appears; type ***−1.5, −2, 4.5*** in the text field in the *Pick Menu*; click on ***OK***.
- A dialog box appears; select ***Cylindrical 1*** in the ***KCS Type of coordinate system*** pull-down menu and type ***−90*** in the ***THYZ Rotation about local X*** text field; click on ***OK***.
- Select the keypoints along the bottom-left hole using the following menu path:

Utility Menu > Select > Entities

- *Select Entities* dialog box appears; choose ***Keypoints*** in first pull-down menu; choose ***By Location*** in the second pull-down menu; type ***0.25/2*** in the ***Min, Max*** text field; click on ***OK***.
- Apply forces on the keypoints using the menu path:

Main Menu > Solution > Define Loads > Apply > Structural > Force/Moment > On Keypoints

- *Pick Menu* appears; click on ***Pick All***.

- In the new dialog box, select ***FZ*** from pull-down menu and enter ***12/4*** for ***Force/moment value***; click on ***OK***.
- Select everything using the following menu path:

Utility Menu > Select > Everything

- Set solution options for the first load step using the following menu path:

Main Menu > Solution > Analysis Type > Sol'n Controls

- *Solution Controls* dialog box appears. By default, the ***Basic*** tab is active.
- On the left side, under ***Time Control***, type ***0.005*** for ***Time at end of loadstep***; select ***On*** from ***Automatic time stepping*** pull-down menu.
- Click on the radio-button for ***Time increment*** and type ***0.0001*** in ***Time step size*** text field.
- On the right side, under ***Write Items to Result File***, click on the radio-button for ***All solution items*** and select ***Write every substep*** from the ***Frequency*** pull-down menu.
- Activate the ***Transient*** tab (by clicking on it).
- Under ***Full Transient Options***, click on the radio-button for ***Ramped loading***; click on ***OK*** to close the *Solution Controls* dialog box.

- Write first *load step file* using the following menu path:

Main Menu > Solution > Load Step Opts > Write LS File

- Type ***1***; click on ***OK***.
- The first load step is written to a file.

- Remove the load at the midpoint node using the following menu path:

Main Menu > Solution > Define Loads > Delete > Structural > Force/Moment > On Keypoints

- Click on ***Pick All***.
- Select ***FZ*** from the pull-down menu; click on ***OK***.

- Set *solution options* for the second load step using the following menu path:

Main Menu > Solution > Analysis Type > Sol'n Controls

- Activate the ***Basic*** tab. On the left side, under ***Time Control***, type ***0.01*** for ***Time at end of loadstep***, leave the other options unchanged.
- Click on ***OK*** to close the *Solution Controls* dialog box.

- Write the second *load step file* using the following menu path:

Main Menu > Solution > Load Step Opts > Write LS File

- Type ***2***; click on ***OK***.
- The second load step is written to a file.

- Set solution options for the third load step using the following menu path:

Main Menu > Solution > Analysis Type > Sol'n Controls

  - On the left side, under ***Time Control***, type ***0.1*** for ***Time at end of loadstep***; leave the other options unchanged.
  - Click on ***OK*** to close the *Solution Controls* dialog box.

- Write the third *load step file* using the following menu path:

Main Menu > Solution > Load Step Opts > Write LS File

  - Type ***3***; click on ***OK***.
  - The third load step is written to a file.

- Obtain the solution using the following menu path:

Main Menu > Solution > Solve > From LS Files

  - Solve *Load Step Files* dialog box appears.
  - Type ***1 for LSMIN Starting LS file number*** and ***3*** for ***LSMAX Ending LS file number***; click on ***OK***.
  - Wait until ANSYS responds with ***Solution is done!***

Postprocessing

- Review time-dependent behaviors of *y*-displacement at the bottom-left and -right corner nodes using the following menu path:

Main Menu > TimeHist Postpro

  - *Time History Variables* dialog box appears.
  - Click on the button with the green plus sign at the top-left to define a variable.
  - *Add Time History Variable* dialog box appears.
  - Successively click on the items ***Nodal Solution***, ***DOF Solution***, and ***Y-component of displacement***; click on ***OK***.
  - Pick Menu appears; pick the bottom-left corner node; click on ***OK***.
  - Note the new variable ***UY_2*** in Time History Variables dialog box.
  - Add new variable for displacement in y-direction at the bottom-right corner node by clicking on the button with the green plus sign and successively clicking on the items ***Nodal Solution***, ***DOF Solution***, and ***Y-component of displacement***; click on ***OK***.
  - Pick the bottom-right corner node; click on ***OK***.
  - Note the new variable ***UY_3*** in Time History Variables dialog box.
  - Highlight the rows ***UY_2*** and ***UY_3*** from the list (by pressing **Ctrl** on the keyboard and clicking on the rows with left mouse button); click on the third from left button to plot the time variation of these displacements.
  - The plot appears in the *Graphics Window*, as shown in Fig. 8.95.
  - Close *Time History Variables* dialog box.

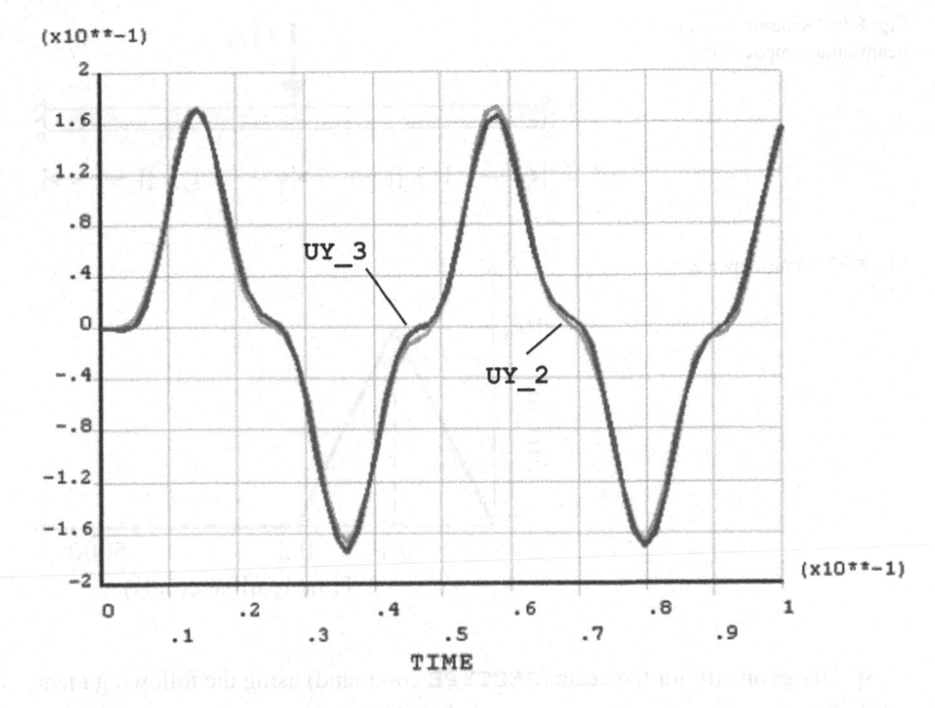

Fig. 8.95 Time variation of displacements in the y-direction at two locations

8.5.3.2 Impact Loading on a Beam

A 3-ft-long steel beam is clamped at both ends and subjected to an impact loading of 100 lbf that lasts 0.0002 s at the midpoint (Fig. 8.96). It is assumed that the loading increases linearly from 0 to 100 lbf in 0.0001 s, and decreases in the same manner, as demonstrated in Fig. 8.97. The beam has a square cross-sectional area (1×1 in) with material properties $E = 29 \times 10^6$ psi and $\nu = 0.32$. The objective is to solve for the time-dependent response of the beam for 0.5 s.

The problem is solved twice: (i) without structural damping, and (ii) with structural damping.

Model Generation

- Define the element type (**ET** command) using the following menu path:

Main Menu > Preprocessor > Element Type > Add/Edit/Delete

- Click on ***Add***.
- Select ***Beam*** immediately below ***Structural Mass*** from the left list and ***2node 188*** from the right list; click on ***OK***.
- Click on ***Close***.

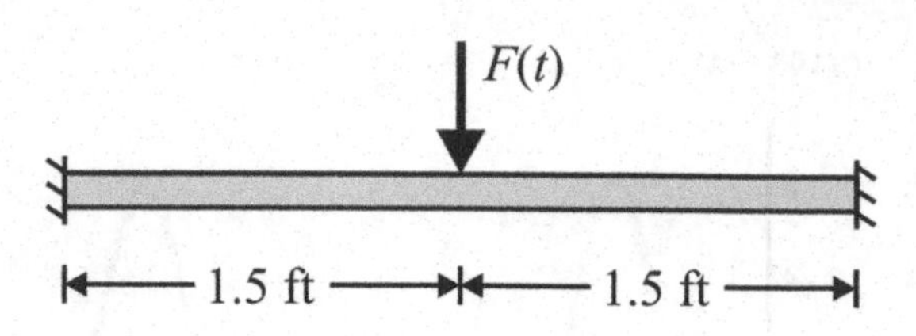

Fig. 8.96 Schematic of the beam under impact loading

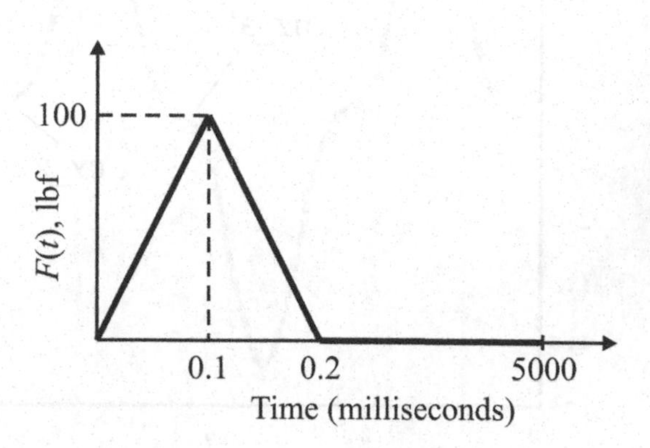

Fig. 8.97 Time-dependent loading

- Specify geometry for the beam (**SECTYPE** command) using the following menu path:

Main Menu > Preprocessor > Sections > Bea > Common Sections

  - *Beam Tool* dialog box appears; enter ***1*** for ***B*** and ***1*** for ***H***; click on ***OK***.
  - Exit from the *Beam Tool* dialog box by clicking on ***OK***.

- Specify material properties (**MP** command) using the following menu path:

Main Menu > Preprocessor > Material Props > Material Models

  - In the *Define Material Model Behavior* dialog box, in the right window, successively left-click on ***Structural***, ***Linear***, ***Elastic***, and, finally, ***Isotropic***, which will bring up another dialog box.
  - Enter ***29e6*** for ***EX*** and ***0.32*** for ***PRXY***; click on ***OK***.
  - In the *Define Material Model Behavior* dialog box, in the right window, left-click on ***Density***, which will bring up another dialog box.
  - Enter ***0.2836*** for ***DENS***; click on ***OK***.
  - Close the *Define Material Model Behavior* dialog box by using the following menu path:

Material > Exit

- Create the solid model. In this problem, the solid model is a line, which is composed of two keypoints. Therefore, first create the keypoints, then the line.
  - Create keypoints using the following menu path:

Main Menu > Preprocessor > Modeling > Create > Keypoints > In Active CS

- In the *Create Keypoints in Active Coordinate System* dialog box, type ***−18*** for ***X*** and ***0*** for ***Y***; click on ***Apply*** (keypoint 1 is created).
- Type ***18*** for ***X*** and ***0*** for ***Y***; click on ***OK*** (keypoint 2 is created)
- Create the line using the following menu path:

Main Menu > Preprocessor > Modeling > Create > Lines > Lines > Straight Line

- Pick keypoints 1 and 2; line is created; click on ***OK*** in the *Pick Menu.*

- Specify the number of elements on the line as **30** divisions using the following menu path:

Main Menu > Preprocessor > Meshing > Size Cntrls > ManualSize > Lines > Picked Lines

- Pick the line; click on ***OK***.
- *Element Sizes on Lines* dialog box appears; type ***30*** in the text field corresponding to ***NDIV*** (the second text field), and uncheck the first checkbox; click on ***Apply***.

- Create the mesh using the following menu path:

Main Menu > Preprocessor > Meshin > Mesh > Lines

- Pick the line; click on ***OK***.

Solution Without Structural Damping

There are three distinct loading regiments: (1) from zero loading to maximum loading, (2) from maximum loading to complete removal of the load, and (3) time period after load removal during which the transient response of the beam is sought. Each of these distinct periods will be defined as a *Load Step*. Each load step will be written to a load step file prior to solution. Once all the load step files are written, the solution will be obtained using these files. Displacement constraints do not change throughout the analysis; they will be specified only once at the beginning.

- Specify the *Analysis Type* as *transient* using the following menu path:

Main Menu > Solution > Analysis Type > New Analysis

- Click on ***Transient***; click on ***OK***.
- A new dialog box appears; click on ***OK***.

- Specify displacement constraints using the following menu path:

Main Menu > Solution > Define Loads > Apply > Structural > Displacement > On Nodes

- Pick the nodes on either end; click on ***OK*** in the *Pick Menu.*

– Highlight ***All DOF***; click on ***OK***.

- Apply the load at the midpoint node (located at the origin for picking conveniently) using the following menu path:

Main Menu > Solution > Define Loads > Apply > Structural > Force/Moment > On Nodes

– Pick the midpoint node; click on ***OK***.
– Select ***FY*** from the pull-down menu and type ***−100*** in the ***VALUE Force/ moment value*** text field; click on ***OK***.

- Set *solution options* for the first *load step* using the following menu path:

Main Menu > Solution > Analysis Type > Sol'n Controls

– *Solution Controls* dialog box appears. By default, the ***Basic*** tab is active.
– On the left side, under ***Time Control***, type ***1e−4*** for ***Time at end of loadstep***, type ***20*** for ***Number of substeps***, and type ***20*** for ***Min no. of substeps***.
– On the right side, under ***Write Items to Results File***, click on the radio-button for ***All solution items*** and select ***Write every substep*** from the ***Frequency*** pull-down menu.
– Activate the ***Transient*** tab (by clicking on it).
– Under ***Full Transient Options***, click on the radio-button for ***Ramped loading***; click on ***OK*** to close the *Solution Controls* dialog box.

- Write the first *load step file* using the following menu path:

Main Menu > Solution > Load Step Opts > Write LS File

– Type ***1***; click on ***OK***.
– The first load step is written to a file.

- Remove the load at the midpoint node using the following menu path:

Main Menu > Solution > Define Loads > Delete > Structural > Force/Moment > On Nodes

– Pick the midpoint node; click on ***OK***.
– Select ***FY*** from the pull-down menu; click on ***OK***.

- Set *solution options* for the second *load step* using the following menu path:

Main Menu > Solution > Analysis Type > Sol'n Controls

– Activate the ***Basic*** tab. On the left side, under ***Time Control***, type ***2e–4*** for ***Time at end of loadstep***; leave ***20*** unchanged for both ***Number of substeps*** and ***Min no. of substeps***.
– On the right side, under ***Write Items to Results File***, click on the radio-button for ***All solution items*** and select ***Write every substep*** from the ***Frequency*** pull-down menu.
– Click on ***OK*** to close the *Solution Controls* dialog box.

- Write the second *load step file* using the following menu path:

Main Menu > Solution > Load Step Opts > Write LS File

  - Type *2*; click on ***OK***.
  - The second *load step* is written to a file.

- Set *solution options* for the third *load step* using the following menu path:

Main Menu > Solution > Analysis Type > Sol'n Controls

  - On the left side, under ***Time Control***, type ***0.5*** for ***Time at end of loadstep*** and type ***200*** for both ***Number of substeps*** and ***Min no. of substeps***.
  - On the right side, under ***Write Items to Results File***, click on the radio-button ***for All solution items*** and select ***Write every substep*** from the ***Frequency*** pull-down menu.
  - Click on ***OK*** to close the Solution Controls dialog box.

- Write the third *load step file* using the following menu path:

Main Menu > Solution > Load Step Opts > Write LS File

  - Type *3*; click on ***OK***.
  - The third *load step* is written to a file.

- Obtain the solution using the following menu path:

Main Menu > Solution > Solve > From LS Files

  - Solve *Load Step Files* dialog box appears.
  - Type ***1*** for ***LSMIN Starting LS file number*** and ***3 for LSMAX Ending LS file number***; click on ***OK***.
  - Wait until ANSYS responds with ***Solution is done!***

Postprocessing

- Review time-dependent behaviors of y-displacement at the midpoint, reaction force in y-direction at the left support, and reaction moment about z-axis at the left support using the following menu path:

Main Menu > TimeHist Postpro

  - *Time History Variables* dialog box appears.
  - Click on the button with the green plus sign at the top-left to define a variable.
  - *Add Time History Variable* dialog box appears.
  - Successively click on the items ***Nodal Solution***, ***DOF Solution***, and ***Y-component of displacement***; click on ***OK***.
  - *Pick Menu* appears; pick the midpoint node; click on ***OK***.
  - Note the new variable ***UY_2*** in *Time History Variables* dialog box.

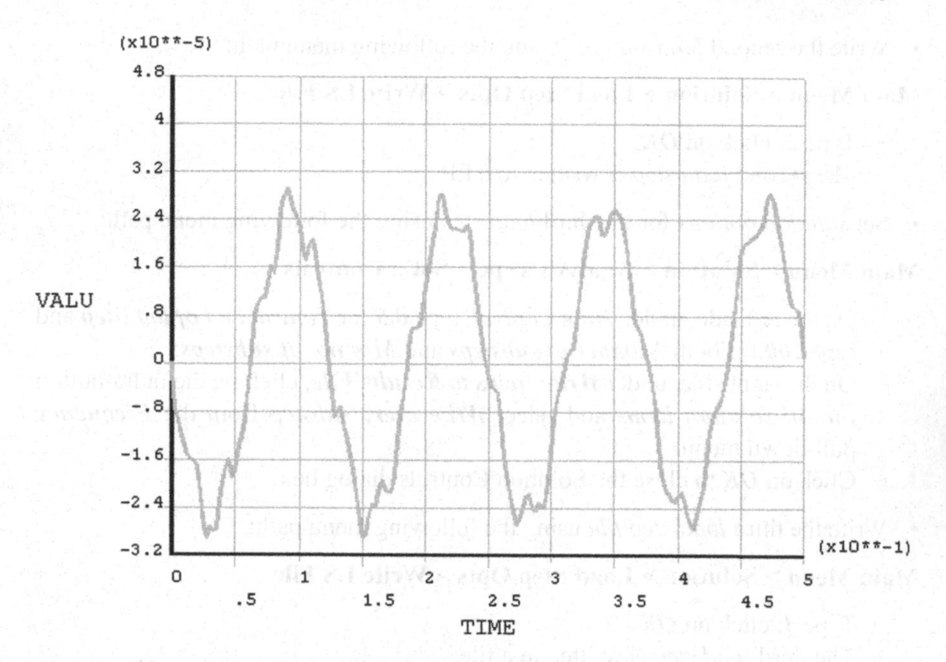

Fig. 8.98 Time variation of the y-displacement at the midpoint

- Highlight the row of ***UY_2*** from the list; click on the third from left button to plot the time variation of this displacement.
- The plot appears in the *Graphics Window*, as shown in Fig. 8.98.
- Note that in this plot the displacement has a periodical behavior with no sign of decay. This is because no damping was specified in the solution. A solution with damping will be obtained in the next section.
- Click on the fourth from left button to list the time variation of this displacement in a separate window.

Solution with Structural Damping

- Repeat the previous solution in its entirety with one difference: when specifying solution options, specify *structural damping* as 0.005. This is accomplished using the following menu path:

Main Menu > Solution > Analysis Type > Sol'n Controls

- Activate the ***Transient*** tab in the *Solution Controls* dialog box.
- Under ***Damping Coefficients***, type 0.005 for ***Stiffness matrix multiplier (BETA)***; click on ***OK*** to close the *Solution Controls* dialog box.
- Note that this needs to be done before each *load step* is written and that all the time and *substep* information needs to be re-entered.

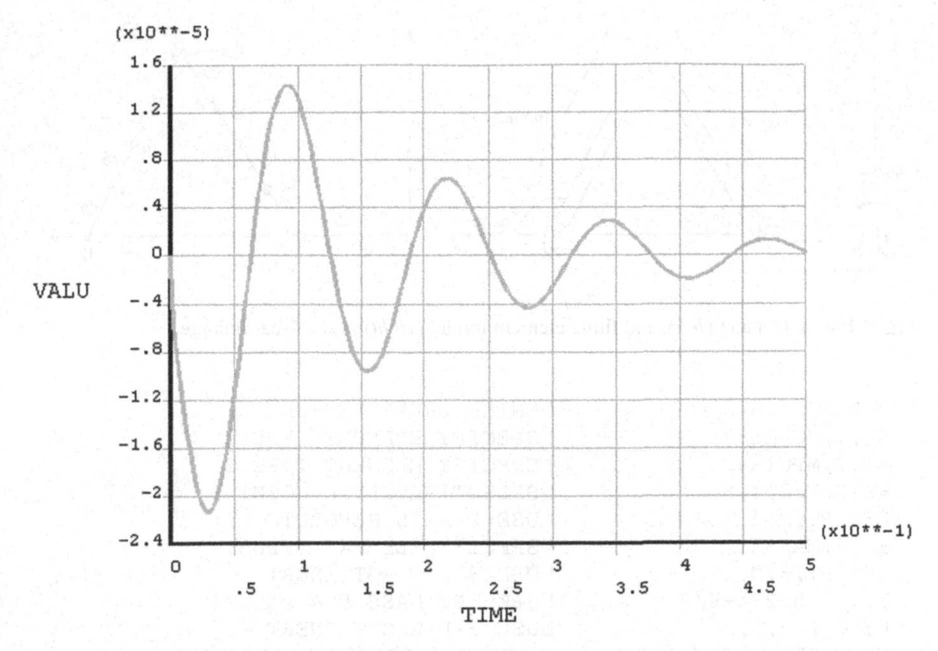

Fig. 8.99 Time variation of the *y*-displacement at the midpoint when damping is included in the analysis

Postprocessing

- Review the time-dependent behavior of the *y*-displacement at the midpoint in the same manner as was done for the case without damping. The corresponding response is given in Fig. 8.99 as it appears in the *Graphics Window*.

8.5.3.3 Dynamic Analysis of a 4-bar Linkage

Consider the 4-bar linkage shown in Fig. 8.100. The far-left node is rotated about the *z*-axis in a clockwise direction. The modeling part of the analysis utilizes direct nodal and elemental definition. Since this model is a flexible kinematics analysis, numerous beam elements are defined for each bar. To reduce the initial vibrations caused by loading, loads are applied in two time steps. A four-revolution displacement is imposed over a time period of 0.6 s. This is equivalent to a 400 rev/min (rpm) displacement.

The following input listing is used for model generation and solution:

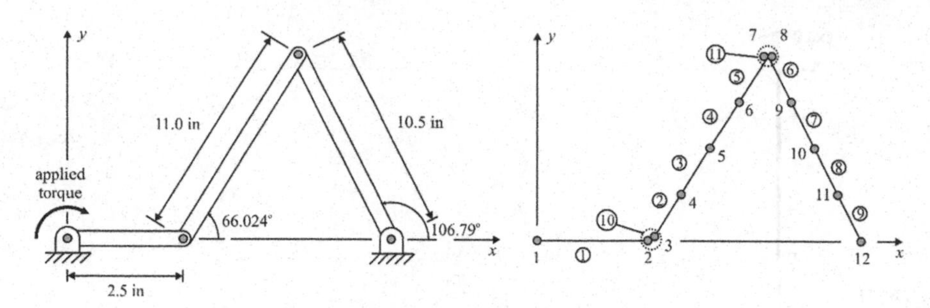

Fig. 8.100 Schematic (*left*), and finite element model (*right*) of the 4-bar linkage

```
/PREP7                        ! ENTER PREPROCESSOR
ET,1,BEAM188                  ! SPECIFY ELEMENT TYPE 1
ET,2,MPC184                   ! SPECIFY ELEMENT TYPE 2
KEYOPT,2,1,6                  ! SELECT REVOLUTE JOINT
KEYOPT,2,4,1                  ! USE Z-AXIS REVOLUTE
ET,3,MASS21                   ! SPECIFY ELEMENT TYPE3
KEYOPT,3,3,2                  ! USE 3-D W ROT INERT
R,1,1.5528e-4/2               ! SPECIFY MASS FOR MASS21
KEYOPT,3,3,2                  ! USE 3-D W ROT INERT
SECTYPE,1,BEAM,RECT           !DEFINE A RECTANGULAR CROSS-
                              !SECTION FOR THE BEAM
SECDATA,0.063,1               !DEFINE SECTION GEOMETRY DATA
MP,EX,1,10e6                  ! ELASTIC MODULUS FOR MATERIAL 1
MP,PRXY,1,0.3                 ! POISSON'S RATIO FOR MATERIAL 1
MP,DENS,1,2.587799e-4         ! DENSITY FOR MATERIAL 1
MP,DAMP,1,0.000139            ! DAMPING MULTIPLIER FOR MATERIAL 1
MP,EX,2,10e6                  ! ELASTIC MODULUS FOR MATERIAL 2
```

```
MP,PRXY,2,0.3               !POISSON'S RATIO FOR MATERIAL 2
MP,DENS,2,2.587799e-4       !DENSITY FOR MATERIAL 2
MP,DAMP,2,0.000278          !DAMPING MULTIPLIER FOR MATERIAL 2
N,1,0,0                     !CREATE NODES
N,2,2.5,0
N,3,2.5,0
CLOCAL,11,0,2.5,0,0,66.024       ! DEFINE LOCAL CS 11
CSYS,11                     !SWITCH TO LOCAL CS
N,4,2.5,0                   !CONTINUE CREATING NODES
N,5,5.5,0
N,6,8.5,0
N,7,11,0
N,8,11,0
CSYS,0                      !SWITCH TO GLOBAL CARTESIAN CS
*AFUN,DEG                   !USE DEGREES FOR TRIG. FUNCTIONS
CLOCAL,12,0,2.5+11*COS(66.024),11*SIN(66.024),0,-73.21
                            !DEFINE LOCAL
                            !COORDINATE SYSTEM (CS) 12
CSYS,12                     !SWITCH TO LOCAL CS 12
N,9,2.5,0                   !CONTINUE CREATING NODES
N,10,5.25,0
N,11,8,0
N,12,10.5,0
E,1,2                       !CREATE BEAM ELEMENTS FOR THE
E,3,4                       !FIRST TWO BARS
E,4,5
E,5,6
E,6,7
MAT,2                       !SWITCH TO MATERIAL 2
E,8,9                       !CREATE BEAM ELEMENTS FOR THE
E,9,10                      !LAST BAR
E,10,11
E,11,12
MAT,3                       !SWITCH TO MATERIAL 3
TB,JOIN,3,1,1,JNS4          !ACTIVATE JOINT MATERIAL MODEL
                            !WITH NONLINEAR ELASTIC STIFFNESS
TBPT,,1E7                   !DEFINE NONLINEAR STIFFNESS
                            !BEHAVIOR IN LOCAL ROTX
TB,JOIN,3,1,1,JNS6          !ACTIVATE JOINT MATERIAL MODEL
                            !WITH NONLINEAR ELASTIC STIFFNESS
TBPT,,1E7                   !DEFINE NONLINEAR STIFFNESS
                            !BEHAVIOR IN LOCAL ROTZ
TYPE,2                      !SWITCH TO ELEMENT TYPE 2
SECTYPE,2,JOINT,REVO        !DEFINE REVOLUTE JOINT AT
SECJOINT,,11,11             !NODES 2 AND 3
SECTYPE,3,JOINT,REVO        !DEFINE REVOLUTE JOINT AT
SECJOINT,,11,11             !NODES 7 AND 8
SECNUM,2                    !SWITCH TO SECTION NUMBER 2
E,2,3                       !CREATE REVOLUTE JOINT ELEMENTS AT
                            !NODES 2 AND 3
SECNUM,3                    !SWITCH TO SECTION NUMBER 3
E,7,8                       !CREATE REVOLUTE JOINT ELEMENTS AT
                            !NODES 7 AND 8
TYPE,3                      !SWITCH TO ELEMENT TYPE 3
E,2                         !DEFINE THE MASS AT NODE 2
```

```
E,3                          ! DEFINE THE MASS AT NODE 3
E,7                          ! DEFINE THE MASS AT NODE 7
E,8                          ! DEFINE THE MASS AT NODE 8
FINISH                       ! FINISH
/SOL                         ! ENTER SOLUTION PROCESSOR
*AFUN,RAD                    ! USE RADIANS FOR TRIG. FUNCTIONS
ANTYPE,TRAN,NEW              ! DEFINE NEW ANALYSIS AS TRANSIENT
NLGEOM,1                     ! INCLUDE LARGE DEFLECTION EFFECTS
D,1,UX,,,12,11,UY            ! SUPPRESS X AND Y DISPLACEMENTS
                             ! AT NODES 1 AND 12
D,ALL,UZ                     ! SUPPRESS Z DISPL. AT ALL NODES
D,1,ROTZ,-0.041888           ! SPECIFY A SMALL ROTATION OF
                             ! -0.041888 RAD (2.4 DEG) AT NODE 1
ACEL,209.4,324.7             ! SPECIFY GRAVITATIONAL
                             ! ACCELERATION IN X AND Y
                             ! DIRECTIONS
TIME,0.001                   ! SPECIFY THE TIME AT THE END OF
                             ! THIS LOAD STEP AS 0.001 SECONDS
NSUBST,1                     ! USE ONLY ONE SUBSTEP
TIMINT,OFF                   ! TURN OFF TRANSIENT EFFECTS
NEQIT,1                      ! USE 1 EQUILIBRIUM ITERATION
                             ! FOR THE SUBSTEP
CNVTOL,F,1,0.001             ! SPECIFY CONVERGENCE TOLERANCE
CNVTOL,M,1,0.001             ! FOR FORCE AND MOMENT
OUTRES,ALL,ALL               ! WRITE RESULTS FOR ALL SUBSTEPS
                             ! TO THE DATABASE FILE
SOLVE                        ! OBTAIN SOL'N FOR THIS LOAD STEP
D,1,ROTZ,-25.13274           ! SPECIFY A ROTATION OF -25.13274
                             ! RAD (1440 DEG) AT NODE 1
TIME,0.6                     ! SPECIFY THE TIME AT THE END OF
                             ! THIS LOAD STEP AS 0.6 SECONDS
NSUBST,299                   ! USE w99 SUBSTEPS
TIMINT,ON                    ! INCLUDE TRANSIENT EFFECTS
NEQIT,100                    ! USE A MAXIMUM OF 100 EQUILIBRIUM
                             ! ITERATIONS FOR EACH SUBSTEP
KBC,0                        ! APPLY THE LOAD (ROTATION) IN A
                             ! RAMPED FASHION
SOLVE                        ! OBTAIN SOL'N FOR THIS LOAD STEP
```

Postprocessing

- Review time-dependent behavior of the torque at node 1 using the following menu path:

Main Menu > TimeHist Postpro

 - *Time History Variables* dialog box appears.
 - Click on the button with the green plus sign at the top-left to define a variable.
 - *Add Time History Variable* dialog box appears.

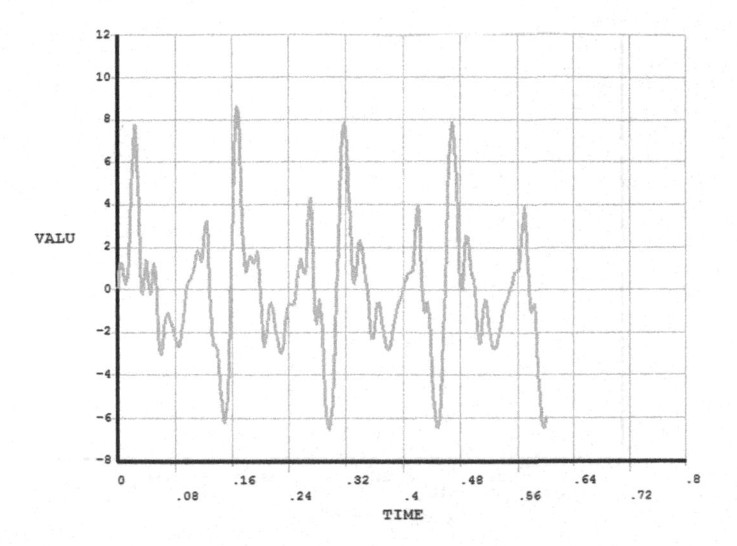

Fig. 8.101 Time variation of the moment at the *left* joint (node 1)

- Successively click on ***Reaction Forces***, ***Structural Moments***, and ***Z-component of moment***; click on ***OK***.
- *Pick Menu* appears; pick node 1; click on ***OK***.
- Note the new variable ***MZ_2*** in the *Time History Variables* dialog box.
- Highlight the row ***MZ_2*** from the list; click on the third from left button to plot the time variation of this displacement.
- The plot appears in the *Graphics Window*, as shown in Fig. 8.101.
- Plot the nodes using the following menu path:

Utility Menu > Plot > Nodes

- Add new variable for displacement in the *y*-direction at node 2 by clicking on the button with the green plus sign and successively clicking on ***Nodal Solution***, ***DOF Solution***, and ***Y-component of displacement***; click on ***OK***.
- *Pick Menu* appears; pick node 2. Since nodes 2 and 3 are coincident, ANSYS asks the user which one to pick. Select node 2, click on ***OK***. Click on ***OK*** in the *Pick Menu*.
- Note the new variable ***UY_3*** in the *Time History Variables* dialog box.
- Highlight the row ***UY_3*** from the list; click on the third from left button to plot the time variation of this displacement.
- The plot appears in the *Graphics Window*, as shown in Fig. 8.102.
- Close the *Time History Variables* dialog box.

• Obtain an animation of the motion of the structure.

  - Enter *General Postprocessor* and read the results using the following menu path:

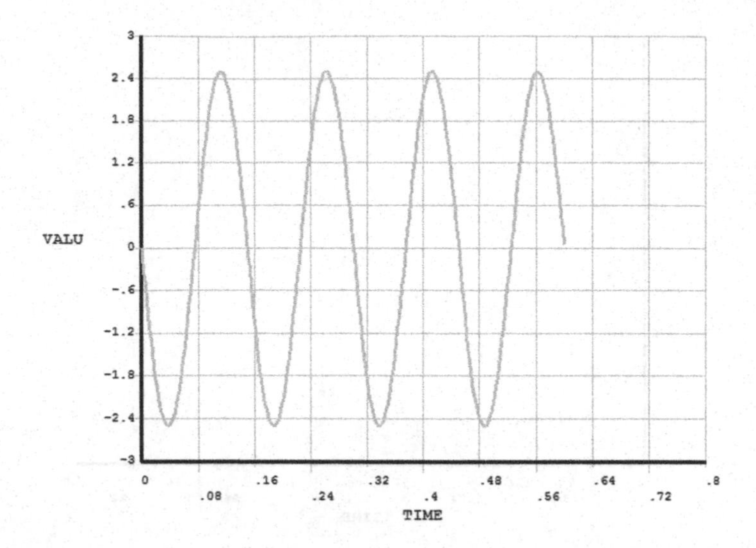

Fig. 8.102 Time variation of the y-displacement at node 2

Main Menu > General Postproc > Read Results > Last Set

– Plot deformed shape using the following menu path:

Main Menu > General Postproc > Plot Results > Deformed Shape

– Select ***Def + undeformed***; click on ***OK***.
– Obtain animation using the following menu path:

Utility Menu > PlotCtrls > Animate > Over Time

– *Animate Over Time* dialog box appears. Enter ***100*** for ***Number of animation frames***. Enter ***0.05*** for ***Animation time delay (sec)***; click on ***OK***.

Chapter 9
Linear Analysis of Field Problems

The field equation,

$$D_x \frac{\partial^2 \phi}{\partial x^2} + D_y \frac{\partial^2 \phi}{\partial y^2} - A\phi + B = 0 \tag{9.1}$$

(where ϕ is the field variable) governs a wide variety of physical problems, referred to as field problems. Several engineering problems, e.g., torsion of noncircular section, ideal irrotational fluid flow, seepage, heat transfer, and electrostatic, are embedded in this equation. For each of these problems, the parameters D_x, D_y, A, and B designate a different physical property. For example, if the problem under consideration involves two-dimensional steady-state heat transfer with no heat generation within the body, then Eq. (9.1) becomes

$$D_x \frac{\partial^2 T}{\partial x^2} + D_y \frac{\partial^2 T}{\partial y^2} = 0 \tag{9.2}$$

where D_x and D_y are the thermal conductivity of the material in the x- and y-directions, respectively, and T represents temperature.

Most of the field problems can be solved using ANSYS software. However, only two types of field problems are considered within the context of this book: (i) heat transfer problems and (ii) moisture diffusion problems.

9.1 Heat Transfer Problems

In certain cases, a thermal analysis is followed by a stress analysis in order to evaluate the structural integrity of the component under the given thermal conditions. In a typical heat transfer problem, the goal is to obtain certain thermal quantities within a body under a specific set of boundary conditions. These quantities include:

The online version of this book (doi: 10.1007/978-1-4939-1007-6_9) contains supplementary material, which is available to authorized users

E. Madenci, I. Guven, *The Finite Element Method and Applications in Engineering Using ANSYS®*, DOI 10.1007/978-1-4899-7550-8_9

temperatures, thermal fluxes and gradients, and the amount of heat dissipated. There are two main types of thermal analyses:

Steady-state heat transfer: Solution is time independent.

Transient heat transfer: Subjected to specific initial conditions, the solution exhibits a time-dependent behavior. If the transient solution is obtained for a sufficiently long time period, the solution is expected to converge to the steady-state solution.

ANSYS accommodates three main heat transfer types: conduction, convection, and radiation. Convection boundary conditions require knowledge of the film coefficient and ambient temperature. There are a few different ways to impose radiation conditions within ANSYS. However, only one of them, the *Radiosity Solver* method, is explained within the context of this book.

Several boundary conditions are available, including specification of temperatures, heat fluxes, convective heat loss/gain, and radiation. As for the body loads, the user can specify heat generation rates within the domain.

In both steady-state and transient analyses, the material properties may be defined as temperature dependent. In the presence of temperature-dependent material properties, the analysis becomes nonlinear, thus requiring an iterative solution.

The most commonly used heat transfer elements are **PLANE55** (two-dimensional plane) and **SOLID70** (three-dimensional brick).

9.1.1 Steady-state Analysis

When the boundary conditions and body loads do not vary with time and there are no specified initial conditions, the solution quantities do not vary with time. In such cases, steady-state solutions are obtained. A steady-state analysis is demonstrated by considering two problems.

9.1.1.1 Analysis of a Tank/Pipe Assembly

A cylindrical tank and a small pipe form a junction, as shown in Fig. 9.1 (only 1/8th of the geometry is shown due to octant-symmetry). Inside the tank, there is fluid at a temperature of 450 °F. A steady flow of a fluid at a temperature of 100 °F is experienced inside the pipe. The film coefficient along the inner surface of the tank is 250 Btu/hr-ft^2-°F whereas the film coefficient along the inner surface of the pipe depends on the surface temperature. The geometric parameters and boundary conditions are given in Table 9.1, and the material properties are summarized in Table 9.2. Note in Tables 9.1 and 9.2 that the length units are in inches and feet. However, in order to obtain a physically correct solution the units must be consistent. Inches are used in this problem, therefore any parameter with the length unit in feet must be converted to inches. The goal is to determine the temperature distribution in the tank.

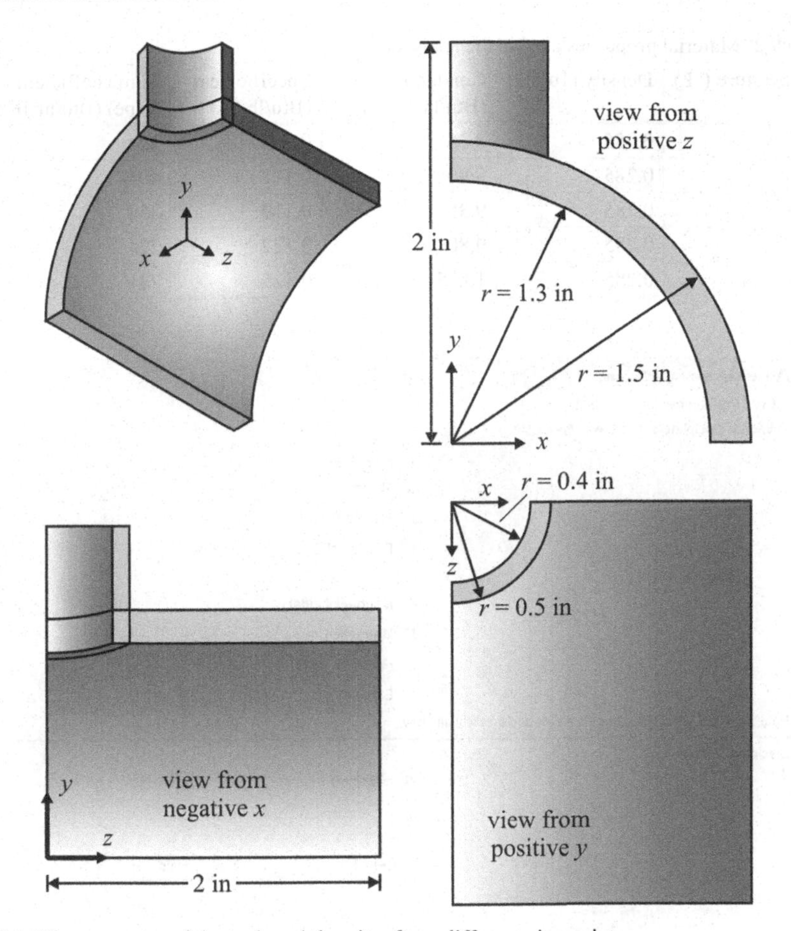

Fig. 9.1 The geometry of the tank and the pipe from different viewpoints

Table 9.1 Geometric parameters and boundary conditions used in the analysis

Parameter	Value
Inside diameter, pipe	0.8 in.
Outside diameter, pipe	1.0 in.
Inside diameter, tank	2.6 in.
Outside diameter, tank	3.0 in.
Inside bulk fluid temperature, tank	450 °F
Inside film coefficient, tank	250 Btu/hr-ft^2-°F
Inside bulk fluid temperature, pipe	100 °F

Table 9.2 Material properties used in the analysis

Temperature (°F)	Density (1b/in3)	Conductivity (Btu/hr-ft-°F)	Specific heat (Btu/lb-°F)	Film coefficient (pipe) (Btu/hr-ft^2-°F)
70	0.285	8.35	0113	426
200	0.285	8.90	0.117	405
300	0.285	9.35	0.119	352
400	0.285	9.80	0.122	275
500	0.285	10.35	0.125	221

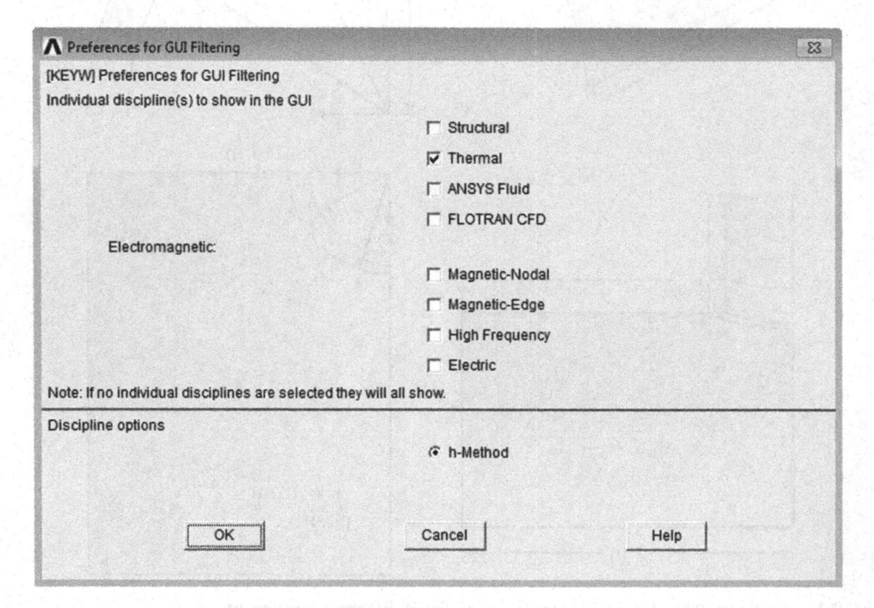

Fig. 9.2 *Preferences for GUI filtering* dialog box

Model Generation

- Declare *Preferences* for *Graphics User Interface* (GUI) filtering (Fig. 9.2) using the following menu path:

Main Menu > Preferences

  – Place a checkmark corresponding to ***Thermal***; click on ***OK***.
  – This action eliminates several menu entries that are not related to thermal analyses, thus resulting in an abridged menu.

- Define the element type (**ET** command) using the following menu path:

Main Menu > Preprocessor > Element Type > Add/Edit/Delete

  – *Element Types* dialog box appears; click on ***Add***.

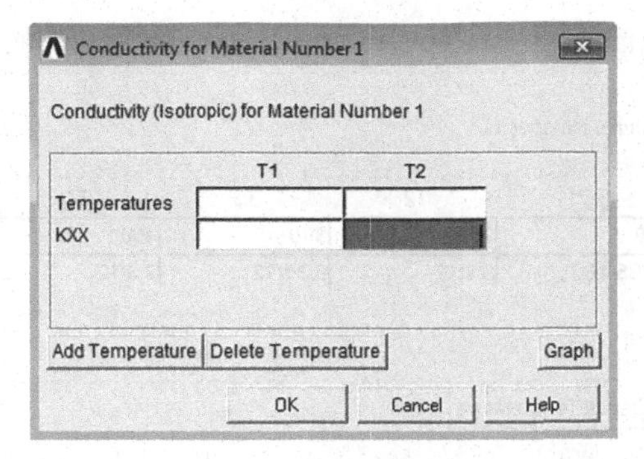

Fig. 9.3 Adding temperatures for temperature-dependent material properties

- Select ***Solid*** immediately below ***Thermal*** in the left list and ***Brick 8 Node 70*** on the right list; click on ***OK***.
- Click on ***Close***.

- Specify the temperature-independent material property, density (**MP** command), using the following menu path:

Main Menu > Preprocessor > Material Props > Material Models

- In the *Define Material Model Behavior* dialog box, in the right window, successively left-click on ***Thermal*** and ***Density***, which will bring up another dialog box.
- Enter ***0.285*** for ***DENS***; click on ***OK***.

- Specify temperature-dependent material properties.

- In the *Define Material Model Behavior* dialog box, in the right window, successively double-click on ***Thermal***, ***Conductivity***, and ***Isotropic***, which will bring up another dialog box.
- Click on the ***Add Temperature*** button four times (so that there are five temperature slots) (Fig. 9.3).
- Enter temperature values (i.e., ***70, …, 500***) in the top row, from left to right in ascending order (Table 9.2).
- Enter the corresponding thermal conductivity values referring to Table 9.2. However, the values given in Table 9.2 are in units of Btu/hr-ft-°F, which must be converted to Btu/hr-in-°F. This can be achieved within the dialog box, as shown in Fig. 9.4. Therefore, enter ***8.35/12, …, 10.35/12***.
- View the variation of thermal conductivity as a function of temperature by clicking on the ***Graph*** button. ***KXX*** versus ***TEMP*** appears in the *Graphics Window* (Fig. 9.5).
- Click on ***OK***.
- Specify specific heat in the same manner as thermal conductivity.

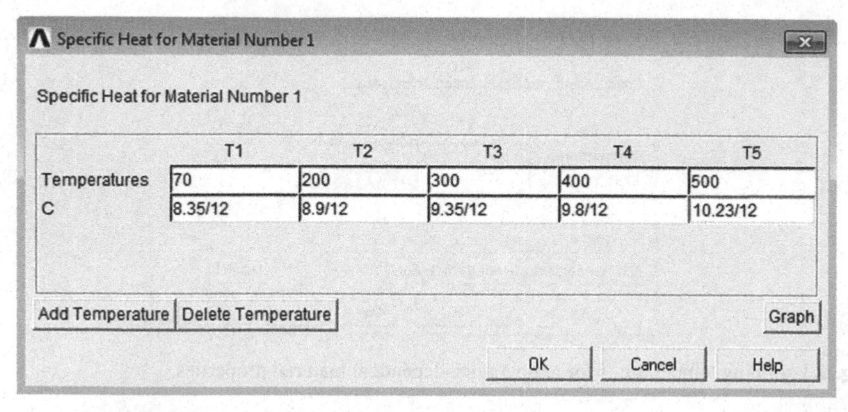

Fig. 9.4 Conductivity values at five different temperatures

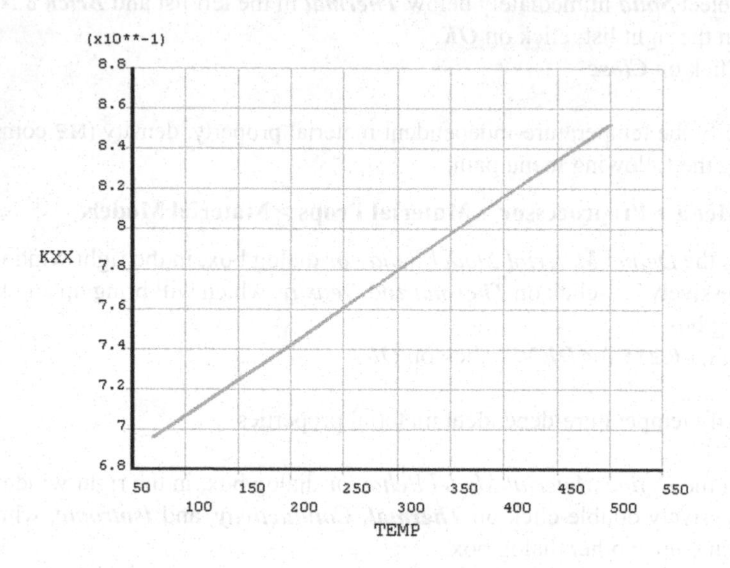

Fig. 9.5 Variation of conductivity with temperature as plotted in ANSYS

- Specify the temperature-dependent film coefficient inside the pipe.
- Although the tank and the pipe are of the same material, their film coefficients are different. Therefore, the film coefficient inside the pipe will be defined as a new material.
- In the *Define Material Model Behavior* dialog box, define a new material reference number using the following menu path:

Material > New Model

- A dialog box appears with the material reference number set to the next available number (in this case 2) (Fig. 9.6); click on OK.

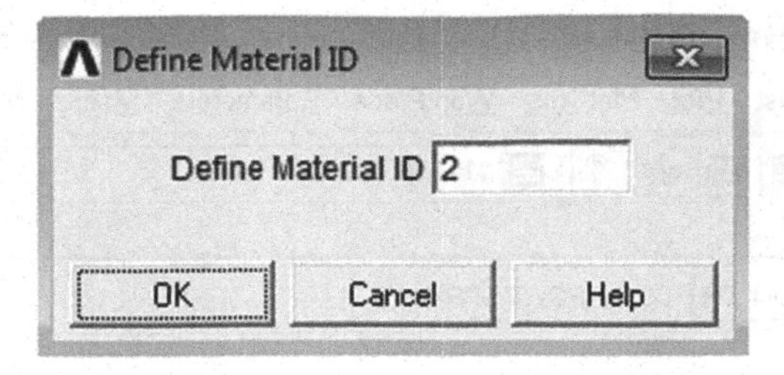

Fig. 9.6 Definition of a new material

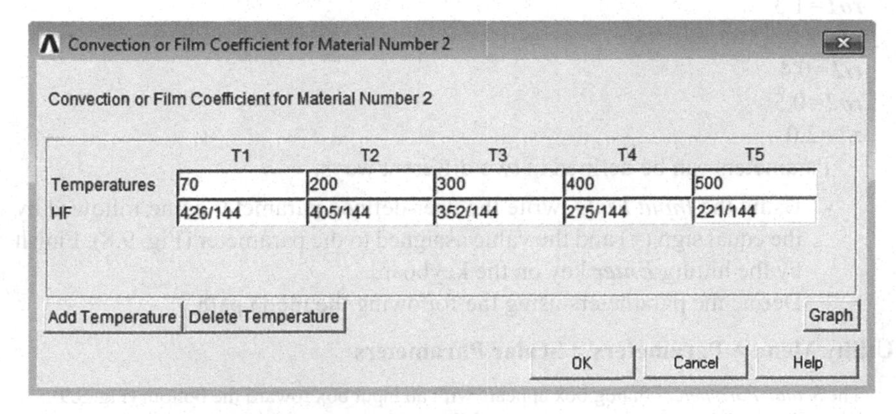

Fig. 9.7 Temperature-dependent film coefficient

- Note the new entry for Material Model Number 2 in the left list in the Define Material Model Behavior dialog box.
- In the right list, left-click on Convection or Film Coef., which brings up a new dialog box.
- Click on the Add Temperature button four times (so that there are five temperature slots); enter temperature values (i.e., 70, … , 500) in the top row, from left to right in ascending order.
- Enter the film coefficient values (division by 144 is required for unit conversion), as shown in Fig. 9.7; click on OK.
- Close the Def*ine Material Model Behavior* dialog box.
- The film coefficient for the inside surface of the tank will be specified later during the application of boundary conditions.

• Define parameters.

- In order to demonstrate the use of parameters in ANSYS the following parameters and their values are defined:

ri1 = 1.3

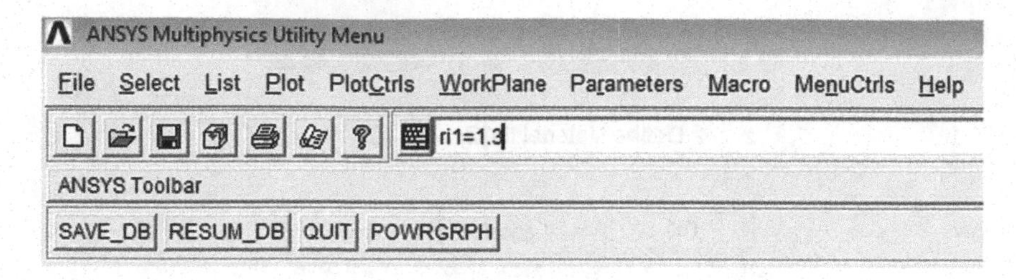

Fig. 9.8 Defining a parameter through the *Input Field*

ro1=1.5
z1=2.0
ri2=0.4
ro2=0.5
z2=2.0

– Parameters can be defined in two different ways:
  - Using the *Input Field*, write the user-defined parameter name followed by the equal sign (=) and the value assigned to the parameter (Fig. 9.8). Finish by the hitting ***Enter*** key on the keyboard.
  - Define the parameters using the following the menu path:

Utility Menu > Parameters > Scalar Parameters

The *Scalar Parameters* dialog box appears with an input box toward the bottom (Fig. 9.9). Write the user-defined parameter name followed by the equal sign (=) and the value assigned to the parameter; click on ***Accept***. Observe that the newly defined parameter appears in the list.

– Define all the parameters using either method.

• Create the volume for the tank—a hollow partial cylinder—(**CYLIND** command) using the following menu path:

Main Menu > Preprocessor > Modeling > Create > Volumes > Cylinder > By Dimensions

– Enter ***ro1*** for ***RAD1 Outer radius***, ***ri1*** for ***RAD2 Optional inner radius***, ***z1*** for the *second* text box in the ***Z1, Z2 Z-coordinates*** row, and ***90*** for ***THETA2 Ending angle (degrees)***; click on ***OK***.
– Observe the hollow partial cylinder created in the Graphics Window.
– View the tank from different angles using the following menu path:

Utility Menu > PlotCtrls > Pan Zoom Rotate

– Click on ***Obliq***.
– Figure 9.10 shows the outcome of this action as it appears in the *Graphics Window*.

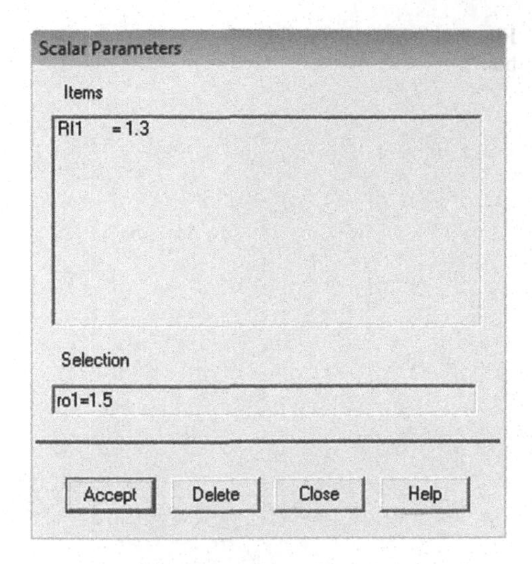

Fig. 9.9 *Scalar Parameters* dialog box

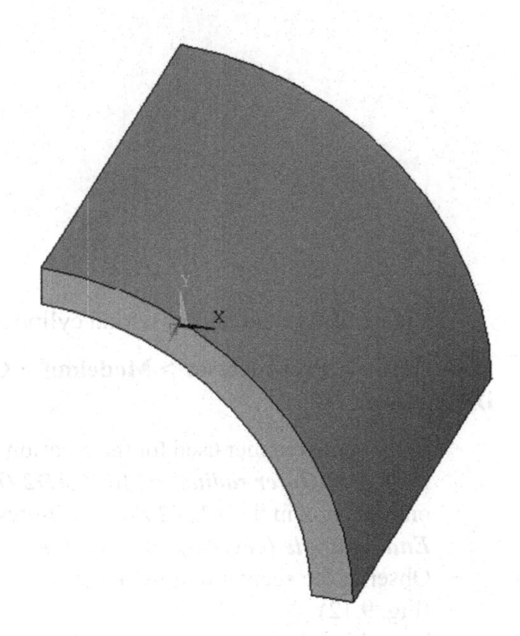

Fig. 9.10 Hollow partial cylinder as it appears in the *Graphics Window*

- Create the volume for the pipe—a second hollow partial cylinder.
  - In order to use the GUI for this action, first the *Working Plane* needs to be rotated using the following menu path:

Utility Menu > WorkPlane > Offset WP by Increments

  - *Offset WP* dialog box appears.
  - Find the *second input field* with the title: ***XY, YZ, ZX Angles***.
  - Enter ***0,−90, 0*** in the second input field; click on ***OK*** (Fig. 9.11).

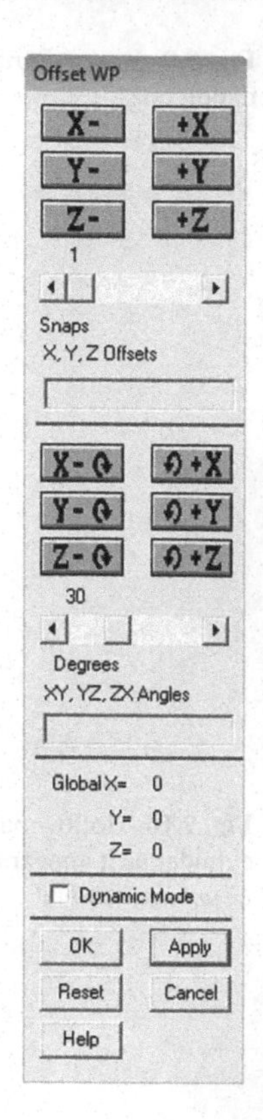

Fig. 9.11 *Offset WP* dialog box

– Create the second hollow partial cylinder using the following menu path:

Main Menu > Preprocessor > Modeling > Create > Volumes > Cylinder > By Dimensions

– In the same manner used for the creation of the cylinder for the tank, enter ***ro2*** for ***RAD1 Outer radius***, ***ri2*** for ***RAD2 Optional inner radius***, ***z2*** for the second text box in the ***Z1, Z2 Z-coordinates*** row, and −***90*** (*not 90*) for ***THETA2 Ending angle (degrees)***; click on ***OK***.
– Observe the second hollow partial cylinder created in the *Graphics Window* (Fig. 9.12).
– Turn the *numbering* on for the volumes using the following menu path:

Utility Menu > PlotCtrls > Numbering

– Place a *checkmark* by clicking on the square box next to ***VOLU Volume numbers***; click on ***OK***.
– In the *Pan Zoom Rotate Window*, click on ***Left*** button.

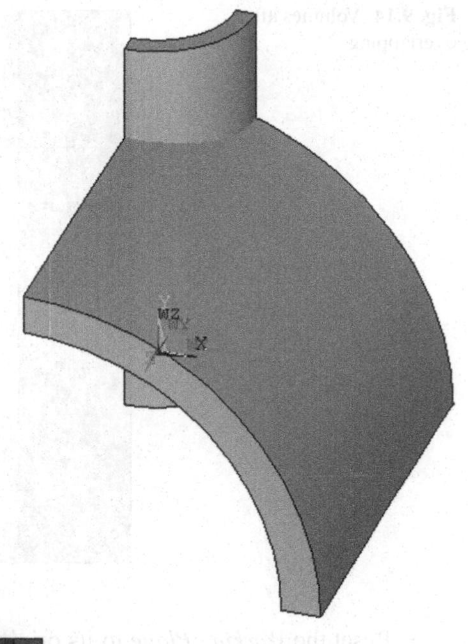

Fig. 9.12 Two hollow partial cylinders as they appear in the *Graphics Window*

Fig. 9.13 Two hollow partial cylinders with volume numbers turned on (*both colors and numbers*)

– Figure 9.13 shows the outcome of this action as it appears in the *Graphics Window*; observe the different colors[1] for each volume and the volume numbers.

---

[1] Colors have not been used in the printed version of the figures. See the accompanying CD-ROM for color versions of the figures.

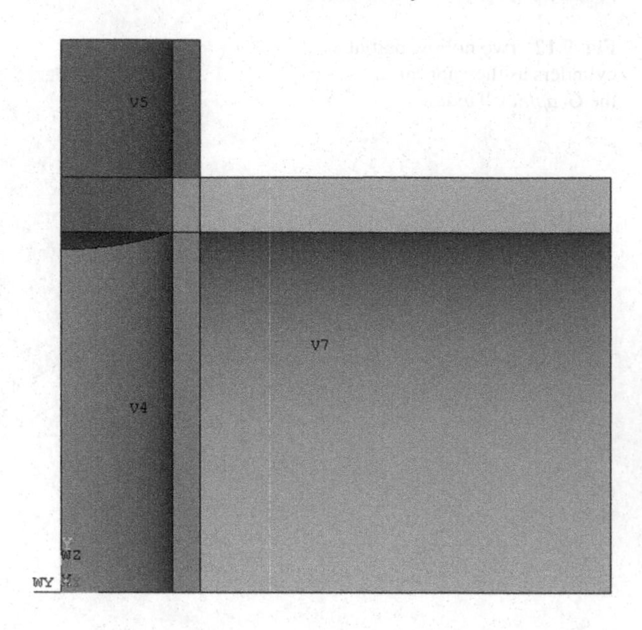

Fig. 9.14 Volumes after overlapping

– Reset the *Working Plane* to its original configuration by using the following menu path:

Utility Menu > WorkPlane > Offset WP by Increments

– *Offset WP* dialog box appears.
– Enter ***0, 90,0*** in the *second input* field; click on ***OK***.

• Use *Boolean Operator Overlap* to define the intersection volume between the two partial cylinders using the following menu path:

Main Menu > Preprocessor > Modeling > Operate > Booleans > Overlap > Volumes

– *Pick Menu* appears; pick the two volumes (or click on ***Pick All*** in the *Pick Menu*); click on ***OK***.
– In the *Pan Zoom Rotate Window*, click on ***Left*** button.
– Figure 9.14 shows the outcome of this action as it appears in the *Graphics Window*.

• Delete the excess volumes using the following menu path:

Main Menu > Preprocessor > Modeling > Delete > Volume and Below

– *Pick Menu* appears; pick the volumes to be deleted (carefully!); the volumes that need to be deleted are **# 3** and **# 4**.
– While picking the volumes, observe the number next to ***Volu No.*** in the *Pick Menu* change as the volumes are picked. This number corresponds to the last volume that is picked. If the wrong volume is picked, *right-click* inside the

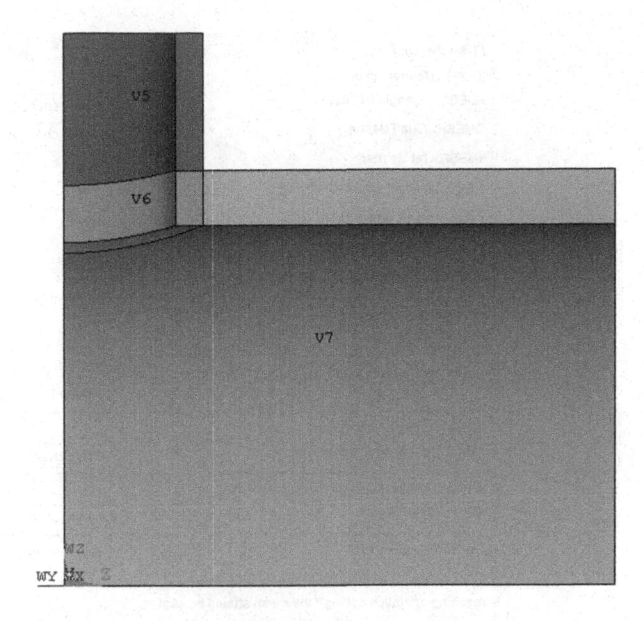

Fig. 9.15 Volumes after deletion

Graphics Window and observe that the mouse pointer changes from *upward arrow* (pick) to *downward arrow* (unpick); then unpick the wrong volume and pick the correct one.

- Once the volumes to be deleted (3 and 4) are picked, click on ***OK*** to finalize the deletion.
- Figure 9.15 shows the left view of the volumes after deletion as they appear in the *Graphics Window.*

• Set the element shape to be used as *hexahedron* (brick shaped) using the following menu path:

Main Menu > Preprocessor > Meshing > Mesher Opts

- In the *Mesher Options* dialog box, click on the ***Mapped*** radio-button (Fig. 9.16); click on ***OK***.
- *Set Element Shape* dialog box appears (Fig. 9.17), click on ***OK***.

• In 3-D, *mapped meshing* can only be performed on volumes with **4** to **6** sides. Examination of the volumes in the current model reveals that volume 7 has 7 sides. In order to be able to perform mapped meshing on this volume, certain areas (2–5 pair) and lines (7–12, 5–10 pairs) must be *concatenated.*

- In order to be able to pick the areas and lines with ease, they should be in plain view. Rotate the model using rotation buttons on the *Pan Zoom Rotate Window* as follows: click on the negative ***X-rotation*** arrow button *twice* and negative ***Y-rotation*** arrow button *once*.

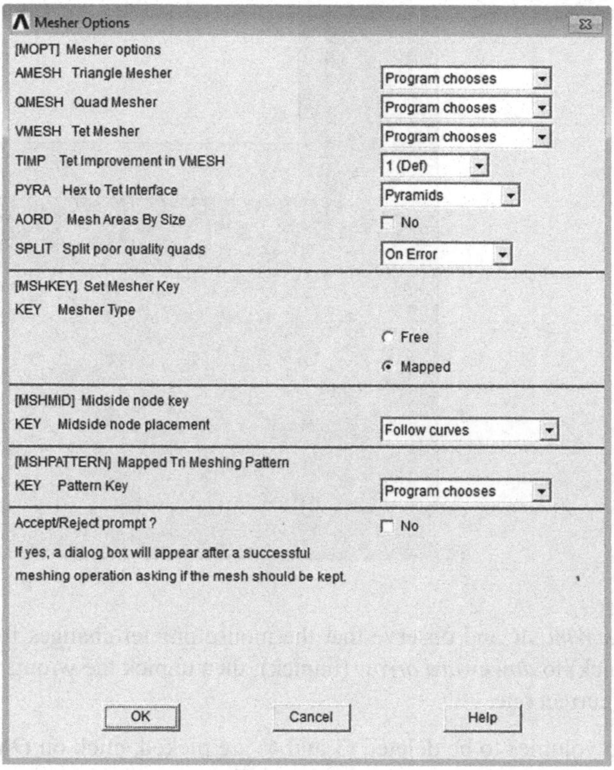

Fig. 9.16 *Mesher Options* dialog box

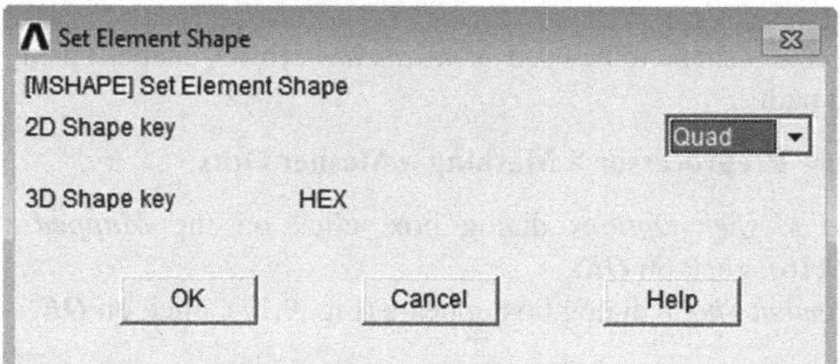

Fig. 9.17 *Set Element Shape* dialog box

– Figure 9.18 shows the outcome of this action as it appears in the *Graphics Window*.
– Turn off volume numbers and turn on area numbers using the following menu path:

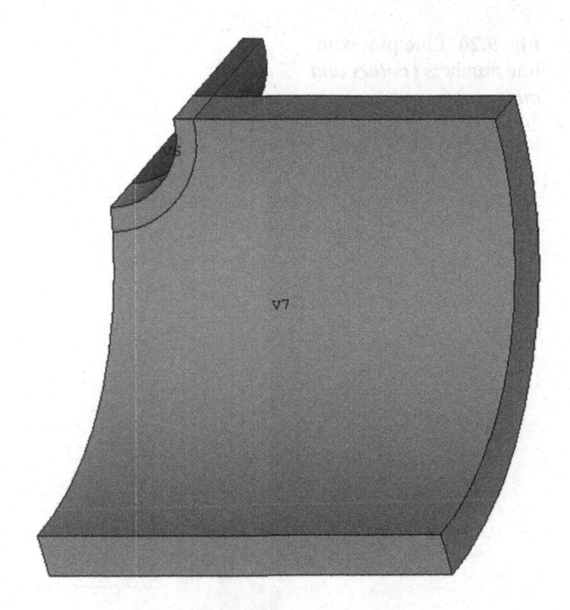

Fig. 9.18 Volumes viewed after rotations

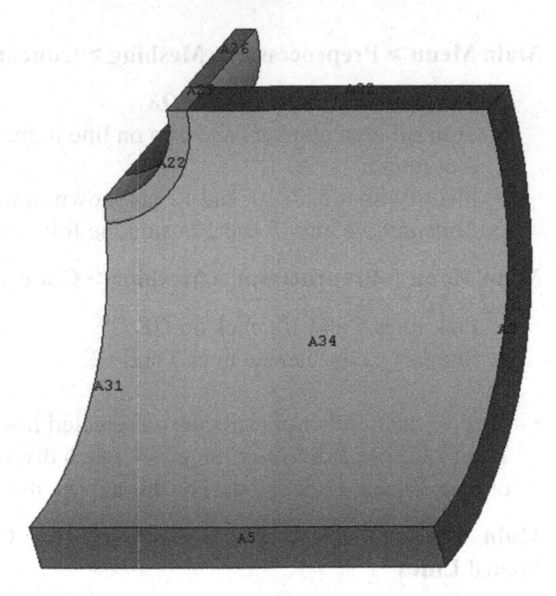

Fig. 9.19 Area plot with area numbers (*colors and numbers*) turned on

Utility Menu > PlotCtrls > Numbering

- Remove the checkmark by clicking on the square box next to ***VOLU Volume numbers*** (turning volume numbers off); place a checkmark on the square box next to ***AREA Area numbers*** (turning area numbers on); click on ***OK***.
- Plot areas using the following menu path:

Utility Menu > Plot > Areas

- Identify areas 2 and 5 (as shown in Fig. 9.19).
- *Concatenate* areas 2 and 5 using the following menu path:

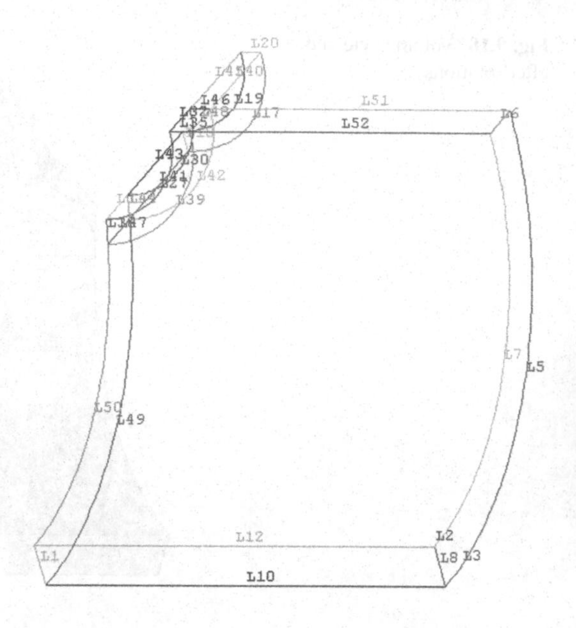

Fig. 9.20 Line plot with line numbers (*colors and numbers*) turned on

Main Menu > Preprocessor > Meshing > Concatenate > Areas

- Pick areas 2 and 5; click on ***OK***.
- Turn off area numbers and turn on line numbers.
- Plot lines.
- Identify lines 5, 7, 10, and 12 (as shown in Fig. 9.20).
- *Concatenate* lines 5 and 10 using the following menu path:

Main Menu > Preprocessor > Meshing > Concatenate > Lines

- Pick lines 5 and 10; click on ***OK***.
- Similarly, concatenate lines 7 and 12.

• Specify the number of elements on selected lines for mapped meshing. On lines 18 and 47, use **2** divisions; on line 7, use **6** divisions; on line 12, use **5** divisions; on line 39, use **11** divisions. For this action, use the following menu path:

Main Menu > Preprocessor > Meshing > Size Cntrls > ManualSize > Lines > Picked Lines

- In the text field of the *Pick Menu*, type ***18*** followed by hitting enter on the keyboard; type ***47*** and hit enter again.
- Click on ***Apply***.
- *Element Sizes on Lines* dialog box appears; type **2** in the text field corresponding to ***NDIV*** (the second text field); uncheck the first check box; click on ***Apply***.
- Repeat this procedure for lines 7, 12, and 39 with their corresponding number of divisions as **6**, **5**, and **11**, respectively. After specifying number of divisions for line 39, click on ***OK*** in the *Element Sizes on Lines* dialog box.

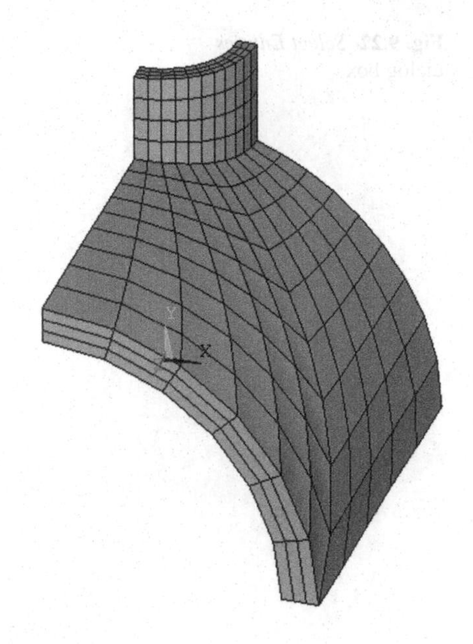

Fig. 9.21 Oblique view of the mesh

- Create the mesh using the following menu path:

Main Menu > Preprocessor > Meshing > Mesh > Volumes > Mapped > 4 to 6 sided

 - In the *Pick Menu*, click on ***Pick All***.
 - Observe the mesh using the different options in the *Pan Zoom Rotate Window*.
 - Figure 9.21 shows the ***oblique*** view of the mesh.

Solution

- Apply uniform temperature to the entire model using the following menu path:

Main Menu > Solution > Define Loads > Apply > Thermal > Temperature > Uniform Temp

 - Dialog box appears; type ***450***; click on ***OK***.

- Apply boundary conditions.

 - Change the active coordinate system to the global cylindrical coordinate system using the following menu path:

Utility Menu > WorkPlane > Change Active CS to > Global Cylindrical

 - Select nodes at the inner surface of the tank for convective boundary conditions using the following menu path:

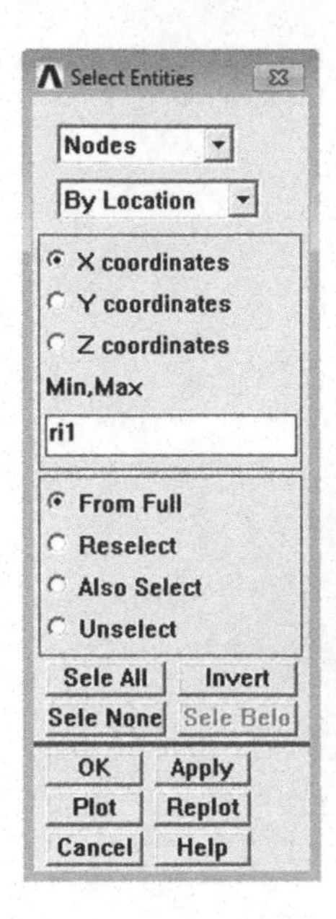

Fig. 9.22 *Select Entities* dialog box

Utility Menu > Select > Entities

- *Select Entities* dialog box appears.
- Choose ***Nodes*** from the first pull-down menu.
- Chose ***By Location*** from the second pull-down menu.
- Click on the radio-button ***X coordinates***. In cylindrical coordinates, x and y correspond to r and θ; therefore, the selection criterion will utilize the r-coordinate as long as the cylindrical coordinate system is active.
- Type ***ri1*** in the text field; click on ***OK*** (Fig. 9.22).
- Activate the *ANSYS Output Window* and examine the lines on the bottom to confirm that several nodes are selected as a result of this action (Fig. 9.23).
- Apply convective boundary conditions on the selected nodes using the following menu path:

Main Menu > Solution > Define Loads > Apply > Thermal > Convection > On Nodes

- Click on ***Pick All***.
- In the dialog box, type ***250/144*** for film coefficient (first text field) and ***450*** for bulk temperature (second text field); click on ***OK*** (Fig. 9.24).

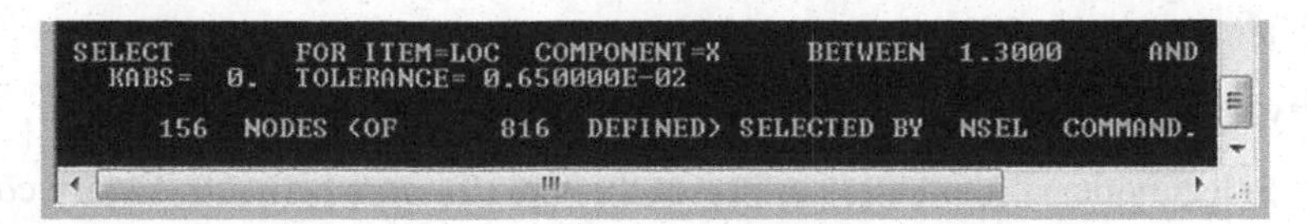

Fig. 9.23 Selection of nodes reported in the *Output Window*

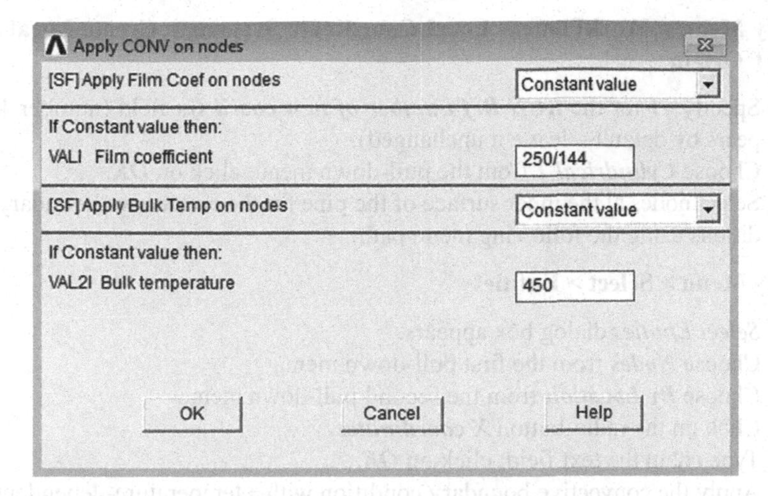

Fig. 9.24 *Apply CONV on nodes* dialog box

- Select nodes at the far edge of the tank for the temperature boundary conditions using the following menu path:

Utility Menu > Select > Entities

- *Select Entities* dialog box appears.
- Choose ***Nodes*** from the first pull-down menu.
- Chose ***By Location*** from the second pull-down menu.
- Click on the radio-button ***Z coordinates***.
- Type ***z1*** in the text field; click on ***OK***.
- Activate the *ANSYS Output Window* and examine the lines on the bottom to confirm that several nodes are selected as a result of this action.
- Apply temperature boundary conditions on the selected nodes using the following menu path:

Main Menu > Solution > Define Loads > Apply > Thermal > Temperature > On Nodes

- Click on ***Pick All***.
- In the dialog box, highlight ***TEMP*** and type ***450*** for ***VALUE Load Temp value***; click on ***OK***.
- Rotate the working plane ***-90°*** (*negative 90°*) about the *x*-axis using the following menu path:

Utility Menu > WorkPlane > Offset WP by Increments

- *Offset WP* dialog box appears.
- Type ***0, -90,0*** in the ***XY, YZ, ZX Angles*** text field (second text field); click on ***OK***.
- Define a local cylindrical coordinate system at the working plane origin using the following menu path:

Utility Menu > WorkPlane > Local Coordinate Systems > Create Local CS > At WP Origin

- Specify ***11*** for the ***KCN Ref number of new coord sys*** field (number 11 appears by default—leave it unchanged).
- Choose ***Cylindrical 1*** from the pull-down menu; click on ***OK***.
- Select nodes at the inner surface of the pipe for the convective boundary conditions using the following menu path:

Utility Menu > Select > Entities

- *Select Entities* dialog box appears.
- Choose ***Nodes*** from the first pull-down menu.
- Choose ***By Location*** from the second pull-down menu.
- Click on the radio-button ***X coordinates***.
- Type ***ri2*** in the text field; click on ***OK***.
- Apply the convective boundary condition with a temperature-dependent film coefficient on the selected nodes using the following menu path:

Main Menu > Solution > Define Loads > Apply > Thermal > Convection > On Nodes

- Click on ***Pick All***.
- In the dialog box, type ***-2*** (*negative 2*) for film coefficient and ***100*** for bulk temperature; click on ***OK***.
- Select everything using the following menu path:

Utility Menu > Select > Everything

- Obtain the solution.
  - Although steady-state (no time dependence), this problem is nonlinear due to the existence of the temperature-dependent film coefficient and thus requires an iterative solution. Therefore, time is introduced as an auxiliary variable, and related parameters are specified.
  - Specify *automatic time stepping* using the following menu path:

Main Menu > Solution > Load Step Opts > Time/Frequenc > Time and Substps

- Type ***50*** in the *second text field* corresponding to ***NSUBST*** (Fig. 9.25).
- Click on the ***ON*** radio-button for *Automatic time stepping*; click on ***OK***.
- Save using the following menu path:

Utility Menu > File > Save as

- Start the solution using the following menu path:

Main Menu > Solution > Solve > Current LS

- *Confirmation Window* appears along with *Status Report Window*.
- Review status; if OK, close the *Status Report Window* and click on ***OK*** in the *Confirmation Window*.
- Wait until ANSYS responds with ***Solution is done!***

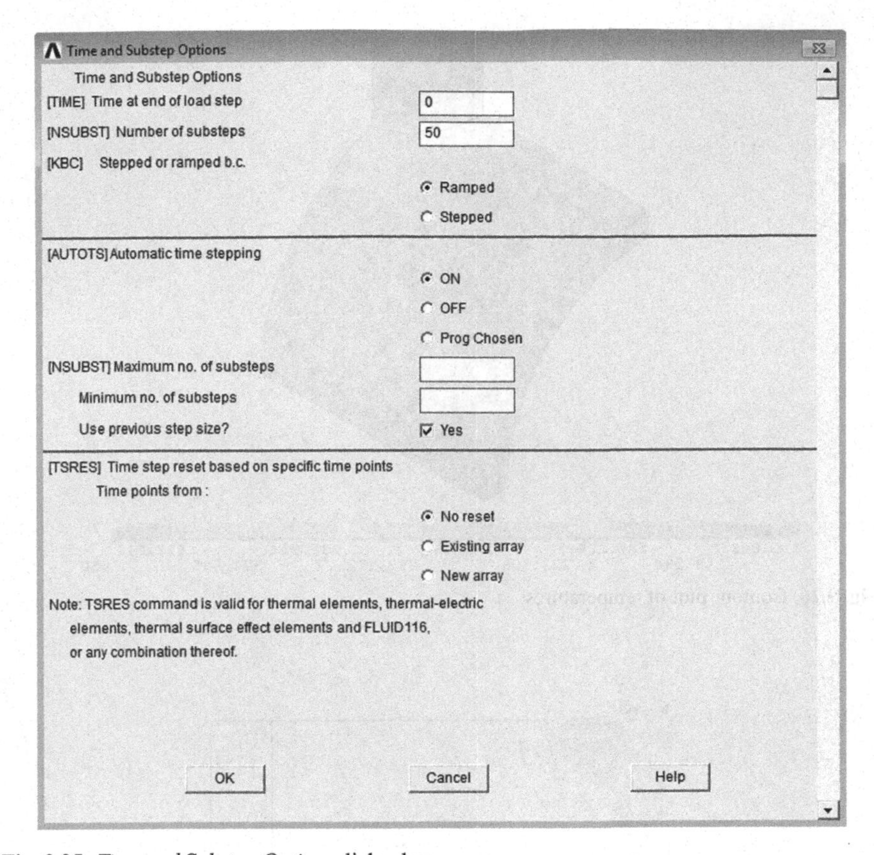

Fig. 9.25 *Time and Substep Options* dialog box

Postprocessing

- Review temperature contours using the following menu path:

Main Menu > General Postproc > Plot Results > Contour Plot > Nodal Solu

  - Select ***DOF Solution*** and ***Nodal Temperature***; click on ***OK***.
  - The temperature contour plot is shown in Fig. 9.26 as it appears in the *Graphics Window*.

- Review thermal flux vectors using the following menu path:

Main Menu > General Postproc > Plot Results > Vector Plot > Predefined

  - Select ***Flux and gradient*** from the left list and ***Thermal flux TF*** from the right list; click on ***OK***.
  - Figures 9.27 and 9.28 show the flux vectors as they appear in the *Graphics Window*, viewed from top and left, respectively.

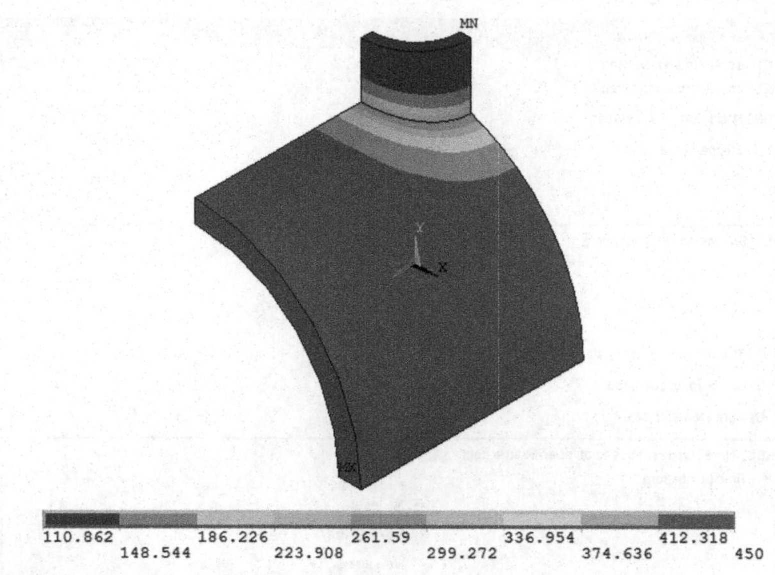

Fig. 9.26 Contour plot of temperatures

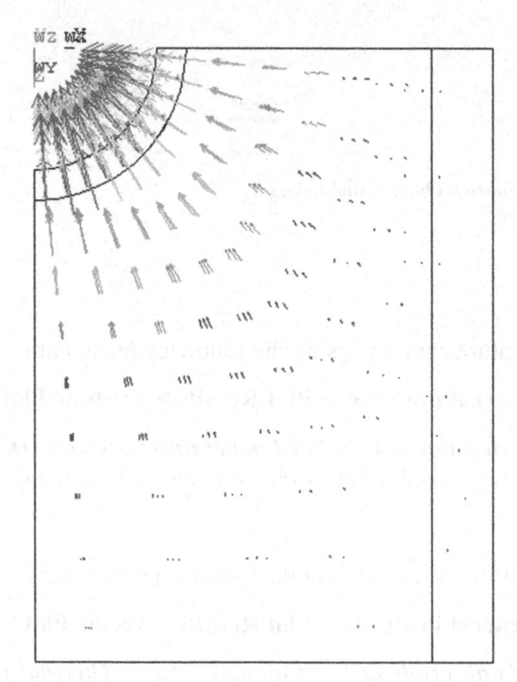

Fig. 9.27 *Top* view of thermal flux vectors

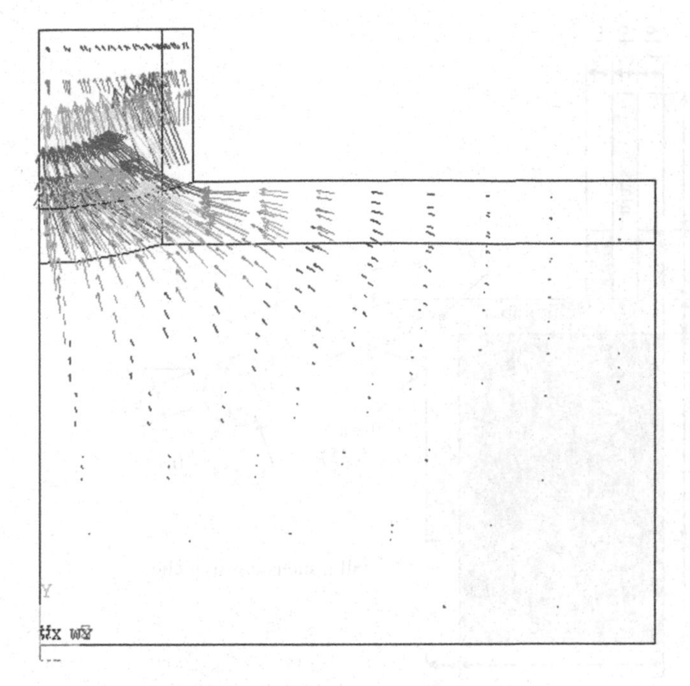

Fig. 9.28 *Left* view of thermal flux vectors

9.1.1.2 Analysis of a Window Assembly

Consider a window assembly composed of concrete, glass, and aluminum, as shown in Fig. 9.29. The inside and outside ambient temperatures are 70 and 0 °F, respectively, as indicated in Fig. 9.30. The film coefficient along the boundaries facing inside is 0.0139 Btu/hr-in^2-°F whereas the film coefficient along the boundaries facing outside is 0.0347 Btu/hr-in^2-°F. The thermal conductivities of glass, concrete, and aluminum are 0.05, 0.06, and 11 Btu/hr-in-°F, respectively. The goal is to determine the steady-state temperature distribution in the system.

Model Generation

- Define the element type (**ET** command) using the following menu path:

Main Menu > Preprocessor > Element Type > Add/Edit/Delete

- *Element Types* dialog box appears; click on ***Add***.
- Select ***Solid*** immediately below ***Thermal*** in the left list and ***Quad 4node 55*** in the right list; click on ***OK***.
- Click on ***Close***.

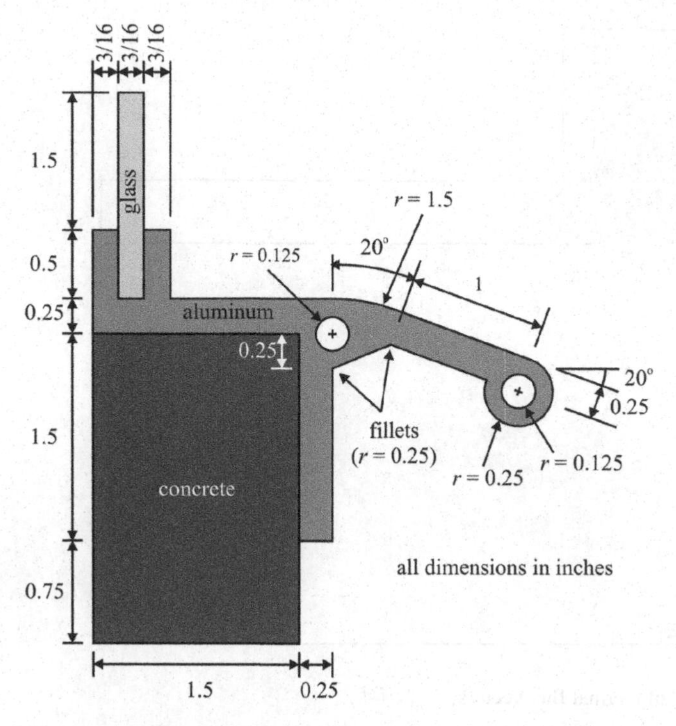

Fig. 9.29 The geometry of the window assembly

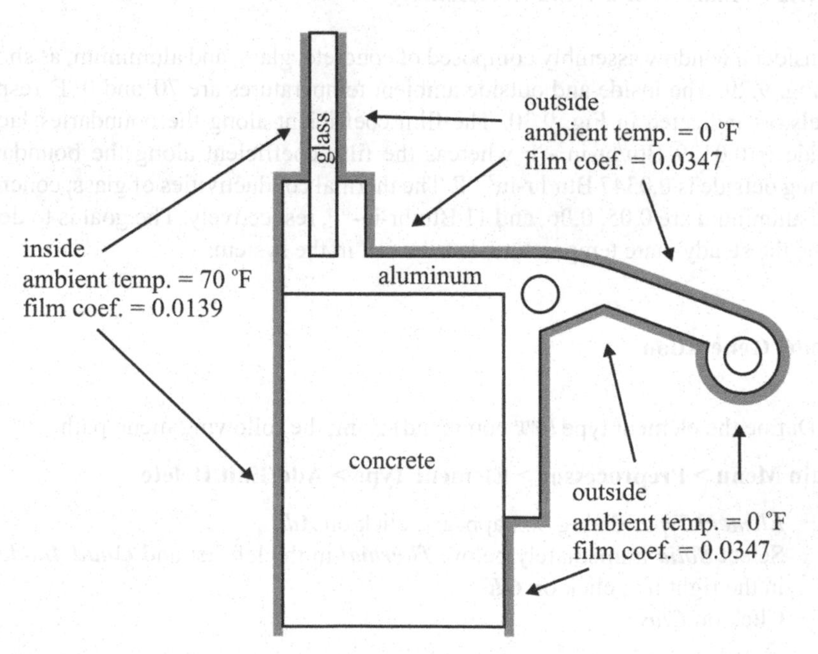

Fig. 9.30 Boundary conditions for the window assembly

Table 9.3 Coordinates defining the rectangles

Area Number	X1	X2	Y1	Y2
	(in.)			
1	0	1.5	0	2.25
2	1.5	1.75	0.75	2.25
3	0	1.75	2.25	2.5
4	0	3/16	2.5	3
5	3/16	3/8	2.5	4
6	3/8	9/16	2.5	3

- Specify thermal conductivities for aluminum (material 1), glass (material 2), and concrete (material 3) (**MP** command) using the following menu path:

Main Menu > Preprocessor > Material Props > Material Models

- In the *Define Material Model Behavior* dialog box, in the right window, successively left-click on ***Thermal***, ***Conductivity***, and ***Isotropic***, which brings up another dialog box.
- Enter ***11*** for ***KXX***; click on ***OK***.
- Add new material for glass. In the *Define Material Model Behavior* dialog box, define a new material reference number using the following menu path:

Material > New Model

- A dialog box appears with the ***material reference number*** set to the next available number (in this case 2); click on ***OK***.
- Note the new entry for ***Material Model Number 2*** in the left list in the *Define Material Model Behavior* dialog box.
- In the right window, successively left-click on ***Thermal***, ***Conductivity***, and ***Isotropic***, which brings up another dialog box.
- Enter ***0.05*** for ***KXX***; click on ***OK***.
- Add new material for concrete and specify the thermal conductivity as 0.06 following the same procedure used for glass above.
- Close the *Define Material Model Behavior* dialog box by using the following menu path:

Material > Exit

- Create the areas (**RECTNG** command) using the following menu path:

Main Menu > Preprocessor > Modeling > Create > Areas > Rectangle > By Dimensions

- *Create Rectangle by Dimensions* dialog box appears; enter ***0*** and ***1.5*** for the ***X-coordinates*** (***X1***, ***X2***) and ***0*** and ***2.25*** for the ***Y-coordinates*** (***Y1***, ***Y2***); click on ***Apply***.
- Referring to Table 9.3, create the remaining six areas. After creating the sixth area, click on ***OK*** (instead of ***Apply***).

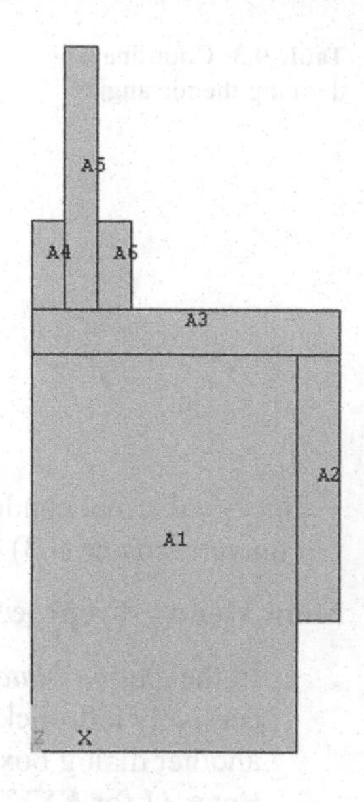

Fig. 9.31 Six areas as they appear in the *Graphics Window*

– Figure 9.31 shows the six areas as they appear in the *Graphics Window*.

- Define a local coordinate system (**LOCAL** command) using the following menu path:

Utility Menu > WorkPlane > Local Coordinate Systems > Create Local CS > At Specified Loc

– *Pick Menu* appears; type ***1.75,1*** in the text box and hit enter (as shown in Fig. 9.32); click on ***OK***.
– *Create Local CS at Specified Location* dialog box appears; select ***Cylindrical 1*** from the ***KCS Type of coordinate system*** pull-down menu; click on ***OK***.
– Observe that the local coordinate system appears in the *Graphics Window*.

- Create a keypoint in the local coordinate system (**K** command) using the following menu path:

Main Menu > Preprocessor > Modeling > Create > Keypoints > In Active CS

– *Create Keypoints in Active Coordinate System* dialog box appears; leave the ***NPT Keypoint number*** field blank (ANSYS issues the lowest available keypoint number to the newly created keypoints) and enter ***1.5*** and ***70*** for ***X, Y, Z Location in active CS*** (leave the ***Z*** field blank). Since the active coordinate system is *cylindrical*, ***X*** corresponds to the radius from the local coordinate system origin and ***Y*** corresponds to the angle measured from the *x*-axis in the counterclockwise direction.
– Upon clicking on ***OK***, ANSYS creates the new keypoint and plots all of the keypoints, as shown in Fig. 9.33. Note that the most recently created keypoint is 25.

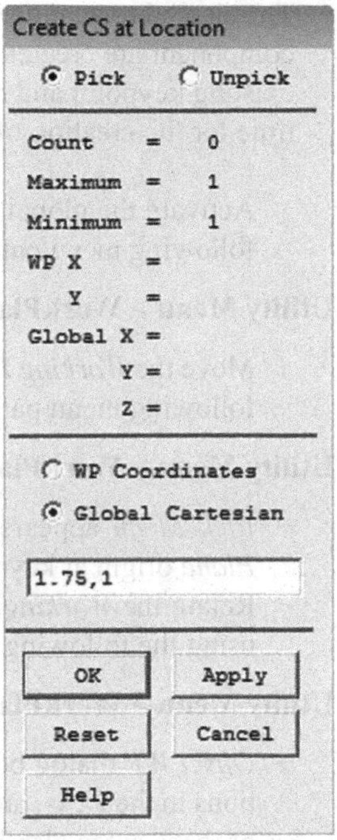

Fig. 9.32 *Pick Menu* for creating a local coordinate system at a location

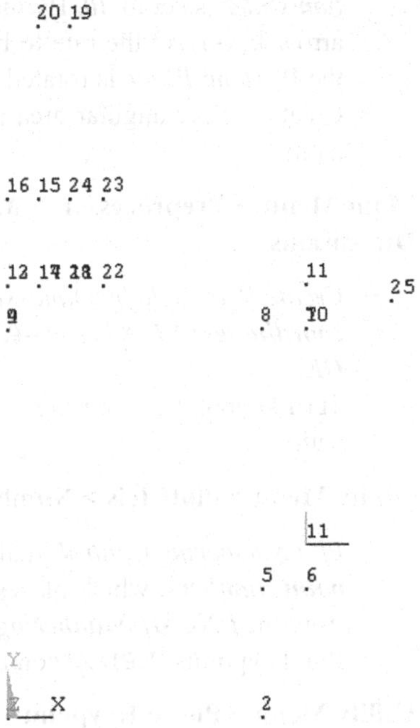

Fig. 9.33 Keypoints as they appear in the *Graphics Window*

- The slanted rectangular area and the circular portion at the end for the aluminum component are created next. For this purpose, the *Working Plane* is moved to an existing keypoint and rotated for the rectangular area. It is then moved one more time for the creation of the circular area.

  – Activate the global Cartesian coordinate system (**CSYS** command) using the following menu path:

Utility Menu > WorkPlane > Change Active CS to > Global Cartesian

  – Move the *Working Plane* origin to keypoint 25 (**KWPAVE** command) using the following menu path:

Utility Menu > WorkPlane > Offset WP to > Keypoints

  – *Pick Menu* appears; pick keypoint 25; click on ***OK***. Observe the *Working Plane* origin at keypoint 25 in the *Graphics Window*.
  – Rotate the *Working Plane* by -20° about the *z*-axis by (**WPROTA** command) using the following menu path:

Utility Menu > WorkPlane > Offset WP by Increments

  – *Offset WP* dialog box appears, which has two main action areas, i.e., translations in the *x*-, *y*-, and *z*-directions, and rotations about *x*-, *y*-, and *z*-axes. Each of these areas also has a *slider* that controls the rate at which translations or rotations are applied. By default, the *translation slider* is set to ***1*** and the *rotation slider* is set to ***30***. By moving the *rotation slider* to the left (or using the arrow keys), set the rate to be ***20°***. Upon clicking on the negative ***Z***-button, the *Working Plane* is rotated −20° about the *z*-axis.
  – Create the rectangular area (**RECTNG** command) using the following menu path:

Main Menu > Preprocessor > Modeling > Create > Areas > Rectangle > By Dimensions

  – *Create Rectangle by Dimensions* dialog box appears; enter ***0*** and ***1*** for the ***X-coordinates*** (***X1***, ***X2***) and ***–0.25*** and ***0*** for the ***Y-coordinates*** (***Y1***, ***Y2***); click on ***OK***.
  – Turn keypoint numbering on (**/PNUM** command) using the following menu path:

Utility Menu > PlotCtrls > Numbering

  – *Plot Numbering Controls* dialog box appears; click on the box next to ***KP Keypoint numbers***, which places a checkmark in the box. Select ***Numbers only*** from the ***[/NUM] Numbering shown with*** pull-down menu; click on ***OK***.
  – Plot keypoints (**KPLOT** command) using the following menu path:

Utility Menu > Plot > Keypoints > Keypoints

  – Move the *Working Plane* origin to keypoint 27 (**KWPAVE** command) using the following menu path:

Utility Menu > WorkPlane > Offset WP to > Keypoints

- *Pick Menu* appears; pick keypoint 27; click on ***OK***. Observe the *Working Plane* origin at keypoint 27 in the *Graphics Window*.
- Create the hollow circular area (**PCIRC** command) using the following menu path:

Main Menu > Preprocessor > Modeling > Create > Areas > Circle > By Dimensions

- *Circular Area by Dimensions* dialog box appears; enter ***0.25*** for ***RAD1 Outer radius*** and ***0.125*** for ***RAD2 Optional inner radius***. Click on ***OK***.
- Turn keypoint numbering off and area numbering on (**/PNUM** command) using the following menu path:

Utility Menu > PlotCtrls > Numbering

- *Plot Numbering Controls* dialog box appears; click on the box next to ***KP Keypoint numbers*** to remove the checkmark (turning keypoint numbering off) and click on the box next to ***AREA Area numbers*** to place a checkmark (turning area numbering on). Click on ***OK***.
- Plot areas (**APLOT** command) using the following menu path:

Utility Menu > Plot > Areas

- Figure 9.34 shows the areas as they appear in the *Graphics Window*.

• Overlap areas 7 and 8 (**AOVLAP** command) using the following menu path:

Main Menu > Preprocessor > Modeling > Operate > Booleans > Overlap > Areas

• *Pick Menu* appears; pick areas 7 and 8; click on ***OK*** in the *Pick Menu*.
• Figure 9.35 shows the areas before and after overlapping.
• Delete area 9 and keypoints and lines attached to it (**ADELE** command) using the following menu path:

Main Menu > Preprocessor > Modeling > Delete > Area and Below

- *Pick Menu* appears; pick area 9; click on ***OK*** in the *Pick Menu*.
- Observe that area 9 disappears in the *Graphics Window*.

• Reset the *Working Plane* to its default configuration (**WPSTYL** command) using the following menu path:

Utility Menu > WorkPlane > Align WP with > Global Cartesian

- *Working Plane* is aligned with global Cartesian coordinate system (both the origin and the orientation).
- Toggle the *Working Plane* display off (**WPSTYL** command) using the following menu path:

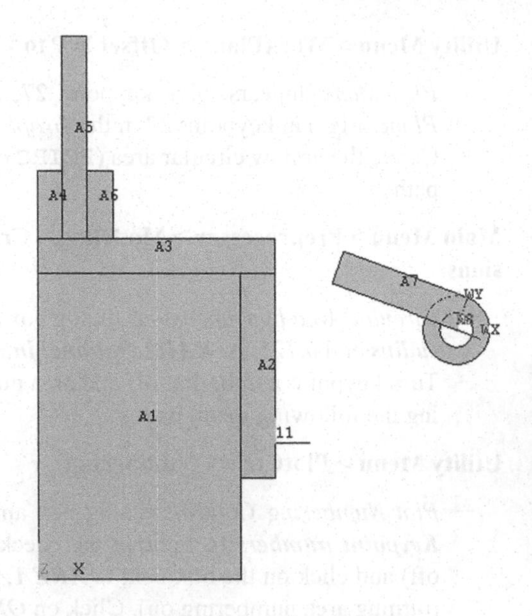

Fig. 9.34 Areas as they appear in the *Graphics Window*

Utility Menu > WorkPlane > Display Working Plane

– If a checkmark appears next to the menu entry ***Display Working Plane***, which means the display is on, click on the menu entry to toggle off the display.

- Create a keypoint (`K` command) using the following menu path:

Main Menu > Preprocessor > Modeling > Create > Keypoints > In Active CS

– *Create Keypoints in Active Coordinate System* dialog box appears; enter ***50*** in the ***NPT Keypoint number*** field and ***1.75*** and ***2*** for ***X, Y, Z Location in active CS*** (leave the **Z** field blank). Click on ***OK***.

- Activate local coordinate system 11 (created earlier) (`CSYS` command) using the following menu path:

Utility Menu > Change Active CS to > Specified Coord Sys

– *Change Active CS to Specified CS* dialog box appears; enter ***11*** for ***KCN Coordinate system number***; click on ***OK***.

- Turn keypoint numbering on and area numbering off. Plot keypoints.
- Create a line between keypoints 11 and 29 (`L` command) using the following menu path:

Main Menu > Preprocessor > Modeling > Create > Lines > Lines > In Active Coord

– *Pick Menu* appears; pick keypoints 11 and 29. Keypoint 29 is coincident with keypoint 25; therefore, when attempting to pick keypoint 29, ANSYS issues

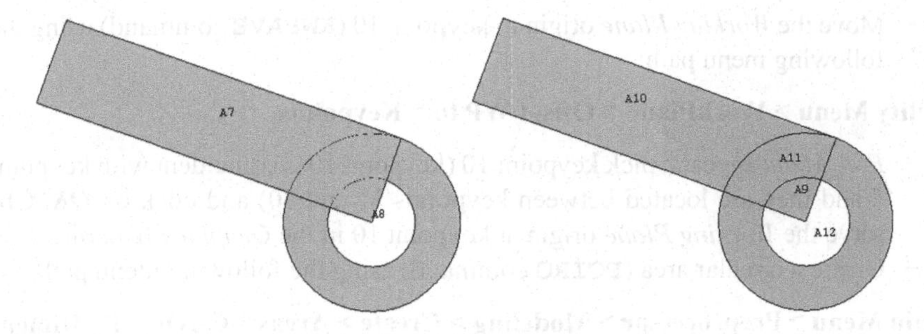

Fig. 9.35 Areas before (*left*) and after (*right*) overlapping

a warning to the user that there are more than one keypoints at the picked location and that the currently picked keypoint is keypoint 25. Click on ***Next*** in this *Multiple_Entities Warning Window*, which picks the next higher numbered keypoint (keypoint 29 in this case) at the coincident location. Click on ***OK*** in the *Multiple_Entities Warning Window*; click on ***OK*** in the *Pick Menu*.

- The line that is just created is a curved line. This is because it is created in the active coordinate system, which is a local cylindrical coordinate system.
- Reset the active coordinate system to be global Cartesian (**CSYS** command) using the following menu path:

Utility Menu > WorkPlane > Change Active CS to > Global Cartesian

- Create an area using keypoints 50, 26, 29, and 11 (**A** command) using the following menu path:

Main Menu > Preprocessor > Modeling > Create > Areas > Arbitrary > Through KPs

- *Pick Menu* appears; pick the keypoints listed above. Note that when picking keypoint 29, since it is coincident with keypoint 25, ANSYS warns the user that there are two keypoints at the picked location. Similar to the previous case, click on ***Next*** to pick keypoint 29. When finished with picking all four keypoints, click on ***OK*** in the *Pick Menu* and observe that area 7 appears in the *Graphics Window*.
- Since a curved line was created between keypoints 11 and 29 previously, and both of these keypoints are used in the definition of area 7, the boundary of the area between keypoints 11 and 29 is a curved one. If the line were not created, then this boundary would have been a straight line.

The circular hole within the aluminum component near the top-right corner of the concrete block is created next. For this purpose, the *Working Plane* origin is moved to keypoint 10. After creating the circle, areas around the circle are overlapped and the ones defining the hole are deleted.

- Plot keypoints (**KPLOT** command) using the following menu path:

Utility Menu > Plot > Keypoints > Keypoints

– Move the *Working Plane* origin to keypoint 10 (**KWPAVE** command) using the following menu path:

Utility Menu > WorkPlane > Offset WP to > Keypoints

– *Pick Menu* appears; pick keypoint 10 (keypoint 10 is coincident with keypoint 7 and they are located between keypoints 11 and 50) and click on ***OK***. Observe the *Working Plane* origin at keypoint 10 in the *Graphics Window*.
– Create a circular area (**PCIRC** command) using the following menu path:

Main Menu > Preprocessor > Modeling > Create > Areas > Circle > By Dimensions

– *Circular Area by Dimensions* dialog box appears; enter ***0.125*** for ***RAD1 Outer radius*** and remove any leftover numbers from the ***RAD2 Optional inner radius*** field. Click on ***OK***.
– Reset the *Working Plane* to its default configuration (**WPSTYL** command) using the following menu path:

Utility Menu > WorkPlane > Align WP with > Global Cartesian

– *Working Plane* is aligned with global Cartesian coordinate system (both the origin and the orientation).
– Toggle the *Working Plane* display off (**WPSTYL** command) using the following menu path:

Utility Menu > WorkPlane > Display Working Plane

– If a checkmark appears next to the menu entry ***Display Working Plane***, which means the display is on, click on the menu entry to toggle off the display.
– Turn area numbering on (**/PNUM** command) and keypoint numbering off using the following menu path:

Utility Menu > PlotCtrls > Numbering

– *Plot Numbering Controls* dialog box appears; click on the box next to ***AREA Area numbers*** to place a checkmark (turning area numbering on). Remove the checkmark for keypoints. Click on ***OK***.
– Plot areas (**APLOT** command) using the following menu path:

Utility Menu > Plot > Areas

– Overlap areas 2, 3, 7, and 8 (**AOVLAP** command) using the following menu path:

Main Menu > Preprocessor > Modeling > Operate > Booleans > Overlap > Areas

– *Pick Menu* appears; pick areas 2, 3, 7, and 8 and click on ***OK*** in the *Pick Menu*.
– Figures 9.36 and 9.37 show the areas before and after overlapping, respectively.

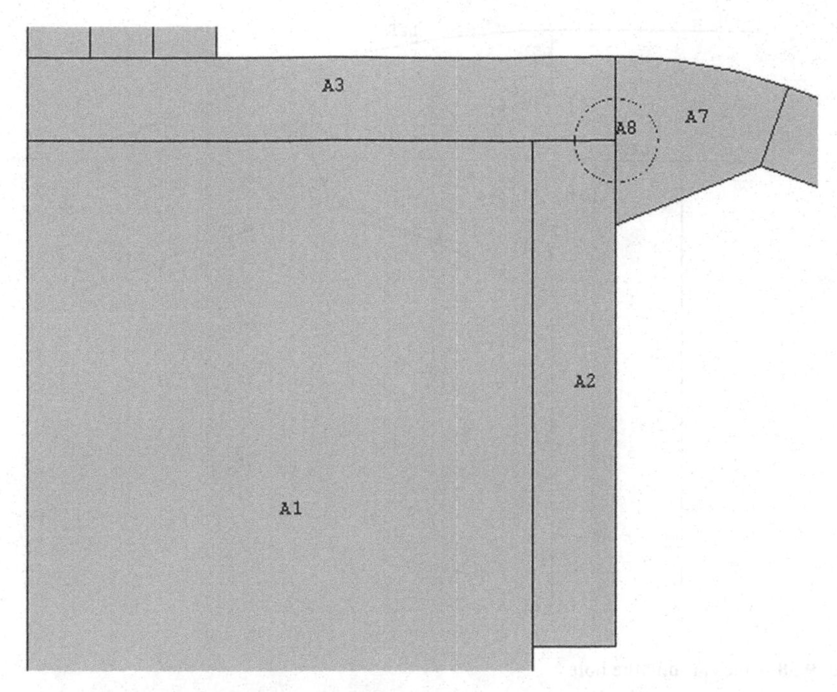

Fig. 9.36 Areas before overlapping

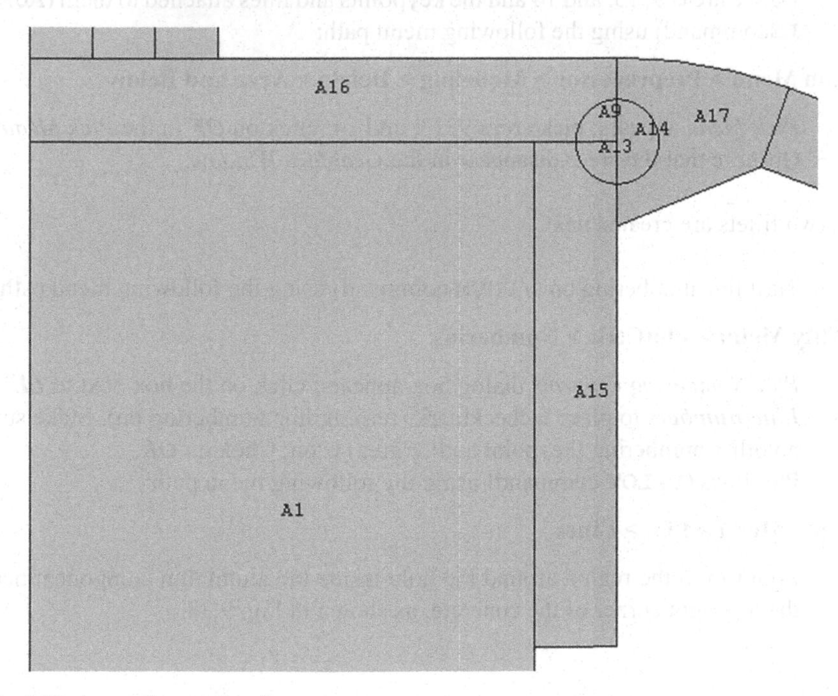

Fig. 9.37 Areas after overlapping

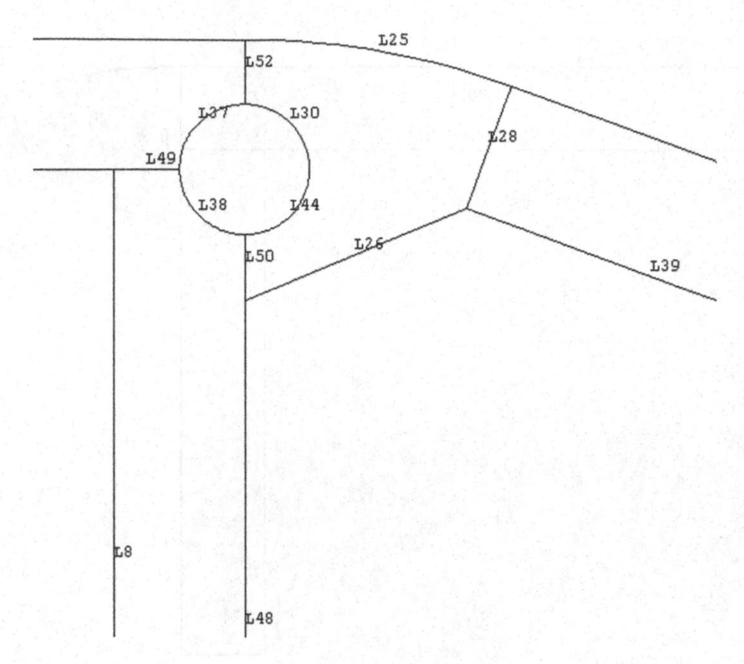

Fig. 9.38 Lines around the hole

- Delete areas 9, 13, and 14 and the keypoints and lines attached to them (**ADE-LE**command) using the following menu path:

Main Menu > Preprocessor > Modeling > Delete > Area and Below

- *Pick Menu* appears; pick areas 9, 13, and 14; click on ***OK*** in the *Pick Menu*.
- Observe that the areas disappear in the *Graphics Window*.

- Two fillets are created next.

- Turn line numbering on (**/PNUM** command) using the following menu path:

Utility Menu > PlotCtrls > Numbering

- *Plot Numbering Controls* dialog box appears; click on the box next to ***LINE Line numbers*** to place a checkmark (turning line numbering on). Make sure no other numbering (keypoint and/or area) is on. Click on ***OK***.
- Plot lines (**LPLOT** command) using the following menu path:

Utility Menu > Plot > Lines

- Zoom-in to the region around the hole inside the aluminum component near the top-right corner of the concrete, as shown in Fig. 9.38.

- Create line fillets between lines 48 and 26 and between lines 26 and 39, both of them having a radius of 0.25 in (**LFILLT** command), using the following menu path:

Main MenuPreprocessor > Modeling > Create > Lines > Line Fillet

- *Pick Menu* appears, prompting the user to pick two intersecting lines. Pick lines ***48*** and ***26***; click on ***OK*** in the *Pick Menu.*
- *Line Fillet* dialog box appears; enter ***0.25*** for ***RAD Fillet radius*** and ***55*** for ***PCENT Number to assign to generated keypoint at fillet center***. Click on ***Apply*** to create the next fillet.
- *Pick Menu* reappears. Repeat the same procedure by picking lines 26 and 39, using the same radius (***0.25***), and using ***60*** for ***PCENT***. When finished, click on ***OK*** in the *Line Fillet* dialog box.
- Turn keypoint numbering on (**/PNUM** command) using the following menu path:

Utility Menu > PlotCtrls > Numbering

- *Plot Numbering Controls* dialog box appears; click on the box next to ***KP Keypoint numbers*** to place a checkmark. Make sure no other numbering (line and/or area) is on. Click on ***OK***.
- Plot keypoints (**KPLOT** command) using the following menu path:

Utility Menu > Plot > Keypoints > Keypoints

- Create areas (for fillets) through keypoints (**A** command) using the following menu path:

Main Menu > Preprocessor > Modeling > Create > Areas > Arbitrary > Through KPs

- *Pick Menu* appears; pick keypoints 7, 10, and 50; click on ***Apply*** in the *Pick Menu.*
- *Pick Menu* remains active; pick keypoints 40, 41, and 26; click on ***OK*** in the *Pick Menu.*
- Turn area numbering on (**/PNUM** command) using the following menu path:

Utility Menu > PlotCtrls > Numbering

- *Plot Numbering Controls* dialog box appears; click on the box next to ***AREA Area numbers*** to place a checkmark (turning area numbering on). Make sure no other numbering (keypoint and/or line) is on. Click on ***OK***.
- Plot areas (**APLOT** command) using the following menu path:

Utility Menu > Plot > Areas

- Areas appear in the *Graphics Window,* as shown in Fig. 9.39.
- Add areas 2, 3, 10, 15, and 17 (**AADD** command) using the following menu path:

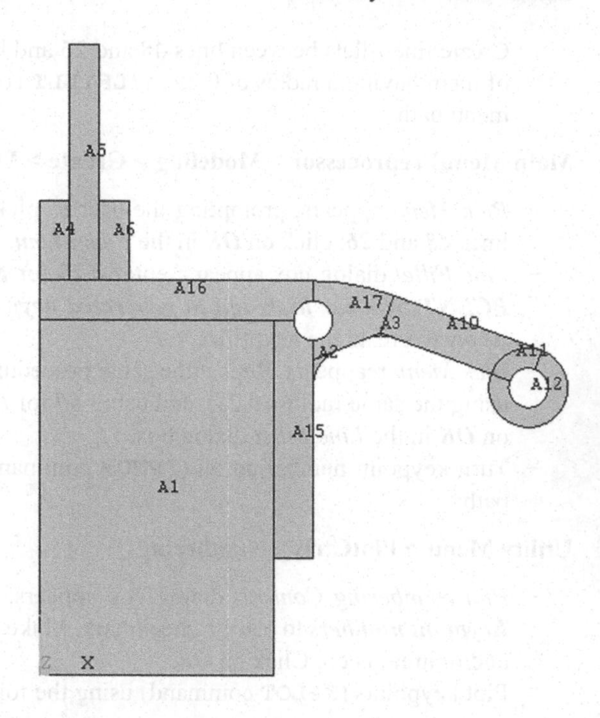

Fig. 9.39 Areas after fillet creation

Main Menu > Preprocessor > Modeling > Booleans > Add > Areas

- *Pick Menu* appears; pick areas 2, 3, 10, 15, and 17; click on ***OK*** in the *Pick Menu*.
- Glue all areas (**AGLUE** command) using the following menu path:

Main Menu > Preprocessor > Modeling > Booleans > Glue > Areas

- *Pick Menu* appears; click on ***Pick All*** in the *Pick Menu*.
- Figure 9.40 shows the areas after adding and gluing operations as they appear in the *Graphics Window*.

• Specify size controls for meshing, compress entity numbers, and create the mesh.

- Specify global element size (**ESIZE** command) usi8.5ng the following menu path:

Main Menu > Preprocessor > Meshing > Size Cntrls > ManualSize > Global > Size

- *Global Element Sizes* dialog box appears; enter ***1/16*** for ***SIZE Element edge length***; click on ***OK***.
- Turn keypoint numbering on (**/PNUM** command) using the following menu path:

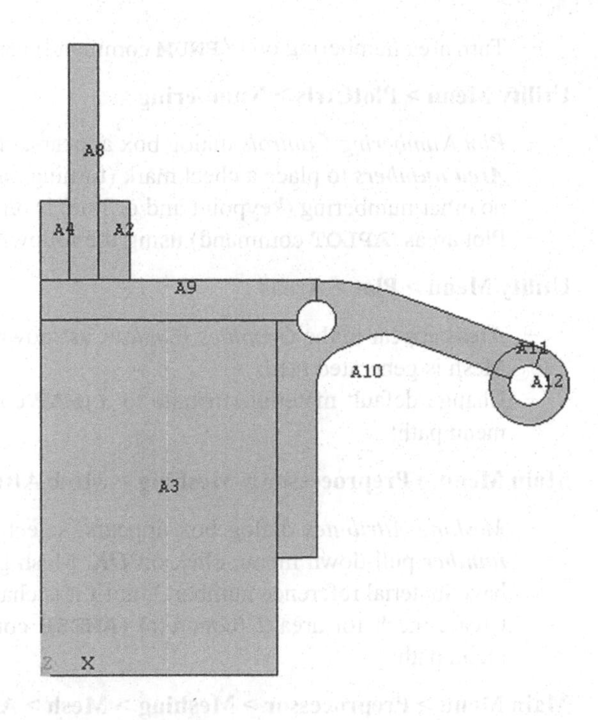

Fig. 9.40 Areas after adding and gluing operations

Utility Menu > PlotCtrls > Numbering

- *Plot Numbering Controls* dialog box appears; click on the box next to ***KP Keypoint numbers*** to place a checkmark. Make sure no other numbering (line and/or area) is on. Click on ***OK***.
- Plot keypoints (**KPLOT** command) using the following menu path:

Utility Menu > Plot > Keypoints > Keypoints

- Specify element size near keypoints 1 and 2 as 0.25 (**KESIZE** command) using the following menu path:

Main Menu > Preprocessor > Meshing > Size Cntrls > ManualSize > Keypoints > Picked KPs

- *Pick Menu* appears; pick keypoints 1 and 2 (these are the bottom corners of the concrete block—keypoint 1 is at the global origin); click on ***OK*** in the *Pick Menu*.
- *Element Size at Picked Keypoints* dialog box appears; enter ***0.25*** for ***SIZE Element edge length***; click on ***OK***.
- Compress entity numbers (**NUMCMP** command) using the following menu path:

Main Menu > Preprocessor > Numbering Ctrls > Compress Numbers

- *Compress Numbers* dialog box appears; select ***All*** from the ***Label Item to be compressed*** pull-down menu; click on ***OK***.

- Turn area numbering on (**/PNUM** command) using the following menu path:

Utility Menu > PlotCtrls > Numbering

- *Plot Numbering Controls* dialog box appears; click on the box next to ***AREA Area numbers*** to place a checkmark (turning area numbering on). Make sure no other numbering (keypoint and/or line) is on. Click on ***OK***.
- Plot areas (**APLOT** command) using the following menu path:

Utility Menu > Plot > Areas

- Areas appear in the *Graphics Window*, as shown in Fig. 9.41.
- Mesh is generated next.
- Change default material attribute to 3 (**MAT** command) using the following menu path:

Main Menu > Preprocessor > Meshing > Mesh Attributes > Default Attribs

- *Meshing Attributes* dialog box appears; select ***3*** from the ***[MAT] Material number*** pull-down menu; click on ***OK***. Mesh generated after this point will have material reference number 3, until it is changed to another number.
- Create mesh for area 2 (concrete) (**AMESH** command) using the following menu path:

Main Menu > Preprocessor > Meshing > Mesh > Areas > Free

- *Pick Menu* appears; pick area 2; click on ***OK***.
- Plot areas (**APLOT** command) using the following menu path:

Utility Menu > Plot > Areas

- Change default material attribute to 2 (**MAT** command) using the following menu path:

Main Menu > Preprocessor > Meshing > Mesh Attributes > Default Attribs

- *Meshing Attributes* dialog box appears; select ***2*** from the ***[MAT] Material number*** pull-down menu; click on ***OK***.
- Create mesh for area 4 (glass) (**AMESH** command) using the following menu path:

Main Menu > Preprocessor > Meshing > Mesh > Areas > Free

- *Pick Menu* appears; pick area 4; click on ***OK***.
- Plot areas (**APLOT** command) using the following menu path:

Utility Menu > Plot > Areas

- Change default material attribute to 1 (**MAT** command) using the following menu path:

Main Menu > Preprocessor > Meshing > Mesh Attributes > Default Attribs

- *Meshing Attributes* dialog box appears; select ***1*** from the ***[MAT] Material number*** pull-down menu; click on ***OK***.

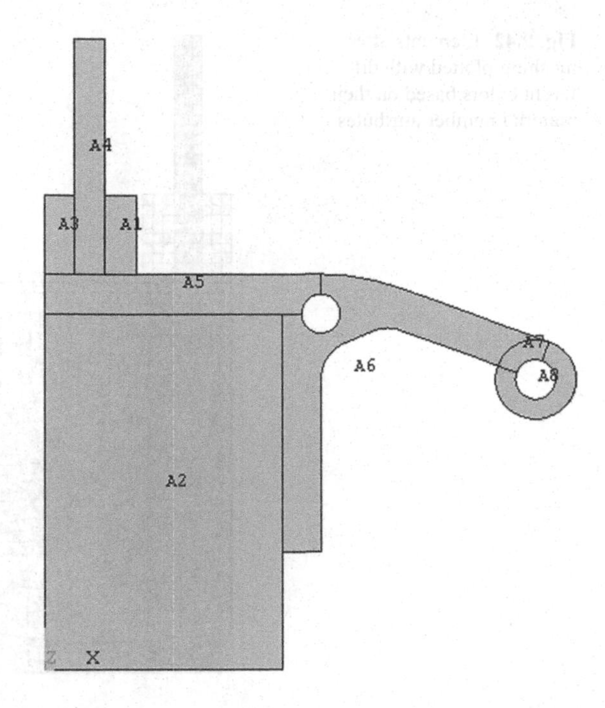

Fig. 9.41 Areas after entity number compression

- Create mesh for the remaining areas (aluminum) (**AMESH** command) using the following menu path:

Main Menu > Preprocessor > Meshing > Mesh > Areas > Free

- *Pick Menu* appears; pick the remaining areas (1, 3, 5–8); click on ***OK***.
- A warning message may appear; if so, close it.
- Turn the element numbering on so that the elements are plotted with different colors based on their material attribute (**/PNUM** command) using the following menu path:

Utility Menu > PlotCtrls > Numbering

- *Plot Numbering Controls* dialog box appears; select ***Material numbers*** from the ***Elem/Attrib numbering*** pull-down menu; select ***Colors only*** from the ***[/NUM] Numbering shown with*** pull-down menu. Click on ***OK***.
- Plot elements (**EPLOT** command) using the following menu path:

Utility Menu > Plot > Elements

- Elements appear in the *Graphics Window*, as shown in Fig. 9.42.

Solution

- Apply convective boundary conditions.

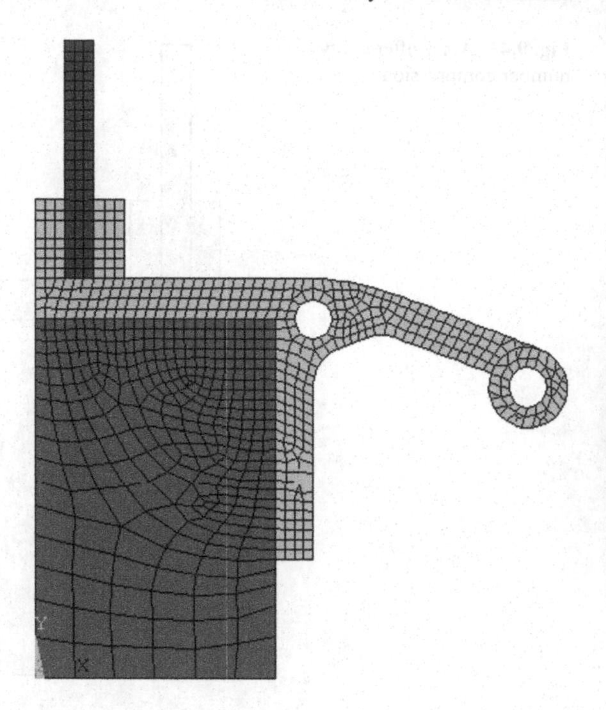

Fig. 9.42 Elements after meshing plotted with different colors based on their material number attributes

- Turn line numbering on (**`/PNUM`** command) using the following menu path:

Utility Menu > PlotCtrls > Numbering

- *Plot Numbering Controls* dialog box appears; click on the box next to ***LINE Line numbers*** to place a checkmark. Select ***No numbering*** from the ***Elem/Attrib numbering*** pull-down menu, and select ***Numbers only*** from the ***[/NUM] Numbering shown with*** pull-down menu. Click on ***OK***.
- Plot lines (**`LPLOT`** command) using the following menu path:

Utility Menu > Plot > Lines

- Apply convective boundary conditions on the lines along the side facing outside (as shown in Fig. 9.30) (**`SFL`** command) using the following menu path:

Main Menu > Solution > Define Loads > Apply > Thermal > Convection > On Lines

- *Pick Menu* appears; pick the lines shown in Fig. 9.43; click on ***OK*** in the *Pick Menu*.
- *Apply CONV on lines* dialog box appears; type ***0.0347*** for ***VALI Film coefficient*** and ***0*** for ***VAL2I Bulk temperature***; click on ***OK***.
- Apply convective boundary conditions on the lines along the side facing inside (as shown in Fig. 9.30) (**`SFL`** command) using the following menu path:

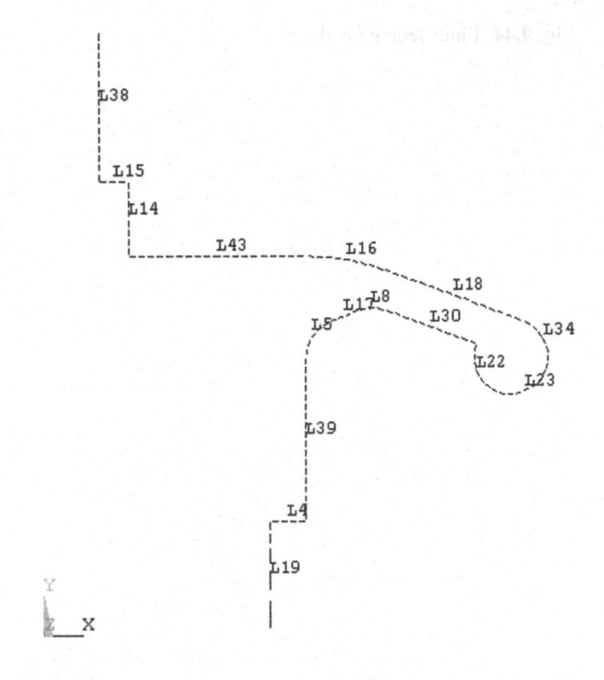

Fig. 9.43 Lines facing outside

Main Menu > Solution > Define Loads > Apply > Thermal > Convection > On Lines

– *Pick Menu* appears; pick the lines shown in Fig. 9.44; click on ***OK*** in the *Pick Menu.*
– *Apply CONV on lines* dialog box appears; type ***0.0139*** for ***VALI Film coefficient*** and ***70*** for ***VAL2I Bulk temperature***; click on ***OK***.

- Obtain the solution (**SOLVE** command) using the following menu path:

Main Menu > Solution > Solve > Current LS

– *Confirmation Window* appears along with *Status Report Window.*
– Review status; if OK, close the *Status Report Window*; click on ***OK*** in the *Confirmation Window.*
– Wait until ANSYS responds with ***Solution is done!***

Postprocessing

- Turn colors on (**/PNUM** command) using the following menu path:

Utility Menu > PlotCtrls > Numbering

– *Plot Numbering Controls* dialog box appears; click on the box next to ***LINE Line numbers*** to remove the checkmark. Select ***Colors only*** from the ***[/NUM] Numbering shown with*** pull-down menu; click on ***OK***.

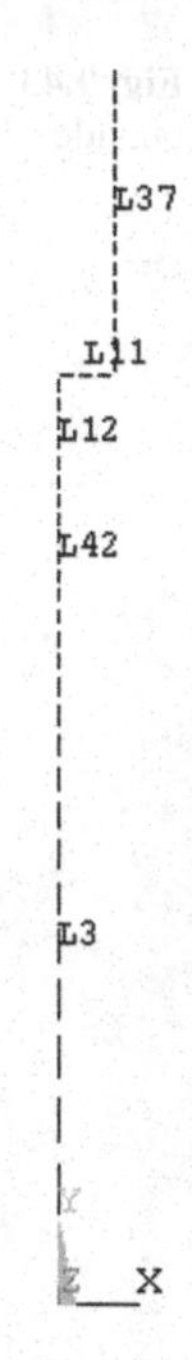

Fig. 9.44 Lines facing inside

- Review temperature contours (**PLNSOL** command) using the following menu path:

Main Menu > General Postproc > Plot Results > Contour Plot > Nodal Solu

  - *Contour Nodal Solution Data* dialog box appears. Successively select ***DOF Solution*** and ***Nodal Temperatue***; click on ***OK***.
  - The temperature contour plot is shown in Fig. 9.45 as it appears in the *Graphics Window*.

- Review thermal flux and gradient vectors using the following menu path:

Main Menu > General Postproc > Plot Results > Vector Plot > Predefined

  - *Vector Plot of Predefined Vectors* dialog box appears. Select ***Flux & gradient*** from the left list and ***Thermal flux TF*** from the right list; click on ***OK***.
  - Figure 9.46 shows the flux vectors as they appear in the *Graphics Window*.
  - Repeat the same procedure for the thermal gradient by selecting ***Thermal grad TG*** from the right list. Figure 9.47 shows the thermal gradient vectors as they appear in the *Graphics Window*.

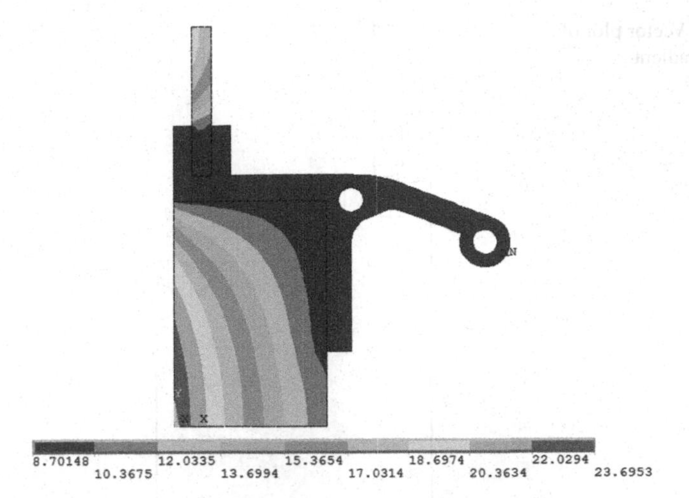

Fig. 9.45 Contour plot of temperatures

Fig. 9.46 Vector plot of thermal flux

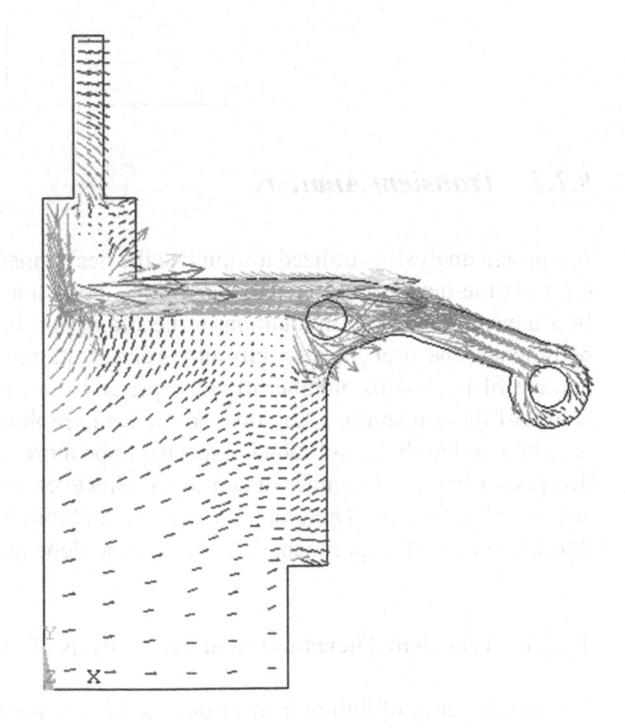

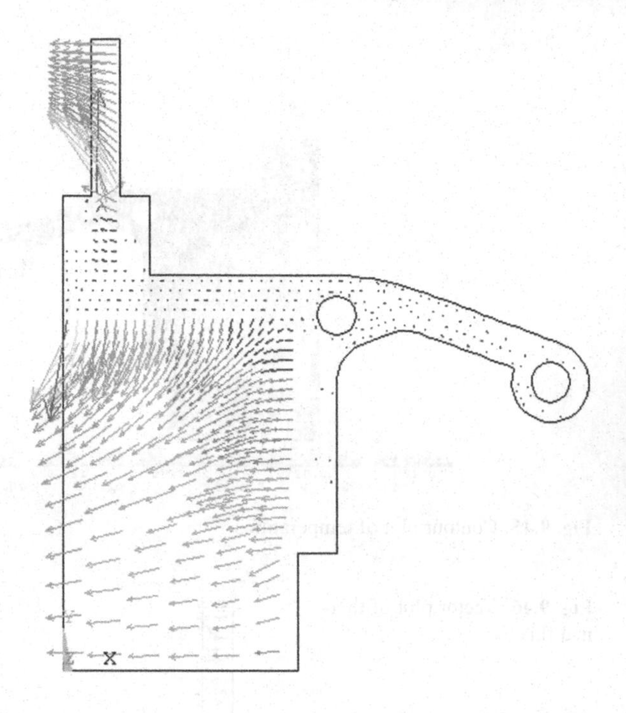

Fig. 9.47 Vector plot of thermal gradient

9.1.2 *Transient Analysis*

A transient analysis is utilized to simulate the heat transfer phenomenon in the presence of time-dependent boundary conditions, body loads, and/or initial conditions. In a transient analysis, in addition to the initial and boundary conditions and the body loads, the user must specify time-related quantities, such as time step size, number of load steps, number of substeps, and the final time. Depending on the values of these quantities, solutions to the same problem may differ considerably. As a rule of thumb, the solution is expected to be more accurate as the time step size decreases (increased number of substeps). However, this may increase the cost of analysis significantly. Therefore, the user has to build a certain level of knowledge through numerical experimentation on the time-dependent parameters.

9.1.2.1 Transient Thermomechanical Analysis of an Electronic Package

A common cause of failure in electronic devices is the thermal stresses at elevated temperatures caused by a coefficient of thermal expansion mismatch. In most thermomechanical analyses, the thermal load is assumed to be uniform. However, a nonuniform temperature distribution may be required for specific cases. In order to obtain a transient thermal stress field, first, a transient heat conduction (diffusion) solution is obtained, followed by a stress analysis utilizing the nonuniform thermal field at a specific time as body loading. ANSYS provides a convenient method for achieving this task. In order to demonstrate this capability, a problem involving

Table 9.4 Properties of the constituent materials in the electronic package

	Substrate	Die-Attach	Silicon	Copper
Material reference no.	1	2	3	4
E (GPa)	22	7.4	163	129
ν	0.39	0.4	0.278	0.344
α (10^{-6}/°C)	18	52	2.6	14.3
κ (W/m·°C)	2	100	150	396
c (J/kg·°C)	840	535	703	384
ρ (kg/m³)	220	6450	2330	8940

Table 9.5 Coordinates defining the areas and the corresponding material reference numbers

Area No.	X1 (mm)	X2	Y1	Y2	Mat. Ref. No.
1	0	5	0	2	1
2	5	7.5	0	2	1
3	7.5	10	0	2	1
4	0	5	2	2.1	2
5	0	5	2.1	3.1	3
6	0	5	3.1	5.1	4

an electronic device is considered. The device contains a silicon die (chip), epoxy die-attach, substrate, and a copper heat spreader, as shown in Fig. 9.48. Thermal and mechanical properties of the constituent materials are given in Table 9.4. The solid model is generated utilizing rectangles with the coordinates given in Table 9.5, along with their associations with material reference numbers.

In this transient thermal analysis, the surrounding air is at a temperature of $T_\infty = 25\,°C$ (ambient temperature). All of the surfaces, except the symmetry line, are subjected to convective heat loss with a heat transfer coefficient of $h=5$ W/(m²·°C). There is no heat transfer through the symmetry line (insulation). In ANSYSthermal analyses, when boundary conditions are not specified along a boundary, insulation is imposed automatically. Heat is generated at the bottom face of the die, which is expressed in terms of a constant heat flux of $q=1000$ W/m. The initial temperature of the device is assumed to be uniform at $T_0=25\,°C$. Heat transfer is simulated for a period of 5 min (300 s), after which the device reaches a steady state.

Once the time-dependent temperature field is obtained, the stress field is calculated based on the temperatures at desired time points.

Transient Thermal Analysis

Model Generation

- Specify the jobname (**`/FILNAM`** command) using the following menu path:

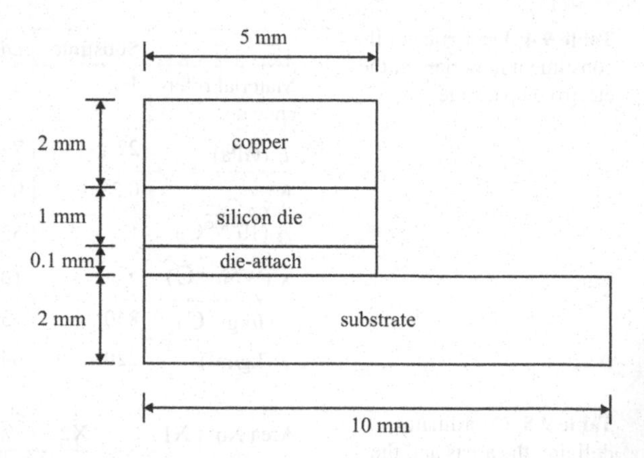

Fig. 9.48 Schematic of the electronic package

Utility Menu > File > Change Jobname

- *Change Jobname* dialog box appears. Type ***TH***; click on ***OK***.

- Define the element type (**ET** command) using the following menu path:

Main Menu > Preprocessor > Element Type > Add/Edit/Delete

- *Element Types* dialog box appears; click on ***Add***.
- Select ***Solid*** immediately below ***Thermal*** in the left list and ***Quad 4 Node 55*** in the right list; click on ***OK***.
- Click on ***Close***.

- Specify material properties (**MP** command) using the following menu path:

Main Menu > Preprocessor > Material Props > Material Models

- *Define Material Model Behavior* dialog box appears. Structural properties are first specified, followed by thermal properties.
- In the right window, successively left-click on ***Structural***, ***Linear***, ***Elastic***, and, finally, ***Isotropic***, which brings up another dialog box
- Referring to Table 9.4, enter ***22E9*** for ***EX*** and ***0.39*** for ***PRXY***; click on ***OK***
- In the right list, successively double-click on ***Structural***, ***Thermal Expansion***, ***Secant Coefficient***, and, finally, ***Isotropic***, which brings up another dialog box.
- Enter ***18E–6*** for ***ALPX***; click on ***OK***.
- Specify density by successively double-clicking on ***Structural*** and ***Density***, which brings up another dialog box. Type ***220***, click on ***OK***.
- In order to specify thermal properties, first successively double-click on ***Thermal***, ***Conductivity***, and ***Isotropic***; then enter ***2*** in the newly appeared dialog box; click on ***OK***. Thermal conductivity is specified.
- Specify specific heat by successively double-clicking on ***Thermal*** and ***Specific Heat***; enter ***840*** in the newly appeared dialog box; click on ***OK***.
- Add new material model using the following menu path:

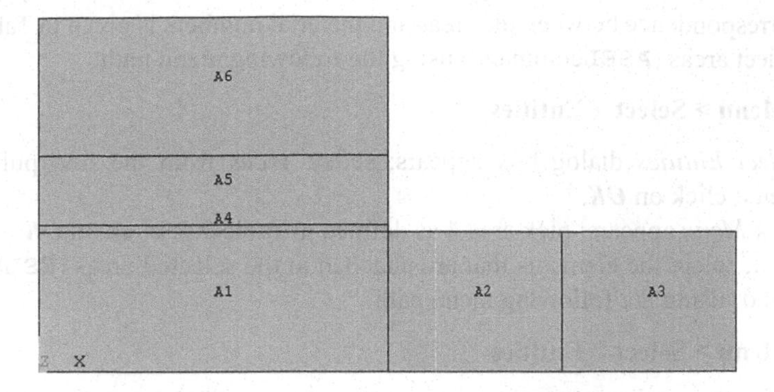

Fig. 9.49 Solid model of the electronic device with area numbers turned on

Material > New Model

– Repeat the procedure for the remaining materials (2 through 4).
– When finished, close the *Define Material Model Behavior* dialog box by using the following menu path:

Material > Exit

- Create rectangles as identified in Fig. 9.49 (**RECTNG** command) using the following menu path:

Main Menu > Preprocessor > Modeling > Create > Areas > Rectangle > By Dimensions

– *Create Rectangle by Dimensions* dialog box appears. Referring to Table 9.5, enter ***0*** and ***5E–3*** for ***X1*** and ***X2*** and ***0*** and ***2E–3*** for ***Y1*** and ***Y2***; click on ***Apply***.
– Repeat the procedure for the remaining areas (2 through 6). When creating Area 6, click on ***OK*** after entering the coordinates.

- Glue the areas (**AGLUE** command) using the following menu path:

Main Menu > Preprocessor > Modeling > Operate > Booleans > Glue > Areas

– *Pick Menu* appears; click on ***Pick All*** button.

- Mesh the areas (**AMESH** command) using the following menu path:

Main Menu > Preprocessor > Meshing > Mesh > Areas > Mapped > 3 or 4 sided

– *Pick Menu* appears; click on ***Pick All.***
– At this point, all the elements have *Material Reference Number* 1. Attributes can be changed after the elements are created. For this purpose, the areas are selected and then the elements that are attached to the selected areas are selected. Finally, elements are modified to have the correct attributes. The

correspondence between the areas and material numbers is given in Table 9.5. Select areas (**ASEL**command) using the following menu path:

Utility Menu > Select > Entities

- *Select Entities* dialog box appears; select ***Areas*** from the first pull-down menu; click on ***OK***.
- *Pick Menu* appears; pick area 4 as defined in Table 9.5; click on ***OK***.
- Now, select the elements that are attached to the selected areas (**ESLA** command) using the following menu path:

Utility Menu > Select > Entities

- *Select Entities* dialog box appears; select ***Elements*** from the first pull-down menu; select ***Attached to*** from the second pull-down menu. Click on the ***Areas*** radio-button; click on ***OK***.
- Modify the attributes of the selected set of elements (**EMODIF** command) using the following menu path:

Main Menu > Preprocessor > Modeling > Move/Modify > Elements > Modify Attrib

- *Pick Menu* appears; click on ***Pick All***, which brings up the *Modify Elem Attributes* dialog box.
- Select ***Material MAT*** from the pull-down menu; enter ***2*** in the ***I1 New attribute number*** field; click on ***OK***.
- Repeat this procedure for area 5 (material reference number 3) and area 6 (material reference number 4).
- When finished, select everything (**ALLSEL** command) using the following menu path:

Utility Menu > Select > Everything

- Save the database (**SAVE** command) using the following menu path:

Utility Menu > Save as Jobname.db

- The database is saved in fi.le ***TH.db***.

Solution

- Declare the new analysis to be a transient analysis (**ANTYPE** command) using the following menu path:

Main Menu > Solution > Analysis Type > New Analysis

- *New Analysis* dialog box appears; click on the ***Transient*** radio-button; click on ***OK***.
- *Transient Analysis* dialog box appears; click on ***OK***.

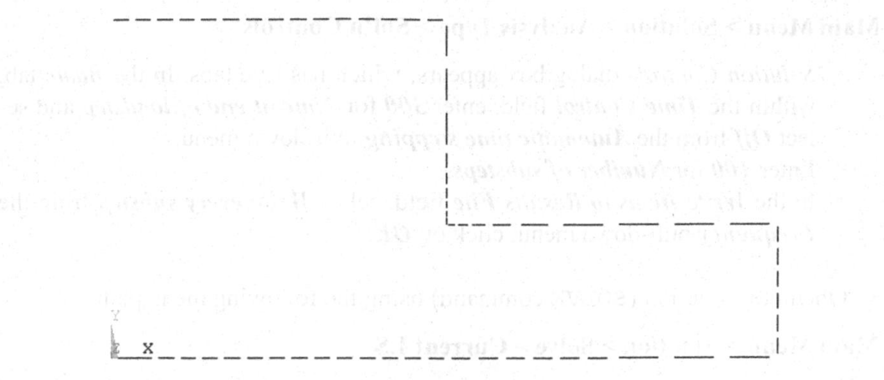

Fig. 9.50 Lines to be selected for convective boundary conditions

- Apply convective boundary conditions on lines (`SFL` command) using the following menu path:

Main Menu > Solution > Define Loads > Apply > Thermal > Convection > On Lines

  – Pick *Menu* appears; pick the exterior lines *except* the ones along $x=0$ (lines to be picked are shown in Fig. 9.50); click on ***OK*** in the *Pick Menu*.
  – Enter ***5*** for ***VALI Film coefficient*** and ***25*** for ***VAL2I Bulk temperature***; click on ***OK***.

- Apply the heat flux condition (`SFL` command) using the following menu path:

Main Menu > Solution > Define Loads > Apply > Thermal > Heat Flux > On Lines

  – *Pick Menu* appears; pick the line between the die and the die-attach; click on ***OK*** in the *Pick Menu*.
  – *Apply HFLUX on lines* dialog box appears; enter ***1000*** for ***VALI Heat flux***; click on ***OK***.

- Apply the initial condition on the nodes (`IC` command) using the following menu path:

Main Menu > Solution > Define Loads > Apply > Initial Condit'n > Define

  – *Pick Menu* appears; click on ***Pick All*** in *Pick Menu*.
  – *Define Initial Conditions* dialog box appears; select ***TEMP*** from the pull-down menu; enter ***25*** for ***VALUE Initial value of DOF***; click on ***OK***.

- Specify solution controls using the following menu path:

Main Menu > Solution > Analysis Type > Sol'n Controls

- *Solution Controls* dialog box appears, which has five tabs. In the ***Basic*** tab, within the ***Time Control*** field, enter ***300*** for ***Time at end of loadstep*** and select ***Off*** from the ***Automatic time stepping*** pull-down menu.
- Enter ***100*** for ***Number of substeps.***
- In the ***Write Items to Results File*** field, select ***Write every substep*** from the ***Frequency*** pull-down menu; click on ***OK***.

• Obtain the solution (**SOLVE** command) using the following menu path:

Main Menu > Solution > Solve > Current LS

- *Confirmation Window* appears along with *Status Report Window*.
- Review status; if OK, close the *Status Report Window*; click on ***OK*** in the Confirmation Window.
- Wait until ANSYS responds with ***Solution is done!***

Postprocessing

• Results are stored in the ***TH.rth*** file in the *Working Directory*. Review the temperature distribution at different time steps using the following menu path:

Main Menu > General Postproc > Read Results > By Load Step

- *Read Results by Load Step Number* dialog box appears; enter ***10*** for ***SBSTEP Substep number***; click on ***OK***.
- Plot temperature contours (**PLNSOL** command) at the selected substep using the following menu path:

Main Menu > General Postproc > Plot Results > Contour Plot > Nodal Solu

- *Contour Nodal Solution Data* dialog box appears. Select ***DOF Solution*** and ***Nodal Temperature***; click on ***OK***. The contour plot appears, as shown in Fig. 9.51.
- Repeat this *procedure* for substeps 50 and 100. Contour plot corresponding to substep 100 (time = 300 s) is shown in Fig. 9.52.

• Stress fields corresponding to substeps 10, 50, and 100 are obtained in the next subsection.

Thermomechanical Analysis

• Clear the database (**/CLEAR** command) using the following menu path:

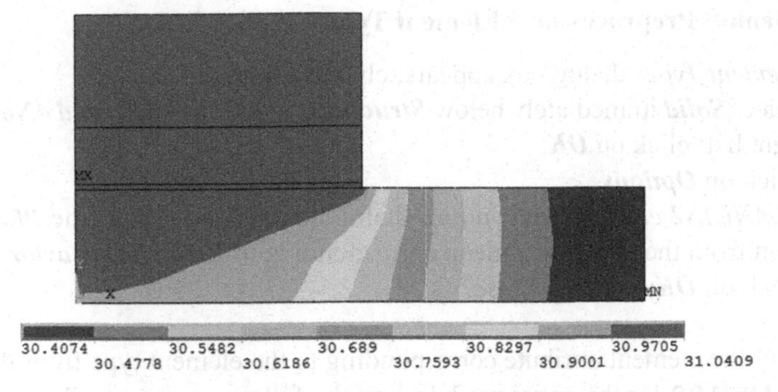

Fig. 9.51 Contour plot of temperature at substep 10 (time = 30 s)

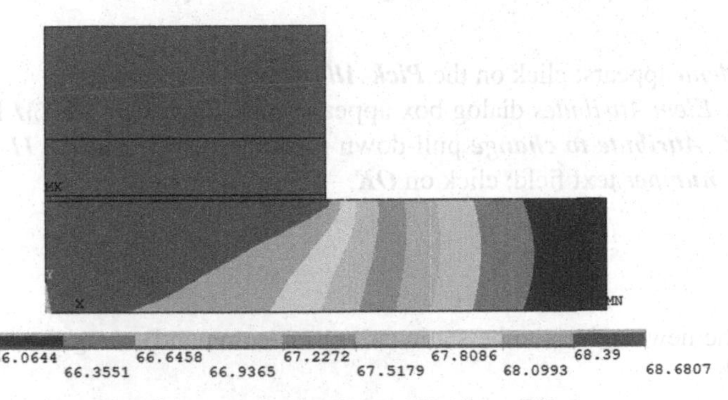

Fig. 9.52 Contour plot of temperature at substep 100 (time = 300 s)

Utility Menu > File > Clear & Start New

- *Clear Database and Start New* dialog box appears; click on ***OK***.
- *Verify* dialog box appears; click on ***Yes***.

- Resume from the ***TH.db*** database file (**`RESUME`** command) using the following menu path:

Utility Menu > File > Resume from

- *Resume Database* dialog box appears. Browse for ***TH.db***; click on ***OK***.

Preprocessor

- In order to perform structural analysis, define a new element type, **`PLANE182`**, using the following menu path:

Main Menu > Preprocessor > Element Type > Add/Edit/Delete

- *Element Types* dialog box appears; click on ***Add***.
- Select ***Solid*** immediately below ***Structural*** in left list and ***Quad 4Node 182*** in right list; click on ***OK***.
- Click on ***Options***.
- *PLANE182 element type options* dialog box appears; select the ***Plane strain*** item from the pull-down menu corresponding to ***Element behavior K3***.
- Click on ***OK***; click on ***Close***.

- Modify the element attribute corresponding to the element type from **`PLANE55`** to **`PLANE182`** for the entire model using the following menu path:

Main Menu > Preprocessor > Modeling > Move/Modify > Elements > Modify Attrib

- *Pick Menu* appears; click on the ***Pick All*** button.
- *Modify Elem Attributes* dialog box appears; pick ***Elem type ELEM*** from the ***STLOC Attribute to change*** pull-down menu and enter ***2*** in the ***I1 New attribute number*** text field; click on ***OK***.

Solution

- Declare the new analysis to be static (**`ANTYPE`** command) using the following menu path:

Main Menu > Solution > Analysis Type > New Analysis

- *New Analysis* dialog box appears; click on the ***Static*** radio-button; click on ***OK***.

- Apply displacement constraints (**`D`** command) using the following menu path:

Main Menu > Solution > Define Loads > Apply > Structural > Displacement > On Nodes

- *Pick Menu* appears; pick the nodes along $x=0$ (y-axis); click on ***OK***.
- *Apply U, ROT on Nodes* dialog box appears; highlight ***UX***; and click on ***Apply***.
- *Pick Menu* reappears; pick the node at $(x, y)=(0, 0)$; click on ***OK***.
- *Apply U, ROT on Nodes* dialog box reappears; highlight ***UX*** and ***UY***; click on ***OK***.

- Specify the stress-free temperature (**`TREF`** command) using the following menu path:

Main Menu > Solution > Define Loads > Settings > Reference Temp

- *Reference Temperature* dialog box appears; enter ***25*** for ***Reference temperature***; click on ***OK***.

- Read nonuniform temperature field as body load (**LDREAD** command) using the following menu path:

Main Menu > Solution > Define Loads > Apply > Structural > Temperature > From Therm Analy

  - *Apply TEMP from Thermal Analysis* dialog box appears; enter ***1*** and ***10*** for ***Load step and substep no***; click on ***Browse.***
  - *Fname Name of results file* dialog box appears; browse for and select the ***TH.rth*** file; click on ***Open***.
  - Click on ***OK*** in the *Apply TEMP from Thermal Analysis* dialog box.

- In real applications in the electronics industry, the copper heat spreader and the die are not in perfect contact; a thermal grease compound is dispensed in between these two components. Therefore, the copper heat spreader will be excluded from the analysis here in order to represent the real situation better. This will be achieved by selecting the nodes associated the substrate, die-attach, and die only (excluding those associated with the heat spreader). Select elements with material numbers 1, 2, and 3 (**ESEL** command) using the following menu path:

Utility Menu > Select > Entities

  - *Select Entities* dialog box appears; select ***Elements*** from the first pull-down menu and ***By Attributes*** from the second pull-down menu. Click on the ***Material num*** radio-button and enter ***1,3,1*** in the ***Min, Max,Inc*** text field. Click on ***Apply***.
  - In the *Select Entities* dialog box, select ***Nodes*** from the pull-down menu and ***Attached to*** from the second pull-down menu. Click on the ***Elements*** radio-button; click on ***OK***.
  - Now, only the nodes associated with the substrate, die-attach, and die are selected (active) and only they will be included in the solution process.

- Obtain the solution (**SOLVE** command) using the following menu path:

Main Menu > Solution > Solve > Current LS

  - *Confirmation Window* appears along with *Status Report Window*.
  - Review status; if OK, close the *Status Report Window*; click on ***OK*** in the Confirmation Window.
  - Wait until ANSYS responds with ***Solution is done!***

Postprocessing

- Review the deformed shape resulting from the nonuniform temperature loading (**PLDISP** command) using the following menu path:

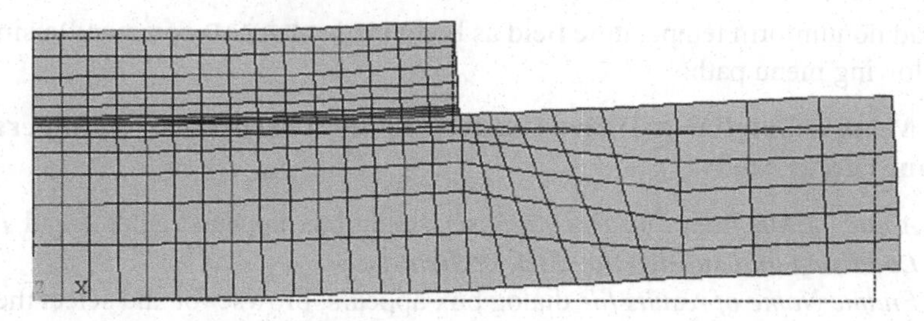

Fig. 9.53 Deformed shape of the electronic package due to thermal loading at load step 1, substep 10

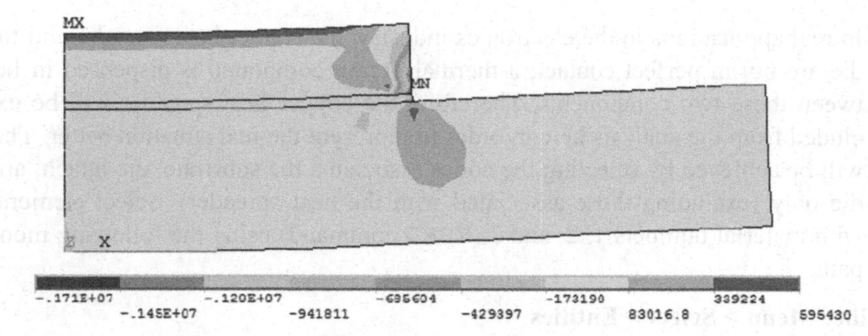

Fig. 9.54 Contour plot of the peeling stress due to thermal loading at load step 1, substep 10

Main Menu > General Postproc > Plot Results > Deformed Shape

- *Plot Deformed Shape* dialog box appears; click on the ***Def+undef edge*** radio-button; click on ***OK***.
- The deformed shape appears, as shown in Fig. 9.53.

• Review normal and shear stress contours (**PLNSOL** command) using the following menu path:

Main Menu > General Postproc > Plot Results > Contour Plot > Nodal Solu

- *Contour Nodal Solution Data* dialog box appears; select ***Stress*** and ***Y- Component of stress***; click on ***OK***.
- The contour plot of normal stress in the *y*-direction (peeling stress) appears, as shown in Fig. 9.54.
- Repeating this procedure for shear stresses, the contour plot appears, as shown in Fig. 9.55.
- The thermomechanical solution may be repeated for any substep of interest. Figs. 9.56 and 9.57 show the contour plots of normal stress in the y-direction and shear stress for substep 100 (last substep, time = 300 s), respectively.

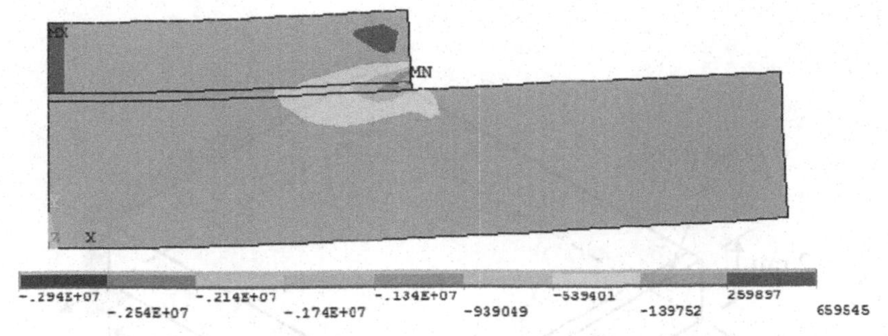

Fig. 9.55 Contour plot of the shearing stress due to thermal loading at load step 1, substep 10

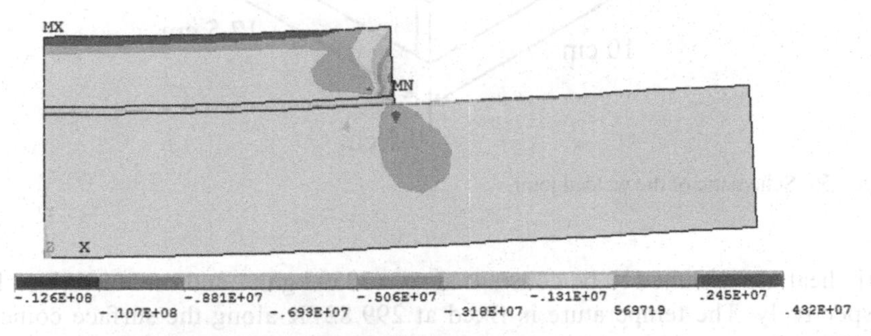

Fig. 9.56 Contour plot of the peeling stress due to thermal loading at load step 1, substep 100

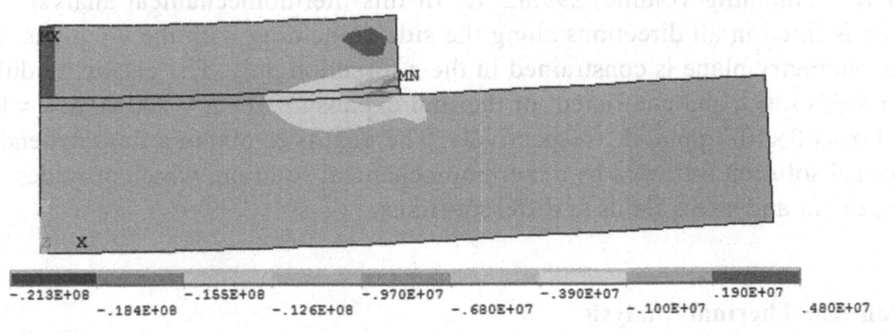

Fig. 9.57 Contour plot of the shearing stress due to thermal loading at load step 1, substep 100

9.1.2.2 Transient Thermomechanical Analysis of a Welded Joint

Consider the welded steel joint shown in Fig. 9.58 (due to symmetry with respect to the plane of the weld pool, only half of the geometry is shown). The plate is 12.5 cm long, 2 cm high, and 10 cm wide. The weld pool is assumed to be 0.5 cm long. In this heat transfer analysis, the surrounding air is at a temperature of $T_o = 229.82$ °K (ambient temperature). The temperature-dependent film coefficients for the top, bottom, and side surfaces are given in Table 9.6, as well as the temperature-dependent thermal conductivity The density and spe-

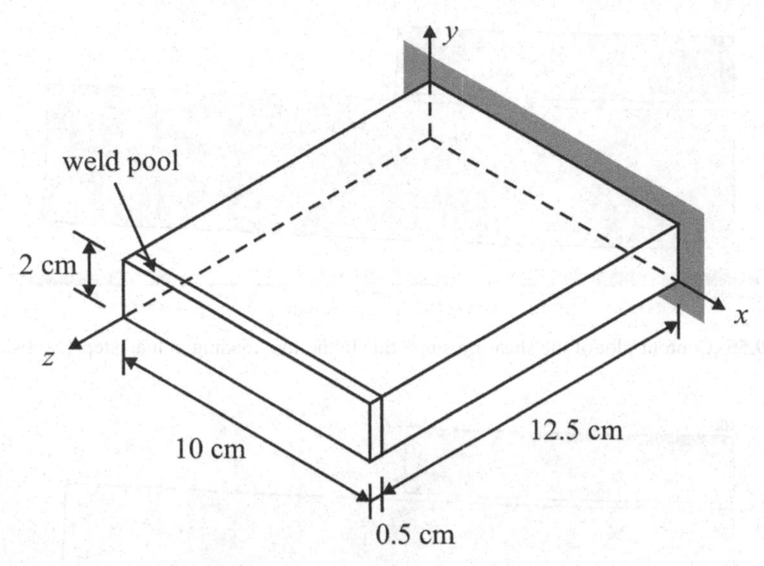

Fig. 9.58 Schematic of the welded joint

cific heat are assumed to be constant at $\rho = 7850$ kg/m^3 and $c = 500$ J/(kg· °K), respectively. The temperature is fixed at 299.82 °K along the surface coinciding with the x-y plane, and there is no heat transfer along the opposite surface (symmetry plane). The initial temperature for the weld pool is 1852.94 °K, and for the remaining volume, 299.82 °K. In this thermomechanical analysis, the plate is fixed in all directions along the side coinciding with the x-y plane, and the symmetry plane is constrained in the z-direction only. The elastic modulus, Poisson's ratio, and coefficient of thermal expansion are $E = 200$ GPa, $\nu = 0.3$, and $\alpha = 60\times10^{-6}$ ppm/^{o}K, respectively. The goal is to obtain a time-dependent thermal solution followed by a thermomechanical solution, which provides displacement and stress fields at different times.

Transient Thermal Analysis

Model Generation

- Specify the jobname (**/FILNAM** command) using the following menu path:

Utility Menu > File > Change Jobname

- *Change Jobname* dialog box appears; type ***WELD***; click on ***OK***.

- Define the element type (**ET** command) using the following menu path:

Main Menu > Preprocessor > Element Type > Add/Edit/Delete

- *Element Types* dialog box appears; click on ***Add***.

Table 9.6 Temperature-dependent properties

Temperature (°K)	Thermal conductivity [W/(m · °K)]	Film coefficient [$W/(m^2 \cdot °K)$]		
		Top surface	Bottom surface	Side surfaces
433.15	30.158	9.08	4.99	8.18
593.15	30.385	11.69	6.05	9.89
753.15	30.611	13.57	6.79	10.96
913.15	30.838	14.96	7.28	11.86
1073.15	31.064	16.19	7.69	12.51
1233.15	31.290	17.25	8.09	13.16
1393.15	31.517	18.15	8.42	13.66
1553.15	31.743	18.97	8.67	14.15
1713.15	31.969	19.79	8.99	14.55
1873.15	25.338	20.52	9.24	14.96

- Select ***Solid*** immediately below ***Thermal*** in the left list and ***Brick 8node 70*** in the right list; click on ***OK***.
- Click on ***Close***.

• Specify material properties (**MP** command) using the following menu path:

Main Menu > Preprocessor > Material Props > Material Models

- *Define Material Model Behavior* dialog box appears. Structural properties are specified first, followed by thermal properties.
- In the right window, successively left-click on ***Structural***, ***Linear***, ***Elastic***, and, finally, ***Isotropic***, which brings up another dialog box.
- Enter ***200E9*** for ***EX*** and ***0.3*** for ***PRXY***; click on ***OK***
- In the right list, successively left-click on ***Structural***, ***Thermal Expansion***, ***Secant Coefficient***, and, finally, ***Isotropic***, which brings up another dialog box.
- Enter ***3.6E–5*** for ***ALPX***; click on ***OK***.
- Specify density by successively clicking on ***Structural*** and ***Density***, which brings up another dialog box. Type ***7850***; click on ***OK***.
- Specify specific heat by successively double-clicking on ***Thermal*** and ***Specific Heat***; enter ***500*** in the newly appeared dialog box; click on ***OK***.
- Specify temperature-dependent thermal conductivity by successively double-clicking on ***Thermal***, ***Conductivity***, and ***Isotropic***.
- *Conductivity for Material Number 1* dialog box appears; click on the ***Add Temperature*** button 9 times (so that there are 10 temperature columns). Referring to Table 9.6, enter temperature values and corresponding thermal conductivity values. When finished, click on ***OK***.

Table 9.7 Coordinates defining the volumes.

Volume number	XI (cm)	X2	Y1	Y2	Z1	Z2
1	0	10	0	2	0	3
2	0	10	0	2	3	8
3	0	10	0	2	8	12
4	0	10	0	2	12	12.5

- Temperature-dependent film coefficient values for top, bottom, and side surfaces are input as having different material reference numbers. Add new material model using the following menu path:

Material > New Model

- In the right list, successively left-click on ***Thermal*** and ***Convection or Film Coef.***, which brings up the *Convection of Film Coefficient for Material 2* dialog box. Click on the ***Add Temperature*** button 9 times (so that there are 10 temperature columns). Referring to Table 9.6, enter temperature values and corresponding film coefficient values at the top surface. When finished, click on ***OK***.
- Repeat the procedure for film coefficient values at the bottom (material 3) and side (material 4) surfaces.
- When finished, close the *Define Material Model Behavior* dialog box by using the following menu path:

Material > Exit

- Create rectangular blocks (**BLOCK** command) using the following menu path:

Main Menu > Preprocessor > Modeling > Create > Volumes > Block > By Dimensions

- *Create Block by Dimensions* dialog box appears. Referring to Table 9.7, enter ***0*** and ***10E–2*** for ***X1*** and ***X2***, ***0*** and ***2E–2*** for ***Y1*** and ***Y2***, and ***0*** and ***3E–2*** for ***Z1*** and ***Z2***; click on ***Apply***.
- Repeat the procedure for the remaining volumes (2 through 4). When creating volume 4, click on ***OK*** after entering the coordinates (instead of ***Apply***).

- Glue the volumes (**VGLUE** command) using the following menu path:

Main Menu > Preprocessor > Modeling > Operate > Booleans > Glue > Volumes

- *Pick Menu* appears; click on ***Pick All*** button.

- Mesh the volumes (**VMESH** command) using the following menu path:

Main Menu > Preprocessor > Meshing > Mesh > Volumes > Mapped > 4 to 6 sided

– *Pick Menu* appears; click on ***Pick All***.

- Save the database (**SAVE** command) using the following menu path:

Utility Menu > Save as Jobname.db

– The database is saved in file ***WELD.db***.

Solution

- Declare the new analysis to be a transient analysis (**ANTYPE** command) using the following menu path:

Main Menu > Solution > Analysis TypeNew Analysis

– *New Analysis* dialog box appears; click on the ***Transient*** radio-button; click on ***OK***.
– *Transient Analysis* dialog box appears; click on ***OK***.

- Specify solution controls using the following menu path:

Main Menu > Solution > Analysis Type > Sol'n Controls

– *Solution Controls* dialog box appears, which has five tabs. In the ***Basic*** tab, within the ***Time Control*** field, enter ***3600*** for ***Time at end of loadstep***.
– Click on the ***Time increment*** radio-button and enter ***36***, ***3.6***, and ***500*** for ***Time step size***, ***Minimum time step***, and ***Maximum time step***, respectively.
– In the ***Write Items to Results File*** field, select ***Write every substep*** from the ***Frequency*** pull-down menu; click on ***OK***.

- Initial conditions are specified next. For this purpose, volumes are selected first, followed by selection of the nodes attached to the selected volumes.

– Select volumes (**VSEL** command) using the following menu path:

Utility Menu > Select > Select Entities

– *Select Entities* dialog box appears; choose ***Volumes*** from the first pull-down menu; click on ***OK***.
– *Pick Menu* appears; pick the volumes indicated in Fig. 9.59; click on ***OK*** in the *Pick Menu*.
– Select nodes attached to the selected volumes (**NSLV** command) using the following menu path:

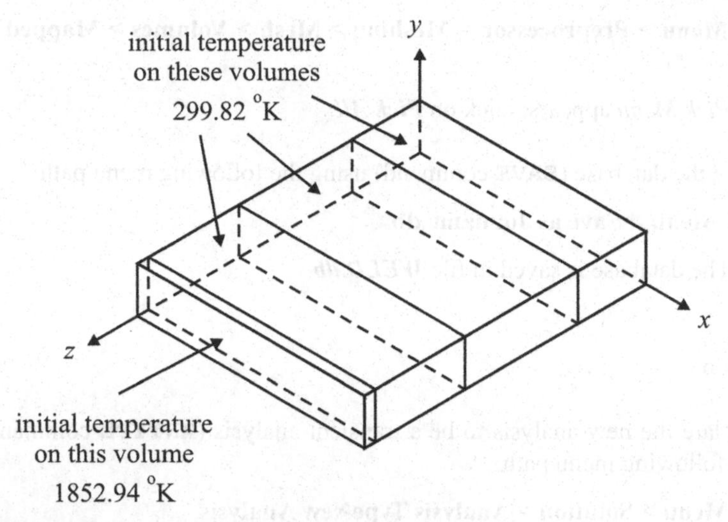

Fig. 9.59 Volumes to be selected for initial temperatures

Utility Menu > Select > Select Entities

- *Select Entities* dialog box appears; choose ***Nodes*** from the first pull-down menu and ***Attached to*** from the second pull-down menu. Click on the ***Volumes, all*** radio-button; click on ***OK***.
- Apply initial conditions on the selected nodes (**IC** command) using the following menu path:

Main Menu > Solution > Define Loads > Apply > Initial Condit'n > Define

- *Pick Menu* appears; click on ***Pick All***.
- *Define Initial Conditions* dialog box appears; select ***TEMP*** from the pull-down menu and enter ***299.8167*** for ***VALUE Initial value of DOF***; click on ***OK***.
- Select everything (**ALLSEL** command) using the following menu path:

Utility Menu > Select > Everything

- Repeat the same procedure to apply initial conditions on nodes attached to the remaining volume corresponding to the welded region. This time, apply a temperature of ***1852.594***°K.
- Select everything (**ALLSEL** command) using the following menu path:

Utility Menu > Select > Everything

- Apply temperature boundary conditions on the selected nodes (**D** command) using the following menu path:

Main Menu > Solution > Define Loads > Apply > Thermal > Temperature > On Nodes

- *Pick Menu* appears; pick the nodes along the $z=0$ surface and click on ***OK*** (in order to pick the nodes conveniently, plot the nodes and obtain a top view).
- *Apply TEMP on Nodes* dialog box appears; highlight ***TEMP*** and type ***299.8167*** for ***VALUE Load Temp value***; click on ***OK***.

- Convective boundary conditions are applied next. Temperature-dependent film coefficients were specified previously for the top (material 2), bottom (material 3), and side (material 4) surfaces.

  - Apply convective boundary conditions on the nodes along the side surfaces (**SF** command) using the following menu path:

Main Menu > Solution > Define Loads > Apply > Thermal > Convection > On Nodes

- *Pick Menu* appears; pick the nodes along the $x = 0$ and $x = 10\times10^{-2}$ surfaces (side surfaces); click on ***OK*** in the *Pick Menu*. For this case, ***Front View*** is the most convenient viewpoint.
- *Apply CONV on nodes* dialog box appears; enter ***−4*** for film coefficient (first text field) and ***299.8167*** for bulk temperature (second text field); click on ***Apply***.
- *Pick Menu* reappears; pick the nodes along the $y = 2\times10^{-2}$ surface (top surface); click on ***OK*** in the *Pick Menu*. ***Front View*** is the most convenient viewpoint for this case, also.
- *Apply CONV on nodes* dialog box appears; enter ***−2*** for film coefficient and ***299.8167*** for bulk temperature; click on ***Apply***.
- *Pick Menu* reappears; pick the nodes along the $y = 0$ surface (bottom surface); click on ***OK*** in the *Pick Menu*. ***Right View*** and ***Left View*** are the most convenient viewpoints for this case.
- *Apply CONV on nodes* dialog box appears; enter ***−3*** for film coefficient and ***299.8167*** for bulk temperature; click on ***OK***.

- Obtain the solution (**SOLVE** command) using the following menu path:

Main Menu > Solution > Solve > Current LS

- *Confirmation Window* appears along with *Status Report Window*.
- Review status; if OK, close the *Status Report Window* and click on ***OK*** in the *Confirmation Window*.
- Wait until ANSYS responds with ***Solution is done!***

Postprocessing

- Results are stored in the ***WELD.rth*** file in the *Working Directory*. Review the temperature distribution at different time points using the following menu path:

Main Menu > General Postproc > Read Results > By Pick

- *Results File: WELD.rth* dialog box appears, in which available solution sets are tabulated. Highlight set ***5***; click on ***Read*** and click on ***Close***.
- Plot temperature contours (`PLNSOL` command) at the selected substep using the following menu path:

Main Menu > General Postproc > Plot Results > Contour Plot > Nodal Solu

- *Contour Nodal Solution Data* dialog box appears. Select ***DOF Solution*** and ***Nodal Temperature***; click on ***OK***. The contour plot appears, as shown in Fig. 9.60.
- Repeat this procedure for substep 22 (last substep). Corresponding contour plot is shown in Fig. 9.61.

- Stress fields corresponding to substeps 10, 50, and 100 are obtained in the next subsection.

Thermomechanical Analysis

- Clear the database (`/CLEAR` command) using the following menu path:

Utility Menu > File > Clear & Start New

- *Clear Database and Start New* dialog box appears; click on ***OK***.
- *Verify* dialog box appears; click on ***Yes***.

- Resume from the ***WELD.db*** database file (`RESUME` command) using the following menu path:

Utility Menu > File > Resume from

- *Resume Database* dialog box appears. Browse for ***WELD.db***; click on ***OK***.

Preprocessor

- In order to perform structural analysis, define a new element type, `SOLID185`, using the following menu path:

Main Menu > Preprocessor > Element Type > Add/Edit/Delete

- *Element Types* dialog box appears; click on ***Add***.

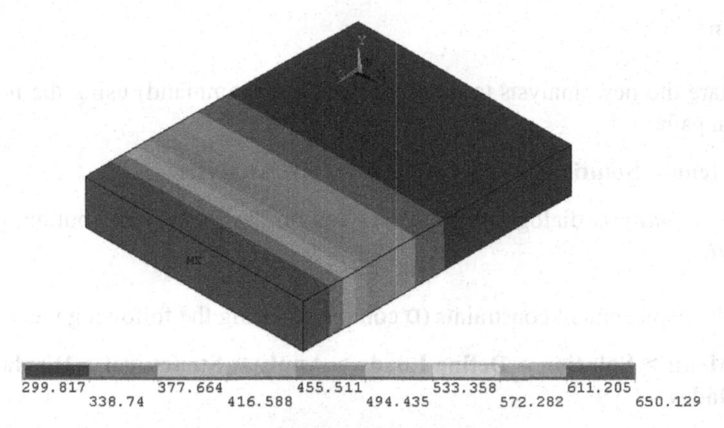

Fig. 9.60 Contour plot of temperature at substep 5 (time = 69.409 s)

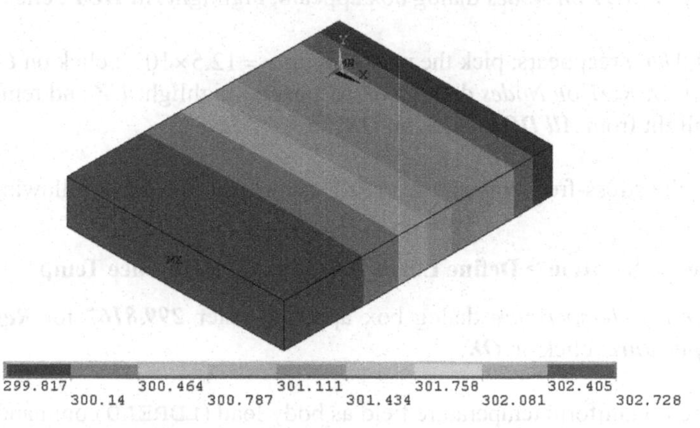

Fig. 9.61 Contour plot of temperature at substep 22 (time = 3600 s)

- Select ***Solid*** immediately below ***Structural*** in left list and ***Brick 8Node 185*** in right list; click on ***OK***.
- Click on ***Close***.

• Modify the element attribute corresponding to element type from `SOLID70` to `SOLID185` for the entire model using the following menu path:

Main Menu > Preprocessor > Modeling > Move/Modify > Elements > Modify Attrib

- *Pick Menu* appears; click on the ***Pick All*** button.
- *Modify Elem Attributes* dialog box appears. Pick ***Elem type ELEM*** from the ***STLOC Attribute to change*** pull-down menu and enter ***2*** in the ***I1 New attribute number*** text field; click on ***OK***.

Solution

- Declare the new analysis to be static (**ANTYPE** command) using the following menu path:

Main Menu > Solution > Analysis Type > New Analysis

  - *New Analysis* dialog box appears; click on the ***Static*** radio-button; click on ***OK***.

- Apply displacement constraints (**D** command) using the following menu path:

Main Menu > Solution > Define Loads > Apply > Structural > Displacement > On Nodes

  - *Pick Menu* appears; pick the nodes along $z=0$; click on ***OK***.
  - *Apply U, ROT on Nodes* dialog box appears; highlight ***All DOF***; click on ***Apply***.
  - *Pick Menu* reappears; pick the nodes along $z=12.5\times10^{-2}$; click on ***OK***.
  - *Apply U, ROT on Nodes* dialog box reappears; highlight ***UZ*** and remove the highlight from ***All DOF***; click on ***OK***.

- Specify the stress-free temperature (**TREF** command) using the following menu path:

Main Menu > Solution > Define Loads > Settings > Reference Temp

  - *Reference Temperature* dialog box appears; enter ***299.8167*** for ***Reference temperature***; click on ***OK***.

- Read the nonuniform temperature field as body load (**LDREAD** command) using the following menu path:

Main Menu > Solution > Define Loads > Apply > Structural > Temperature > From Therm Analy

  - *Apply TEMP from Thermal Analysis* dialog box appears; enter ***1*** and ***5*** for ***Load step and substep no.***; click on ***Browse***.
  - *Fname Name of results file* dialog box appears; search for and select ***WELD.rth*** file; click on ***OK***.
  - Click on ***OK*** in the *Apply TEMP from Thermal Analysis* dialog box.

- Obtain the solution (**SOLVE** command) using the following menu path:

Main Menu > Solution > Solve > Current LS

  - *Confirmation Window* appears along with *Status Report Window*
  - Review status; if OK, close the *Status Report Window*; click on ***OK*** in the *Confirmation Window*.
  - Wait until ANSYS responds with ***Solution is done!***

Postprocessing

- Review contours for normal stresses in the *z*-direction (**PLNSOL** command) using the following menu path:

Main Menu > General Postproc > Plot Results > Contour Plot > Nodal Solu

  - *Contour Nodal Solution Data* dialog box appears; select ***Stress*** and ***Z-Component of stress***; click on ***OK***.
  - The contour plot of normal stresses in the *z*-direction appears, as shown in Fig. 9.62.
  - The thermomechanical solution may be repeated for any substep of interest. Figure 9.63 shows the contour plots of normal stress in the *z*-direction for substep 22 (last substep).

9.1.3 Radiation Analysis

Radiation is the transfer of thermal energy between two surfaces through electromagnetic waves. No medium is required for radiation heat transfer to take place. The transfer of thermal energy between two surfaces depends on the difference between the fourth powers of absolute temperatures along the surfaces. When radiation conditions are present, the problem is certain to be nonlinear. A radiation analysis requires specific knowledge of concepts that are not covered in this book. Therefore, it is highly recommended that the user have a good understanding of the radiation phenomenon before using ANSYS for radiation analyses.

Radiation analyses in ANSYS require the Stefan-Boltzmann constant, as well as the emissivity, view factor, and space temperature values for the surfaces involved. A radiation analysis is demonstrated by solving a simple problem using the *Radiosity Solver* method.

Two partial hollow circular regions are radiating to each other, as shown in Fig. 9.64. The emissivity of the outer surface of the inner region and inner surface of the outer region are 0.9 and 0.7, respectively, while the inner surface of the inner region and outer surface of the outer region are maintained at temperatures of 1500 °F and 100 °F, respectively. The space temperature is 70 °F. Both regions have a thermal conductivity of 0.1. The goal is to obtain the steady-state temperature and heat flux variations.

Model Generation

- Define the element type (**ET** command) using the following menu path:

Main Menu > Preprocessor > Element Type > Add/Edit/Delete

  - *Element Types* dialog box appears; click on ***Add***.
  - Select ***Solid*** immediately below ***Thermal*** in the left list and ***Quad 4 Node 55*** in the right list; click on ***OK***.

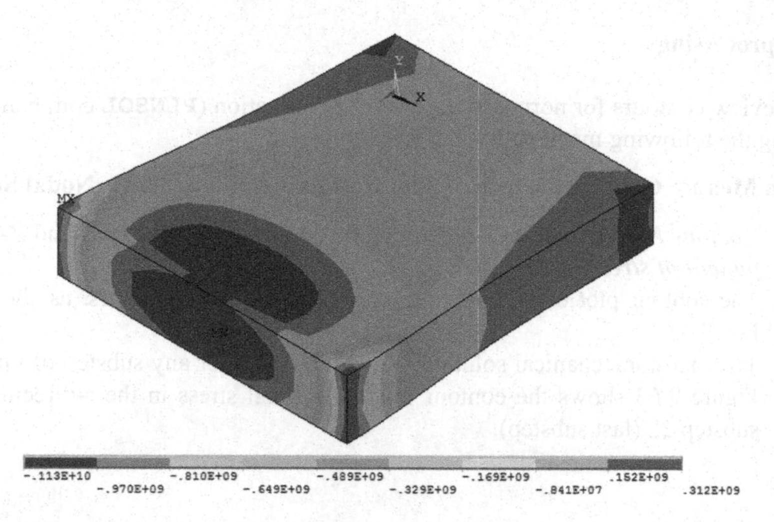

Fig. 9.62 Contour plot of the stresses in the *z*-direction due to thermal loading at load step 1, substep 5

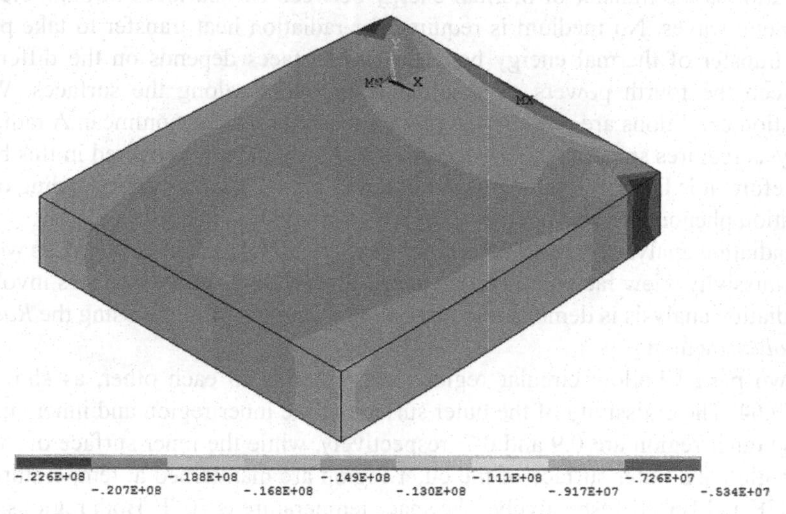

Fig. 9.63 Contour plot of the stresses in the *z*-direction due to thermal loading at load step 1, substep 22

- Exit from the *Element Types* dialog box by clicking on ***Close***.

- Specify material properties (**MP** command) using the following menu path:

Main Menu > Preprocessor > Material Props > Material Models

- *Define Material Model Behavior* dialog box appears. In the right window, successively left-click on ***Thermal***, ***Conductivity***, and ***Isotropic***, which brings up another dialog box.

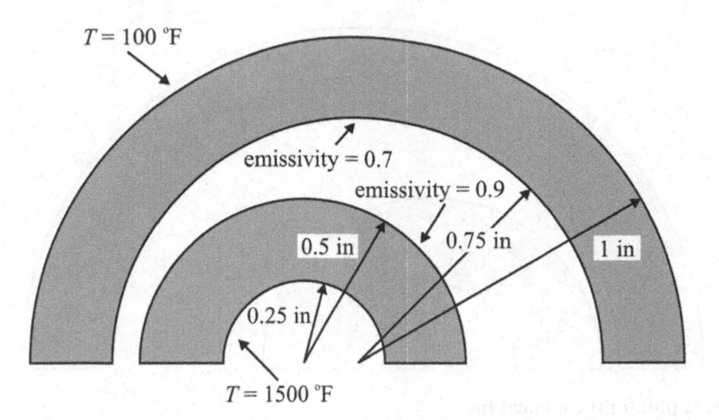

Fig. 9.64 Schematic of the conical fin cross section

- Enter ***0.1*** for ***KXX***; click on ***OK***.
- Close the *Define Material Model Behavior* dialog box by using the following menu path:

Material > Exit

- Create partial hollow circular areas (**PCIRC** command) using the following menu path:

Main Menu > Preprocessor > Modeling > Create > Areas > Circle > By Dimensions

- *Circular Area by Dimensions* dialog box appears. Referring to Fig. 9.64, enter ***0.5*** and ***0.25*** for ***RAD1*** and ***RAD2***, and ***0*** and ***180*** for ***THETA1*** and ***THETA2***; click on ***OK***.
- Move the center of the *Working Plane* by 0.2 in the *x*-direction using the following menu path:

Utility Menu > WorkPlane > Offset WP by Increments

- *Offset WP* menu appears; click on the ***+X*** button four times and observe the *Working Plane* triad move in the *Graphics Window*. Also observe, toward the bottom of the *Offset WP* menu, that ***0.2*** appears next to ***Global X=***.
- Exit from the *Offset WP* menu by clicking on ***OK***.
- Create the remaining partial hollow circular area (**PCIRC** command) using the following menu path:

Main Menu > Preprocessor > Modeling > Create > Areas > Circle > By Dimensions

- *Circular Area by Dimensions* dialog box appears. Referring to Fig. 9.64, enter ***1*** and ***0.75*** for ***RAD1*** and ***RAD2*** and ***0*** and ***180*** for ***THETA1*** and ***THETA2***; click on ***OK***.
- The areas appear in the *Graphics Window*, as shown in Fig. 9.65.

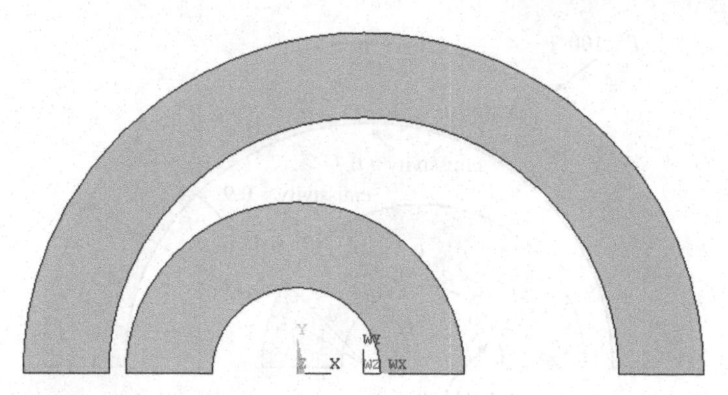

Fig. 9.65 Area plot for the conical fin

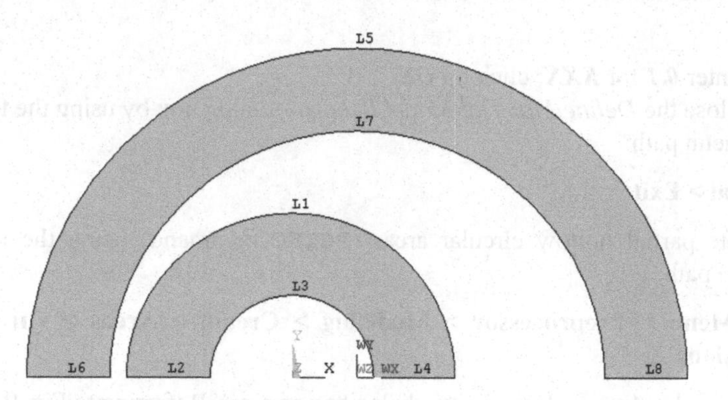

Fig. 9.66 Area plot with line numbers turned on

- Turn line numbering on using the following menu path:

Utility Menu > PlotCtrls > Numbering

  - *Plot Numbering Controls* dialog box appears; click on the square box next to ***LINE Line numbers*** so that a checkmark appears. Click on ***OK***. Line numbers appear, as shown in Fig. 9.66.

- Specify the number of elements along selected lines (**`LESIZE`** command) using the following menu path:

Main Menu > Preprocessor > Meshing > Size Cntrls > ManualSize > Lines > Picked Lines

  - *Pick Menu* appears; pick lines 1, 3, 5, and 7; click on ***OK***.
  - *Element Sizes on Picked Lines* dialog box appears; enter ***40*** in the text field for ***NDIV*** and remove the checkmark next to ***KYNDIV***. Click on ***Apply***.
  - *Pick Menu* reappears; pick lines 2, 4, 6, and 8; click on ***OK***.
  - In the *Element Sizes on Picked Lines* dialog box, enter ***10*** in the text field for ***NDIV***; click on ***OK***.

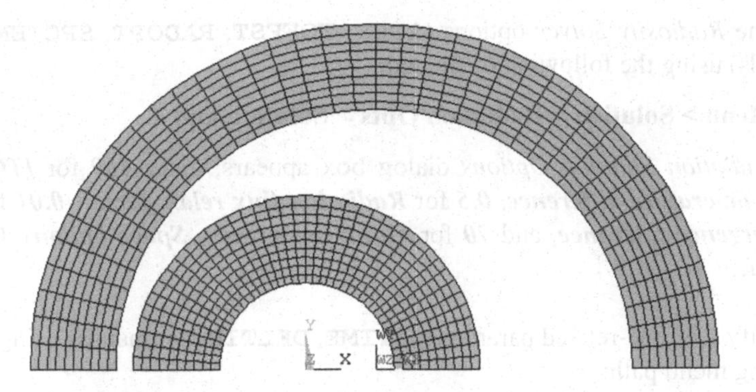

Fig. 9.67 Mesh representing the conical fin

- Mesh the areas (**AMESH** command) using the following menu path:

Main Menu > Preprocessor > Meshing > Mesh > Areas > Mapped > 3 or 4 sided

- *Pick Menu* appears; click on ***Pick All***.
- The mesh appears in the *Graphics Window*, as shown in Fig. 9.67.

Solution

- Apply radiation conditions on lines (**SFL** command) using the following menu path:

Main Menu > Solution > Define Loads > Apply > Thermal > Radiation > On Lines

- *Pick Menu* appears; pick line 1 (refer to Fig. 9.66); click on ***OK*** in the *Pick Menu*.
- *Apply RDSF on Lines* dialog box appears; enter ***0.9*** for ***VALUE Emissivity*** and ***1*** for ***VALUE2 Enclosure number***. Click on ***Apply***.
- *Pick Menu* reappears; pick line 7; click on ***OK*** in the *Pick Menu*.
- In the *Apply RDSF on Lines* dialog box, enter ***0.7*** for ***VALUE Emissivity*** and ***1*** for ***VALUE2 Enclosure number***. Click on ***OK***.

- Specify the temperatures on lines (**DL** command) using the following menu path:

Main Menu > Solution > Define Loads > Apply > Thermal > Temperature > On Lines

- *Pick Menu* appears; pick line 3; click on ***OK*** in the *Pick Menu*.
- *Apply TEMP on Lines* dialog box appears; highlight ***TEMP*** from the list and enter ***1500*** for ***VALUE Load TEMP value***; click on ***Apply***.
- *Pick Menu* reappears; pick line 5; and click on ***OK*** in the *Pick Menu*.
- In the *Apply TEMP on Lines* dialog box enter ***100*** for ***VALUE Load TEMP value***; click on ***OK***.

- Define *Radiosity Solver* options (**STEF**, **TOFFST**, **RADOPT**, **SPCTEMP** commands) using the following menu path:

Main Menu > Solution > Radiation Opts > Solution Opt

  - *Radiation Solution Options* dialog box appears; enter ***460*** for ***[TOFFST] Temperature difference, 0.5*** for ***Radiation flux relax. factor, 0.01*** for ***Convergence tolerance***, and ***70*** for ***Value*** (underneath ***Space option***). Click on ***OK***.

- Specify the time-related parameters (**TIME**, **DELTIM** commands) using the following menu path:

Main Menu > Solution > Load Step Opts > Time/Frequenc > Time—Time Step

  - *Time and Time Step Options* dialog box appears. Enter *1* for ***[TIME] Time at end of load step***, *0.5* for ***[DELTIM] Time step size, 0.1*** for ***[DELTIM] Minimum time step size***, and *1* for ***Maximum time step size***. Click on ***OK***.

- Specify the maximum number of equilibrium iterations (**NEQIT** command) using the following menu path:

Main Menu > Solution > Load Step Opts > Nonlinear > Equilibrium Iter

  - *Equilibrium Iterations* dialog box appears. Enter ***1000*** for ***[NEQIT] No. of equilibrium iter***; click on ***OK***.

- Obtain the solution (**SOLVE** command) using the following menu path:

Main Menu > Solution > Solve > Current LS

  - *Confirmation Window* appears along with *Status Report Window*.
  - Review status; if OK, close the *Status Report Window*; click on ***OK*** in the *Confirmation Window*.

- Wait until ANSYS responds with ***Solution is done!***

Postprocessing

- Review the temperature distribution (**PLNSOL** command) using the following menu path:

Main Menu > General Postproc > Plot Results > Contour Plot > Nodal Solu

  - *Contour Nodal Solution Data* dialog box appears; select ***DOF Solution*** and ***Nodal Temperature***; click on ***OK***.
  - Temperature contours appear in the *Graphics Window*, as shown in Fig. 9.68.

- Review flux vectors (**PLVECT** command) using the following menu path:

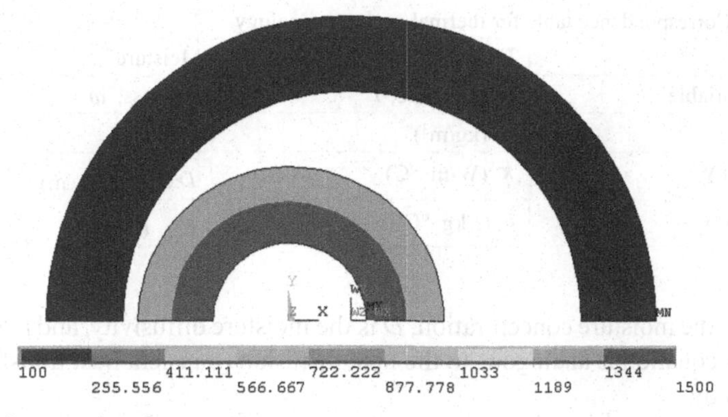

Fig. 9.68 Contour plot of temperature

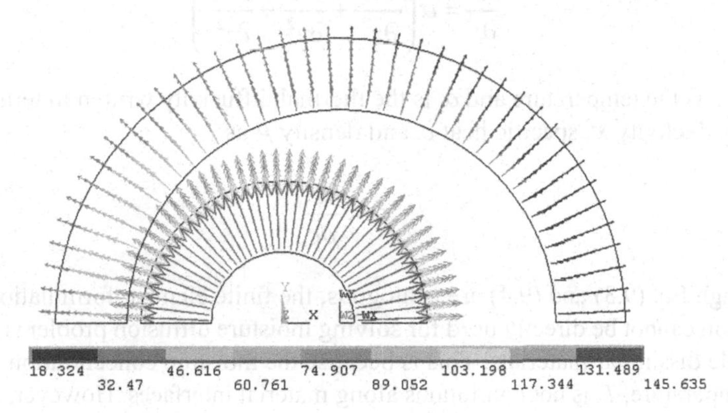

Fig. 9.69 Vector plot of heat flux

Main Menu > General Postproc > Plot Results > Vector Plot > Predefined

- *Vector Plot of Predefined Vectors* dialog box appears; select ***Flux & gradient*** from the left list and ***Thermal flux TF*** from the right list; click on ***OK***.
- Flux vectors appear in the *Graphics Window*, as shown in Fig. 9.69.

9.2 Moisture Diffusion

The moisture diffusion phenomenon is an important issue because polymeric materials are being used more extensively in engineering applications. The moisture diffusion into a water permeable medium is governed by

$$\frac{\partial C}{\partial t} = D\left(\frac{\partial^2 C}{\partial x^2} + \frac{\partial^2 C}{\partial y^2} + \frac{\partial^2 C}{\partial z^2}\right) \tag{9.3}$$

Table 9.8 Correspondence table for thermal/moisture analogy

Property	Thermal	Moisture
Primary variable	Temperature, T	Wetness, ω
Density	ρ (kg/m^3)	1
Conductivity	κ (W/m· °C)	$D{\cdot}C_{sat}$ (kg/s· m)
Specific heat	c (J/kg· °C)	C_{sat} (kg/m^3)

where C is the moisture concentration, D is the moisture diffusivity, and t designates time. This equation is analogous to the heat diffusion (transient heat transfer) equation given by

$$\frac{\partial T}{\partial t} = \alpha\left(\frac{\partial^2 T}{\partial x^2} + \frac{\partial^2 T}{\partial y^2} + \frac{\partial^2 T}{\partial z^2}\right) \tag{9.4}$$

where T is the temperature and α is the thermal diffusivity written in terms of thermal conductivity κ, specific heat c, and density ρ as

$$\alpha = \frac{\kappa}{\rho c} \tag{9.5}$$

Although Eq. (9.3) and (9.4) are analogous, the finite element formulation for heat diffusion cannot be directly used for solving moisture diffusion problems involving multiple dissimilar materials. This is because the moisture concentration, C, unlike the temperature, T, is not continuous along material interfaces. However, when C is normalized with respect to the saturated moisture concentration, C_{sat}, this incompatibility is removed and the finite element formulation for heat diffusion can now be used for solving moisture diffusion problems. The normalized moisture concentration is called the *wetness* parameter and is written as

$$w = \frac{C}{C_{sat}} \tag{9.6}$$

As is clear from Eq. (9.4), thermal conductivity, κ, specific heat, c, and density, ρ, need to be known to solve a heat diffusion problem. After the introduction of the wetness parameter, in order to utilize the finite element formulation for heat diffusion for solving moisture diffusion problems, the user must use the correspondence table given in Table 9.8.

Once the solution is obtained, the weight of the moisture absorbed by an element, $W^{(e)}$, is computed by the product of average moisture concentration and the volume of the element in the form

$$W^{(e)} = \left(\frac{1}{N}\sum_{i=1}^{N} \omega_i \cdot C_{sat}\right) \cdot V^{(e)} \tag{9.7}$$

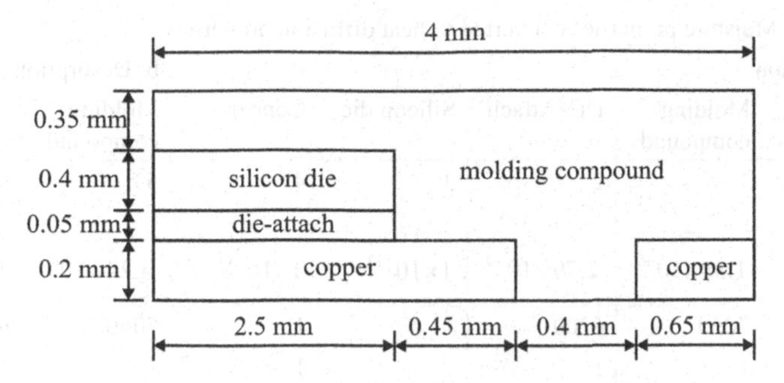

Fig. 9.70 Schematic of the electronic device undergoing moisture diffusion

Table 9.9 Moisture diffusion related material properties of the constituent materials in the electronic package

	Molding compound	Die-Attach
D (mm²/sec) absorption	7.43×10^{-7}	1.25×10^{-5}
D (mm²/sec) desorption	0.18	0.35
C_{sat} (mg/mm³)	7.06×10^{-3}	6.20×10^{-3}

in which N is the number of nodes per element and $V^{(e)}$ represents the volume of the element. The wetness at the ith node is indicated by ω_i. The total weight of the absorbed moisture can then be found by simply summing the weight of the moisture absorbed by each element.

When simulating absorption conditions, the exposed surfaces of the moisture-absorbing materials are subjected to "*wet*" conditions, i.e., $T = 1$. Similarly, in desorption simulations, those surfaces are subjected to "*dry*" conditions, $T = 0$.

Once the solution is complete, the analyst can use Eq. (9.6) to obtain the moisture concentration, C, in each material.

The moisture diffusion simulation using ANSYS is demonstrated by considering a typical example from the electronics industry: an analysis of the moisture diffusion of an electronic package.

Consider an electronic package consisting of a silicon die, die-attach, copper, and molding compound, as shown in Fig. 9.70. The package is preconditioned at 85 % relative humidity and a temperature of 85 °C for one week (168 h) before being placed in a reflow oven at 225 °C for 5 min. The moisture diffusivity during absorption and desorption and the saturated moisture concentration values of the molding compound and the die-attach are tabulated in Table 9.9. After applying the substitutions required for a thermal-moisture analogy, the values given in Table 9.10 are obtained. These are the material properties to be used in the ANSYS analysis. As noted in Table 9.11, the silicon die and copper do not absorb moisture and so appropriate material properties are used for these components. The geometry of the problem possesses half-symmetry; thus only half the package is simulated, with insulation along the symmetry axis.

Table 9.10 Moisture properties converted to heat diffusion properties

	a. Absorption				b. Desorption	
	Molding compound	Die-Attach	Silicon die	Copper	Molding compound	Die-Attach
Material ref. no.	1	2	3	4	5	6
κ (kg/hr·m)	1.88×10^{-5}	2.79×10^{-4}	1×10^{-11}	1×10^{-11}	4.24×10^{-3}	9.32×10^{-3}
c (kg/m^3)	7060	6200	1	1	7060	6200
ρ	1	1	1	1	1	1

Table 9.11 Coordinates defining the areas

Area number	X1	X2	Y1	Y2
	(mm)			
1	0	2.50	0	1.00
2	0	2.95	0	1.00
3	0	3.35	0	1.00
4	0	4.00	0	1.00
5	0	4.00	0	0.20
6	0	4.00	0	0.25
7	0	4.00	0	0.65

The solution has two stages. First, absorption is simulated. During absorption, all the external surfaces of the moisture-absorbing materials are subjected to $T = 1$. After the absorption simulation is complete, desorption is simulated, with boundary conditions $T = 0$ along the same surfaces.

Model Generation

- Define the element type (**ET** command) using the following menu path:

Main Menu > Preprocessor > Element Type > Add/Edit/Delete

 – *Element Types* dialog box appears; click on ***Add***.
 – Select ***Solid*** immediately below ***Thermal*** in the left list and ***Quad 4 Node 55*** in the right list; click on ***OK***.
 – Click on ***Close***.

- Specify material properties (**MP** command) using the following menu path:

Main Menu > Preprocessor > Material Props > Material Models

 – *Define Material Model Behavior* dialog box appears. Specify conductivity by successively clicking on ***Thermal***, ***Conductivity***, and ***Isotropic***; enter ***1.88E–5*** in the newly appeared dialog box; click on ***OK***.

- Specify density by successively clicking on ***Thermal*** and ***Specific Heat***, which brings up another dialog box. Type ***7060***, click on ***OK***.
- Specify specific heat by successively clicking on ***Thermal*** and ***Density***; enter ***1*** in the newly appeared dialog box; click on ***OK***.
- Add new material model using the following menu path:

Material > New Model

- Repeat the procedure for the remaining materials (2 through 6) by referring to Table 9.10.
- When finished, close the *Define Material Model Behavior* dialog box by using the following menu path:

Material > Exit

- Create rectangular areas (**RECTNG** command) using the following menu path:

Main Menu > Preprocessor > Modeling > Create > Areas > Rectangle > By Dimensions

- *Create Rectangle by Dimensions* dialog box appears. Referring to Table 9.11, enter ***0*** and ***2.5E–3*** for ***X1*** and ***X2*** and ***0*** and ***1E–3*** for ***Y1*** and ***Y2***; click on ***Apply***. Note that the units used in the analysis are in meters while the values given in Table 9.11 are in millimeters.
- Repeat the procedure for the remaining areas (2 through 7). When creating Area 7, click on ***OK*** after entering the coordinates.

- Overlap the areas (**AOVLAP** command) using the following menu path:

Main Menu > Preprocessor > Modeling > Operate > Booleans > Overlap > Areas

Pick Menu appears; click on ***Pick All*** button.

- Compress entity numbers (**NUMCMP** command) using the following menu path:

Main Menu > Preprocessor > Numbering Ctrls > Compress Numbers

- *Compress Numbers* dialog box appears; select ***All*** from the pull-down menu; click on ***OK***.

- Specify the global element size (**ESIZE** command) using the following menu path:

Main Menu > Preprocessor > Meshing > Size Cntrls > ManualSize > Global > Size

- *Global Element Sizes* dialog box appears; enter ***4E–5*** for ***SIZE Element edge length***; click on ***OK***.

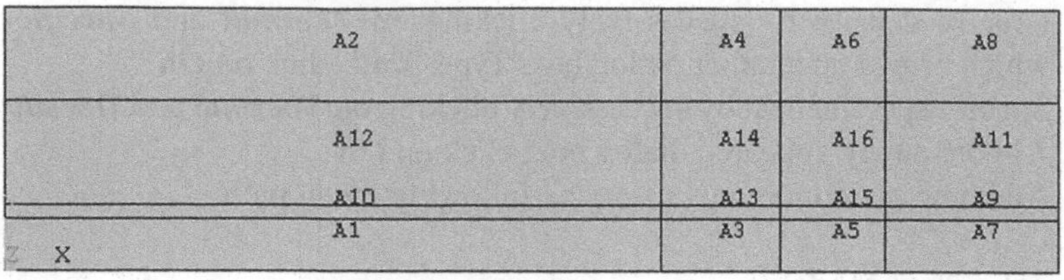

Fig. 9.71 Area plot with area numbers turned on

Table 9.12 Areas corresponding to specific materials

Material	Area numbers
Molding compound	2, 4, 5, 6, 8, 9, 11, 13, 14, 15, 16
Die-attach	10
Silicon die	12
Copper	1, 3, 7

- Mesh the areas (**AMESH** command) using the following menu path:

Main Menu > Preprocessor > Meshing > Mesh > Areas > Mapped > 3 or 4 sided

- *Pick Menu* appears; click on ***Pick All***.
- At this point, all the elements have *Material Reference Number* 1. Attributes can be changed after the elements are created. For this purpose, areas are selected and then the elements that are attached to the selected areas are selected. Finally, elements are modified to have the correct attributes. The correspondence between the areas and material numbers are given in Table 9.12. Figure 9.71 shows the area plot with area numbers printed. Plot areas (**APLOT** command) using the following menu path:

Utility Menu > Plot > Areas

- Turn area numbers on (**/PNUM** command) using the following menu path:

Utility Menu > PlotCtrls > Numbering

- *Plot Numbering Controls* dialog box appears; place a checkmark next to ***AREA Area numbers*** by clicking on the box; select ***Numbers only*** from the pull-down menu next to ***[/NUM] Numbering shown with***. Click on ***OK***. Areas appear with their numbers printed, as shown in Fig. 9.71.
- Select areas (**ASEL** command) using the following menu path:

Utility Menu > Select > Entities

- *Select Entities* dialog box appears; select ***Areas*** from the first pull-down menu; click on ***OK***.

- *Pick Menu* appears; pick area 10; click on ***OK***.
- Now, select the elements that are attached to the selected areas (**ESLA** command) using the following menu path:

Utility Menu > Select > Entities

- *Select Entities* dialog box appears; select ***Elements*** from the first pull-down menu; select ***Attached to*** from the second pull-down menu. Click on the ***Areas*** radio-button; click on ***OK***.
- Modify the attributes of the selected set of elements (**EMODIF** command) using the following menu path:

Main Menu > Preprocessor > Modeling > Move/Modify > ElementsModify Attrib

- *Pick Menu* appears; click on ***Pick All***, which brings up the *Modify Elem Attributes* dialog box.
- Select ***Material MAT*** from the pull-down menu and enter ***2*** in the ***I1 New attribute number*** field; click on ***OK***.
- Repeat this procedure for area 12 (material reference number 3), and areas 1, 3, and 7 (material reference number 4).
- When finished, select everything (**ALLSEL** command) using the following menu path:

Utility Menu > Select > Everything

- Plot the elements (**EPLOT** command) using the following menu path:

Utility Menu > Plot > Elements

- Plot the elements with different colors based on their material reference numbers (**/PNUM** command) using the following menu path:

Utility Menu > PlotCtrls > Numbering

- *Plot Numbering Controls* dialog box appears; remove the checkmark next to ***AREA Area numbers*** by clicking on the box. Select ***Material Numbers*** for ***Elem/ Attrib numbering*** pull-down menu; select ***Colors only*** from the pull-down menu next to ***[/NUM] Numbering shown with***; click on ***OK***. When elements are plotted, they appear in different colors based on their material reference numbers, as shown in Fig. 9.72.

Solution

- Declare the new analysis to be a transient analysis (**ANTYPE** command) using the following menu path:

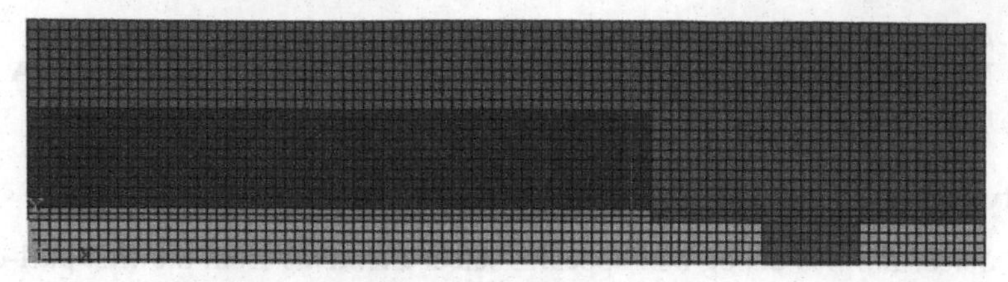

Fig. 9.72 *Mesh* representing the electronic device

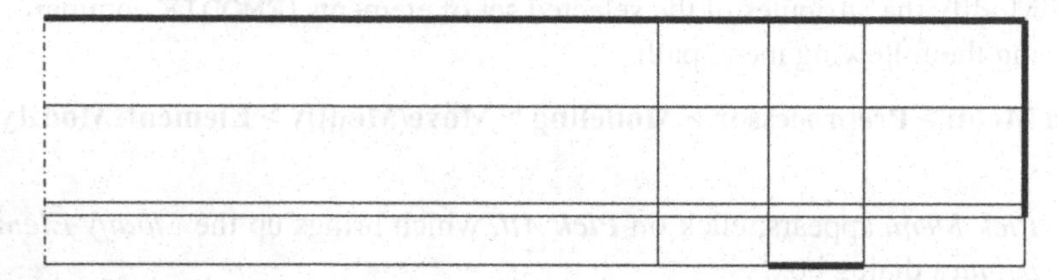

Fig. 9.73 *Bold lines* represent the surfaces of moisture diffusion

Main Menu > Solution > Analysis Type > New Analysis

- *New Analysis* dialog box appears; click on the ***Transient*** radio-button; click on ***OK***.
- *Transient Analysis* dialog box appears; click on ***OK***.

- Apply the initial condition on the nodes (`IC` command) using the following menu path:

Main Menu > Solution > Define Loads > Apply > Initial Condit'n > Define

- *Pick Menu* appears; click on ***Pick All*** in *Pick Menu*.
- *Define Initial Conditions* dialog box appears; select ***TEMP*** from the pull-down menu; enter ***0*** for ***VALUE Initial value of DOF***; click on ***OK***.

- Apply temperature (wetness) boundary conditions on the lines (`SFL` command) using the following menu path:

Main Menu > Solution > Define Loads > Apply > Thermal > Temperature > On Lines

- *Pick Menu* appears; pick the exterior lines that are shown in bold in Fig. 9.73; click on ***OK*** in the *Pick Menu*.
- *Apply TEMP on Lines* dialog box appears; select TEMP from the list and enter ***1*** for ***VALUE Load TEMP value***; click on ***OK***.

- Specify solution controls using the following menu path:

Main Menu > Solution > Analysis Type > Sol'n Controls

- *Solution Controls* dialog box appears, which has five tabs. In the ***Basic*** tab, within the ***Time Control*** field, enter ***168*** for ***Time at end of loadstep***.
- Click on the ***Time increment*** radio-button; enter ***1*** for ***Time step size***.
- In the ***Write Items to Results File*** field, select ***Write every substep*** from the ***Frequency*** pull-down menu; click on ***OK***.

• Obtain the solution (**SOLVE** command) using the following menu path:

Main Menu > Solution > Solve > Current LS

- *Confirmation Window* appears along with *Status Report Window*.
- Review status; if OK, close the *Status Report Window* and click on ***OK*** in the *Confirmation Window*.
- Wait until ANSYS responds with ***Solution is done!***

The solution for pre-conditioning is complete. Next, the solution for the reflow process is obtained. The moisture diffusivity values for the molding compound and the die-attach are different for pre-conditioning (absorption) and reflow (desorption). For this purpose, the material attributes of the molding compound and die-attach need to be modified to be 5 and 6, respectively.

• Select elements corresponding to the molding compound (**ESEL** command) using the following menu path:

Utility Menu > Select Entities

- *Select Entities* dialog box appears; select ***Elements*** from the first pull-down menu; select ***By Attributes*** from the second pull-down menu. Click on the ***Material num*** radio-button and type ***1*** in the test field; click on ***OK***.

• Modify the attributes of the selected set of elements (**EMODIF** command) using the following menu path:

Main Menu > Solution > Other > Change Mat Props > Change Mat Num

- *Change Material Number* dialog box appears; type ***5*** in the ***Mat New material number*** text field and type ***ALL*** in the ***ELEM Element no. to be modified*** text field; click on ***OK***.

• Repeat the same procedure for die-attach elements; first select elements with material attribute number 2; then modify their material attribute number to be 6.
• When finished, select everything (**ALLSEL** command) using the following menu path:

Utility Menu > Select > Everything

- Apply temperature (wetness) boundary conditions on lines (**SFL** command) using the following menu path:

Main Menu > Solution > Define Loads > Apply > Thermal > Temperature > On Lines

- *Pick Menu* appears; pick the exterior lines shown in Fig. 9.73; click on ***OK*** in the *Pick Menu*.
- *Apply TEMP on Lines* dialog box appears; select TEMP from the list and enter ***0*** for ***VALUE Load TEMP value***; click on ***OK***.

- Specify solution controls using the following menu path:

Main Menu > Solution > Analysis Type > Sol'n Controls

- *Solution Controls* dialog box appears, which has five tabs. In the ***Basic*** tab, within the ***Time Control*** field, enter ***168+5/60*** for ***Time at end of loadstep***.
- Click on the ***Time increment*** radio-button; enter ***5/60/20*** for ***Time step size***.
- In the ***Write Items to Results File*** field, select ***Write every substep*** from the ***Frequency*** pull-down menu; click on ***OK***.

- Obtain the solution (**SOLVE** command) using the following menu path:

Main Menu > Solution > Solve > Current LS

- *Confirmation Window* appears along with *Status Report Window*.
- Review status; if OK, close the *Status Report Window* and click on ***OK*** in *Confirmation Window*.
- Wait until ANSYS responds with ***Solution is done!***

Postprocessing

- Read the results for the last time point using the following menu path:

Main Menu > General Postproc > Read Results > Last Set

- Plot temperature (wetness) contours (**PLNSOL** command) at the last time point using the following menu path:

Main Menu > General Postproc > Plot Results > Contour Plot > Nodal Solu

- *Contour Nodal Solution Data* dialog box appears. Select ***DOF Solution*** and ***Nodal Temperature***; click on ***OK***. The contour plot appears, as shown in Fig. 9.74.
- Plot flux vector contours (**PLVECT** command) at the last time point using the following menu path:

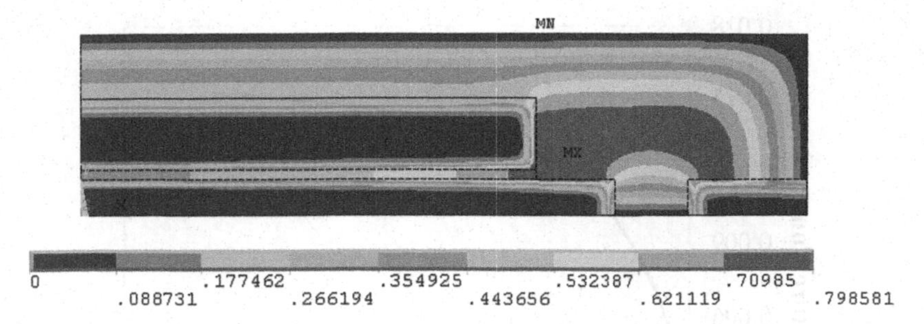

Fig. 9.74 Contour plot of wetness

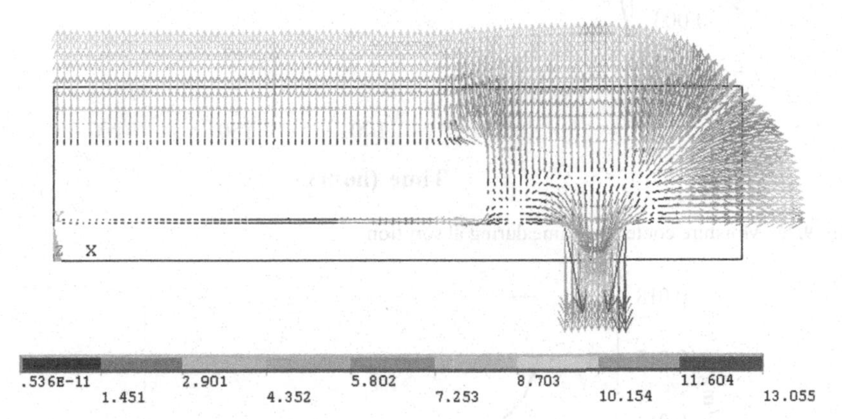

Fig. 9.75 Vector plot of flux

Main Menu > General Postproc > Plot Results > Vector Plot > Predefined

– *Vector Plot of Predefined Vectors* dialog box appears; click on ***OK***. The vector plot appears, as shown in Fig. 9.75.

- Obtain an animation of the temperature (wetness) over time using the following menu path:

Utility Menu > PlotCtrls > Animate > Over Time

- *Animate Over Time* dialog box appears; enter ***100*** for ***Number of animation frames***, select ***Load Step Range*** radio-button and enter ***1*** and ***2*** for ***Range Minimum, Maximum***. Click on ***OK***; wait until the *Animation Controls Window* appears.
- Obtain the moisture content as a function of time. This portion of the postprocessing is performed using the ANSYS Parametric Design Language (APDL) as discussed in substantial detail in Chap. 7. It is assumed that the results are stored in a file named ***WET.RTH***. The input file (written using APDL) is given below. There are two `DO` loops: one over the substeps and the other over the elements. During each of the loops over the elements, each element's area, saturated

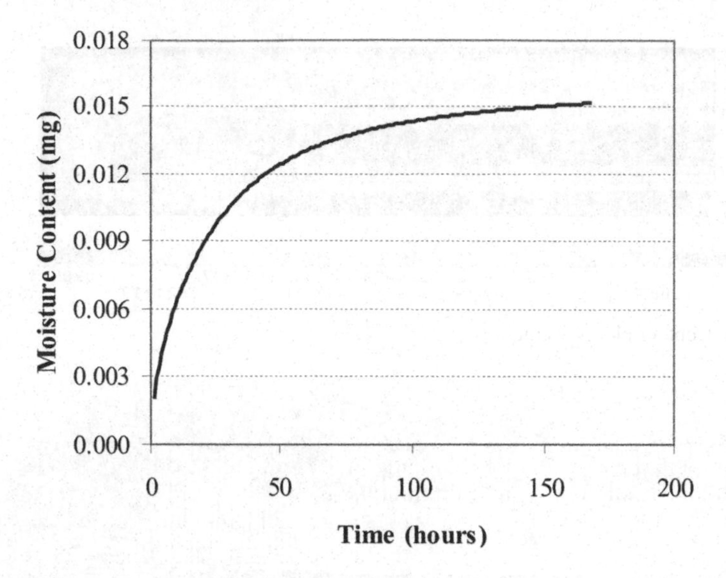

Fig. 9.76 Moisture content vs. time during absorption

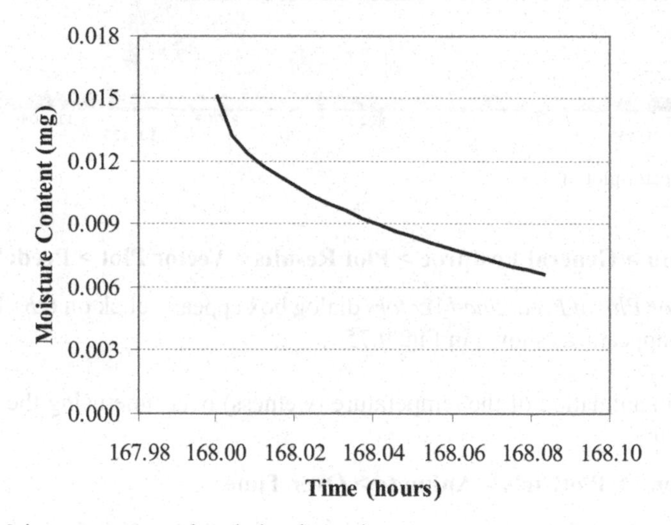

Fig. 9.77 Moisture content vs. time during desorption

moisture concentration, and wetness parameter are used to calculate the element's moisture content in milligrams. Since this is a two-dimensional idealization, the package is assumed to have a unit thickness in the third direction. The results (time and total moisture content in the package) are written to two different text files, i.e., ***ABS.OUT*** for absorption and ***DES.OUT*** for desorption. Figs. 9.76 and 9.77 show the variation of total moisture content in the package against time during absorption and desorption, respectively.

```
/POST1                          ! ENTER POSTPROCESSOR
FILE,'WET','RTH','.'            ! DECLARE RESULTS FILE
CSAT5=7.06E-3                   ! C_sat FOR MOLDING COMPOUND
CSAT6=6.20E-3                   ! C_sat FOR DIE-ATTACH
CSAT3=0                         ! C_sat FOR DIE
CSAT4=0                         ! C_sat FOR COPPER
*GET,NUMEL,ELEM,0,COUNT         ! OBTAIN NUMBER OF ELEMENTS
SET,FIRST                       ! READ 1ST RESULTS SET
                                ! (SUBSTEP 1)
*DO,J,1,188                     ! LOOP OVER SUBSTEPS (TOTAL
                                ! 188)
ETABLE,,TEMP                    ! STORE WETNESS IN ELEMENT
                                ! TABLE
*GET,MINEL,ELEM,0,NUM,MIN       ! OBTAIN MINIMUM ELEMENT
                                ! NUMBER
*GET,TIM,ACTIVE,0,SET,TIME      ! OBTAIN CURRENT TIME
SUM=0                           ! INITIALIZE TOTAL MOISTURE
                                ! MASS
*DO,I,1,NUMEL                   ! LOOP OVER ELEMENTS
*GET,EA,ELEM,MINEL,AREA         ! OBTAIN ELEMENT AREA
*GET,EM,ELEM,MINEL,ATTR,MAT     ! OBTAIN ELEMENT MATERIAL
                                ! NUMBER
*GET,EW,ETAB,1,ELEM,MINEL       ! OBTAIN ELEMENT WETNESS
EC=1E6*EA*EW*CSAT%EM%           ! CALCULATE MOISTURE MASS (mg)
SUM=SUM+EC                      ! UPDATE TOTAL MOISTURE MASS
MINEL=ELNEXT(MINEL)             ! MOVE ON TO NEXT ELEMENT
*ENDDO                          ! END LOOP OVER ELEMENTS
*IF,J,LE,168,THEN               ! IF TRUE - ABSORPTION
/OUTPUT,ABS,OUT,,APPEND         ! REDIRECT OUTPUT TO FILE
                                ! ABS.OUT

*VWRITE,TIM,SUM                 ! WRITE TIME AND MOISTURE MASS



                                ! TO FILE
(E16.8,5X,E16.8)                ! FORMAT FOR *VWRITE COMMAND
/OUTPUT                         ! REDIRECT OUTPUT TO OUTPUT
                                ! WINDOW
*ELSEIF,J,GT,168,THEN           ! IF TRUE - DESORPTION
/OUTPUT,DES,OUT,,APPEND         ! REDIRECT OUTPUT TO FILE
                                ! DES.OUT
*VWRITE,TIM,SUM                 ! WRITE TIME AND MOISTURE MASS
                                ! TO FILE
(E16.8,5X,E16.8)                ! FORMAT FOR *VWRITE COMMAND
/OUTPUT                         ! REDIRECT OUTPUT TO OUTPUT
                                ! WINDOW
*ENDIF                          !
*IF,J,LT,188,THEN               !
SET,,,,,,,J+1                   ! READ NEXT RESULTS SET
*ENDIF                          !
*ENDDO                          ! END LOOP OVER SUBSTEPS
```

Chapter 10
Nonlinear Structural Analysis

The nonlinear load-displacement relationship—the stress-strain relationship with a nonlinear function of stress, strain, and/or time; changes in geometry due to large displacements; irreversible structural behavior upon removal of the external loads; change in boundary conditions such as a change in the contact area and the influence of loading sequence on the behavior of the structure—requires a nonlinear structural analysis. The structural nonlinearities can be classified as geometric nonlinearity, material nonlinearity, and contact or boundary nonlinearity.

The governing equations concerning large deformations are nonlinear with respect to displacements and velocities. The material behavior can be linear or nonlinear, and the boundary conditions can also exhibit nonlinearity. Geometrical nonlinearity arising from large deformations is associated with the necessity to distinguish between the coordinates of the initial and final states of deformation, and also with the necessity to use the complete expressions for the strain components. The material can exhibit either time-dependent or time-independent nonlinear behavior. Nonlinearity due to boundary conditions emerges from a nonlinear relationship between the external forces and the boundary displacements. The presence of contact conditions also leads to a nonlinear structural analysis because the extent of the contact region and the contact stresses are not known a priori.

The solution to the nonlinear governing equations can be achieved through an incremental approach. The incremental form of the governing equations can be written as

$$\mathbf{K}(\mathbf{u})\Delta\mathbf{u} = \Delta\mathbf{P} \tag{10.1}$$

in which $\Delta\mathbf{u}$ and $\Delta\mathbf{P}$ represent the unknown incremental displacement vector and the known incremental applied load vector, respectively. The solution is constructed by taking a series of linear steps in the appropriate direction in order to closely approximate the exact solution. Depending on the nature of the nonlinearity, the magnitude of each step and its direction may involve several iterations. The computational algorithms and the associated parameters must be chosen with extreme care. The solution to nonlinear problems may not be unique.

The online version of this book (doi: 10.1007/978-1-4939-1007-6_10) contains supplementary material, which is available to authorized users

E. Madenci, I. Guven, *The Finite Element Method and Applications in Engineering Using ANSYS®*, DOI 10.1007/978-1-4899-7550-8_10

When solving nonlinear problems, ANSYS uses the Newton-Raphson (N-R) method, which involves an iterative procedure. This method starts with a trial (assumed) solution, $\mathbf{u} = \mathbf{u}_i$, to determine the magnitude of the next step (increment), $\Delta\mathbf{u}_i = \mathbf{K}^{-1}(\mathbf{u}_i)\Delta\mathbf{P}$, and the corresponding *out-of-balance load vector*, $\Delta\mathbf{R}_i = \Delta\mathbf{P} - \mathbf{K}(\mathbf{u}_i)\Delta\mathbf{u}_i$, which is the difference between the applied loads and the loads evaluated based on the assumed solution. In order to satisfy the equilibrium conditions exactly, *the out-of-balance load vector* must be zero. However, as the nonlinear equilibrium conditions are solved approximately, a tolerance is introduced for the out-of-balance load vector in order to terminate the solution procedure. In each iteration, the N-R method computes the out-of-balance load vector and checks for convergence based on the specified tolerance. If the convergence criterion is not satisfied, the trial solution is updated as $\mathbf{u}_{i+1} = \mathbf{u}_i + \Delta\mathbf{u}_i$ based on the calculated incremental displacements, and the next incremental solution vector is determined as $\Delta\mathbf{u}_{i+1} = \mathbf{K}^{-1}(\mathbf{u}_{i+1})\Delta\mathbf{P}$ leading to the computation of the new out-of-balance load vector $\Delta\mathbf{R}_{i+1} = \Delta\mathbf{P} - \mathbf{K}(\mathbf{u}_{i+1})\Delta\mathbf{u}_{i+1}$; this procedure is repeated until convergence is accomplished.

Several methods for improving the convergence (or convergence rate) are available in ANSYS. These include automatic time stepping, a bisection method, and line search algorithms. The user may choose to have full control or let ANSYS choose the options.

In a nonlinear solution in ANSYS, there are three distinct levels: (1) *Load Steps*, (2) *Substeps*, and (3) *Equilibrium Iterations*. The number of load steps is specified by the user. Different load steps must be used if the loading on the structure changes abruptly. The use of load steps also becomes necessary if the response of the structure at specific points in time is desired. A solution within each load step is obtained by applying the load incrementally in substeps. Within each substep, several equilibrium iterations are performed until convergence is accomplished, after which ANSYS proceeds to the next substep. As the number of substeps used increases, the accuracy of the solution improves. However, this also means that more computational time is being used. ANSYS offers the *Automatic Time Stepping* feature to optimize the task of obtaining a solution with acceptable accuracy in a reasonable amount of time. The automatic time stepping feature decides on the number and size of substeps within load steps. When using automatic time stepping, if a solution fails to converge within a substep, the *bisection* method is activated, which restarts the solution from the last converged substep.

The ANSYS program has default values for all of the nonlinear *solution controls*, including the convergence options. The **SOLCONTROL** command is used to turn these defaults on or off. The help page for the **SOLCONTROL** command provides a comprehensive list of the default values of nonlinear analysis settings when solution controls are on (**SOLCONTROL, ON**), which is the default setting. It is also possible to modify specific controls while leaving the rest for ANSYS to assign. Some of the commonly used commands for modifying/specifying nonlinear analysis settings with brief descriptions are:

AUTOTS Command: Turns automatic time stepping on or off.

DELTIM Command: Specifies time step size and/or minimum and maximum time step sizes to be used within a load step.

NSUBST Command: Specifies number of substeps and/or minimum and maximum number of substeps to be used within a load step.

NEQIT Command: Specifies maximum number of equilibrium iterations within a substep. If this number is reached with no converged solution, and if automatic time stepping is on, then ANSYS employs the *bisection* method to achieve convergence. Otherwise, the solution is terminated.

KBC Command: Specifies whether the loads are interpolated (ramped) for each substep from the values of the previous load step to the values of the current load step.

EQSLV Command: Specifies the type of solver to be used to solve the matrix system of equations. By default, it is the *Sparse Solver*; however, there are several other solvers available that may be more efficient for the particular problem being solved.

CNVTOL Command: Specifies convergence tolerance values for the nonlinear analyses.

NROPT Command: Specifies which type of Newton-Raphson method is used in the solution.

LNSRCH Command: Specifies whether a line search is to be used with the Newton-Raphson method in the solution.

PRED Command: Specifies whether a predictor algorithm is to be used in the solution.

ARCLEN Command: Toggles the *arc-length* method on or off.

SSTIF Command: The stiffness of certain materials increases with the increased stress levels within the structure (e.g., cables and membranes). This command toggles the *stress stiffening* effects on or off.

TIMINT Command: Toggles the transient effects on or off.

OUTRES Command: Specifies the amount and frequency of the data saved in the results file. By default, results associated with the last substep of each load step are written in the results file.

The commands described above require special attention, which may be crucial to the success of the analysis. There are no "golden standard" values for time step sizes, the number of equilibrium iterations, or the number of substeps. The user accumulates knowledge on the use of these features with every new analysis. It is highly recommended that the user consult the ANSYS Help pages on nonlinear analysis and individual commands, which provide detailed guidelines. Some general suggestions on achieving success in nonlinear analyses are as follows:

Nonlinear analyses require more computational time. Therefore, when solving nonlinear static problems, it may be helpful to solve a preliminary version of the problem with no nonlinearities. The results from the *linear solution* may indicate mistakes in modeling, meshing, and application of boundary conditions in a shorter time frame. Also, the linear solution provides information about the regions where high stress gradients are expected, thus guiding the user to modify the mesh (make it more refined) in those regions.

In nonlinear analyses, it is important to utilize all possible simplifications in order to improve convergence and reduce the computational cost. For example, if the problem can be simplified as a plane stress or plane strain idealization, then the user should take advantage of this opportunity.

Reading the contents of the *Output Window* and the *Error File* (`jobname.err`) is crucial in finding the specific reason why the solution does not converge.

Another important consideration in dealing with nonlinear problems is the *path dependency* of the solution. When all the materials in a problem exhibit linear behavior, the order in which the loads are applied does not make any difference in the results. However, when a nonlinear material behavior is present, results obtained by applying the same set of loads in different orders may differ from each other.

Detailed step-by-step instructions for numerous example problems are provided in Chaps. 6, 8, and 9. The command line equivalents of each of these example problems, written in the ANSYS Parametric Design Language (APDL), are included in the accompanying CD-ROM. The use of the APDL is described in Chap. 7.

In this chapter, the nonlinear structural analyses arising from (i) geometric nonlinearity, (ii) material nonlinearity, and (iii) contact conditions are considered in order to demonstrate the nonlinear features of ANSYS. However, APDL is chosen to be the main method of interacting with ANSYS because of its versatility and efficiency. Explanations are included in each command line after an exclamation mark (`!`). Step-by-step instructions are sporadically given when they are considered to be beneficial to the user. It is highly recommended that the user have a good understanding of APDL before delving into this chapter.

10.1 Geometric Nonlinearity

Geometric nonlinearities arise from the presence of large strain, small strains but finite displacements and/or rotations, and loss of structural stability. Large strains, over 5 % may occur in rubber structures and metal forming. Slender structures such as bars and thin plates may experience large displacements and rotations with small strains. Initially stressed structures with small strains and displacements may undergo a loss of stability by buckling.

Two problems are considered. The first problem involves a thin cantilever plate subjected to a point load at one of the free corners. Because the plate is thin, the resulting displacement components are in comparable order to its geometric dimensions, thus the geometry changes. This requires the stiffness matrix to be modified by accounting for the changes in the geometry. Results are compared to the solution of the same problem obtained by disregarding the nonlinear geometry effects.

The second problem involves a composite plate with a circular hole subjected to compression. As the applied loading increases, the plate is expected to buckle. Eigenvalue Buckling Analysis is one of the methods that could be used to solve this problem (as explained in Chap. 8). However, Eigenvalue Buckling Analysis evaluates only the buckling load; it does not solve for the events after buckling occurs. Alternatively, nonlinear geometry effects are turned on and post-buckling behavior is evaluated along with the buckling load.

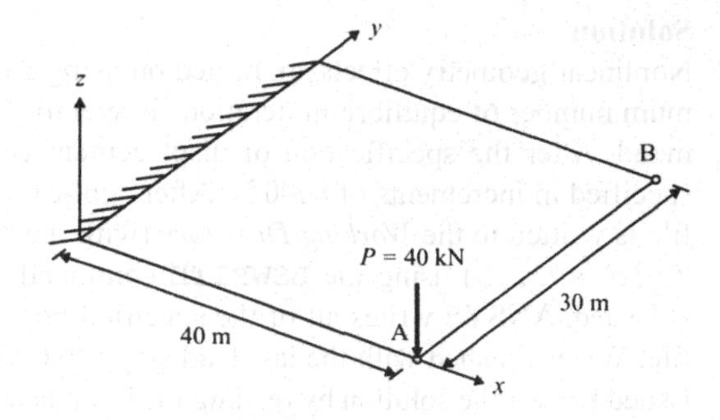

Fig. 10.1 Cantilever plate with a transverse force at one corner

10.1.1 Large Deformation Analysis of a Plate

Consider the cantilever plate with a transverse force at one corner shown in Fig. 10.1. The plate has a length, width, and thickness of 40, 30, and 0.4 m, respectively. Its elastic modulus is 120 MPa, and the Poisson's ratio is 0.3. The maximum applied load of 40 kN is reached in five equal increments. The nonlinear geometry option is used in the ANSYS solution. This is achieved by writing *Load Step Files* for each increment, and obtaining the solution from these files (**LSSOLVE** command). The goal is to find the displacement components as the applied load increases, at points A and B shown in Fig. 10.1.

Model Generation

Element type **SHELL181** is used in the analysis. The number of element divisions on all of the lines is specified as 20, and mapped meshing is utilized. The following command input is used for model generation:

```
/PREP7                ! ENTER PREPROCESSOR
ET,1,181              ! SPECIFY ELEMENT TYPE AS SHELL 181
MP,EX,1,1.2E8         ! SPECIFY ELASTIC MODULUS
MP,NUXY,1,0.3         ! SPECIFY POISSON'S RATIO
SECT,1,SHELL          ! SPECIFY PLATE THICKNESS
SECDATA,0.4,1         ! DEFINE THICKNESS FOR THE PLATE
K,1,0                 ! CREATE KEYPOINTS
K,2,40                !
K,3,40,30             !
K,4,0,30              !
A,1,2,3,4             ! CREATE AREA THROUGH KEYPOINTS
LESIZE,ALL,,,20       ! SPECIFY NO. OF DIVISIONS ON LINES
MSHKEY,1              ! ENFORCE MAPPED MESHING
AMESH,ALL             ! MESH THE AREA
FINISH                ! EXIT PREPROCESSOR
```

Solution

Nonlinear geometry effects are turned on using the **NLGEOM** command. The maximum number of equilibrium iterations is specified as 1000 using the **NEQIT** command. After the specification of displacement constraints, the transverse load is specified in increments of 8000 N. After application of each increment, a load step file is written to the *Working Directory* (with the naming convention: `file.s01`, `file.s02`, ...) using the **LSWRITE** command. When the **LSWRITE** command is issued, ANSYS writes all of the specified boundary conditions to the load step file. When finished with the last load step (40,000 N), the **LSSOLVE** command is issued to start the solution by reading the boundary conditions from load step files 1 through 5. The following command input is used for the solution:

```
/SOLU                      ! ENTER SOLUTION PROCESSOR
ANTYPE,STATIC              ! SPECIFY ANALYSIS TO BE STATIC
NLGEOM,ON                  ! TURN NONLINEAR GEOMETRY
                           ! EFFECTS ON
NEQIT,1000                 ! SPECIFY MAX. # OF EQUILIBRIUM
                           ! ITERATIONS
OUTRES,ALL,ALL             ! WRITE ALL SOLUTION ITEMS FOR
                           ! EVERY SUBSTEP
NSEL,S,LOC,X,0             ! SELECT NODES AT X = 0
D,ALL,ALL                  ! CONSTRAIN ALL DOFS AT SELECTED
                           ! NODES
ALLSEL                     ! SELECT EVERYTHING
                           ! START FIRST LOAD INCREMENT
NSEL,S,LOC,X,40            ! SELECT NODES AT X = 40
NSEL,R,LOC,Y,0             ! SELECT NODES AT Y = 0 FROM THE
                           ! SELECTED SET
F,ALL,FZ,-8000             ! SPECIFY FORCE IN NEG. Z-DIR AT
                           ! SELECTED NODES
ALLSEL                     ! SELECT EVERYTHING
LSWRITE                    ! WRITE LOAD STEP FILE 1
                           ! SECOND LOAD INCREMENT
NSEL,S,LOC,X,40            ! SELECT NODES AT X = 40
NSEL,R,LOC,Y,0             ! SELECT NODES AT Y = 0 FROM THE
                           ! SELECTED SET
F,ALL,FZ,-16000            ! SPECIFY FORCE IN NEG. Z-DIR AT
                           ! SELECTED NODES
ALLSEL                     ! SELECT EVERYTHING
LSWRITE                    ! WRITE LOAD STEP FILE 2
                           ! THIRD LOAD INCREMENT
NSEL,S,LOC,X,40            ! SELECT NODES AT X = 40
NSEL,R,LOC,Y,0             ! SELECT NODES AT Y = 0 FROM THE
                           ! SELECTED SET
F,ALL,FZ,-24000            ! SPECIFY FORCE IN NEG. Z-DIR AT
                           ! SELECTED NODES
ALLSEL                     ! SELECT EVERYTHING
LSWRITE                    ! WRITE LOAD STEP FILE 3
                           ! FOURTH LOAD INCREMENT
NSEL,S,LOC,X,40            ! SELECT NODES AT X = 40
```

```
NSEL,R,LOC,Y,0                  ! SELECT NODES AT Y = 0 FROM THE
                                ! SELECTED SET
F,ALL,FZ,-32000                 ! SPECIFY FORCE IN NEG. Z-DIR AT
                                ! SELECTED NODES
ALLSEL                          ! SELECT EVERYTHING
LSWRITE                         ! WRITE LOAD STEP FILE 4
                                ! FIFTH LOAD INCREMENT
NSEL,S,LOC,X,40                 ! SELECT NODES AT X = 40
NSEL,R,LOC,Y,0                  ! SELECT NODES AT Y = 0 FROM THE
                                ! SELECTED SET
F,ALL,FZ,-40000                 ! SPECIFY FORCE IN NEG. Z-DIR AT
                                ! SELECTED NODES
ALLSEL                          ! SELECT EVERYTHING
LSWRITE                         ! WRITE LOAD STEP FILE 5
LSSOLVE,1,5,1                   ! SOLVE FROM LOAD STEP FILES
FINISH                          ! EXIT SOLUTION PROCESSOR
```

Postprocessing
In the postprocessing, node numbers for the nodes at points A and B are stored in parameters **NA** and **NB**, respectively. A do loop is used for extraction of the results data at different results sets corresponding to the load steps. The commands **/OUTPUT** and ***VWRITE** are used for redirecting output and writing parameters to external files. The use of shortcuts for ***GET** functions **UX(*N*)**, **UY(*N*)**, and **UZ(*N*)**, is also demonstrated. These functions retrieve the *x*-, *y*-, and *z*-displacements of node ***N***. The following command input is used for postprocessing:

```
/POST1                            !ENTER POSTPROCESSOR
NSEL,S,LOC,X,40                   !SELECT NODES AT X = 40
NSEL,R,LOC,Y,0                    !RESELECT NODE AT Y = 0
*GET,NA,NODE,0,NUM,MIN            !STORE NODE # OF PT A
                                  !INTO NA
NSEL,S,LOC,X,40                   !SELECT NODES AT X = 40
NSEL,R,LOC,Y,30                   !RESELECT NODE AT Y = 30
*GET,NB,NODE,0,NUM,MIN            !STORE NODE # OF PT B
                                  !INTO NB
ALLSEL                            !SELECT EVERYTHING
*DO,I,1,5                         !START DO LOOP
SET,I                             !SET RESULTS TO LOAD STEP I
/OUTPUT,ADISP,OUT,,APPEND         !REDIRECT OUTPUT TO FILE
                                  !ADISP.OUT
*VWRITE,UX(NA),UY(NA),UZ(NA)      !WRITE DISPLACEMENTS TO FILE
(E16.8,5X,E16.8,5X,E16.8,5X)      !FORMAT STATEMENT
/OUTPUT                           !REDIRECT OUTPUT TO OUTPUT
                                  !WINDOW
/OUTPUT,BDISP,OUT,,APPEND         !REDIRECT OUTPUT TO FILE
                                  !BDISP.OUT
*VWRITE,UX(NB),UY(NB),UZ(NB)      !WRITE DISPLACEMENTS TO FILE
(E16.8,5X,E16.8,5X,E16.8,5X)      !FORMAT STATEMENT
/OUTPUT                           !REDIRECT OUTPUT TO
                                  !OUTPUT WINDOW
*ENDDO                            !END DO LOOP
```

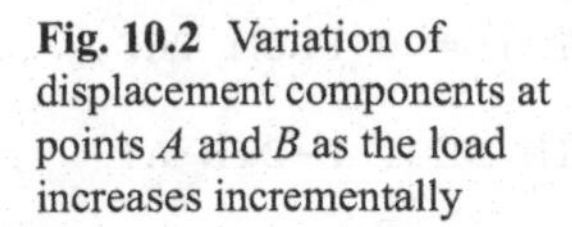

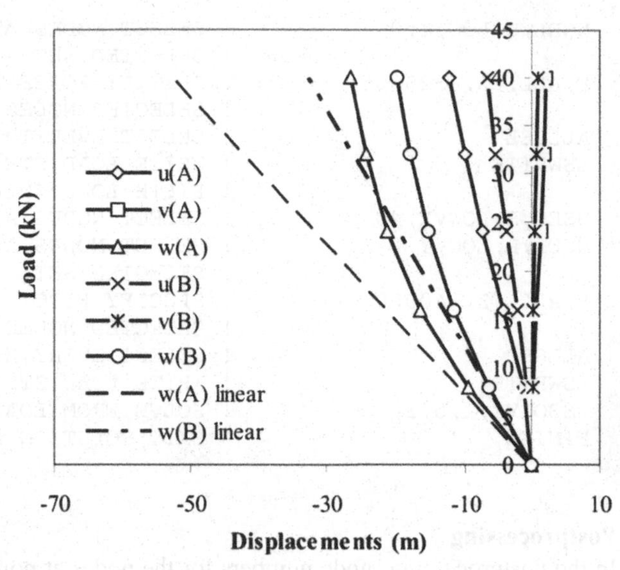

Fig. 10.2 Variation of displacement components at points *A* and *B* as the load increases incrementally

After execution of the command input given above, the x-, y-, and z-displacements at points A and B are written to files *ADISP.OUT* and *BDISP.OUT*, respectively. Figure 10.2 shows the variation of displacement components as the load increases incrementally. Also plotted in Fig. 10.2 are the z-displacements of points A and B obtained by disregarding geometric nonlinearity (indicated in the legend as "linear"). Displacements in the x- and y-directions are identically zero for this case.

10.1.2 Post-buckling Analysis of a Plate with a Hole

Consider a 9.5-in-square composite plate with a circular hole of radius 1.5 in, as illustrated in Fig. 10.3. The laminate lay-up is $[\pm 30]_{3s}$, with a total of 12 layers, symmetric with respect to the mid-plane, and the orientation of the layers alternates between 30 and -30. Each layer has moduli of $E_L = 18.5\times10^6$ psi, $E_T = 1.6\times10^6$ psi, and $G_{LT} = 0.832\times10^6$ psi, and a Poisson's ratio of $\nu_{LT} = 0.35$. Each layer is 0.01 in. thick, resulting in a total laminate thickness of 0.12 in.

Along the right edge of the laminate, an axial concentrated load of 12,000 lb is introduced through a rigid end. This type of load introduction, requiring the x-displacement to be uniform, is enforced by coupled degrees of freedom, as explained in Chap. 11. All degrees of freedom (displacements and rotations) are constrained along the left edge. Along the horizontal edges, in-plane displacements and rotations about these edges are permitted. In order to trigger the nonlinear response, a

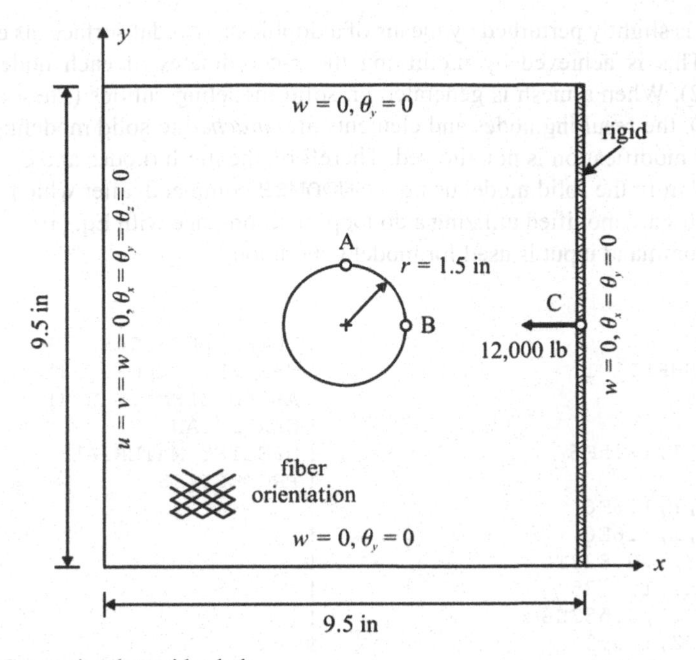

Fig. 10.3 Composite plate with a hole

sinusoidal imperfection with an amplitude of 1 % of the total laminate thickness is used as follows:

$$z = 0.012 \sin\left(\frac{\pi x}{9.5}\right)\sin\left(\frac{\pi y}{9.5}\right) \tag{10.2}$$

The goal is to obtain the variations of the z-displacement at points A (4.75, 6.25, 0) and B (6.25, 4.75, 0), and the x-displacement at point C (9.5, 4.75, 0) as the applied load increases (points A, B, and C are indicated in Fig. 10.3).

Model Generation

The element type used in the analysis is **SHELL181**, which is a layered element especially useful for modeling composite plates. By default, the element coordinate system is derived from the local geometry of each element. Although this may be convenient in certain cases, in this problem it is required that the element coordinate system for each element be aligned with the global Cartesian system. This is achieved by using **ESYS** the command. After specifying real constants (thickness and layer information) and orthotropic material properties, and creating the solid model, the mesh is generated. The mesh has all of its nodes and elements on the x-y plane, i.e., $z = 0$. However, in order to capture the buckling behavior, the flatness of the plate is slightly perturbed by means of a double-sinusoidal surface, as explained earlier. This is achieved by modifying the *z*-coordinates of each node through Eq. (10.2). When a mesh is generated on solid modeling entities (lines, areas and volumes), the resulting nodes and elements are *attached* to solid modeling entities and their modification is not allowed. Therefore, the mesh (nodes and elements) is *detached* from the solid model using the **MODMSH** command, after which the nodal coordinates are modified utilizing a do loop in accordance with Eq. (10.2). The following command input is used for model generation:

```
/PREP7                              ! ENTER PREPROCESSOR
ET,1,SHELL181                       ! SPECIFY ELEMENT TYPE
ESYS,0                              ! ALIGN ELEM CS WITH GLOBAL
                                    ! CARTESIAN
MP,EX,1,18.5E6                      ! SPECIFY MATERIAL
                                    ! PROPERTIES
MP,EY,1,1.6E6                       !
MP,EZ,1,1.6E6                       !
MP,GXY,1,0.832E6                    !
MP,PRXY,1,0.35                      !
MP,GYZ,1,0.533E6                    !
MP,PRYZ,1,0.5                       !
MP,GXZ,1,0.832E6                    !
MP,PRXZ,1,0.35                      !
SECT,1,SHELL                        ! SPECIFY REAL CONSTANTS
                                    ! DEFINE THICKNESS AND
                                    ! ORIENTATION FOR LAYER 1-12
SECDATA,0.01,1,30                   !
SECDATA,0.01,1,-30                  !
SECDATA,0.01,1,30                   !
SECDATA,0.01,1,-30                  !
SECDATA,0.01,1,30                   !
SECDATA,0.01,1,-30                  !
SECDATA,0.01,1,30                   !
SECDATA,0.01,1,-30                  !
SECDATA,0.01,1,30                   !
SECDATA,0.01,1,-30                  !
SECDATA,0.01,1,30                   !
SECDATA,0.01,1,-30                  !
K,1,0                               ! CREATE KEYPOINTS
K,2,9.5                             !
K,3,9.5,9.5                         !
K,4,0,9.5                           !
A,1,2,3,4                           ! CREATE SQUARE AREA
CYL4,4.75,4.75,1.5                  ! CREATE CIRCULAR AREA
ASBA,1,2                            ! SUBTRACT AREAS
NUMCMP,ALL                          ! COMPRESS ENTITY NUMBERS
LSEL,S,,,1,4                        ! SELECT LINES
```

```
LESIZE,ALL,,,10                  ! SPECIFY # OF DIVISIONS
LSEL,S,,,5,8                     ! SELECT LINES
LESIZE,ALL,,,5                   ! SPECIFY # OF DIVISIONS
AMESH,ALL                        ! MESH AREA
MODMSH,DETACH                    ! DETACH MESH FROM SOLID
                                 ! MODEL
*GET,NNUMBER,NODE,0,NUM,MAX      ! GET MAXIMUM NODE # (ALSO #
                                 ! OF NODES AS NUMBERS ARE
                                 ! COMPRESSED)
PI=4*ATAN(1)                     ! DEFINE PI (3.1415...)
*DO,I,1,NNUMBER                  ! START DO LOOP ON NODES
*GET,TMPX,NODE,I,LOC,X           ! GET X-COORD OF CURRENT
                                 ! NODE
*GET,TMPY,NODE,I,LOC,Y           ! GET Y-COORD OF CURRENT
                                 ! NODE
TMPZ=SIN(PI*NX(I)/9.5)*SIN(PI*NY(I)/9.5)*0.012
                                 ! EQ. (10.2)
N,I,NX(I),NY(I),TMPZ             ! REDEFINE CURRENT NODE
*ENDDO                           ! END DO LOOP
FINISH                           ! EXIT FROM PREPROCESSOR
```

Solution

In the solution phase of this problem, first, nonlinear geometry effects are turned on (**NLGEOM** command). Results associated with each substep are written to the results file using the **OUTRES** command. In order to improve accuracy and convergence, the number of substeps is specified as 100 using the **NSUBST** command. Displacement constraints are then applied using the **D** command. The condition that the x-displacement be uniform along the right boundary ($x = 9.5$) is imposed by selecting all of the nodes along that boundary and issuing the CP command. This command defines a set of coupled degrees of freedom (DOF) of the selected nodes (in this case the x-displacement) and enforces these DOF to be equal. The concept of coupled DOF is described as a separate topic in Sect. 11.1. The following command input is used for the solution:

```
/SOLU                  ! ENTER SOLUTION PROCESSOR
ANTYPE,STATIC          ! SPECIFY STATIC SOLUTION
NLGEOM,ON              ! TURN NONLINEAR GEOMETRY ON
OUTRES,ALL,ALL         ! WRITE RESULTS FOR EVERY SUBSTEP
NSUBST,100             ! SET NUMBER OF SUBSTEPS TO BE 100
NSEL,S,LOC,X,0         ! SELECT NODES AT X = 0
D,ALL,ALL              ! CONSTRAIN ALL DOFS
NSEL,S,LOC,Y,0         ! SELECT NODES AT Y = 0
NSEL,A,LOC,Y,9.5       ! ADD TO SELECTION NODES AT Y = 9.5
D,ALL,UZ,0             ! CONSTRAIN Z-DISPLACEMENTS
D,ALL,ROTY,0           ! CONSTRAIN ROTATIONS ABOUT Y-AXIS
NSEL,S,LOC,X,9.5       ! SELECT NODES AT X = 9.5
D,ALL,UZ,0             ! CONSTRAIN Z-DISPLACEMENTS
D,ALL,ROTX,0           ! CONSTRAIN ROTATIONS ABOUT X-AXIS
D,ALL,ROTY,0           ! CONSTRAIN ROTATIONS ABOUT Y-AXIS
```

```
ALLSEL               ! SELECT EVERYTHING
D,ALL,ROTZ,0         ! CONSTRAIN ROTATIONS ABOUT Z-AXIS
NSEL,S,LOC,X,9.5     ! SELECT NODES AT X = 9.5
CP,1,UX,ALL          ! COUPLE X-DISPL. OF SELECTED NODES
NSEL,R,LOC,Y,4.75    ! RESELECT NODE AT Y = 4.75
F,ALL,FX,-12000      ! APPLY TOTAL LOAD IN X-DIRECTION
ALLSEL               ! SELECT EVERYTHING
SOLVE                ! OBTAIN SOLUTION
FINISH               ! EXIT SOLUTION PROCESSOR
```

Postprocessing

Because the loading is applied slow enough that there are no dynamic effects within the structure, referred to as quasi-static, the nonlinear nature of the problem requires that the load is incrementally increased.

Turning on the nonlinear geometry effects using the **NLGEOM** command leads to the problem being solved in small increments (substeps), even though the analysis type is declared as static. ANSYS automatically assigns the value ***1*** as the time at the end of the load step. In nonlinear static solutions, time is a measure of the fraction of the total load applied at the current substep. For example, if time has a value of ***0.2*** at a particular substep, this means that the load applied during solution at that substep is 20 % of the total load. In the following command input, the node numbers of points A, B, and C are stored in parameters **NA**, **NB**, and **NC**, respectively, followed by extraction of the total number of substeps (parameter **SB**). A do loop is set up so that the results associated with every substep can be retrieved and written to external files sequentially. Within the do loop, the time associated with the current substep is extracted (parameter **TT**) for subsequent scaling of the applied load (parameter **TF**), and z-displacements at points A and B and the absolute value of the x-displacement at point C are written to files *AZD.OUT*, *BZD.OUT*, and *CXD.OUT*, respectively. Figure 10.4 shows the variation of these displacements with incremental load. In Fig. 10.4, z-displacements at points A and B are denoted by w(A) and w(B), respectively, and u(C) designates the absolute value of the x-displacement at point C.

```
/POST1                     ! ENTER POSTPROCESSOR
NSEL,S,LOC,X,4.75          ! SELECT NODES AT X = 4.75
NSEL,R,LOC,Y,6.25          ! RESELECT NODE AT Y = 6.25
*GET,NA,NODE,0,NUM,MIN     ! STORE NODE # OF PT A INTO NA
NSEL,S,LOC,X,6.25          ! SELECT NODES AT X = 6.25
NSEL,R,LOC,Y,4.75          ! RESELECT NODE AT Y = 4.75
*GET,NB,NODE,0,NUM,MIN     ! STORE NODE # OF PT B INTO NB
NSEL,S,LOC,X,9.5           ! SELECT NODES AT X = 9.5
NSEL,R,LOC,Y,4.75          ! RESELECT NODE AT Y = 4.75
*GET,NC,NODE,0,NUM,MIN     ! STORE NODE # OF PT C INTO NC
ALLSEL                     ! SELECT EVERYTHING
```

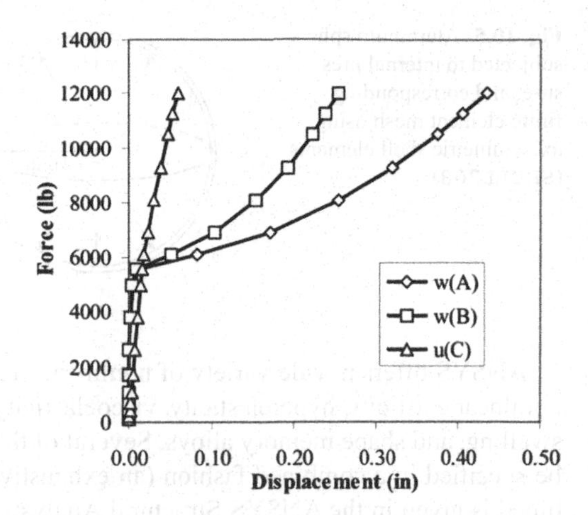

Fig. 10.4 Variation of displacement components at points *A*, *B*, and *C* as the load increases incrementally

```
SET,LAST                          ! SET RESULTS TO LAST SUBSTEP
*GET,SB,ACTIVE,0,SET,SBST         ! STORE # OF SUBSTEPS TO SB
*DO,I,1,SB                        ! START DO LOOP
SET,1,I                           ! SET RESULTS TO SUBSTEP I
*GET,TT,ACTIVE,0,SET,TIME         ! STORE CURRENT TIME TO TT
TF=TT*12000                       ! FIND CURRENT FORCE
/OUTPUT,AZD,OUT,,APPEND           ! REDIRECT OUTPUT TO FILE AZD.OUT
*VWRITE,TF,UZ(NA)                 ! WRITE TIME & DISP TO FILE
(E16.8,5X,E16.8)                  ! FORMAT STATEMENT
/OUTPUT                           ! REDIRECT OUTPUT TO
                                  ! OUTPUT WINDOW
/OUTPUT,BZD,OUT,,APPEND           ! REDIRECT OUTPUT TO FILE BZD.OUT
*VWRITE,TF,UZ(NB)                 ! WRITE TIME & DISP TO FILE
(E16.8,5X,E16.8)                  ! FORMAT STATEMENT
/OUTPUT                           ! REDIRECT OUTPUT TO
                                  ! OUTPUT WINDOW
/OUTPUT,CXD,OUT,,APPEND           ! REDIRECT OUTPUT TO FILE CXD.OUT
*VWRITE,TF,ABS(UX(NC))            ! WRITE TIME & DISP TO FILE
(E16.8,5X,E16.8)                  ! FORMAT STATEMENT
/OUTPUT                           ! REDIRECT OUTPUT TO
                                  ! OUTPUT WINDOW
*ENDDO                            ! END DO LOOP
```

10.2 Material Nonlinearity

Material nonlinearities arise from the presence of time-independent behavior, such as plasticity, time-dependent behavior such as creep, and viscoelastic/viscoplastic behavior where both plasticity and creep effects occur simultaneously. They may result in load sequence dependence and energy dissipation (irreversible structural behavior).

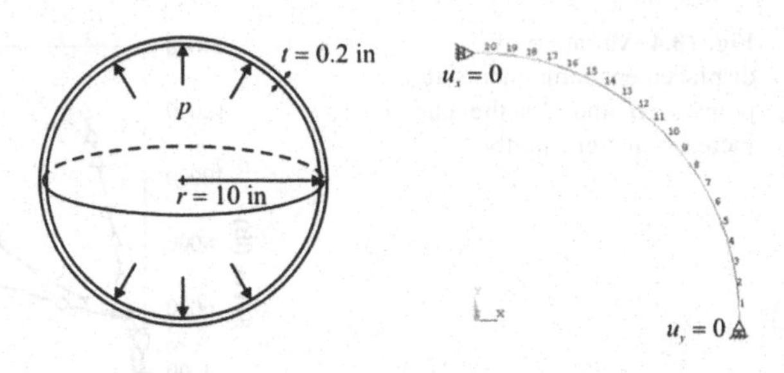

Fig. 10.5 Aluminum sphere subjected to internal pressure, and corresponding finite element mesh using axisymmetric shell elements (**SHELL208**)

ANSYS offers a wide variety of nonlinear material behavior models, including nonlinear elasticity, hyperelasticity, viscoelasticity, plasticity, viscoplasticity, creep, swelling, and shape memory alloys. Several of these nonlinear material models can be specified in a combined fashion (an exhaustive list of models that can be combined is given in the ANSYS Structural Analysis Guide). In the following subsections, four problems are considered that demonstrate the solution methods involving plasticity with isotropic hardening, viscoelasticity, viscoplasticity with Anand's model, and combined plasticity and creep.

10.2.1 Plastic Deformation of an Aluminum Sphere

Consider a thin-walled aluminum sphere with a radius of $r = 10$ in and a thickness of $t = 0.2$ in, as shown in Fig. 10.5. The sphere is subjected to an internal pressure of $p_0 = 1600$psi. The elastic modulus and Poisson's ratio of the shell are $E = 10^7$ psi and $\nu = 0.3$, respectively. The plastic behavior of aluminum is governed by

$$\sigma_e = 30000 + 136000(\varepsilon_p)^{1/2} \tag{10.3}$$

in which σ_e is the effective stress and ε_p designates the plastic strain. Figure 10.6 shows the stress vs. total strain ($\varepsilon = \varepsilon_e + \varepsilon_p$, ε_e: elastic strain) curve based on Eq. (10.3). This curve is input in ANSYS by means of a data table for nonlinear material behavior, which is given through data points (see Table 10.1). The goal is to obtain the radial displacements, as well as the strain field.

Model Generation
In order to model the thin-walled aluminum sphere, a 2-noded axisymmetric shell element (**SHELL208**) is used. For the nonlinear material behavior, the multiple-point isotropic hardening rule is chosen (**TB** command with **MISO** option). Twenty data points for strain and stress values are entered using the **TBPT** command [the data point (0, 0) is implied]. The stress value of the first data point defines the yield stress, i.e., $\sigma_{ys} = 30000$psi. Due to the symmetry conditions, only a quarter circle is modeled.

```
/PREP7                        ! ENTER PREPROCESSOR
ET,1,208                      ! USE SHELL208 ELEMENT
MP,EX,1,1E7                   ! SPECIFY ELASTIC MODULUS
MP,NUXY,1,0.3                 ! SPECIFY POISSON'S RATIO
SECT,1,SHELL                  ! SPECIFY SECTION TYPE
SECDATA,0.2,1                 ! DEFINE THICKNESS FOR SHELL
TB,MISO,1,1,20,               ! SPECIFY MULTILINEAR ISOTROPIC
                              ! HARDENING
TBPT,,0.00300,30000           ! ENTER STRAIN VS STRESS DATA
                              ! POINTS
TBPT,,0.00350,33041           !
TBPT,,0.00400,34300           !
TBPT,,0.00450,35267           !
TBPT,,0.00500,36082           !
TBPT,,0.00550,36800           !
TBPT,,0.00600,37449           !
TBPT,,0.00650,38045           !
TBPT,,0.00700,38601           !
TBPT,,0.00750,39123           !
TBPT,,0.00800,39616           !
TBPT,,0.00850,40086           !
TBPT,,0.00875,40312           !
TBPT,,0.00900,40534           !
TBPT,,0.00925,40751           !
TBPT,,0.00950,40964           !
TBPT,,0.00975,41173           !
TBPT,,0.02400,49708           !
TBPT,,0.04000,56160           !
TBPT,,0.06300,63313           !
K,1,0,0                       ! CREATE KEYPOINTS
K,2,10,0                      !
K,3,0,10                      !
LARC,2,3,1,10                 ! CREATE ARC
LESIZE,ALL,,,20               ! SPECIFY # OF ELEMENTS ON LINE
LMESH,ALL                     ! MESH LINE
FINISH                        ! EXIT PREPROCESSOR
```

Solution

The solution phase of this problem is rather straightforward and involves application of boundary conditions and the internal pressure. No specific solution controls options are specified; ANSYS uses default settings for nonlinear solution.

```
/SOLU                    ! ENTER SOLUTION PROCESSOR
P0=1600                  ! DEFINE PARAMETER FOR INTERNAL
                         ! PRESSURE
NSEL,S,LOC,Y,0           ! SELECT NODES AT Y = 0
D,ALL,UY,0               ! CONSTRAIN Y-DISPL AT SELECTED NODES
NSEL,S,LOC,Z,0           ! SELECT NODES AT Z = 0
SFE,ALL,1,PRES,,P0       ! SPECIFY INTERNAL PRESSURE ALONG ELEMS
SOLVE                    ! OBTAIN SOLUTION
FINISH                   ! EXIT SOLUTION PROCESSOR
```

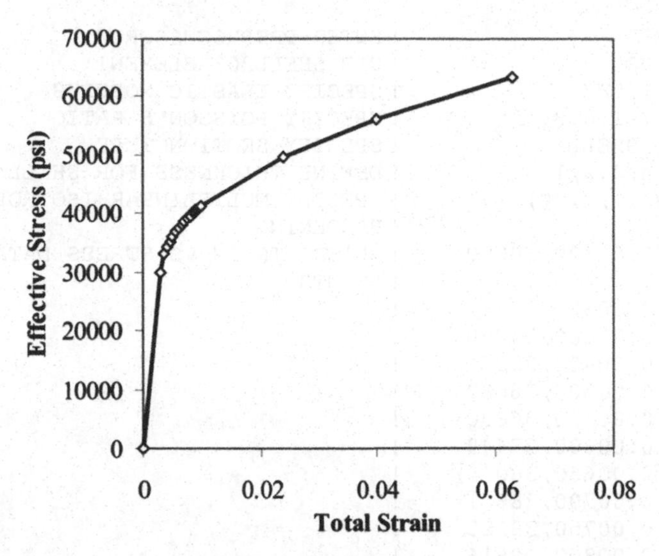

Fig. 10.6 Nonlinear stress-strain behavior of aluminum

Postprocessing

The elastic and plastic strains along the inner surface of the sphere are stored in element tables using the **ETABLE** command. Three strain components are considered, i.e., meridional, through-the-thickness, and hoop strains. Finally, total strains are obtained by adding elastic and plastic strains for each component using the **SADD** command, which adds columns to the element table. The element tables are listed in the *Output Window* using the **PRETAB** command. Figure 10.7 shows the listing of total strains at elements as they appear in the *Output Window*.

```
/POST1                        ! ENTER POSTPROCESSOR
RSYS,2                        ! ACTIVATE GLOBAL SPHERICAL
                              ! COORDINATE SYSTEM
ETABLE,EPELX,EPEL,X           ! STORE ELASTIC
                              ! STRAINS IN ELEMENT
ETABLE,EPELY,EPEL,Y           ! TABLE
ETABLE,EPELZ,EPEL,Z           !
ETABLE,EPPLX,EPPL,X           ! STORE PLASTIC
                              ! STRAINS IN ELEMENT
ETABLE,EPPLY,EPPL,Y           ! TABLE
ETABLE,EPPLZ,EPPL,Z           !
```

```
SADD,T_THK_TP,EPELX,EPPLX,1,1              ! ADD ELEMENT TABLE
SADD,T_MER_TP,EPELY,EPPLY,1,1              ! ITEMS TO FIND TOTAL
SADD,T_HOP_TP,EPELZ,EPPLZ,1,1              ! STRAINS
PRETAB,EPELX,EPELY,EPELZ                   ! LIST ELASTIC
                                           ! STRAINS
PRETAB,EPPLX,EPPLY,EPPLZ                   ! LIST PLASTIC
                                           ! STRAINS
PRETAB,T_MER_TP,T_THK_TP,T_HOP_TP          ! LIST TOTAL STRAINS
```

10.2.2 Plastic Deformation of an Aluminum Cylinder

Consider a thin-walled aluminum cylinder with a radius of $r = 20$ in., height of $r = 72$ in., a thickness of $t = 0.5$ in., and an extremely stiff panel on one end, as shown in Fig. 10.8. The cylinder and the stiff panel are first subjected to an internal pressure of $p_0 = 1500$ psi. With the internal pressure in place, the stiff panel is subjected to four tangential forces of 10^5 lb, as illustrated in Fig. 10.8. All displacements and rotations at the cylinder's end opposite to the stiff panel are constrained. The elastic modulus and Poisson's ratio of the aluminum shell are $E_{al} = 10^7$ psi and

Table 10.1 Data points for nonlinear stress-strain behavior of aluminum

ε	σ_e (psi)
0	0
0.00300	30,000
0.00350	33,041
0.00400	34,300
0.00450	35,267
0.00500	36,082
0.00550	36,800
0.00600	37,449
0.00650	38,045
0.00700	38,601
0.00750	39,123
0.00800	39,616
0.00850	40,086
0.00875	40,312
0.00900	40,534
0.00925	40,751
0.00950	40,964
0.00975	41,173
0.02400	49,708
0.04000	56,160
0.06300	63,313

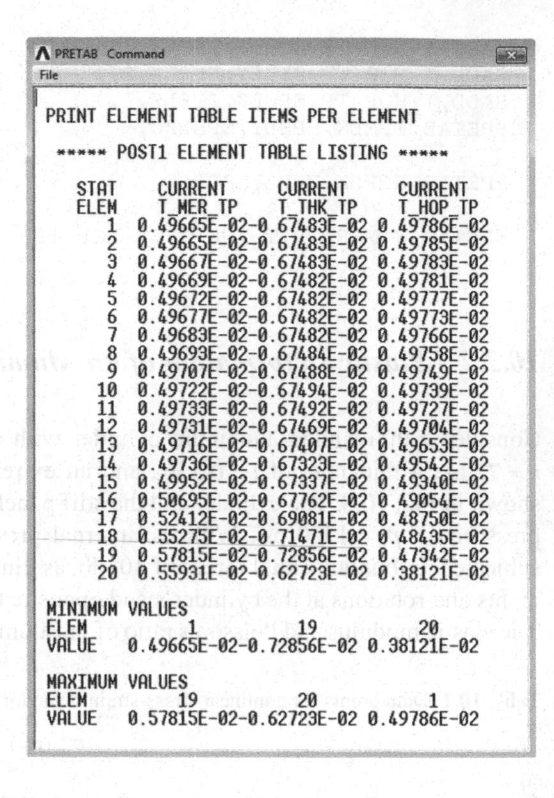

```
PRETAB Command
File

PRINT ELEMENT TABLE ITEMS PER ELEMENT

 ***** POST1 ELEMENT TABLE LISTING *****

   STAT      CURRENT      CURRENT      CURRENT
   ELEM      T_MER_TP     T_THK_TP     T_HOP_TP
      1   0.49665E-02-0.67483E-02 0.49786E-02
      2   0.49665E-02-0.67483E-02 0.49785E-02
      3   0.49667E-02-0.67483E-02 0.49783E-02
      4   0.49669E-02-0.67482E-02 0.49781E-02
      5   0.49672E-02-0.67482E-02 0.49777E-02
      6   0.49677E-02-0.67482E-02 0.49773E-02
      7   0.49683E-02-0.67482E-02 0.49766E-02
      8   0.49693E-02-0.67484E-02 0.49758E-02
      9   0.49707E-02-0.67488E-02 0.49749E-02
     10   0.49722E-02-0.67494E-02 0.49739E-02
     11   0.49733E-02-0.67492E-02 0.49727E-02
     12   0.49731E-02-0.67469E-02 0.49704E-02
     13   0.49716E-02-0.67407E-02 0.49653E-02
     14   0.49736E-02-0.67323E-02 0.49542E-02
     15   0.49952E-02-0.67337E-02 0.49340E-02
     16   0.50695E-02-0.67762E-02 0.49054E-02
     17   0.52412E-02-0.69081E-02 0.48750E-02
     18   0.55275E-02-0.71471E-02 0.48435E-02
     19   0.57815E-02-0.72856E-02 0.47342E-02
     20   0.55931E-02-0.62723E-02 0.38121E-02

MINIMUM VALUES
ELEM            1            19           20
VALUE    0.49665E-02-0.72856E-02 0.38121E-02

MAXIMUM VALUES
ELEM           19            20            1
VALUE    0.57815E-02-0.62723E-02 0.49786E-02
```

Fig. 10.7 Listing of element table items as they appear in the *Output Window*

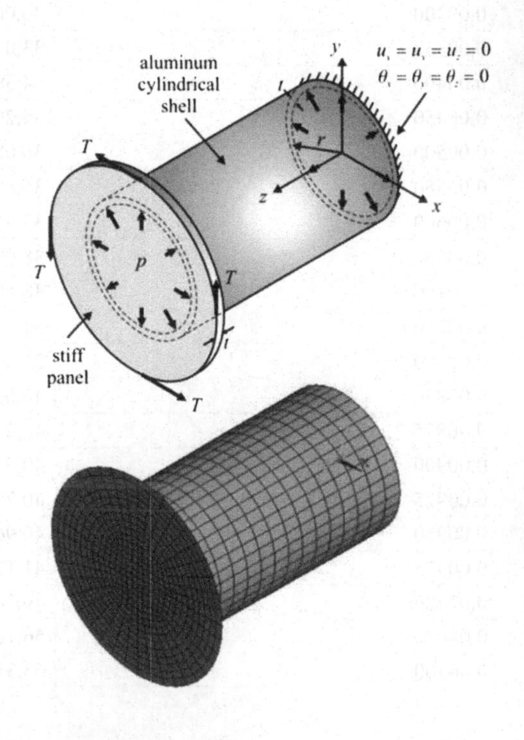

Fig. 10.8 Geometry (*top*) and the corresponding mesh (*bottom*) of the aluminum cylinder and the stiff panel

$\nu_{al} = 0.3$, respectively. The same set of properties corresponding to the stiff panel are $E_{st} = 10^{11}$ psi and $\nu_{st} = 0$. The plastic behavior of aluminum is governed by

Equation (10.3). The data points for the stress vs. strain are plotted in Fig. 10.6 and tabulated in Table 10.1. The goal is to obtain the plastic strain field resulting from the internal pressure and added torsion.

Model Generation

Both the aluminum cylinder and the stiff panel are modeled using an 8-noded shell element, **SHELL281**. For the nonlinear material behavior, the multiple-point isotropic hardening rule is chosen (**TB** command with **MISO** option). Although the geometry is axisymmetric, the loading is not. Therefore, the entire geometry is modeled.

```
/PREP7                    ! ENTER PREPROCESSOR
R=20                      ! PARAMETER FOR RADIUS
H=72                      ! PARAMETER FOR HEIGHT
T=R/40                    ! PARAMETER FOR THICKNESS
NDIV1=10                  ! # OF DIVISIONS IN RADIAL
                          ! DIRECTION
NDIV2=15                  ! # OF DIVISIONS IN HEIGHT
                          ! DIRECTION
P=1500                    ! INTERNAL PRESSURE
F=100E3                   ! TANGENTIAL FORCE
ET,1,281                  ! USE SHELL281 ELEMENT TYPE
KEYOPT,1,4,0              ! NO USER SUBROUTINE FOR ELEMENT CS
KEYOPT,1,8,2              ! STORE DATA FOR TOP, BOTTOM & MID
                          ! SURFACES
MP,EX,1,1E7               ! SPECIFY ELASTIC MODULUS FOR
                          ! ALUMINUM
MP,NUXY,1,0.3             ! SPECIFY POISSON'S RATIO FOR
                          ! ALUMINUM
SECT,1,SHELL              ! SPECIFY SECTION TYPE
SECDATA,T,1               ! SPECIFY THICKNESS FOR ALUMINUM
TB,MISO,1,1,20,           ! MULTILINEAR ISOTROPIC HARDENING
                          ! PLASTICITY
TBPT,,0.00300,30000       ! ENTER STRAIN VS STRESS DATA
                          ! POINTS
TBPT,,0.00350,33041       !
TBPT,,0.00400,34300       !
TBPT,,0.00450,35267       !
TBPT,,0.00500,36082       !
TBPT,,0.00550,36800       !
TBPT,,0.00600,37449       !
TBPT,,0.00650,38045       !
TBPT,,0.00700,38601       !
TBPT,,0.00750,39123       !
TBPT,,0.00800,39616       !
TBPT,,0.00850,40086       !
TBPT,,0.00875,40312       !
TBPT,,0.00900,40534       !
TBPT,,0.00925,40751       !
TBPT,,0.00950,40964       !
TBPT,,0.00975,41173       !
TBPT,,0.02400,49708       !
```

```
TBPT,,0.04000,56160         !
TBPT,,0.06300,63313         !
MP,EX,2,1E11                ! SPECIFY ELASTIC MODULUS FOR STIFF
                            ! PANEL
MP,NUXY,2,0                 ! SPECIFY POISSON'S RATIO FOR STIFF
                            ! PANEL
SECT,2,SHELL                ! SPECIFY SECTION TYPE
SECDATA,T,2                 ! SPECIFY THICKNESS FOR STIFF PANEL
CYLIND,R,0,0,H              ! CREATE CYLINDRICAL VOLUME
VDEL,ALL                    ! DELETE VOLUME (WITHOUT DELETING
                            ! AREAS)
ASEL,S,LOC,Z,0              ! SELECT AREAS AT Z = 0
ASEL,A,LOC,Z,H              ! ADD AREAS AT Z = H TO SELECTION
ADEL,ALL                    ! DELETE SELECTED AREAS
ALLSEL                      ! SELECT EVERYTHING
LSEL,S,LOC,Z,H/2            ! SELECT LINES AT Z = H/2
LESIZE,ALL,,,NDIV2          ! SPECIFY # OF DIVISIONS ON
                            ! SELECTED LINES
ALLSEL                      ! SELECT EVERYTHING
LESIZE,ALL,,,NDIV1          ! SPECIFY # OF DIVISIONS ON ALL
                            ! LINES
LSEL,S,LOC,Z,0              ! SELECT LINES AT Z = 0
LSEL,R,LOC,Y,0,R            ! RESELECT LINES BETWEEN Y = 0 AND
                            ! Y = R
LCCAT,ALL                   ! CONCATENATE LINES FOR MAPPED
                            ! MESHING
LSEL,S,LOC,Z,0              ! SELECT LINES AT Z = 0
LSEL,R,LOC,Y,0,-R           ! RESELECT LINES BETWEEN Y = 0 AND
                            ! Y = -R
LCCAT,ALL                   ! CONCATENATE LINES FOR MAPPED
                            ! MESHING
LSEL,S,LOC,Z,H              ! SELECT LINES AT Z = H
LSEL,R,LOC,Y,0,R            ! RESELECT LINES BETWEEN Y = 0 AND
                            ! Y = R
LCCAT,ALL                   ! CONCATENATE LINES FOR MAPPED
                            ! MESHING
LSEL,S,LOC,Z,H              ! SELECT LINES AT Z = H
LSEL,R,LOC,Y,0,-R           ! RESELECT LINES BETWEEN Y = 0 AND
                            ! Y = -R
LCCAT,ALL                   ! CONCATENATE LINES FOR MAPPED
                            ! MESHING
ALLSEL                      ! SELECT EVERYTHING
MSHKEY,1                    ! ENFORCE MAPPED MESHING
MAT,1                       ! SWITCH MATERIAL ATTRIBUTE TO
                            ! MAT # 1
SECNUM,1                    ! SWITCH SECTION ATTRIBUTE TO
                            ! SEC # 1
AMESH,ALL                   ! MESH ALL AREAS
WPOFFS,0,0,H                ! OFFSET WORKING PLANE
PCIRC,R,0,0,90              ! CREATE CIRCLE
PCIRC,1.5*R,R,0,90          ! CREATE HOLLOW CIRCLE
ASEL,S,LOC,Z,H              ! SELECT AREAS AT Z = H
AGLUE,ALL                   ! GLUE SELECTED AREAS
CSYS,1                      ! SWITCH TO GLOBAL CYLINDRICAL CS
```

```
LSLA,S                     !SELECT LINES ATTACHED TO SELECTED
                           !AREAS
LSEL,U,LOC,X,1.25*R        !UNSELECT LINES AT X = 1.25*R
LESIZE,ALL,,,NDIV1         !SPECIFY # OF DIVS ON SELECTED
                           !LINES
LSLA,S                     !SELECT LINES ATTACHED TO SELECTED
                           !AREAS
LSEL,R,LOC,X,1.25*R        !RESELECT LINES AT X = 1.25*R
LESIZE,ALL,,,NDIV1/2       !SPECIFY # OF DIVS ON SELECTED
                           !LINES
MAT,2                      !SWITCH MATERIAL ATTRIBUTE TO MAT
                           !# 2
SECNUM,2                   !SWITCH SECTION ATTRIBUTE TO SEC
                           !# 2
AMESH,ALL                  !MESH SELECTED AREAS
CSYS                       !SWITCH TO GLOBAL CARTESIAN CS
ARSYM,X,ALL                !REFLECT SELECTED AREAS ABOUT Y-Z
                           !PLANE
ARSYM,Y,ALL                !REFLECT SELECTED AREAS ABOUT X-Z
                           !PLANE
ALLSEL                     !SELECT EVERYTHING
NUMMRG,ALL                 !MERGE DUPLICATE ENTITIES
FINISH                     !EXIT PREPROCESSOR
```

Solution

The solution is obtained in two steps. First, the internal pressure is applied and the first load step file is generated (**LSWRITE** command). Then, the concentrated loads are applied while the internal pressure is still present, which constitutes the second load step file. The solution is obtained sequentially from these load step files using the **LSSOLVE** command.

```
/SOLU                      !ENTER SOLUTION PROCESSOR
NSEL,S,LOC,Z,0             !SELECT NODES AT Z = 0
D,ALL,ALL                  !CONSTRAIN ALL DOFS (DISPL &
                           !ROTATIONS)
ALLSEL                     !SELECT EVERYTHING
ESEL,S,MAT,,1              !SELECT ELEMENTS WITH MAT # 1
SFE,ALL,1,PRES,,P          !APPLY INTERNAL PRES. ON SELECTED
                           !ELEMS
CSYS,1                     !SWITCH TO GLOBAL CYLINDRICAL CS
ASEL,S,LOC,Z,H             !SELECT AREAS AT Z = H
ASEL,R,LOC,X,-R,R          !RESELECT AREAS BETWEEN X = -R &
                           !X = R
CSYS                       !SWITCH TO GLOBAL CARTESIAN CS
ESLA,S                     !SELECT ELEMENTS ATTACHED TO
                           !SELECTED AREAS
SFE,ALL,1,PRES,,P          !APPLY INTERNAL PRES. ON SELECTED
                           !ELEMS
ALLSEL                     !SELECT EVERYTHING
LSWRITE,1                  !WRITE LOAD STEP FILE # 1
N=NODE(1.5*R,0,H)          !STORE NODE # IN N FOR GIVEN
                           !COORDS.
```

```
F,N,FY,F                     !APPLY CONCENTRATED LOAD ON NODE N
N=NODE(-1.5*R,0,H)           !STORE NODE # IN N FOR GIVEN
                             !COORDS.
F,N,FY,-F                    !APPLY CONCENTRATED LOAD ON NODE N
N=NODE(0,1.5*R,H)            !STORE NODE # IN N FOR GIVEN
                             !COORDS.
F,N,FX,-F                    !APPLY CONCENTRATED LOAD ON NODE N
N=NODE(0,-1.5*R,H)           !STORE NODE # IN N FOR GIVEN
                             !COORDS.
F,N,FX,F                     !APPLY CONCENTRATED LOAD ON NODE N
ALLSEL                       !SELECT EVERYTHING
LSWRITE,2                    !WRITE LOAD STEP FILE # 1
LSSOLVE,1,2                  !OBTAIN SOLUTION FROM LOAD STEPS
                             !FILES
FINISH                       !EXIT SOLUTION PROCESSOR
```

Postprocessing
Once the solution is complete, the quantities of interest, i.e., radial, circumferential, and longitudinal displacements and circumferential plastic strains, are retrieved at a specific node and written to an external text file for both load steps. The channeling of data to the external text file, referred to as the *Command File*, is achieved by using the ***CFOPEN** command (detailed description is given in Sect. 11.5.2.2). After issuing the ***CFOPEN** command, quantities of interest are written to the file **PLASTIC OUT** using the ***VWRITE** command. The aforementioned quantities are reviewed at a single node, which is chosen along the middle of the cylinder surface in the z-direction (i.e., $z = H/2$). Plastic strains are calculated for elements. Therefore, plastic strains in the circumferential direction are stored in an element table using the **ETABLE** command. There are two elements attached to the selected single node. Therefore, the plastic strains at these two elements are averaged to find its value at the shared node. The results associated with specific load steps are read using the **SET** command.

```
/POST1                       !ENTER GENERAL POSTPROCESSOR
*CFOPEN,PLASTIC,OUT          !OPEN "COMMAND FILE"
CSYS,1                       !SWITCH TO GLOBAL CYLINDRICAL
                             !CS
RSYS,1                       !USE CYLINCRICAL CS FOR
                             !RESULTS
SET,1                        !READ RESULTS OF LOAD STEP 1
ETABLE,EPT,EPPL,Y            !STORE PLASTIC STRAINS IN
                             !THETA DIRECTION IN ELEMENT
                             !TABLE
NSEL,S,LOC,Z,H/2             !SELECT NODES AT Z = H/2
NSEL,R,LOC,Y,0               !RESELECT NODES AT THETA = 0
ESLN,S                       !SELECT ELEMENTS ATTACHED TO
                             !SELECTED NODE
*GET,E1,ELEM,0,NUM,MAX       !STORE MAX ELEM # IN PARAMETER
                             !E1
*GET,E2,ELEM,0,NUM,MIN       !STORE MIN ELEM # IN PARAMETER
                             !E2
```

```
*GET,EPT1,ETAB,1,ELEM,E1    ! STORE ELEM TABLE ITEMS FOR E1
*GET,EPT2,ETAB,1,ELEM,E2    ! & E2 IN PARAMETERS EPT1 &
                            ! EPT2
EPTAV=(EPT1+EPT2)/2         ! FIND AVERAGE OF EPT1 & EPT2
*GET,NODE1,NODE,0,NUM,MAX   ! RETRIEVE SELECTED NODE NUMBER
*GET,UR1,NODE,NODE1,U,X     ! RETRIEVE RADIAL DISPLACEMENT
*GET,UT1,NODE,NODE1,U,Y     ! RETRIEVE CIRCUMF.
                            ! DISPLACEMENT
*GET,UZ1,NODE,NODE1,U,Z     ! RETRIEVE Z-DISPLACEMENT
*VWRITE,'TIME = 1'          ! START WRITING TO COMMAND FILE
(A8)                        ! FORMAT STATEMENT
*VWRITE,'UR = ',UR1         ! WRITE RADIAL DISPLACEMENT
(A9,E14.5)                  ! FORMAT STATEMENT
*VWRITE,'UTHETA = ',UT1     ! WRITE CIRCUMF. DISPLACEMENT
(A9,E14.5)                  ! FORMAT STATEMENT
*VWRITE,'UZ = ',UZ1         ! WRITE Z-DISPLACEMENT
(A9,E14.5)                  ! FORMAT STATEMENT
*VWRITE,'EPLTH = ',EPTAV    ! WRITE AVERAGED PLASTIC STRAIN
(A9,E14.5)                  ! FORMAT STATEMENT
SET,NEXT                    ! READ RESULTS OF LOAD STEP 2
ETABLE,REFL                 ! UPDATE ELEMENT TABLE
*GET,EPT1,ETAB,1,ELEM,E1    ! STORE ELEM TABLE ITEMS FOR E1
*GET,EPT2,ETAB,1,ELEM,E2    ! & E2 IN PARAMETERS EPT1 &
                            ! EPT2
EPTAV=(EPT1+EPT2)/2         ! FIND AVERAGE OF EPT1 & EPT2
*GET,UR1,NODE,NODE1,U,X     ! RETRIEVE RADIAL DISPLACEMENT
*GET,UT1,NODE,NODE1,U,Y     ! RETRIEVE CIRCUMF.
                            ! DISPLACEMENT
*GET,UZ1,NODE,NODE1,U,Z     ! RETRIEVE Z-DISPLACEMENT
*VWRITE,'TIME = 2'          ! RESUME WRITING TO COMMAND
                            ! FILE
(A8)                        ! FORMAT STATEMENT
*VWRITE,'UR = ',UR1         ! WRITE RADIAL DISPLACEMENT
(A9,E14.5)                  ! FORMAT STATEMENT
*VWRITE,'UTHETA = ',UT1     ! WRITE CIRCUMF. DISPLACEMENT
(A9,E14.5)                  ! FORMAT STATEMENT
*VWRITE,'UZ = ',UZ1         ! WRITE Z-DISPLACEMENT
(A9,E14.5)                  ! FORMAT STATEMENT
*VWRITE,'EPLTH = ',EPTAV    ! WRITE AVERAGED PLASTIC STRAIN
(A9,E14.5)                  ! FORMAT STATEMENT
*CFCLOS                     ! CLOSE "COMMAND FILE"
FINISH                      ! EXIT GENERAL POSTPROCESSOR
```

After the command input segment given above is executed, the contents of the file PLASTIC OUT is given as

```
TIME = 1
 UR      =    0.52429E+00
 UTHETA  =    0.46792E-09
 UZ      =    0.60653E-01
 EPLTH   =    0.20618E-01
TIME = 2
 UR      =    0.62531E+00
 UTHETA  =    0.19920E+00
 UZ      =    0.64957E-01
 EPLTH   =    0.25581E-01
```

10.2.3 Stress Analysis of a Reinforced Viscoelastic Cylinder

Consider a hollow viscoelastic cylinder reinforced by an elastic material along the outer periphery subjected to internal pressure, as shown in Fig. 10.9. The cylinder is long in the out-of-plane direction. The inner and outer radii of the cylinder are $a = 2$ in. and $b = 4$ in., respectively; the thickness of the reinforcing layer is $h = 4/33$ in. The elastic modulus and Poisson's ratio of the reinforcement are $E_r = 3\times10^7$ psi and $\nu_r = 1/\sqrt{11}$.

The shear and bulk moduli of the viscoelastic cylinder behave as

$$G(t) = G_0 e^{-t} \tag{10.4a}$$

$$K(t) = K_\infty H(t) \tag{10.4b}$$

in which G_0 and K_∞ are defined as $G(0) = G_0 = E_c/(1+2\nu_c)$ and $K_\infty = E_c/[3(1-2\nu_c)]$. The elastic modulus and Poisson's ratio of the cylinder are $E_c = 10^5\, psi$ and $\nu_c = 1/3$.

Within ANSYS, the viscoelastic material behavior is specified as

$$G(t) = G_0\left[\frac{G_\infty}{G_0} + \frac{G_1}{G_0} e^{\left(-t/\tau^G\right)}\right] \tag{10.5a}$$

$$K(t) = K_0\left[\frac{K_\infty}{K_0} + \frac{K_1}{K_0} e^{\left(-t/\tau^K\right)}\right] \tag{10.5b}$$

In accordance with Eq. (10.4), the parameters in Eq. (10.5) take the following values: $G(0) = G_0 = G_1$, $\lim_{t\to\infty} G(t) = G_\infty = 0$, $K_0 = K_\infty$, and $K_1 = 0$. Therefore, the relative shear and bulk moduli, (G_1/G_0) and (K_1/K_0), respectively, are assigned the values of 1.0 and 0.0. The parameters τ^G and τ^K represent relative times for shear and bulk behavior, respectively; both of them are assumed to be 1.0. The goal is to find the time-dependent behavior of radial and hoop stresses.

Model Generation

The problem is solved using two-dimensional 8-noded **PLANE183** elements with the axisymmetric option. A Prony series representation of viscoelasticity is specified **TB** using and **TBDATA** commands. The model is a rectangular cross section of the cylinder and the reinforcement layer with a height of 1 in.

```
/PREP7                    ! ENTER PREPROCESSOR
ET,1,PLANE183,,,1         ! USE PLANE183 ELEM WITH
                          ! AXISYMMETRY
MP,EX,1,1.0E5             ! ELASTIC MODULUS OF MAT 1
MP,NUXY,1,1/3             ! POISSON'S RATIO OF MAT 1
TB,PRONY,1,,1,SHEAR       ! USE PRONY SERIES VISCOELASTICITY
                          ! FOR SHEAR
TBDATA,1,1.0,1.0          ! RELATIVE MODULUS AND TIME
TB,PRONY,1,,1,BULK        ! USE PRONY SERIES VISCOELASTICITY
                          ! FOR BULK
TBDATA,1,0.0,1.0          ! RELATIVE MODULUS AND TIME
MP,EX,2,3.0E7             ! ELASTIC MODULUS OF MAT 2
MP,NUXY,2,1/SQRT(11)      ! POISSON'S RATIO OF MAT 2
R1=2                      ! PARAMETER FOR INNER RADIUS OF
                          ! CYLINDER
R2=4                      ! PARAMETER FOR OUTER RADIUS OF
                          ! CYLINDER
R3=R2+4/33                ! PARAMETER FOR OUTER RADIUS OF
                          ! REINFORC.
K,1,R1                    ! CREATE KEYPOINTS
K,2,R2                    !
K,3,R3                    !
KGEN,2,1,3,1,,1           ! GENERATE KPS FROM EXISTING
                          ! PATTERN
L,1,2,5                   ! CREATE LINES
L,2,3,1                   !
A,4,1,2,5                 ! CREATE AREAS
A,5,2,3,6                 !
LESIZE,1,,,10             ! SPECIFY # OF LINE DIVISIONS
LESIZE,2,,,2              !
ESIZE,0.5                 ! SPECIFY ELEMENT SIZE
MSHKEY,1                  ! ENFORCE MAPPED MESHING
MSHAPE,0,2D               ! ENFORCE QUADRILATERAL ELEMENT
                          ! SHAPE
MAT,1                     ! SWITCH TO MATERIAL 1
```

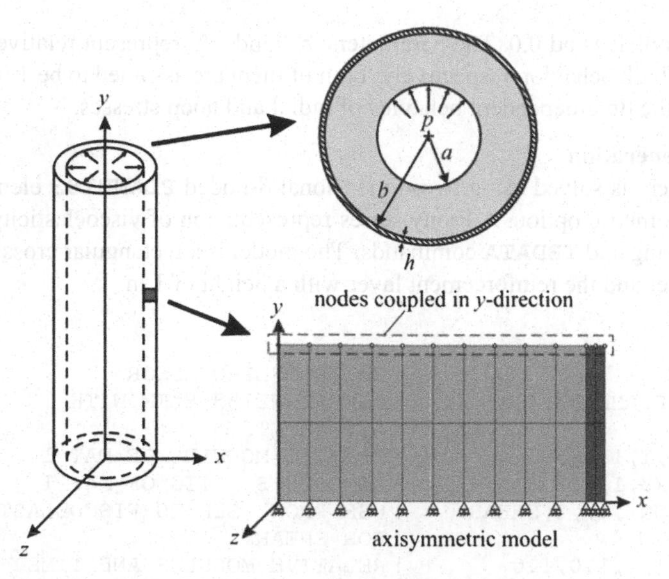

Fig. 10.9 Hollow viscoelastic cylinder with external reinforcement

```
AMESH,1                     ! MESH AREA 1
MAT,2                       ! SWITCH TO MATERIAL 2
AMESH,2                     ! MESH AREA 2
FINISH                      ! EXIT PREPROCESSOR
```

Solution

The solution is obtained in two load steps. The first load step, which spans an extremely small duration of 0.00001 s, is used to set up the initial conditions for the problem. As shown in Fig. 10.9, the nodes along the bottom row are constrained to move in the vertical direction (*y*-direction). In order to capture the plane strain characteristic of the problem, displacements are enforced to be the same on the top row of nodes. This is accomplished by coupling the degree of freedom in the *y*-direction of those nodes using the **CP** command. The solution controls option is turned off in this analysis, and values of the specific nonlinear solution controls items are specified. In order to reduce the size of the results file, only a limited amount of data is written (**OUTRES** command). The accuracy of the solution is improved by specifying a small convergence tolerance value using the **CNVTOL** command. Finally, the time step size is specified to be 0.1 s using the **DELTIM** command.

```
/SOLU                         ! ENTER SOLUTION PROCESSOR
 ANTYPE,STATIC                ! SPECIFY ANALYSIS TYPE
 NSEL,S,LOC,Y,0               ! SELECT NODES AT Y = 0
 D,ALL,UY                     ! CONSTRAIN Y-DISP AT SELECTED
                              ! NODES
 NSEL,S,LOC,Y,1               ! SELECT NODES AT Y = 1
 CP,1,UY,ALL                  ! COUPLE Y-DISP OF SELECTED NODES
 ALLSEL                       ! SELECT EVERYTHING
 NSEL,S,LOC,X,R1              ! SELECT NODES AT X = R1
 PI0 = 1                      ! PARAMETER FOR INNER PRESSURE
 SF,,PRES,PI0                 ! APPLY PRESSURE BC AT SELECTED
                              ! NODES
 ALLSEL                       ! SELECT EVERYTHING
 SOLCONTROL,0                 ! TURN SOLUTION CONTROLS OFF
 OUTRES,BASIC,ALL             ! SAVE BASIC OUTPUT AT EVERY
                              ! SUBSTEP
 CNVTOL,F,,,,1E-7             ! SMALL CONVERGENCE TOLER. ENFORCED
 TIME,0.00001                 ! TIME AT THE END OF 1ST LOAD STEP
 SOLVE                        ! OBTAIN SOLUTION FOR 1ST LOAD STEP
 TIME,10                      ! TIME AT THE END OF 2ND LOAD STEP
 DELTIM,0.1                   ! SPECIFY TIME STEP SIZE
 SOLVE                        ! OBTAIN SOLUTION FOR 2ND LOAD STEP
```

Postprocessing
Once the solution is obtained, a component of nodes (the center row of nodes) is created using the **CM** command. The radial and hoop stresses are written to an external file, which is opened using the ***CFOPEN** command and closed using the ***CFCLOSE** command. Between these two commands, desired quantities are written to the external file using the ***VWRITE** command followed by a format

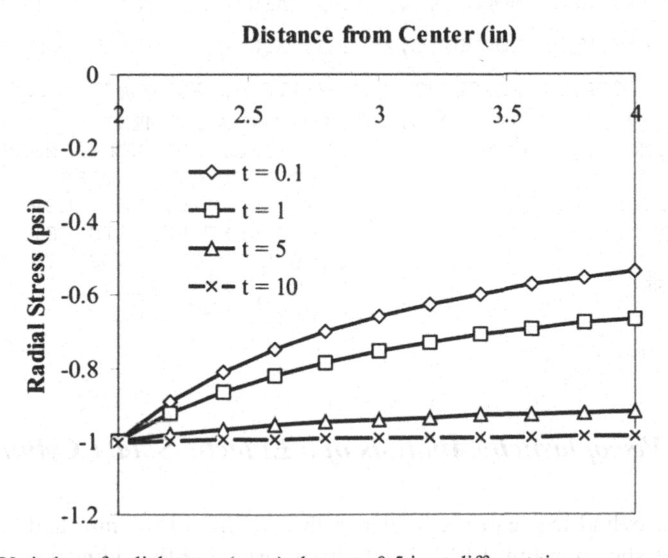

Fig. 10.10 Variation of radial stress (σ_{rr}) along y=0.5 in at different times

statement (written in FORTRAN syntax). Figs. 10.10 and 10.11 show the variation of radial (σ_{rr}) and hoop ($\sigma_{\theta\theta}$) stresses, respectively, along $y = 0.5$ in for times $t = 0.1, 1, 5$, and 10.

```
/POST1                            ! ENTER POSTPROCESSOR
ESEL,S,MAT,,1                     ! SELECT ELEMENTS WITH MAT 1
NSLE,S,CORNER                     ! SELECT CORNER NODES ATTACHED
                                  ! TO THE SELECTED ELEMENTS
NSEL,R,LOC,Y,0.5                  ! RESELECT NODES AT Y = 0.5
CM,NLIST,NODE                     ! CREATE A COMPONENT OF
                                  ! NODES NAMED "NLIST"
*CFOPEN,'STRS','OUT'              ! OPEN DATA FILE "STRS.OUT"
SET,FIRST                         ! READ RESULTS FOR 1ST LOAD
                                  ! STEP
*DO,J,1,1000                      ! LOOP OVER RESULTS SETS
ALLSEL                            ! SELECT EVERYTHING
*GET,TIM,ACTIVE,0,SET,TIME        ! OBTAIN TIME AT CURRENT
                                  ! RESULTS SET
*VWRITE,'********'                ! WRITE A SEPARATOR ROW TO
                                  ! FILE
(A8)                              ! FORMAT STATEMENT
*VWRITE,'TIME = ',TIM             ! WRITE CURRENT TIME TO FILE
(A7,E10.3)                        ! FORMAT STATEMENT
NSEL,S,NODE,,NLIST                ! SELECT COMPONENT "NLIST"
*GET,NCOUNT,NODE,0,COUNT          ! OBTAIN NUMBER OF NODES
*DO,I,1,NCOUNT,1                  ! LOOP OVER SELECTED NODES
NODNUM = NODE(0.0,0.0,0.0)        ! OBTAIN NODE # CLOSEST
                                  ! TO THE CENTER
LOCA=NX(NODNUM)                   ! OBTAIN X-COORD OF
                                  ! THE NODE "NODNUM"
*GET,SIGRR,NODE,NODNUM,S,X        ! OBTAIN σrr AT NODE "NODNUM"
*GET,SIGTT,NODE,NODNUM,S,Z        ! OBTAIN σθθ AT NODE "NODNUM"
*VWRITE,LOCA,SIGRR,SIGTT          ! WRITE σrr AND σθθ TO FILE
(E10.3,3X,E10.3,3X,E10.3)         ! FORMAT STATEMENT
NSEL,U,NODE,,NODNUM               ! UNSELECT THE NODE "NODNUM"
*ENDDO                            ! END LOOP OVER NODES
*IF,TIM,GE,10.0,*EXIT             ! IF TIM = 10, END LOOP
SET,NEXT                          ! READ NEXT RESULTS SET
*ENDDO                            ! END LOOP OVER RESULTS SETS
*CFCLOSE                          ! CLOSE DATA FILE
ALLSEL                            ! SELECT EVERYTHING
```

10.2.4 Viscoplasticity Analysis of a Eutectic Solder Cylinder

Consider a cylindrical eutectic solder with a radius of 10 mm and a height of 100 mm, as shown in Fig. 10.12. The bottom surface of the cylinder is constrained

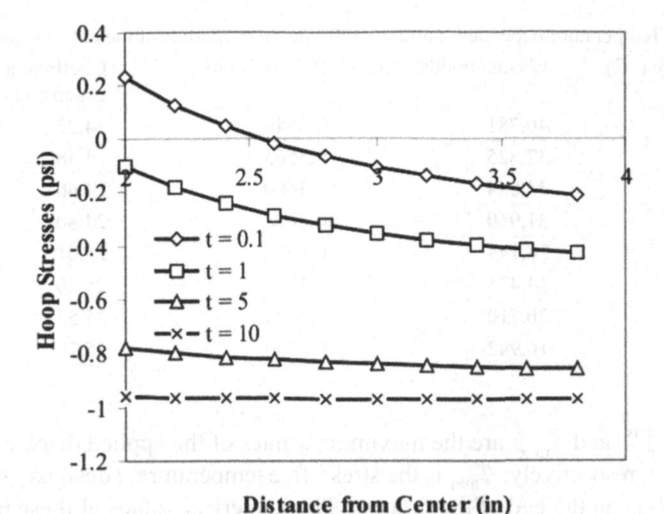

Fig. 10.11 Variation of hoop stresses ($\sigma_{\theta\theta}$) along $y=0.5$ in at different times

Fig. 10.12 Cylindrical eutectic solder

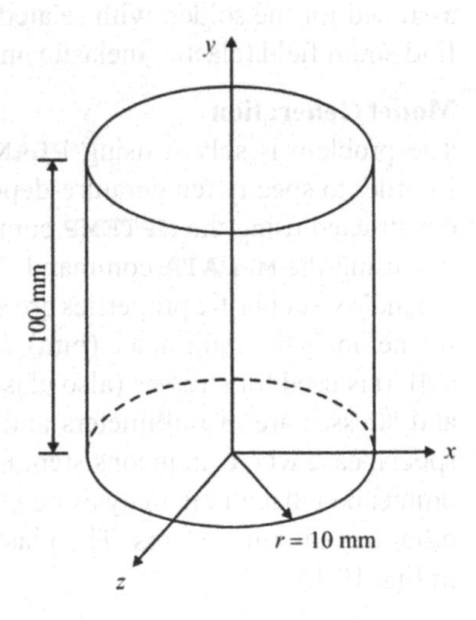

in all directions while the top surface is subjected to a prescribed displacement in the *y*-direction as a sinusoidal function of time. In addition, the eutectic solder is exposed to temperature, which exhibits the same time-dependent behavior as the prescribed displacement.

These loading conditions are given by

$$u_y(x, y=100, z, t) = u_y^{\max} \sin\left(\frac{\pi t}{t_{fin}}\right)$$

$$T(x, y, z, t) = T_{\min} + (T_{\max} - T_{\min}) \sin\left(\frac{\pi t}{t_{fin}}\right) \tag{10.6}$$

Table 10.2 Temperature-dependent variation of material properties of the eutectic solder

Temperature (°C)	Elastic modulus (MPa)	Poisson's ratio	Coefficient of thermal expansion (ppm/°C)
−35	40,781	0.3540	24.27
−15	37,825	0.3565	24.48
5	34,884	0.3600	24.66
25	31,910	0.3628	24.80
50	28,149	0.3650	25.01
75	24,425	0.3700	25.26
100	20,710	0.3774	25.52
125	16,942	0.3839	25.79

in which $u_y^{\max}$ and $T_{\max}$ are the maximum values of the applied displacement and temperature, respectively; $T_{\min}$ is the stress-free temperature; t designates time; and t_{fin} is the time at the end of the process. The numerical values of these parameters are given as $u_y^{\max} = 0.5$ mm, $T_{\min} = 0°$ C, $T_{\max} = 125°$ C, and $t_{fin} = 60$s . The material properties of the solder vary with temperature, as tabulated in Table 10.2. In addition to the elastic material properties, Anand's viscoplastic material behavior is assumed for the solder, with related parameters listed in Table 10.3. The goal is to find strain field (elastic, inelastic and total strains) at different times.

Model Generation

The problem is solved using **PLANE182** element with the axisymmetric option. In order to specify temperature-dependent elastic properties, a temperature table is constructed using the **MPTEMP** command, followed by the specification of properties using the **MPDATA** command. Note that temperatures are specified in Kelvin. Anand's viscoplastic properties are specified using the **TB** and **TBDATA** commands. In the analysis, millimeter (mm) is used for length dimensions and megapascal (MPa) is used for stresses (also elastic modulus). Thus, the resulting displacements and stresses are in millimeters and megapascals, respectively. However, this is a special case where an inconsistent unit system works correctly, and it is highly recommended that a unit analysis be performed before applying similar approaches to other mixed-unit systems. The plastic behavior at different temperatures is shown in Fig. 10.13.

Table 10.3 Numerical values of parameters used in Anand's material model for the eutectic solder

Parameter	Description	Value
S_0	Initial deformation resistance	12.41 (MPa)
Q/R	Ratio of activation energy to universal gas constant	9400 (1/°K)
A	Pre-exponential factor	$4\times10^6 (1/\text{sec})$
ξ	Stress multiplier	1.5
m	Strain rate sensitivity of stress	0.303
h_0	Hardening/softening constant	1379 (MPa)
$\hat{s}$	Coefficient for deformation resistance saturation value	13.79 (MPa)
n	Strain rate sensitivity of saturation value	0.07
a	Strain rate sensitivity of hardening or softening	1.3

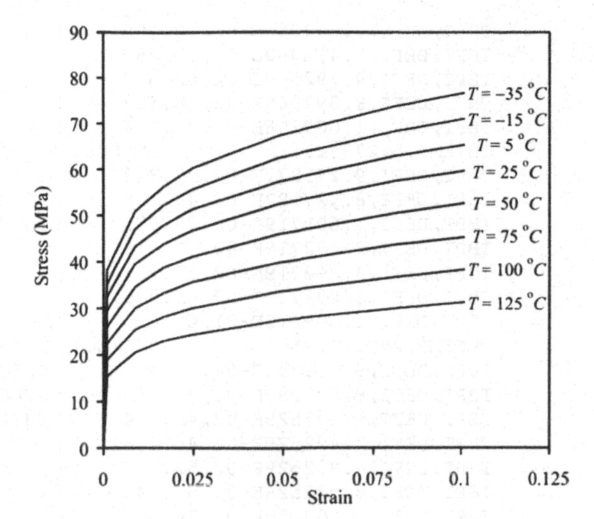

Fig. 10.13 Nonlinear stress-strain behavior of solder at different temperatures

```
/PREP7                        ! ENTER PREPROCESSOR
ET,1,PLANE182                 ! USE PLANE182 ELEMENT
KEYOPT,1,3,1                  ! SPECIFY AXISYMMETRY
MPTEMP                        ! INITIALIZE MATERIAL TABLE
                              ! SPECIFY TEMPERATURE POINTS
MPTEMP,1,238.15,258.15,278.15
MPTEMP,4,298.15,323.15,348.15,373.15,398.15
                              ! SPECIFY ELASTIC MODULUS
MPDATA,EX,1,1,40781,37825,34884
MPDATA,EX,1,4,31910,28149,24425,20710,16942
                              ! SPECIFY POISSON'S RATIO
MPDATA,NUXY,1,1,0.354,0.3565,0.36
MPDATA,NUXY,1,4,0.3628,0.365,0.37,0.3774,0.3839
                              ! SPECIFY COEFFICIENT OF THERMAL
                              ! EXPANSION
MPDATA,ALPX,1,1,2.427e-5,2.448e-5,2.466e-5
MPDATA,ALPX,1,4,2.48e-5,2.501e-5,2.526e-5,2.552e-5,
                                                     ➥2.579e-5
TB,RATE,1,1,9,9               ! SPECIFY ANAND'S VISCOPLASTICITY
                              ! SPECIFY ANAND'S PARAMETERS
TBDATA,,12.41,9400,4e6,1.5,0.303,1379
TBDATA,,13.79,0.07,1.3
TB,MISO,1,8,7,                   ! SPECIFY ISOTROPIC HARDENING
TBTEMP,-35+273.15                ! SPECIFY TEMPERATURE AS -35 C
TBPT,DEFI,9.313161E-04,37.980      ! SPECIFY STRAIN VS
TBPT,DEFI,8.931316E-03,50.677      ! STRESS DATA POINTS AT
TBPT,DEFI,1.693132E-02,56.307      ! THIS TEMPERATURE
TBPT,DEFI,2.493132E-02,60.142      !
TBPT,DEFI,4.893132E-02,67.620      !
TBPT,DEFI,8.093132E-02,73.889      !
TBPT,DEFI,1.009313E-01,76.836      !
TBTEMP,-15+273.15                ! SPECIFY TEMPERATURE AS -15 C
TBPT,DEFI,9.295968E-04,35.162      ! SPECIFY STRAIN VS
TBPT,DEFI,8.929597E-03,46.917      ! STRESS DATA POINTS AT
```

```
TBPT,DEFI,1.692960E-02,52.129     !THIS TEMPERATURE
TBPT,DEFI,2.492960E-02,55.680     !
TBPT,DEFI,4.892960E-02,62.603     !
TBPT,DEFI,8.092960E-02,68.407     !
TBPT,DEFI,1.009296E-01,71.136     !
TBTEMP,5+273.15                 !SPECIFY TEMPERATURE AS 5 C
TBPT,DEFI,9.271872E-04,32.344     !SPECIFY STRAIN VS
TBPT,DEFI,8.927187E-03,43.158     !STRESS DATA POINTS AT
TBPT,DEFI,1.692719E-02,47.952     !THIS TEMPERATURE
TBPT,DEFI,2.492719E-02,51.218     !
TBPT,DEFI,4.892719E-02,57.586     !
TBPT,DEFI,8.092719E-02,62.925     !
TBPT,DEFI,1.009272E-01,65.435     !
TBTEMP,25+273.15                !SPECIFY TEMPERATURE AS 25 C
TBPT,DEFI,9.252899E-04,29.526     !SPECIFY STRAIN VS
TBPT,DEFI,8.925290E-03,39.398     !STRESS DATA POINTS AT
TBPT,DEFI,1.692529E-02,43.774     !THIS TEMPERATURE
TBPT,DEFI,2.492529E-02,46.756     !
TBPT,DEFI,4.892529E-02,52.569     !
TBPT,DEFI,8.092529E-02,57.443     !
TBPT,DEFI,1.009253E-01,59.734     !
TBTEMP,50+273.15                !SPECIFY TEMPERATURE AS 50 C
TBPT,DEFI,9.237984E-04,26.004     !SPECIFY STRAIN VS
TBPT,DEFI,8.923798E-03,34.698     !STRESS DATA POINTS AT
TBPT,DEFI,1.692380E-02,38.552     !THIS TEMPERATURE
TBPT,DEFI,2.492380E-02,41.178     !
TBPT,DEFI,4.892380E-02,46.298     !
TBPT,DEFI,8.092380E-02,50.590     !
TBPT,DEFI,1.009238E-01,52.608     !
TBTEMP,75+273.15                !SPECIFY TEMPERATURE AS 75 C
TBPT,DEFI,9.204504E-04,22.482     !SPECIFY STRAIN VS
TBPT,DEFI,8.920450E-03,29.998     !STRESS DATA POINTS AT
TBPT,DEFI,1.692045E-02,33.330     !THIS TEMPERATURE
TBPT,DEFI,2.492045E-02,35.601     !
TBPT,DEFI,4.892045E-02,40.027     !
TBPT,DEFI,8.092045E-02,43.738     !
TBPT,DEFI,1.009205E-01,45.482     !
TBTEMP,100+273.15               !SPECIFY TEMPERATURE AS 100 C
TBPT,DEFI,9.154515E-04,18.959     !SPECIFY STRAIN VS
TBPT,DEFI,8.915451E-03,25.298     !STRESS DATA POINTS AT
TBPT,DEFI,1.691545E-02,28.108     !THIS TEMPERATURE
TBPT,DEFI,2.491545E-02,30.023     !
TBPT,DEFI,4.891545E-02,33.756     !
TBPT,DEFI,8.091545E-02,36.885     !
TBPT,DEFI,1.009155E-01,38.357     !
TBTEMP,125+273.15               !SPECIFY TEMPERATURE AS 125 C
TBPT,DEFI,9.111675E-04,15.437     !SPECIFY STRAIN VS
TBPT,DEFI,8.911168E-03,20.598     !STRESS DATA POINTS AT
TBPT,DEFI,1.691117E-02,22.886     !THIS TEMPERATURE
TBPT,DEFI,2.491117E-02,24.445     !
TBPT,DEFI,4.891117E-02,27.485     !
TBPT,DEFI,8.091117E-02,30.033     !
TBPT,DEFI,1.009112E-01,31.231     !
RECTNG,0,10,0,100           !CREATE RECTANGLE
ESIZE,2                     !SPECIFY ELEMENT SIZE
```

```
MSHKEY,1                            ! ENFORCE MAPPED MESHING
AMESH,ALL                           ! MESH AREA
FINISH                              ! EXIT PREPROCESSOR
```

Solution

The solution is obtained in twenty load steps. Nonlinear geometry effects must be turned on when using **PLANE182** elements. Automatic time stepping is turned on (**AUTOTS** command) so that ANSYS can adjust the time step size values for the substeps within each load step. However, the starting time step size and the minimum and maximum time step size to be used in the analysis are specified using the **DELTIM** command. A do loop is utilized for creating load step files, in which the applied displacement and temperature conditions are calculated and written to load step files. Finally, the solution is obtained using the **LSSOLVE** command.

```
/SOLU                               ! ENTER SOLUTION PROCESSOR
TMIN=273.15                         ! PARAMETER FOR MIN
                                    ! TEMPERATURE
TMAX=398.15                         ! PARAMETER FOR MAX
                                    ! TEMPERATURE
DELT=TMAX-TMIN                      ! TEMPERATURE DIFFERENCE
VMAX=0.5                            ! MAXIMUM APPLIED
                                    ! DISPLACEMENT
NLS=20                              ! NUMBER OF LOAD STEPS
TMX=60                              ! FINAL TIME (IN SECONDS)
PI=4*ATAN(1)                        ! PARAMETER FOR π
ANTYPE,STATIC                       ! DECLARE ANALYSIS TYPE
NLGEOM,ON                           ! TURN NONLINEAR GEOMETRY ON
KBC,0                               ! APPLY RAMPED LOADING
AUTOTS,ON                           ! TURN AUTOMATIC TIME
                                    ! STEPPING ON
TREF,273.15                         ! STRESS-FREE TEMPERATURE
DELTIM,1E-2,1E-3,0.3                ! SPECIFY TIME STEP SIZE
NSEL,S,LOC,Y,0                      ! SELECT NODES AT Y = 0
D,ALL,ALL                           ! CONSTRAIN ALL DOFS
ALLSEL                              ! SELECT EVERYTHING
TIM=TMX/NLS                         ! FIND TIME AT CURRENT LOAD
                                    ! STEP
*DO,I,1,NLS                         ! BEGIN DO LOOP ON LOAD STEPS
TIME,TIM                            ! SPECIFY TIME AT END OF LOAD
                                    ! STEP
VC=VMAX*SIN(PI*TIM/TMX)             ! FIND CURRENT DISPLACEMENT
                                    ! BC
TC=TMIN+DELT*SIN(PI*TIM/TMX)        ! FIND CURRENT TEMPERATURE
NSEL,S,LOC,Y,100                    ! SELECT NODES AT Y = 100
D,ALL,UY,VC                         ! APPLY CURRENT DISPLACEMENT
                                    ! BC
ALLSEL                              ! SELECT EVERYTHING
BFUNIF,TEMP,TC                      ! APPLY CURRENT TEMPERATURE
                                    ! LOAD
LSWRITE,I                           ! WRITE LOAD STEP FILE
```

```
TIM=TIM+TMX/NLS                 ! UPDATE TIME AT END OF LOAD
                                ! STEP
*ENDDO                          ! END DO LOOP ON LOAD STEPS
LSSOLVE,1,NLS                   ! SOLVE FROM LOAD STEP FILES
FINISH                          ! EXIT SOLUTION PROCESSOR
```

Postprocessing
Once the solution is obtained, the user can review a multitude of results items, including elastic, plastic, and total strain components and stress components. Also available is the plastic work. The command input segment given below obtains contour plots for a few of the items mentioned above, in two different load steps. The results for different load steps are read using the **SET** command. For clarity in the contour plots, the triad and the symbols for minimum and maximum quantities are turned off using the **/TRAID** and **/PLOPTS** commands. The two load steps considered here are the 10th and the last load steps. The 10th load step corresponds to the maximum displacement and temperature conditions while the last load step brings the structure to its initial configuration. Although initially stress free, after applying time-dependent displacement and temperature loads and returning to a no-load state, the structure experiences residual stresses and strains. Figs. 10.14 and 10.15 show the contour plots of plastic work at load steps 10 and 20, respectively.

```
/POST1                    ! ENTER POSTPROCESSOR
/TRIAD,OFF                ! TURN THE TRIAD OFF
/PLOPTS,MINM,0            ! TURN MIN & MAX SYMBOLS OFF
SET,10                    ! READ RESULTS AT LOAD STEP 10
PLNSOL,S,EQV              ! PLOT EQUIVALENT STRESS CONTOURS
PLNSOL,EPPL,Y             ! PLOT CONTOURS FOR PLASTIC STRAIN
                          ! IN Y-DIR.
PLNSOL,NL,PLWK            ! PLOT PLASTIC WORK CONTOURS
SET,20                    ! READ RESULTS AT LOAD STEP 20
PLNSOL,S,EQV              ! PLOT EQUIVALENT STRESS CONTOURS
PLNSOL,EPPL,Y             ! PLOT CONTOURS FOR PLASTIC STRAIN
                          ! IN Y-DIR.
PLNSOL,NL,PLWK            ! PLOT PLASTIC WORK CONTOURS
```

10.2.5 Combined Plasticity and Creep

In this subsection, the cylindrical eutectic solder column considered in the previous subsection is reconsidered, this time with nonlinear material properties by means of rate-independent plasticity combined with creep (Fig. 10.12). Both top and bottom surfaces of the cylinder are constrained in all directions. Temperature within the cylinder varies as a sinusoidal function of time as given by

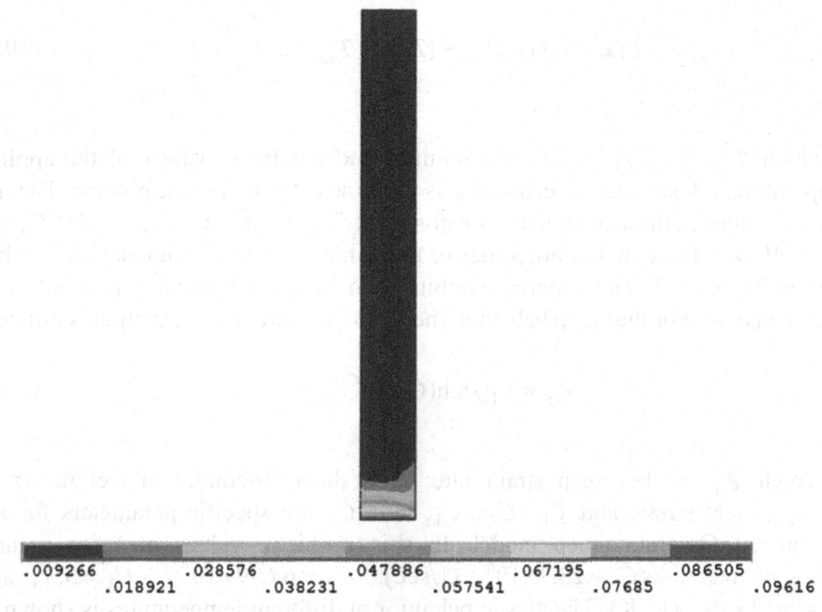

Fig. 10.14 Plastic work contours at load step 10 (t=30 s)

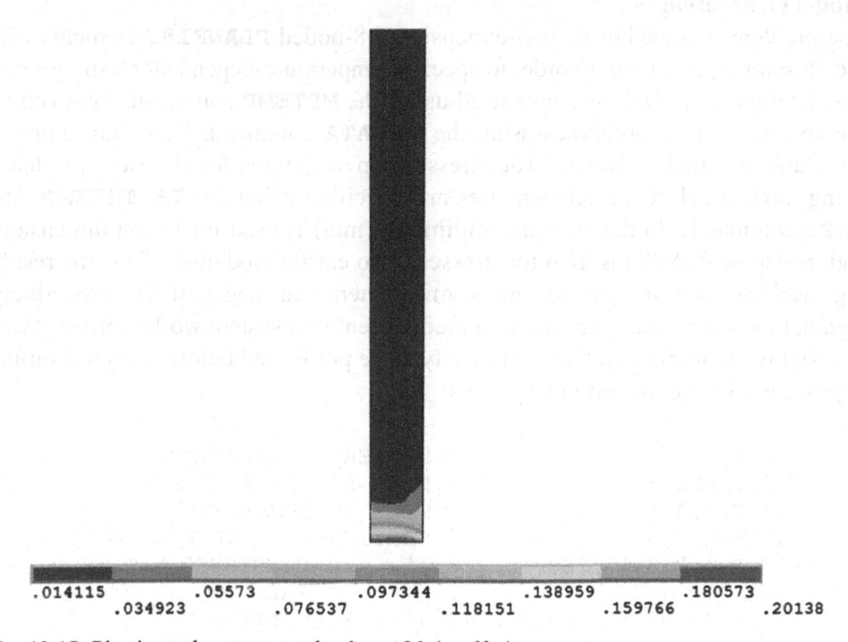

Fig. 10.15 Plastic work contours at load step 20 (t=60 s)

$$T(x,y,z,t) = T_{\min} + (T_{\max} - T_{\min})\sin\left(\frac{\pi t}{t_{fin}}\right) \quad (10.7)$$

in which $T_{\min}$ and $T_{\max}$ are the minimum and maximum values of the applied temperature, t designates time, and t_{fin} is the time at the end of the process. The numerical values of these parameters are given as: $T_{\min} = -35\ °C$, $T_{\max} = 125\ °C$, and $t_{fin} = 60$ sec. The material properties of the solder vary with temperature, as tabulated in Table 10.2. The material exhibits both isotropic hardening plasticity and creep behavior. For the creep behavior, the following strain rate equation is utilized:

$$\dot{\varepsilon}_{cr} = C_1[\sinh(C_2\sigma)]^{C_3} e^{-C_4/T} \quad (10.8)$$

in which $\dot{\varepsilon}_{cr}$ is the creep strain rate, T is the temperature in Kelvin, σ is the equivalent stress, and C_1, C_2, C_3, and C_4 are specific parameters for the generalized Garofalo creep model. In this problem, values of these parameters are taken as $C_1 = 23.3\times10^6$ (1/sec), $C_2 = 6.699\times10^{-2}$, $C_3 = 3.3$, and $C_4 = 8.12\times10^3$ (1/°K). The plastic behavior at different temperatures is shown in Fig. 10.13. The goal is to find strain fields, including plastic and creep strains, at different times.

Model Generation

The problem is solved using two-dimensional 8-noded **PLANE183** elements with the axisymmetric option. In order to specify temperature-dependent elastic properties, a temperature table is constructed using the **MPTEMP** command, followed by the specification of properties using the **MPDATA** command. Note that temperatures are specified in Kelvin. The stress-strain variations for the isotropic hardening model at different temperatures are specified using the **TB**, **TBTEMP**, and **TBPT** commands. In the analysis, millimeter (mm) is used for length dimensions and megapascal (MPa) is used for stresses (also elastic modulus). Thus, the resulting displacements and stresses are in millimeters and megapascals, respectively. Again, this is a special case where an inconsistent unit system works correctly, and it is highly recommended that a unit analysis be performed before applying similar approaches to other mixed-unit systems.

```
/PREP7                              ! ENTER PREPROCESSOR
ET,1,PLANE183                       ! USE PLANE183 ELEMENT
KEYOPT,1,3,1                        ! USE AXISYMMETRY
MPTEMP                              ! INITIALIZE MAT TEMP TABLE
MPTEMP,1,238.15,258.15,278.15          ! CONSTRUCT MAT TEMP
                                       ! TABLE
MPTEMP,4,298.15,323.15,348.15,373.15,398.15
                                    ! SPECIFY TEMPERATURE
                                    ! DEPENDENT ELASTIC MODULUS
MPDATA,EX,1,1,40781,37825,34884
```

```
MPDATA,EX,1,4,31910,28149,24425,20710,16942
                              ! SPECIFY TEMPERATURE
                              ! DEPENDENT POISSON'S RATIO
MPDATA,NUXY,1,1,0.354,0.3565,0.36
MPDATA,NUXY,1,4,0.3628,0.365,0.37,0.3774,0.3839
                              ! SPECIFY TEMPERATURE
                              ! DEPENDENT CTE
MPDATA,ALPX,1,1,2.427E-5,2.448E-5,2.466E-5
MPDATA,ALPX,1,4,2.48E-5,2.501E-5,2.526E-5,
                                      ➥2.552E-5,2.579E-5
MPTEMP                        ! INITIALIZE MAT TEMP TABLE
TB,MISO,1,8,7,                ! SPECIFY ISOTROPIC HARDENING
TBTEMP,-35+273.15             ! SPECIFY TEMPERATURE AS -35 C
TBPT,DEFI,9.313161E-04,37.980     ! SPECIFY STRAIN VS
TBPT,DEFI,8.931316E-03,50.677     ! STRESS DATA POINTS AT
TBPT,DEFI,1.693132E-02,56.307     ! THIS TEMPERATURE
TBPT,DEFI,2.493132E-02,60.142     !
TBPT,DEFI,4.893132E-02,67.620     !
TBPT,DEFI,8.093132E-02,73.889     !
TBPT,DEFI,1.009313E-01,76.836     !
TBTEMP,-15+273.15             ! SPECIFY TEMPERATURE AS -15 C
TBPT,DEFI,9.295968E-04,35.162     ! SPECIFY STRAIN VS
TBPT,DEFI,8.929597E-03,46.917     ! STRESS DATA POINTS AT
TBPT,DEFI,1.692960E-02,52.129     ! THIS TEMPERATURE
TBPT,DEFI,2.492960E-02,55.680     !
TBPT,DEFI,4.892960E-02,62.603     !
TBPT,DEFI,8.092960E-02,68.407     !
TBPT,DEFI,1.009296E-01,71.136     !
TBTEMP,5+273.15               ! SPECIFY TEMPERATURE AS 5 C
TBPT,DEFI,9.271872E-04,32.344     ! SPECIFY STRAIN VS
TBPT,DEFI,8.927187E-03,43.158     ! STRESS DATA POINTS AT
TBPT,DEFI,1.692719E-02,47.952     ! THIS TEMPERATURE
TBPT,DEFI,2.492719E-02,51.218     !
TBPT,DEFI,4.892719E-02,57.586     !
TBPT,DEFI,8.092719E-02,62.925     !
TBPT,DEFI,1.009272E-01,65.435     !
TBTEMP,25+273.15              ! SPECIFY TEMPERATURE AS 25 C
TBPT,DEFI,9.252899E-04,29.526     ! SPECIFY STRAIN VS
TBPT,DEFI,8.925290E-03,39.398     ! STRESS DATA POINTS AT
TBPT,DEFI,1.692529E-02,43.774     ! THIS TEMPERATURE
TBPT,DEFI,2.492529E-02,46.756     !
TBPT,DEFI,4.892529E-02,52.569     !
TBPT,DEFI,8.092529E-02,57.443     !
TBPT,DEFI,1.009253E-01,59.734     !
TBTEMP,50+273.15              ! SPECIFY TEMPERATURE AS 50 C
TBPT,DEFI,9.237984E-04,26.004     ! SPECIFY STRAIN VS
TBPT,DEFI,8.923798E-03,34.698     ! STRESS DATA POINTS AT
TBPT,DEFI,1.692380E-02,38.552     ! THIS TEMPERATURE
TBPT,DEFI,2.492380E-02,41.178     !
TBPT,DEFI,4.892380E-02,46.298     !
TBPT,DEFI,8.092380E-02,50.590     !
TBPT,DEFI,1.009238E-01,52.608     !
TBTEMP,75+273.15              ! SPECIFY TEMPERATURE AS 75 C
TBPT,DEFI,9.204504E-04,22.482     ! SPECIFY STRAIN VS
TBPT,DEFI,8.920450E-03,29.998     ! STRESS DATA POINTS AT
TBPT,DEFI,1.692045E-02,33.330     ! THIS TEMPERATURE
```

```
TBPT,DEFI,2.492045E-02,35.601     !
TBPT,DEFI,4.892045E-02,40.027     !
TBPT,DEFI,8.092045E-02,43.738     !
TBPT,DEFI,1.009205E-01,45.482     !
TBTEMP,100+273.15              !SPECIFY TEMPERATURE AS 100 C
TBPT,DEFI,9.154515E-04,18.959     !SPECIFY STRAIN VS
TBPT,DEFI,8.915451E-03,25.298     !STRESS DATA POINTS AT
TBPT,DEFI,1.691545E-02,28.108     !THIS TEMPERATURE
TBPT,DEFI,2.491545E-02,30.023     !
TBPT,DEFI,4.891545E-02,33.756     !
TBPT,DEFI,8.091545E-02,36.885     !
TBPT,DEFI,1.009155E-01,38.357     !
TBTEMP,125+273.15              !SPECIFY TEMPERATURE AS 125 C
TBPT,DEFI,9.111675E-04,15.437     !SPECIFY STRAIN VS
TBPT,DEFI,8.911168E-03,20.598     !STRESS DATA POINTS AT
TBPT,DEFI,1.691117E-02,22.886     !THIS TEMPERATURE
TBPT,DEFI,2.491117E-02,24.445     !
TBPT,DEFI,4.891117E-02,27.485     !
TBPT,DEFI,8.091117E-02,30.033     !
TBPT,DEFI,1.009112E-01,31.231     !
C1 = 23.3E6                    !DEFINE CREEP MODEL
C2 = 6.6990E-2                 !PARAMETERS
C3 = 3.300                     !
C4 = 6.7515E4/8.314            !
TB,CREEP,1,,,8                 !SPECIFY GENERALIZED
                               !GAROFALO CREEP MODEL
TBDATA,1,C1,C2,C3,C4           !SPECIFY CREEP MODEL PAR.
RATE,1                         !USE CREEP STRAIN RATE EFFECT
RECTNG,0,10,0,100              !CREATE RECTANGLE
ESIZE,2                        !SPECIFY ELEMENT SIZE
MSHKEY,1                       !ENFORCE MAPPED MESHING
AMESH,ALL                      !MESH AREA
FINISH                         !EXIT PREPROCESSOR
```

Solution

The solution is obtained in twenty load steps. Nonlinear geometry effects are turned on using the **NLGEOM** command. Automatic time stepping is turned on (**AUTOTS** command) so that ANSYS can adjust the time step size values for the substeps within each load step. The loads (temperature change) are interpolated for each substep from the values of the previous load step to the values of the current load step using the **KBC** command. A do loop is utilized for creating load step files, in which the temperature conditions are calculated and applied (**BFUNIF** command) before they are written to load step files (**LSWRITE** command). Finally, the solution is obtained using the **LSSOLVE** command.

```
/SOLU                    !ENTER SOLUTION PROCESSOR
TMIN=238.15              !DEFINE PARAMETER FOR MIN
                         !TEMPERATURE
TMAX=398.15              !DEFINE PARAMETER FOR MAX
                         !TEMPERATURE
```

```
NLS=20                       ! DEFINE PARAMETER FOR # OF LOAD
                             ! STEPS
TMX=60                       ! DEFINE PARAMETER FOR FINAL TIME
PI=4*ATAN(1)                 ! DEFINE PARAMETER FOR π
ANTYPE,STATIC                ! SPECIFY ANALYSIS TYPE AS STATIC
NLGEOM,ON                    ! TURN NONLINEAR GEOMETRY EFFECTS
KBC,0                        ! ENFORCE RAMPED LOADING
AUTOTS,ON                    ! TURN AUTOMATIC TIME STEPPING ON
TREF,273.15                  ! SPECIFY STRESS-FREE TEMPERATURE
DELTIM,1E-2,1E-3,3           ! SPECIFY TIME STEP SIZE PARAMETERS
NSEL,S,LOC,Y,0               ! SELECT NODES AT Y = 0
D,ALL,ALL                    ! CONSTRAIN ALL DOFS
ALLSEL                       ! SELECT EVERYTHING
NSEL,S,LOC,Y,100             ! SELECT NODES AT Y = 100
D,ALL,ALL                    ! CONSTRAIN ALL DOFS
ALLSEL                       ! SELECT EVERYTHING
TIM=TMX/NLS                  ! FIND END TIME FOR CURRENT LOAD
                             ! STEP
*DO,I,1,NLS                  ! LOOP OVER LOAD STEPS
TIME,TIM                     ! SPECIFY TIME FOR CURRENT LOAD
                             ! STEP
TC=TMIN+(TMAX-TMIN)*SIN(PI*TIM/TMX)      ! FIND TEMP FOR
                                         ! CURRENT LS
BFUNIF,TEMP,TC               ! SPECIFY TEMPERATURE FOR
                             ! CURRENT LS
LSWRITE,I                    ! WRITE LOAD STEP FILE
TIM=TIM+TMX/NLS              ! FIND END TIME FOR NEXT LOAD STEP
*ENDDO                       ! END LOOP OVER LOAD STEPS
LSSOLVE,1,NLS                ! SOLVE FROM LOAD STEPS
FINISH                       ! EXIT SOLUTION PROCESSOR
```

Postprocessing

Once the solution is obtained, the user can review a multitude of results items, including elastic, plastic, creep, and total strain components, as well as the stress components. The command input segment given below obtains contour plots for a few of the items mentioned above, in two different load steps. The results for different load steps are read using the **SET** command. For clarity in the contour plots, the triad and the symbols for minimum and maximum quantities are turned off. Using the **/TRIAD** and **/PLOPTS** commands. The two load steps considered here are the 10th and the last load steps. The 10th load step corresponds to the maximum temperature condition while the last load step brings the structure to its initial configuration. Although initially stress free, after applying a time-dependent temperature load and returning to the initial temperature, the structure experiences residual stresses and strains. Figure 10.16 shows the contour plot of equivalent stress at load step 20, which exhibits a significant amount of stress. Figures 10.17 and 10.18 show contour plots of plastic and creep strains at load step 20, respectively.

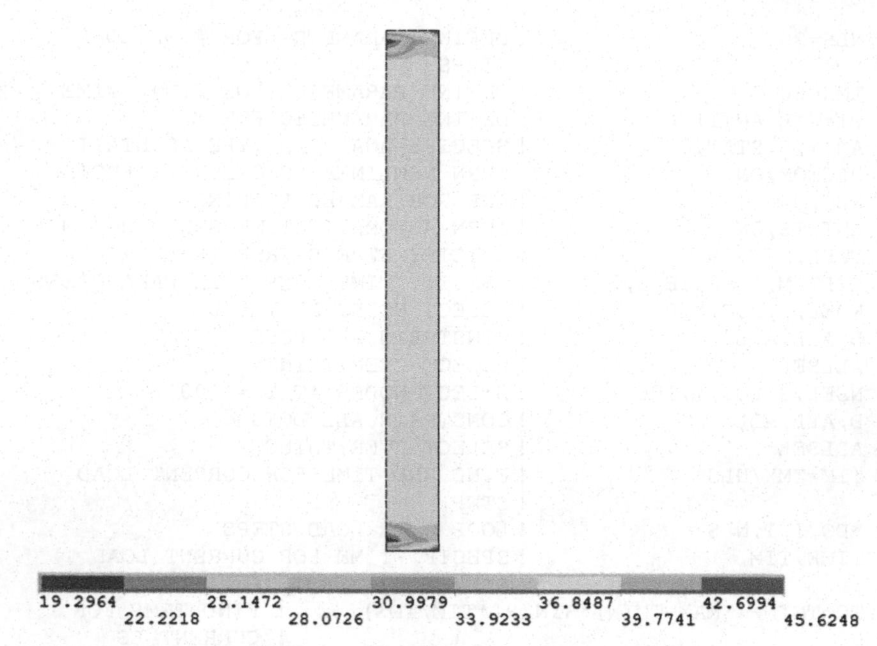

Fig. 10.16 Contours of residual equivalent stress at load step 20 (t=60 s)

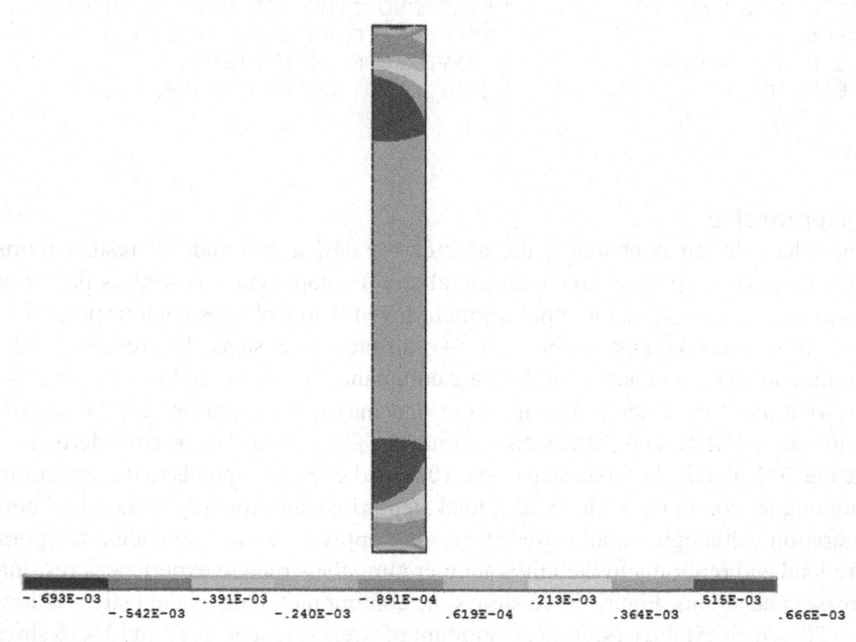

Fig. 10.17 Contours of plastic strain in the y-direction at load step 20 (t=60 s)

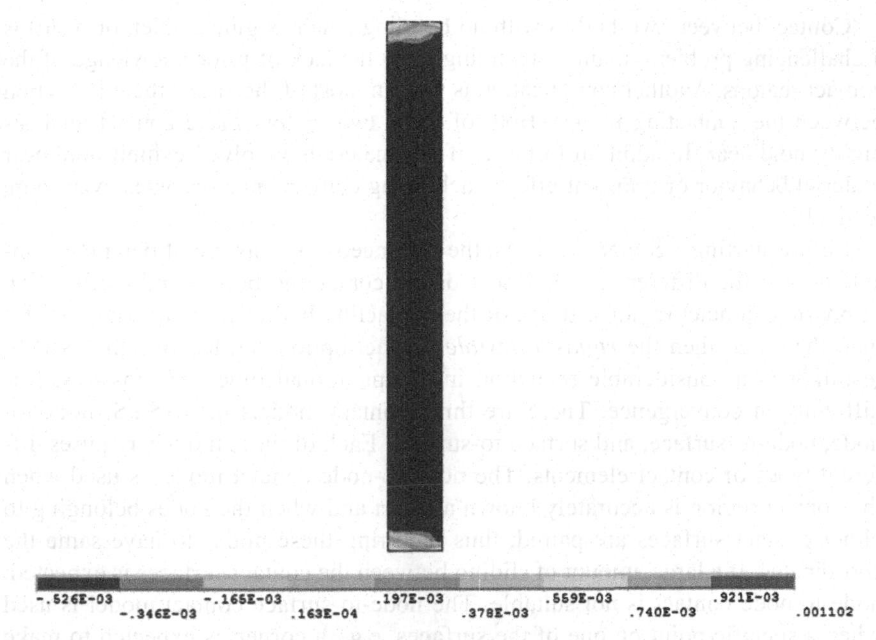

Fig. 10.18 Contours of creep strain in the y-direction at load step 20 (t=60 s)

```
/POST1                    ! ENTER GENERAL POSTPROCESSOR
/TRIAD,OFF                ! TURN THE TRIAD OFF
/PLOPTS,MINM,0            ! TURN MIN & MAX SYMBOLS OFF
SET,10                    ! READ RESULTS AT LOAD STEP 10
PLNSOL,NL,PLWK            ! PLOT PLASTIC WORK CONTOURS
PLNSOL,S,EQV              ! PLOT EQUIVALENT STRESS CONTOURS
PLNSOL,EPPL,Y             ! PLOT CONTOURS FOR PLASTIC STRAIN
                          ! IN Y-DIR.
PLNSOL,EPCR,Y             ! PLOT CONTOURS FOR CREEP STRAIN IN
                          ! Y-DIR.
SET,20                    ! READ RESULTS AT LOAD STEP 20
PLNSOL,NL,PLWK            ! PLOT PLASTIC WORK CONTOURS
PLNSOL,S,EQV              ! PLOT EQUIVALENT STRESS CONTOURS
PLNSOL,EPPL,Y             ! PLOT CONTOURS FOR PLASTIC STRAIN
                          ! IN Y-DIR.
PLNSOL,EPCR,Y             ! PLOT CONTOURS FOR CREEP STRAIN IN
                          ! Y-DIR.
```

10.3 Contact

Nonlinearity due to contact conditions arises because the prescribed displacements on the boundary depend on the deformation of the structure. Furthermore, no-interpenetration conditions are enforced while the extent of the contact area is unknown.

Contact between two bodies with no bonding (such as glue, solder, or weld) is a challenging problem, mainly stemming from the lack of prior knowledge of the contact regions. Another complication is that, in most of the cases, there is friction between the contacting bodies. Both of these two factors make contact analysis highly nonlinear. In addition to these, if the materials involved exhibit nonlinear material behavior or transient effects, achieving convergence becomes even more difficult.

Before starting a contact analysis, the user needs to be aware of two main considerations: the difference in stiffness of the contacting bodies and the location of possible contact regions. If one of the contacting bodies is significantly stiffer than the other, then the *rigid-to-flexible* contact option can be used in ANSYS, resulting in a considerable reduction in computational time and, possibly, less difficulty in convergence. There are three contact models in ANSYS: node-to-node, node-to-surface, and surface-to-surface. Each of these models requires different types of contact elements. The node-to-node contact model is used when the contact region is accurately known a priori and when the nodes belonging to either contact surfaces are paired, thus requiring these nodes to have same the coordinates. If a large amount of sliding between the contact surfaces is expected, node-to-node contact is not suitable. The node-to-surface contact model is used when a specific point on one of the surfaces, e.g., a corner, is expected to make contact with a rather smooth surface. This model does not require accurate a priori knowledge of the contact region, and the mesh pattern on either surface does not need to be compatible. The surface-to-surface contact model is used when contact regions are not known accurately and a significant amount of sliding is expected. In this model, one of the surfaces is called the *contact surface* and the other, the *target surface*.

In a typical ANSYS contact analysis, the model is first meshed with conventional elements (beam, plane, or solid), and then the contact elements are created along the potential contact regions. Different element types are used for different contact models.

A contact analysis can be highly complicated, requiring the user to have a good understanding beforehand. It is recommended that the user read the section entitled "Contact" in the ANSYS Structural Analysis Guide as it provides a highly detailed description on how to perform a contact analysis in ANSYS, as well as several helpful hints to achieve converged solutions.

In the following subsections, two contact problems are used to demonstrate a contact analysis within ANSYS. In the first problem, the node-to-surface contact model is used in the simulation of a block dropping on a beam. Thus, the contact analysis is combined with dynamics, leading to time-dependent results.

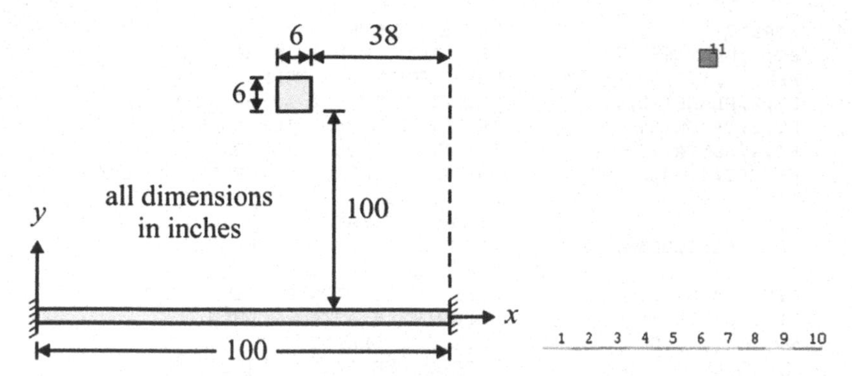

Fig. 10.19 Schematic of the block and beam (*left*), and the corresponding finite element mesh (*right*)

The second problem is the simulation of a nano-indentation test of a thin film deposited on a hard substrate. The problem is solved using surface-to-surface contact elements, and the thin film exhibits an elastic-perfectly plastic material behavior.

10.3.1 Contact Analysis of a Block Dropping on a Beam

Consider a 6-in × 6-in × 1-in block free falling onto a 100-in-long beam, as shown in Fig. 10.19. The block and the beam are made of the same material, with elastic modulus $E = 1\times10^6$ psi and density $\rho = 0.001$ lb/in^3. The beam has a cross-sectional area of 0.5 in^2, an area moment of inertia of $I_{yy} = 0.05$ in^4, and a height of 1 in. The block is initially at 100 in above the beam, with its center point 9 in to the right of the mid-point of the beam. The goal is to obtain the time-dependent response of the beam and the block.

Model Generation

The problem is solved using two-dimensional elements for both the beam (**BEAM188**) and the block (**PLANE182**). In modeling the block, plane stress with thickness idealization is used. Between the block (contact surface) and the beam (target surface), contact elements are defined using element types **CONTAC172** and **TARGE169**. This is achieved by first defining two *components*, one containing the nodes of the target surface and the other containing the nodes of the contact surface.

```
/PREP7                      ! ENTER PREPROCESSOR
ET,1,BEAM188                ! ELEMENT TYPE 1 IS BEAM188
KEYOPT,1,3,3                ! SPECIFY CUBIC FORM
ET,2,PLANE182,,,3           ! ELEMENT TYPE 2 IS PLANE182
ET,3,TARGE169               ! ELEMENT TYPE 1 IS TARGE169
ET,4,CONTA172               ! ELEMENT TYPE 2 IS CONTA172
KEYOPT,4,10,2               ! UPDATE CONTACT STIFFNESS AT EACH
                            ! ITERATION
R,1,1                       ! REAL CONSTANT SET 2 FOR PLANE182
SECTYPE,1,BEAM,RECT         ! DEFINE A RECTANGULAR CROSS-
                            ! SECTION FOR BEAM
SECDATA,0.5,1               ! DEFINE SECTION GEOMETRY DATA
MP,EX,1,1E6                 ! SPECIFY ELASTIC MODULUS
MP,DENS,1,.001              ! SPECIFY DENSITY
K,1,0,0                     ! CREATE KEYPOINT 1 (LEFT END OF
                            ! BEAM)
K,2,100,0                   ! CREATE KEYPOINT 2 (RIGHT END OF
                            ! BEAM)
L,1,2                       ! CREATE LINE FOR BEAM
ESIZE,,10                   ! USE 10 ELEMENTS PER LINE
LMESH,1                     ! MESH THE LINE WITH BEAM3 ELEMENTS
RECTNG,56,62,100,106        ! CREATE RECTANGULAR AREA
ESIZE,,1                    ! USE 1 ELEMENT PER LINE
TYPE,2                      ! SWITCH TO ELEMENT TYPE 2
                            ! (PLANE182)
REAL,1                      ! SWITCH TO REAL CONSTANT SET 1
AMESH,ALL                   ! MESH THE RECTANGULAR AREA
R,2                         ! REAL CONSTANT SET 2 FOR CONTACT
                            ! PAIRS
REAL,2                      ! SWITCH TO REAL CONSTANT SET 2
                            ! GENERATE THE TARGET SURFACE
LSEL,S,,,1                  ! SELECT LINE 1
CM,TARGET,LINE              ! DEFINE COMPONENT NAMED "TARGET"
TYPE,3                      ! SWITCH TO ELEMENT TYPE 3
                            ! (TARGE169)
NSLL,S,1                    ! SELECT NODES ASSOCIATED WITH THE
                            ! SELECTED LINE
ESLN,S,0                    ! SELECT THE ELEMENTS ATTACHED TO
                            ! THOSE NODES
ESURF                       ! GENERATE ELEMENTS OVERLAID ON
                            ! THE FREE FACES OF EXISTING
                            ! SELECTED ELEMENTS
                            ! GENERATE THE CONTACT SURFACE
LSEL,S,,,2,5                ! SELECT LINES 2,3,4, AND 5
CM,CONTACT,LINE             ! DEFINE COMPONENT NAMED "TARGET"
TYPE,4                      ! SWITCH TO ELEMENT TYPE 4
                            ! (CONTA172)
NSLL,S,1                    ! SELECT NODES ASSOCIATED WITH THE
                            ! SELECTED LINES
ESLN,S,0                    ! SELECT THE ELEMENTS ATTACHED TO
                            ! THOSE NODES
ESURF                       ! GENERATE ELEMENTS OVERLAID ON
                            ! THE FREE FACES OF EXISTING
                            ! SELECTED ELEMENTS
ALLSEL                      ! SELECT EVERYTHING
```

Solution

The transient solution is obtained in two load steps. The first load step encompasses a very short duration (0.002 s) and is solved using two substeps without time integration. This is done in order to set up the initial conditions for the transient solution. In the second load step, time integration is turned on (**TIMINT** command) and so is the automatic time stepping option (**AUTOTS** command). Also, the predictor option is turned on before the solution is obtained for the second load step using the **PRED** command. This option allows ANSYS to make a prediction of the displacements at the beginning of each substep, thus improving convergence. Structural damping is specified in terms of viscous damping, which produces a damping matrix in the form $[\mathbf{C}] = \alpha\,[\mathbf{M}] + \beta\,[\mathbf{K}]$, in which $[\mathbf{C}]$, $[\mathbf{M}]$, and $[\mathbf{K}]$ are the damping, mass, and stiffness matrices, respectively. In this case, only the damping through stiffness matrix is enforced by specifying the stiffness matrix multiplier β using the **BETAD** command.

```
/SOLU                     ! ENTER SOLUTION PROCESSOR
ANTYPE,TRANS              ! DECLARE ANALYSIS TYPE A
                          ! TRANSIENT
NLGEOM,ON                 ! TURN NONLINEAR GEOMETRY EFFECTS
                          ! ON
LUMPM,ON                  ! USE LUMPED MASS ASSUMPTION
KSEL,S,KP,,1,2            ! SELECT KEYPOINTS 1 AND 2
NSLK,S                    ! SELECT NODES ATTACHED TO
                          ! KEYPOINTS
D,ALL,ALL,0               ! CONSTRAIN ALL DOFS AT SELECTED
                          ! NODES
ALLSEL,ALL                ! SELECT EVERYTHING
ESEL,S,ENAME,,182         ! SELECT ELEMENTS OF TYPE PLANE182
NSLE,S                    ! SELECT NODES ATTACHED TO SELECTED
                          ! ELEMENTS
D,ALL,ALL,0               ! CONSTRAIN ALL DOFS AT SELECTED
                          ! NODES
ALLSEL,ALL                ! SELECT EVERYTHING
ACEL,,386                 ! DEFINE GRAVITATIONAL ACCELERATION
TIME,.0002                ! SPECIFY TIME AT THE END OF 1ST
                          ! LOAD STEP
DELTIM,.0001              ! SPECIFY TIME STEP SIZE
KBC,1                     ! ENFORCE STEPPED LOADING
BETAD,.000318             ! STIFFNESS MATRIX MULTIPLIER FOR
                          ! DAMPING
TIMINT,OFF                ! TURN OFF TIME INTEGRATION
CNVTOL,F,,.00001          ! SPECIFY FORCE CONVERGENCE
                          ! TOLERANCE
OUTRES,ALL,LAST           ! SAVE ONLY THE LAST RESULTS SET
SOLVE                     ! OBTAIN SOLUTION FOR LOAD STEP 1
ESEL,S,ENAME,,182         ! SELECT ELEMENTS OF TYPE PLANE182
NSLE,S                    ! SELECT NODES ATTACHED TO SELECTED
                          ! ELEMENTS
DDELE,ALL,ALL             ! DELETE DOF CONSTRAINTS ON
```

```
                        ! SELECTED NODES
ALLSEL,ALL              ! SELECT EVERYTHING
TIME,3                  ! TIME AT THE END OF 2ND LS AS
                        ! 3 SEC
DELTIM,.02,.0002,.02    ! USE INITIAL TIME STEP SIZE 0.02
                        ! WITH MINIMUM 0.0002 AND MAXIMUM
                        ! 0.02
AUTOTS,ON               ! TURN ON AUTOMATIC TIME STEPPING
TIMINT,ON               ! TURN ON TIME INTEGRATION
CNVTOL,F                ! RESET FORCE CONVERGENCE TOLERANCE
PRED,ON                 ! ACTIVATE PREDICTOR FOR INITIAL
                        ! GUESS
OUTRES,ALL,ALL          ! SAVE RESULTS FOR EVERY SUBSTEP
SOLVE                   ! OBTAIN SOLUTION
FINISH                  ! EXIT SOLUTION PROCESSOR
```

Postprocessing

After the solution is obtained, results are reviewed in both the *General Postprocessor* (results associated with the whole structure at a specific time) and *Time History Postprocessor* (results associated with a specific node in the structure along the entire time domain). First, in the *Time History Postprocessor*, the displacements in the *y*-direction at the mid-point of the beam (parameter **MID**) and at four corners of the block (parameters **BL**, **BR**, **TR**, and **TL**) are stored (**NSOL** command) in user-defined parameters, and they are plotted against time (**PLVAR** command), as shown in Fig. 10.20. Similarly, the reaction forces in the *y*-direction at both ends of the beam are stored (**RFOR** command) in user-defined parameters (parameters **FL** and **FR**), and they are plotted against time (Fig. 10.21). In both of these graphs,

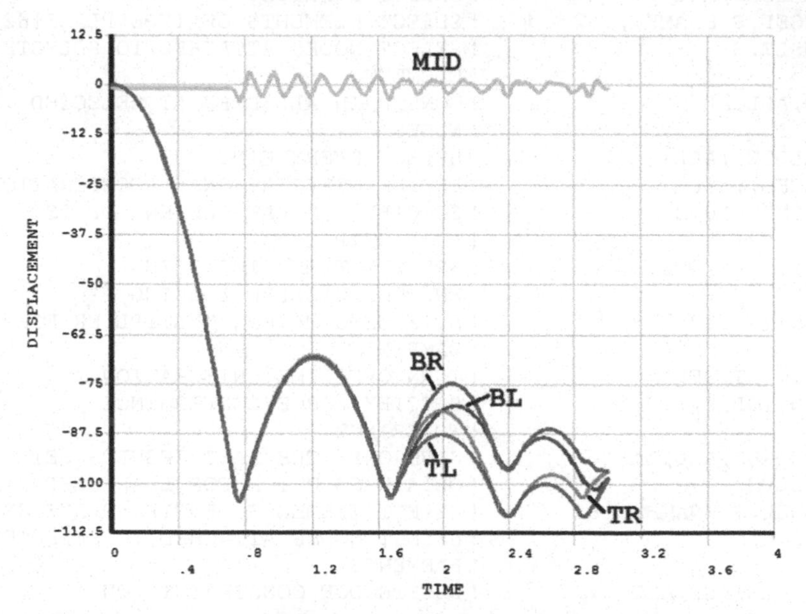

Fig. 10.20 Time variation of *y*-displacements at four corners of the block and at the midpoint of the beam

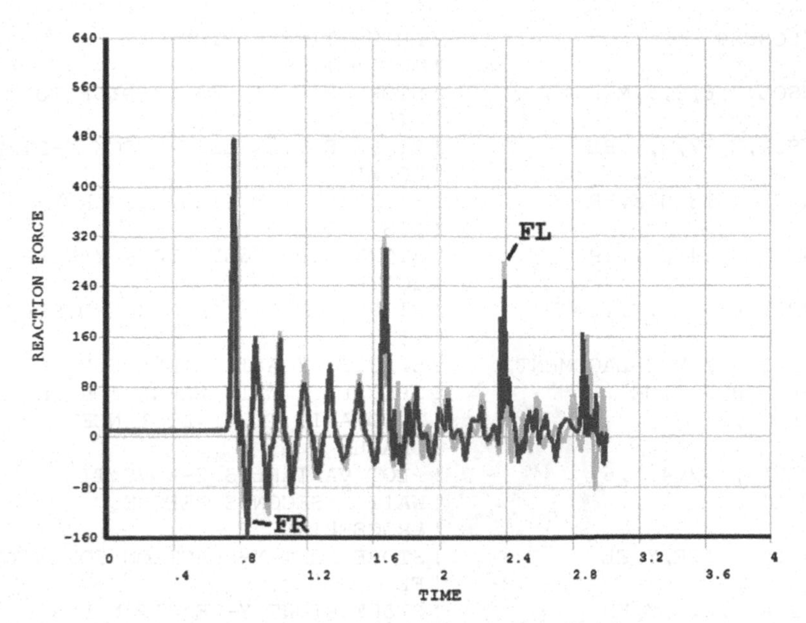

Fig. 10.21 Time variation of reaction forces in the beam in the y-direction

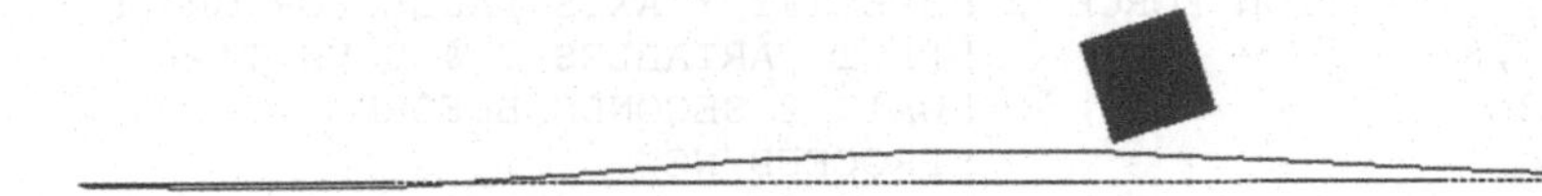

Fig. 10.22 Deformed configuration at time 1.742 s

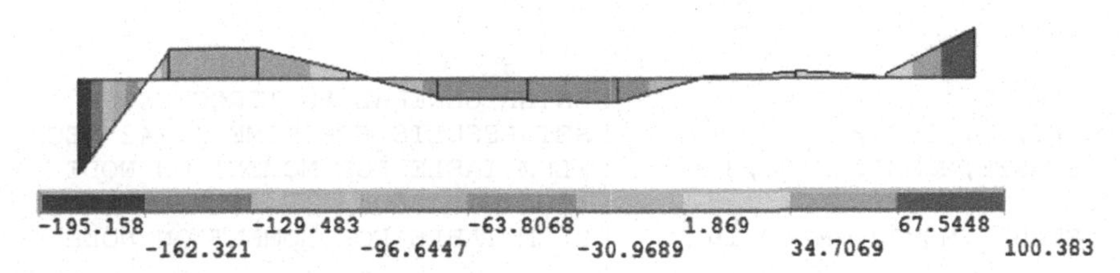

Fig. 10.23 Moment diagram of the beam at time 1.742 s

labels and ranges for the x- and y-axes are modified using the **/AXLAB**, **/AUTO**, **/XRANGE**, and **/YRANGE** commands. After reviewing the responses of specific nodes as functions of time, the deformed shape and moment diagram at a specific time (1.742 s) are reviewed in the *General Postprocessor*. The results associated with time 1.742 s are read using the **SET** command, and the deformed shape is obtained using the **PLDISP** command (Fig. 10.22). The moment results associated with nodes **I** and **J** of each element are stored in element table items **MOMZI** and **MOMZJ**, respectively, using the **ETABLE** command. Finally, the **PLLS** command is used for plotting the moment diagram (Fig. 10.23). The command input block below includes a **/WAIT** command after each plot command (**PLVAR** and **PLDISP** commands), which causes ANSYS to suspend operations for a specified duration (2 s in this case).

```
/POST26                          !ENTER TIME HISTORY
                                 !POSTPROCESSOR
NSOL,2,6,U,Y,MID                 !STORE MID-PT BEAM Y-DISP TO
                                 !MID
NSOL,3,12,U,Y,BL                 !STORE BOTTOM LEFT BLOCK Y-DISP
                                 !TO BL
NSOL,4,13,U,Y,BR                 !STORE BOTTOM RIGHT BLOCK Y-
                                 !DISP TO BR
NSOL,5,14,U,Y,TR                 !STORE TOP RIGHT BLOCK Y-DISP
                                 !TO TR
NSOL,6,15,U,Y,TL                 !STORE TOP LEFT BLOCK Y-DISP TO
                                 !TL
/AXLAB,Y,DISPLACEMENT            !SPECIFY Y AXIS LABEL FOR GRAPH
/AXLAB,X,TIME                    !SPECIFY X AXIS LABEL FOR GRAPH
/AUTO,1                          !AUTOMATIC FOCUS/DISTANCE FOR
                                 !GRAPH
PLVAR,2,3,4,5,6                  !PLOT VARIABLES 2-6 VS TIME
/WAIT,2                          !WAIT 2 SECONDS BEFORE
                                 !PROCEEDING
RFOR,7,1,F,Y,FL                  !STORE LEFT Y-REACTION FORCE TO
                                 !FL
RFOR,8,2,F,Y,FR                  !STORE RIGHT Y-REACTION FORCE
                                 !TO FR
/YRANGE                          !RESET Y AXIS RANGE
/AXLAB,Y,REACTION FORCE          !SPECIFY Y AXIS LABEL FOR GRAPH
PLVAR,7,8                        !PLOT VARIABLES 7 & 8 VS TIME
/WAIT,2                          !WAIT 2 SECONDS BEFORE
                                 !PROCEEDING
FINISH                           !EXIT TIME HISTORY
                                 !POSTPROCESSOR
```

```
/POST1                           !ENTER GENERAL POSTPROCESSOR
SET,,,,,1.742                    !SET RESULTS FOR TIME 1.742 SEC
ETABLE,MOMZI,SMISC,3             !ELEM TABLE FOR MOMENT ON NODE
                                 !I
ETABLE,MOMZJ,SMISC,16            !ELEM TABLE FOR MOMENT ON NODE
                                 !J
/DSCALE,1,1                      !DO NOT SCALE DISPLACEMENTS
/TRIAD,OFF                       !TURN OFF THE TRIAD
PLDISP,1                         !PLOT DEFORMED SHAPE
/WAIT,2                          !WAIT 2 SECONDS BEFORE
                                 !PROCEEDING
ESEL,S,TYPE,,1                   !SELECT BEAM ELEMENTS (TYPE 1)
NSLE,S                           !SELECT NODES ATTACHED TO
                                 !ELEMENTS
PLLS,MOMZI,MOMZJ                 !PLOT ELEM TABLE ITEMS ON MESH
ALLSEL,ALL                       !SELECT EVERYTHING
FINISH                           !EXIT GENERAL POSTPROCESSOR
```

In order to obtain an animation of the time-dependent displacement response of the system:

- Read the results associated with load step 2 (**SET** command) using the following menu path:

Main Menu > General Postproc > Read Results > Last Set

- Obtain the deformed shape (**PLDISP** command) using the following menu path:

Main Menu > General Postproc > Plot Results > Deformed Shape

- *Plot Deformed Shape* dialog box appears; select ***Def + undef edge*** radio-button; click on ***OK***.
- Create animation using the following menu path:

Utility Menu > PlotCtrls > Animate > Over Time

- *Animate Over Time* dialog box appears; enter ***100*** for ***Number of animation frames***. Click on ***OK*** and wait until *Animation Controls Window* appears.

10.3.2 Simulation of a Nano-Indentation Test

Nano-indentation tests are commonly used for evaluation of the response of thin films. Consider a sol-gel layer deposited on a glass substrate, which is indented by means of a conical diamond indenter as shown in Fig. 10.24. The thickness values for the sol-gel film and the glass substrate are $2\ \mu m$ and $6\ \mu m$, respectively. The indenter has an angle of 68° measured from the axis of rotation. In order to correctly simulate the indentation phenomenon, a contact analysis is utilized. For this purpose, target elements (**TARGE170**) are placed along the top surface of the film and contact elements (**CONTA174**) are used along the bottom surface of the indenter. The indentation is simulated by applying displacement boundary conditions in the *y*-direction to the nodes along the bottom surface of the substrate. Consequently, the top surface of the film is pressed against the bottom surface of the indenter, thus exerting the contact elements against the target elements. The contact is assumed to be frictionless. The indentation is performed using several displacement steps, each of which is written to a load step file. Both loading and unloading are simulated. Displacement step sizes for loading and unloading are $0.04\ \mu m$ and $0.03\ \mu m$, respectively. All three materials are modeled using **SOLID185** elements. Since the problem possesses symmetry with respect to the *y*-axis, only one octant ($1/8^{th}$) of the geometry is modeled. Normal displacements are constrained along the symmetry planes, and *x*- and *z*-displacements are constrained along the axis of rotation. The top surface of the indenter is constrained in all directions. The sol-gel film exhibits elastic-perfectly plastic behavior with yield stress of $700\ MPa$ while the diamond indenter and the glass substrate are both elastic. The elastic modulus and Poisson's ratio values for the constituent materials are given in Table 10.4. The goal is to obtain indentation vs. force response for the film.

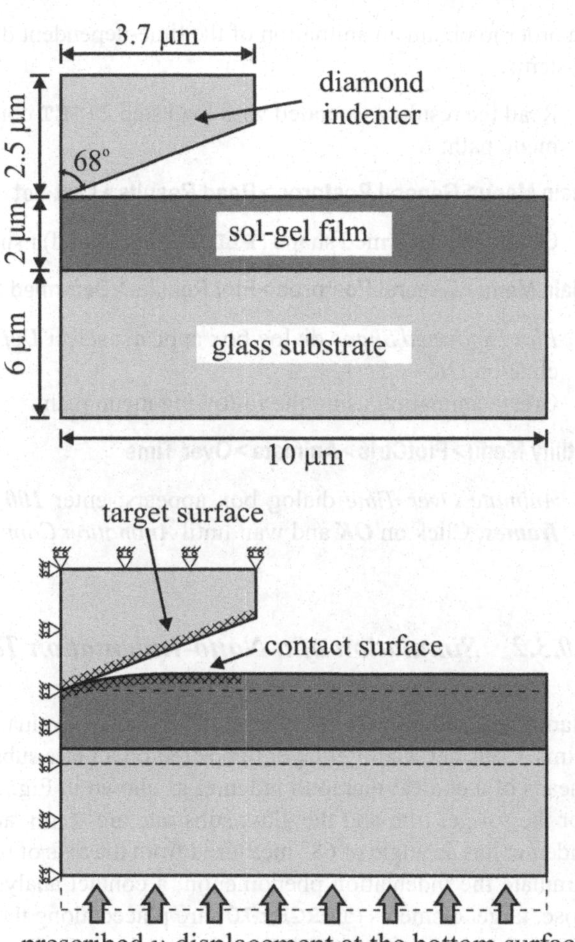

Fig. 10.24 Sol-gel film on glass substrate, and diamond indenter

Model Generation

In the solid modeling phase, a *bottom-up approach* is used. The solid model is created, starting with keypoints, then lines, areas, and volumes. The elastic-perfectly plastic behavior of the thin film is incorporated using the bilinear isotropic hardening rule (**TB** command with **BISO** option) with a tangent modulus of zero. The model is first generated in the *x-y* plane and meshed using two-dimensional elements (**PLANE182**), after which the **VROTAT** command is used to generate

Table 10.4 Material properties used in nano-indentation simulation

	Elastic modulus (GPa)	Poisson's ratio	Material reference number
Glass substrate	75	0.23	1
Sol-gel film	4.5	0.35	2
Diamond indenter	1141	0.07	3

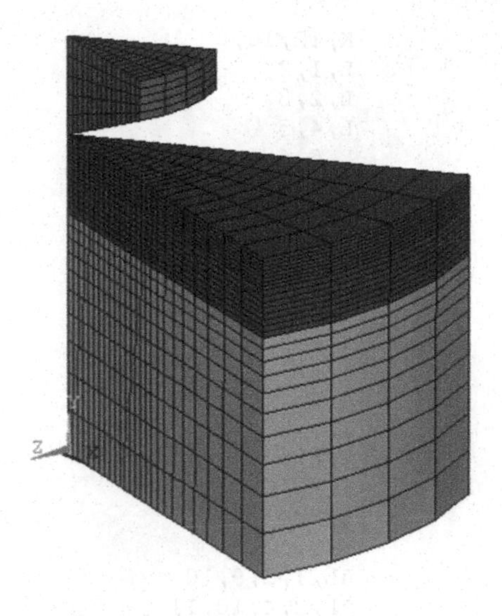

Fig. 10.25 Isometric view of the finite element mesh used for the nano-indentation simulation

the three-dimensional mesh by rotating the meshed areas about the *y*-axis by 45°. Before issuing the **VROTAT** command, the default element type attribute must be changed to the one for three-dimensional elements (element type *2* for **SOLID185** in this case). Figure 10.25 shows an isometric view of the mesh used in this analysis.

```
/FILNAM,NANO              ! SPECIFY JOBNAME
/PREP7                    ! ENTER PREPROCESSOR
ET,1,182                  ! ELEMENT TYPE 1 IS PLANE182
ET,2,185                  ! ELEMENT TYPE 2 IS SOLID185
ET,3,TARGE170             ! ELEMENT TYPE 3 IS TARGE170
ET,4,CONTA174             ! ELEMENT TYPE 4 IS CONTA174
MP,EX,1,75000             ! GLASS SUBSTRATE MAT PROPS
MP,NUXY,1,0.23            !
MP,EX,2,4500              ! SOL GEL FILM MAT PROPS
MP,NUXY,2,0.35            !
TB,BISO,2,1               ! BILINEAR ISOTROPIC HARDENING RULE
TBTEMP,0                  !
TBDATA,1,700,0            ! YIELD STRENGTH
MP,EX,3,1141000           ! DIAMOND INDENTER MAT PROPS
MP,NUXY,3,0.07            !
K,1                       ! CREATE KEYPOINTS
K,2,4
K,3,10
K,4,,6
K,5,4,6
K,6,10,6
K,7,,7.2
K,8,4,7.2
K,9,10,7.2
K,10,,8
K,11,4,8
```

```
K,12,10,8
L,1,2                     ! CREATE LINES
L,2,3
L,4,5
L,5,6
L,7,8
L,8,9
L,10,11
L,11,12
L,1,4
L,2,5
L,3,6
L,4,7
L,5,8
L,6,9
L,7,10
L,8,11
L,9,12
AL,1,3,9,10               ! CREATE AREAS
AL,2,4,10,11
AL,3,5,12,13
AL,4,6,13,14
AL,5,7,15,16
AL,6,8,16,17
LESIZE,7,,,15             ! SPECIFY LINE DIVISIONS
LESIZE,5,,,15
LESIZE,3,,,15
LESIZE,1,,,15
LESIZE,8,,,10,4
LESIZE,6,,,10,4
LESIZE,4,,,10,4
LESIZE,2,,,10,4
LESIZE,9,,,10,1/4
```

```
LESIZE,10,,,10,1/4
LESIZE,11,,,10,1/4
LESIZE,12,,,8
LESIZE,13,,,8
LESIZE,14,,,8
LESIZE,15,,,8
LESIZE,16,,,8
LESIZE,17,,,8
MSHKEY,1                      ! ENFORCE MAPPED MESHING
TYPE,1                        ! SWITCH TO ET 1
MAT,1                         ! SWITCH TO MATERIAL 1
AMESH,1,6                     ! CREATE MESH
TYPE,2                        ! SWITCH TO ET 2
ESIZE,,4                      ! SPECIFY # OF ELEMS FOR SWEEP
VROTAT,1,2,3,4,5,6,1,10,45       ! SWEEP AREAS TO CREATE
                                 ! VOLUME
NSEL,S,LOC,Y,6,8              ! SELECT NODES AT 6 µm ≤ y ≤ 8 µm
ESLN,S,1                      ! SELECT ELEMENTS WITH SELECTED
                              ! NODES
EMODIF,ALL,MAT,2              ! MODIFY ELEMS TO BE GLASS
TYPE,4                        ! SWITCH TO ET 4
AMESH,22                      ! CREATE CONTACT AREA MESH
ESEL,S,TYPE,,4                ! SELECT ELEMENTS WITH ET 4
ESURF,,REVE                   ! ADJUST OUTWARD NORMAL OF ELEMENTS
                              ! GENERATE SOLID MODEL FOR INDENTER
K,23,,8                       ! CREATE KEYPOINTS
K,26,,10.5
K,24,3.7,9.5
K,25,3.7,10.5
L,23,24                       ! CREATE LINES
L,24,25
L,25,26
L,26,23
AL,40,41,42,43                ! CREATE AREA
LESIZE,41,,,5                 ! SPECIFY LINE DIVISIONS
LESIZE,43,,,5
LESIZE,40,,,15
LESIZE,42,,,15
TYPE,1                        ! SWITCH TO ET 1
MAT,3                         ! SWITCH TO MATERIAL 3
AMESH,27                      ! CREATE MESH FOR INDENTER
TYPE,2                        ! SWITCH TO ET 2
ESIZE,,4                      ! SPECIFY # OF ELEMS FOR SWEEP
VROTAT,27,,,,,,23,26,45       ! SWEEP AREAS TO CREATE VOLUME
R,1                           ! REAL CONSTANT SET 1 FOR CONTACT
                              !PAIR
REAL,1                        ! SWITCH TO REAL CONSTANT SET 1
                              ! GENERATE THE TARGET SURFACE
ASEL,S,,,22                   ! SELECT AREA 22
CM,TARGET,AREA                ! DEFINE COMPONENT NAMED "TARGET"
TYPE,3                        ! SWITCH TO ET 3
NSLA,S,1                      ! SELECT NODES ASSOCIATED WITH THE
                              !SELECTED AREA
ESLN,S,0                      ! SELECT THE ELEMENTS ATTACHED TO
                              ! THOSE NODES
ESURF                         ! GENERATE ELEMENTS OVERLAID ON
                              ! THE FREE FACES OF EXISTING
```

```
                              ! SELECTED ELEMENTS
                              ! GENERATE THE CONTACT SURFACE
ASEL,S,,,28                   ! SELECT LINES 2,3,4, AND 5
CM,CONTACT,AREA               ! DEFINE COMPONENT NAMED "TARGET"
TYPE,4                        ! SWITCH TO ET 4
NSLA,S,1                      ! SELECT NODES ASSOCIATED WITH THE
ESLN,S,0                      ! SELECT THE ELEMENTS ATTACHED TO
                              ! THOSE NODES
ESURF                         ! GENERATE ELEMENTS OVERLAID ON
                              ! THE FREE FACES OF EXISTING
                              ! SELECTED ELEMENTS
ALLSEL                        ! SELECT EVERYTHING
FINISH                        ! EXIT PREPROCESSOR
```

Solution

As mentioned previously, the top surface of the diamond indenter is constrained in all directions. In addition, the vertical planes are not allowed to move in their normal direction due to the symmetry boundary conditions imposed by using the **DSYM** command. The indentation is simulated by prescribing a uniform *y*-displacement along the bottom surface of the glass substrate, which causes the film to contact the indenter, starting the indentation process. This displacement is applied in increments of 0.04 μm for the loading phase, and its values are stored in the array parameter **DIS**. The array parameter is created/initialized using the ***DIM** command, with an array size of 57. The first 37 entries of this array correspond to the loading phase displacements and the following 20 entries are for the unloading phase displacements. Once all the displacement values are stored in the array **DIS**, 57 load step files are written (**LSWRITE** command) utilizing a do loop. Nonlinear geometry effects are turned on (**NLGEOM** command), as well as automatic time stepping (**AUTOTS** command). A full Newton-Raphson method is utilized without the *adaptive descent* option (**NROPT** command). The maximum number of equilibrium iterations in each substep is set to 100 using the **NEQIT** command. Prior to initiating the solution using the **LSSOLVE** command, the two-dimensional elements must be unselected, as the sole purpose of their existence is for meshing.

```
/SOLU                         ! ENTER SOLUTION PROCESSOR
SOLCONTROL,0                  ! TURN OFF SOLUTION CONTROLS
ANTYPE,STATIC                 ! SPECIFY ANALYSIS TYPE AS STATIC
NSEL,S,LOC,Z                  ! SELECT NODES AT Z = 0
DSYM,SYMM,Z,0                 ! APPLY SYMMETRY BC ABOUT X-Y PLANE
CLOCAL,11,,,,,,,45            ! DEFINE LOCAL CS
NSEL,S,LOC,Z                  ! SELECT NODES AT Z = 0 (IN LOCAL
                              ! CS)
DSYM,SYMM,Z,11                ! APPLY SYMMETRY BC IN LOCAL CS
CSYS,0                        ! SWITCH TO GLOBAL CARTESIAN CS
NSEL,S,LOC,Y,10.5             ! SELECT NODES AT Y = 10.5
D,ALL,UY,0                    ! CONSTRAIN Y-DISP AT SELECTED
```

```
                              ! NODES
NSEL,S,LOC,X                  ! SELECT NODES AT X = 0
NSEL,R,LOC,Z                  ! RESELECT NODES AT Z = 0
D,ALL,UX                      ! CONSTRAIN X-DISP AT SELECTED
                              ! NODES
D,ALL,UZ                      ! CONSTRAIN Z-DISP AT SELECTED
                              ! NODES
*DIM,DIS,ARRAY,57             ! INITIALIZE ARRAY PARAMETER DIS
A=0                           ! INITIALIZE PARAMETER A
*DO,I,1,36                    ! START DO LOOP FOR LOADING
DIS(I)=A                      ! STORE DISP VALUE IN DIS FOR
                              ! CURRENT LS
A=A+0.04                      ! UPDATE PARAMETER A
*ENDDO                        ! END DO LOOP ON LOADING
B=0.03                        ! INITIALIZE PARAMETER B
*DO,I,37,57                   ! START DO LOOP FOR UNLOADING
DIS(I)=A-B                    ! STORE DISP VALUE IN DIS FOR
                              ! CURRENT LS
B=B+0.03                      ! UPDATE PARAMETER B
*ENDDO                        ! END DO LOOP ON UNLOADING
NLGEOM,ON                     ! TURN ON NONLINEAR GEOMETRY
                              ! EFFECTS
AUTOTS,ON                     ! TURN ON AUTOMATIC TIME STEPPING
OUTRES,,1                     ! SAVE RESULTS FOR LAST SUBSTEP OF
                              ! EACH LS
NROPT,FULL,,OFF               ! USE FULL NEWTON-RAPHSON WITH NO
                              ! ADAPTIVE DESCENT
NEQIT,100                     ! USE MAXIMUM 100 EQUILIBRIUM
                              ! ITERATIONS
ALLSEL                        ! SELECT EVERYTHING
*DO,I,1,57                    ! START DO LOOP FOR WRITING LOAD
                              ! STEPS
NSEL,S,LOC,Y                  ! SELECT NODES AT Y = 0
D,ALL,UX                      ! CONSTRAIN X-DISP AT SELECTED
                              ! NODES
D,ALL,UZ                      ! CONSTRAIN Z-DISP AT SELECTED
                              ! NODES
D,ALL,UY,DIS(I)               ! SPECIFY Y-DISP ALONG THE BOTTOM
                              ! SURFACE OF THE SUBSTRATE
ALLSEL                        ! SELECT EVERYTHING
LSWRITE,I                     ! WRITE LOAD STEP FILE
*ENDDO                        ! END DO LOOP ON WRITING LOAD STEPS
ESEL,U,TYPE,,1                ! UNSELECT ELEMENT TYPE 1
LSSOLVE,1,57                  ! SOLVE FROM LS FILES (1 TO 57)
FINISH                        ! EXIT SOLUTION PROCESSOR
```

Postprocessing

With the goal of obtaining the load vs. indentation depth response of the sol-gel thin film deposited on a glass substrate, the following command input segment is used. Load and indentation depth values for each load step are extracted within a do loop,

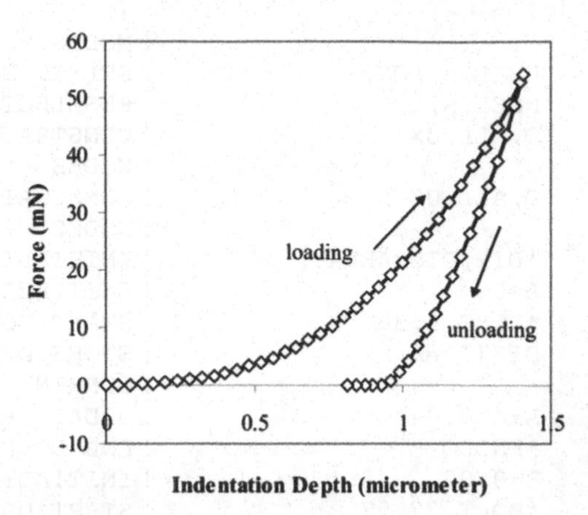

Fig. 10.26 Load vs. indentation depth for sol-gel film on glass substrate

in which the results associated with the current load step are read using the **SET** command. The parameter **SUM** is the total reaction force in the *y*-direction along the bottom surface of the glass substrate. This quantity must be identical to the force exerted by the indenter to the top surface of the film because of equilibrium of forces. The command ***GET** is used on numerous occasions to retrieve model and results information about the nodes. For each load step, the applied indentation depth from array parameter **DIS**, and the corresponding total reaction force stored in parameter **SUM** are written to the file **NANO_RF_D.OUT** using a combination of **/OUTPUT *VWRITE** and commands. Figure 10.26 shows the resulting load vs. indentation depth response.

```
/POST1                              ! ENTER GENERAL POSTPROCESSOR
*DO,J,1,57                          ! LOOP OVER LOAD STEPS
SET,J                               ! READ RESULTS SET
NSEL,S,LOC,Y                        ! SELECT NODES AT Y = 0
*GET,NUMNOD,NODE,0,COUNT            ! STORE # OF NODES IN NUMNOD
*GET,CURNOD,NODE,0,NUM,MIN          ! STORE MIN NODE # TO CURNOD
SUM=0                               ! INITIALIZE TOTAL REACTION F
*DO,I,1,NUMNOD                      ! LOOP OVER SELECTED NODES
*GET,RFY,NODE,CURNOD,RF,FY          ! STORE Y REACT. FORCE IN RFY
SUM=SUM+RFY                         ! UPDATE TOTAL REACTION FORCE
CURNOD=NDNEXT(CURNOD)               ! UPDATE CURRENT NODE NUMBER
*ENDDO                              ! END LOOP OVER SELECTED NODES
DISJ=DIS(J)                         ! STORE CURRENT INDENTATION
                                    ! DEPTH TO PARAMETER DISJ
SUM=SUM/100                         ! CONVERT TO MILLINEWTONS
/OUTPUT,NANO_RF_D,OUT,,APPEND! REDIRECT OUTPUT TO FILE
*VWRITE,DISJ,SUM                    ! WRITE DISJ AND SUM TO FILE
(E16.8,5X,E16.8)                    ! FORMAT STATEMENT
/OUTPUT                             ! REDIRECT OUTPUT TO OUTPUT W
*ENDDO                              ! END LOOP OVER LOAD STEPS
```

Chapter 11
Advanced Topics in ANSYS

Selected advanced topics are discussed in this chapter. First, coupled degrees of freedom and constraint equations are explained. Discussions on submodeling and superelements are included next, with one example problem for each. Finally, a brief section on how to interact with external files from within the ANSYS environment is followed by a section on modification of the ANSYS GUI.

11.1 Coupled Degrees of Freedom

In certain engineering problems, the behavior of some of the unknown degrees of freedoms may be known. For example, certain points (nodes) may be expected to have the same displacement in a certain direction. One can take advantage of this behavior and enforce it in order to achieve an accurate solution with minimum computational resources. If a particular degree of freedom at several nodes is expected to have the same unknown value, these degrees of freedoms can be coupled. The use of coupled degrees of freedom is clarified with the following example: a long, hollow cylinder subjected to internal pressure (see Fig. 11.1)

This problem can be solved using two different approaches: (i) plane strain idealization and (ii) axisymmetry. Use of plane strain idealization involves meshing on the *x-y* plane and does not require imposition of additional constraints. Figure 11.2 shows the plane strain model with applied boundary conditions (quarter-symmetry is used).

In order to utilize axisymmetry, the mesh is generated on the *x-z* plane, as shown in Fig. 11.3. Along the left vertical boundary, the inner pressure is specified while the right vertical boundary is traction free. Based on the problem definition (long in the *z*-direction), it is known that the *z*-displacement on the *x-y* plane is uniform but its value is unknown. This condition is enforced by constraining the *z*-displacements along the bottom boundary and coupling the *z*-displacements of the nodes along the top surface.

The online version of this book (doi: 10.1007/978-1-4939-1007-6_11) contains supplementary material, which is available to authorized users

E. Madenci, I. Guven, *The Finite Element Method and Applications in Engineering Using ANSYS®*, DOI 10.1007/978-1-4899-7550-8_11

Fig.11.1 A long, hollow cylinder subjected to internal pressure

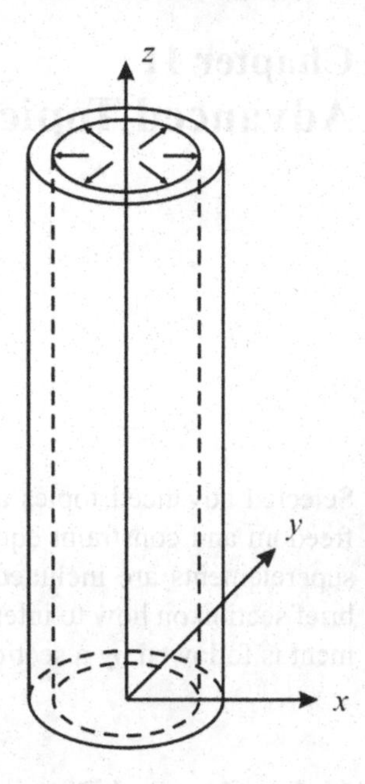

Both plane strain and axisymmetric models are valid. The solution obtained using plane strain idealization produces identical results in the circumferential direction (when viewed in the cylindrical coordinate system). Although both models may have a similar number of nodes, the accuracy of the results in the radial direction provided by the axisymmetric solution is much higher than that of the plane strain solution.

A set of coupled degrees of freedom is defined using the **CP** command, which has the following syntax:

```
CP, NSET, Lab, NODE1, NODE2,..., NODE17
```

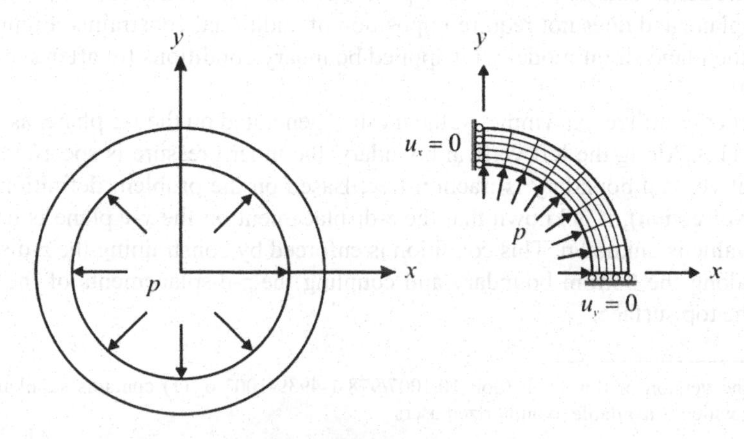

Fig. 11.2 Plane strain idealization with applied boundary conditions

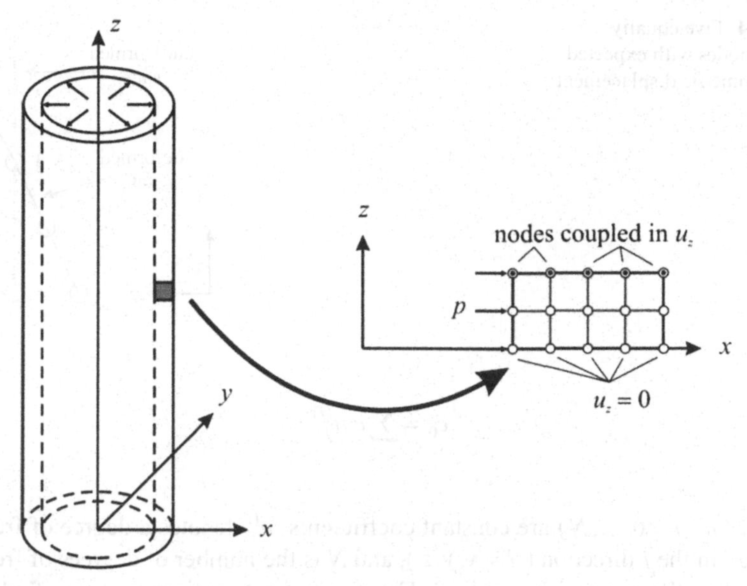

Fig. 11.3 Axisymmetric model with applied boundary conditions

in which **NSET** is the coupled degree of freedom set number, **Lab** is the degree of freedom label (e.g., **UX**, **UY**, etc.) and **NODE1** through **NODE17** are the nodes to be coupled. It is common to first select the nodes to be coupled and then assign **ALL** for the argument NODE1. For example, the z-displacements of all nodes along $y = 1$ are coupled, using the following command input:

```
NSEL,S,LOC,Y,1
CP,1,UZ,ALL
```

The coupled degrees of freedom set must not include any degrees of freedom that are specified as boundary conditions; they must be unknown.

11.2 Constraint Equations

The coupled degrees of freedom feature discussed in the previous section is a subset of the constraint equations. When coupling two degrees of freedom, say x-displacements at nodes 2 and 4, the following constraint equation is enforced:

$$u_x^{(2)} - u_x^{(4)} = 0 \tag{11.1}$$

in ANSYS, it is also possible to specify more general constraint equations. The general form of these equations is

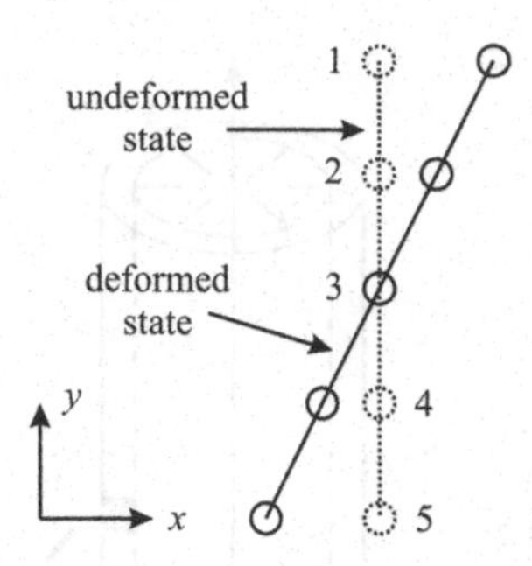

Fig. 11.4 Five equally spaced nodes with expected anti-symmetric displacement

$$c_0 = \sum_{i=1}^{N} c_i u_j^{(i)} \tag{11.2}$$

in which c_i ($i = 0,\ldots,N$) are constant coefficients, $u_j^{(i)}$ denotes a degree of freedom at node i in the j-direction ($j = x, y, z$), and N is the number of degrees of freedom involved in the constraint equation. The constraint equations are specified using the **CE** command with the following syntax:

```
CE,NEQN,CONST,NODE1,Lab1,C1,NODE2,Lab2,C2,NODE3,Lab3,C3
```

in which **NEQN** is the constraint equation reference number; **CONST**, **C1**, **C2**, and **C3** are the coefficients c_i of Eq. (11.2); **NODE1**, **NODE2**, and **NODE3** denote the node numbers involved in the constraint equation; and, finally, **Lab1**, **Lab2** and **Lab3** are the degree of freedom labels. The use of this command is explained through an example.

Consider five equally spaced nodes, as shown in Fig. 11.4. Assume that the x-displacements of these nodes are expected to exhibit anti-symmetric behavior. This implies two constraint equations, which are written as

$$\begin{aligned} 0 &= 1 \times u_x^{(1)} + (-1) \times u_x^{(5)} \\ 0 &= 1 \times u_x^{(2)} + (-1) \times u_x^{(4)} \end{aligned} \tag{11.3}$$

The equivalent command line input using the **CE** command is given by

```
CE,1,0,1,UX,1,5,UX,-1
CE,2,0,2,UX,1,4,UX,-1
```

Constraint equations are particularly useful when a rigid connection between two separate regions is needed. For example, the modeling of a composite sandwich panel where the core is enclosed by two thin facesheets on the top and bottom surfaces is ideal for using rigid constraints. In this case, the facesheets are modeled using shell elements while solid brick elements are used for the core, as shown in

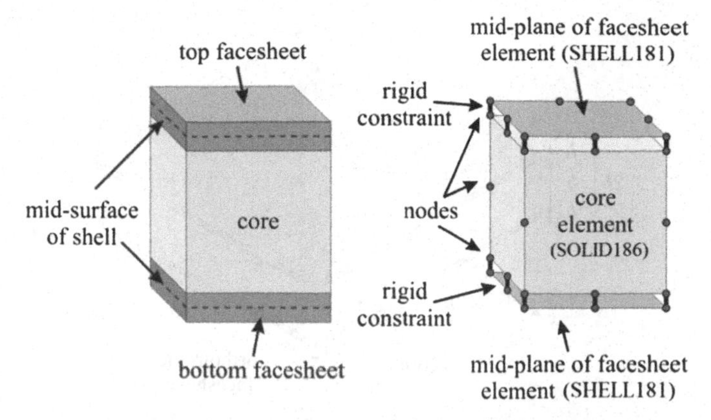

Fig. 11.5 A sandwich panel and corresponding finite element model utilizing rigid constraints between the shell and solid elements

Fig. 11.5. However, the shell elements, by default, place the nodes at the mid-plane of the plate. If the nodes of the shell elements were shared with the solid brick elements, half of the facesheet element in the thickness direction would penetrate into the solid brick element, thus creating a wrong representation of the actual structure. The use of shell elements whose mid-plane is offset by means of rigid constraints removes this problem.

Rigid constraints are specified using the **CERIG** command with the following syntax:

```
CERIG,MASTE,SLAVE,Ldof,Ldof2,Ldof3,Ldof4,Ldof5
```

in which MASTE and SLAVE are the node numbers for the *master* and *slave* nodes, respectively, and Ldof,…,Ldof5 are the degree of freedom labels. Once the constraint is specified, only one of these two nodes is retained in the matrix system of equations, which is called the *master node*, while the remaining node is called the *slave node*. Generally, only the argument Ldof is used as it provides three labels, ALL, UXYZ, and RXYZ, in addition to the usual degree of freedom labels (UX, UY, UZ, ROTX, ROTY, and ROTZ). The use of the CERIG command is demonstrated by an analysis of a sandwich composite panel under transverse pressure.

The construction of a sandwich panel with facesheets and a core is illustrated in Fig. 11.6. The mid-plane of the sandwich panel coincides with the (x,y) plane of the reference coordinate system. As shown in Fig. 11.6, the panel has a uniform thickness of 2 h. The thicknesses of the core and facesheets are denoted by h_c and h_f, respectively. The panel has a square planar geometry described by $2a = 2b = 6$ in. The facesheets at the bottom and top of the core are of equal thickness, $h_f = 0.01$ in, and the core thickness is $h_c = 0.5$ in. The facesheets are homogeneous, elastic, and orthotropic material layers. The elastic modulus and Poisson's ratio values are defined with

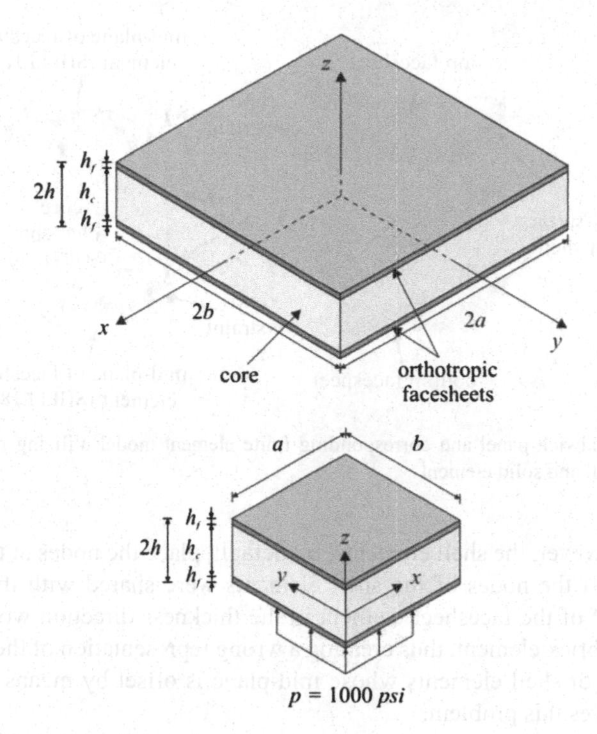

Fig. 11.6 Sandwich composite panel under transverse pressure

respect to the reference frame as $E_{xx} = E_{yy} = 9.58 \times 10^6$ psi, $E_{zz} = 1.66 \times 10^6$ psi, $G_{xy} = 709 \times 10^3$ psi, $G_{yz} = G_{xz} = 731 \times 10^3$ psi, $\nu_{xy} = 0.033$, and $\nu_{yz} = \nu_{xz} = 0.352$. The core is isotropic with Young's modulus of $E_c = 18.5 \times 10^3$ psi and Poisson's ratio of $\nu_c = 0.3$.

The panel is on roller supports along all edges and is subjected to a square step load of 1000 psi in the positive z-direction, applied over a square area (2×2 in^2) centered along the bottom surface of the panel. Due to the quarter-symmetry of the problem, symmetry boundary conditions are applied along the $x = 0$ and $y = 0$ planes.

```
/PREP7                   ! ENTER PREPROCESSOR
ET,1,181                 ! USE SHELL181 AS ELEM TYPE 1
KEYOPT,1,8,1             ! STORE DATA FOR ALL LAYERS
ET,2,186                 ! USE SOLID186 AS ELEM TYPE 2
UIMP,1,EX,EY,EZ,9.582E6,9.582E6,1.66E6, ! SPECIFY MAT
                                        ! PROPS
UIMP,1,PRXY,PRYZ,PRXZ,0.03192,0.35231,0.35231, !      FOR
UIMP,1,GXY,GYZ,GXZ,709E3,731E3,731E3,     !     FACESHEETS
UIMP,2,EX, , ,18.51E3,   ! SPECIFY MAT PROPS FOR CORE
UIMP,2,NUXY, , ,.3,      !
SECT,1,SHELL             ! SPECIFY SECTION TYPE
SECDATA,0.01,1           ! DEFINE THICKNESS FOR SHELL
LOCAL,11,0               ! DEFINE A LOCAL CS
ESYS,11                  ! ALIGN ELEMENT CS WITH CS 11
W=6                      ! PARAMETER FOR X-LENGTH OF PANEL
H=6                      ! PARAMETER FOR Y-LENGTH OF PANEL
D=1                      ! PARAMETER FOR SUBDIVISIONS
TF1=0.01                 ! TOP FACESHEET THICKNESS
TF2=0.01                 ! BOTTOM FACESHEET THICKNESS
TC=0.5                   ! CORE THICKNESS
P=1000                   ! APPLIED PRESSURE PSI
DIV1=10                  ! PARAMETERS FOR NUMBER OF
DIV2=7                   ! DIVISIONS AT VARIOUS LOCATIONS
DIV3=5                   !
DIV4=DIV1/2              !
RECTNG,,W/2,,H/2,        ! CREATE RECTANGLES
RECTNG,,D,,H/2           !
RECTNG,,W/2,,D           !
RECTNG,,2*D,,H/2         !
RECTNG,,W/2,,2*D         !
AOVLAP,ALL               ! OVERLAP ALL AREAS
ALLSEL                   ! SELECT EVERYTHING
LSEL,S,LOC,X,D/2         ! SELECT LINES AT X = D/2
LSEL,A,LOC,Y,D/2         ! ALSO SELECT LINES AT Y = D/2
LESIZE,ALL,,,DIV1        ! SPECIFY # OF DIVISIONS
LSEL,S,LOC,X,3*D/2       ! SELECT LINES AT X = 3D/2
LSEL,A,LOC,Y,3*D/2       ! ALSO SELECT LINES AT Y = 3D/2
LESIZE,ALL,,,DIV2        ! SPECIFY # OF DIVISIONS
LSEL,S,LOC,X,5*D/2       ! SELECT LINES AT X = 5D/2
LSEL,A,LOC,Y,5*D/2       ! ALSO SELECT LINES AT Y = 5D/2
LESIZE,ALL,,,DIV3        ! SPECIFY # OF DIVISIONS
TYPE,1                   ! USE SHELL181 ELEMENTS
MAT,1                    ! SWITCH TO MATERIAL 1
SECNUM,1                 ! SWITCH TO SECTION NUMBER 1
```

```
MSHKEY,1                 !ENFORCE MAPPED MESHING
AMESH,ALL                !MESH ALL AREAS
LSEL,S,LCCA              !SELECT CONCATENATED LINES
LDEL,ALL                 !DELETE SELECTED LINES
ALLSEL                   !SELECT EVERYTHING
AGEN,2,ALL,,,,-TF1/2,,0    ! GENERATE AREAS FROM EXISTING
                           ! PATTERN
TYPE, 2                  !USE SOLID186 ELEMENTS
MAT,2                    !SWITCH TO MATERIAL 2
EXTOPT,ESIZE,6,0,        !SPECIFY OPTIONS FOR EXTRUDING
ASEL,S,LOC,Z,-TF1/2      !SELECT AREAS AT Z = -TF1/2
VEXT,ALL,,,0,0,-TC       !EXTRUDE AREAS IN Z-DIR
ALLSEL                   !SELECT EVERYTHING
ASEL,S,LOC,Z,-TC-TF1/2 !SELECT AREAS AT Z = -TC-TF1/2
AGEN,2,ALL,,,,-TF1/2,,1    ! GENERATE AREAS FROM EXISTING
                           ! PATTERN
TYPE,1                   !USE SHELL181 ELEMENTS
MAT,1                    !SWITCH TO MATERIAL 1
SECNUM,1                 !SWITCH TO SECTION NUMBER 1
AMESH,ALL                !MESH SELECTED AREAS
MODMSH, DETACH           !DETACH MODEL FROM MESH
NSEL,S,LOC,Z,0           !SELECT NODES AT Z = 0
NSEL,A,LOC,Z,-TC-TF1     !ADD NODES TO SELECTION
ESLN                     !SELECT ELEMENTS ATTACHED TO NODES
ESEL,INVE                !INVERT ELEMENT SELECTION
ESEL,U,TYPE,,2           !UNSELECT TYPE 2 ELEMENTS
NSLE                     !SELECT NODES ATTACHED TO ELEMENTS
EDEL,ALL                 !DELETE SELECTED ELEMENTS
NDEL,ALL                 !DELETE SELECTED NODES
ALLSEL                   !SELECT EVERYTHING
NSEL,S,LOC,Z,-TF1/2      !SELECT TOP FACESHEET NODES
*GET,NUM1,NODE,0,COUNT !RETRIEVE # OF NODES
*GET,MIN1,NODE,0,NUM,MIN   ! RETRIEVE MINIMUM NODE #
CURNOD=MIN1              !INITIALIZE CURRENT NODE #
*DO,I,1,NUM1             !BEGIN LOOP ON TOP FACESHEET NODES
ALLSEL                   !SELECT EVERYTHING
SLAVE=NODE(NX(CURNOD),NY(CURNOD),0)      ! STORE NODE # OF
                         !CORRESPONDING CORE NODE
CERIG,CURNOD,SLAVE,UXYZ  ! APPLY RIGID CONSTRAINT
NSEL,S,LOC,Z,-TF1/2      !SELECT TOP FACESHEET NODES
CURNOD=NDNEXT(CURNOD)    !UPDATE CURRENT NODE #
*ENDDO                   !END LOOP ON TOP FACESHEET NODES
ALLSEL                   !SELECT EVERYTHING
NSEL,S,LOC,Z,-TC-TF1/2 !SELECT BOTTOM FACESHEET NODES
*GET,NUM1,NODE,0,COUNT !RETRIEVE # OF NODES
*GET,MIN1,NODE,0,NUM,MIN   !      RETRIEVE MINUMUM NODE #
CURNOD=MIN1              !INITIALIZE CURRENT NODE #
*DO,I,1,NUM1             !BEGIN LOOP ON BOTTOM FACESHEET
                         !NODES
ALLSEL                   !SELECT EVERYTHING

SLAVE=NODE(NX(CURNOD),NY(CURNOD),-TC-TF1)   ! STORE NODE #
                         !OF CORRESPONDING CORE NODE
CERIG,CURNOD,SLAVE,UXYZ  ! APPLY RIGID CONSTRAINT
NSEL,S,LOC,Z,-TC-TF1/2 !SELECT BOTTOM FACESHEET NODES
```

```
CURNOD=NDNEXT(CURNOD)      ! UPDATE CURRENT NODE #
*ENDDO                     ! END LOOP ON BOTTOM FACESHEET
                           ! NODES
FINISH                     ! EXIT PREPROCESSOR
/SOLU                      ! ENTER SOLUTION PROCESSOR
NSEL,S,LOC,X,0             ! SELECT NODES AT X = 0
D,ALL,UX                   ! CONSTRAIN UX AT SELECTED NODES
NSEL,S,LOC,Y,0             ! SELECT NODES AT Y = 0
D,ALL,UY                   ! CONSTRAIN UY AT SELECTED NODES
NSEL,S,LOC,Y,H/2-1E-2      ! SELECT NODES AT Y = H/2-1E-2
D,ALL,UX                   ! CONSTRAIN UX AT SELECTED NODES
D,ALL,UZ                   ! CONSTRAIN UZ AT SELECTED NODES
NSEL,S,LOC,X,W/2-1E-2      ! SELECT NODES AT X = H/2-1E-2
D,ALL,UY                   ! CONSTRAIN UY AT SELECTED NODES
D,ALL,UZ                   ! CONSTRAIN UZ AT SELECTED NODES
ALLSEL                     ! SELECT EVERYTHING
NSEL,S,LOC,Z,-TC-TF1       ! SELECT NODES AT Z = -TC-TF1
NSEL,R,LOC,X,0,1           ! RESELECT NODES AT X = 0 & 1
NSEL,R,LOC,Y,0,1           ! RESELECT NODES AT Y = 0 & 1
SF,ALL,PRES,P              ! APPLY PRESSURE ON SELECTED NODES
ALLSEL                     ! SELECT EVERYTHING
/NERR,5,100000             ! INCREASE ALLOWED # OF ERROR MSG
SOLVE                      ! OBTAIN SOLUTION
SAVE                       ! SAVE MODEL
FINISH                     ! EXIT SOLUTION PROCESSOR
/POST1                     ! ENTER GENERAL POSTPROCESSOR
PLNSOL,S,EQV               ! OBTAIN EQUIVALENT STRESS CONTOURS
PLNSOL,S,Z                 ! OBTAIN STRESS CONTOURS IN Z-DIR
```

Figures 11.7 and 11.8 show the contours for equivalent stress and stress in the z-direction, respectively, obtained through the command input given above.

11.3 Submodeling

Submodeling is a method used for obtaining more accurate results in a specific region of the domain. It requires an existing model, which is called the *global* model, and the corresponding solution. The global model uses a coarse mesh to minimize computational cost while the submodel has a much finer mesh to enhance (capture) accuracy. The solution to the global model provides the boundary conditions for the submodel. Consider the gear shown in Fig. 11.9. If the primary interest is the level of stress intensity around the region indicated in Fig. 11.9, then a separate finite element model (submodel) of that region is generated with a refined mesh (also shown in Fig. 11.9). The boundaries of the submodel that are not external boundaries are called *cut boundaries*. The degree of freedom results from the global model are interpolated and applied as boundary conditions along the cut boundaries. The boundary conditions along the external boundaries of the submodel must be the

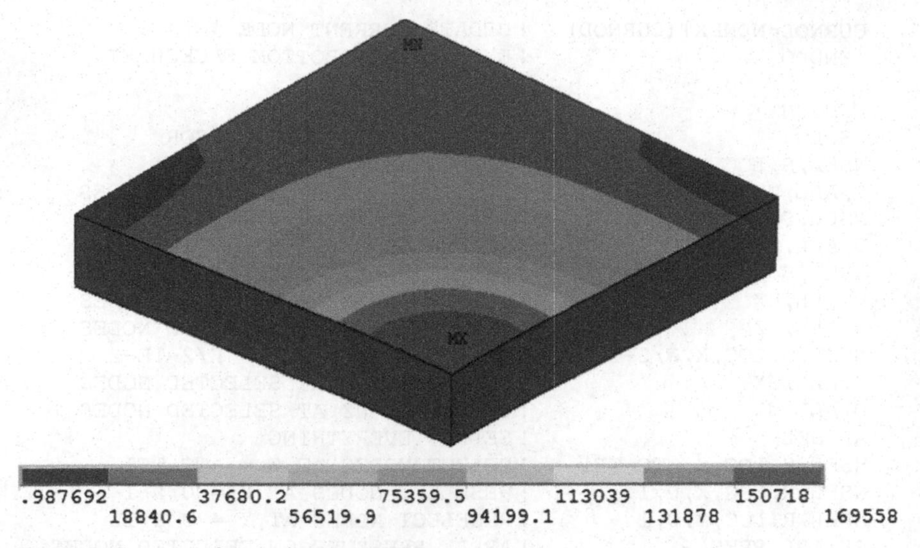

Fig. 11.7 Contours of equivalent stress

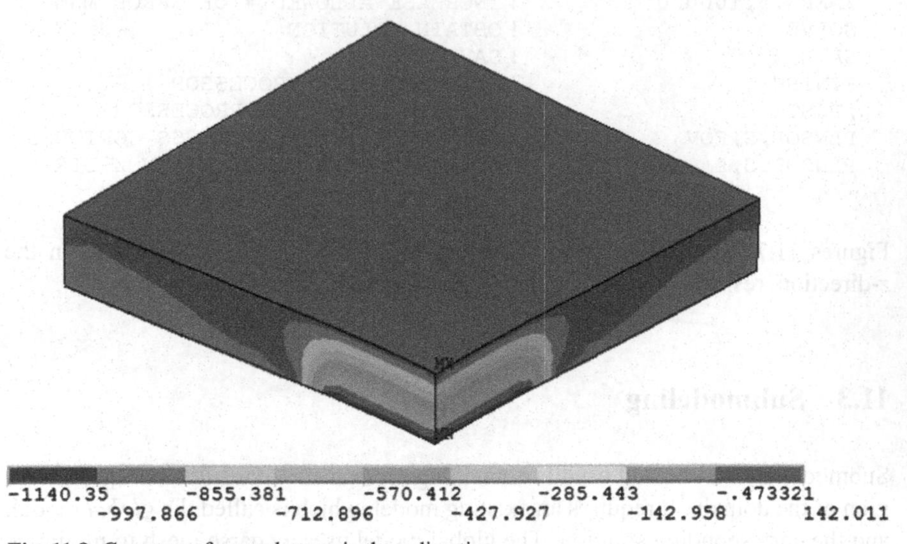

Fig. 11.8 Contours of normal stress in the *z*-direction

same as those of the global model. Cut boundaries do not need to coincide with the global element boundaries.

The steps required in a typical submodeling analysis are given as follows:

- Create global model and obtain solution.
- Save global model and solution.
- Create submodel and save.
- Select nodes along cut boundaries and write to a file.

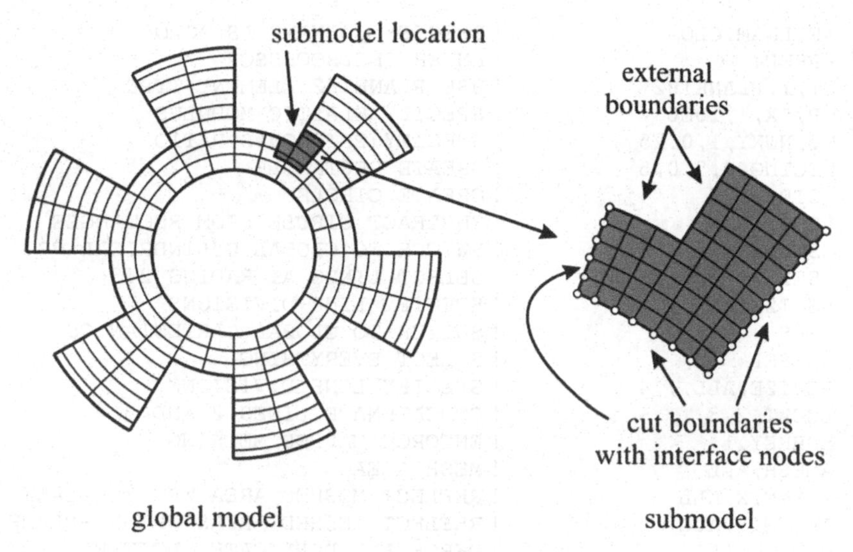

Fig. 11.9 Meshes for the global model and submodel

- Resume from the global model.
- Retrieve global solution.
- Perform cut boundary interpolation (writes boundary conditions to cut boundary condition file).
- Resume from the submodel.
- Read cut boundary condition file.
- Apply remaining loads (body loads and/or conditions along external boundaries).
- Obtain submodel solution.

The following example—submodeling analysis of a plate with a circular hole—demonstrates how the submodeling method is used in an analysis.

A thin, square plate ($10\times10\text{in}^2$) with a circular hole (radius r=2.5 in) is subjected to a uniformly distributed tensile loading (1000 psi) in the vertical direction along its top edge while being fixed along the bottom edge (Fig. 11.10). Because the plate is thin and subjected to in-plane loads, plane stress idealization is valid. The material properties are elastic modulus, $E = 10\times10^6$ psi, and Poisson's ratio $\nu = 0.25$. The goal is to perform a submodeling analysis around the hole to obtain more accurate stress results.

Global Model Analysis

The global model analysis is straightforward. However, in order for the submodeling analysis to be successful, it is crucial that distinct *jobnames* for both the global model and the submodel be specified (**/FILNAM** command). The *jobname* for the global model is specified as **GLO** in this example. Once the global model solution is obtained, the **SAVE** command is issued, which saves both the model and the solution.

```
/FILNAM,GLO              ! SPECIFY JOBNAME AS 'GLO'
/PREP7                   ! ENTER PREPROCESSOR
ET,1,PLANE182            ! USE PLANE182 ELEMENT TYPE
MP,EX,1,10E6             ! SPECIFY ELASTIC MODULUS
MP,NUXY,1,0.25           ! SPECIFY POISSON'S RATIO
RECTNG,0,5,0,5           ! CREATE RECTANGLE
PCIRC,2.5                ! CREATE CIRCLE
ASBA,1,2                 ! SUBTRACT CIRCLE FROM RECTANGLE
CSYS,1                   ! SWITCH TO GLOBAL CYLINDRICAL CS
LSEL,S,LOC,X,2.5         ! SELECT LINES AT RADIUS 2.5
LESIZE,ALL,,,8           ! SPECIFY LINE DIVISIONS
CSYS                     ! SWITCH TO GLOBAL CARTESIAN CS
ALLSEL                   ! SELECT EVERYTHING
LESIZE,ALL,,,4           ! SPECIFY LINE DIVISIONS
LCCAT,2,3                ! CONCATENATE LINES 2 AND 3
MSHKEY,1                 ! ENFORCE MAPPED MESHING
AMESH,ALL                ! MESH AREA
ARSYM,X,ALL              ! REFLECT MESHED AREA WRT YZ-PLANE
ARSYM,Y,ALL              ! REFLECT MESHED AREAS WRT XZ-PLANE
NUMMRG,ALL               ! MERGE ALL DUPLICATE ENTITIES
FINISH                   ! EXIT PREPROCESSOR

/SOLU                    ! ENTER SOLUTION PROCESSOR
NSEL,S,LOC,Y,-5          ! SELECT NODES AT Y = -5
D,ALL,ALL                ! CONSTRAIN ALL DOFS AT SELECTED
                         ! NODES
NSEL,S,LOC,Y,5           ! SELECT NODES AT Y = 5
SF,ALL,PRES,-1000        ! APPLY PRESSURE AT SELECTED NODES
ALLSEL                   ! SELECT EVERYTHING
SOLVE                    ! OBTAIN GLOBAL MODEL SOLUTION
SAVE                     ! SAVE MODEL AND RESULTS
FINISH                   ! EXIT SOLUTION PROCESSOR

/POST1                   ! ENTER GENERAL POSTPROCESSOR
PLNSOL,S,Y               ! OBTAIN STRESS CONTOURS IN Y-
                         ! DIRECTION
FINISH                   ! EXIT GENERAL POSTPROCESSOR
```

Submodel Analysis

As part of the submodeling analysis, the mesh for the area enclosed by dashed lines indicated in Fig. 11.10 is created; the submodel encompasses this area. Figure 11.11 shows the mesh for the global model and the region where the submodel is created. The refined mesh for the submodel is shown in Fig. 11.12. The *jobname* for the submodel is specified as **SUB** in this example. All of the lines except the circular hole boundary are the cut boundaries. After creating the mesh for the submodel, nodes along the cut boundaries are selected and written to an external ASCII file (**SBS.NOD**) using the **NWRITE** command, after which the submodel is saved (**SAVE** command). The cut boundary interpolation can be performed using the cut boundary nodes in the SBS.NOD file and the global model solution in the results file

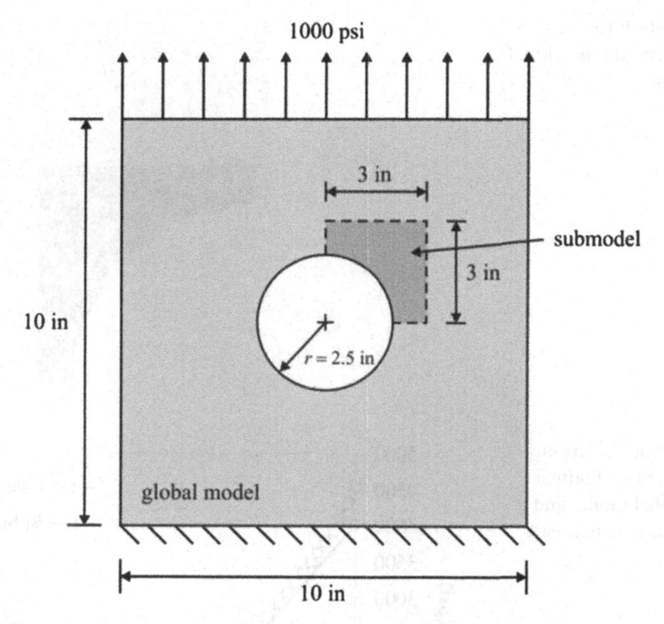

Fig. 11.10 Plate with a circular hole subjected to uniform tensile loading

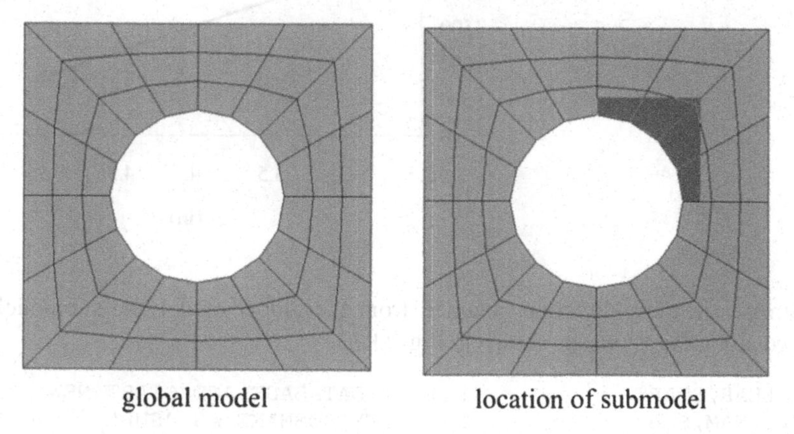

Fig. 11.11 Mesh for the global model (*left*), and the location of the submodel (*right*)

(`GLO.RST`). For this purpose, the global model is loaded (**`RESUME`** command), and results are read from the results file **`GLO.RST`**.

Using the **`CBDOF`** command, the cut boundary interpolation is performed, which writes the interpolated results in the cut boundary condition file (**`SBS.CBD`** in this example). The submodel is then loaded and the **`SBS.CBD`** file is read as input.

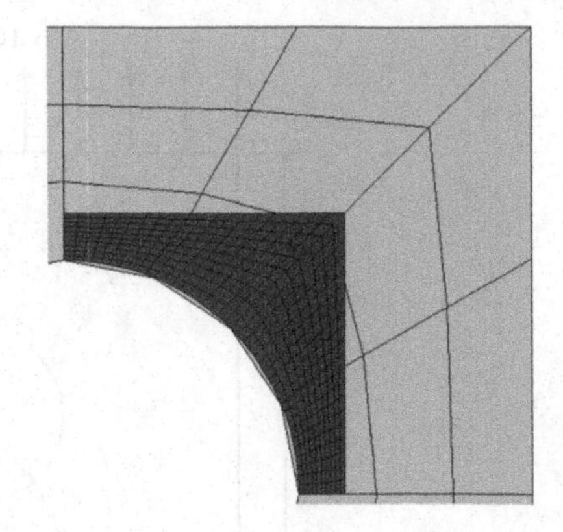

Fig. 11.12 Mesh for the submodel (overlaid on the global model mesh)

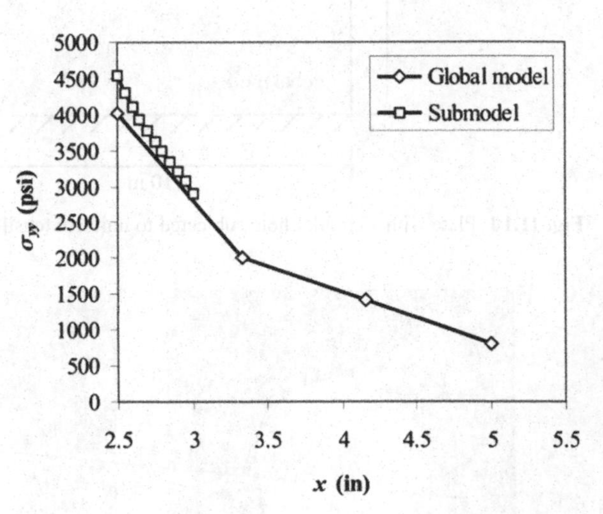

Fig. 11.13 Normal stress in the y-direction obtained from the global model and submodel, plotted against x

Stresses in the y-direction obtained from the global model and submodel are plotted along the x-axis, as shown in Fig. 11.13.

```
/CLEAR,START                  ! CLEAR DATABASE AND START NEW
/FILNAM,SUB                   ! SPECIFY JOBNAME AS 'SUB'
/PREP7                        ! ENTER PREPROCESSOR
ET,1,PLANE182                 ! USE PLANE182 ELEMENT TYPE
MP,EX,1,10E6                  ! SPECIFY ELASTIC MODULUS
MP,NUXY,1,0.25                ! SPECIFY POISSON'S RATIO
RECTNG,0,3,0,3                ! CREATE RECTANGLE
PCIRC,2.5                     ! CREATE CIRCLE
```

```
ASBA,1,2                   ! SUBTRACT CIRCLE FROM RECTANGLE
LESIZE,9,,,10              ! SPECIFY LINE DIVISIONS
LESIZE,10,,,10             !
LESIZE,5,,,40              !
LESIZE,2,,,20              !
LESIZE,3,,,20              !
LCCAT,2,3                  ! CONCATENATE LINES 2 AND 3
MSHKEY,1                   ! ENFORCE MAPPED MESHING
AMESH,ALL                  ! MESH AREAS
NSEL,S,LOC,Y,0             ! SELECT NODES AT Y = 0
NSEL,A,LOC,X,0             ! ADD NODES AT X = 0 TO SELECTION
NSEL,A,LOC,Y,3             ! ADD NODES AT Y = 3 TO SELECTION
NSEL,A,LOC,X,3             ! ADD NODES AT X = 3 TO SELECTION
NWRITE,SBS,NOD             ! WRITE NODES TO FILE 'SBS.NOD'
ALLSEL                     ! SELECT EVERYTHING
SAVE                       ! SAVE MODEL
FINISH                     ! EXIT PREPROCESSOR

RESUME,GLO,DB              ! RESUME FROM 'GLO.DB' FILE
/POST1                     ! ENTER GENERAL POSTPROCESSOR
FILE,GLO,RST               ! DECLARE RESULTS FILE
CBDOF,SBS,NOD,,SBS,CBD     ! PERFORM CUT BOUNDARY
                           ! INTERPOLATION
FINISH                     ! EXIT GENERAL POSTPROCESSOR

RESUME,SUB,DB              ! RESUME FROM 'SUB.DB' FILE
/SOLU                      ! ENTER SOLUTION PROCESSOR
/INPUT,SBS,CBD             ! READ CUT BOUNDARY CONDITIONS
SOLVE                      ! OBTAIN SOLUTION
FINISH                     ! EXIT SOLUTION PROCESSOR

/POST1                     ! ENTER GENERAL POSTPROCESSOR
PLNSOL,S,Y                 ! OBTAIN STRESS CONTOURS IN Y-
                           ! DIRECTION
```

11.4 Substructuring: Superelements

Consider the aircraft shown in Fig. 11.14. A detailed finite element model of an aircraft may contain several hundred thousand degrees of freedom. In order to reduce computational cost without losing accuracy, the matrix equations of the tail and wing sections can be condensed in terms of the corresponding interface nodes (*Master DOFs*) identified in Fig. 11.14. Thus, during the analysis, the size of the matrix system of equations becomes the number of unknown degrees of freedom of the fuselage section (*main structure*). Once the solution is obtained, all degrees of freedom belonging to the nodes of the fuselage are known, including the *Master DOFs*. Finally, the solution for the tail and wing sections is obtained by using the known *Master DOFs* as the boundary conditions.

The linear matrix equations representing a group of finite elements (and nodes) can be condensed into a smaller matrix system corresponding to a number of selected degrees of freedom (DOF). This procedure is commonly known as **Static Condensation or Substructuring.**

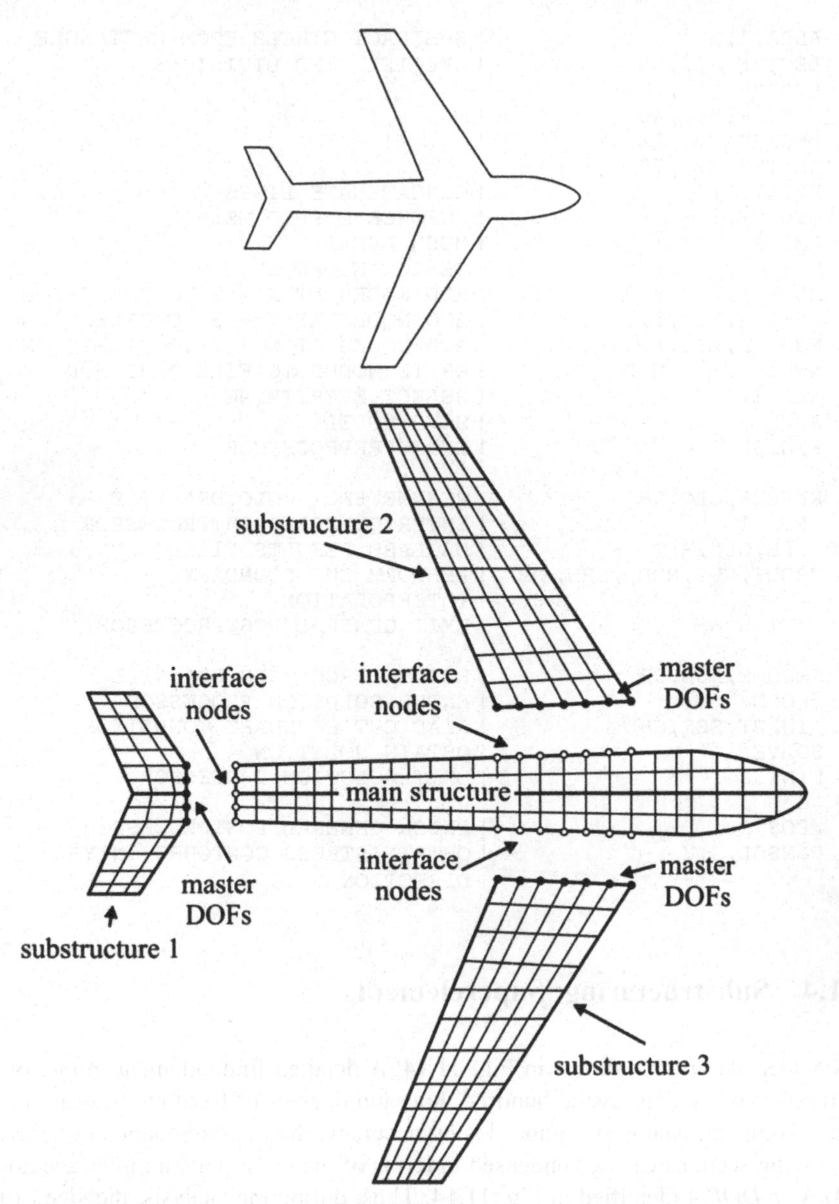

Fig. 11.14 An aircraft structure and its finite element model divided into four parts (three superelements)

The DOF selected in the analysis is referred to as the *Master DOF*. The resulting matrix may be considered as the element equations of a special element, which is called a *Substructure* or *Superelement*. Superelements are attached to the main structure through the *Master DOF*. The main advantage of substructuring is that it allows for the solution of large complex structures at a reduced computational cost (both memory and CPU time). Another advantage is that if the problem involves

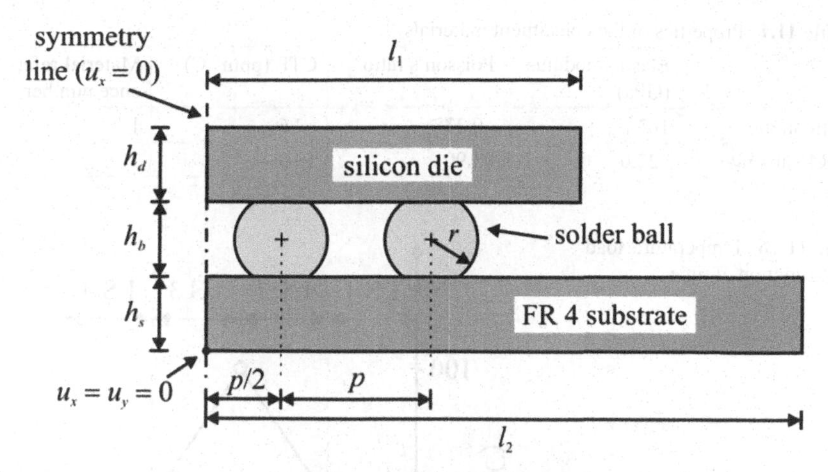

Fig. 11.15 Electronic package with FR4 substrate, solder balls, and silicon die

a combination of materials with linear and nonlinear material behavior, the linear portion can be condensed into a superelement so that only the matrices corresponding to the nonlinear portion are recalculated during the iterations of the nonlinear solution.

A substructuring analysis involves three main steps:

Generation Pass: Superelements are generated during this step. It involves selection of the master degrees of freedom, which are located along the interface between the superelement and the main structure. At the end of this step, matrices corresponding to each superelement are written to separate files in the *Working Directory*.
Use Pass: The main structure is generated in this step. Also, the superelements are introduced to the main structure. At the end of this step, the solution for the main structure is obtained, which includes the solution for the master degrees of freedom of each superelement.
Expansion Pass: During this step, complete solutions for the superelements are obtained by *expanding* the master degree of freedom solution available from the *use pass*.

Substructuring is demonstrated by considering an analysis of an electronic package with linear and nonlinear material behaviors.

The electronic package shown in Fig. 11.15 is composed of an FR4 substrate, two solder balls, and a silicon die. As indicated in the figure, thicknesses of the substrate and die are h_s and h_d, respectively, and the stand-off height is h_b. The lengths of the die and substrate are l_1 and l_2, respectively. The pitch for the solder balls is p, and the balls are assumed to have circular geometry with radius r. Material properties of the die and the substrate are tabulated in Table 11.1. The solder behavior is viscoplastic using Anand's model, having properties given in Fig. 10.13, Tables 10.2, and 10.3. The following numerical values of the parameters are used in the analysis: $h_s = 2$ mm, $h_b = 0.7$ mm, $h_d = 1$ mm, $p = 2$ mm, and $r = 0.5$ mm. The temperature profile as a function of time is shown in Fig. 11.16, which is implemented in the ANSYS solution using four load steps. The stress-free temperature is taken as 0 °C. The temperature of the entire structure is first increased to 60 °C

Table 11.1 Properties of the constituent materials

	Elastic modulus (GPa)	Poisson's ratio	CTE (ppm/°C)	Material reference number
Silicon die	163	0.278	2.6	1
FR4 substrate	22.0	03.90	18.0	3

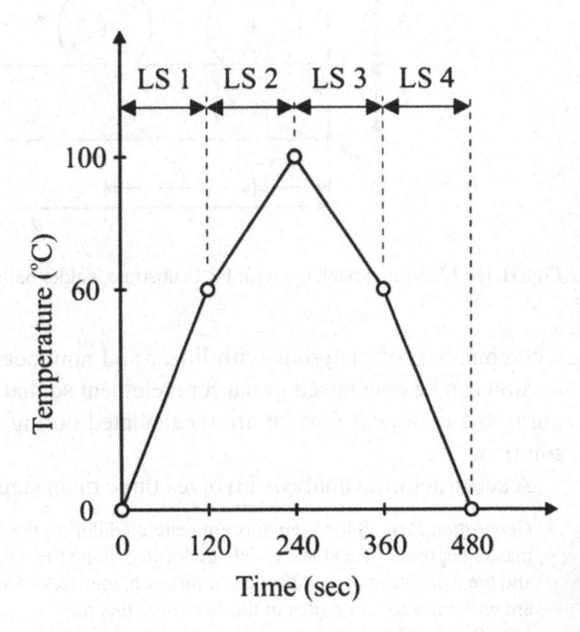

Fig. 11.16 Temperature load as a function of time

over a period of 120 s, followed by an increase to 100 °C over 120 s. Finally, the package is cooled down, following the same path along the temperature vs. time curve. The symmetry plane is constrained in the x-direction. Also, the y-direction displacement is suppressed at the bottom-left corner in order to prevent rigid-body translations. The goal is to solve this problem utilizing the substructuring technique. Figure 11.17 shows the superelement and main structure, and a close-up of the master degrees of freedom is shown in Fig. 11.18.

11.4.1 Generation Pass

A *Generation Pass* is composed of two main parts: (i) model generation and (ii) superelement generation. In this analysis, the solder balls are the main model and the substrate and the die constitute the superelement.

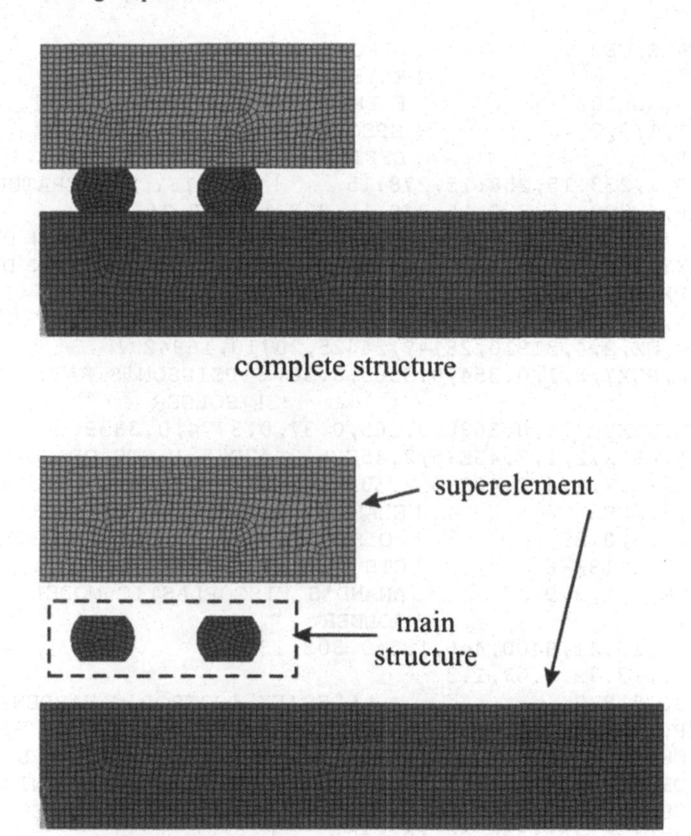

Fig. 11.17 Superelement and main structure

11.4.1.1 Model Generation

The solder material is modeled using **PLANE182** elements. In order to specify temperature-dependent elastic properties, a temperature table is constructed using the **MPTEMP** command, followed by the specification of properties using the MPDATA command. Note that temperatures are specified in Kelvin. Anand's viscoplastic properties are specified using the **TB** and **TBDATA** commands. In order for the main model and superelement interface nodes to have the same numbers (for continuity), the full model is generated (shown in Fig. 11.17) and saved in a file named **SUB1.** DB. After this operation, the elements (and attached nodes) corresponding to the solder are deleted (**EDEL** and **NDEL** commands), leaving the mesh for only the superelement.

```
/FILNAME,GEN                  !SPECIFY JOBNAME AS GEN
/PREP7                        !ENTER PREPROCESSOR
ET,1,PLANE182                 !ELEMENT TYPE 1 IS PLANE182
KEYOPT,1,3,2                  ! SPECIFY PLANE STRAIN
MPTEMP                        !DEFINE TEMPERATURE TABLE
MPTEMP,1,238.15,258.15,278.15     ! SPECIFY TEMPERATURES
MPTEMP,4,298.15,323.15,348.15,373.15,398.15 !
MP,EX,1,163E3                 !ELASTIC MODULUS OF SILICON DIE
MP,NUXY,1,0.278               !POISSON'S RATIO OF SILICON DIE
MP,ALPX,1,2.6E-6              !CTE OF SILICON DIE
MPDATA,EX,2,1,40781,37825,34884  ! ELASTIC MOD. OF SOLDER
MPDATA,EX,2,4,31910,28149,24425,20710,16942 !
MPDATA,NUXY,2,1,0.354,0.3565,0.36  ! POISSON'S RAT. OF
                                   ! SOLDER
MPDATA,NUXY,2,4,0.3628,0.365,0.37,0.3774,0.3839 !
MPDATA,ALPX,2,1,2.43E-5,2.45E-5,2.47E-5 ! CTE OF SOLDER
MPDATA,ALPX,2,4,2.48E-5,2.50E-5,2.53E-5,2.55E-5,2.58E-5 !
MP,EX,3,22E3                  !ELASTIC MODULUS OF FR4 SUBSTRATE
MP,NUXY,3,0.39                !POISSON'S RATIO OF FR4 SUBSTRATE
MP,ALPX,3,18E-6               !CTE OF FR4 SUBSTRATE
TB,RATE,2,1,9,9               !ANAND'S VISCOPLASTIC MODEL FOR
                              !SOLDER
TBDATA,,12.41,9400,4e6,1.5,0.303,1379
TBDATA,,13.79,0.07,1.3
TB,MISO,2,8,7,                   !SPECIFY ISOTROPIC HARDENING
TBTEMP,-35+273.15                !SPECIFY TEMPERATURE AS -35 C
TBPT,DEFI,9.313161E-04,37.980      !SPECIFY STRAIN VS
TBPT,DEFI,8.931316E-03,50.677      !STRESS DATA POINTS AT
TBPT,DEFI,1.693132E-02,56.307      !THIS TEMPERATURE
TBPT,DEFI,2.493132E-02,60.142      !
TBPT,DEFI,4.893132E-02,67.620      !
TBPT,DEFI,8.093132E-02,73.889      !
TBPT,DEFI,1.009313E-01,76.836      !
TBTEMP,-15+273.15                !SPECIFY TEMPERATURE AS -15 C
TBPT,DEFI,9.295968E-04,35.162      !SPECIFY STRAIN VS
TBPT,DEFI,8.929597E-03,46.917      !STRESS DATA POINTS AT
TBPT,DEFI,1.692960E-02,52.129      !THIS TEMPERATURE
TBPT,DEFI,2.492960E-02,55.680      !
TBPT,DEFI,4.892960E-02,62.603      !
TBPT,DEFI,8.092960E-02,68.407      !
TBPT,DEFI,1.009296E-01,71.136      !
TBTEMP,5+273.15                  !SPECIFY TEMPERATURE AS 5 C
TBPT,DEFI,9.271872E-04,32.344      !SPECIFY STRAIN VS
TBPT,DEFI,8.927187E-03,43.158      !STRESS DATA POINTS AT
```

```
TBPT,DEFI,1.692719E-02,47.952     !THIS TEMPERATURE
TBPT,DEFI,2.492719E-02,51.218     !
TBPT,DEFI,4.892719E-02,57.586     !
TBPT,DEFI,8.092719E-02,62.925     !
TBPT,DEFI,1.009272E-01,65.435     !
TBTEMP,25+273.15              !SPECIFY TEMPERATURE AS 25 C
TBPT,DEFI,9.252899E-04,29.526     !SPECIFY STRAIN VS
TBPT,DEFI,8.925290E-03,39.398     !STRESS DATA POINTS AT
TBPT,DEFI,1.692529E-02,43.774     !THIS TEMPERATURE
TBPT,DEFI,2.492529E-02,46.756     !
TBPT,DEFI,4.892529E-02,52.569     !
TBPT,DEFI,8.092529E-02,57.443     !
TBPT,DEFI,1.009253E-01,59.734     !
TBTEMP,50+273.15              !SPECIFY TEMPERATURE AS 50 C
TBPT,DEFI,9.237984E-04,26.004     !SPECIFY STRAIN VS
TBPT,DEFI,8.923798E-03,34.698     !STRESS DATA POINTS AT
TBPT,DEFI,1.692380E-02,38.552     !THIS TEMPERATURE
TBPT,DEFI,2.492380E-02,41.178     !
TBPT,DEFI,4.892380E-02,46.298     !
TBPT,DEFI,8.092380E-02,50.590     !
TBPT,DEFI,1.009238E-01,52.608     !
TBTEMP,75+273.15              !SPECIFY TEMPERATURE AS 75 C
TBPT,DEFI,9.204504E-04,22.482     !SPECIFY STRAIN VS
TBPT,DEFI,8.920450E-03,29.998     !STRESS DATA POINTS AT
TBPT,DEFI,1.692045E-02,33.330     !THIS TEMPERATURE
TBPT,DEFI,2.492045E-02,35.601     !
TBPT,DEFI,4.892045E-02,40.027     !
TBPT,DEFI,8.092045E-02,43.738     !
TBPT,DEFI,1.009205E-01,45.482     !
TBTEMP,100+273.15             !SPECIFY TEMPERATURE AS 100 C
TBPT,DEFI,9.154515E-04,18.959     !SPECIFY STRAIN VS
TBPT,DEFI,8.915451E-03,25.298     !STRESS DATA POINTS AT
TBPT,DEFI,1.691545E-02,28.108     !THIS TEMPERATURE
TBPT,DEFI,2.491545E-02,30.023     !
TBPT,DEFI,4.891545E-02,33.756     !
TBPT,DEFI,8.091545E-02,36.885     !
TBPT,DEFI,1.009155E-01,38.357     !
TBTEMP,125+273.15             !SPECIFY TEMPERATURE AS 125 C
TBPT,DEFI,9.111675E-04,15.437     !SPECIFY STRAIN VS
TBPT,DEFI,8.911168E-03,20.598     !STRESS DATA POINTS AT
TBPT,DEFI,1.691117E-02,22.886     !THIS TEMPERATURE
TBPT,DEFI,2.491117E-02,24.445     !
TBPT,DEFI,4.891117E-02,27.485     !
TBPT,DEFI,8.091117E-02,30.033     !
TBPT,DEFI,1.009112E-01,31.231     !
HS=2                          !SUBSTRATE THICKNESS
HB=0.7                        !STAND-OFF HEIGHT
HD=1                          !THICKNESS OF DIE
R=0.5                         !RADIUS OF SOLDER BALLS
P=2                           !PITCH
L1=5                          !LENGTH OF DIE
L2=10                         !LENGTH OF SUBSTRATE
R1=SQRT(R**2-(HB/2)**2)       !HALF-LENGTH OF SOLDER AT
                              !TRUNCATION PT
CYL4,P/2,HS+HB/2,R            !CREATE 1ST SOLDER BALL
```

```
CYL4,3*P/2,HS+HB/2,R     ! CREATE 2ND SOLDER BALL
BLC4,0,0,L2,HS           ! CREATE SUBSTRATE
BLC4,0,HS+HB,L1,HS       ! CREATE DIE
ASEL,S,,,1,2             ! SELECT AREAS FOR SOLDER BALLS
CM,A1,AREA               ! CREATE COMPONENT OF AREAS
ASEL,S,,,3,4             ! SELECT AREAS FOR DIE AND
                         ! SUBSTRATE
CM,A2,AREA               ! CREATE COMPONENT OF AREAS
ALLSEL                   ! SELECT EVERYTHING
ASBA,A1,A2,,,KEEP        ! SUBTRACT AREAS (KEEP SUBTRACTED
                         ! AREAS)
AGLUE,ALL                ! GLUE ALL AREAS
ESIZE,0.1                ! SPECIFY ELEMENT SIZE
AMESH,ALL                ! MESH AREAS
ASEL,S,LOC,Y,HS/2        ! SELECT AREAS AT Y = HS/2
ESLA,S                   ! SELECT ELEMENTS ATTACHED TO AREAS
EMODIF,ALL,MAT,3         ! MODIFY MATERIAL ATTRIBUTE
ASEL,S,LOC,Y,HS+HB/2     ! SELECT AREAS AT Y = HS+HB/2
ESLA,S                   ! SELECT ELEMENTS ATTACHED TO AREAS
EMODIF,ALL,MAT,2         ! MODIFY MATERIAL ATTRIBUTE
EMODIF,ALL,TYPE,2        ! MODIFY ELEMENT TYPE ATTRIBUTE
ALLSEL                   ! SELECT EVERYTHING
MODMSH,DETACH            ! DETACH SOLID MODEL FROM MESH
SAVE,SUB1,DB             ! SAVE AS SUB1.DB
ESEL,S,MAT,,2            ! SELECT SOLDER ELEMENTS
EDEL,ALL                 ! DELETE SELECTED ELEMENTS
NDEL,ALL                 ! DELETE FREE NODES
ALLSEL                   ! SELECT EVERYTHING
FINISH                   ! EXIT PREPROCESSOR
```

11.4.1.2 Superelement Generation

The analysis type is specified as *Substructuring*. This makes sure that, when the **SOLVE** command is issued, the superelement file (**GEN.SUB**) is generated in the *Working Directory*. The superelement name is specified using the **SEOPT** command. The interface nodes between the main model and the superelement are defined as the master degrees of freedom using the M command (Fig. 11.18). After applying the displacement constraints (D command) and thermal load (**BFUNIF** command), the substructure is saved in the **GEN.DB** file, which is necessary for expanding the solution for the superelement after the solution for the main model is obtained. Figure 11.19 shows the superelement with the interface nodes indicated. Although the superelement contains only materials with linear behavior, time values consistent with the applied thermal load profile (shown in Fig. 11.16) must be given. This is performed using the **TIME** command. The stress-free temperature is specified using the **TREF** command. Load vectors corresponding to each load step are obtained by using the **SOLVE** command repeatedly while changing the time value and thermal load.

```
CM,N1,NODE                          ! COMPONENT 1
NSEL,S,LOC,Y,HS                     !
NSEL,R,LOC,X,P/2-R1,P/2+R1          !
CM,N2,NODE                          ! COMPONENT 2
NSEL,S,LOC,Y,HS+HB                  !
NSEL,R,LOC,X,3*P/2-R1,3*P/2+R1      !
CM,N3,NODE                          ! COMPONENT 3
NSEL,S,LOC,Y,HS                     !
NSEL,R,LOC,X,3*P/2-R1,3*P/2+R1      !
CM,N4,NODE                          ! COMPONENT 4
ALLSEL                              ! SELECT EVERYTHING
CMSEL,S,N1,NODE                     ! SELECT COMPONENT N1
CMSEL,A,N2,NODE                     ! ADD COMPONENT N2 TO THE
                                    ! SELECTION
CMSEL,A,N3,NODE                     ! ADD COMPONENT N3 TO THE
                                    ! SELECTION
CMSEL,A,N4,NODE                     ! ADD COMPONENT N4 TO THE
                                    ! SELECTION
M,ALL,ALL                           ! DEFINE MASTER DEGREES OF
                                    ! FREEDOM
ALLSEL                              ! SELECT EVERYTHING
NSEL,S,LOC,X,0                      ! SELECT NODES ALONG X = 0
D,ALL,UX,0                          ! CONSTRAIN X-DISPL AT
                                    ! SELECTED NODES
NSEL,R,LOC,Y,0                      ! RESELECT NODE AT Y = 0
D,ALL,ALL                           ! CONSTRAIN ALL DOFS
ALLSEL                              ! SELECT EVERYTHING
TREF,273.15                         ! STRESS-FREE TEMPERATURE
                                    ! IS 273.15 K
BFUNIF,TEMP,333.15                  ! UNIFORM TEMPERATURE
                                    ! INCREASE TO 333.15 K
TIME,120                            ! TIME AT THE END OF LOAD
                                    ! STEP = 120 SEC
SAVE                                ! SAVE DATABASE TO GEN.DB
                                    ! OR EXPANSION PASS
SOLVE                               ! OBTAIN SOLUTION -
                                    ! CREATES GEN.SUB FILE
                                    ! WHICH CONTAINS
                                    ! SUPERELEMENT MATRIX AND
                                    ! LOAD VECTOR FOR 1ST LOAD
                                    ! STEP
BFUNIF,TEMP,373.15                  ! UNIFORM TEMPERATURE
                                    ! INCREASE TO 373.15 K
TIME,240                            ! TIME AT THE END OF LOAD
                                    ! STEP = 240 SEC
SOLVE                               ! OBTAIN LOAD VECTOR FOR
                                    ! SECOND LOAD STEP
BFUNIF,TEMP,333.15                  ! UNIFORM TEMPERATURE
                                    ! DECREASE TO 333.15 K
TIME,360                            ! TIME AT THE END OF LOAD
                                    ! STEP = 360 SEC
SOLVE                               ! OBTAIN LOAD VECTOR FOR
                                    ! THIRD LOAD STEP
BFUNIF,TEMP,273.15                  ! UNIFORM TEMPERATURE
                                    ! DECREASE TO 273.15 K
TIME,480                            ! TIME AT THE END OF LOAD
                                    ! STEP = 480 SEC
SOLVE                               ! OBTAIN LOAD VECTOR FOR
                                    ! FOURTH LOAD STEP
FINISH                              ! EXIT SOLUTION PROCESSOR
```

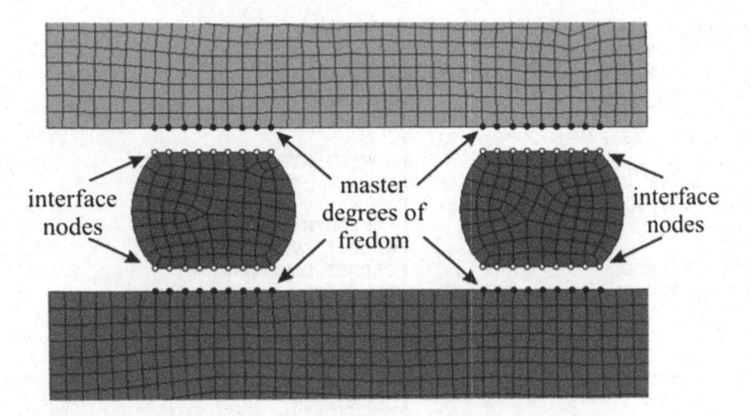

Fig. 11.18 Close-up of master degrees of freedom

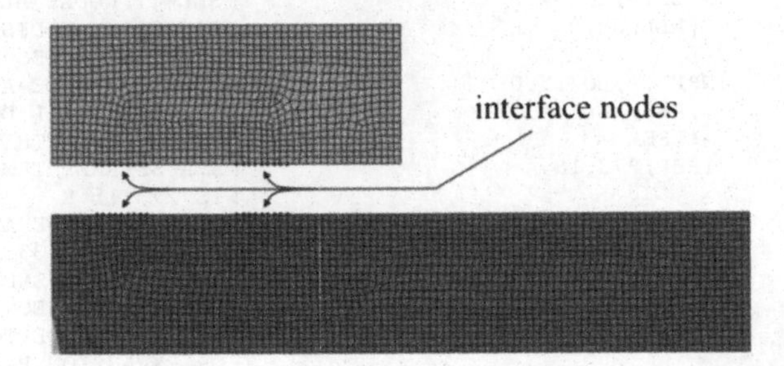

Fig. 11.19 Superelement mesh with interface nodes indicated

11.4.2 Use Pass

At this point, the *Working Directory* includes the superelement file **GEN.SUB**. The main model is created conveniently by resuming from the **SUB1.DB** file and eliminating the elements (and attached nodes) corresponding to the substrate and the die. Element type 3 is defined as **MATRIX50**, which is used for superelements. The superelement file **GEN.SUB** is read by using the SE command, after which an element plot is obtained (**EPLOT** command) to verify the location of the superelement. In the solution phase of the *Use Pass*, thermal loading for the main model is applied, followed by reading the superelement load vector corresponding to the first load step using the **SFE** command with the **SELV** label. Finally, the solution for the main model is obtained. This procedure is repeated for load steps 2 through 4. Figure 11.20 shows the main model with the interface nodes indicated.

```
/CLEAR                          ! CLEAR DATABASE
/FILNAME,USE                    ! SPECIFY JOBNAME AS USE
RESUME,SUB1,DB                  ! RESUME FROM SUB1.DB
/PREP7                          ! ENTER PREPROCESSOR
ET,2,MATRIX50                   ! MATRIX50 IS ELEMENT TYPE 2
                                ! (FOR SUPERELEMENT)
ESEL,U,MAT,,2                   ! UNSELECT SOLDER ELEMENTS
```

```
EDEL,ALL                      ! DELETE ALL SELECTED ELEMENTS
NDEL,ALL                      ! DELETE ALL FREE NODES
ALLSEL                        ! SELECT EVERYTHING
TYPE,2                        ! SWITCH TO ELEMENT TYPE 2
SE,GEN                        ! READ IN SUPERELEMENT
EPLOT                         ! VERIFY LOCATION OF
                              ! SUPERELEMENT
ALLSEL                        ! SELECT EVERYTHING
FINISH                        ! ENTER PREPROCESSOR
/SOLU                         ! ENTER SOLUTION PROCESSOR
NLGEOM,ON                     ! TURN NONLINEAR GEOMETRY
                              ! EFFECTS ON
KBC,0                         ! INTERPOLATE LOADS BETWEEN
                              ! LOAD STEPS
AUTOTS,ON                     ! TURN AUTOMATIC TIME
                              ! STEPPING ON
ESEL,S,TYPE,,2                ! SELECT SUPERELEMENT
*GET,SENUM,ELEM,0,NUM,MAX     ! STORE ELEM # INTO
                              ! PARAMETER SENUM
ALLSEL                        ! SELECT EVERYTHING
TREF,273.15                   ! STRESS-FREE TEMP IS 273.15 K
BFUNIF,TEMP,333.15            ! UNIFORM TEMP. INCREASE TO
                              ! 333.15 K
SFE,SENUM,1,SELV,,1           ! READ IN SUPERELEMENT LOAD
                              ! VECTOR 1
TIME,120                      ! TIME AT THE END OF
                              ! LS = 120 SEC
DELTIM,10,1,10                ! TIME STEP SIZE SETTINGS
SAVE                          ! SAVE DATABASE TO USE.DB
SOLVE                         ! OBTAIN SOLUTION FOR
                              ! LOAD STEP 1
BFUNIF,TEMP,373.15            ! UNIFORM TEMP. INCREASE TO
                              ! 373.15 K
TIME,240                      ! TIME AT THE END OF
                              ! LS = 240 SEC
SFE,SENUM,2,SELV,,1           ! READ IN SUPERELEMENT LOAD
                              ! VECTOR 2
SOLVE                         ! OBTAIN SOLUTION FOR
                              ! LOAD STEP 2
BFUNIF,TEMP,333.15            ! UNIFORM TEMP. DECREASE TO
                              ! 333.15 K
TIME,360                      ! TIME AT THE END OF
                              ! LS = 360 SEC
SFE,SENUM,3,SELV,,1           ! READ IN SUPERELEMENT LOAD
                              ! VECTOR 3
SOLVE                         ! OBTAIN SOLUTION FOR
                              ! LOAD STEP 3
BFUNIF,TEMP,273.15            ! UNIFORM TEMP. DECREASE TO
                              ! 273.15 K
TIME,480                      ! TIME AT THE END OF
                              ! LS = 480 SEC
SFE,SENUM,4,SELV,,1           ! READ IN SUPERELEMENT LOAD
                              ! VECTOR 4
SOLVE                         ! OBTAIN SOLUTION FOR
                              ! LOAD STEP 4
FINISH                        ! EXIT SOLUTION PROCESSOR
/POST1                        ! ENTER GENERAL POSTPROCESSOR
PLNSOL,NL,PLWK                ! PLOT CONTOURS OF PLASTIC WORK
FINISH                        ! EXIT GENERAL POSTPROCESSOR
```

The solution contains all results items for the main model, including the degree of freedom results for the nodes along the interface between the main model and the superelement. The superelement degree-of-freedom solution is also stored in the

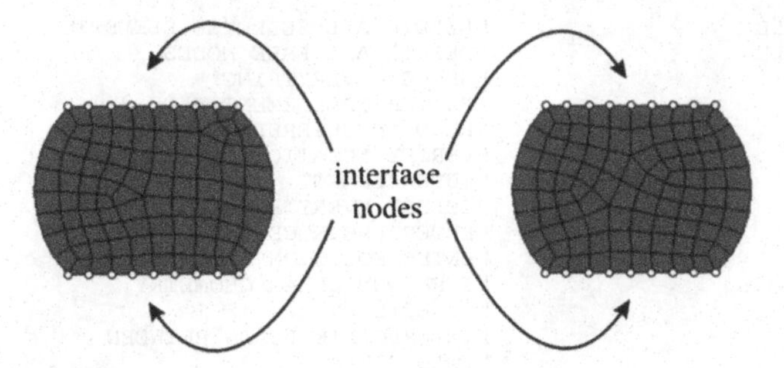

Fig. 11.20 Main model mesh with interface nodes indicated

file USE.DSUB, which is used in the *Expansion Pass*. Figure 11.21 shows the contour plots of plastic work at the end of the fourth load step.

11.4.3 Expansion Pass

In order to obtain a complete a solution for the superelement, the database is first cleared, followed by resuming (**RESUME** command) from the **GEN.DB** file. *Expansion Pass* is turned on using the **EXPASS** command. Names of the superelement file (**GEN.SUB**) and the file containing the superelement solution (**USE.DSUB**) are declared using the **SEEXP** command. The command **EXPSOL** is used for obtaining a solution at different time values (load steps).

```
/CLEAR                    ! CLEAR DATABASE
/FILNAME,GEN              ! SPECIFY JOBNAME AS GEN
RESUME                    ! RESUME FROM GEN.DB
/SOLU                     ! ENTER SOLUTION PROCESSOR
EXPASS,ON                 ! ACTIVATE EXPANSION PASS
SEEXP,GEN,USE             ! SUPPLY SUPERELEMENT NAME TO BE
                          ! EXPANDED
EXPSOL,,,120              ! TIME OF INTEREST IS 120 SEC
SOLVE                     ! OBTAIN EXPANSION PASS SOLUTION
EXPSOL,,,240              ! TIME OF INTEREST IS 240 SEC
SOLVE                     ! OBTAIN EXPANSION PASS SOLUTION
EXPSOL,,,360              ! TIME OF INTEREST IS 360 SEC
SOLVE                     ! OBTAIN EXPANSION PASS SOLUTION
EXPSOL,,,480              ! TIME OF INTEREST IS 480 SEC
SOLVE                     ! OBTAIN EXPANSION PASS SOLUTION
FINISH                    ! EXIT SOLUTION PROCESSOR
/POST1                    ! ENTER GENERAL POSTPROCESSOR
SET,LAST                  ! SET RESULTS TO LAST LOAD STEP
PLNSOL,S,Y,2              ! PLOT CONTOURS OF STRESS IN Y-
                          ! DIRECTION
PLNSOL,S,XY,2             ! PLOT CONTOURS OF STRESS IN XY-
                          ! DIRECTION
FINISH                    ! EXIT GENERAL POSTPROCESSOR
```

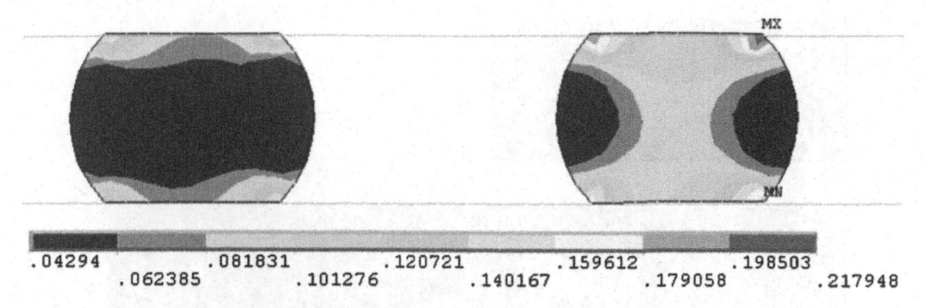

Fig. 11.21 Contour plot of plastic work at the end of the fourth load step

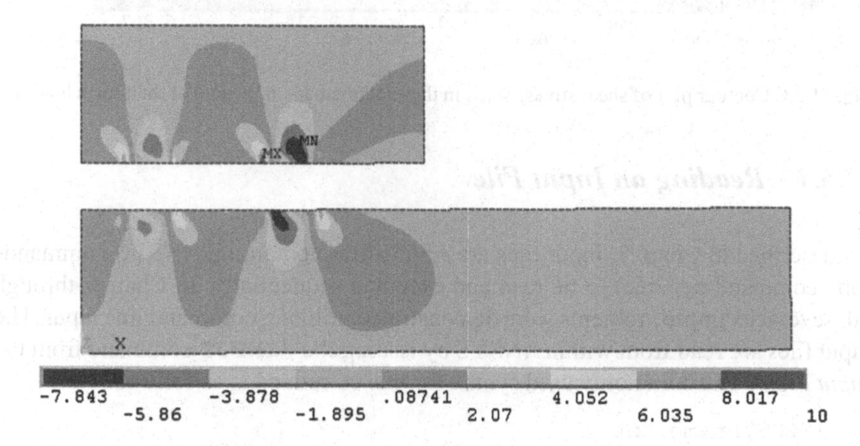

Fig. 11.22 Contour plot of normal stress, σ_{yy}, in the superelement at the end of the fourth load step

Figures 11.22 and 11.23 show the contour plots of the normal and shear stresses, σ_{yy} and σ_{xy}, respectively, in the superelement at the end of fourth load step.

11.5 Interacting with External Files

Interaction with external files makes ANSYS a very versatile software. The most common such interactions include:

Reading an input file.
Writing data to ASCII (plain text) files.
Reading data from ASCII (plain text) files.
Passing a command string to the operating system.

These topics are briefly discussed in this section. However, the descriptions of commands are not complete; only their most common usage is described. Therefore, it is recommended that the user consult the ANSYS *Help System* for complete instructions.

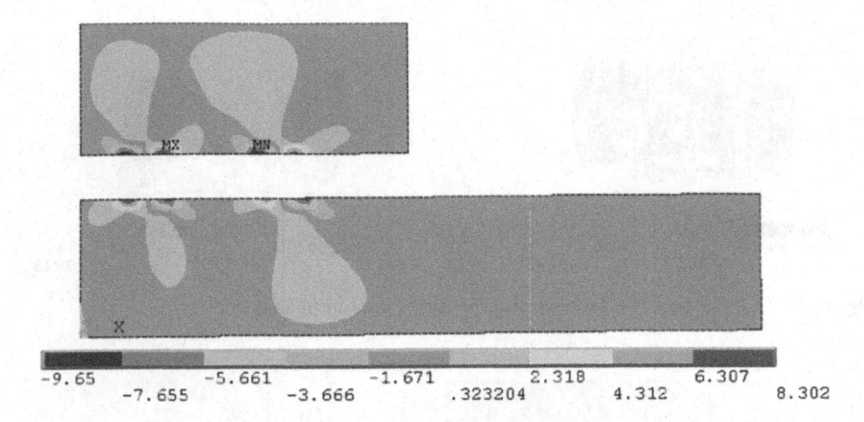

Fig. 11.23 Contour plot of shear stress, σ_{xy}, in the superelement at the end of the fourth load step

11.5.1 Reading an Input File

As described in Chap. 7, input files are ASCII files containing ANSYS commands (one command per line) to be read and executed sequentially. In Chap. 7 through 10, several example problems were demonstrated utilizing command line input. The input files are read from within ANSYS by issuing the **/INPUT** command from the *Input Field*. The commonly used syntax for this command is as follows:

```
/INPUT,Fname,Ext
```

which **Fname** and **Ext** are the path with the file name and extension of the input file, respectively. If the path is not specified, i.e., the argument **Fname** includes only the file name, then the file is searched for in the *Working Directory*.
in

11.5.2 Writing Data to External ASCII Files

There are two main motivations in writing data to files:

ANSYS software may be used as a pre- and postprocessor while the actual computations are performed by an external executable file. In such cases, the executable file almost always requires input data. Such data can be written to ASCII files, with a specific format, to be read by the executable file as input.
Although ANSYS provides powerful contour and line plot options, it may not always offer the most convenient environment for specific plotting formats. In such cases, it may be more efficient to write the required data to an external ASCII file; this file can then be processed with specialized graphics or spreadsheet software to produce the desired plots.

There are two main commands that enable writing data to external files: /OUTPUT and *CFOPEN. Once either of these commands is issued, data are written to an external file using the *VWRITE command. These commands are discussed in the following subsections.

11.5.2.1 The /OUTPUT Command

The **/OUTPUT** command redirects text output to a specified file. The syntax for the **/OUTPUT** command is as follows:

```
/OUTPUT,Fname,Ext, ,Loc
```

in which **Fname** and **Ext** are the path (file name and extension of the file, respectively) and Loc determines whether to write starting from the top of the file (effectively deleting the contents of the file) or appending at the bottom of the file. Note that the third argument is not used. If the **Loc** field is left blank, then ANSYS writes data starting from the top of the file. In order to append an existing file, the phrase **APPEND** is entered in the **Loc** field. After issuing the **/OUTPUT** command with its arguments, text that normally appears in the *Output Window* is written to the external file. Therefore, immediately after writing the data to the file, the text output needs to be redirected to the *Output Window* by issuing the **/OUTPUT** command without any arguments. An example is shown below that writes the parameters in an external file named **DATA.OUT** in the *Working Directory*:

```
/OUTPUT,DATA,OUT, ,APPEND
*VWRITE,CH1,NUM1,NUM2
(A8,2X,F12.8,2X,E13.6)
/OUTPUT
```

11.5.2.2 The *CFOPEN and *CFCLOSE Commands

A better way of writing data to an external file is to open a "command" file using the ***CFOPEN** command. It has a syntax similar to the **/OUTPUT** command, with the same argument definitions:

```
*CFOPEN,Fname,Ext, ,Loc
```

After issuing this command with a file name and extension, anytime the ***VWRITE** command is issued, the data are written to the file. Once finished with the data transfer, the file can be closed using the ***CFCLOSE** command, which has no arguments.

11.5.2.3 The *VWRITE Command

The syntax for the *VWRITE command is as follows:

```
*VWRITE,Par1,Par2,...,Par19
```

in which **Par1, ..., Par19** are user-defined parameters. The ***VWRITE** command line must be followed by a format statement enclosed in parenthesis. The format statement follows FORTRAN programming language syntax. There are five distinct format descriptors addressing real numbers (three different descriptors), characters, and blank spaces. Real number format descriptors are **F, E**, and **D**, which have the following syntax

```
Fw.d
Ew.d
Dw.d
```

in which is the total number of digits allocated for the number in the file (total width) while denotes the number of decimal places (number of digits to the right of the decimal point). The information given herein is common to FORTRAN compilers used in MS Windows operating systems. For compilers used under different operating systems, small changes may be expected, but the general concept remains the same. The descriptor **F** writes the real number with no exponent while descriptors **E** and **D** write with E and D exponents, respectively. The number must account for the sign of the number, decimal point and the exponent field. The exponent field, when using descriptors **E** and **D**, takes up four digits: one digit for the exponent character **E** or **D**, one digit for the sign of the exponent, and two digits for the exponent itself. Therefore, in format statements using **E** and **D** descriptors, the difference $(w - d)$ must be at least seven.

In order to specify the format for characters, the descriptor **A** is used with the following syntax:

```
Aw
```

where w is the total width. The maximum width allowed by ANSYS is eight digits. Finally, blank spaces are inserted using the **X** descriptor with the following syntax:

```
wX
```

in which is the total number of blank spaces.

The use of descriptors **F, E, D, A**, and **X** is demonstrated next.

Consider the numbers 152.67328199 and -3.251667×10^6, which are stored in parameters **NUM1** and **NUM2**, respectively, and the character string **RESULT1** stored in parameter **CH1** in ANSYS. They can be written to an external ASCII file using the following commands lines:

```
*VWRITE,CH1,NUM1,NUM2
(A8,2X,F12.8,2X,E13.6)
```

which produces the following output in the ASCII file

RESULT1 152.67328199 -0.3251667E+07

8 digits 2 digits 12 digits 2 digits 13 digits

There are two observations to be made: (i) when the character string is given more digits than its length, ANSYS inserts as many blank spaces as needed to make the field the same width as that specified and justifies the string to the left, and (ii) when using **E** or **D** descriptors, ANSYS places a 0 immediately to the left of the decimal.

When the format statement above is replaced with

```
(A2,1X,F11.8,1X,E13.2)
```

The following output is written in the ASCII file:

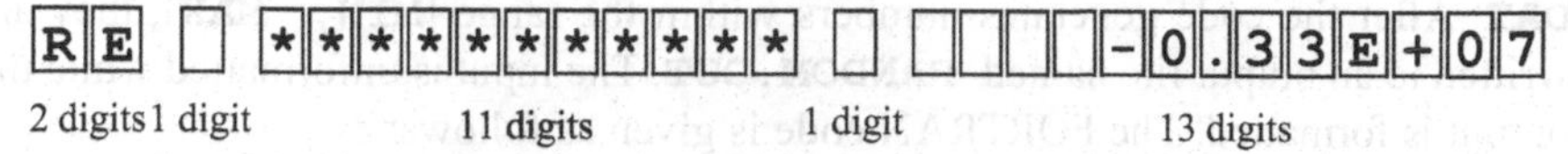

Observations:

When the number of digits allocated by the descriptor **A** is less than the actual length of the character string, ANSYS truncates the character string.
When the number of digits allocated by the descriptor **F** (or **E** or **D**) is less than the actual length of the real number, ANSYS prints the character * as many times as the specified number digits.

When the format statement is replaced with

```
(A7,1X,E11.4,1X,D14.7)
```

the following output is written in the ASCII file:

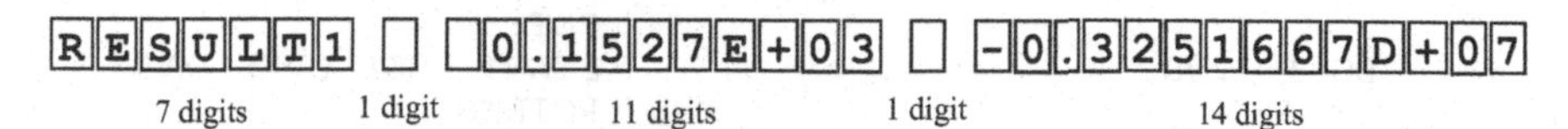

11.5.3 Executing an External File

In certain cases, executing an external file may be required. This could be software performing an independent analysis. The command **/SYS**, which passes a command string to the operating system, is used in such cases. Its syntax is written as follows:

```
/SYS,String
```

in which the argument String is the command string. When used for the aforementioned purpose, the argument String is the name of the executable file. For example, if the name of the executable file is **ROOTS.EXE**, then the following command is issued to execute this file:

```
/SYS,ROOTS.EXE
```

Upon execution of this command, the **ROOTS.EXE** file runs, and ANSYS does not allow any action to be taken during this time.

As an example, we consider the generation of a time-dependent random thermal load. In this example, an external executable file, **RANDOM.EXE**, generating random numbers is created using the FORTRAN programming language. The code requires values for five input parameters:

TOT Total time the load is applied.
ND Number of data points (random numbers).
MIN Minimum value of load.
MAX Maximum value of load.
SEED Seeding number for random number generation.

The values of these parameters are provided through an input file named **RANDOM.DAT**. After the code generates numbers within the range [**MIN, MAX**], they are written to an output file named **RANDOM.OUT**. The input is unformatted while the output is formatted. The FORTRAN code is given as follows:

```
      REAL(8) NUM,TOT,DT,T,MIN,MAX       ! REAL VARIABLES
      INTEGER(4) ND,F,I,SEED             ! INTEGER VARIABLES
      OPEN(1,FILE='RANDOM.DAT')          ! DECLARE INPUT FILE
      OPEN(2,FILE='RANDOM.OUT')          ! DECLARE OUTPUT FILE
      READ(1,*) TOT,ND,MIN,MAX,SEED      ! READ INPUT
      DT=TOT/ND                          ! FIND TIME STEP SIZE
      T=DT                               ! INITIALIZE TIME
      F=0                                ! INITIALIZE RANDOM
                                         ! NUMBER
                                         ! GENERATOR FLAG
      AMP=MAX-MIN                        ! FIND LOAD RANGE
      NUM=DRAND(SEED)                    ! GENERATE RANDOM
                                         ! NUMBER
                                         ! SET
      DO I=1,ND                          ! START LOOP OVER DATA
                                         ! POINTS
      NUM=MIN+AMP*DRAND(F)               ! FIND LOAD VALUE
      WRITE(2,100) T,NUM                 ! WRITE TIME AND LOAD
                                         ! VALUES
      T=T+DT                             ! UPDATE TIME
      ENDDO                              ! END LOOP OVER DATA
                                         ! POINTS
  100 FORMAT (2(E16.8,2X))               ! FORMAT STATEMENT
      END                                ! END OF PROGRAM
```

In order to create the *executable* file, one must compile and link the code above using a FORTRAN compiler.

The following command input segment, executed from within ANSYS, writes the required input to file **RANDOM.DAT**, executes the file **RANDOM.EXE** (resides in the *Working Directory*), reads the time vs. load values from file **RANDOM.OUT**, and, finally, plots the variation of load vs. time. After the file **RANDOM.EXE** is executed, three array parameters are defined (***SET** command) and given dimensions

(***DIM** command). The time and load values are stored in one large array using the ***VREAD** command. Finally, axis labels are specified using the **/AXLAB** command and the plot is obtained using the ***VPLOT** command.

```
/PREP7                          ! ENTER PREPROCESSOR
TOT=8                           ! DEFINE PARAMETERS
ND=160                          !
MN=0                            !
MX=100                          !
SEED=12                         !
/OUTPUT,RANDOM,DAT              ! REDIRECT OUTPUT TO RANDOM.DAT
*VWRITE,TOT,ND,MN,MX,SEED       ! WRITE PARAMETERS TO FILE
(E16.8,2X,F8.0,2X,E10.2,2X,E10.2,2X,F8.0) ! FORMAT
                                          ! STATEMENT
/OUTPUT                         ! REDIRECT OUTPUT TO OUTPUT
                                ! WIN.
/SYS,RANDOM.EXE                 ! EXECUTE RANDOM.EXE
*SET,ARY                        ! INITIALIZE PARAMETER ARY
*SET,TIM                        ! INITIALIZE PARAMETER TIM
*SET,LOAD                       ! INITIALIZE PARAMETER LOAD
*DIM,ARY,ARRAY,2*ND             ! DIMENSION PARAMETER ARY
*DIM,TIM,TABLE,ND               ! DIMENSION PARAMETER TIM
*DIM,LOAD,TABLE,ND              ! DIMENSION PARAMETER LOAD
*VREAD,ARY(1),RANDOM,OUT        ! FILL ARY ARRAY FROM
                                ! RANDOM.OUT
(2(E16.8,2X))                   ! FORMAT STATEMENT
*DO,I,1,ND                      ! LOOP OVER DATA POINTS
TIM(I)=ARY(2*I-1)               ! SEPARATE ARRAY ARY INTO TIM
LOAD(I)=ARY(2*I)                ! AND LOAD ARRAYS
*ENDDO                          ! END LOOP OVER DATA POINTS
/AXLAB,X,TIME                   ! DEFINE X-AXIS LABEL
/AXLAB,Y,LOAD                   ! DEFINE Y-AXIS LABEL
*VPLOT,TIM(1),LOAD(1)           ! PLOT TWO ARRAYS AGAINST EACH
                                ! OTHER
```

When the command input segment above, with the given values of parameters, is read as an input file from within ANSYS, the plot shown in Fig. 11.24 is produced in the *Graphics Window*.

11.5.4 Modifying ANSYS Results

ANSYS software can be used as a postprocessor for results calculated by an external program utilizing the **DNSOL** or **DESOL** commands, which impose those results on the ANSYS mesh. The syntax for **DNSOL** and **DESOL** commands are:

```
DNSOL,NODE,Item,Comp,V1,V2,V3,V4,V5,V6
DESOL,ELEM,NODE,Item,Comp,V1,V2,V3,V4,V5,V6
```

in which **NODE** and **ELEM** are the node and elements numbers, respectively, at which the results are specified, **Item** and **Comp** are labels for the specific degree

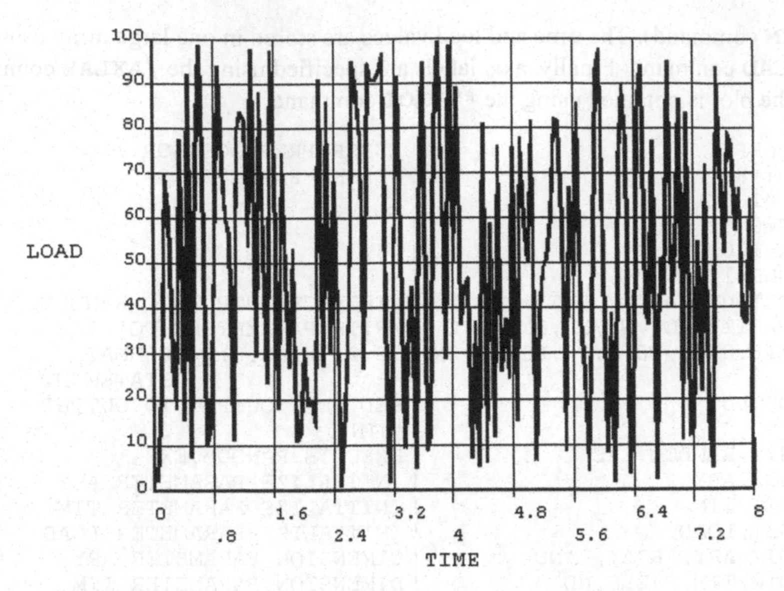

Fig. 11.24 Variation of load with time as read from the RANDOM.OUT file

of freedom, and **V1** through **V6** are the values of the results. If the *y*-displacement (**Item** is **U** and **Comp** is **Y**) at node 12 is modified to be 2.5, the following command input is used:

```
DNSOL,12,U,Y,2.5
```

Similarly, when shear stress in the *yz*-direction (**Item** is **S** and **Comp** is **YZ**) at node 212 of element 23 is modified to be 1200, the following command input is used:

```
DESOL,23,212,S,YZ,1200
```

11.6 Modifying the ANSYS GUI

ANSYS allows the user to modify/customize its Graphical User Interface (GUI). This capability is very advantageous, particularly when the user performs similar types of analyses over and over again. However, it is not a straightforward task. The main resource for learning how to customize the GUI is the *ANSYS UIDL Programmer's Guide*, in which *UIDL* stands for *User Interface Design Language*. This guide is not included in the *ANSYS Help System*; it must be purchased separately.

This section does not provide exhaustive information on UIDL. It is written with the intention of giving the reader a few examples on how the UIDL is modified. First, a brief summary of steps involved in modifying the UIDL are discussed, after which two simple examples are given for demonstration purposes. The first example involves customized dialog boxes, and the second involves a customized pick menu.

The following steps are taken to customize the ANSYS GUI:

- Locate the *Control Files* **UIFUNC1.GRN**, **UIFUNC2.GRN**, and **UIMENU.GRN**, and the file **menulist**.ans**. The characters ****** correspond to the ANSYS version being used. For example if ANSYS Release 14.0 is being used, then the file name is **menulist140.ans**. In later ANSYS releases, these files are located in the same directory. In a PC using a Windows operating system, a possible location is:

```
\Program Files\ANSYS Inc\v140\ansys\gui\en-us\UIDL
```

- Create a temporary directory, e.g.:

```
\user\temp
```

- Copy the control files and the file **menulist140.ans** into the temporary directory.
- Using parts of the existing control files located in the temporary directory, create new control files (this point is explained in detail later).
- Copy the new control files and the **menulist140.ans** file to a *Working Directory* and add the names of the new control files, including their path, in the menulist140.ans file.
- Start ANSYS with the directory containing the new control files and the **menulist140.ans** file as the *Working Directory* and verify that the modified GUI works properly.
- When finished, copy the **menulist140.ans** file to additional *Working Directories*; the modified GUI will be available when ANSYS is started using those directories as *Working Directory*.

There are three control files: **UIFUNC1.GRN**, **UIFUNC2.GRN**, and **UIMENU.GRN**. The first two contain function blocks and the last one is for the menu blocks. The general logic of menu and function blocks is explained considering existing blocks. First, menu blocks are discussed, followed by a discussion on function blocks.

At the beginning of the **UIMENU.GRN** file, the following *Control File Header* is placed:

```
:F UIMENU.GRN
:D Modified on %E%, Revision (SID) = %I%
:I    403,  100064,  110164
:!
```

in which the **:F** command defines the name of the control file, **:D** gives a description of the control file, and the **:I** command stores internally generated index numbers. At the beginning of every control file, these three commands must appear in the sequence given above. In a new control file, the **:I** command must have comma-separated zeros in columns 9, 18 and 27. Once ANSYS runs and reads the file, it assigns new numbers and modifies the control files. The *Control File Header* must be separated from the menu blocks by a line containing :! (the fourth line above).

The *Main Menu* in ANSYS is created by the following menu block, taken from **UIMENU.GRN** file:

```
:N MenuRoot
:S    488,     76,     405
:T Menu
:A Main Menu
:D ANSYS ROOT MENU
Men_UVBA_Main_T1
Men_UVBA_Main_T2
Men_UVBA_Main_T3
Fnc_Preferences
Sep_
Men_Preproc
Men_Solution
Men_GenlPost
Men_TimePost
K_LN(DROPTEST)
Sep_
K_LN(DROPTEST)
Men_DropTest
Sep_
Men_Topo
Men_ROM
Men_DesXplorerVT
K_LN(alpha)
Men_DesOpt
K_LN(ALPHA)
Men_DesOpt_al
Men_ProbDesign
Men_Aux12
Men_RunStat
Sep_
Fnc_UNDO
Sep_
Fnc_FINISH
K_LN(UTILMENU)
Men_UtilMenu
Men_UVBA_Main_B1
Men_UVBA_Main_B2
Men_UVBA_Main_B3
:E END
:!
```

which is composed of following three distinct sections:

Block Header Section: Commands belonging to this section always start with a colon (**:**). The first command must be **:N**, which defines a unique internal name given to the block.

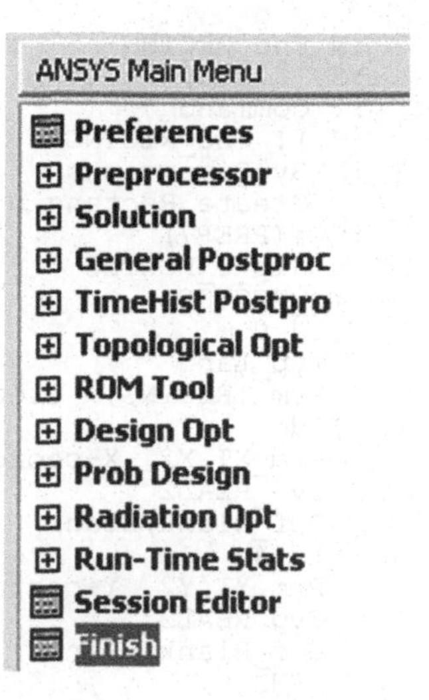

Fig. 11.25 ANSYS *Main Menu* with functions and submenus

The next command must be the **:S** command, which stores internally generated index numbers. In a new control file, the **:S** command must have comma-separated zeros in columns 9, 16, and 23. Once ANSYS runs and reads the file, it assigns new numbers and modifies the control files. The command **:T** identifies the block as a menu block (**Menu**) or a command block (**Cmd**). The command **:T** defines the heading of the block as it appears in the menu system. There are optional commands that can be used in the *Block Header Section*, such as **:D**, which is the description of the block.

Data Controls Section: This section contains four distinct data control commands: **Men_, Fnc_, Sep_,** and **K_LN.** The control commands **Men_** and **Fnc_** are calls to existing menu and function blocks, respectively. For example, the line **Men_Preproc** is a call to the menu block for the ANSYS*Preprocessor* in the same control file (**UIMENU.GRN**), which has the internal name **Men_Preproc** defined by the :N command. The function blocks are stored in one of two files: **UIFUNC1.GRN** or **UIFUNC2.GRN**. Thus, the line Fnc_Preferences calls the function block that is in UIFUNC2.GRN file. Figure 11.25 shows the ANSYS Release 14.0 *Main Menu*, with Preferences (**Fnc_Preferences**) and Preprocessor (Men Preproc). The command **Sep_** is used for inserting a separator bar between menu items. However, in newer releases of ANSYS, the whole menu is in one window with a tree-like structure (as opposed to separate windows for menus). Therefore, the separators in menus are no longer visible. The command **K_LN** is related to ANSYS *keyword logic*, which decides to show or not show certain menu items; this topic is not discussed within the context of this book.

Ending Section: Every menu block must end with these two lines:

```
:E END
:!
```

Function blocks are contained in the control files **UIFUNC1.GRN** and **UIFUNC2.GRN**. They execute ANSYS commands. As an example for function blocks, the creation of rectangles by dimensions function block, as given below, is considered (dialog box shown in Fig. 11.26). This block executes the **RECTNG** command. The function block is taken from the **UIFUNC2.GRN** file.

```
:N Fnc_RECTNG
:S   355,    145,    203
:T Command
:C )! Fnc_RECTNG
:A By Dimensions
:D Create Rectangle by Dimensions
:K #(PREP7)
:H Hlp_C_RECTNG
Cmd_RECTNG
 Fld_0
  Typ_Lab
  Prm_[RECTNG]  Create Rectangle by Dimensions
 Fld_2
  Prm_X1,X2  X-coordinates
  Typ_REAL2
  Def_Blank,Blank
 Fld_4
  Prm_Y1,Y2  Y-coordinates
  Typ_REAL2
  Def_Blank,Blank
:E END
:!
```

Function blocks are also composed of three distinct sections:

Block Header Section: Definitions of commands belonging to this section are the same as those for the menu blocks. In addition to the commands described under menu blocks, there is another important command::C. This command is used for executing ANSYS commands before the actual function is executed. An exclamation mark designates, as in the APDL, comments.

Data Controls Section: Many more commands can be used within this section under function blocks than can be used under menu blocks. For a complete list of available commands, the reader should consult the ANSYS *UIDL Programmer's Guide*. A partial list is given in Table 11.2.

In the example above, the *Data Controls Section* starts with the line containing **Cmd_RECTNG**. The command **Cmd_** initiates a command sequence, in this case the **RECTNG** command. The **Fld** command defines the field number, which is associated with the command being used. If no data are collected in a field (such as a text block in the dialog box), then 0 is assigned to the **Fld_** command, i.e., **Fld_0**. Each field has a type defined by the **Typ** command. In this case, the text block **[RECTNG] Create Rectangle by Dimensions**, in the dialog box (Fig. 11.26) is created using the **Typ_Lab** (stands for type label) command. Finally the label is printed using the **Prm_** command, which is immediately followed by the text block.

If data are collected in a field, then non-zero, sequential field numbers must be assigned. The field number for the first data item is 2, i.e., **Fld_2**. In this particular example, **X1** and **X2** coordinates are collected, which constitute field numbers 2 and 3. Hence, the next two data items, **Y1** and **Y2**, are assigned field numbers 4 and 5. All of the data collected in this case are real numbers. Also, two data items per row are collected, which is achieved using the command **Typ_Real2**. Single and three real numbers per row are collected using **Typ_Real** and **Typ_Real3** commands. Similarly, integer data can be collected using commands **Typ_Int**, **Typ_Int2**, and **Typ_Int3**. Default values for the data can be supplied by using the **Def_**

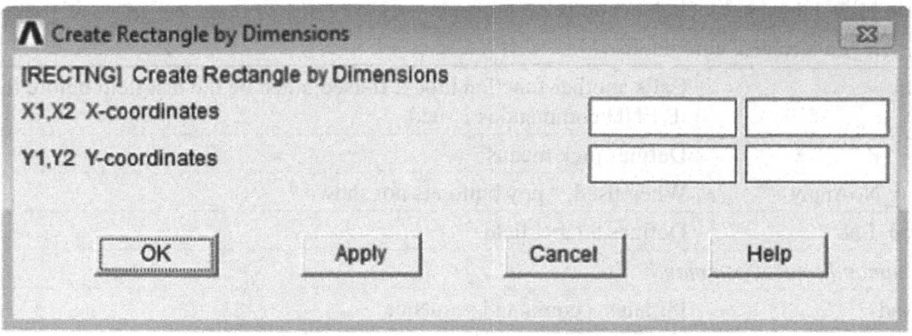

Fig. 11.26 Dialog box produced by the Fnc_RECTNG function block

command. In this case, blank data fields are enforced, i.e., **Def_Blank** is used. If, for example, the default values of X1 and X2 are 0.1 and 0.4, respectively, then **Def_0.1,0.4** would be used.

Ending Section: Every function block must end with the following two lines:

```
:E END
:!
```

ANSYS dialog boxes offer numerous modes of interaction with the user. A partial list of these interactions is:

- *Single Selection List*: Only one selection can be made.
- *Text Option Button List*: Creates a pull-down menu.
- *Radio-button List*: Creates several radio-buttons (only one can be selected).
- *Multiple Selection List*: Multiple items can be selected.
- *Side-by-side Parent Child List*: Creates two lists, one on the left (parent) and the other on the right (child). Every item in the parent list has at least one child in the child list.
- *Data Field*: Creates data fields for real and integer numbers, characters, and yes/no types of interactions.

Each of these interactions passes values corresponding to the selection made to ANSYS commands (or user macro files). In the case of *Data Fields*, the value that is passed is specified in the text boxes.

GUI modification using UIDL is demonstrated by considering three examples. The first example demonstrates how to create function blocks utilizing the different modes of interaction listed above. In the second example, data are collected through a dialog box used as input for an external code (**RANDOM.EXE** of Sect. 11.5.3), which is subsequently executed. A menu item that reads and plots the output generated by the external executable file is also generated. Finally, a simple example utilizing a customized picking menu is included.

Table 11.2 Function block data control commands

General data control commands	
Cal_	Calls another function block. If used, must be the last item before the :E.END command is issued
Inp_P	Defines pick menus
Inp_NoApply	When used, Appy button is not shown
Typ_Lab	Defines a label field
Command control commands	
Cmd_	Initiates a command sequence
General field control commands	
Def_	Sets default values for the data to be shown in dialog boxes
Fld_	Assigns numbers to fields
Prm_	Prints text in dialog boxes
Picking control commands	
Max_	Sets maximum number of items to be picked
Min_	Sets minimum number of items to be picked
Sel_	Defines type of selection allowed
Typ_Entity	Retrieves the numbers of the specified entities (NODE, ELEM, KEYP, LINE,AREA, VOLU)
Typ_XYZ	Retrieves the coordinates of a point
Listing control commands	
Typ_Idx	Creates scrolling lists from subsequent Idx_ commands
Typ_Lis_OptionB	Creates option button lists
Typ_Lis_RadioB	Creates radio-button lists
Typ_Lis	Creates single scrolling lists
Typ_MLis	Creates multiple scrolling lists
Numerical control commands	
Typ_Int	Creates single input field for integers
Typ_Int2	Creates two input fields for integers
Typ_Int3	Creates three input fields for integers
Typ_Real	Creates single input field for real numbers
Typ_Real2	Creates two input fields for real numbers
Typ_Real3	Creates three input fields for real numbers
Character control commands	
Typ_Char	Creates single input field for characters
Logical control commands	
Typ_Logi	Creates toggle buttons

11.6.1 GUI Development Demonstration

In this example, a menu item under the ANSYS *Preprocessor* menu is generated. Its name appears as **DEMO**. There are three function blocks under the **DEMO** menu: **F1**, **F2**, and **F3**. The function block F1 involves a *Single Selection List*, a *Text Option Button List* and a *Radio-button List*. In function block **F2**, creation of a *Multiple Selection List* and a *Side-by-side Parent Child List* is demonstrated. Finally, function block **F3** utilizes various *Data Fields*. The following is a complete step-by-step list of actions taken to achieve this task.

Step 1 Copy the original control files **UIFUNC1.GRN**, **UIFUNC2.GRN,** and **UIMENU.GRN**, and the file menulist140.ans file to the temporary directory

```
\user\temp.
```

Step 2 Create two temporary files with names **zero1.dat** and **zero2.dat** in the temporary directory. Open both files using a text editor. In the **zero1.dat** file, enter the following phrase:

```
:I      0,       0,       0
```

making sure the zeros are placed in columns 9, 18, and 27. In the **zero2.dat** file, enter the following phrase:

```
:S      0,     0,     0
```

This time, the zeros must be placed in columns 9, 16, and 23. Save both files and leave them open.

Step 3 Create a new control file in the temporary directory with a user-defined name, say **USERM.GRN**. Open both the **USERM.GRN** and **UIMENU.GRN** files using a text editor. Copy the first four lines from the **UIMENU.GRN** file and paste them in the file **USERM.GRN**. Modify the first two lines in the **USERM.GRN** file to be:

```
:F USERM.GRN
:D User Defined Menu
```

Copy the single line in the **zero1.dat** file and paste it to replace the third line in the **USERM.GRN** file. At this point, the **USERM.GRN** file looks like:

```
:F USERM.GRN
:D User Defined Menu
:I      0,       0,       0
:!
```

Step 4 Because the new menu item is going to be under the ANSYS Preprocessor, find the menu block corresponding to the ANSYS Preprocessor. It was observed earlier that the internal name for the Preprocessor is **Preproc**. Perform a word search of the **UIMENU.GRN** file for the phrase **Men_ Preproc**, which produces several results. Find the location with the phrase:

```
:N Men_Preproc
```

Copy the menu block **Preproc** from file **UIMENU.GRN** and paste it in the file USERM.GRN, after the first four lines modified earlier. In the **USERM.GRN** file, find the line starting with the **: S** command (sixth line). Copy the single line in the zero2.dat file and paste it to replace the sixth line in the **USERM.GRN** file.

The **DEMO** menu is going to be the first item available under the ANSYS *Preprocessor*. So, in the **USERM.GRN** file, first find the line with the phrase

```
Men_ElemType
```

then insert the following line above it:

```
Men_DEMO,
```

which declares that there is a menu block with an internal name **DEMO**.

Create the menu block for **DEMO** at the bottom of the **USERM.GRN** file as follows:

```
:N Men_DEMO
:S      0,      0,      0
:T Menu
:A Demo Menu
:D Demo Menu
Fnc_F1
Fnc_F2
Fnc_F3
:E END
:!
```

Make sure the menu block for the *Preprocessor* is separated from the **DEMO** menu block by the following line:

```
:!
```

Creation of the customized menu is complete. Save the **USERM.GRN** file.

Step 5 Three function blocks are created next. For this purpose, create a new control file in the temporary directory with a user-defined name, say **USERF.GRN**. Open the newly created **USERF.GRN** file using a text editor. Copy the first four lines from the **USERM.GRN** file and paste them in the file **USERF.GRN**. Modify the first two lines in the **USERF.GRN** file to be

```
:F USERF.GRN
:D User Defined Functions
```

The first function block utilizes a *Single Selection List*, a *Text Option Button List*, and a *Radio-button List*. Enter the following text in the **USERF.GRN** file, starting from line 5:

```
:N Fnc_F1
:S      0,      0,      0
:T Command
:A Function F1
:D Function F1 Heading
Cmd_F1
 Fld_0
  Typ_Lab
  Prm_Type the description of function F1 here
 Fld_0
  Typ_Sep
 Fld_0
  Typ_Lab
  Prm_Typ_Lis command is used to generate the list
 Fld_2
  Typ_Lis
   Lis_Item 1,1
   Lis_Item 2,2
   Lis_Item 3,3
   Lis_Item 4,4
   Lis_Item 5,5
  Prm_Selected item appears in the box
 Fld_0
  Typ_Sep
 Fld_0
  Typ_Lab
  Prm_Typ_Lis_OptionB command is for pull-down menus
 Fld_3
  Prm_Choose one option
  Typ_Lis_OptionB
   Lis_Option 1,10
   Lis_Option 2,20
   Lis_Option 3,35
   Lis_Option 4,89
 Fld_0
  Typ_Sep
 Fld_0
  Typ_Lab
  Prm_Typ_Lis_RadioB command is for radio-buttons
 Fld_4
  Prm_Choose one of the buttons
  Typ_Lis_RadioB
   Lis_Button 1,4
   Lis_Button 2,7
   Lis_Button 3,23
:E END
:!
```

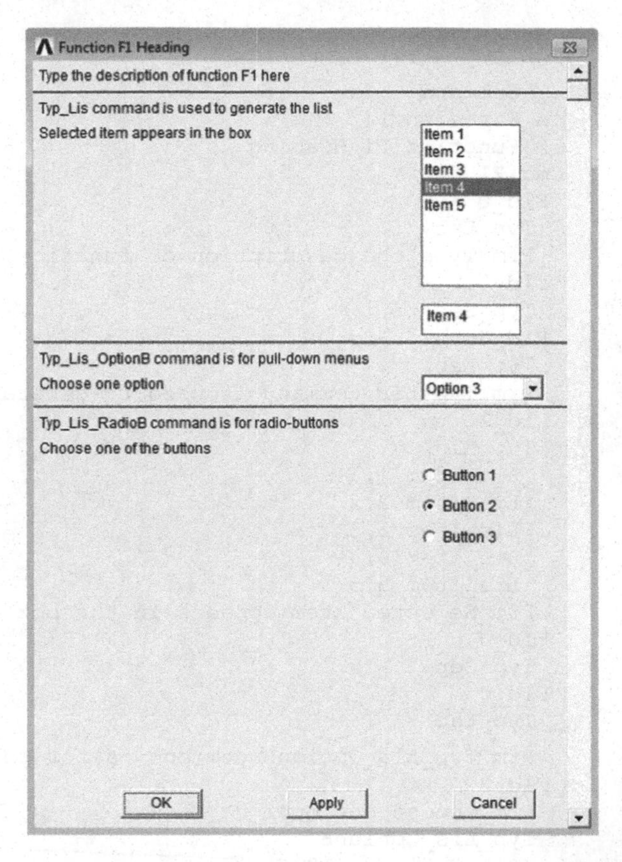

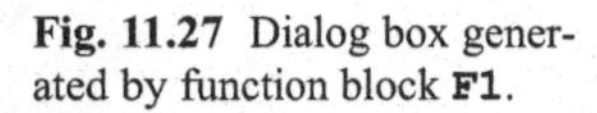
Fig. 11.27 Dialog box generated by function block **F1**.

The function block above produces the dialog box shown in Fig. 11.27. There are three data fields (fields 2, 3, and 4). After making the selections and clicking **OK**, ANSYS issues the command **F1** with three arguments (which is not a standard ANSYS command). For example, if the user selects ***Item 4*** from the single selection list, ***Option 3*** from the pull-down menu, and ***Button 2*** from the radio-buttons, then ANSYS issues the following command input:

```
F1,4,35,7
```

The arguments for each list item are specified in the UIDL segment for **F1** given above. For example, selecting ***Button 3*** passes 23 as the third argument of the command **F1**, enforced by the line

```
Lis_Button 3,23
```

The second function block demonstrates *Multiple Selection* and *Side-by-side Parent Child Lists*. Append the following text in the **USERF.GRN** file, after the function block for **F1**:

```
:N Fnc_F2
:S      0,      0,      0
:T Command
:A Function F2
:D Function F2 Heading
Cmd_F2
 Fld_0
  Typ_Lab
  Prm_Type the description of function F2 here
 Fld_0
  Typ_Sep
 Fld_0
  Typ_Lab
  Prm_Typ_MLis command is used to generate this list
 Fld_0
  Typ_Lab
  Prm_Bnd_ command is used to set limits
 Fld_2
  Prm_Select a maximum of 5 items
  Typ_MLis
Bnd_1.0,5.0
   Lis_Multi Item 1,1
   Lis_Multi Item 2,2
   Lis_Multi Item 3,3
   Lis_Multi Item 4,4
   Lis_Multi Item 5,5
   Lis_Multi Item 6,6
   Lis_Multi Item 7,7
   Lis_Multi Item 8,8
   Lis_Multi Item 9,9
   Lis_Multi Item 10,10
Fld_0
  Typ_Sep
 Fld_0
  Typ_Lab
  Prm_Typ_Idx command is for parent-child lists
 Fld_3
  Prm_Select one item from each list
  Typ_Idx
   Idx_Parent 1,Child 1 of Parent 1,1
   Idx_Parent 1,Child 2 of Parent 1,2
   Idx_Parent 1,Child 3 of Parent 1,3
   Idx_Parent 1,Child 4 of Parent 1,4
   Idx_Parent 1,Child 5 of Parent 1,5
   Idx_Parent 2,Child 1 of Parent 2,10
   Idx_Parent 2,Child 2 of Parent 2,15
   Idx_Parent 3,Child 1 of Parent 3,17
   Idx_Parent 4,Child 1 of Parent 4,-1
   Idx_Parent 4,Child 2 of Parent 4,0
   Idx_Parent 4,Child 3 of Parent 4,-21
   Idx_Parent 4,Child 4 of Parent 4,212
   Idx_Parent 4,Child 5 of Parent 4,18
   Idx_Parent 4,Child 6 of Parent 4,19
   Idx_Parent 4,Child 7 of Parent 4,20
:E END
:!
```

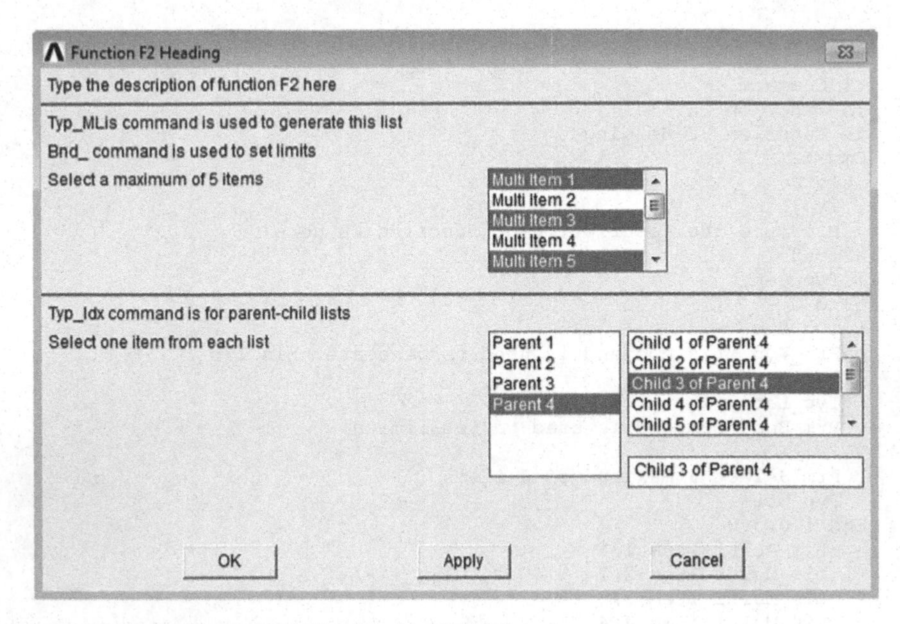

Fig. 11.28 Dialog box generated by function block F2.

This function block for **F2** produces the dialog box shown in Fig. 11.28. Similar to function block **F1**, clicking on **OK** in the dialog box issues the command **F2** with appropriate arguments. In this case, however, the number of arguments is not predetermined due to the presence of the *Multiple Selection List*. For example, if items 1, 3, and 5 from the *Multiple Selection List* and **Parent 4/Child 3** from the *Parent Child List* are selected, then ANSYS issues the following command input:

```
F2,1,3,5,-21
```

On the other hand, if only item 3 from the *Multiple Selection List* and **Parent 4/Child 3** from the *Parent Child List* are selected, then ANSYS issues the following command input:

```
F2,3,-21
```

Note that there are a minimum and maximum number of items that can be selected from the *Multiple Selection List*. In this case, their values are set to be 1 and 5, respectively, using the Bnd_ command.

Finally, function block **F3** is created. Append the following text in the **USERF.GRN** file, after the function block for **F2**:

```
:N Fnc_F3
:S      0,      0,      0
:T Command
:A Function F3
:D Function F3 Heading
Cmd_F3
 Fld_0
  Typ_Lab
  Prm_Type the description of function F3 here
 Fld_0
  Typ_Sep
 Fld_0
  Typ_Lab
  Prm_Typ_Int command is used for integer input fields
 Fld_2
  Typ_Int
  Prm_Typ_Int
 Fld_3
  Typ_Int2
  Prm_Typ_Int2
 Fld_5
  Typ_Int3
  Prm_Typ_Int3
 Fld_0
  Typ_Sep
 Fld_0
  Typ_Lab
  Prm_Typ_Real command is used for real input fields
 Fld_8
  Typ_Real
  Prm_Typ_Real
 Fld_9
  Typ_Real2
  Prm_Typ_Real2
 Fld_11
  Typ_Real3
  Prm_Typ_Real3
 Fld_0
  Typ_Sep
Fld_0
  Typ_Lab
  Prm_Typ_Char command is used for character input fields
 Fld_14
  Typ_Char
  Prm_Typ_Char
Fld_0
  Typ_Lab
  Prm_Typ_Logi command is used for logical input fields
 Fld_15
  Typ_Logi,No,Yes
  Prm_Typ_Logi
:E END
:!
```

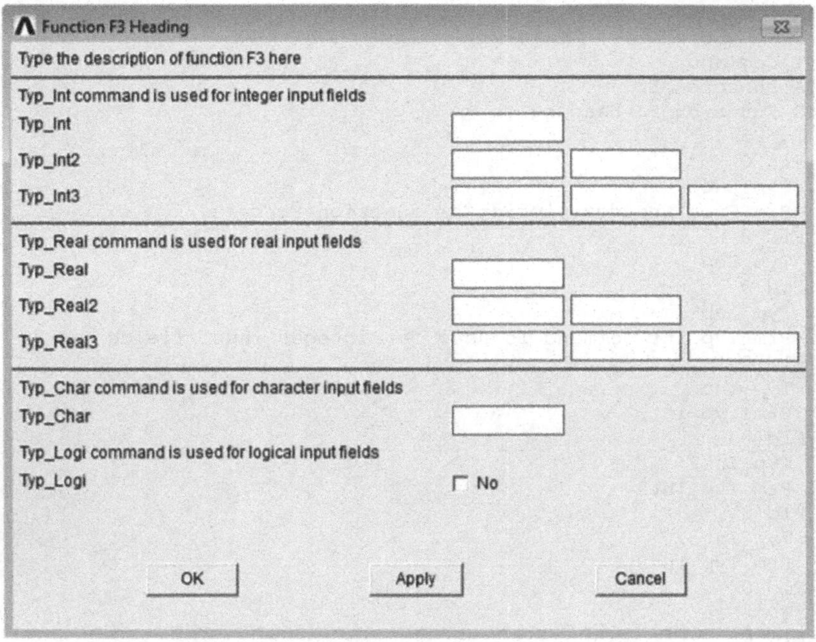

Function F3 Heading
Type the description of function F3 here
Typ_Int command is used for integer input fields
Typ_Int
Typ_Int2
Typ_Int3
Typ_Real command is used for real input fields
Typ_Real
Typ_Real2
Typ_Real3
Typ_Char command is used for character input fields
Typ_Char
Typ_Logi command is used for logical input fields
Typ_Logi No
OK Apply Cancel

Fig. 11.29 Dialog box generated by function block F3

This function block produces the dialog box shown in Fig. 11.29. When the logical field, created using the **`Typ_Logi`** command, is unchecked, the corresponding argument has the value 0; otherwise (when checked), the value 1 is assigned.

Step 6 Copy the new control files, **`USERM.GRN`** and **`USERF.GRN`**, located in the temporary directory and paste them in the *Working Directory* where they will be accessed and processed by ANSYS. In this example, the following directory is considered to be the *Working Directory*: **`C:\user\`**.

Step 7 Add the file names of the new control files, **`USERM.GRN`** and **`USERF.GRN`**, to the **`menulist140.ans`** file. Copy the **`menulist140.ans`** file located in the temporary directory, paste it in the *Working Directory*, and, finally, open it using a text editor. Its contents should look similar to the following:

```
C:\Program Files\ANSYS Inc\v140\ansys\gui\en-us\UIDL\UIMENU.GRN
C:\Program Files\ANSYS Inc\v140\ansys\gui\en-us\UIDL\UIFUNC1.GRN
C:\Program Files\ANSYS Inc\v140\ansys\gui\en-us\UIDL\UIFUNC2.GRN
C:\Program Files\ANSYS Inc\v140\ansys\gui\en-us\UIDL\MECHTOOL.AUI
```

Add the following two lines at the bottom of the file:

```
C:\user\USERM.GRN
C:\user\USERF.GRN
```

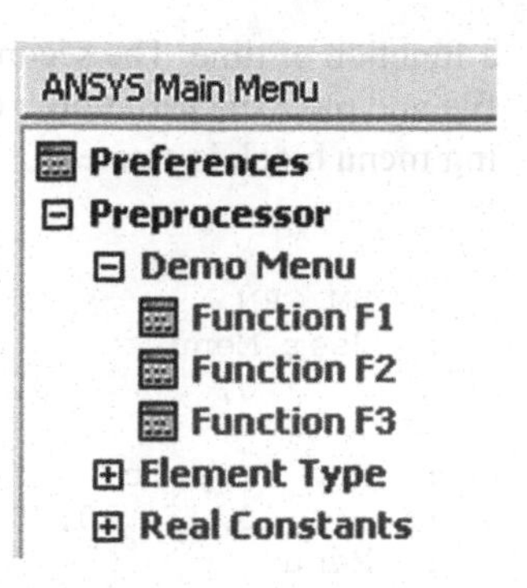

Fig. 11.30 Customized menu structure

Step 8 Start ANSYS with **C:\user** as the *Working Directory* and verify that the customized GUI is working properly. If there are problems, edit the files in the temporary directory, then copy and paste them in the **C:\user** directory and restart ANSYS for verification.

Step 9 Once completed, the **menulist140.ans** file can be copied to additional *Working Directories*; the customized GUI will be available when ANSYS is started using those directories as the *Working Directory*. Figure 11.30 shows a portion of the newly created menu structure.

Step 10 Although the menu structure and the dialog boxes are working properly, the data collected through them have not been processed, i.e., **F1**, **F2**, and **F3** are not commands. ANSYS *Macros* are used for processing the data collected from customized dialog boxes. These files are written in the *ANSYS Parametric Design Language* (APDL). In this case, three macro files are necessary: **F1.MAC**, **F2.MAC**, and **F3.MAC**. Macro files must have their extensions as **MAC**. If they reside in the *Working Directory*, they are automatically recognized by ANSYS. Otherwise, the **/PSEARCH** command is used to point to their location. Macro files corresponding to customized dialog boxes expect arguments in the sequence they are entered in the dialog boxes. The first number in the dialog box data set corresponds to ARG1 in the macro file. A maximum of 19 scalar parameters can be passed on to each macro file from dialog boxes, i.e., **ARG1, ARG2, … , ARG9, AR10, AR11, … , AR19**. In the following subsection, this point is clarified.

11.6.2 GUI Modification for Obtaining a Random Load Profile

In this subsection, ANSYS GUI is customized to have an additional menu item with two functions under the *Preprocessor*. The first function collects data to use the external file **RANDOM.EXE** of Sect. 11.5.3 for generation of random loading as a function of time. The second function reads the random loading from an external file and plots the variation. Following the steps outlined in Sect. 11.6.1, the following menu block is created:

```
:F UM.GRN
:D User Menu
:I      0,        0,        0
:!
:N Men_Preproc
:S      0,      0,      0
:T Menu
:A Preprocessor
:C )/nopr
:C )*get,_z1,active,,routin
:C )!*IF,_z1,NE,17,THEN
:C /PREP7
:C )!*ENDIF
:C )/go
:D Preprocessor (PREP7)
Men_UVBA_P7_T1
Men_UVBA_P7_T2
Men_UVBA_P7_T3
Men_Random
Men_ElemType
Men_Real_copy
Fnc_R_too
Men_Material
Fnc_DYNA_Shell
Men_Section_Defs
Sep_
Men_Modeling
Sep_
Men_Meshing
K_LN(lsdyna)
Sep_
Men_Trefftz_Dom
K_LN(ELECSTAT)
Sep_
Men_CheckCtrl
Men_NumCtrl
Men_Archive
Sep_
```

```
Men_CoupCeqn
K_LN(lsdyna+flotran)
Sep_
Men_FLOTRAN
K_LN(LSDYNA)
Sep_
Men_DYNAPREP
K_LN(lsdyna)
Sep_
K_LN(lsdyna)
Men_FSI
K_LN(lsdyna)
Men_MFSET
K_LN(lsdyna+ROMES)
Sep_
K_LN(lsdyna+ROMES)
Men_ROM_Tool
K_LN(lsdyna)
Sep_
Men_Radiosity
Sep_
K_LN(ALPHA)
Men_QA_Test
K_LN(lsdyna)
Men_Loads
K_LN(lsdyna)
Sep_
Men_MultiPhys
Men_CrePath
Men_UVBA_P7_B1
Men_UVBA_P7_B2
Men_UVBA_P7_B3
:E END
:!
:N Men_Random
:S       0,      0,      0
:T Menu
:A Random Loading
:D Random Loading
Fnc_Generate
Fnc_ReadPlot
:E END
:!
```

As indicated in the first line of the listing, the control file for menus is named **UM.GRN** and the internal name for the customized menu block is **Random**, which is included at the bottom of the listing. The control file for customized functions is expected to have two function blocks: one for **Generate** and the other for **ReadPlot**. The contents of the control file, **UF.GRN**, for customized functions are as follows:

```
:F UF.GRN
:D User Defined Functions
:I      0,        0,       0
:!
:N Fnc_Generate
:S      0,      0,      0
:T Command
:A Generate Load
:D Generate Random Loading
Cmd_F3
 Fld_0
  Typ_Lab
  Prm_[GENERATE] Generates random loading
 Fld_0
  Typ_Sep
 Fld_2
  Typ_Real
  Prm_Duration of loading
 Fld_3
  Typ_Int
  Prm_Number of data points
 Fld_4
  Typ_Real2
  Prm_Range of load (MIN & MAX)
 Fld_0
  Typ_Lab
  Prm_Seed number for random number generation
 Fld_6
  Typ_Int
  Prm_Seed number
:E END
:!
:N Fnc_ReadPlot
:S      0,      0,      0
:T Command
:A Read/Plot
:D Read and Plot Random Loading
Cmd_F3
 Fld_0
  Typ_Lab
  Prm_[ReadPlot] Reads and plots random loading data
 Fld_0
  Typ_Sep
 Fld_3
  Prm_Select one option
  Typ_Lis_OptionB
   Lis_Read,1
   Lis_Read and Plot,2
:E END
:!
```

The resulting menu structure appears in the ANSYS GUI, as shown in Fig. 11.31. The dialog boxes corresponding to function blocks **Generate** and **ReadPlot** are shown in Fig. 11.32 and 11.33, respectively.

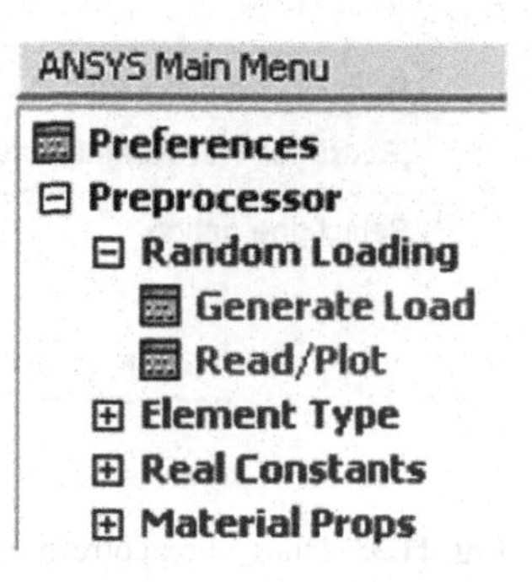

Fig. 11.31 Customized menu structure for random loading data generation.

Two macro files corresponding to the function blocks are created and stored in the *Working Directory*. The first macro file is **`Generate.MAC`**:

```
TOT=ARG1                            ! STORE ARGUMENTS IN
                                    ! PARAMETERS
ND=ARG2                             !
MN=ARG3                             !
MX=ARG4                             !
SEED=ARG5                           !
/OUTPUT,RANDOM,DAT                  ! REDIRECT OUTPUT TO
                                    ! RANDOM.DAT
*VWRITE,TOT,ND,MN,MX,SEED           ! WRITE PARAMETERS TO FILE
(E16.8,2X,F8.0,2X,E10.2,2X,E10.2,2X,F8.0)
/OUTPUT                             ! REDIRECT OUTPUT TO OUTPUT
                                    ! WINDOW
/SYS,RANDOM.EXE                     ! EXECUTE RANDOM.EXE
```

Generate Random Loading
[GENERATE] Generates random loading
Duration of loading
Number of data points
Range of load (MIN ,MAX)
Seed number for random number generation
Seed number
OK Apply Cancel

Fig. 11.32 Dialog box corresponding to function Generate.

Read and Plot Random Loading
[ReadPlot] Reads and plots random loading data
Select one option Read and Plot
OK Apply Cancel

Fig. 11.33 Dialog box corresponding to function ReadPlot.

The macro file **ReadPlot.MAC** is given as

```
*SET,ARY                        ! INITIALIZE PARAMETER ARY
*SET,TIM                        ! INITIALIZE PARAMETER TIM
*SET,LOAD                       ! INITIALIZE PARAMETER LOAD
*DIM,ARY,ARRAY,2*ND             ! DIMENSION PARAMETER ARY
*DIM,TIM,TABLE,ND               ! DIMENSION PARAMETER TIM
*DIM,LOAD,TABLE,ND              ! DIMENSION PARAMETER LOAD
*VREAD,ARY(1),RANDOM,OUT        ! FILL ARY ARRAY FROM RANDOM.OUT
(2(E16.8,2X))                   ! FORMAT STATEMENT
*DO,I,1,ND                      ! LOOP OVER DATA POINTS
TIM(I)=ARY(2*I-1)               ! SEPARATE ARRAY ARY INTO TIM
LOAD(I)=ARY(2*I)                ! AND LOAD ARRAYS
*ENDDO                          ! END LOOP OVER DATA POINTS
*IF,ARG1,EQ,2,THEN
/AXLAB,X,TIME                   ! DEFINE X-AXIS LABEL
/AXLAB,Y,LOAD                   ! DEFINE Y-AXIS LABEL
*VPLOT,TIM(1),LOAD(1)           ! PLOT TWO ARRAYS AGAINST EACH
                                ! OTHER
*ENDIF
```

Finally, a copy of the external executable file **RANDOM.EXE** is stored in the *Working Directory*. The input shown in Fig. 11.34 produces the random loading shown in Fig. 11.35, which is obtained using the ReadPlot function.

11.6.3 Function Block for Selecting Elements Using a Pick Menu

The final example for demonstrating the ANSYS GUI environment involves a simple selection function. The function block is called **Selel** and is given below.

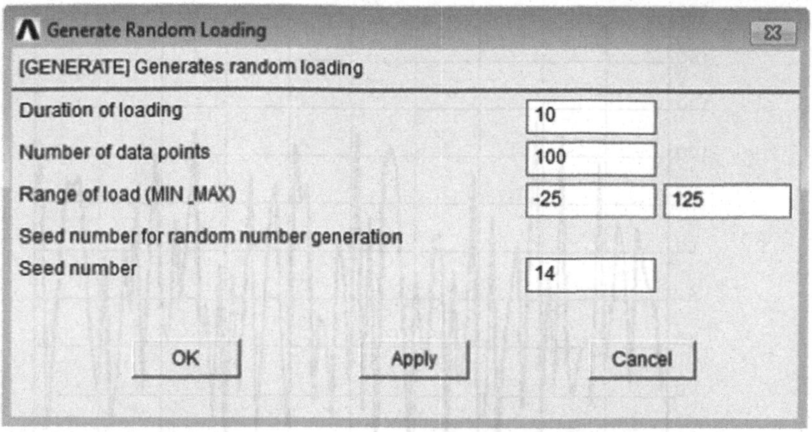

Generate Random Loading
[GENERATE] Generates random loading
Duration of loading 10
Number of data points 100
Range of load (MIN_MAX) -25 125
Seed number for random number generation
Seed number 14
OK Apply Cancel

Fig. 11.34 Input for random load generation through a dialog box.

```
:N Fnc_Selel
:S 0, 0, 0
:T Cmd_P
:A Element select
:D Pick elements
Inp_P
Cmd_ESEL
 Fld_2
 Typ_Def_S
 Fld_5
Prm_Pick elements within the global element
 Typ_ELEM
 Min_1
 Cnt_s
 PCN_1
 PFM_1
Cal_Fnc_Store
:!
```

The expression in the third line,:**T Cmd_P**, declares that this function block is for creating a *Pick Menu*. Similarly, the line **Inp_P** is used for picking, and it must be the first line in the data controls block. The line **Typ_ELEM** states that elements are to be picked. A minimum of 1 element must be picked, based on the specification **Min**_1. The line **Cnt_s** specifies that the selected entities produce an unordered list and that there is no limit on how many entities can be picked. The command

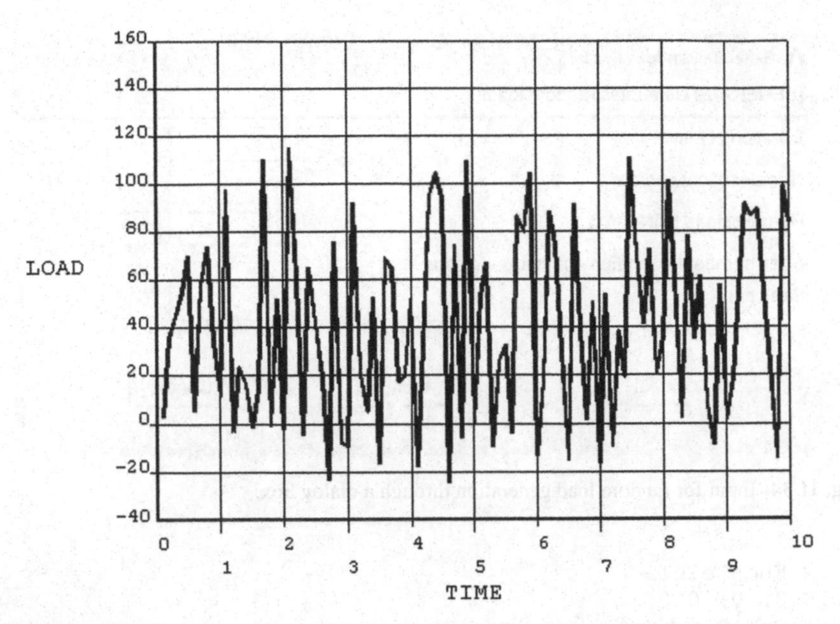

Fig.11.35 Variation of load vs. time generated using ReadPlot function block.

PCN_1 specifies that a set of picked entities should not be ordered for retrieval picking. The command **PFM_1** forces ANSYS to use commands **FITEM** and **FLST** for processing of the picked items. Finally, a call to another function block, named **Store**, is made in the line **Cal_Fnc_Store**.

Erratum to: The Finite Element Method and Applications in Engineering Using ANSYS®

Erdogan Madenci, Ibrahim Guven

Erratum to:
The Finite Element Method and Applications in Engineering Using ANSYS®
DOI 10.1007/978-1-4899-7550-8

Electronic supplementary material for using ANSYS® can be found at http://link.springer.com/book/10.1007/978-1-4899-7550-8

The online version of the original book can be found under
DOI 10.1007/978-1-4899-7550-8

Erdogan Madenci (✉)
Department of Aerospace and Mechanical Engineering
The University of Arizona
Tucson, Arizona
USA
e-mail: madenci@email.arizona.edu

Ibrahim Guven
Department of Mechanical and Nuclear Engineering
Virginia Commonwealth University
Richmond, Virginia
USA
e-mail: iguven@vcu.edu

E. Madenci, I. Guven, *The Finite Element Method and Applications in Engineering Using ANSYS®*, DOI 10.1007/978-1-4899-7550-8_12

References

Abramowitz, M., & Stegun, I. A. (1972). *Handbook of mathematical functions*. New York: Dover.
Barsoum, R. S. (1976). On the use of isoparametric finite elements in linear fracture mechanics. *International Journal for Numerical Methods in Engineering, 10,* 25–37.
Barsoum, R. S. (1977). Triangular quarter-point elements as elastic and perfectly plastic crack tip elements. *International Journal for Numerical Methods in Engineering, 11,* 85–98.
Bathe, K.-J. (1996). *Finite element procedures*. Englewood Cliffs: Prentice Hall.
Bathe, K.-J., & Wilson, C. L. (1976). *Numerical methods in finite element analysis*. Englewood Cliffs: Prentice Hall.
Carslaw, H. S., & Jaeger, J. C. (1959). *Conduction of heat in solids*. London: Oxford University Press.
Cook, R. D. (1981). *Concepts and application of finite element analysis* (2nd ed.). New York: Wiley.
Desai, C., & Abel, J. (1971). *Introduction to the finite element method*. Reinhold: Van Nostrand.
Dym, C. L., & Shames, I. H. (1973). *Solid mechanics: A variational approach*. New York: McGraw-Hill.
Ergatoudis, J., Irons, B., & Zienkiewicz, O. C. (1968). Curved isoparametric quadrilateral elements for finite element analysis. *International Journal of Solids and Structures, 4,* 31–42.
Finlayson, B. A. (1972). *The method of weighted residuals and variational principles*. New York: Academic Press.
Gallagher, R. H. (1975). *Finite element analysis: Fundamentals*. Englewood Cliffs: Prentice Hall.
Henshell, R. D., & Shaw, K. G. (1975). Crack tip finite elements are unnecessary. *International Journal for Numerical Methods in Engineering, 9,* 495–507.
Huebner, K. H. (1975). *The finite element method for engineers*. New York: Wiley.
Huebner, K. H., Dewhirst, D. L., Smith, D. E., & Byrom, T. G. (2001). *The finite element method for engineers* (4th ed.). New York: Wiley.
Oden, J. T. (1972). *Finite elements of nonlinear continua*. New York: McGraw-Hill.
Pu, S. L., Hussain, M. A., & Lorenson, W. E. (1978). The collapsed cubic isoparametric element as singular element for crack problems. *International Journal for Numerical Methods in Engineering, 12,* 1727–1742.
Stroud, A. H., & Secrest, D. (1966). *Gaussian quadrature formulas*. Englewood Cliffs: Prentice Hall.
Turner, M. J., Clough, R. W., Martin, H. C., & Topp, L. J. (1956). Stiffness and deflection analysis of complex structures. *Journal of the Aeronautical Sciences, 23,* 805–823.
Washizu, K. (1982). *Variational methods in elasticity and plasticity* (3rd ed.). New York: Pergamon.
Zienkiewicz, O. C. (1977). *The finite element method* (3rd ed.). New York: McGraw-Hill.

E. Madenci, I. Guven, *The Finite Element Method and Applications in Engineering Using ANSYS®*, DOI 10.1007/978-1-4899-7550-8

Index

E. Madenci, I. Guven, *The Finite Element Method and Applications in Engineering Using ANSYS®*, DOI 10.1007/978-1-4899-7550-8

Printed in the United States
By Bookmasters